Introduction to

Aircraft Performance, Selection, and Design

Introduction to
Aircraft Performance, Selection, and Design

Francis J. Hale
North Carolina State University

John Wiley & Sons
New York Chichester Brisbane Toronto Singapore

Library of Congress Cataloging in Publication Data:

Hale, Francis J.
Introduction to aircraft performance, selection,
and design.

Includes indexes.
1. Aeronautics. I. Title.
TL546.H28 1984 629.13 83-16995
ISBN 0-471-07885-9

Printed in the United States of America

10 9 8 7 6 5 4 3 2

Preface

This is a teaching text for an introductory course in aeronautical or aerospace engineering. The course has no prerequisites, contrary to the usual practice of requiring one or more aerodynamics courses plus propulsion and structures courses prior to a performance course. Such prerequisites not only eliminate nonaerospace engineering students but also delay the exposure of aero students to the reasons for the configuration and performance of various types of aircraft and deprive them of the motivation and rationale for the supporting technical courses. This course has been taught many times here at North Carolina State University as well as at The Middle East Technical University in Ankara, Turkey and at the United States Military Academy at West Point.

The major objective of the course and of this book is to impart an understanding of why conventional aircraft look and fly as they do. By the end of the course, the student should be able to take the physical characteristics of any existing aircraft (estimating those data that might be missing; manufacturers have a habit of not providing everything we would like) and determine its performance, namely, its range, flight regime, rate of climb, turning rate, etc.

If it is possible to determine the operational performance of an aircraft with a set of specified physical characteristics, it seems logical to be able to do the converse, that is, to design an aircraft that will meet a set of operational requirements that specify such things as the desired range, cruise airspeed, and payload capability. Such a design, using the techniques of this book, is a conceptual or feasibility design, yielding the major characteristics such as the maximum gross weight, the wing area and span, the drag polar, the thrust required, and the fuel consumption and load. Skill in performance and design implies the capability to evaluate and compare competing aircraft, thus the chapter on selection and the inclusion of the word in the title. Selection is an important process to the person, organization, or country faced with the problem of picking the best aircraft despite myriad, often conflicting details and claims.

Since the emphasis throughout the book is on the determination of the key parameters and a physical appreciation of their influence on the performance and design of an aircraft, analytic expressions and closed-form solutions are essential. Consequently, the techniques developed and used are based on assumptions and idealizations (with varying effects on the preciseness of the numerical values obtained) that are dependent on the flight regime. In other words, this book does not presume to treat all possible flight conditions or all aircraft configurations with equal accuracy of results.

After the development of the relevant equations of motion and subsystem characteristics, the first portion of the book is devoted to the examination of the performance of conventional subsonic aircraft with an idealized turbojet propulsion system. The mathematics for the turbojet is simpler, more straightforward, and easier to visualize physically than that for the piston-prop aircraft. With an understanding of the turbojet, it is much easier to evaluate and visualize the differences in the performance, flight conditions, and design criteria of piston-prop aircraft. Turbojet and piston-prop aircraft establish the boundaries for the performance and design of fixed-wing aircraft. Since both the turbofan and the turboprop combine to varying degrees the characteristics of turbojets and piston-props, their performance falls somewhere between these boundaries and is difficult to analyze quantitatively. Consequently, only the qualitative aspects of the performance and design of these aircraft will be discussed in this book.

Although the design examples appear to be restricted to a single chapter at the end of the book, they in fact will occur throughout the book, either by referral or as chapter problems, which may be used as illustrative examples during a lecture or worked outside of the classroom. The important thing to remember is that analysis and design are a combined process rather than two separate and isolated processes.

The courses that this book is designed to support are for students in any curriculum. For the aero engineer it can be either an introductory course to be followed by the technical courses, which is the sequence that I favor, or it can be a capstone course that ties the technical courses together. The nonaero or non-engineering student may well take this as a terminal course with the sole objective of understanding a mode of transportation that is characteristic of our modern society. This student, or reader, will know why there are no transcontinental helicopters, for example, and why on long-range flights a jet aircraft flies at high altitudes and increases that altitude as fuel is consumed.

The mathematical skills used in this book are minimal, namely, the ability to solve a quadratic equation, take a first derivative, and perform a single integration. Since the results of these operations are always given as closed-form expressions, these skills need be exercised only if the reader wants to verify the solutions. Since the English system of units is still used in aircraft operations and is generally more familiar to the average person than the SI system, this book uses the English system of units. In the early chapters, the corresponding SI units and values are shown in parentheses following the English values and units to give the reader a feel for the equivalence of numbers and units. In this book an inconsistency with aircraft operational practice exists since miles per hour and statute miles are used rather than knots and nautical miles.

This book covers more material than can possibly be dealt with in a one-semester course. I have found over the years that the most satisfactory solution is to cover the first seven chapters as written. It is possible, and may even be desirable, to minimize the discussion of the alternate flight programs and emphasize only the principal ones. Then cover the first three sections of Chap. 8, which summarize the differences between turbojets and piston-props and discuss the turbofan and turboprop. A brief discussion on the remainder of Chap. 8, a few words on the effects of wind, and the semester is over. I use simple design problems, such as those at the end of some of the chapters, as take-home exercises.

Finally, since the book is self-contained, with no prerequisites or prior knowledge necessary, and the required mathematical skills are minimal, it should be suitable for self-teaching and study. I recommend it to anyone who is interested in aircraft, in their design and operation, and particularly to pilots who would like to have a better understanding of the rationale behind their operating manuals and procedures and who might wish to sharpen and improve their individual procedures and techniques. As a former military pilot, I know that, in developing this book and teaching a corresponding course, my own knowledge and understanding of aircraft have greatly increased. Professional pilots who have taken the course from time to time seem to agree.

I would like to thank Leroy S. Fletcher, Texas A & M University and Charles Libove, Syracuse University for their comments and suggestions made while reviewing the the text.

Francis J. Hale

Contents

Introduction to

Aircraft Performance, Selection, and Design

Chapter One

Introduction

1-1 AIRCRAFT FLIGHT BEHAVIOR

The flight path and behavior of an aircraft are determined by the interaction between the characteristics of the aircraft itself and the environment in which it is flying. The aircraft characteristics can be categorized as the physical characteristics, such as the shape, mass, volume, and surface area; the characteristics of the subsystems, such as the propulsion, guidance, and control subsystems; and the structural characteristics, such as the loading and temperature limitations and the stiffness or rigidity of the structure.

The environment affects the flight of an aircraft through the field forces and the surface forces. The only field force that we need consider for an aircraft is gravity, which appears as the weight and is a function of the mass of the aircraft. The surface forces are the aerodynamic forces (the lift, the drag, and the side force), which are very strongly dependent upon the shape and the surface area of the aircraft, especially of the wings, and upon the properties of the atmosphere. In addition, consideration must be given to the inertia forces. They are the consequence of nonequilibrium processes and play an important role in dynamic and stress analyses of aircraft.

There are two basic and fundamental modes of aircraft behavior, particularly when the aircraft is assumed to be a rigid body. The first mode is the *translational mode* in which we treat the aircraft as a point mass that has 3 degrees of freedom and can move up or down, frontwards or backwards, and sideways. The force equations of motion are sufficient to determine and describe the translational mode. The second mode is the *rotational mode* with 3 additional degrees of freedom comprising angular motion about three mutually orthogonal axes, whose origin is normally located at the center of mass (the center of gravity) of the aircraft. The moment equations, as well as the force equations, are required to determine and describe the rotational mode.

In this book, we shall limit ourselves to rigid aircraft even though most modern aircraft have varying amounts of elasticity arising from the performance and design objectives of reducing structural weight to the minimum. Fortunately, aeroelastic effects usually need not be considered in performance analyses or preliminary configuration designs. In addition, we can limit ourselves to the force equations because we shall be treating the aircraft as a point mass in our performance analyses and feasibility designs. In Chap. 11, when we take a brief look

at stability and control and its implications with respect to performance are discussed, the moment equations will be introduced. Finally, we shall consider only quasi-steady-state flight; that is, we shall assume that velocities and other flight-path parameters are either constant or are changing so slowly that their rates of change can be neglected. Consequently, with the exception of the centrifugal force in turning flight, the inertia forces can be neglected.

1-2 THE PERFORMANCE ANALYSIS

Strictly speaking, in a performance analysis, a person takes an existing set of physical characteristics of a particular aircraft and determines such things as how fast and how high it can fly and how far it can travel with a specified amount of fuel. Performance also means examining different ways of flying a mission so as to exploit the characteristics and capabilities of the aircraft in question. For example, two major objectives of commercial aircraft are to minimize both the fuel consumption and the flight time. These objectives can be met by a combination of proper design and appropriate operational procedures.

The aircraft undergoing a performance analysis may be an existing aircraft, a real aircraft. On the other hand, it may be a paper aircraft, one that is being studied for possible adoption and manufacture. As a result of the performance analysis, it may be rejected for any further consideration or some of its characteristics may be modified and another performance analysis carried out. And then possibly another modification and another analysis might follow. Thus, the performance analysis and the preliminary design process are often intertwined in a series of iterations, and it may be hard at times to distinguish between the two.

There are two basic approaches to a performance analysis. The first is primarily graphical, and the other is primarily analytical. The latter will be used in this book with emphasis on classifying aircraft on a broader basis than the former approach. For example, the graphical approach might describe an aircraft in terms of its gross weight, whereas the analytical approach would use the wing loading. In fact, we shall find that a knowledge of the wing loading, the maximum lift-to-drag ratio, and the type of propulsion system will give us insight as to the design mission and performance of an aircraft.

This is not an aerodynamics book nor is it a propulsion book. The coverage of these two subjects is limited to what is necessary or helpful in understanding the significance and importance of the various aerodynamic and propulsive parameters. In spite of the overriding importance of structural weight and integrity in the design of aircraft and to the initial and operating costs, the complexity of the subject precludes anything more than a continued awareness of its importance.

The absence of any computer programs or mention of computational methods is deliberate but is not to be interpreted as a rejection of the computer for use in performance and design. To the contrary, the computer in its many forms is a most useful tool. However, computers come in many hardware and software configurations, and it would not be possible to treat them adequately in this book.

It is left to the individual to decide how the computer can best be used.

After a chapter of background material, which can serve as a review for the more knowledgeable person, level flight of turbojet aircraft is our first introduction to performance. The mathematics is relatively simple and straightforward and the analytical results quite satisfying. It is of some interest to discover that jets fly high and fast not necessarily because they want to but because they have to in order to be competitive.

With some understanding of the performance of turbojet aircraft, it is easier to relate to the performance of piston-props, which is surprisingly different. The two types of aircraft fly in different ways to exploit their respective advantages. Then we look briefly at turbofans and turboprops which combine the features of turbojets and piston-props to varying degrees.

Although we shall reduce the performance problem to a set of simple, two-dimensional statics problems, it is always nice to be aware of their origins. Therefore, this chapter concludes with a brief section on the fundamental equations of motion and coordinate systems.

1-3 EQUATIONS OF MOTION AND COORDINATE SYSTEMS

In its most general form, Newton's law governing the linear momentum of a continuous system can be written in vector form as

$$\mathbf{F} = \int \mathbf{a}\, dm \tag{1-1}$$

where $\mathbf{F}$ is the vector sum of the external forces, $\mathbf{a}$ is the acceleration of a particle mass of the system with respect to inertial space (a nonrotating and nonaccelerating reference frame). For aircraft performance analyses, the Earth can be taken to be the inertial space and we can neglect the rotation and curvature of the Earth. Consequently, with a flat, nonrotating Earth as the inertial reference, the vector equation of motion of a rigid aircraft can now be written as

$$\mathbf{F} = m\mathbf{a} = \frac{W}{g}\frac{d\mathbf{V}}{dt} \tag{1-2}$$

where m and W are the mass and weight of the aircraft, $\mathbf{g}$ is the acceleration due to Earth's gravity (32.2 ft/s or 9.8 m/s^2), $\mathbf{a}$ is the acceleration of the center of gravity (the cg) of the aircraft with respect to the Earth, and $\mathbf{V}$ is the velocity of the cg of the aircraft with respect to the Earth.

For flight over a flat Earth, there are three rectilinear and right-handed coordinate systems of interest: the *ground-axes system* $EXYZ$; the *local-horizon system* $Ox_hy_hz_h$: and the *wind-axes system* $Ox_wy_wz_w$. We now define these systems with the assumption that the aircraft has a plane of symmetry, as do all aircraft now in operation. As a matter of historical interest, the Germans did build and fly an experimental asymmetrical aircraft during World War II. Also, the skewed-wing concept proposed for low-supersonic aircraft would have varying degrees of asymmetry as a function of the wing angle. Returning to the plane of symmetry, with an aircraft sitting on the ground, it is the vertical plane passing through the center line of the fuselage that divides the aircraft into two parts that are mirror images of each other.

The *ground-axes system* is fixed to the surface of the Earth. Its origin is any point on the Earth's surface, usually the starting point for a flight. The X axis and the Y axis are in the horizontal plane and are used to measure distances on the surface of the Earth; for flight in the vertical plane, the X axis is located in the direction of flight. The Z axis is vertical and positive downward so that the three axes form a right-handed cartesian system.

The *local-horizon system* has its origin at the center of gravity of the aircraft, which is in the plane of symmetry. The axes are all parallel to the corresponding axes of the ground system and do not rotate but do translate with the aircraft. The x and y axes form a plane that is always parallel to the surface of the Earth, thus forming what is known as the local horizon, and that is carried along with the aircraft.

In defining the *wind-axis system*, we assume that the atmosphere is at rest with respect to the Earth, a "no-wind" assumption. The wind-axes system has the same origin as the local-horizon system; i.e., the cg of the aircraft. The x axis is tangent to the flight path, which means that it lies along the aircraft velocity vector (in the direction of the *relative wind*) and is positive in the forward direction of flight. The z axis is perpendicular to the x axis, is in the plane of symmetry, and is positive downward for a normal aircraft attitude. The y axis is perpendicular to the xz plane and forms a right-handed set of cartesian coordinates. The wind axes are *not body axes*; that is, they are not fixed to the aircraft other than at the cg. A change in the direction of flight can change x without changing the attitude of the aircraft; conversely, x can remain stationary, say, in the horizontal plane, while the aircraft attitude changes.

Since the ground and local-horizon axes systems are always parallel, their corresponding unit vectors are identical, and there are no angular relationships among them to worry about. However, this is not the case with the orientation of the wind-axis system with respect to the local-horizon axis system. This orientation is described in terms of three successive rotations, always made in the same sequence and with clockwise rotation taken as positive. The three standard angles used, in order of their application, are the yaw angle χ, the flight-path angle γ, and the bank or roll angle ϕ. As a side note, these are not necessarily identical to the Euler angles that are used in Chapter 11 to relate the local-horizon system to a set of body axes. The three successive rotations use two intermediate coordinate systems. The first rotation is about the local-horizon x axis and through the yaw angle. The second rotation is about the first intermediate y axis and through the flight-path angle, and the last rotation is about the second intermediate x axis and through the roll angle. When these three rotations are completed in the correct sequence, the relationship among the unit vectors can be expressed by the vector equation

$$\mathbf{1}_w = \mathbf{A}\mathbf{1}_h \tag{1-3}$$

where $\mathbf{1}_w$ is the column vector of the wind-axes unit vectors and $\mathbf{1}_h$ is the column vector of the local-horizon systems, and $\mathbf{A}$ is the transformation matrix, which is orthogonal; that is, $\mathbf{A}\cdot\mathbf{A}^T = \mathbf{I}$. The transformation from the wind axes to the local-horizon axes is given by

$$\mathbf{1}_h = \mathbf{A}^{-1}\mathbf{1}_w \tag{1-4}$$

Fortunately, we need not concern ourselves with either the details of the rotation or the composition of the transformation matrix. These can be found in any standard text on dynamics, if you really want to know more about coordinate transformations. For our purposes, we can think of the flight-path angle γ as describing the direction of flight in the vertical plane, of the yaw angle χ as describing the direction of flight in the horizontal plane, and of the roll angle ϕ as the angle between the wings and the horizontal plane.

Chapter Two

Aircraft Forces and Subsystems

2-1 INTRODUCTION

The principal forces acting on a rigid symmetrical aircraft flying over a flat, non-rotating Earth are the thrust produced by the propulsion subsystem; the major aerodynamic forces, namely, the lift and drag of the wing (we shall neglect the lift of the other components, such as the tail, fuselage, and engine nacelles and include their drag in that of the wing); and the weight of the aircraft, which is the vertical force resulting from the acceleration due to Earth's gravity. Since the thrust and aerodynamic forces are strongly influenced by the properties of the atmosphere, these properties will be defined, in terms of our flight regime of interest, in the next section. Then there will be brief descriptions of the aerodynamic and propulsive forces, and the chapter will conclude with a discussion of structural and weight influences and considerations.

2-2 THE ATMOSPHERE

The atmospheric model that we shall use is the International Standard Atmosphere, which is based on the ARDC Model Atmosphere of 1959. Although this model has seven concentric layers, we shall limit ourselves to the first two layers of the lower atmosphere. The layer next to the surface of the Earth (starting at sea level) is called the *troposphere* and is characterized by a decreasing ambient temperature. At 36,089 ft (11,000 m) above mean sea level, the temperature becomes constant and remains so until an altitude of 82,021 ft (25,000 m) above mean sea level. This second layer is called the *stratosphere*, and the separating altitude of 36,089 ft (11,000 m) is known as the *tropopause*. The sea-level properties with symbols and with the subscript SL denoting sea-level values for a standard day are:

Ambient temperature: $\Theta_{SL} = 59$ deg F (15 deg C)

Pressure: $P_{SL} = 2{,}116$ lb/ft^2 (1.013×10^5 N/m^2)

Density: $\rho_{SL} = 23.769 \times 10^{-4}$ lb-s^2/ft^4 (1.226 kg/m^3)

Speed of sound: $a_{SL} = 1{,}116$ fps (340 m/s)

Gas constant: $R = 1.7165 \times 10^3$ ft^2/s^2-deg R (0.287 kJ/kg-K)

Ratio of specific heats: $k = 1.4$

Acceleration of gravity: $g_{SL} = 32.174$ ft/s^2 (9.8 m/s^2)

We shall often use values with fewer significant figures in our calculations. Using these sea-level values, the value of a property at any altitude, which is given the symbol h, can be found from the property-ratio tables in Appendix A.

Although the values from the tables are more accurate and normally should be used, there are times when it is desirable to have a mathematical expression for the density ratio in terms of the altitude. One such expression is

$$\sigma_1 = \frac{\rho_1}{\rho_{SL}} = \exp\left(\frac{-h_1}{\beta}\right) \tag{2-1}$$

where ρ_1 is the atmospheric density at altitude h_1 on a standard day. If the density ratios at two altitudes are known, the difference in altitude can be obtained from

$$\Delta h = h_2 - h_1 = \beta \ln\left(\frac{\sigma_1}{\sigma_2}\right) \tag{2-2}$$

We shall use two values for β. A value of 30,500 ft (9,296 m) gives very good results for both Eqs. 2-1 and 2-2 within the troposphere (altitudes below 36,000 ft) but introduces large errors at higher altitudes. For altitudes above the tropopause (36,000 ft) and up to approximately 250,000 ft, a value of 23,800 ft (7254 m) gives good results for the difference in altitude but when used in Eq. 2-1 does introduce errors of the order of 20 percent for altitudes between 35,000 and 55,000 ft; the error decreases with altitude and at 80,000 ft and higher is of the order of 5 percent or less.

To summarize the preceding paragraph, use Eq. 2-1 *only* when a mathematical relationship is required. Equation 2-2 can be used with the appropriate value of β when only reasonable accuracy is needed for an altitude difference.

Returning to the sea-level properties for a standard day, the atmosphere can be assumed to satisfy the equation of state for an ideal gas; i.e.,

$$P = \rho R \Theta \tag{2-3}$$

where P is the pressure, R is the gas constant, and Θ is the air temperature expressed as an absolute temperature (Rankine or Kelvin). Consequently, an increase in the ambient temperature over the standard will decrease the density, as will a decrease in the pressure. In other words, on a hot day and/or with a below-standard barometric pressure, the sea-level density will be lower than standard and will affect the performance of the aircraft. With a decreased atmospheric density, the aircraft thinks and performs as though it were at a higher altitude than it actually is. We shall assume a standard day in this book, unless specifically stated otherwise, but we should be aware of the possible effects of any deviations.

2-3 AERODYNAMIC FORCES

The resultant or vector aerodynamic force produced by the motion of the aircraft through the atmosphere is resolved into components along the wind-axes. The component along the x axis is called the *drag* and given the symbol D; it is in the opposite direction to the velocity and resists the motion of the aircraft. The component along the z axis (perpendicular to the aircraft velocity) is called the *lift* and given the symbol L; the lift is normally in an upward direction and its function is to counteract the weight of the aircraft. It is the lift that keeps the aircraft in the air. The third component, along the y axis, is a side force that appears only with asymmetrical aircraft or when the velocity vector of a symmetrical aircraft is not in the plane of symmetry; i.e., when there is a side-slip angle. The latter case, which is called "uncoordinated flight," is a normally undesirable flight condition and will not be considered in this book. Consequently, we shall concern ourselves only with the lift and drag forces.

Although all parts of an aircraft generate lift and drag, with a well-designed aircraft, the wing is the major source of the aerodynamic forces. We need at this point to define some of the wing parameters important to our analyses. The *wing span* b is the distance from wingtip to wingtip. The *wing area* S is the surface area of one side of the wing, to include the area occupied by the fuselage. The *chord* is the straight-line distance from the front (the leading edge) of an airfoil section to the back (the trailing edge). The chord length will vary in value in the span-wise direction if there is any wing taper. The arithmetic mean of the chord values is known as the *mean geometric chord* or the *average chord*, for which we will use the symbol $\bar{c}$. The ratio of the wing span to the average chord is called the *aspect ratio* and given the symbol AR. The following relationships exist among the wing characteristics:

$$\text{AR} = \frac{b}{\bar{c}} = \frac{b^2}{S} \tag{2-4}$$

where

$$S = b\bar{c} \tag{2-5}$$

One other wing characteristic (of importance in determining wing drag) is the *thickness ratio* ($t/\bar{c}$), which is the ratio of the maximum thickness (from top to bottom) of the wing to the average chord length (see Fig. 2-1).

The cross section of a wing is called an *airfoil* and if it is *symmetrical* with respect to the chord, we speak of a symmetrical wing. Many current airfoils are symmetrical or nearly so. *Asymmetrical airfoils* (with positive or negative camber), such as those in Fig. 2-2, are generally used when stability considerations or special operational considerations are more important than range or speed.

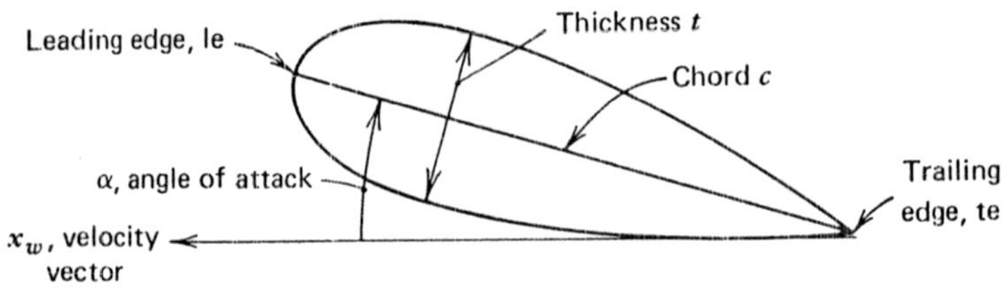

FIGURE 2-1

A symmetrical wing cross section (airfoil).

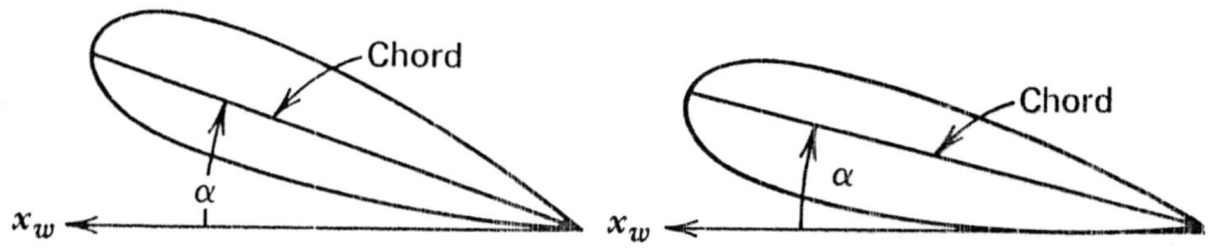

FIGURE 2-2

Examples of cambered airfoils: (*a*) positive camber; (*b*) negative camber.

It has been determined from dimensional analyses and experiments that the lift and drag forces can be found from the following expressions:

$$L = qSC_L = \tfrac{1}{2}\rho V^2 SC_L \tag{2-6a}$$

$$D = qSC_D = \tfrac{1}{2}\rho V^2 SC_D \tag{2-6b}$$

In these equations, q is called the dynamic pressure, has the dimensions of lb/ft^2 (N/m^2), and is given by

$$q = \tfrac{1}{2}\rho V^2 \tag{2-7}$$

where ρ is the atmospheric density which can be expressed as $\rho_{SL}\sigma$, V (the true airspeed) is the speed of the aircraft through the atmosphere, and S is the wing area.

In Eqs. 2-5 and 2-6, C_L and C_D are the dimensionless *lift* and *drag coefficients*, respectively. How these coefficients are evaluated is beyond the scope of this course; however, they do have the following functional relationships:

$$C_L = C_L(\alpha, M, \text{Re, shape}) \tag{2-8a}$$

$$C_D = C_D(\alpha, M, \text{Re, shape}) \tag{2-8b}$$

where α is the *angle of attack* of the aircraft (see Fig. 2-1) and is the angle between the velocity vector (the x wind axis) and the wing chord; M is the *Mach number*, which is the ratio of the airspeed to the speed of sound; Re is the *Reynolds number*, which we will ignore since its influence is strongest at high angles of attack; and "shape" refers to the wing shape to include the airfoil section, the taper, the sweep angle, and the aspect ratio. Figure 2-3 shows the variation of the lift coefficient as a function of the angle of attack for a typical subsonic wing for a given Mach number and Reynolds number for three airfoils of different camber. Notice that the C_L versus α relationship is almost linear for low values of α. At point S, the air flow separates from the wing and C_L decreases, usually sharply. This is called the *stall point*, and conventional aircraft do not (and usually cannot) fly beyond this point.

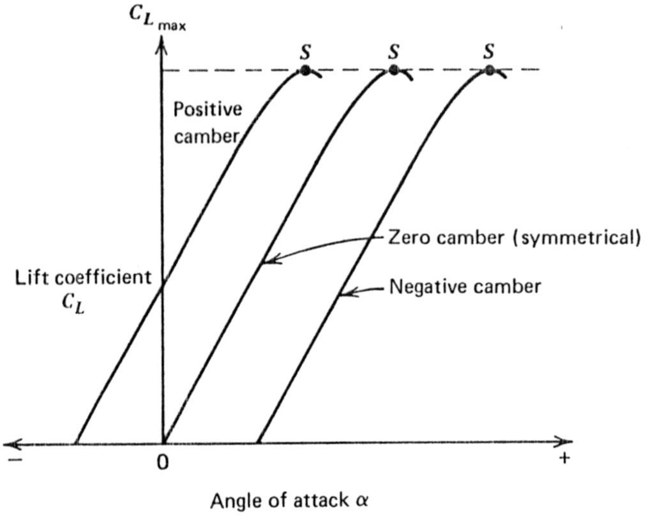

FIGURE 2-3
Wing lift coefficient as a function of the angle of attack for various cambers for a given Mach number and Reynolds number.

An expression that can be used to obtain an approximate value for the slope of the lift curve is

$$\frac{dC_L}{d\alpha}=\frac{\pi(\mathrm{AR})}{1+\left[1+\left(\dfrac{\mathrm{AR}}{2\cos\Lambda}\right)^2\right]^{1/2}} \tag{2-9}$$

where the slope is the change per radian (divide by 57.3 to get the change per degree) and Λ is the sweep angle of the wing. Although it does not include the effects of the Mach number, it is a useful relationship and does show that the slope decreases as the aspect ratio decreases and as the sweep angle increases.

Figures 2-4*a* and 2-4*b* are typical lift and drag curves for a symmetrical wing, where C_L and C_D are shown as functions of the angle of attack. If α is eliminated from the functional relationships of Eq. 2-8, we obtain what is known as the *drag polar*, where C_D is expressed as a function of C_L so that

$$C_D=C_D(C_L, M, \mathrm{Re}) \tag{2-10}$$

as is shown in Fig. 2-4*c*. The drag polar is very important to performance analyses and can often be very difficult to obtain from an aircraft manufacturer. For our purposes, the overall drag coefficient can be divided into two components and written as

$$C_D=C_{D0}+C_{Di} \tag{2-11}$$

where C_{DO} is the *zero-lift drag coefficient* (i.e., the drag coefficient when the lift coefficient is equal to zero) and C_{Di} is called the *drag-due-to-lift coefficient* (sometimes called the *induced-drag coefficient*). See Fig. 2-4c. Since C_{Di} arises from the generation of lift, it can be expressed as

$$C_{Di} = KC_L^x \tag{2-12}$$

so that Eq. 2-11 becomes

$$C_D = C_{DO} + KC_L^x \tag{2-13}$$

Equation 2-13 is known as the *generalized drag polar*; C_{DO}, K, and the exponent x are all functions not only of the wing shape but also of the Mach and Reynolds numbers. Fortunately, there are many efficient subsonic and thin-winged supersonic configurations and flight regimes where these parameters can be considered constant and x can be set equal to 2. Consequently, Eq. 2-13 can be written as

$$C_D = C_{DO} + KC_L^2 \tag{2-14}$$

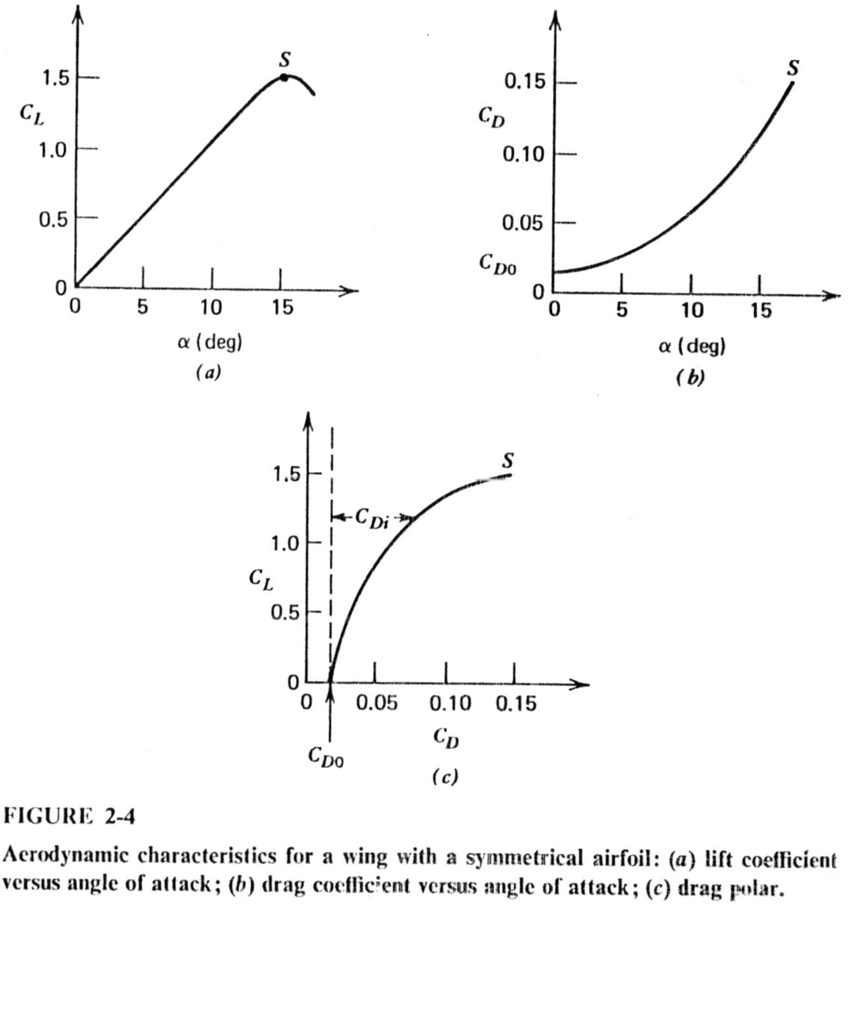

FIGURE 2-4

Aerodynamic characteristics for a wing with a symmetrical airfoil: (*a*) lift coefficient versus angle of attack; (*b*) drag coefficient versus angle of attack; (*c*) drag polar.

This particular drag polar is known as the *parabolic drag polar* and is the one used throughout this book.

The value of K is strongly dependent upon the aspect ratio, as can be seen from the expression

$$K = \frac{1}{\pi(\text{AR})e} \tag{2-15}$$

where e is the *Oswald span efficiency factor*. An idealized wing of infinite span, flying under idealized conditions, would have a value of unity for e. Practical values of e range from 0.6 to 0.9 (in general, decreasing with a decrease in the aspect ratio) and are difficult to determine for a particular aircraft-wing combination. We shall normally use compromise values of 0.8 or 0.85 in our determination of K. Incidentally, the addition of properly designed wingtip tanks, end plates, or winglets (and the ground effect) will increase the span efficiency, even beyond the value of unity.

If a wing is symmetrical, C_{D0} is also the minimum-drag coefficient. If the wing is cambered, the drag polar will still be parabolic but the minimum-drag coefficient will be located at a lift coefficient other than zero, as shown in Fig. 2-5 for a wing with some positive camber. In such cases, the parabolic drag polar can be represented by the expression

$$C_D = C_{D\text{min}} + K(C_L - C_{L0})^2$$

with the minimum-drag coefficient sometimes called the parasite drag coefficient. In the interest of simplicity, we shall limit ourselves to the parabolic drag polar of Eq. 2-14, with the understanding that C_{D0} represents the minimum-drag coefficient of the entire aircraft, with the wing of a well-designed and reasonably clean aircraft

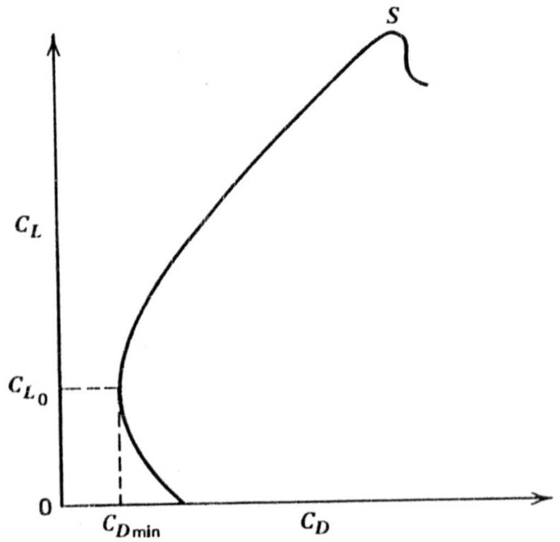

FIGURE 2-5
Parabolic drag polar for a wing with positive camber.

making the largest contribution. Typical subsonic values of C_{DO} are of the order of 0.011 for a clean fighter aircraft (at subsonic airspeeds), 0.016 for a jet transport, and 0.025 for a propeller-driven aircraft.

An important performance and design parameter of an aircraft is the *lift-to-drag ratio*, or *aerodynamic efficiency*. Using the symbol E, it is defined as

$$E = \frac{L}{D} = \frac{C_L}{C_D} \tag{2-16}$$

and for a given shape is a function of the angle of attack (and thus of the lift coefficient). However, there is a maximum value for each aircraft, E_m, that can be found by setting the derivative of E with respect to C_L equal to zero. For a parabolic drag polar, substituting Eq. 2-14 into Eq. 2-16 yields

$$E = \frac{C_L}{C_{DO} + KC_L^2} \tag{2-17}$$

so that setting dE/dC_L equal to zero and solving produces the condition that C_{DO} is equal to KC_L^2, which is also equal to C_{Di}. This is the flight condition at which the zero-lift and induced drags are equal and

$$C_{L,E_m} = \left(\frac{C_{DO}}{K}\right)^{1/2} \tag{2-18}$$

Substitution of Eq. 2-18 back into Eq. 2-17 leads to an expression for the maximum lift-to-drag ratio, namely,

$$E_m = \frac{1}{2(KC_{DO})^{1/2}} \tag{2-19}$$

It should be realized that the maximum lift-to-drag ratio is a design characteristic of an aircraft, because K and C_{DO} are design, not flight, parameters. Each aircraft has its own value that it cannot exceed, although it can fly at lower values. Some order of magnitude values of E_m for several classes of aircraft are:

Sailplanes	35
M 0.8 transports	18
Subsonic fighters	10
Supersonic aircraft	7
Helicopters	3

Since the maximum range of a flight vehicle, for a given fuel load, is directly proportional to the maximum lift-to-drag ratio, we can understand why there are no transcontinental helicopters as yet and why there are doubters as to the economic soundness of an SST.

The parabolic drag polar, with a constant C_{DO} and K, can be considered valid for airspeeds up to those approaching the *drag-rise Mach number* M_{DR}, at which point the drag starts to rise and the maximum lift-to-drag ratio begins to fall off. This increase in drag is due to the *wave drag* associated with the shock waves forming on the wing as the velocity of the air flowing over the wing becomes sonic and supersonic even though the aircraft itself is flying at a subsonic airspeed. The aircraft (free-stream) Mach number at which the first local shock forms is known as the *critical Mach number* M_{cr}; it is lower than the drag-rise Mach number, which can be considered to be the start of the transonic region.

Obviously, it is desirable to have M_{cr}, and thus M_{DR}, as large as possible. This can be done by reducing the thickness ratio ($t/\bar{c}$) of the wing, thereby reducing the increase in the velocity of the air as it flows over the wing. Since only the component of the free-stream airflow perpendicular to the leading edge of the wing affects the local velocity of the flow over the wing, sweeping the wing (either forward or backward) reduces the magnitude of this component (see Fig. 2-6). This reduction delays the onset of the local shock waves and increases the critical and drag-rise Mach numbers. There are disadvantages to reducing the thickness of a wing and sweeping it. A thin wing, especially one with a high aspect ratio, presents structural problems and, surprisingly enough, has a higher weight than a thicker wing. Wing sweep has such other disadvantages as a reduction in the slope of the lift curve and in $C_{L_{max}}$ as well as a susceptibility to tip stall.

The use of an airfoil shape that is called a *supercritical airfoil* does not change the critical Mach number for a given thickness ratio or sweep angle, but it will increase the drag-rise Mach number for that configuration. If the drag-rise Mach number is kept constant, a supercritical wing can be thicker or have less sweep than a wing with a conventional airfoil.

The parabolic drag polars used in this book will, in general, be considered to be valid for airspeeds up to the order of M 0.85, although we should realize that the drag-rise Mach number can be less than that if the wing is not designed for that speed range. When analyzing the performance of an existing high-subsonic aircraft, the manufacturer's stated best-range cruise Mach number is often very close to the drag-rise Mach number.

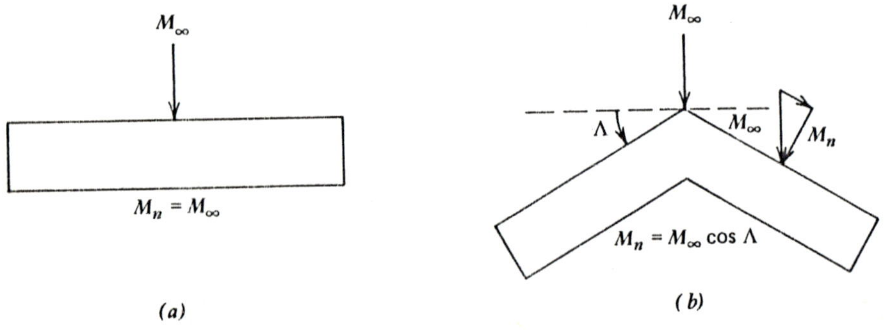

FIGURE 2-6

Effect of wing sweep on component of air velocity normal to wing: (*a*) unswept wing; (*b*) swept wing.

In the parabolic drag polar, the zero-lift drag coefficient is the minimum value of the drag coefficient and includes the drag contribution of not only the wing but also of the other components of the aircraft, such as the fuselage, engine nacelles and the empennage (the tail section), which are all designed so as to minimize their contribution to this minimum or zero-lift drag coefficient. With modern aircraft, the wing is the major lift-producing device and is considered, particularly in such analyses as we shall perform, to be the only contributor to the drag-due-to-lift.

In the *transonic region*, which is defined as starting at the drag-rise Mach number and ending at a Mach number of the order of M 1.1, C_{D0} rises sharply and then decreases to approach a reasonably constant value in the supersonic region. K, on the other hand, increases almost linearly in the transonic region, leveling off in the supersonic region. The maximum lift-to-drag ratio drops off sharply in the transonic region and levels off in the supersonic region. These effects are shown qualitatively in Fig. 2-7. As mentioned previously, the upper limit for a subsonic drag polar is in the vicinity of M_{DR}. There is a lower limit also, which is established by large lift coefficients that approach the maximum. A parabolic drag polar can also be used for analyses of the supersonic performance of certain wing configurations in certain flight regimes if the appropriate values for C_{D0} and K are used.

The shape and size of the wings of an aircraft often indicate the speed, range, and purpose for which the aircraft was designed. Large, straight, wide (where width refers to the chord length), and thick wings, often cambered, indicate a low-

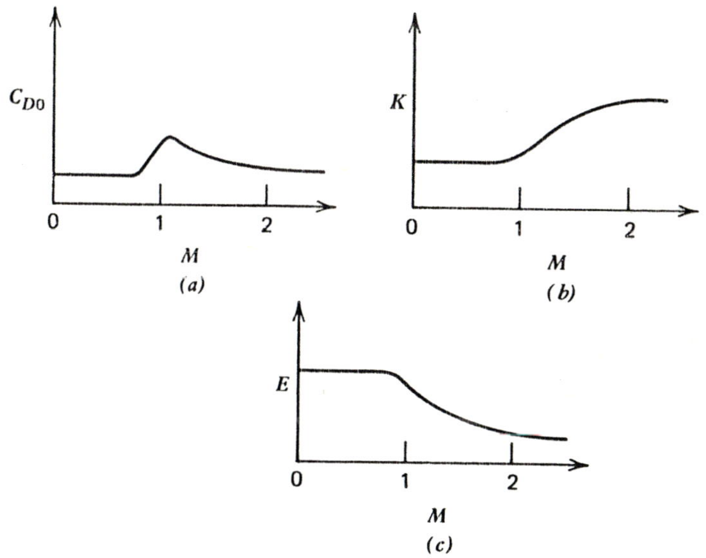

FIGURE 2-7

Qualitative variation of aerodynamic characteristics as a function of the Mach number: (*a*) zero-lift (minimum) drag coefficient; (*b*) induced drag parameter; (c) lift-to-drag ratio.

speed, short-range aircraft with emphasis on low stall speeds and short take-offs and landing runs. As the speed and range of the aircraft increase, the wings become smaller, narrower, and thinner, and the camber decreases. This trend continues as the cruise Mach number increases, and at a Mach number of the order of *M* 0.7 wing sweep begins to appear, reaching an angle of the order of 35 deg for an *M* 0.85 subsonic transport or bomber. As yet, no aircraft is designed to operate in the transonic region, and so the next increase in airspeed results in a supersonic aircraft. The wings become thinner, the leading edges much sharper, and the sweep more pronounced, and now the aspect ratio begins to decrease (the wings become stubbier). The combination of increased sweep and reduced aspect ratio often results in a delta wing. An aircraft designed to operate primarily in the high supersonic region may well have straight stubby wings with a very thin, symmetrical airfoil section and with sharp leading edges. There are several aircraft that employ variable sweep wings, so as to tailor the wing configuration to the airspeed at which it is operating.

2-4 THE PROPULSION SUBSYSTEM

In order to conduct performance analyses or perform preliminary design studies, we need functional relationships for the propulsive force (the *thrust* T) and for the fuel consumption rate for the various types of aircraft engines that might be used currently or in the near future. These are all classified as air-breathers in that they use the oxygen from the atmosphere to burn with a petroleum-product fuel, either gasoline or a form of kerosene, which is commonly referred to as jet fuel. Fortunately, for the purposes of this book, we need not concern ourselves with the inner workings or internal design of an aircraft engine but can instead treat the engine as a "black box," which in turn can be represented by a few operational and design parameters.

The four types of aircraft propulsion systems to be considered are the pure *turbojet*, the reciprocating engine and propeller combination (the *piston-prop*), the *turboprop*, and the *turbofan*. We shall assume that the engine has been properly selected and matched to both the aircraft and the operating regime for which the aircraft has been designed. Consequently, any values given in this book represent design-point values, that is, best values.

The *turbojet* engine produces thrust by expanding hot combustion gases through a nozzle. Referring to the schematic diagram of Fig. 2-8*a*, the air enters the diffuser section at the true airspeed of the aircraft and is decelerated and partially compressed. The air then passes through a mechanical compressor section where the pressure is increased to that desired for combustion, which occurs in the burners. The hot combustion products are partially expanded through the turbine section to provide power for driving the compressor(s) and for operating aircraft accessories, and then is fully expanded through the nozzle (the tail pipe) to produce thrust. This thrust is a function of the altitude and airspeed of the aircraft and of

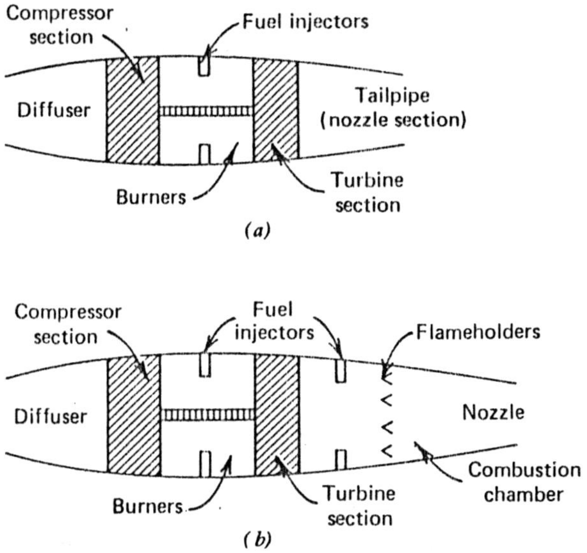

FIGURE 2-8

A turbojet schematic schematic: (*a*) without afterburner; (*b*) with afterburner.

the engine control setting and can be expressed in a generalized form as

$$T = T(h, V, \Pi) \tag{2-20}$$

where Π is the customary symbol for the control setting, which for the sake of consistency will be called the *throttle setting*. Since the throttle setting symbol never appears in any of our equations, we need not worry about any possible confusion with the symbol for pi.

For turbojet engines, the fuel consumption rate is described by the *thrust specific fuel consumption* tsfc, or simply c, which is defined as the fuel weight flow rate per hour per pound of thrust and has the dimensions of *lb/h/lb* (*N/h/N*), sometimes expressed as inverse hours with English units. The specific fuel consumption is an engine characteristic and has the functional relationships

$$c = \frac{dW_f/dt}{T} = c(h, V, \Pi) \tag{2-21}$$

From its definition and Eq. 2-21, we see that lowering the specific fuel consumption of an engine decreases the amount of fuel required to generate a pound (Newton) of thrust.

It is possible to simplify further the functional relationships of Eqs. 2-20 and 2-21 by approximations and assumptions. First of all, the thrust of a turbojet engine that is properly designed and matched to the aircraft in which it is installed can be considered, to a first approximation, to be independent of the airspeed, so that

$$T = T(h, \Pi) \tag{2-22}$$

The thrust of a turbojet for a given throttle setting is directly proportional to the mass flow rate of the air through the engine. Consequently, as the density of the atmosphere decreases with an increase in altitude, so does the available thrust. The thrust T_1 at any given altitude h_1 can be expressed in terms of its sea-level value by the relationship

$$\frac{T_1}{T_{SL}} = \left(\frac{\rho_1}{\rho_{SL}}\right)^x = \sigma_1^x \tag{2-23}$$

In the troposphere, the exponent x takes on values of the order of 0.7, whereas in the isothermal stratosphere, it can be set equal to unity, which yields the simple relationship that

$$\frac{T_1}{T_{SL}} = \frac{\rho_1}{\rho_{SL}} = \sigma_1 \tag{2-24}$$

We shall use the simpler relationship of Eq. 2-24 that says that the thrust of a turbojet is directly proportional to the atmospheric density, in both the troposphere and the atmosphere, even though the actual decrease in thrust with altitude will be somewhat less than that indicated.

With respect to the specific fuel consumption, it is less affected by the altitude than is the thrust. The minimum value of the specific fuel consumption occurs at the tropopause. Within the troposphere, it decreases as the 0.2 power of the density ratio and increases even more slowly within the stratosphere. We shall neglect these variations due to altitude in our analyses other than to remember that the lowest specific fuel consumption does occur at the tropopause. In addition, in consonance with our earlier assumption that the aircraft is operating in the region of its engine-airframe design point, we shall neglect any variations with changes in either the airspeed or the throttle setting. As a consequence, we shall assume that *the specific fuel consumption will be constant for all flight conditions.*

Turbojet engines come in all sizes these days, from 50 lb of thrust up to the order of 25,000 lb. The uninstalled thrust-to-engine weight ratio is constantly increasing and is of the order of 4 to 6 lb of thrust for every pound of uninstalled engine weight. Some turbojet engines contain an *afterburner* (see Fig. 2-8*b*) to take advantage of the fact that the gaseous mixture leaving the turbine contains an excess of unburned air in order to keep the temperature below the maximum allowable for safe turbine operation. Injecting and burning additional fuel behind the turbine will greatly increase the thrust with the only temperature limitation being that of the tailpipe wall (the wall can withstand higher temperatures than can the turbine blades). Although the thrust is approximately doubled, the increase in the thrust (the *thrust augmentation*) is accompanied by a large increase in the specific fuel consumption (of the order of two to three times as much). Consequently, an afterburner is installed only when the operational requirements demand one, such as for an interceptor or for a supersonic capability, and then is used sparingly.

Let us leave the turbojet engine and discuss the *piston-prop*. An internal combustion reciprocating engine burns air and gasoline (there are no diesel aviation engines as yet) and produces power rather than thrust. The power output of the engine is commonly measured in units of horsepower (hp), is essentially independent of the airspeed, and is a function of the altitude and throttle setting. The fuel consumption rate is proportional to the horsepower (HP), so that

$$\frac{dW_f}{dt}=\hat{c}(\mathrm{HP}) \tag{2-25}$$

where $\hat{c}$ is the *horsepower specific fuel consumption* (hpsfc), which is defined as the fuel flow rate per horsepower and has the units of lb/h/hp (N/h/kW).

The engine shaft power is converted into thrust power by the propeller. The *thrust power* P is equal to the product of the thrust and the airspeed of the aircraft, that is,

$$P=TV \tag{2-26}$$

where the thrust is in pounds (Newtons) and the airspeed is expressed in either ft/s (m/s) or mi/h (km/h), whichever is more convenient to use at the time. The engine horsepower and the propeller thrust power are related by the expression

$$P=TV=k\eta_p(\mathrm{HP}) \tag{2-27}$$

where η_p is the propeller efficiency and is of the order of 80 to 85 percent and k is a conversion factor with a value of 375 lb-mi/h/hp when V is in mph and a value of 550 ft-lb/s/hp when V is in fps. Notice that, for a given horsepower, the available thrust is inversely proportional to the airspeed, decreasing as the airspeed increases, whereas the thrust of a turbojet is constant and the thrust power increases as the airspeed increases. Since the horsepower specific fuel consumption has the same variations with altitude and airspeed as does the thrust specific fuel consumption, it will also be assumed to be constant in our analyses. Although the two specific fuel consumptions are "constant" for their respective power plants, they are related to each other through the following expression:

$$\hat{c}=\frac{k\eta_p c}{V} \tag{2-28}$$

This equation will be used to develop "equivalent" specific fuel consumptions for turboprops and turbofans in Chap. 8.

The altitude variation of the power produced depends upon whether or not the engine is supercharged. If not, we speak of an *aspirated engine*, and the variation of the output power with altitude is essentially that of the thrust of a turbojet engine (see Eqs. 2-23 and 2-24). The superchargers of today use a turbine driven by the exhaust gases to increase the density of the air entering the cylinders and are called *turbochargers*. With a constant throttle setting, the power will remain constant up to the *critical altitude*, which has a maximum value of the order of 20,000 ft. Above the critical altitude, the power of a *turbocharged engine* decreases with altitude in the same manner as an aspirated engine. The penalties of a turbocharger

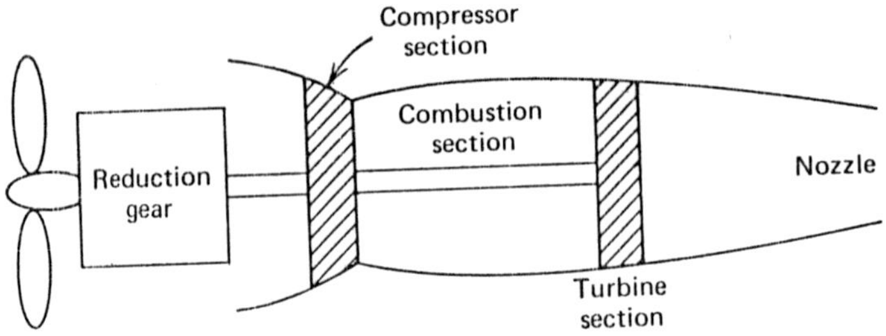

FIGURE 2-9

A turboprop schematic.

are a slight decrease in the useful power output of the engine, a slight increase in the engine weight, and a slight increase in cost.

Current piston-prop engines are relatively small, ranging in size from around 50 hp to the order of 600 hp, because they are the heaviest of all the engines. The uninstalled horsepower-to-engine weight ratio is of the order of 0.5 hp/lb of engine weight.

Turboprop engines and turbofan engines are basically turbojet engines in which the combustion gases are more fully expanded in the turbine section to develop more power than is needed to drive the compressor and accessories. This excess power is then used to drive either a propeller, in the case of a turboprop (see Fig. 2-9), or a multibladed ducted fan, in the case of a turbofan (see Fig. 2-10), to produce thrust power. Any energy remaining in the gaseous mixture leaving the drive turbines is then expanded in a nozzle to produce what is known as jet thrust. This jet thrust, obviously, is considerably less than that produced by a comparable turbojet.

In the *turboprop* engine, the residual jet thrust is converted into an equivalent horsepower at some design airspeed, and the engine is then described in the

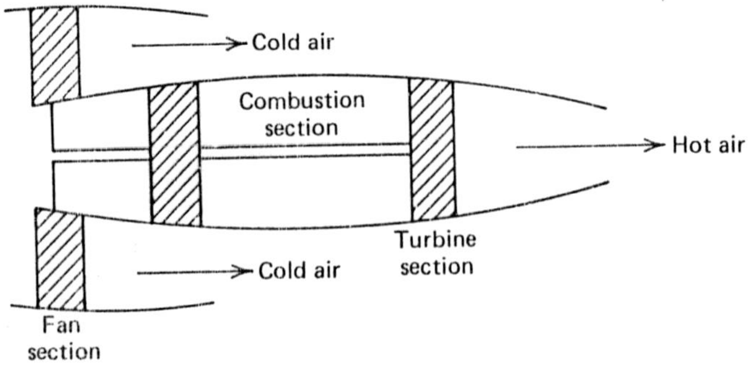

FIGURE 2-10

A front turbofan schematic.

terminology of the piston-prop, i.e., an equivalent shaft horsepower (eshp) and an equivalent shaft horsepower specific fuel consumption (eshp sfc) that are somewhat sensitive to variations in the airspeed of the aircraft. Since the jet thrust power of a turboprop engine is of the order of 15 to 20 percent of the total power, it is reasonable to treat the turboprop in the same manner as an aspirated (no turbocharger) piston-prop. The turboprop has a slightly higher specific fuel consumption than does the piston-prop but the weight of the engine is considerably less, the heaviest item being the propeller gearbox. Horsepower-to-engine-weight ratios are of the order of 2 hp/lb of engine weight, and the largest engine in current operational use is of the order of 6,000 eshp.

Although a *turbofan* engine is described as though it were a turbojet, its characteristics are determined by the *bypass ratio*, which is the ratio of the mass of the "cold air" passing through the fan to the mass of the "hot air" passing through the burners and turbine section. If the bypass ratio is zero, then the turbofan is obviously a pure turbojet. As the bypass ratio is increased, the percentage of jet thrust decreases and the turbofan more and more resembles a turboprop. For example, with a bypass ratio of 10, the theoretical jet thrust is of the order of 17 percent. Current maximum bypass ratios are of the order of 5 to 6 and thrust specific fuel consumptions, for a specified airspeed, are of the order of 0.7 lb/h/lb. Although the frontal area of a turbofan increases rapidly with an increase in the bypass ratio, the length decreases; consequently, there is less of an increase in the engine drag and weight than might be expected. Thrust-to-engine-weight ratios are of the order of 5 to 6 lb of thrust per pound of engine weight. The maximum thrust of an individual engine is increasing and currently is of the order of 60,000 lb.

The specific fuel consumption is an extremely important performance parameter. Some typical values, all expressed as an equivalent thrust specific fuel consumption in lb/h/lb, are:

Rocket engines	10
Ramjets	3
Turbojets (afterburner)	2.5
Turbojets	1
High bypass turbofans	0.6–0.8
Turboprops	0.5–0.6
Piston-props	0.4–0.5

It is interesting to note that, for various reasons, this list also serves to indicate the relative airspeed regime of the flight vehicles in which these various types of engines are used. For example, piston-prop engines are used in aircraft with airspeeds of the order of 250 mph or less; turboprop engines at the higher airspeeds up to approximately M 0.7; turbofan engines for airspeeds up to M 0.85; and turbojet engines and very low bypass ratio engines in supersonic aircraft. The ramjet is suitable for flight vehicles flying at M 3.0 and higher, and rocket engines are used in ballistic missiles and space boosters. Furthermore, the piston-prop is the cheapest and the heaviest of the engines with the weight decreasing and the cost increasing as we move up the list.

2-5 WEIGHT FRACTIONS

Our intuition is correct when it tells us that weight is an important consideration in the performance and design of an aircraft. In actuality, it may well be the most important consideration, and design experience has shown that the lowest-weight design is also the lowest-cost and most-efficient design. Every extra pound of weight is accompanied by an increase in the wing area, thrust, fuel, etc., all leading to a further increase in the aircraft weight and adversely affecting the performance and costs (both initial and operational) of the aircraft. *Weight fractions* are very useful in performance and design analyses and are obtained by expressing the gross (total) weight of an aircraft as the sum of the weights of the individual components and subsystems and then dividing through by the gross weight.

Although the "bookkeeping" process can be, and in final design is, very detailed, we shall consider the total weight of an aircraft to be made up of the structural weight W_s, the engine weight W_e, the payload weight W_{PL}, and the fuel weight W_f. The structural weight will include not only the weight of the structure itself but also the weight of everything not included in the other categories. It will include all the equipment, for example, and even the weight of the flight and cabin crew. Usually, aircraft manufacturers will lump the structural and engine weights together and call the sum the *empty weight* or the *operational empty weight*, OEW, and combine the fuel and payload weights into the *useful load*.

With our subdivisions and knowing that the weight of the whole aircraft is the sum of the individual weights, we can write the *gross weight* W_0 of the aircraft as

$$W_0 = W_s + W_e + W_{PL} + W_f \tag{2-29}$$

If we divide through by the gross weight, we obtain

$$1 = \frac{W_s}{W_0} + \frac{W_e}{W_0} + \frac{W_{PL}}{W_0} + \frac{W_f}{W_0} \tag{2-30}$$

and we establish these *weight fractions*: the *structural weight fraction*, the *engine weight fraction*, the *payload weight fraction*, and the *fuel weight fraction*. The sum of these individual weight fractions *must always* be unity. As an aside, the engine weight fraction for turbojets and turbofans can be found from the relationship that

$$\frac{W_e}{W_0} = \frac{T_{max}/W_0}{T_{max}/W_e} \tag{2-31}$$

where the numerator is the ratio of the maximum engine thrust to the gross weight of the aircraft and the denominator is the ratio of the maximum engine thrust to the weight of the engine. A similar relationship, using horsepower rather than thrust, is used with the propeller-driven aircraft.

The analysis and design of aircraft structures and the determination of component and subsystem weights are obviously beyond the scope of this book. We shall work instead with order of magnitude values for the structural weight fraction. As the gross weight of an aircraft increases, its structural weight fraction decreases. There are minimum volume requirements for cabin and cargo space, and the weights of fixed equipment (such as radios and navigation equipment) have a larger impact on the structural weight fraction of the lower gross weight aircraft. Large subsonic transports, such as the C-5 and 747, have structural weight fractions of the order of 0.4 to 0.45 whereas smaller aircraft, such as general aviation and fighter aircraft, can have structural weight fractions in excess of 0.5. Remember that the structural weight fraction does *not* include the weight of the engines.

We should never forget the importance and significance of the weight fractions. For example, a payload weight fraction of 0.05 means that 20 lb of aircraft weight is required for each pound of payload. Therefore, a payload of 2,000 lb calls for a 40,000-lb aircraft. Look at the problems at the end of this chapter for some examples of the impact of the engine and structural weight fractions upon the gross weight, the range, and the payload capability of an aircraft. The weight fractions can be used to dramatically demonstrate the range and payload limitations associated with the increased engine weight required for a true vertical take-off and landing (VTOL) aircraft as compared with a conventional take-off and landing aircraft (CTOL).

A word of caution is in order with respect to the structural weight fraction. In attempts to reduce weight, the structural designer must not go so far as to introduce undesirable aeroelastic effects or to produce structural members that cannot handle the aerodynamic loads that might be encountered.

2-6 MISCELLANY

The symbol V without a subscript will always be used to denote the *true airspeed*, which is the actual speed of the aircraft through the atmosphere. The basic instrument for measuring airspeed is the pitot tube, which measures the stagnation pressure (P_0) of the air, in combination with a static source to measure the local (ambient) pressure P. If the air is assumed to be incompressible, then Bernoulli's equation can be used to determine the true airspeed from

$$V = \left[\frac{2(P_0 - P)}{\rho}\right]^{1/2} \tag{2-32}$$

where $P_0 - P$ is the pressure difference measured by the pitot tube–static source combination. Since the ambient atmospheric density is difficult to measure, the standard practice is to use the standard day sea-level value and call the result the *calibrated airspeed* (CAS), where

$$\mathrm{CAS} = \left[\frac{2(P_0 - P)}{\rho_{SL}}\right]^{1/2} \tag{2-33}$$

Equations 2-32 and 2-33 provide the relationships between the true and calibrated airspeeds, namely,

$$V = \frac{\mathrm{CAS}}{\sigma^{1/2}} \tag{2-34a}$$

$$\mathrm{CAS} = V\sigma^{1/2} \tag{2-34b}$$

The *indicated airspeed* (IAS) is what the pilot reads on his airspeed indicator in the cockpit and is the calibrated airspeed with any errors arising from the construction of the instrument itself and from the installation and location of the pitot tube and the static source. We shall assume the IAS and the CAS to be identical.

At sea level on a standard day, the values of the calibrated airspeed and of the true airspeed are equal, but this is the only time and place where they are. If the CAS is held constant at 250 mph, the true airspeed will increase with altitude as shown in the following table:

h (1000 ft)	σ (ρ/ρ_{SL})	CAS (mph)	V (mph)
SL	1.000	250	250
5	0.862	250	269
10	0.738	250	291
15	0.629	250	315
20	0.533	250	342
30	0.374	250	409
40	0.185	250	476
50	0.152	250	541

As the true airspeed increases beyond a Mach number of the order of 0.3, the assumption of the incompressibility of the air begins to lose its validity and the effect of the compressibility upon the measurement of the true airspeed should be taken into account. The calibrated airspeed corrected for compressibility is often called the *equivalent airspeed.* At high airspeeds, the airspeed indicator is replaced by or complemented with the Machmeter, which measures the *Mach number M*, where

$$M = \frac{V}{a} \tag{2-35}$$

where V is the true airspeed and a, the local speed of sound in air (the acoustic velocity), is a function of the local air temperature and is given by

$$a = (kR\Theta)^{1/2} \tag{2-36}$$

In Eq. 2-36, k is the ratio of specific heats (1.4 for air) and R is the gas constant for air. For a constant true airspeed, the Mach number increases with altitude until the tropopause is reached; it then remains constant in the isothermal stratosphere. For a constant true airspeed of 250 mph, its relationships with the CAS and the Mach number are as follows:

h (1,000 ft)	V (mph)	M	CAS (mph)
SL	250	0.238	250
10	250	0.336	215
20	250	0.353	182
30	250	0.368	153
40	250	0.378	124
50	250	0.378	97

Notice that there is considerably less change in the Mach number with altitude than in the CAS. If the Mach number is held constant, then the true airspeed will decrease with altitude until the tropopause is reached.

The *stall speed* V_s of an aircraft is the true airspeed at the stalling point of the wing, which occurs at the maximum lift coefficient ($C_{L_{max}}$) of the wing. In level flight, the lift must be equal to the weight so that

$$L = W = \tfrac{1}{2}\rho_{SL}\sigma V^2 S C_L \tag{2-37}$$

Solving for the level-flight true airspeed yields

$$V = \left[\frac{2(W/S)}{\rho_{SL}\sigma C_L}\right]^{1/2} \tag{2-38}$$

where W/S is called the *wing loading* of the aircraft and has the units of lb/ft^2 (N/m^2). To determine the stall speed, set the lift coefficient equal to its maximum value to obtain

$$V_s = \left[\frac{2(W/S)}{\rho_{SL}\sigma C_{L_{max}}}\right]^{1/2} \tag{2-39}$$

The stall speed increases with an increase in the wing loading and decreases with an increase in the maximum lift coefficient. For a given wing loading and lift coefficient, the stall speed increases with altitude such that

$$V_s = \frac{V_{s,SL}}{\sigma^{1/2}} \tag{2-40}$$

If the stall speed is expressed in terms of the calibrated airspeed rather than the true airspeed, Eq. 2-34 can be used to show that the calibrated stall speed is independent of the altitude, which is convenient for a pilot who only has to remember the sea-level stall speeds.

Since the value of the maximum lift coefficient increases with the use of flaps (and other high-lift devices) and decreases with the landing gear down, it is necessary to define the flap and gear configuration as well as the weight (the wing loading) in specifying the stall speed. It is customary to describe the level-flight stall speed in terms of the calibrated airspeed, a given weight (wing loading), various flap settings, and whether the gear is up or down.

Equation 2-38 shows that, for a given wing loading and altitude, the true airspeed is inversely proportional to the square root of the lift coefficient at which the aircraft is flying. In other words, as the lift coefficient (and thus the angle of attack) is increased, the true airspeed required to maintain level flight decreases. Figure 2-11 shows the variation of the flight lift-to-drag ratio E and the sea-level true airspeed V for a particular aircraft with a wing loading of 100 lb/ft^2 and the parabolic drag polar, $C_D = 0.015 + 0.06\,C_L^2$, as a function of the lift coefficient. In addition to the general shape of the curves, the thing to notice is that, although the value of the lift coefficient that maximizes the lift-to-drag ratio is obvious, the value that maximizes the product of the airspeed and the lift-to-drag ratio is not. The significance of this observation will become apparent in subsequent sections dealing with the best range of aircraft with turbojet and turbofan engines.

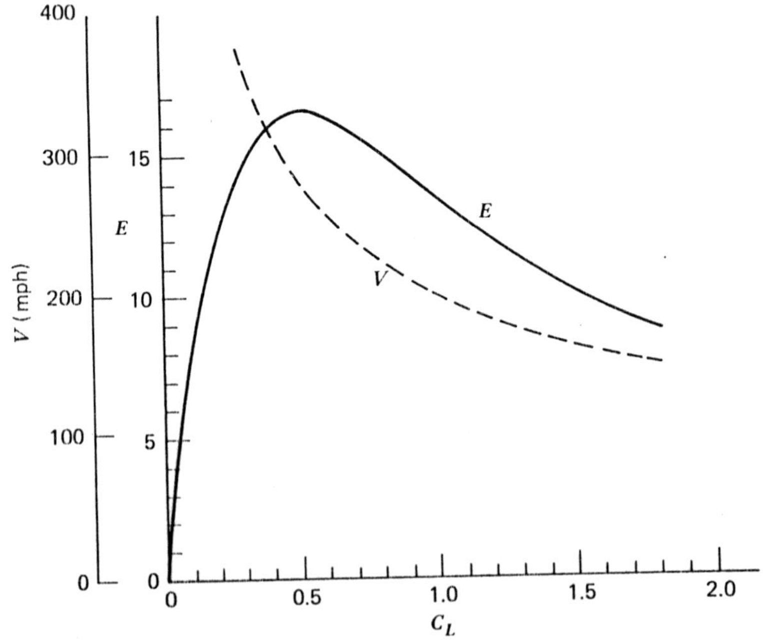

FIGURE 2-11

A typical variation of the lift-to-drag ratio and true airspeed as a function of the lift coefficient.

PROBLEMS

2-1. Find the altitude in feet for each of the density ratios listed below, using Table A-1 and Eq. 2-2 with $\beta = 30{,}500$ ft and then with $\beta = 23{,}800$ ft:

a. 0.310 b. 0.862

c. 0.490 d. 0.120

e. 0.225 f. 0.060

g. 0.533 h. 0.374

2-2. Find the difference in altitude in feet for each of the density ratio pairs listed below, using Table A-1 and Eq. 2-2 with the two values of β, that is, 30,500 and 23,800 ft:

a. 0.862 and 0.152 b. 0.862 and 0.310

c. 0.629 and 0.194 d. 0.310 and 0.152

e. 0.194 and 0.120 f. 0.533 and 0.297

g. 0.374 and 0.246 h. 0.246 and 0.152

2-3. For each of the altitudes listed below, find the density ratio, using Table A-1 and then Eq. 2-1 with $\beta = 23{,}800$ ft:

a. 10,000 ft b. 82,000 ft

c. 15,000 ft d. 30,000 ft

e. 20,000 ft f. 45,000 ft

g. 70,000 ft h. 36,000 ft

2-4. Do Prob. 2-3 using Table A-1 and a value of 30,500 ft for β.

2-5. Using the equation of state for air (Eq. 2-3) along with relevant values from Table A-1, find the pressure ratio for each of the altitudes listed below and compare it with the value listed in Table A-1:

a. 10,000 ft b. 15,000 ft

c. 30,000 ft d. 82,000 ft

e. 25,000 ft f. 50,000 ft

g. 35,000 ft h. 5,000 ft

2-6. Redo Prob. 2-5 to find the temperature ratio and the temperature itself (in degrees Rankine and in degrees Fahrenheit).

2-7. Redo Prob. 2-5 to find the density ratio at each listed altitude.

2-8. An aircraft has the drag polar, $C_D = 0.015 + 0.05\, C_L^2$.

a. Find the maximum lift-to-drag ratio and the associated lift coefficient.

b. Find the AR required to maintain a constant value of 0.05 for K for each of the following values of the Oswald span efficiency: 0.75, 0.8, 0.85, and 0.95.

c. For a lift coefficient of 0.5, find the value of the drag coefficient and of the lift-to-drag ratio.

d. Sketch the lift-to-drag ratio as a function of the lift coefficient.

2-9. Do Prob. 2-8 for the parabolic drag polar, $C_D = 0.025 + 0.04\, C_L^2$.

2-10. An aircraft with a symmetrical airfoil wing has the following aerodynamic data:

α(deg)	0	3	5	10	15	20	25
C_L	0	0.3	0.5	1.0	1.5	1.8	1.7
C_D	0.02	0.024	0.032	0.07	0.13	0.2	0.32

a. Plot the lift and drag coefficients as a function of the angle of attack.

b. Plot the actual drag polar, that is, C_D versus C_L.

c. If the drag polar can be approximated by $C_D = 0.02 + 0.05\, C_L^2$, superimpose a plot of this parabolic drag polar on that of (b), above.

2-11. Find the dynamic pressure q in lb/ft^2 from sea level to 40,000 ft at 10,000-ft intervals, for each of the following true airspeeds:

a. 150 mph b. 200 mph

c. 300 mph d. 350 mph

e. 400 mph f. 500 mph

2-12. For a wing area of 1,400 ft^2, a lift coefficient of 0.5, and a lift-to-drag ratio of 12, find the lift and drag at sea level, 20,000 ft, and 30,000 ft for each of the airspeeds of Prob. 2-11.

2-13. a. For the aircraft of Prob. 2-12, find the airspeed in mph required to achieve a level-flight lift of 140,000 lb (the weight of the aircraft) at sea level, 20,000 ft, and 30,000 ft.

b. Find the corresponding drag and lift-to-drag ratio.

2-14. Do Prob. 2-13 for a level-flight lift of 70,000 lb.

2-15. An aircraft has a wing with an aspect ratio of 4 and an angle of attack at stall of 20 deg. Find the slope of the lift curve ($a_w = dC_L/d\alpha$) per radian and per degree for each of the following wing sweep angles:

a. 0 deg b. 10 deg

c. 20 deg d. 30 deg

e. 35 deg f. 40 deg

2-16. Do Prob. 2-15 for an aspect ratio of 8.

2-17. Do Prob. 2-15 for an aspect ratio of 12.

2-18. A supersonic transport (SST) has an aspect ratio of 1.8 and a sweep angle of 40 deg.

a. Find the slope of the lift curve per degree.

b. If the wing airfoil section is symmetrical and the maximum lift coefficient for landing is 1.2, find the landing angle of attack.

c. Does the answer of (c) above indicate why the nose of the Concorde SST is drooped for landing?

2-19. A turbojet has a maximum sea-level thrust of 50,000 lb. Using the approximation of Eq. 2-24, find the maximum thrust at:

a. 10,000 ft
b. 20,000 ft
c. 30,000 ft
d. 36,000 ft
e. 40,000 ft
f. 50,000 ft

2-20. Do Prob. 2-19 using Eq. 2-23 with $x=0.7$ in the troposphere and $x=1.0$ in the stratosphere.

2-21. A turbojet has a tsfc of 0.9 lb/h/lb. Find the fuel consumption rate in both lb/h and gal/h for the following thrust levels:

a. 10,000 lb
b. 20,000 lb
c. 30,000 lb
d. 40,000 lb
e. 50,000 lb
f. 60,000 lb

2-22. With the assumption that the thrust of a turbojet is independent of the airspeed, plot the thrust power in horsepower (hp) as a function of the true airspeed in mph from zero to 500 mph (at 100-mph intervals) for the following thrust levels:

a. 10,000 lb
b. 20,000 lb
c. 30,000 lb
d. 50,000 lb

2-23. An aspirated piston-prop engine has a maximum sea-level horsepower of 500 hp. Find the maximum available horsepower, using the approximation, at the following altitudes:

a. 10,000 ft
b. 20,000 ft
c. 30,000 ft
d. 40,000 ft

2-24. If the engine of Prob. 2-23 is turbocharged with a critical altitude of 15,000 ft, do Prob. 2-23 to include the 15,000-ft altitude.

2-25. Assuming that the horsepower of a piston-prop engine is independent of the airspeed, sketch the thrust as a function of the true airspeed for 1,000 hp and a propeller efficiency of 85 percent.

2-26. Do Prob. 2-25 for 500 hp and a propeller efficiency of 80 percent.

2-27. A 350-hp piston-prop engine has a hpsfc of 0.4 lb/h/hp and a propeller efficiency of 80 percent. Find the fuel consumption rate in lb/h and in gal/h.

2-28. Do Prob. 2-27 for 1,000 hp, a hpsfc of 0.45, and a propeller efficiency of 85 percent.

2-29. An aircraft has the following weight allocations:

Structure	8,000 lb
Engines	1,000 lb
Fuel	6,000 lb
Payload	5,000 lb

a. Find the gross weight, the operational empty weight, and the useful load.

b. Find the weight fractions: structural, engine, fuel, payload, and operational empty weight.

2-30. A CTOL (conventional take-off and landing) turbojet aircraft is designed for a 20,000-lb payload. The structural weight fraction is 0.44, and the design fuel weight fraction is 0.3. If the thrust-to-weight ratio of the aircraft is 0.25 and the thrust-to-engine weight is 5:

a. Find the payload weight fraction and the gross weight of the aircraft.

b. Find the structural, engine, and fuel weights.

2-31. In order to transform the aircraft of Prob. 2-30 from CTOL to VTOL (vertical take-off and landing), the thrust-to-weight ratio of the aircraft is increased to 1.1, the thrust-to-engine weight ratio remaining at 5.

a. Find the payload weight fraction and the new gross weight.

b. If the gross weight is held constant at the value obtained in Prob. 2-30 for the CTOL aircraft, as are the structural and fuel weight fractions, find the payload weight and compare it with the design value.

c. Do (b), letting only the fuel weight fraction change. Find the value of the new fuel weight fraction and compare it with the design value.

2-32. A piston-prop aircraft has a hp-to-aircraft weight ratio of 0.1 and a hp-to-engine weight ratio of 0.5. The structural weight fraction is 0.45, and the design fuel weight fraction is 0.1.

a. If the payload is 5,000 lb, find the payload weight fraction and the gross weight of the aircraft.

b. If the structural weight fraction can be reduced to 0.40 with the use of composites, find the new payload weight fraction and gross weight.

c. For (b), increase the aircraft HP/W ratio to 0.2 (in order to increase the maximum airspeed), keeping the payload weight and other fractions constant. Find the new payload weight fraction and gross weight.

2-33. For each of the following combinations of calibrated airspeed and altitude, find the true airspeed in mph and the Mach number:

a. 250 mph and sea level
b. 250 mph and 20,000 ft
c. 400 mph and 15,000 ft
d. 250 mph and 40,000 ft
e. 300 mph and 30,000 ft
f. 200 mph and 50,000 ft

2-34. a. For a Mach number of 0.7, find the true airspeed and the calibrated airspeed (both in mph) at sea level, 20,000 ft, and 40,000 ft.

b. Do (a) above for a Mach number of 0.55.

c. Do (a) above for a Mach number of 0.84.

2-35. An aircraft that weighs 100,000 lb has a wing area of 1,000 ft^2. Find the lift coefficient required for a level-flight true airspeed of 450 mph for the following altitudes:

a. Sea level
b. 10,000 ft
c. 20,000 ft
d. 30,000 ft
e. 36,000 ft
f. 45,000 ft

2-36. Do Prob. 2-35 for a true airspeed of 250 mph.

2-37. For the aircraft and altitudes of Prob. 2-35, if the maximum lift coefficient is 2.0, find the respective stall speeds, both the true airspeed (TAS) and the calibrated airspeed (CAS).

2-38. For a maximum lift coefficient of 1.7, sketch the level-flight stall speed (TAS) at sea level as a function of the wing loading (W/S), ranging in values from 10 to 150 lb/ft^2.

Chapter Three

Level Flight in the Vertical Plane: Turbojets

3-1 GOVERNING EQUATIONS

Our performance analyses will be limited to rigid, symmetrical aircraft flying in a quasi-steady-state condition over a flat, nonrotating earth.

Quasi-steady-state flight implies that any accelerations or rates of change of any of the key variables are sufficiently small in magnitude or duration so that they may be neglected. Furthermore, since we are not concerned with the attitude of the aircraft, we need not consider the rotational modes of the aircraft. Consequently, the Newtonian equations of motion of interest are reduced to the single static force vector equation, namely,

$$\sum \mathbf{F}_{\text{ext}} = 0 \tag{3-1}$$

Equation 3-1 simply means that the sum of all the external forces, with due regard for their direction, must be equal to zero. Substituting the external forces on an aircraft, Eq. 3-1 becomes

$$\mathbf{T} + \mathbf{A} + \mathbf{W} = 0 \tag{3-2}$$

where $\mathbf{T}$ is the thrust vector, $\mathbf{A}$ is the aerodynamic force vector, and $\mathbf{W}$ is the weight vector, which is equal to $m\mathbf{g}$ (the product of the aircraft mass and the gravity vector).

The vertical plane is defined as the plane containing the X and Z axes of the ground axis system. We shall assume that the aircraft will fly in a straight line along the X axis with its wings level and with no side forces. The "no side forces" assumption means that the sideslip angle is zero (that the aircraft is in what is known as *coordinated flight* with its velocity vector in the plane of symmetry) and that the plane of symmetry is in the vertical plane. We shall also assume that the thrust vector always coincides with the velocity vector (is along the wind x axis), which means that the engines are movable (which is not the case with conventional aircraft) or that the angle between the thrust and velocity vectors is sufficiently small to be neglected (a valid assumption for reasonable angles of attack). With no side forces, the aerodynamic force vector $\mathbf{A}$ can be resolved into two components, both in the plane of symmetry. They are the drag vector $\mathbf{D}$, which lies along the negative wind x axis, and the lift vector $\mathbf{L}$, which lies along the negative wind z axis.

Resolving the external forces along the x and z wind axes (see Fig. 3-1) and applying the conditions of Eq. 3-2 yields the two scalar equations

$$T - D - W \sin \gamma = 0 \tag{3-3}$$

and

$$L - W \cos \gamma = 0 \tag{3-4}$$

In these two equations, γ is the *flight-path angle* and is the angle between the airspeed velocity vector (the direction of travel) and the local horizon.

The motion of the aircraft with respect to the earth is described by the two kinematic equations

$$\frac{dV}{dt} = V_g \cos \gamma_g \tag{3-5}$$

and

$$\frac{dh}{dt} = V_g \sin \gamma_g \tag{3-6}$$

where h is the altitude above the mean sea level (MSL), V_g is the *ground speed* (the speed of the aircraft with respect to the surface of the earth), and γ_g is the angle

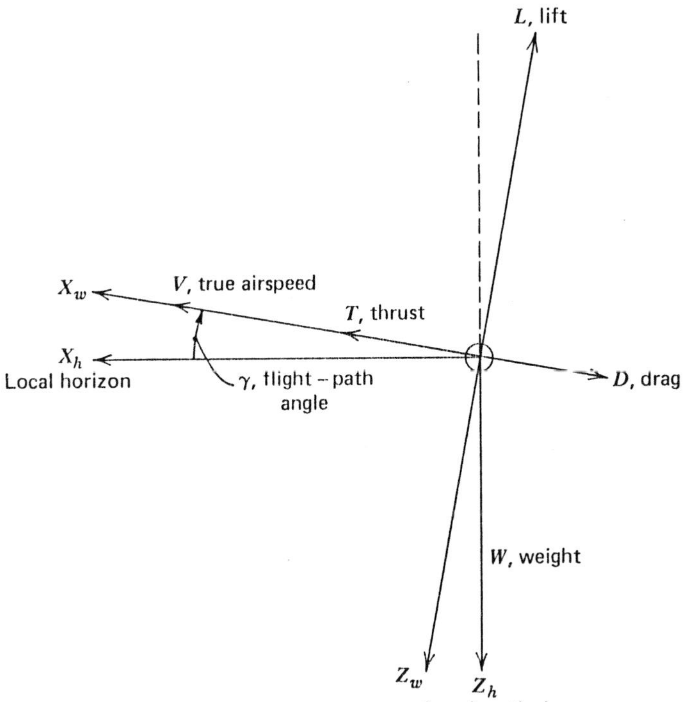

FIGURE 3-1

Steady-state forces in the vertical plane.

between the ground speed velocity vector and the horizon. The ground speed is the sum of the true airspeed plus or minus the x component of the wind velocity V_w, so that

$$V_g = V \pm V_w \tag{3-7}$$

where the plus sign is associated with a tailwind and the minus sign with a headwind. We shall assume a no-wind condition (a rare occasion) and replace V_g by V (the airspeed) and γ_g by γ in Eqs. 3-5 and 3-6. The effects of wind upon flight conditions and performance will be discussed in Chap. 10.

Since the weight of the aircraft is decreasing with time as fuel is being used, an additional relationship is needed. From the definition of the thrust specific fuel consumption and Eq. 2-21, we obtain the expression that

$$\frac{-dW}{dt} = cT \tag{3-8}$$

where W is the instantaneous gross weight of the aircraft, c is the thrust specific fuel consumption in lb/h/lb (N/h/N), and T is the actual thrust in pounds (Newtons). This weight balance equation accounts only for the fuel consumption. There may be discrete weight changes, such as the dropping of bombs and cargo, or other continuous weight changes if, for example, chemicals are sprayed.

The five equations that we shall use in examining the flight of turbojet aircraft are repeated here and renumbered for the sake of convenience.

$$T - D - W \sin \gamma = 0 \tag{3-9}$$

$$L - W \cos \gamma = 0 \tag{3-10}$$

$$\frac{dX}{dt} = V \cos \gamma \tag{3-11}$$

$$\frac{dh}{dt} = V \sin \gamma \tag{3-12}$$

$$\frac{-dW}{dt} = cT \tag{3-13}$$

Keep in mind the no-wind assumption and the fact that the thrust is a function of both the altitude and the throttle setting and that the drag is a function of the altitude, the airspeed, and the lift. Since there are more variables than equations, it is necessary to treat some of the variables as parameters, as will be seen.

3-2 LEVEL FLIGHT

For level flight in the vertical plane, the flight-path angle is zero and the altitude remains constant. Furthermore, the x and z axes of the wind and local horizon axes will be coincident. The governing flight equations of the preceding section reduce, for level flight, to

$$T = D \tag{3-14}$$

$$L = W \tag{3-15}$$

$$\frac{dV}{dt} = V \tag{3-16}$$

$$\frac{dh}{dt} = 0 \tag{3-17}$$

$$\frac{-dW}{dt} = cT \tag{3-18}$$

where

$$T = T(h, \Pi) \tag{3-19}$$

and, since $L = W$,

$$D = D(h, V, L) = D(h, V, W) \tag{3-20}$$

This series of simple equations shows that, in level flight, sufficient lift must be generated to balance the weight of the aircraft; that is, keep the aircraft in the air. This lift is developed by moving the wing through the air; this movement is resisted by the drag, which must in turn be balanced by the thrust of the engine(s).

With the introduction of the parabolic drag polar, the drag can be expressed as

$$D = qSC_D = qSC_{DO} + qSKC_L^2 \tag{3-21}$$

where q is the dynamic pressure, as defined by

$$q = \tfrac{1}{2}\rho V^2 = \tfrac{1}{2}\rho_{SL}\sigma V^2 \tag{3-22}$$

Whether q or its equivalent expression is used is a matter of convenience.

The lift coefficient in Eq. 3-21 can be expressed in terms of the wing loading by combining the lift equation with Eq. 3-15 to obtain

$$C_L = \frac{W/S}{q} = \frac{2(W/S)}{\rho_{SL}\sigma V^2} \tag{3-23}$$

Substituting Eq. 3-23 into Eq. 3-21 yields two drag expressions that will be used quite often, one expression is for the drag as a function of the weight and the other is for the drag-to-weight ratio (which is also the level-flight thrust-to-weight ratio) in terms of the wing loading. These expressions are:

$$D = qSC_{DO} + \frac{KW^2}{qS} = \tfrac{1}{2}\rho_{SL}\sigma V^2 SC_{DO} + \frac{2KW^2}{\rho_{SL}\sigma V^2} \tag{3-24}$$

$$\frac{D}{W} = \frac{qC_{DO}}{W/S} + \frac{K(W/S)}{q} = \frac{\rho_{SL}\sigma V^2 C_{DO}}{2(W/S)} + \frac{2K(W/S)}{\rho_{SL}\sigma V^2} \tag{3-25}$$

Since, from Eq. 3-14, the thrust produced by the engines must be equal to the drag, either of these equations can be written as a quadratic equation in q, namely,

$$q^2 - \frac{T/S}{2C_{DO}} q + \frac{K(W/S)^2}{C_{DO}} = 0 \tag{3-26}$$

which can easily be solved to obtain an expression for the level-flight dynamic pressure,

$$q = \frac{T/S}{2C_{DO}} \left\{ 1 \pm \left[1 - \frac{4KC_{DO}}{(T/W)^2} \right]^{1/2} \right\} \tag{3-27}$$

Then, from the definition of the dynamic pressure, the level-flight airspeed is found to be

$$V = \left(\frac{T/S}{\rho_{SL}\sigma C_{DO}} \left\{ 1 \pm \left[1 - \frac{4KC_{DO}}{(T/W)^2} \right]^{1/2} \right\} \right)^{1/2} \tag{3-28}$$

The presence of the $\pm$ sign in Eq. 3-28 indicates that mathematically there are two possible values for the level-flight airspeed for a given aircraft with a specified throttle setting and altitude. There is a high-speed solution V_1, associated with the plus sign, and a low-speed solution V_2, associated with the minus sign. The physical significance of these two possible solutions will be discussed after an examination of the radicand in Eq. 3-28.

The radicand $1-4KC_{DO}(W/T)^2$ can be negative, positive, or equal to zero. If the radicand is negative, then

$$1 - 4KC_{DO}\left(\frac{W}{T}\right)^2 < 0 \tag{3-29}$$

and there are no real solutions. Consequently, steady-state level flight is *not* possible when

$$\frac{T}{W} < 2(KC_{DO})^{1/2} \tag{3-30}$$

Since, for a parabolic drag polar, $E_m = 1/[2(KC_{DO})^{1/2}]$ (see Eq. 2-19), there can be no level flight if

$$\frac{T}{W} < \frac{1}{E_m} \tag{3-31}$$

If the radicand is positive, there will be two level-flight solutions with the condition that

$$\frac{T}{W} > \frac{1}{E_m} \tag{3-32}$$

The third condition arises when the radicand is identically equal to zero, so that

$$\frac{T}{W} = \frac{1}{E_m} \tag{3-33}$$

With the substitution of Eq. 3-33, Eq. 3-28 yields only one level-flight solution, which will subsequently be shown to be the absolute ceiling condition. This examination of the radicand in Eq. 3-28 can be summarized by stating that a *necessary condition for steady-state level flight is that the thrust-to-weight ratio be greater than or equal to the reciprocal of the maximum lift-to-drag ratio*, i.e.,

$$\frac{T}{W} \geqslant \frac{1}{E_m} \tag{3-34}$$

where T is the actual, instantaneous thrust, which is a function of the throttle setting and altitude, and where W is the actual, instantaneous gross weight of the aircraft. Remember that E_m is a design characteristic of the aircraft.

Let us now look at the physical significance of the level-flight solutions and conditions by writing Eq. 3-14 in functional form as

$$T(h, V, \Pi) - D(h, V, W) = 0 \tag{3-35}$$

There are four variables but only one equation. Let us specify our aircraft and select three of the variables (h, W, and Π) as parameters, so that Eq. 3-35 becomes

$$T(V) - D(V) = 0 \tag{3-36}$$

If we sketch, as a function of the airspeed, a typical available thrust-to-weight ratio (T_1/W) for a given throttle setting, weight, and altitude, as in Fig. 3-2, we see that the thrust is, to a first approximation, independent of the airspeed. The total drag-to-weight ratio, however, is strongly dependent upon the airspeed, being very large at low airspeeds (where the drag-due-to-lift is dominant since the lift coefficient is large), decreasing with increasing airspeed to reach a minimum value,

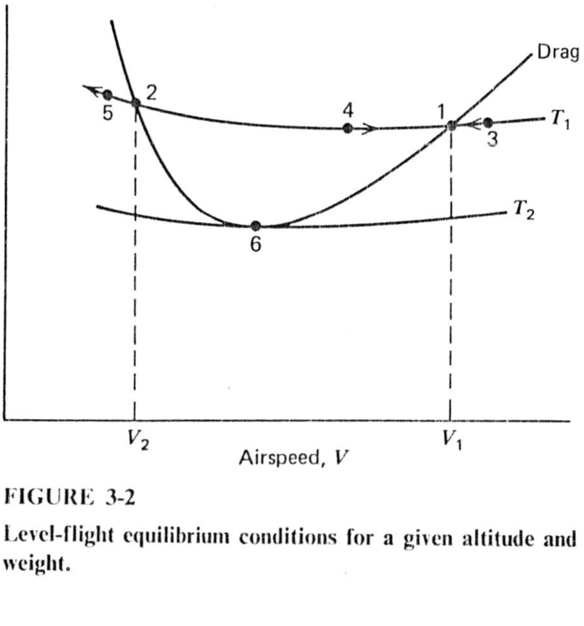

FIGURE 3-2

Level-flight equilibrium conditions for a given altitude and weight.

and then becoming very large again at high airspeeds (where the zero-lift drag is dominant). For the throttle setting for T_1/W, there are two points of intersection of the available thrust and drag (which is the thrust required) curves; these points are the graphical solutions of Eqs. 3-14 and 3-24. Point 1 is the high-speed solution (V_1) and point 2 is the low-speed solution (V_2), both for steady-state level flight.

The static stability of the airspeed at the equilibrium points, 1 and 2, can be determined by examining off-point conditions. If V is greater than V_1, such as at point 3, the thrust is less than the drag and the aircraft will decelerate until point 1 is reached. If V is between V_1 and V_2, as at point 4, the thrust is greater than the drag and the aircraft will accelerate away from point 2 until point 1 is again reached. Thus, point 1 and V_1 represent a *statically stable* equilibrim condition.

If, however, the airspeed is less than V_2, such as at point 5, the thrust is less than the drag and the aircraft will decelerate rapidly and move away from point 2. Point 2 and V_2, therefore, represent a *statically unstable* equilibrium condition. The region to the left of point 6 is often referred to as the *back side of the power curve* and is a region to be avoided, particularly on take-off or at landing. If the airspeed is allowed to drop below V_6, it is not always possible to increase the thrust (by increasing the throttle setting) either sufficiently or rapidly enough to prevent the aircraft from stalling.

Remember that V_1 and V_2 may not be physically realizable: V_1 if it exceeds the upper limits of the valid range of the parabolic drag polar (the drag-rise Mach number) and V_2 if it is less than the stall speed. If the available thrust is reduced by coming back on the throttle or by increasing the altitude, the two airspeed solutions approach each other, coming together at the point where the available thrust curve is tangent to the drag curve, such as at point 6. This point represents the *absolute ceiling* of the aircraft for this thrust. If the thrust is further reduced, the available thrust will be less than the drag. The airspeed will then begin to drop off, decreasing the lift, and the aircraft will lose altitude and descend until the available thrust again becomes equal to the drag.

If the throttle setting and weight are kept constant, then a theoretical flight envelope that shows the region of possible airspeeds as a function of the altitude can be constructed, as shown in Fig. 3-3. The solid line represents a variation in the thrust that is proportional to the density ratio raised to the 0.7th power in the troposphere, whereas the dashed line represents our assumption that the thrust is directly proportional to the density ratio in both the troposphere and the stratosphere. The flight envelope can be further refined by superimposing lines for the stall speed and the drag-divergence Mach number as functions of the altitude.

Let us return to Eq. 3-28, which can be written as

$$V = \left[\frac{T/S}{\rho_{SL}\sigma C_{D0}}\left(1 \pm \left\{1 - \frac{1}{[E_m(T/W)]^2}\right\}^{1/2}\right)\right]^{1/2} \tag{3-37}$$

where T, the available thrust is determined by the throttle setting and the altitude (and, of course, by the size of the engines). This equation shows that a high airspeed calls for a large T/S and/or a small C_{D0}. The thrust-to-wing-area ratio (T/S) is not considered a primary performance parameter because it is the product of the

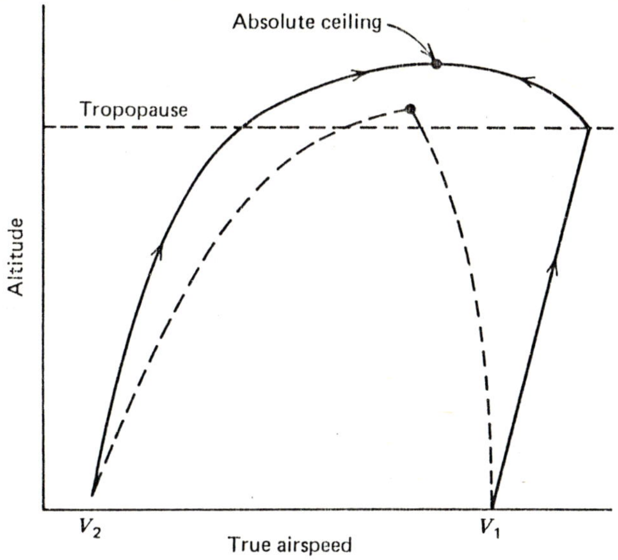

FIGURE 3-3

Level-flight airspeed-altitude envelope for a given weight and constant throttle setting.

thrust-to-weight ratio and the wing loading, which are primary parameters. Since it does appear at times, we shall call it the *thrust loading* and find its value from the relationship that

$$\frac{T}{S}=\left(\frac{T}{W}\right)\left(\frac{W}{S}\right) \tag{3-38}$$

The use of Eq. 3-37 to find the maximum airspeed of a high-speed aircraft with a maximum throttle setting often results in unrealistically high values that exceed the drag-rise Mach number, and may well be supersonic, inasmuch as the maximum available thrust-to-weight ratio is usually larger than that required for level flight, being determined by such operational constraints as minimum take-off run, rate of climb, and minimum ceiling. Furthermore, the parabolic drag polar assumption does not provide for the rapid rise in the drag in the transonic region.

Do not be overly concerned if at this point you feel confused. The main objective of this section has been to introduce you to the level-flight relationships among the flight parameters, such as airspeed and altitude, and the design characteristics, such as the wing loading, thrust-to-weight ratio, maximum lift-to-drag ratio, and the drag polar. As these relationships are further developed in subsequent sections, your familiarity will increase and your sense of discomfort should disappear.

3-3 CEILING CONDITIONS

The condition for steady-state level flight is that the flight thrust-to-weight ratio be greater than or equal to the reciprocal of the maximum lift-to-drag ratio of the aircraft. When, with a given throttle setting and thrust level, the thrust-to-weight ratio is exactly equal to the reciprocal of the maximum lift-to-drag ratio, the aircraft is flying at the *absolute ceiling for that throttle setting*, with the absolute ceiling defined as the highest altitude at which the aircraft can maintain steady-state level flight. If the altitude is increased beyond that of the absolute ceiling, the thrust will decrease, with the result that the thrust-to-weight ratio will be less than the reciprocal of the maximum lift-to-drag ratio, and level flight is no longer possible. If the original thrust is less than the maximum available, the throttle setting can be increased so as to establish the new altitude as the absolute ceiling for the new thrust level. Once the thrust reaches the maximum available, that altitude becomes the *absolute ceiling of the aircraft*. Obviously, an aircraft can fly above its absolute ceiling for a limited period of time by zooming and exchanging kinetic energy for potential energy. Eventually, however, the aircraft must descend to its absolute ceiling altitude.

The ceiling condition can be used to determine the altitude of the absolute ceiling of an aircraft by first writing the ceiling condition as

$$T_c = T_m = \frac{W_c}{E_m} \tag{3-39}$$

where the subscript c denotes the ceiling conditions. The ceiling thrust can be related to the maximum sea-level thrust by making use of Eq. 2-23 to write

$$\frac{T_c}{T_{m,\mathrm{SL}}} = \sigma_c{}^x \tag{3-40}$$

Combining Eqs. 3-38 and 3-40 yields an expression for the density ratio at the ceiling, namely,

$$\sigma_c{}^x = \frac{W_c}{E_m T_{m,\mathrm{SL}}} \tag{3-41}$$

where x takes on a value of 0.7 in the troposphere and becomes unity in the stratosphere. We see from this expression that as fuel is used, W_c will decrease and so will σ_c, so that the ceiling altitude will gradually increase with time.

Equation 3-41 can be simplified to obtain an approximate value for the *minimum absolute ceiling of an aircraft* by letting x be equal to unity in the troposphere as well as in the stratosphere and by assuming that the aircraft weights at the ceiling and at sea level are the same. The simplified expression becomes

$$\sigma_c = \frac{1}{E_m(T_m/W)_{\mathrm{SL}}} \tag{3-42a}$$

and shows that the maximum lift-to-drag ratio and the sea-level maximum thrust-to-weight ratios are the only design characteristics that affect the ceiling of a turbojet. In a subsequent paragraph, we shall see that the ceiling airspeed is also affected by the wing loading and the zero-lift-drag coefficient.

If a turbojet has a maximum lift-to-drag ratio of 18 and a maximum sea-level thrust-to-weight ratio of 0.25, Eq. 3-42*a* yields a value of 0.222 for the density ratio at the ceiling. Referring to Table A-1 (A-2) in the Appendix and interpolating, we find that the absolute ceiling is approximately 42,000 ft (12,800 m) above mean sea level.

If the more precise relationship of Eq. 3-41 is used with x equal to 0.7 in the troposphere and to unity in the stratosphere, the density ratio is found to be 0.154, corresponding to an absolute ceiling altitude of approximately 50,000 ft (15,240 m).

The *service ceiling* is commonly used as a performance and design specification and is defined as the altitude at which the maximum rate of climb is 100 fpm. The *cruise ceiling* is defined as the altitude at which the maximum rate of climb is 300 fpm. The service and cruise ceilings are obviously lower than the absolute ceiling and are in the vicinity of the minimum absolute ceiling obtained from Eq. 3-42*a*, which will be adequate for our use.

In all cases, the ceilings will increase as fuel is consumed and the weight of the aircraft decreases. We should be aware that an aircraft with a high wing loading may not always be able to achieve these calculated ceilings because the required airspeeds may exceed the drag-rise Mach number. The airspeed at the ceiling can be found from Eq. 3-28 by setting the radicand equal to zero. Doing so, several expressions for the ceiling airspeed are:

$$V_c = \left(\frac{T}{\rho S C_{DO}}\right)^{1/2} = \left[\frac{(T_{SL}/S)}{\rho_{SL} C_{DO}}\right]^{1/2} = \left[\frac{(T_{SL}/W)(W/S)}{\rho_{SL} C_{DO}}\right]^{1/2} \tag{3-42b}$$

Notice the influence of the wing loading and the zero-lift-drag coefficient. If our turbojet with an approximate ceiling of 42,000 ft has a wing loading of 100 lb/ft^2 and a zero-lift-drag coefficient of 0.016, the required airspeed at the ceiling is 811 fps, or M 0.84, which is in the vicinity of the typical drag-rise Mach number.

3-4 CRUISE RANGE

Cruise flight starts at the completion of the climb phase, when the airspeed has changed from the climb speed to the cruise speed, and ends when the descent phase begins. The *cruise range* is the horizontal distance X traveled with respect to the surface of the Earth and does not include any distances traveled during climb or descent. Unless otherwise stated, range will refer to the cruise range only.

In this section, we shall examine the influence of the design and flight parameters upon the range of a turbojet aircraft. The applicable level-flight equations (with the flight-path angle equal to zero) are repeated here for convenience.

$$T = D \tag{3-43}$$

$$L = W \tag{3-44}$$

$$\frac{dX}{dt} = V \tag{3-45}$$

$$\frac{-dW}{dt} = cT \tag{3-46}$$

If Eq. 3-43 is divided by Eq. 3-44, we obtain an expression that shows that the required thrust-to-weight ratio is identically equal to the drag-to-weight ratio and, therefore, equal to the reciprocal of the flight lift-to-drag ratio, i.e.,

$$\frac{T}{W} = \frac{D}{W} = \frac{1}{E} \tag{3-47}$$

where E is the *flight* or instantaneous lift-to-drag ratio, *not* the maximum lift-to-drag ratio, and is a function of both the design characteristics and the flight conditions. The value of the flight lift-to-drag ratio is established by the lift coefficient required to maintain level flight at a specified airspeed and altitude. Obviously, the minimum thrust required from the engines occurs when the aircraft is flying at the maximum value of the lift-to-drag ratio, and just as obviously, the larger the maximum lift-to-drag ratio (the aerodynamic efficiency), the lower the minimum required thrust.

Dividing Eq. 3-45 by Eq. 3-46, and making use of Eq. 3-43, results in the relationships that

$$\frac{dX}{-dW} = \frac{V}{cT} = \frac{V}{cD} = \frac{VE}{cW} \tag{3-48}$$

where $dX/-dW$ is often referred to as the *instantaneous range* or the *specific range* and is the exchange ratio between range and fuel. It has the units of miles per pound of fuel (km/kg) and is analogous to the *mileage* of an automobile, especially when it is expressed in miles per gallon (kilometers/liter). *A gallon of jet fuel weighs approximately 6.75 lb.*

The instantaneous range is one measure of what is known as the point performance of an aircraft, i.e., the performance at a specified point on the flight path or at a specified instant of time. Our interest, however, primarily lies in determining the overall flight performance; namely, how far can a particular aircraft fly with a given amount of fuel or conversely, how much fuel is required to fly a specified range. This can be done by integrating the point performance over the interval between specified initial and final points, usually the start and end of cruise. Conditions at the start of cruise will be identified by the subscript 1 and at the end of cruise by the subscript 2.

Before integrating Eq. 3-48, let us define the mass ratio, MR, and the cruise-fuel weight fraction ζ. These are important design and performance parameters. The *mass ratio* is the ratio of the total weight of the aircraft at the start of cruise (W_1) to its total weight at the end of cruise (W_2). The *cruise-fuel weight fraction* is defined as the ratio of the weight of fuel consumed during cruise (ΔW_f) to the total weight of the aircraft at the start of cruise, i.e.,

$$\zeta = \frac{\Delta W_f}{W_1} \tag{3-49}$$

The relationships among the mass ratio, the cruise–fuel weight fraction, and the aircraft weights are:

$$MR = \frac{W_1}{W_2} = \frac{W_1}{W_1 - \Delta W_f} = \frac{1}{1-\zeta} \tag{3-50}$$

$$W_2 = W_1(1-\zeta) \tag{3-51}$$

$$\zeta = 1 - \frac{1}{MR} = \frac{MR-1}{MR} \tag{3-52}$$

With the assumption that the specific fuel consumption is constant, Eq. 3-48 can be partially integrated to obtain the integral range equation:

$$X = \frac{-1}{c} \int_1^2 \frac{V}{D}\, dW \tag{3-53}$$

Before Eq. 3-53 can be further integrated, it is necessary to define the flight programs to be considered.

Of the many possible cruise flight programs only three will be examined; in each case, two flight parameters will be held constant throughout cruise. The three flight programs of interest are:

1. Constant altitude-constant lift coefficient flight
2. Constant airspeed-constant lift coefficient flight
3. Constant altitude-constant airspeed flight

For each flight program, the integral equation will be set up and then only the final range equation will be shown and discussed. The details of the integrations are given in Appendix B for anyone who is interested.

The first flight program to be examined is the *constant altitude-constant lift coefficient flight program.* Since the lift coefficient is held constant throughout cruise, the flight-lift-to-drag ratio E will also be constant. It is convenient, therefore, to express the instantaneous drag as the ratio of the instantaneous weight to the instantaneous lift-to-drag ratio and rewrite Eq. 3-53 as

$$X_{h,C_L} = -\frac{E}{c} \int_1^2 \frac{V}{W}\, dW \tag{3-54}$$

Performing the integration yields the range equation

$$X_{h,C_L} = \frac{2EV_1}{c}\left[1 - (1-\zeta)^{1/2}\right] \tag{3-55}$$

where V_1, the initial airspeed, is given by

$$V_1 = \left[\frac{2(W_1/S)}{\rho_{SL}\sigma C_L}\right]^{1/2} \tag{3-56}$$

where W_1 is the gross weight of the aircraft at the start of cruise. It can be seen from Eq. 3-56 that the airspeed must be decreased as fuel is used if C_L is to be kept constant as the weight decreases along the flight path. It can also be shown that the final airspeed V_2 is

$$V_2 = V_1(1-\zeta)^{1/2} \tag{3-57}$$

since

$$W_2 = W_1(1-\zeta) \tag{3-58}$$

Since E is held constant, Eq. 3-47 shows that the thrust must constantly be decreased (by coming back on the throttle) as the fuel is used and the gross weight decreases. The variations in the flight parameters for this flight program are shown in Fig. 3-4 as functions of the cruise-fuel weight fraction, which is a measure of the range flown. There are three drawbacks to this flight program. The first is the need to continuously compute the airspeed along the flight path and to reduce the throttle setting accordingly. The second is that reducing the airspeed increases the flight and block times. The third is the fact that air traffic control rules require a "constant" true airspeed for cruise flight, currently constant is ± 10 knots.

The second flight program to be examined is the *constant airspeed-constant lift coefficient flight program*, which is commonly referred to as *cruise-climb flight*.

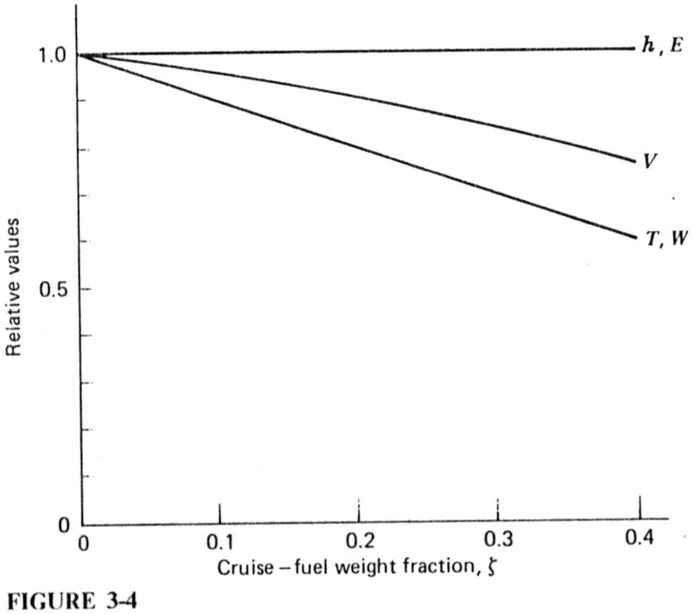

FIGURE 3-4

Variation of flight parameters along the flight path for constant-altitude constant lift coefficient flight.

With both V and E constant, Eq. 3-54 can be written as

$$X_{V,C_L} = \frac{-EV}{c} \int_1^2 \frac{dW}{W} \tag{3-59}$$

The resulting range equation is

$$X_{V,C_L} = \frac{EV}{c} \ln \text{MR} = \frac{EV}{c} \ln \left(\frac{1}{1-\zeta} \right) \tag{3-60}$$

This is the general form of what is known as the *Breguet range equation.* In order to keep *both* the airspeed and the lift coefficient constant as the weight of the aircraft decreases, Eq. 3-56 shows that ρ must decrease in a similar manner so as to keep the ratio of the weight to the atmospheric density (W/ρ) constant. The only way that this can be done is to increase the altitude in an appropriate manner. Consequently, the aircraft will be in a continuous climb (thus, the name cruise-climb), which appears to violate the level-flight condition of a zero flight-path angle. It will be shown in a subsequent section that the cruise-climb flight-path angle is sufficiently small so as to justify the use of the level-flight equations and solutions for cruise-climb. The thrust required will decrease along the flight path in such a manner that in the stratosphere the available thrust will decrease in an identical manner. Therefore, cruise-climb flight in the stratosphere requires no computations or efforts by the pilot. After establishing the desired cruise airspeed, the pilot simply engages the Mach-hold mode (or constant-airspeed mode) on the autopilot and the aircraft will slowly climb at the desired flight-path angle as the fuel is burned. The variations in the flight parameters along the flight path are shown in Fig. 3-5. Only under certain limited conditions will cruise-climb flight be allowed by air traffic control.

Generally, when flight is conducted under the jurisdiction of flight traffic control regulations, the accepted flight program is the *constant altitude-constant airspeed flight program.* The integral range equation for this flight program can be written as

$$X_{h,V} = \frac{-V}{c} \int_1^2 \frac{dW}{D} \tag{3-61}$$

The integration over the cruise interval is a bit messy (see Appendix B) but does yield the following range equation, which is also a bit messy:

$$X_{h,V} = \frac{2E_m V}{c} \arctan \left[\frac{\zeta E_1}{2E_m(1 - KC_{L_1} E_1 \zeta)} \right] \tag{3-62}$$

where E_1 and C_{L_1} are the values of the flight lift-to-drag ratio and of the lift coefficient at the start of cruise; these two flight parameters will decrease along the flight path. The thrust (throttle setting) must be reduced along the flight path so as to maintain a constant airspeed; this can be done manually or by a combination of altitude and airspeed (Mach number) hold modes. The variations of the flight parameters along the flight path of this program are sketched in Fig. 3-6. This flight program is the one used the most, but many performance analyses use the

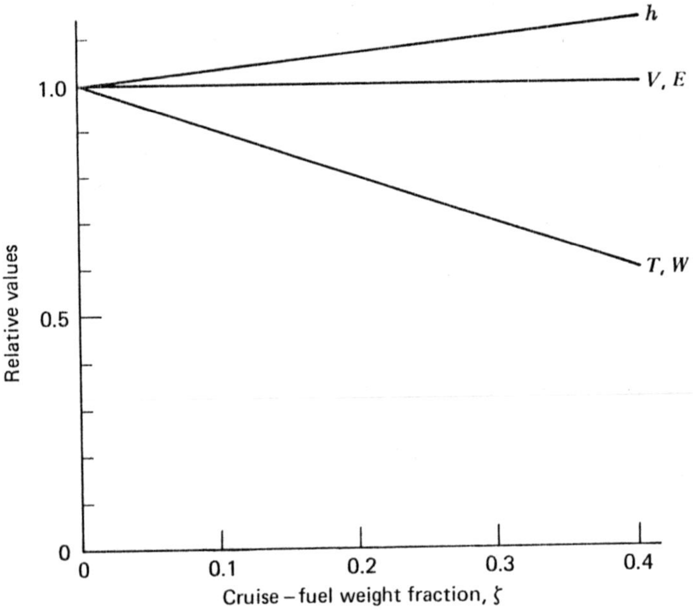

FIGURE 3-5

Variation of flight parameters along the flight path for constant airspeed–constant lift coefficient flight.

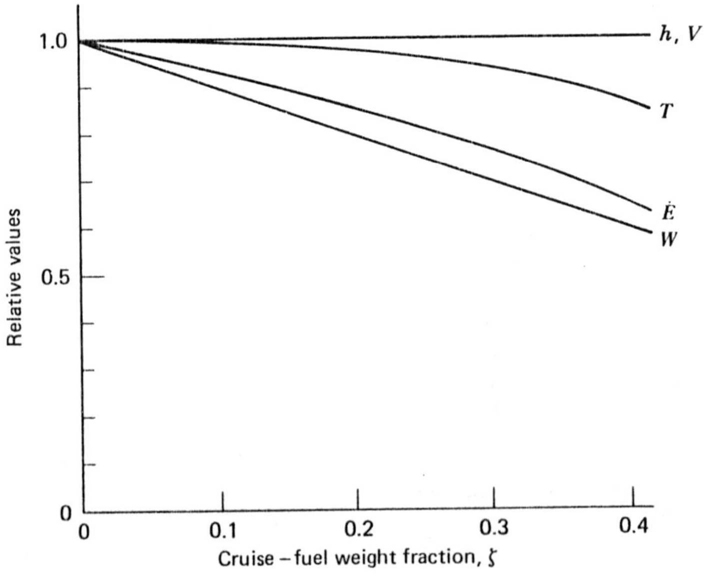

FIGURE 3-6

Variation of flight parameters along the flight path for constant altitude–constant airspeed flight.

Breguet range equation of the cruise-climb program because the mathematics is simpler and the errors are normally not significant.

In order to compare the ranges attainable with each of these flight programs, consider an aircraft that starts its cruise at30,000 ft ($\sigma = 0.374$) at 550 fps (167.6 m/s) or 375 mph (603 km/h) and has a wing loading of 60 lb/ft^2 (2873 N/m^2), a parabolic drag polar with $C_D = 0.016 + 0.065\ C_L^2$, and a sfc of 0.8 lb/h/lb. The relevant conditions at the start of cruise are:

$$q_1 = \tfrac{1}{2}\rho_{SL}\sigma_1 V_1^2 = 134.4 \text{ lb/ft}^2 \ (6435 \text{ N/m}^2)$$

$$C_{L_1} = \frac{W_1/S}{q_1} = 0.446$$

$$C_{D_1} = 0.016 + 0.065\ C_{L1}^2 = 0.0289$$

$$E_1 = \frac{C_{L_1}}{C_{D_1}} = 15.4$$

Note that

$$Em = \frac{1}{2(KC_{DO})^{1/2}} = 15.5$$

Only the cruise-fuel weight fraction remains to be specified. Obviously, the larger that ζ is, the more the available fuel and the greater the range. Let us set ζ equal to 0.3 for a reasonably long-range flight.

For the constant altitude-constant lift coefficient flight, substitute the appropriate values in Eq. 3-55 to obtain

$$X_{h,C_L} = \frac{2 \times 15.4 \times 375}{0.8}[1 - (1 - 0.3)^{1/2}]$$

so that

$$X_{h,C_L} = 2{,}358 \text{ mi } (3{,}794 \text{ km})$$

Watch your units. Since c is expressed in lb/h/lb and since we are interested in the range in statute miles, V in the range equation must be expressed in mph. A convenient, easy-to-remember, conversion relationship is the fact that 60 *mph is equal to* 88 *fps*. Although nautical miles (nmi) and knots (kt) are commonly used operationally for range and airspeed, (statute) miles and miles per hour will be the basic units used in this book; 1 nmi = 1.1515 mi and 1 kt = 1.1515 mph.

Using Eq. 3-60, the cruise-climb range is found to be

$$X_{V,C_L} = \frac{15.4 \times 375}{0.8} \ln\left(\frac{1}{1 - 0.3}\right)$$

or

$$X_{V,C_L} = 2{,}575 \text{ mi } (4143 \text{ km})$$

The constant altitude-constant airspeed range is found, from Eq. 3-62, to be

$$X_{h,V} = \frac{2 \times 15.5 \times 375}{0.8} \arctan\left[\frac{0.3 \times 15.4}{2 \times 15.5(1 - 0.065 \times 0.446 \times 15.4 \times 0.3)}\right]$$

or $\quad X_{h,V} = 2{,}476 \text{ mi } (3{,}984 \text{ km})$

We see that, for this aircraft and for this set of initial cruise conditions, the cruise-climb flight program yields about 4 percent more range than the constant altitude-constant airspeed program and approximately 9 percent more than the constant altitude-constant lift coefficient program. These ranges, however, are not necessarily the maximum ranges that can be achieved by this aircraft using each of these programs. The next section will be devoted to determining the conditions for the maximum, or best range, for each flight program and for the aircraft itself.

3-5 BEST (MAXIMUM) RANGE

Range is an important performance and design criterion for all aircraft. It is primary for transports and secondary for fighter and special-purpose aircraft. The conditions for the maximum range, which will henceforth be called the *best range*, can be determined by maximizing either the individual range equations or by maximizing the instantaneous range. Both approaches will arrive at the same conclusion, and we shall choose the latter approach. We shall maximize the instantaneous range with respect to the airspeed by setting the first derivative with respect to the airspeed equal to zero and solving for the best-range airspeed. Using Eq. 3-48, the maximization process starts with

$$\frac{d}{dV}\left(\frac{dX}{-dW}\right)=\frac{d}{dV}\left(\frac{V}{cD}\right)=0 \qquad (3\text{-}63)$$

Performing the differentiation indicated yields the equation

$$\frac{-V}{cD^2}\frac{dD}{dV}-\frac{V}{c^2D}\frac{dc}{dD}+\frac{1}{cD}=0 \qquad (3\text{-}64)$$

Since the variation in the thrust specific fuel consumption with the airspeed is small and can be assumed equal to zero, the *condition for best instantaneous range* becomes

$$\frac{dD}{dV}=\frac{D}{V} \qquad (3\text{-}65)$$

Although Eq. 3-65 is valid for any drag polar, the parabolic drag polar is now introduced in order to obtain analytic and closed-form expressions and solutions. By making use of the drag expression of Eq. 3-24, Eq. 3-65 can be solved for the best-range dynamic pressure, which is found to be

$$q_{\text{br}}=\left(\frac{W}{S}\right)\left(\frac{3K}{C_{DO}}\right)^{1/2} \qquad (3\text{-}66)$$

so that

$$V_{\text{br}}=\left[\frac{2(W/S)}{\rho_{\text{SL}}\sigma}\right]^{1/2}\left[\frac{3K}{C_{DO}}\right]^{1/4} \qquad (3\text{-}67)$$

With the substitution of Eq. 3-66 into the parabolic drag expressions, the best-range drag is

$$D_{br}=\frac{4W(KC_{D0})^{1/2}}{3^{1/2}}=\frac{1.155W}{E_m}=T_{br} \tag{3-68}$$

and

$$E_{br}=\frac{W}{D_{br}}=0.866E_m \tag{3-69}$$

With appropriate substitutions into Eq. 3-48, the expression for the *best instantaneous range* can be written as

$$\frac{dX_{br}}{-dW}=\frac{0.866E_m V_{br}}{cW} \tag{3-70}$$

where the value of V_{br} can be found from Eq. 3-67. From Eq. 3-70, we see that, for "good mileage," we want a large maximum lift-to-drag ratio, a high best-range airspeed, a low specific fuel consumption, and a low aircraft gross weight. The maximum lift-to-drag ratio and the specific fuel consumption are design characteristics whereas the best-range airspeed and the gross weight are a combination of design characteristics and operational considerations. Since the gross weight will decrease along the flight path, the best instantaneous range will increase along the flight path and the best mileage will be at the end of the flight when the aircraft is the lightest.

The best-range lift coefficient is

$$C_{L,br}=\frac{(W/S)}{q_{br}}=\left(\frac{C_{D0}}{3K}\right)^{1/2}=0.577C_{L,Em} \tag{3-71}$$

Equation 3-71 tells us that, in order to maximize the instantaneous range at all points along the flight path, the lift coefficient must be kept constant at all times and be equal to the best-range lift coefficient, which is a design characteristic. Therefore, these best-range conditions can be applied only to the two constant lift coefficient flight programs of the preceding section.

Introducing the best-range conditions into Eq. 3-55, the *best-range equation for constant altitude-constant lift coefficient flight* can be written as

$$X_{br;h,C_L}=\frac{1.732E_m V_{br1}}{c}\left[1-(1-\zeta)^{1/2}\right] \tag{3-72}$$

where V_{br1} is the best-range airspeed at the start of cruise. Remember that the airspeed must be appropriately reduced during cruise.

From Eq. 3-60 and the best-range conditions, the *best-range equation for cruise-climb* is

$$X_{br;V,C_L}=\frac{0.866E_m V_{br}}{c}\ln\left(\frac{1}{1-\zeta}\right) \tag{3-73}$$

The term in parentheses with ζ can be, and often is, replaced by its equivalent, which is simply the mass ratio MR.

Although the best-range condition of a constant lift coefficient cannot be satisfied by the constant altitude-constant airspeed flight program, the range can be maximized to a first approximation by setting the initial cruise conditions equal to those for best range. By so doing, the "*best-range*" *equation for constant altitude-constant airspeed cruise* can be written as

$$X_{br;h,V} \cong \frac{2E_m V_{br}}{c} \arctan\left(\frac{0.433\zeta}{1-0.25\zeta}\right) \tag{3-74}$$

where the angle represented by the arc tangent term *must* be expressed in radians. The exact conditions for best range for this flight program can be obtained by maximizing the range equation of Eq. 3-63 and are a function of the cruise-fuel weight fraction as well as of the design and flight characteristics. The difference in range between the exact solution and that of Eq. 3-74 is small for small values of the cruise-fuel weight fraction. Although the difference does increase with increasing values of ζ, long-range flights at constant altitude and constant airspeed are normally broken down into smaller segments (with low values of ζ), which are flown at successively increasing altitudes. Such a flight program is known as *stepped-altitude flight*; it approaches cruise-climb flight in the limit as the number of altitude steps increases and will be discussed in the next section. Consequently, Eq. 3-74 can be used as the "best-range" equation for constant altitude-constant airspeed (sometimes called hard altitude-hard airspeed) flight with little loss of accuracy.

At this point, let us return to the aircraft of the preceding section and determine the best ranges for each flight program. The aircraft has a W/S of 60 lb/ft^2 (2873 N/m^2), $C_D=0.016+0.065\,C_L^2$, $c=0.8$ lb/h/lb (0.81 kg/h/N), and $E_m=15.5$. Cruise will start at 30,000 ft (9,144 m), and ζ will be 0.3, as before. The value of V_{br} at start of cruise is found, from Eq. 3-67, to be

$$V_{br}=\left(\frac{2\times 60}{\rho_{SL}\times 0.374}\right)^{1/2}\left(\frac{3\times 0.065}{0.016}\right)^{1/4}=686.5 \text{ fps}$$

$$V_{br}=468 \text{ mph}=M\,0.69$$

In SI units, $V_{br}=209$ m/s $=753$ km/h $=M\,0.69$. The best-range equations yield the following values:

$$X_{br;h,C_L}=2{,}565 \text{ mi } (4{,}127 \text{ km})$$

$$X_{br;V,C_L}=2{,}800 \text{ mi } (4{,}505 \text{ km})$$

$$X_{br;h,V}=2{,}530 \text{ mi } (4{,}070 \text{ km})$$

Comparing these values to those obtained for a specified airspeed of 375 mph shows not only that the maximum range of the aircraft has been increased by 9 percent but also that the cruise-climb program is the best of the three, producing 9 percent more range than the constant altitude-constant lift coefficient program, which in turn is about 1.3 percent better than cruise at constant altitude and airspeed.

The differences among the ranges for the various flight programs increase as the ranges increase, i.e., as the cruise-fuel fraction ζ increases. For example, the ratio of the range for cruise-climb flight to that for constant altitude-constant lift coefficient flight is

$$\frac{X_{br;V,C_L}}{X_{br;h,C_L}}=\frac{\ln\left[1/(1-\zeta)\right]}{2\left[1-(1-\zeta)^{1/2}\right]} \tag{3-75}$$

and the ratio of ranges for cruise-climb and constant altitude-constant airspeed flight is

$$\frac{X_{br;V,C_L}}{X_{br;h,V}}=\frac{0.433\ln\left[1/(1-\zeta)\right]}{\arctan\left(0.433\zeta/1-0.25\zeta\right)} \tag{3-76}$$

These relative ranges are plotted in Fig. 3-7 and show, as we might expect, that the differences in range among the various flight programs are not as important for short-range flight as for long-range flight.

Let us now examine Eq. 3-67 with a view to maximizing the best-range airspeed as a means of increasing the best range. We see first that V_{br} is inversely proportional to the square root of the density ratio and thus increases with altitude. The influence of the altitude upon the airspeed and range is strong, as evidenced by the fact that the range of a particular aircraft is 60 percent greater at 30,000 ft than at sea level. We next see that V_{br} increases in direct proportion to the square root of the wing loading. A doubling of the wing loading results in a 14 percent increase in range.

Typical values of the wing loading are of the order of 100 to 120 lb/ft^2 (4,788 to 5,746 N/m^2) for long-range subsonic transports, of the order of 50 lb/ft^2 (2,400

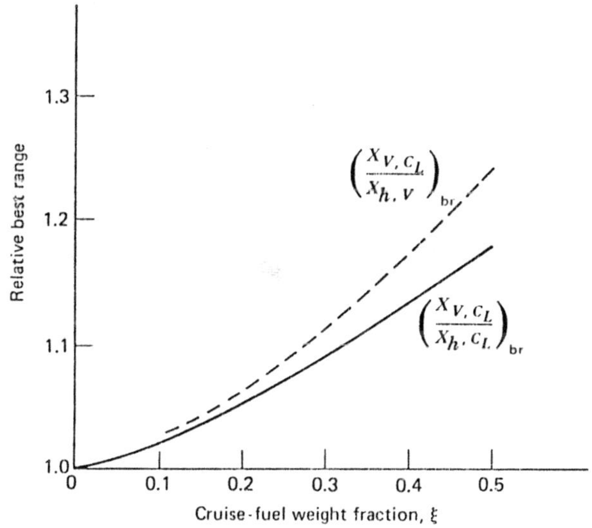

FIGURE 3-7

Relative best range as a function of the range, i.e., the cruise-fuel weight fraction.

N/m²) for short-range transports and fighter aircraft, and of the order of 15 to 20 lb/ft² (718 to 958 N/m²) for light aircraft. The lower values for the last two classes of aircraft arise from operational requirements more pressing than range, generally a short take-off run for the smaller aircraft and maneuverability for the fighters.

Two cautionary comments are in order with respect to increasing the best-range airspeed. First of all, do not attempt to increase the wing loading by increasing the weight, reduce the wing area instead. Although V_{br} is directly proportional to the square root of the gross weight, the instantaneous range is inversely proportional to the weight itself. Secondly, do not try to increase V_{br} by increasing K (by decreasing the aspect ratio) because the range penalty associated with the accompanying decrease in the maximum lift-to-drag ratio will be predominant. On the other hand, reducing the zero-lift drag coefficient will increase not only V_{br} but also E_m, both salutory. Furthermore, a reduction in C_{D0} improves all other aspects of performance and is a goal to be pursued vigorously, as is a reduction in weight.

The cruise-fuel weight fraction is a measure of the fuel available for cruise. Obviously, all of the fuel loaded aboard an aircraft on the ramp is not available for cruise because some of it is used for taxiing, take-off, and climb to the cruise altitude. In addition, there must be fuel remaining at the end of cruise in case the winds and fuel consumption were not as planned and for descent, landing, and taxiing. If the flight is conducted in weather (under instrument flight rules), there must be sufficient fuel to fly to an alternate airport. This remaining fuel is called the *reserve*, either VFR or IFR, depending on whether the flight is conducted under visual flight rules or instrument flight rules.

For large aircraft, it is reasonable to assume that $0.1W_f$ (W_f is the total fuel loaded aboard the aircraft) is consumed prior to the start of cruise and that the reserve is also $0.1W_f$, leaving a ΔW_f available for cruise of $0.8W_f$. For small aircraft, assume $0.2W_f$ to reach cruise altitude and the same reserve so that the ΔW_f available for cruise is $0.7W_f$. With these assumptions, W_1, the initial cruise weight, for large aircraft is equal to $W_0 - 0.1W_f$ and for small aircraft is $W_0 - 0.2W_f$. If all the fuel available for cruise is not needed and not used, it can be shown that the performance of the aircraft has been penalized by carrying the fuel as dead weight.

From either Eq. 3-47 or Eq. 3-68, we see that the thrust required for best-range flight is

$$\frac{T_{br}}{W} = \frac{1}{E_{br}} = \frac{1.155}{E_m} \tag{3-77}$$

We see, first of all, that the thrust-to-weight ratio is greater than the reciprocal of the maximum lift-to-drag ratio, as is required for level flight, but not much more, indicating that the altitude for best-range flight lies below but close to the absolute ceiling for the associated throttle setting. By using Eq. 2-2 and the ceiling thrust-to-weight ratio, it can be shown that the best-range altitude is approximately 3,500 ft below the instantaneous absolute ceiling for that particular throttle setting.

We also see that the thrust required is independent of the altitude itself, being dependent only upon the instantaneous weight of the aircraft and upon E_m. In other words, the thrust required at sea level is identical (as is the drag) to that required at high altitudes. We fly at altitude to obtain the range benefits (and the secondary benefits of flying above the weather and surface turbulence) accruing to the increase in the airspeed necessary (with the reduced air density) to generate the lift required to counteract the weight and keep the airplane in the air. As we increase the altitude, however, the available thrust does decrease (it's the required thrust that remains constant) and we must be sure that there is sufficient available thrust at the cruise altitude to satisfy Eq. 3-77. In addition to worrying about the thrust available versus that required, we must also check to be sure that the Mach number associated with the best-range airspeed does not exceed the drag-rise Mach number for our aircraft.

The instantaneous fuel consumption rate is also of interest; measured in lb/h, it is displayed to the pilot as part of his instrument array. For best-range flight,

$$\left(\frac{dW_f}{dt}\right)_{\text{br}} = cT_{\text{br}} = \frac{1.155cW}{E_m} \tag{3-78}$$

If the specific fuel consumption is assumed to be independent of the altitude, then the best-range fuel consumption rate is also independent of the altitude. This means that *for a given fuel load, a turbojet aircraft can stay in the air just as long at sea level as at altitude.* It is the increased airspeed required at the higher altitudes that results in the greater range. In actuality, however, the specific fuel consumption does decrease slowly with altitude, reaching its minimum value at the tropopause and then increasing even more slowly in the stratosphere. Consequently, there is a slight advantage to flying in the vicinity of the tropopause, all other things being equal.

The time required for any of the best-range, constant lift coefficient flight programs can be determined by inverting Eq. 3-78 and integrating over the flight path from W_1 to W_2 to obtain

$$t_{\text{br};C_L} = \frac{0.866E_m}{c}\ln\left(\frac{1}{1-\zeta}\right) \tag{3-79}$$

The time of flight is the same for all best-range, constant lift coefficient flight programs at any altitude; i.e., it is independent of the actual range flown.

The cruise flight time for the constant altitude-constant airspeed flight program can easily be found by dividing the range by the airspeed. The same procedure can also be used for the cruise-climb flight time.

3-6 CRUISE-CLIMB AND STEPPED-ALTITUDE FLIGHT

In the preceding section, the level-flight equations were used to evaluate cruise-climb flight with the statement that the flight-path (climb) angle was sufficiently small to justify this action. The basic operating condition for cruise-climb flight is that the ratio of the instantaneous weight of the aircraft to the air density be kept constant along the flight path. Therefore,

$$\frac{W}{\rho}=\frac{W_1}{\rho_1}=\frac{W_2}{\rho_2} \tag{3-80}$$

so that, with Eq. 3-52, the density ratio at the end of cruise-climb flight can be expressed in terms of both the initial density ratio and the cruise-fuel weight fraction, i.e.,

$$\sigma_2=\sigma_1(1-\zeta) \tag{3-81}$$

With σ_2 known, Table A-1 can be used to determine the final altitude. The change in altitude can also be determined mathematically by making use of Eq. 2-2 to write the expression

$$\Delta h=23{,}800 \ln\left(\frac{1}{1-\zeta}\right) \tag{3-82}$$

where Δh is in feet and the value of 23,800 implies flight in or near the stratosphere. Both Eq. 3-81 and Eq. 3-82 show that the increase in altitude during a cruise-climb flight is a function only of the cruise-fuel weight fraction, which is a measure of the range flown.

Returning to the best-range example of the preceding section, the cruise-fuel weight fraction was 0.3, so that the increase in altitude, as determined from Eq. 3-82, would be approximately 8,500 ft. This altitude change, in conjunction with the range of 2,800 miles, represents an average flight-path angle of 5.7×10^{-4} rad, or 0.033 deg, a small angle indeed.

Since the flight-path angle is small, the tangent can be replaced by the angle itself, expressed in radians. Therefore,

$$\gamma_{V,C_L}=\frac{\Delta h}{5{,}280\,X}=\frac{23{,}800c}{5{,}280VE} \tag{3-83}$$

for any cruise-climb flight, with V in mph and γ in radians. For best-range cruise-climb flight, Eq. 3-83 becomes

$$\gamma_{\mathrm{br};V,C_L}=\frac{5.2c}{V_{\mathrm{br}}E_m} \tag{3-84}$$

In Eqs. 3-83 and 3-84, the rate of change of the flight-path angle is essentially zero and the flight-path angle itself can be assumed to be constant.

Even though the best-range flight-path angle is small, it is not zero and it will have some effect on the value of the actual range. In order to determine the errors in the range associated with the level-flight solution, an exact solution has been obtained elsewhere using the non-level-flight equations. The conclusions are that, for aircraft that are apt to use cruise-climb flight, the errors in range from using the "level-flight" (Breguet) range equation are of the order of 1 percent or less.

Although cruise-climb flight can greatly increase the range of an aircraft for long-range flights, it does involve a continuous increase of altitude that is not compatible with safe flight when the presence of other aircraft must be considered. Consequently, the opportunity to use cruise-climb flight is limited. It can be approximated, however, on long-range flights by the use of *stepped-altitude flight*, which is a series of constant altitude-constant airspeed flight segments conducted at different altitudes. Stepped-altitude flight is consistently used on long-range flights, such as transcontinental and transoceanic flights, in order to reduce the total fuel consumption, an important consideration in these days of fuel scarcity and/or price increases.

Although analytic relationships have been developed for stepped-altitude flight, they are somewhat awkward and it is easier to show the use and benefits of such flight by example. The situation is sketched in Fig. 3-8*a*, where the problem is to fly a prescribed range with a minimum expenditure of fuel at a constant airspeed and at a constant altitude. If the altitude remains unchanged throughout the

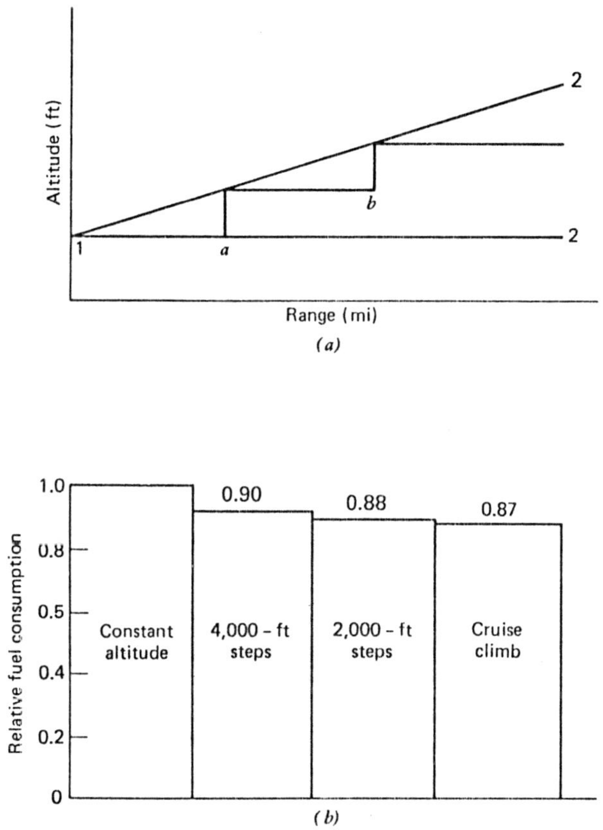

FIGURE 3-8

Comparison of stepped-altitude flight with constant altitude and cruise-climb flight: (*a*) flight profiles; (*b*) relative fuel consumption.

entire flight, the theoretical best-range airspeed will decrease below the actual airspeed and the range will suffer. If cruise-climb is used, the altitude will be continually increasing in order to keep the best-range airspeed equal to the actual airspeed. The key to stepped-altitude flight is to increase the altitude in steps that are compatible with air traffic control regulations and of such magnitude and timing so as to rematch the actual airspeed to the theoretical best-range airspeed at regular intervals. As the number of steps increases, the stepped-altitude flight path obviously becomes closer to the cruise-climb flight path.

Since air traffic control assigns odd-numbered altitudes to traffic going in one direction and even-numbered altitudes to traffic in the opposite direction, the steps in altitude must be multiples of 2,000 ft. At present, air traffic control limits such steps to 4,000 ft although the airlines would like 2,000-ft steps. If ζ_i is used to designate the cruise-fuel weight fraction for a constant-altitude segment, Eq. 3-82 can be solved to obtain

$$\zeta_i = 1 - \exp\left(\frac{-\Delta h}{23{,}800}\right) \tag{3-85}$$

where

$$\zeta_i = \frac{\Delta W_{fi}}{W_i} \tag{3-86}$$

In Eq. 3-86, ΔW_{fi} is the cruise-fuel used during the ith segment and W_i is the gross weight at the start of the ith segment. Returning to Eq. 3-85, the segment ζ_i associated with a 4,000-ft change is 0.1547 and with a 2,000-ft change is 0.0806.

Using the illustrative aircraft of the preceding sections, the minimum fuel consumption for four flight programs will be compared for a specified range of 4,000 mi and for an aircraft weight at the start of cruise of 200,000 lb. In review, this aircraft has a wing loading of 60 lb/ft^2, $C_D = 0.016 + 0.065\,C_L^2$, $c = 0.8$ lb/h/lb, $E_m = 15.5$, and a V_{br} for 30,000 ft of 468 mph.

The "best-range" constant altitude-constant airspeed cruise-fuel weight fraction can be calculated by substituting into Eq. 3-74 and solving to obtain a value for ζ of 0.4585. This is an unrealistically high value that is primarily the consequence of a wing loading that is too low for such a long range. With this value and an initial aircraft weight of 200,000 lb, the fuel consumption will be 91,700 lb. This flight program with this fuel consumption and a flight time of 8.55 hours, will be the baseline used in the comparisons. This particular flight program starts and ends at 30,000 ft.

The fuel consumption for the stepped-altitude programs requires a little more effort to determine. The 4,000-ft step case will be treated first and the fuel consumption for each segment will be calculated in turn; points of altitude change will be denoted by $a, b, c, \ldots$. Referring to Fig. 3-8a, the first segment starts at 1 and ends at a, the second starts at b and ends at c, and so on. In general, there will be n segments of identical length and ζ_i plus a final segment with a different length and a different ζ_i.

The first action is to use Eq. 3-74 to calculate the length of the first segment using a ζ_1 of 0.1547, the proper value for a 4,000-ft altitude change. The first segment

10257 0471078859 10257

Location: Row G5 Bay 8

GWSVM.5B4C

Title:	Introduction to Aircraft Performance, Selection, and Design
Cond:	Acceptable
User:	sv_dharsie
Station:	IT-00975
Date:	2024-11-14 15:13:24 (UTC)
Account:	Goodwill of Silicon Valley
Orig Loc:	Row G5 Bay 8
mSKU:	GWSVM.5B4C
vSKU:	GWSVV.0471078859.A
Seq#:	10257
unit_id:	20748038
width:	0.93 in
rank:	2,332,498

GWSVV.0471078859.A

delist unit# 20748038

xxxxx

length is 1,261.6 mi. With an overall range of 4,000 mi, there can be either three segments of 1,261.6 mi each and a final segment of only 215.2 mi with three altitude changes or two segments of 1,261.6 mi each and a final segment of 1,476.8 mi with two altitude changes. The latter case will be the only one examined since it is the more practical of the two. With the length and number of the segments established, we can proceed with the calculation of the fuel consumption.

For the first segment, ζ_i is 0.1547 and is equal to $\Delta W_{f1}/W_1$, so that ΔW_{f1} is 30,940 lb. For the second segment, ζ_2 is identical in value to ζ_1 but is now equal to $\Delta W_{f2}/W_a$, where $W_a = W_1(1-\zeta_1)$, or 169,060 lb. Therefore, ΔW_{f2} is equal to 26,154 lb.

For the third and final segment, ζ_3 must be calculated for a range of 1,476.8 mi. It has a value of 0.17935 and is equal to $\Delta W_{f3}/W_b$, where $W_b = W_a(1-\zeta_2)$ or 142,906 lb. There, ΔW_{f3} is 25,630 lb and the total amount of fuel used during cruise is 82,724 lb. This represents a 9.8 percent fuel savings over single-altitude cruise and amounts to an actual savings of 8,976 lb (1,330 gal). The overall cruise-fuel weight fraction is 0.4136, the final altitude is 38,000 ft, and the altitude changes are made every 2.7 hours.

The procedure followed in the case of 2,000-ft altitude changes is the same. There will be five segments of 645.64 mi each with a ζ_i of 0.0806 and a final segment of 771.8 mi with a ζ_6 of 0.09598 and five altitude changes. The total amount of fuel used is 81,224 lb for an 11.4 percent savings, which amounts to 10,746 lb (1,552 gal). The overall cruise-fuel weight fraction is 0.406, the final altitude is 40,000 ft, and altitude changes are made every 1.4 hours.

The cruise-fuel weight fraction for best-range, cruise-climb flight can easily be found from Eq. 3-73 to be 0.3991 so that ΔW_f is 79,828 lb. This represents the maximum fuel savings of 13 percent or 11,872 lb (1,759 gal). The altitude at the completion of cruise-climb is approximately 42,000 ft.

The relative fuel consumption of each of the flight profiles is shown in Fig. 3-8*b*. Notice that the most dramatic reduction is in going from single-altitude flight to 4,000-ft steps. The fuel consumption for stepped-altitude flight will be slightly increased by the climb requirements. There is also an extra burden placed on air traffic control to ensure safe clearance from any traffic at the altitudes crossed during climb and then at the new cruise altitudes.

3-7 BEST RANGE WITH A SPECIFIED OR RESTRICTED AIRSPEED

The best-range flight conditions in the preceding sections were determined for a given wing loading with the tacit assumption that the calculated best-range airspeed does not exceed that for the drag-rise Mach number. Under these conditions, the best range occurs when the turbojet is flying at a lift-to-drag ratio less than the maximum value, specifically at 86.6 percent of the maximum value. If the cruise airspeed is specified or if the theoretical best-range airspeed exceeds that for the drag-rise Mach number, then the conditions for best range will be different.

If the design airspeed is specified for operational reasons or is limited by the drag-rise Mach number, then, obviously, it cannot be varied in order to maximize either the instantaneous or the flight-path range. If the instantaneous range is written as

$$\frac{dX}{-dW}=\frac{VE}{cW} \tag{3-87}$$

then for a given weight, specific fuel consumption, and airspeed, the instantaneous range is maximized when the flight lift-to-drag ratio is equal to the maximum lift-to-drag ratio of the aircraft. However, V and E are not independent of each other; specifying one for a given aircraft specifies the other. The implications of these statements will be explored by examples. Only the cruise-climb (Breguet) range equation in its general form will be used, i.e.,

$$X=\frac{EV}{c}\ln\left(\frac{1}{1-\zeta}\right) \tag{3-88}$$

It is also apparent from this equation that if V is fixed, then the maximum range will occur when E is as large as possible, remaining compatible with V.

Let us first consider a turbojet aircraft that has a wing loading of 115 lb/ft^2; a drag polar, $C_D=0.0154+0.05\ C_L^2\ (E_m=18)$; and a drag-rise Mach number of 0.85 These are typical values for current long-range, wide-body, subsonic transports. The initial cruise altitude is to be 35,000 ft ($\sigma=0.310$). The theoretical best-range airspeed, from Eq. 3-67, is

$$V_{\text{br}}=\left(\frac{2\times 115}{\rho_{\text{SL}}\times 0.310}\right)^{1/2}\left(\frac{3\times 0.05}{0.0154}\right)^{1/4}=987\ \text{fps}=673\ \text{mph}$$

Since, from Table A-1, the speed of sound at 35,000 ft is 972 fps, this best-range airspeed corresponds to a slightly supersonic Mach number, specifically M 1.01. This is an unrealistic Mach number for this aircraft and the aircraft should cruise at M 0.85, which is 826 fps or 563 mph. This decision establishes the airspeed to be used in Eq. 3-88, and this airspeed establishes the lift coefficient as

$$C_L=\frac{2\times 115}{\rho_{\text{SL}}\times 0.310\times(826)^2}=0.4575$$

The drag coefficient can now be found from the drag polar to be 0.0259, so that the cruise lift-to-drag ratio is 17.7. This value is less than the maximum value of 18 but greater than the theoretical best-range value of 15.6. If the specific fuel consumption is 0.8 lb/h/lb and the cruise-fuel weight fraction is 0.3, then Eq. 3-88 shows a maximum range for this aircraft of 4,443 mi.

If, somehow, it were possible to keep the airspeed at M 0.85 and increase the lift-to-drag value to its maximum value of 18, the range could be increased to 4,518 mi, an increase of 1.7 percent. Although this is probably not enough of an increase for us to concern ourselves with, let us pursue its attainment for the sake of illustrating a point and procedure. Since the lift coefficient for the maximum lift-to-drag ratio is equal to $(C_{D0}/K)^{1/2}$ or 0.555, then use of the level-flight lift equation provides the relationship between the flight loading and density ratio that

$$\frac{W/S}{\sigma}=\frac{\rho_{SL}V^2}{2}\left(\frac{C_{DO}}{K}\right)^{1/2}=450\ \text{lb/ft}^2 \tag{3-89}$$

If the initial cruise altitude is to be kept at 35,000 ft, then the wing loading must be increased to 139.5 lb/ft^2 by reducing the wing area, with due consideration given to the effects on other aspects of performance, principally the take-off run. If the wing loading is to be kept at 115 lb/ft^2, then the altitude must be increased to 38,000 ft ($\sigma=0.255$) with due consideration given to the ceiling (T/W) constraints and the fuel required to climb to the higher altitude.

If an aircraft is to be designed to cruise in or near the stratosphere *at a specified airspeed, then the best-range condition is to fly at the maximum lift-to-drag ratio,* using this condition to establish an exchange ratio between the wing loading and the initial cruise altitude. For example, if the specified cruise airspeed in the stratosphere of this illustrative aircraft is to be M 0.8 (774 fps or 528 mph), then Eq. 3-89 yields the exchange (trade-off) ratio

$$\frac{W/S}{\sigma}=395\ \text{lb/ft}^2 \tag{3-90}$$

For an initial cruise altitude of 35,000 ft, the wing loading should be 122.5 lb/ft^2; for 40,000 ft, it need only be 97 lb/ft^2.

If we examine the Breguet (cruise-climb) range equation,

$$X=\frac{VE}{c}\ln\left(\frac{1}{1-\zeta}\right) \tag{3-91}$$

we see that, for a given fuel load and a given specific fuel consumption, the range is maximized by maximizing the product of the airspeed and the lift-to-drag ratio (VE). If there are no constraints on V, the product is maximized when V is such that E is equal to $0.866E_m$. If, however, V is constrained to a specified value, then E is also specified and the value of their product VE is established. The product VE can now be increased only by increasing E to its maximum attainable value, namely, the maximum lift-to-drag ratio for the aircraft. This can be accomplished either by the operational technique of increasing the cruise altitude, if such is possible, or by the redesign of the aircraft so as to increase the wing loading.

3-8 MAXIMUM ENDURANCE

Endurance is the length of time that an aircraft can remain airborne for a given expenditure of fuel and for a specified set of flight conditions. The instantaneous endurance, or exchange ratio between time and fuel, is the inverse of the fuel consumption rate, and for level or cruise-climb flight it can be written as

$$\frac{dt}{-dW}=\frac{1}{cD}=\frac{E}{cW} \tag{3-92}$$

In a preceding section, this expression was integrated, using the best-range conditions, to obtain the flight time for constant lift coefficient flight.

Maximum endurance is a primary design and operational criterion for aircraft with such special missions as patrol, antisubmarine warfare, observation, and command and control and is of interest for all aircraft during the loiter phase of any flight. *Loiter* is defined as flight where endurance is paramount and range is either secondary or of no importance at all. An intercontinental bomber, such as the B-52 or B-1, loiters while on airborne alert. A fighter on a combat air patrol (CAP) loiters while awaiting the assignment of targets or the sighting of intruding aircraft. All aircraft should loiter, if so permitted, in holding patterns while awaiting clearance for further flight to their destinations or for let-down and landing.

Inspection of Eq. 3-92 shows that the instantaneous endurance is at a maximum when both the specific fuel consumption and the drag are at a minimum. The minimum-drag condition (also shown by Eq. 3-92) occurs when the aircraft weight is as low as possible and the lift-to-drag ratio is at its maximum. Therefore, the endurance is maximized by the airspeed (ground speed does not affect endurance) that produces the lowest drag and thus requires the lowest amount of thrust, which in turn minimizes the fuel consumption rate. With a parabolic drag polar, the maximum endurance (minimum-drag) conditions are:

$$C_{L;t_{\max}} = C_{L,E_m} = \left(\frac{C_{D0}}{K}\right)^{1/2} \tag{3-93a}$$

$$E_{t_{\max}} = E_m \tag{3-93b}$$

$$V_{t_{\max}} = V_{md} = \left[\frac{2(W/S)}{\rho_{SL}\sigma}\right]^{1/2}\left[\frac{K}{C_{D0}}\right]^{1/4} \tag{3-93c}$$

$$\left(\frac{T}{W}\right)_{t_{\max}} = \frac{1}{E_m} \tag{3-93d}$$

Comparison with the best-range flight conditions for a given aircraft shows that maximum-endurance flight is slower (24 percent for identical altitudes), and thus at a higher lift coefficient, and that the flight altitude is the absolute ceiling for the throttle setting that produces the thrust required to maintain the airspeed specified by Eq. 3-93*c*.

If the flight lift-to-drag ratio is set at the maximum value for the aircraft and is kept constant along the flight path, then Eq. 3-92 can easily be integrated to yield

$$t_{\max;C_L} = \frac{E_m}{c}\ln\left(\frac{1}{1-\zeta}\right) = \frac{E_m}{c}\ln \mathrm{MR} \tag{3-94}$$

Equation 3-94 shows that with a given weight of fuel, maximum endurance calls for a large E_m and a low gross weight (as does best range), but that the endurance does not increase with an increase in the value of the minimum-drag airspeed. Consequently, the maximum endurance is independent of the wing loading and of the altitude. An aircraft flying at 40,000 ft at its minimum-drag airspeed will not remain airborne any longer (with the specific fuel consumption assumed to be independent of altitude) than if it flew at sea level at the lower sea-level minimum-drag airspeed. The aircraft will, however, fly farther at altitude because the airspeed is higher. It is also interesting to note that a comparison of the expressions for flight times shows that the maximum-endurance flight time is always 15.5 percent longer than the best-range flight time, no matter how long the range flown is.

Since Eq. 3-94 is valid only for constant lift coefficient flight, maintaining a constant airspeed requires cruise-climb flight, and maintaining a constant altitude requires a continuous decrease in the airspeed. If both altitude and airspeed are to be kept constant, then the expression for maximum endurance can be approximated by

$$t_{\max;h,V} = \frac{2E_m}{c} \arctan\left(\frac{0.5\zeta}{1-0.5\zeta}\right) \tag{3-95}$$

and the ratio of the times for such flight is

$$\left(\frac{t_{\max}}{t_{\text{br}}}\right)_{h,V} = \frac{\arctan\,[0.5\zeta/(1-0.5)]}{\arctan\,[0.433\zeta/(1-0.25\zeta)]} \cong \frac{1.155(1-0.25\zeta)}{1-0.5\zeta} \tag{3-96}$$

Equation 3-96 shows that the increase in the flight time (the endurance) by flying at the maximum lift-to-drag ratio, rather than at $0.866E_m$, increases with ζ, and that this increase in endurance is greater than the corresponding increase for constant lift coefficient flight.

PROBLEMS

The three aircraft whose major characteristics are given below will be used in many of the problems of this chapter and of the two subsequent chapters.

	Aircraft *A*	**Aircraft *B***	**Aircraft *C***
Type	**Executive**	**Medium Range**	**Long Range**
Gross weight (lb)	24,000	140,000	600,000
Wing area (ft^2)	600	2,333	5,128
Maximum thrust (lb)	6,000	37,800	180,000
C_{D0}	0.02	0.018	0.017
K	0.056	0.048	0.042
M_{DR}	0.72	0.8	0.85
$C_{L_{\max}}$	1.8	2.0	2.2

3-1. For Aircraft *A*, flying at sea level:

a. Plot the drag-to-weight ratio, D/W, and the maximum thrust-to-weight ratio, T/W, in the same figure as a function of the true airspeed in mph.

b. From (a) above, find the maximum and minimum airspeeds in mph and as a Mach number. Are these airspeeds physically attainable? Justify your answer.

c. Calculate the maximum and minimum airspeeds and the associated lift coefficients and lift-to-drag ratios.

d. Calculate the maximum lift-to-drag ratio and compare its value with that obtained from the plots of a above.

3-2. Do Prob. 3-1 for Aircraft *A* but at an altitude of 20,000 ft.

3-3. Do Prob. 3-1 for Aircraft *A* but at an altitude of 30,000 ft.

3-4. Do Prob. 3-1 for Aircraft *A* but at an altitude of 40,000 ft.

3-5. For Aircraft *A*:

a. Find the maximum absolute ceiling using the approximation of Eq. 3-42*a*.

b. Find the airspeed (mph) and Mach number at this ceiling. If the latter is greater than the drag-rise Mach number, determine the actual ceiling.

c. Do (a) above, using the more exact relationship of Eq. 3-41 and compare with the results of (a).

3-6. Do Prob. 3-1 for Aircraft *B*.

3-7. Do Prob. 3-2 for Aircraft *B*.

3-8. Do Prob. 3-3 for Aircraft *B*.

3-9. Do Prob. 3-4 for Aircraft *B*.

3-10. Do Prob. 3-5 for Aircraft *B*.

3-11. Do Prob. 3-1 for Aircraft *C*.

3-12. Do Prob. 3-2 for Aircraft *C*.

3-13. Do Prob. 3-3 for Aircraft *C*.

3-14. Do Prob. 3-4 for Aircraft *C*.

3-15. Do Prob. 3-5 for Aircraft *C*.

3-16. Find the density ratio and altitude (ft) at the absolute ceiling for each of the following T/W and maximum lift-to-drag combinations:

a. 1.2; 12   b. 0.25; 14
c. 0.25; 20   d. 0.7; 10
e. 0.3; 18   f. 0.28; 16

3-17. Aircraft *A* is to cruise at sea level with a tsfc of 0.95 lb/h/lb, a cruise-fuel weight of 4,800 lb, and a cruise Mach number of 0.4.

a. Find the T/W and L/D ratios at the start of cruise along with the mileage in mi/lb and mi/gal and the fuel flow rate (lb/h).

b. For a constant altitude-constant lift coefficient flight program, find the range along with the airspeed, Mach number, and mileage (mi/gal) at the end of cruise.

c. For a cruise-climb flight program, find the range along with the altitude and mileage (mi/gal) at the end of cruise and the average flight-path angle.

d. For a constant altitude-constant airspeed flight program, find the range along with the mileage (mi/gal) at the end of cruise.

3-18. Do Prob. 3-17 for Aircraft *A* but at a cruise altitude of 35,000 ft.

3-19. Do Prob. 3-17 but for Aircraft *B* with a tsfc of 0.9 lb/h/lb, a cruise-fuel weight of 35,000 lb, and a cruise Mach number of 0.6.

3-20. Do Prob. 3-18 for Aircraft *B* at a cruise altitude of 35,000 ft.

3-21. Do Prob. 3-17 but for Aircraft *C* with a tsfc of 0.85 lb/h/lb, a cruise-fuel weight of 180,000 lb, and a cruise Mach number of 0.7.

3-22. Do Prob. 3-21 for Aircraft *C* at a cruise altitude of 35,000 ft.

3-23. Do Prob. 3-17 for Aircraft *A* at a cruise altitude of 35,000 ft, but this time cruise at the best-range Mach number or at the drag-rise Mach number, whichever is lower. Find the best-range airspeed and Mach number and then do parts (a) through (d).

3-24. Do Prob. 3-23 in conjunction with Prob. 3-19 for Aircraft *B*.

3-25. Do Prob. 3-23 in conjunction with Prob. 3-21 for Aircraft *C*.

3-26. Find the highest best-range cruise altitudes for Aircraft *A*, *B*, and *C*, using maximum thrust and assuming that the ceiling weight is the gross weight.

3-27. Aircraft *A* is scheduled to fly 1,200 mi under best-range conditions, starting at 30,000 ft with a tsfc of 0.95 lb/h/lb.

a. Find the fuel required (lb and gal) and the cruise-fuel weight fraction for a constant altitude-constant airspeed flight program.

b. Find the fuel required and the fuel fraction for stepped-altitude flight with 4,000-ft steps. What is the altitude at the end of cruise?

c. Do (b) above, using 2,000-ft steps.

d. Find the fuel required and the fuel fraction for cruise climb along with the altitude at the end of cruise.

3-28. Do the stepped-altitude problem of Prob. 3-27, using Aircraft *B* with a tsfc of 0.9 lb/h/lb and for a range of 2,000 mi.

3-29. Do the stepped-altitude problem of Prob. 3-27, using Aircraft *C* with a tsfc of 0.8 lb/h/lb and for a range of 4,000 mi.

3-30. Aircraft *A* has a cruise-fuel weight fraction of 0.2 and a tsfc of 0.95 lb/h/lb.

a. At what altitude will the best-range airspeed be equal to the drag-rise Mach number? What will the cruise-climb range be? Is there enough thrust available to fly at this altitude?

b. At what altitude will the drag-rise Mach number be equal to the minimum-drag (maximum lift-to-drag ratio) airspeed? What will the cruise-climb range be? Is there enough thrust available to fly this program?

3-31. Do Prob. 3-30 with Aircraft *B* with a cruise-fuel weight fraction of 0.25 and a tfsc of 0.9 lb/h/lb.

3-32. Do Prob. 3-30 with Aircraft *C* with a cruise-fuel weight fraction of 0.3 and a tsfc of 0.8 lb/h/lb.

3-33. Aircraft *A* has a cruise-fuel weight of 4,800 lb and a tfsc of 0.95 lb/h/lb.

a. Find its maximum-endurance airspeed, range, and flight time at sea level.

b. Do (a) above, for 30,000 ft.

3-34. Do Prob. 3-33 for Aircraft *B* with a cruise-fuel weight of 35,000 lb and a tfsc of 0.9 lb/h/lb.

3-35. Do Prob. 3-33 for Aircraft *C* with a cruise-fuel weight of 150,000 lb and a tfsc of 0.85 lb/h/lb.

3-36. You are to do the preliminary sizing of the lowest weight aircraft to meet the following operational requirements: a payload of 20 passengers (at 200 lb each) plus 1,500 lb of cargo, a cruise range of 1,800 mi at 30,000 ft and at an airspeed of 440 mph, and an absolute ceiling of 40,000 ft. Based on your experience with this type of aircraft, you choose the following tentative values for your first try: AR $=7$, $C_{D0}=0.019$, $e=0.9$, tsfc $=$ 0.8 lb/h/lb, a structural weight fraction of 0.44, and a thrust-to-engine weight of 5.

a. Find the cruise lift coefficient and then the wing loading.

b. Find the cruise-fuel weight fraction. Then with the assumption that the cruise fuel is 80 percent of the total fuel loaded aboard the aircraft, find the total fuel weight fraction.

c. Find the payload weight fraction and the gross weight of the aircraft.

d. Using a minimum of two engines, determine the number and size of the engines.

e. Find the remaining characteristics of the aircraft, such as the wing area, wing span, engine weight, average chord, operational empty weight.

3-37. Unused fuel is excess weight and penalizes the performance of an aircraft. Aircraft *A* is loaded with 4,800 lb of cruise fuel for a cruise-climb range of 1,000 mi, starting at 35,000 ft. Assume a tsfc of 0.9 lb/h/lb.

a. Find the best-range airspeed and the fuel required (lb) for this cruise range. How much excess cruise fuel is there?

b. Remove this excess fuel, thus reducing the gross weight and the wing loading of the aircraft. Keeping the cruise airspeed equal to that of (a) above (remember that the lift-to-drag ratio will change), find the fuel now required for this cruise range and compare with the results of (a).

c. Find the new best-range airspeed with the reduced wing loading and do (b) with this airspeed and its associated lift-to-drag ratio.

3-38. Aircraft B is loaded with 35,000 lb of fuel for cruise for a cruise-climb range of 2,000 mi, starting at 35,000 ft. Assume a tsfc of 0.85 lb/h/lb. Do Prob. 3-37 but do not exceed the drag-rise Mach number.

3-39. Aircraft C is loaded with 180,000 lb of cruise fuel for a cruise-climb range of 3,000 mi, starting at 35,000 ft. Assume a tsfc of 0.8 lb/h/lb, and do not exceed the drag-rise Mach number. Do Prob. 3-37.

3-40. Do Prob. 3-37 with the cruise range reduced to 500 mi.

3-41. Do Prob. 3-38 with the cruise range reduced to 1,000 mi.

3-42. Do Prob. 3-39 with the cruise range reduced to 2,000 mi.

Chapter Four

Other Flight in the Vertical Plane: Turbojets

4-1 TAKE-OFF AND LANDING

The ground-run requirement for an aircraft, though not a flight condition, is important for performance and operational reasons and often may be the determining factor in the selection of certain design and subsystem characteristics, such as the thrust-to-weight ratio and the wing loading.

The equation of motion governing the take-off ground run of a conventional aircraft can be written as

$$T-D-\mu(W-L)=\frac{W}{g}\frac{dV}{dt} \tag{4-1}$$

where μ, the coefficient of rolling friction, is of the order of 0.02 for a dry, hard runway. Rather than use numerical or iterative procedures to obtain a solution of this equation, we shall obtain a closed-form expression by assuming that the thrust is constant throughout the run and by neglecting the drag and friction forces, which account for 10 to 20 percent of the energy expended. With these assumptions, Eq. 4-1 can now be written as

$$T=\frac{W}{g}\frac{dV}{dX}\frac{dX}{dt}=\frac{WV}{g}\frac{dV}{dX} \tag{4-2}$$

Separating the variables and integrating from the start of the run to lift-off (lift-off conditions will be denoted by the subscript LO) yields the expression

$$d=\frac{V_{\text{LO}}^2}{2g(T/W)} \tag{4-3}$$

The lift-off airspeed (V_{LO}) is generally set equal to $1.2V_s$, where V_s is the stall speed. These two airspeeds can be found from

$$V_s=\left[\frac{2(W/S)}{\rho_{\text{SL}}\sigma C_{L_{\max,\text{TO}}}}\right]^{1/2} \tag{4-4}$$

$$V_{\text{LO}}=\left[\frac{2(W/S)}{\rho_{\text{SL}}\sigma C_{L,\text{LO}}}\right]^{1/2}=\left[\frac{2.88(W/S)}{\rho_{\text{SL}}\sigma C_{L_{\max}\ \text{TO}}}\right]^{1/2} \tag{4-5}$$

where $C_{L,\text{LO}}=0.694C_{L_{\max},\ \text{TO}}$.

With the substitution of Eq. 4-5 into Eq. 4-3, we obtain

$$d=\frac{1.44(W/S)}{\rho_{SL}\sigma g C_{L_{max},TO}(T/W)}=\frac{1.44(W/S)}{\rho_{SL}\sigma^2 g C_{L_{max},TO}(T/W)_{SL}} \tag{4-6}$$

This expression gives overly optimistic values for the take-off run (of the order of 10 to 20 percent too low) but is valuable for its insights into the parameters affecting the run. We see that to decrease the take-off run, we need to decrease the wing loading or increase the thrust-to-weight ratio, the maximum lift coefficient, or the atmospheric density.

Decreasing the wing loading, however, decreases the best-range airspeed and thus the cruise range for a given cruise-fuel weight fraction. Increasing the thrust-to-weight ratio beyond that required for cruise or for special mission requirements (e.g., ceilings) increases the engine weight with accompanying increases in the gross weight of the aircraft and in the weight of the fuel required for a specified range. Increasing the maximum lift coefficient by means of flaps (Fowler-type flaps also decrease the wing loading by increasing the effective wing area during the take-off) adds weight and complexity and increases the zero-lift drag coefficient during the ground run. The atmospheric density cannot be controlled by man, but its effect on take-off performance cannot be neglected. Obviously, an airfield located 5,000 ft above mean sea level will require a longer ground run than one at sea level. In addition, a hot day and/or a low pressure area will also decrease the air density, thus increasing the ground run.

We can develop an approximate, and also somewhat optimistic, expression for the time required for lift-off by recognizing that the thrust-to-weight ratio, although appearing dimensionless, actually is an acceleration expressed in g's. Neglecting all other forces, Newton's law can be written as

$$T=ma=mg\left(\frac{a}{g}\right)=W\left(\frac{a}{g}\right)$$

so that

$$\frac{T}{W}=\frac{a}{g} \tag{4-7}$$

Consequently, a T/W ratio of 0.25 represents an acceleration of 0.25 g's or 8.05 ft/s^2. Since

$$d=\tfrac{1}{2}at^2=\frac{g(T/W)t^2}{2}$$

we can solve for

$$t=\left[\frac{0.062d}{\sigma(T/W)_{SL}}\right]^{1/2}=\frac{V_{LO}}{g\sigma(T/W)_{SL}} \tag{4-8}$$

where d, the ground run in feet, can be obtained from Eq. 4-6.

Let us look at several typical classes of aircraft. A long-range high-subsonic transport with a wing loading of 100 lb/ft^2, a thrust-to-weight ratio of 0.25, and a maximum take-off lift coefficient of 1.8 would require, according to Eqs. 4-6 and 4-8, a sea-level ground run of 4,200 ft with a lift-off airspeed of 177 mph and an elapsed time of 32 seconds. Remembering that these are optimistic figures and noting that provisions must be made for nonstandard day conditions, clearance of a 50-ft obstacle, factors of safety, take-off aborts, etc., such a ground-run value implies the need for runway lengths of the order of 8,000 to 10,000 ft for such an aircraft.

Short-range transports used for travel between and from smaller cities cannot count on finding 10,000-ft runways. Consequently, their design characteristics must be modified so as to reduce the ground run, generally by reducing the wing loading and increasing the maximum lift coefficient for take-off. The thrust-to-weight ratio is left unchanged, if possible, for weight and operating economy reasons. Reducing W/S from 100 to 50 lb/ft^2 and increasing the lift coefficient from 1.8 to 2.0, but leaving T/W equal to 0.25, results in a ground run of approximately 1,900 ft with a lift-off airspeed of 119 mph and an elapsed time of 22 seconds. This ground-run value implies a need for runway lengths of the order of 4,000 to 5,000 ft.

A STOL (Short Take-Offs and Landing) aircraft is designed for exceptionally short ground runs as well as steep climbs and descents. It is characterized by a low wing loading, a high lift coefficient, and a large thrust-to-weight ratio. For example, if W/S is 40 lb/ft^2, $C_{L_{max}}$ is 2.6, and T/W is 0.6, then the ground run becomes of the order of 500 ft with a lift-off airspeed of 93 mph and an elapsed time of 7 seconds.

Air-to-air fighter aircraft, particularly those equipped with guns, are characterized by a lower wing loading than transports for reasons of maneuverability, higher thrust-to-weight ratios, as might be expected, and lower values of the maximum lift coefficient (to keep the wing thin and clean so as to minimize C_{D0} and the wave drag). An advanced fighter with a T/W of 1.3 (without afterburners), a W/S of 66 lb/ft^2, and a maximum C_L of 1.2 would have a ground run of the order of 800 ft and an elapsed ground-run time of 6 seconds. The sea-level stall and lift-off airspeeds for such an aircraft would be 215 fps (147 mph) and 258 fps (176 mph), respectively.

The landing requirements must also be considered. They have increased in importance as wing loadings (and thus approach and touchdown airspeeds) have increased and as the drag has decreased. The landing maneuver comprises the final approach, the landing flare, the touchdown (to include getting all the wheels on the ground), and the ground run. A simple but crude expression for the landing-run distance, in feet, is

$$d \cong \tfrac{1}{2} f_a V_s^2 \tag{4-9}$$

where f_a, the deceleration factor, is of the order of 0.4 for a conventional jet transport and V_s is the stall speed at landing. This expression is crude because of the variety and complexity of retardation devices on modern aircraft (such as thrust reversers and spoilers) and the differences between wet and dry runways. If our long-range transport has been on a ζ equal to 0.4 mission and has a landing lift coefficient of 2.0, then the landing wing loading will be equal to $100(1-\zeta)$ or

60 lb/ft^2. Consequently, V_s will be 159 fps (108 mph); the final approach airspeed is normally 20 percent higher than the stall speed. With f_d equal to 0.4, the landing run will be of the order of 5,000 ft. So we can see that stopping a clean, high-speed aircraft is neither a trivial nor unimportant task.

4-2 CLIMBING FLIGHT

In order to examine climbing flight, we shall return to the dynamic and kinematic equations of Sec. 3-1, which are repeated here along with the weight balance equation for a turbojet aircraft:

$$T - D - W \sin \gamma = 0 \tag{4-10}$$

$$L - W \cos \gamma = 0 \tag{4-11}$$

$$\frac{dX}{dt} = V \cos \gamma \tag{4-12}$$

$$\frac{dh}{dt} = V \sin \gamma \tag{4-13}$$

$$\frac{-dW}{dt} = cT \tag{4-14}$$

From Eq. 4-10, we find that

$$\sin \gamma = \frac{T - D}{W} \tag{4-15}$$

which shows that the climb angle is determined by the excess thrust per unit weight, where excess thrust refers to that thrust not required to counteract the drag. Substituting Eq. 4-15 into Eq. 4-13 gives an expression for dh/dt, which we shall call the rate of climb and write as R/C. This expression, i.e.,

$$\text{R/C} = V \sin \gamma = \frac{TV - DV}{W} \tag{4-16}$$

tells us that the rate of climb is determined by the excess thrust power per unit weight.

Returning to Eq. 4-15 and rewriting it as

$$\sin \gamma = \frac{T}{W} - \frac{D}{W} = \frac{T}{W} - \frac{\cos \gamma}{E} \tag{4-17}$$

we see that the largest value that the sine of the climb angle can assume is equal to the maximum thrust-to-weight ratio. Thus, we can obtain a qualitative feel for the magnitude of the maximum climb angle. If the maximum T/W is 0.25, a typical value for a subsonic jet transport, then the maximum climb angle will be less than 15 deg at sea level; if the maximum T/W is 0.6, the maximum climb angle will be less than 37 deg. As the altitude increases during the climb, the available T/W will decrease (if we assume W to be constant) and the maximum climb angle will, therefore, decrease also, going to zero at the absolute ceiling of the aircraft, as might be expected.

For an aircraft with a parabolic drag polar, that is, $C_D = C_D + KC_L^2$, the climbing-flight drag function can be written as

$$D = qSC_{DO} + \frac{KW^2 \cos^2 \gamma}{qS} \tag{4-18}$$

since

$$C_L = \frac{L}{qS} = \frac{W \cos \gamma}{qS} \tag{4-19}$$

The first term of Eq. 4-18 represents the zero-lift drag, and the second term the drag-due-to-lift (induced drag). In general, for climbing flight, the induced drag will be less than the zero-lift drag and will decrease as the climb angle increases, going to zero for a 90 deg climb angle (flight straight up). For such flight, the drag will be entirely zero-lift drag and will be equal to qSC_{DO}; obviously, the thrust-to-weight ratio must be greater than unity. For our analyses, we shall ignore the effects of the climb angle upon the drag function and use the approximation that

$$D = qSC_{DO} + \frac{KW^2}{qS} \tag{4-20}$$

This level-flight drag approximation can be used with reasonable accuracy for climb angles up to the order of 30 deg or so.

In order to find the rate of climb for a specified airspeed at a specified altitude, use Eq. 4-15 in conjunction with Eq. 4-20 to find the sine of the climb angle and then use Eq. 4-16 to obtain the rate of climb. We shall assume the gross weight of the aircraft to be constant throughout the climb because the weight of the fuel burned is a small portion of the gross weight of a well-designed aircraft climbing in an efficient manner. The largest rate of climb for a given throttle setting (thrust) will occur at sea level, and the rate of climb will decrease to zero at the absolute ceiling.

To illustrate the use of the climbing equations, let us consider our jet transport with a W/S of 100 lb/ft^2, a maximum T/W of 0.25, and the drag polar, $C_D = 0.015 + 0.06C_L^2$. We plan to climb at a constant airspeed of 400 mph (587 fps) and want to find the rate of climb at sea level and at 30,000 ft. At sea level, the dynamic pressure q is equal to 409.5 lb/ft^2. Since we do not know either W or S, let us rewrite Eq. 4-20 as the drag-to-weight ratio,

$$\frac{D}{W} = \frac{qC_{DO}}{W/S} + \frac{K(W/S)}{q} \tag{4-21}$$

From Eq. 4-21, the sea-level D/W is

$$\left(\frac{D}{W}\right)_{SL} = 0.0614 + 0.0146 = 0.076$$

From Eq. 4-17, with the maximum T/W of 0.25, we find that

$$\sin \gamma_{SL} = 0.25 - 0.076 = 0.174$$

so that the sea-level climb angle is 10 deg. Substituting into Eq. 4-16 results in

$$(R/C)_{SL} = 587 \times 0.174 = 102.14 \text{ fps} = 6{,}128 \text{ fpm}$$

At 30,000 ft, the dynamic pressure is 153 lb/ft^2, and the maximum thrust-to-weight ratio is 0.0935 (0.25 × 0.374), so that the D/W ratio is

$$\left(\frac{D}{W}\right)_{30K} = 0.0203 + 0.039 = 0.062$$

and the sine of the climb angle is

$$\sin \gamma = 0.0935 - 0.062 = 0.0315$$

The climb angle is only 1.8 deg and the R/C is

$$(RIC)_{30K} = 587 \times 0.0315 = 18.5 \text{ fps} = 1{,}110 \text{ fpm}$$

From this set of calculations, we see that for a constant climb airspeed of 400 mph, the climb angle and rate of climb decreased from their sea-level values of 10 deg and 6,128 fpm, respectively, to 1.8 deg and 1,110 fpm at 30,000 ft.

There are three climbing-flight conditions of special interest. They are steepest climb γ_{max}, fastest climb $(R/C)_{max}$, and most economical climb $(-dW/dh)_{min}$. Steepest climb will be described and discussed in the next section, and the other two will be treated in the sections following.

4-3 STEEPEST CLIMB

Steepest climb is climbing flight at the maximum climb angle. It is of interest in clearing obstacles, such as mountains or trees and buildings at the end of a runway, and in establishing the upper limit for the climb angle. If Eq. 4-17 is rewritten here for convenience,

$$\sin \gamma = \frac{T}{W} - \frac{D}{W} = \frac{T}{W} - \frac{\cos \gamma}{E} \tag{4-22}$$

then the maximum climb angle obviously is obtained when the excess thrust is maximized by maximizing the thrust-to-weight ratio and by minimizing the ratio of the cosine of the climb angle to the lift-to-drag ratio. With the constant-weight assumption and the level-flight drag approximation, the maximum climb angle can be found from

$$\sin \gamma_{max} = \frac{T_m}{W} - \frac{1}{E_m} \tag{4-23}$$

This is an approximate solution, as are the steepest-climb conditions that follow

$$V_{\gamma max} = \left[\frac{2(W/S)}{\rho_{SL}\sigma}\right]^{1/2}\left[\frac{K}{C_{DO}}\right]^{1/4} \tag{4-24}$$

$$E_{\gamma max} = E_m \tag{4-25}$$

$$(R/C)_{\gamma max} = V_{\gamma max} \sin \gamma_{max} \tag{4-26}$$

Although approximate, they do provide useful values of suitable accuracy for most conventional aircraft. Special high-performance aircraft, such as interceptor aircraft, should be given special consideration to include acceleration along the flight path. The energy-state approximation of Sec. 8-6 is one technique for examining the climbing performance of such aircraft.

For the turbojet transport of the preceding section, the steepest-climb conditions at sea level are:

$$V_{\gamma max} = \left(\frac{2 \times 100}{\rho_{SL}}\right)^{1/2}\left(\frac{0.06}{0.015}\right)^{1/4} = 410 \text{ fps} = 280 \text{ mph}$$

$$\sin \gamma_{max} = 0.25 - \frac{1}{16.67} = 0.190$$

$$\gamma_{max} = 11 \text{ deg}$$

$$(R/C)_{\gamma max} = 410 \times 0.190 = 77.9 \text{ fps} = 4{,}674 \text{ fpm}$$

At 30,000 ft, the steepest-climb conditions are:

$$V_{\gamma max} = \frac{V_{\gamma max, SL}}{\sigma^{1/2}} = \frac{410}{(0.374)^{1/2}} = 670 \text{ fps} = 457 \text{ mph}$$

$$\sin \gamma_{max} = 0.25\sigma - \frac{1}{16.67} = 0.0335$$

$$\gamma_{max} = 1.92 \text{ deg}$$

$$(R/C)_{\gamma max} = 670 \times 0.0335 = 22.4 \text{ fps} = 1{,}347 \text{ fpm}$$

At the absolute ceiling of the aircraft, the steepest-climb angle is zero, as can be seen from

$$\sin \gamma_{max} = \frac{T_m \sigma_c}{W} - \frac{1}{E_m} = 0$$

Examination of Fig. 4-1, which is a sketch of the drag (required thrust) and of the available thrust for a given weight and altitude indicates that the maximum excess thrust occurs in the vicinity of the minimum-drag airspeed. The minimum-drag airspeed is the airspeed where the thrust-to-weight ratio is a maximum.

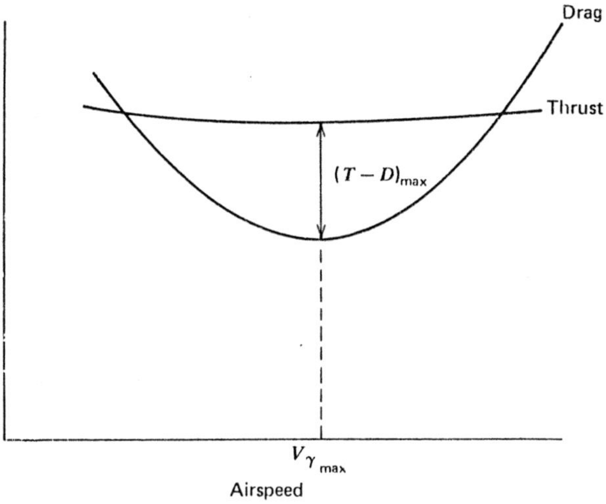

FIGURE 4-1

Graphical representation of steepest-climb condition for a turbojet for a given weight, altitude, and throttle setting.

4-4 FASTEST CLIMB

Fastest climb is synonymous with climb at the maximum rate of climb and is of much greater interest to us than steepest climb. Fastest climb requires the minimum time to climb to a specified altitude, which is of importance to air traffic control who must keep the intervening air space clear of traffic. Furthermore, to a first approximation, fastest climb requires the smallest amount of fuel, thus increasing the amount available for cruise.

The rate of climb is proportional to the excess power per unit weight (see Eq. 4-16 and Fig. 4-2) and can be maximized by setting its first derivative with respect to the true airspeed equal to zero, i.e.,

$$\frac{d(\mathrm{R/C})}{dV}=\frac{d}{dV}\left(\frac{TV-DV}{W}\right)=0 \tag{4-27}$$

Carrying out this differentiation, with the aircraft weight assumed constant, produces the condition for the fastest climb of a turbojet that

$$T-D-V\frac{dD}{dV}=0 \tag{4-28}$$

With the realization that the climb angle will be less than that for steepest climb, substitution of the drag function of Eq. 4-20 into Eq. 4-28 and solving for the dynamic pressure yields the expression

$$q_{\mathrm{FC}}=\frac{T/S}{6C_{DO}}\left(1\pm\left\{1+\frac{3}{[E_m(T/W)]^2}\right\}^{1/2}\right) \tag{4-29}$$

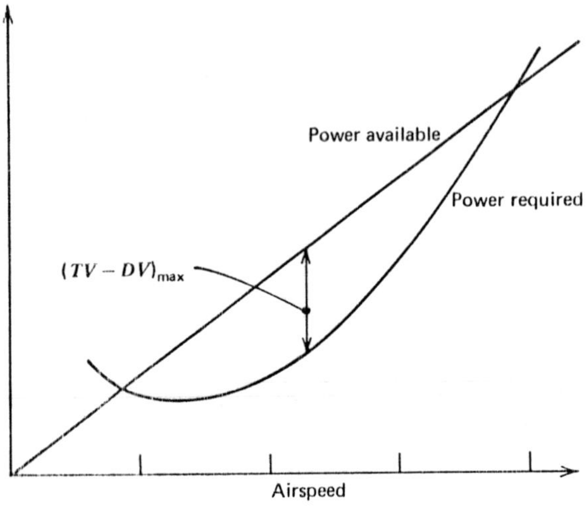

FIGURE 4-2

Graphical representation of fastest-climb condition for a turbojet for a given weight, altitude, and power setting.

where the subscript FC denotes fastest-climb conditions. The first thing to notice is that the dynamic pressure is not necessarily independent of the altitude inasmuch as the thrust available for a given throttle setting decreases with altitude. The second thing to notice is that the minus sign within the larger parenthesis results in negative values for q, an unrealistic situation.

Dropping the minus sign, Eq. 4-29 and subsequent expressions can be simplified by giving the term within the larger parentheses the symbol Γ, where

$$\Gamma = 1 + \left\{1 + \frac{3}{[E_m(T/W)]^2}\right\}^{1/2} \tag{4-30}$$

At the absolute ceiling for a given throttle setting, the thrust-to-weight ratio is equal to the reciprocal of the maximum lift-to-drag ratio, so that Γ at the ceiling is equal to three. Since the T/W ratio has its largest values at sea level, the sea-level value of Γ will approach a value of 2 for large values of the product of E_m and T/W. Consequently, Γ must take on values between 2 and 3.

With the substitution of Γ, Eq. 4-29 can now be written as

$$q_{FC} = \frac{(T/S)\Gamma}{6C_{DO}} \tag{4-31}$$

so that

$$V_{FC} = \left[\frac{(T/S)\Gamma}{3\rho_{SL}\sigma C_{DO}}\right]^{1/2} = \left[\frac{(T_{SL}/S)\Gamma}{3\rho_{SL} C_{DO}}\right]^{1/2} \tag{4-32}$$

The remaining flight conditions for fastest climb are:

$$\left(\frac{D}{W}\right)_{FC} = \frac{(T/W)\Gamma}{6} + \frac{3}{2\Gamma E_m^2(T/W)} \tag{4-33}$$

$$\sin \gamma_{FC} = \frac{T}{W}\left(1 - \frac{\Gamma}{6}\right) - \frac{3}{2\Gamma E_m^2(T/W)} \tag{4-34}$$

$$(R/C)_{max} = (R/C)_{FC} = V_{FC} \sin \gamma_{FC} \tag{4-35}$$

If there are no flight constraints or limitations, the fastest climb of an aircraft will obviously occur when the maximum available thrust is used. Examination of the fastest-climb relationships shows that a turbojet aircraft with a high rate of climb is characterized by one or more of the following: a large thrust loading (T/S), a large thrust-to-weight ratio (T/W), a large maximum lift-to-drag ratio (E_m), and a small zero-lift drag coefficient. The maximum rate of climb occurs at sea level and decreases with altitude, going to zero at the absolute ceiling.

In calculating the maximum rate of climb at any specified altitude, the simplest procedure is to calculate Γ for that altitude, use that value to determine the corresponding airspeed and the sine of the climb angle, and then multiply the last two together to obtain the fastest-climb rate. Remember that the airspeed will be in fps as will the climb rate.

Our illustrative turbojet has a maximum T/W of 0.25, a W/S of 100 lb/ft^2, a C_{D0} of 0.015, and an E_m of 16.67; the corresponding maximum T/S is 25 lb/ft^2. Using maximum thrust throughout the climb, the maximum rate of climb will be calculated at sea level, at 30,000 ft, and at the absolute ceiling of the aircraft.

At sea level: $$\Gamma = 1 + \left[1 + \frac{3}{(16.67 \times 0.25)^2}\right]^{1/2} = 2.083$$

$$V_{FC} = \left(\frac{25 \times 2.083}{3\rho_{SL} \times 0.015}\right)^{1/2} = 698 \text{ fps} = 476 \text{ mph} = M\ 0.63$$

$$\sin \gamma_{FC} = 0.25\left(1 - \frac{2.083}{6}\right) - \frac{3}{2 \times 2.083 \times (16.67)^2 \times 0.25} = 0.153$$

$$\gamma_{FC} = 8.8 \text{ deg}$$

$$(R/C)_{FC} = 698 \times 0.153 = 106.8 \text{ fps} = 6{,}408 \text{ fpm}$$

Notice that the climb angle of 8.8 deg is less than the value of 11 deg obtained for steepest climb.

At 30,000 ft, where $\sigma = 0.374$, the T/W is 0.0935 (0.25×0.347) and the T/S, therefore, is 9.35 lb/ft^2, so that:

$$\Gamma = 1 + \left[1 + \frac{3}{(16.67 \times 0.0935)^2}\right]^{1/2} = 2.495$$

$$V_{FC} = \left(\frac{9.35 \times 2.495}{3\rho_{SL} \times 0.374 \times 0.015}\right)^{1/2} = 764 \text{ fps} = 520 \text{ mph} = M\ 0.78$$

$$\sin \gamma_{FC} = 0.0935\left(1 - \frac{2.495}{6}\right) - \frac{3}{2 \times 2.495 \times (16.67)^2 \times 0.935} = 0.0315$$

$$\gamma_{FC} = 1.8 \text{ deg}$$

$$(R/C)_{FC} = 764 \times 0.0315 = 24 \text{ fps} = 1{,}440 \text{ fpm}$$

Notice the increase in the airspeed and the reduction in both the climb angle and the rate of climb.

At the absolute ceiling of the aircraft, the product of the thrust-to weight ratio and the maximum lift-to-drag ratio becomes equal to unity so that Γ becomes equal to 3 and

$$\frac{T_c}{W} = \frac{1}{E_m}$$

$$\sin \gamma_{FC} = \frac{1}{E_m}\left(1 - \frac{3}{6}\right) - \frac{3E_m}{2 \times 3E_m{}^2} = 0$$

$$\gamma_{FC} = 0$$

$$(R/C)_{FC} = 0$$

It can be shown that the airspeeds for steepest climb, fastest climb, and the minimum-drag airspeed are all identically equal at the absolute ceiling. Since, at the ceiling, $\sigma = 0.24$ and $T/S = 6.0$, the true airspeed is 837.4 fps or 571 mph or M 0.865, which is probably in excess of the drag-rise Mach number. Therefore, this represents a case where the theoretical absolute ceiling cannot actually be reached.

In order to show the variations with altitude of the airspeed, the climb angle, and the rate of climb for the fastest-climb program from sea level to the absolute ceiling, the values at 5,000-ft intervals were calculated and used to construct Table 4-1, which is also sketched in Fig. 4-3.

The actual rate of climb is of less interest than the length of time required to climb to a particular altitude and the amount of fuel consumed during climb. Rather than attempt to integrate Eq. 4-35 and its auxiliary equations, average values will be used to determine the increment of time for a specific altitude interval using the expression

$$\Delta t = \frac{\Delta h}{(R/C)_{FC,\text{ave}}} \tag{4-36}$$

Similar expressions for the fuel consumed and the ground distance covered during an increment of time are

$$\frac{\Delta W_f}{W} = \frac{c(T/W)_{\text{ave}}\,\Delta t}{3{,}600} \tag{4-37}$$

$$\Delta X = \frac{V_{\text{ave}} \cos \gamma_{\text{ave}}\,\Delta t}{5{,}280} \tag{4-38}$$

TABLE 4-1

Fastest-Climb Values

$T_{max}/W = 0.25$; $W/S = 100$ lb/ft^2; $E_{max} = 16.67$; $C_{D_0} = 0.015$; sfc = 0.8 lb/h/lb

h (1,000 ft)	T_{max}/W (g's)	Γ	V (fps)	$\sin\gamma$	$(R/C)_{max}$ (fps)	γ (deg)
SL	0.25	2.083	698	0.153	106.7	8.8
5	0.216	2.11	702	0.128	89.7	7.35
10	0.184	2.148	708	0.105	74.0	6.00
15	0.157	2.20	717	0.099	71.3	5.7
20	0.133	2.27	728	0.065	47.2	3.72
25	0.112	2.36	743	0.047	35.3	2.72
30	0.0935	2.495	763	0.0315	24.0	1.8
35	0.0775	2.672	790.4	0.0169	13.4	0.97
40	0.0615	2.9631	832.3	0.00149	1.244	0.08
40.58	0.06	3.0	837.4	0	0	0

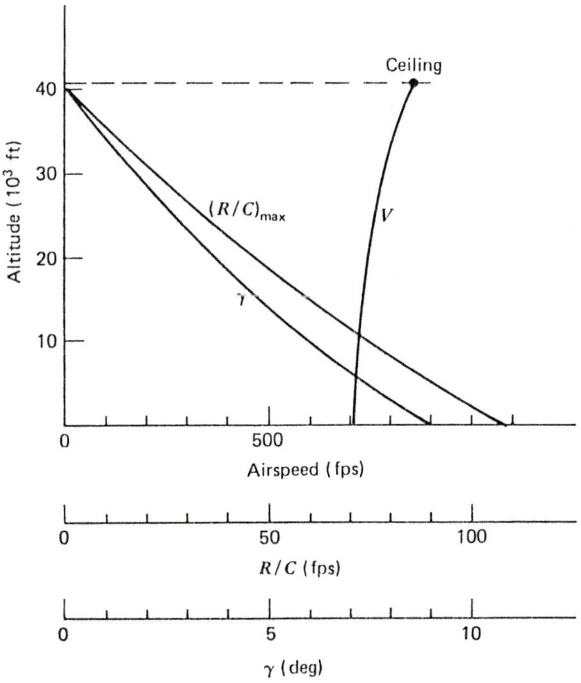

FIGURE 4-3

Fastest-climb conditions for the illustrative turbojet.

with Δt in seconds, V in fps, and ΔX in miles. Equation 4-37 comes from the weight balance equation with both sides divided by the aircraft weight to avoid the need to specify the actual weight of the aircraft. $\Delta W_f/W$ represents an increment of climb-fuel weight fraction. The average and cumulative values are tabulated in Table 4-2 and sketched in Fig. 4-4 for our illustrative turbojet with a specific fuel consumption of 0.8 lb/h/lb.

TABLE 4-2
Additional Fastest-Climb Values

h (ft)	$(R/C)_{max}$ (fps)	Δt (s)	t (s)	$(T/W)_{ave}$	$\Delta W_f/W$ ($\times 10^3$)	$\sum \Delta W_f/W$ ($\times 10^3$)	V_{ave} (fps)	ΔX (mi)	X (mi)
SL–5,000	98.2	50.9	50.9	0.233	2.63	2.63	700.0	6.7	6.7
5–10,000	81.85	61.1	112.0	0.200	2.71	5.34	705.0	8.1	14.8
10–15,000	72.65	68.8	180.8	0.1705	2.61	7.95	712.5	9.2	24.0
15–20,000	59.25	84.4	265.2	0.145	2.72	10.67	722.5	11.5	35.5
20–25,000	41.25	121.2	386.4	0.1225	3.30	13.97	735.5	16.9	52.4
25–30,000	29.65	168.6	555.0	0.1027	3.80	17.77	753.0	17.3	69.7
30–35,000	18.7	267.4	822.4	0.0855	5.00	22.77	776.7	39.3	149.0
35–40,000	7.3	682.9	1,505.3	0.0695	10.51	33.28	811.4	104.9	213.9
40–40,576	0.6	833.3	2,338.6	0.0607	11.12	44.40	834.8	131.7	345.6

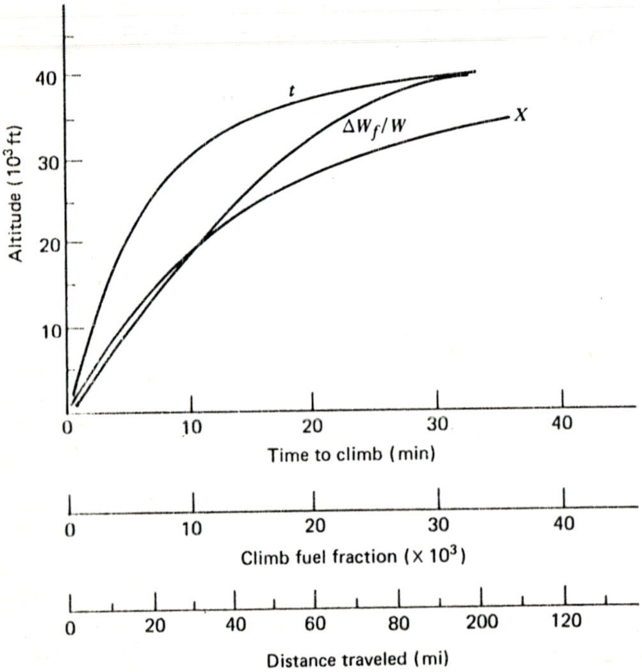

FIGURE 4-4
Fastest-climb values for the illustrative turbojet.

From Table 4-2, the total time (which is the minimum time) to climb from sea level to 30,000 ft is 555 seconds or 9.25 minutes, for an average climb rate of 3,243 fpm. Note that it takes almost four times as long to climb the last 5,000 ft (from 25,000 to 30,000 ft) as it does to climb the first 5,000 ft. The total W_f/W ratio to 30,000 ft is 0.0178, which means that the amount of fuel used during the climb is 1.78 percent of the total weight of the aircraft at the start of climb. If the aircraft weighs 100,000 lb, 1,780 lb of fuel will be used during climb. If the climb to 30,000 ft is made in the direction of the desired flight, the aircraft will be 69.7 miles closer to its destination.

Notice how nonlinear the climb time, the fuel consumption, and the range are as the absolute ceiling is approached.

Tabulating values and averaging to determine time to climb, fuel consumed, and distance traveled is tedious and time consuming. It would be nice to have closed-form expressions that would give reasonable values. If we assume that Γ is constant and assign it a value of 2, then

$$V_{\mathrm{FC}} \cong \left[\frac{2(T/S)}{3\rho_{\mathrm{SL}}\sigma C_{DO}}\right]^{1/2} \tag{4-39}$$

This expression gives values at the higher altitudes that are too low. In addition,

$$\sin \gamma_{\mathrm{FC}} \cong \frac{2(T/W)}{3} - \frac{3}{4E_m^{\,2}(T/W)} \tag{4-40}$$

which gives values at the higher altitudes that are too high. This expression can be further simplified, with additional loss of accuracy, to

$$\sin \gamma_{\mathrm{FC}} \cong \frac{2(T/W)}{3} \tag{4-41}$$

Combining Eqs. 4-40 and 4-41 results in a single expression for the maximum rate of climb (fastest climb), namely,

$$(\mathrm{R/C})_{\mathrm{FC}} \cong \frac{2(T/W)}{3}\left[\frac{2(T/S)}{3\rho_{\mathrm{SL}}\sigma C_{DO}}\right]^{1/2} = \left[\frac{8(T/W)^3(W/S)}{27\rho_{\mathrm{SL}}\sigma C_{DO}}\right]^{1/2} \tag{4-42}$$

Equation 4-42 shows that a high rate of climb calls for a large thrust-to-weight ratio, a large wing loading, and a low zero-lift coefficient. The magnitude of *the thrust itself is the most important parameter for fast climb.* The effects of changes in the thrust upon the fastest-climb rate and airspeed of our illustrative jet are sketched in Fig. 4-5.

These simplified expressions *should not* be used to determine climb performance at a specified altitude; instead, use the more exact expressions. The simplified model of Eq. 4-42 can be used, however, to develop closed-form expressions that can be used to calculate directly the minimum time to climb to a particular altitude, along with the fuel used and distance traveled, without the need to construct tables or graphs. Since the rate of climb is equal to dh/dt, Eq. 4-42 can be inverted,

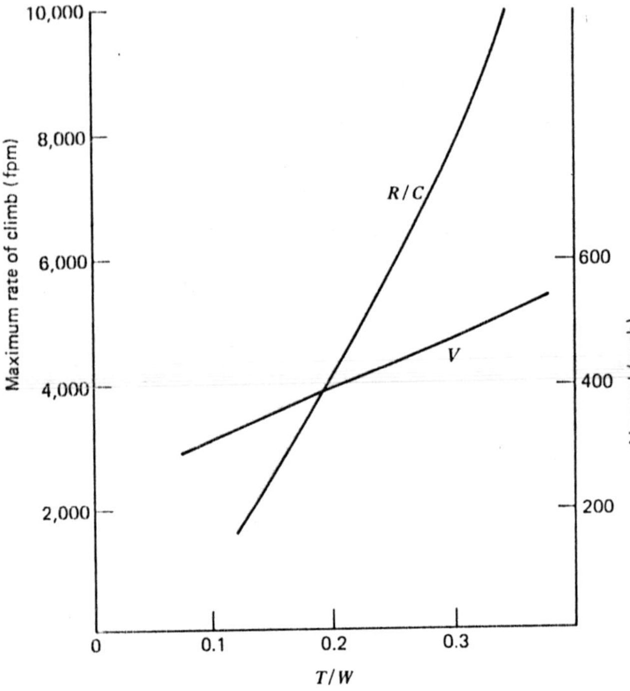

FIGURE 4-5
Effect of changes in T/W on fastest-climb rates and airspeeds for illustrative turbojet.

partially integrated, and written as

$$t_{FC} = t_{min} = \int_1^2 \frac{1}{T/W}\left[\frac{27\rho_{SL}\sigma C_{DO}}{8(T/S)}\right]^{1/2} dh \tag{4-43}$$

An expression for the fuel consumed during fastest climb can be obtained by first establishing the fuel-altitude exchange ratio:

$$\frac{-dW}{dh} = \frac{-dW/dt}{dh/dt} = \frac{cT}{(R/C)_{FC}} \tag{4-44}$$

With the substitution of Eq. 4-42, Eq. 4-44 becomes

$$-dW = \frac{c}{3{,}600}\left[\frac{27\rho_{SL}\sigma C_{DO}}{8(T/S)}\right]^{1/2} dh \tag{4-45}$$

which is partially integrated to yield

$$\ln MR_{FC} = \frac{c}{3{,}600}\int_1^2 \left[\frac{27\rho_{SL}\sigma C_{DO}}{8(T/S)}\right]^{1/2} dh \tag{4-46}$$

which can be further reduced to

$$\left(\frac{\Delta W_f}{W}\right)_{FC} = 1 - \exp\left\{-\frac{c}{3{,}600}\int_1^2\left[\frac{27\rho_{SL}\sigma C_{DO}}{8(T/S)}\right]^{1/2} dh\right\} \tag{4-47}$$

The expression for the distance traveled during climb is developed from the range-altitude exchange ratio:

$$\frac{dX}{dh} = \frac{dX/dt}{dh/dt} = \frac{V\cos\gamma}{(\text{R/C})_{FC}} = \left(\frac{\cos\gamma}{\sin\gamma}\right)_{FC} \tag{4-48}$$

With the assumption of a "small" climb angle, so that the cosine of the climb angle can be set equal to unity, Eq. 4-48 can be simplified by the substitution of Eq. 4-41, rearranged, and partially integrated to obtain

$$X_{FC} = \frac{1.5}{5{,}280}\int_1^2 \frac{dh}{T/W} \tag{4-49}$$

The climb program must now be established before the remaining integrations can be carried out. Of the many possible climb programs, only the *constant throttle setting climb* will be examined here as it gives the minimum climb time to altitude for quasi-steady-state climb. With a constant throttle setting, the simple relationship that T/ρ is constant can be introduced into Eq. 4-43 to obtain

$$t_{FC} = t_{min} = \frac{1}{(T/W)_{SL}}\left[\frac{27\rho_{SL}C_{DO}}{8(T/S)_{SL}}\right]^{1/2}\int_1^2\frac{dh}{\sigma} \tag{4-50}$$

If the exponential approximation of the density ratio (Eq. 2-1) is used with $\beta =$ 23,800, then

$$t_{FC} = t_{min} = \frac{23{,}800}{(T/W)_{SL}}\left[\frac{27\rho_{SL}C_{DO}}{8(T/S)_{SL}}\right]^{1/2}\left(e^{h_2/23{,}800} - e^{h_1/23{,}800}\right) \tag{4-51}$$

Using the same relationships, the other two expressions can be integrated to give

$$\left(\frac{\Delta W_f}{W}\right)_{FC} = 1 - \exp\left\{\frac{-c}{3{,}600}\left[\frac{27\rho_{SL}C_{DO}}{8(T/S)_{SL}}\right]^{1/2}(h_2 - h_1)\right\} \tag{4.52}$$

and

$$X_{FC} = \frac{6.76}{(T/W)_{SL}}\left(e^{h_2/23{,}800} - e^{h_1/23{,}800}\right) \tag{4-53}$$

These closed-form expressions will now be used to calculate the values for the *maximum-thrust climb* of our illustrative turbojet transport to 30,000 ft.

$$t_{min} = \frac{23{,}800}{0.25}\left(\frac{27\rho_{SL}\times 0.015}{8\times 25}\right)^{1/2}\left(e^{30{,}000/23{,}800} - 1\right) = 528\ \text{s} = 8.8\ \text{min}$$

$$\left(\frac{\Delta W_f}{W}\right)_{FC} = 1 - \exp\left[\frac{-0.8}{3{,}600}\left(\frac{27\rho_{SL}\times 0.015}{8\times 25}\right)^{1/2}\times 30{,}000\right] = 0.0145$$

$$X_{FC} = \frac{6.76}{0.25}\left(e^{30{,}000/23{,}800} - 1\right) = 68\ \text{mi}$$

Comparison with the tabulated values shows a calculated climb time of 528 seconds versus 555 seconds, for a 3.8 percent error on the low side; 0.0145 versus 0.0178, for a 17 percent error on the low side; and 68 mi versus 69.7 mi. For lower destination altitudes, the correlation is much better and for higher altitudes much worse. As the final altitude approaches the absolute ceiling of the aircraft, these expressions should not be used.

In these climbing examples, the maximum available thrust was used. Aircraft normally do not climb at maximum thrust, not only for consideration of engine life but also for other reasons such as passenger comfort, structural limitations, or air traffic control rules.

The next chapter on turning flight will discuss the increase in the loads exerted on the passengers and on the aircraft structure by virtue of the lift force being greater than the weight of the aircraft. This condition is described by the lift-to-weight ratio L/W, which is called the *load factor*, given the symbol n, and has the units of g's. In level flight, the lift is equal to the weight and the load factor is unity (one-g flight); in climbing flight, $L = W \cos \gamma$ and n is less than one. However, it is possible in both level and climbing flight to increase the magnitude of the load factor by suddenly pulling up or nosing down.

The maximum steady-state load factor possible, with aerodynamic considerations only, is equal to the product of the thrust-to-weight ratio and the maximum lift-to-drag ratio, i.e.,

$$n_m = \left(\frac{T}{W}\right) E_m \tag{4-54}$$

There is also a maximum allowable load factor that is imposed to avoid over-stressing the aircraft structure; this allowable load factor is usually less than the possible maximum. It is normal practice to keep the flight thrust-to-weight ratio below the value that could inadvertently cause the instantaneous load factor to exceed the allowable limit. As an example, let us assume that our illustrative turbo-jet is limited to a maximum allowable load factor of 2.5 g's even though the maximum possible is 4.17 g's (0.25×16.67). Equation 4-54 tells us that the climbing thrust-to-weight ratio should be limited to 0.15 to avoid the possibility of exceeding the maximum allowable load factor. Consequently, this value should be used in the fastest-climb expressions rather than the value of 0.25. If we do so, the sea-level rate of climb is reduced from 6,408 to 2,620 fpm and the sea-level climb speed is reduced from 476 to 380 mph.

Returning to Eq. 4-30, it can be seen that Γ can also be expressed in terms of n_m for use in the more exact expressions, so that

$$\Gamma = 1 + \left(1 + \frac{3}{n_m^2}\right)^{1/2} \tag{4-55}$$

If the T/W ratio is kept less than the maximum T/W at the start of climb, it is now possible to climb at a constant thrust, by increasing the throttle setting, until the available thrust becomes equal to maximum. Climb beyond that point will be with a constant throttle setting. For the constant-thrust portion of the climb,

Eqs. 4-43, 4-47, and 4-49 can be used in conjunction with the exponential approximation of the density ratio.

If the climb is to be made in an air traffic control area, the climb airspeed is limited to a maximum calibrated airspeed of 250 knots for altitudes under 10,000 ft. The exact expressions and tabulated results are best suited for handling this climb program.

Returning to the fastest-climb program using the maximum thrust available, examination of the tabulated values in Table 4-1 shows that the required climb airspeed increases from 476 mph at sea level to 520 mph at 30,000 ft, indicating a tangential acceleration along the flight path that we have ignored. By doing so, we have also ignored the propulsive energy used to generate the increase in the kinetic energy of the aircraft. By adding the tangential acceleration term ($m\, dV/dt$) to the right-hand side of Eq. 4-10, an *acceleration correction factor* can be developed that relates the accelerated rate of climb to our quasi-steady-state rate of climb. This correction factor is

$$C = \frac{1}{1 + \dfrac{V}{g}\dfrac{dV}{dh}} \tag{4-56}$$

For the tabulated example, a feeling for the magnitude of this correction factor can be obtained by approximating dV/dh by a $\Delta V/\Delta h$ equal to 2.17×10^{-3} ft/s/ft and using an average airspeed of 731 fps; these values were obtained from Table 4-1. Substitution into Eq. 4-56 yields a correction factor of 0.953, which means that the actual rate of climb is of the order of 95.3 percent of the value we obtained from the steady-state expressions. As the climb airspeed and acceleration along the climb path increase, the correction factor decreases, reducing the actual rate of climb and thus increasing the time to climb, the fuel consumed, and the distance traveled. These acceleration effects should be considered for high-speed and high-thrust aircraft. Such aircraft are discussed in Sec. 8-6, which looks at the energy-state approach to accelerated climbs.

4-5 MOST-ECONOMICAL CLIMB

The third climb program of interest is the *most-economical climb*, the climb that uses the smallest amount of fuel. The fuel-altitude exchange ratio ($-dW/dh$) is obtained by dividing Eq. 4-14 by Eq. 4-16 and can be written as

$$\frac{-dW}{dh} = \frac{cTW}{TV - DV} \tag{4-57}$$

which represents the fuel flow rate per unit of excess climb power per unit of aircraft weight. For most-economical climb, we wish to minimize this value, which can be done by maximizing its reciprocal, $dh/-dW$. With our idealized turbojet assumptions that the thrust is independent of the airspeed and that the specific fuel consumption is constant, we obtain the same conditions for most-economical climb that we did for fastest flight. In other words, with our assumptions and approximations, fastest climb is the most-economical climb.

When the actual variations in the thrust and specific fuel consumption are considered along with the compressibility effects of high speed upon the performance of the engine, the airspeed for most-economical climb is lower than that for fastest climb but is much closer to the fastest-climb airspeed than to the steepest-climb airspeed. It is so close that most aircraft do not distinguish between the two programs and fly the fastest-climb program whenever possible. In any event, for preliminary performance and design analyses, fastest climb is also considered to be the most-economical climb.

The relationships among the airspeeds for level flight and for the three climb programs are sketched in Fig. 4-6.

4-6 UNPOWERED FLIGHT

Unpowered flight occurs when a single-engine aircraft has engine failure or runs out of fuel, when a multiengine aircraft runs out of fuel, and when an aircraft has no propulsion system (a glider or a sailplane).

The equations of motion for unpowered flight in the vertical plane are obtained by setting the thrust equal to zero in the set of general equations for quasi-steady flight, listed in Sec. 3-1. The glide angle (the flight-path angle) is sufficiently small, but not zero, so that its cosine may be replaced by unity and its sine replaced by the

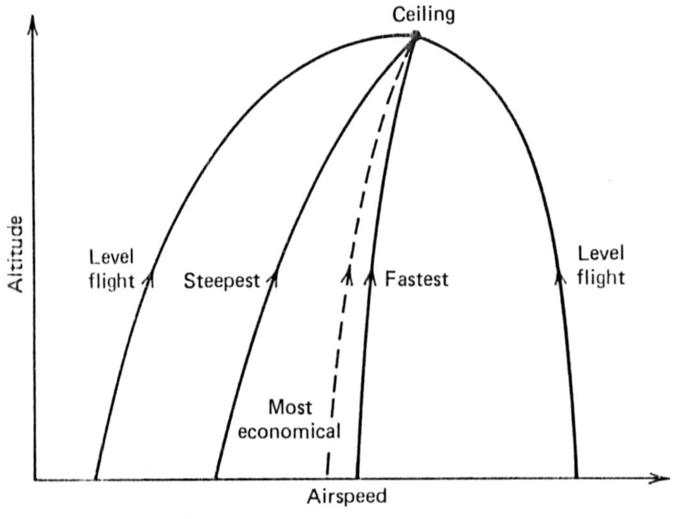

FIGURE 4-6

Relationships among level-flight and climb airspeeds.

angle itself, always expressed in radians. The relevant equations are:

$$D - W\gamma = 0 \tag{4-58}$$

$$L = W \tag{4-59}$$

$$\frac{dX}{dt} = V \tag{4-60}$$

$$\text{R/D} = \frac{dh}{dt} = V\gamma \tag{4-61}$$

There are two unpowered flight programs of special interest: maximum (best) range and maximum endurance (minimum rate of descent). The first is of particular importance to the pilot with engine failure or fuel starvation and in the design of a military glider. The second is the key to the design of sailplanes for soaring.

The various relationships of interest, as derived from the equations of motion above, are:

$$\gamma = \frac{-D}{W} = \frac{-D}{L} = \frac{-1}{E} \tag{4-62}$$

$$\frac{dX}{dh} = \frac{1}{\gamma} = -E \tag{4-63}$$

$$\text{R/D} = \frac{dh}{dt} = \frac{-V}{E} = \frac{-DV}{W} \tag{4-64}$$

where Eq. 4-63 is the range-altitude exchange ratio or instantaneous range.

The maximum (best) range conditions will be determined first. By inspection of Eq. 4-63, the instantaneous range is maximized by maximizing the lift-to-drag ratio, so that Eq. 4-63 can be written as

$$dX_{\text{br}} = -E_m dh \tag{4-65}$$

Integrating from the start of the glide to its completion, usually touchdown, yields the best-range equation

$$X_{\text{br}} = E_m(h_1 - h_2) \tag{4-66}$$

The best range is affected only by the maximum lift-to-drag ratio of the aircraft and by the altitude. The associated best-range flight conditions are:

$$\gamma_{\text{br}} = \frac{-1}{E_m} \tag{4-67}$$

$$V_{\text{br}} = V_{\text{md}} = \left[\frac{2(W/S)}{\rho_{\text{SL}}\sigma}\right]^{1/2} \left(\frac{K}{C_{D0}}\right)^{1/4} = \frac{V_{\text{br,SL}}}{\sigma^{1/2}} \tag{4-68}$$

$$(\text{R/D})_{\text{br}} = \frac{-V_{\text{br}}}{E_m} = \frac{-V_{\text{br,SL}}}{\sigma^{1/2} E_m} \tag{4-69}$$

$$t_{br} = -E_m \int_1^2 \frac{dh}{V_{br}} = \frac{-E_m}{V_{br,SL}} \int_1^2 \sigma^{1/2}\, dh \tag{4-70}$$

Using the exponential approximation for the density ratio and a value of 23,800 for β, Eq. 4-70 can be integrated to give

$$t_{br} = \frac{47{,}600 E_m}{V_{br,SL}} \left(e^{-h_2/47{,}600} - e^{-h_1/47{,}600}\right) \tag{4-71}$$

In the equations above, note that the glide angle is constant and represents the smallest possible glide angle; this glide program is sometimes referred to as *flattest glide*, or *minimum-angle glide*. Note also that pilots must control their true airspeed in accordance with Eq. 4-68, decreasing it as they lose altitude so as to maintain a constant lift coefficient. Their task is simplified by the fact that the calibrated airspeed remains constant throughout the descent. When the *glide ratio* of an aircraft is given, it is the maximum lift-to-drag ratio of the aircraft in the glide configuration.

Assume that our illustrative turbojet transport ($W/S = 100$ lb/ft^2, $C_D = 0.015 + 0.06\, C_L^2$, $E_m = 16.67$) has run out of fuel at 40,000 ft over the Atlantic Ocean and the pilot wants to glide as far as possible before ditching.

$$X_{br} = 16.67 \times 40{,}000 = 6.67 \times 10^5 \text{ ft} = 126.3 \text{ mi}$$

$$\gamma_{br} = \frac{-1}{16.67} = -0.06 \text{ rad} = -3.44 \text{ deg}$$

Notice that the glide angle is indeed small.

$$V_{br,SL} = \left(\frac{2 \times 100}{\rho_{SL}}\right)^{1/2} \left(\frac{0.06}{0.015}\right)^{1/4} = 410 \text{ fps} = 280 \text{ mph}$$

$$V_{br,40K} = \frac{V_{br,SL}}{\sigma^{1/2}} = \frac{410}{(0.246)^{1/2}} = 827 \text{ fps} = 564 \text{ mph} = \text{M } 0.855$$

Notice that the initial airspeed is quite high, much higher than one's intuition might lead one to expect. The time to ditching is

$$t_{br} = \frac{47{,}600 \times 16.67}{410} \left(1 - e^{-40{,}000/47{,}600}\right) = 1{,}100 \text{ s} = 18.3 \text{ min}$$

and the initial and final rates of descent are

$$(\text{R/D})_{40K} = \frac{-V_{br,40K}}{E_m} = -49.6 \text{ fps} = -2{,}973 \text{ fpm}$$

$$(\text{R/D})_{SL} = \frac{-V_{br,SL}}{E_m} = -24.6 \text{ fps} = -1{,}476 \text{ fpm}$$

These figures support the warning to the pilot who is striving for maximum power-off range (to reach the end of the runway, for example) to "Never stretch a glide."

To minimize the rate of descent (and thus maximize the endurance), the derivative of Eq. 4-64 with respect to the airspeed is set equal to zero, i.e.,

$$\frac{d}{dV}\left(\frac{dh}{dt}\right)=\frac{d}{dV}\left(\frac{-DV}{W}\right)=0 \tag{4-72}$$

or

$$\frac{dD}{dV}=\frac{-D}{V} \tag{4-73}$$

With the introduction of the parabolic drag polar, Eq. 4-73 can be solved for the maximum-endurance dynamic pressure and airspeed.

$$q_{t_{\max}}=\left(\frac{W}{S}\right)\left(\frac{K}{3C_{DO}}\right)^{1/2} \tag{4-74}$$

$$V_{t_{\max}}=\left[\frac{2(W/S)}{\rho_{SL}\sigma}\right]^{1/2}\left(\frac{K}{3C_{DO}}\right)^{1/4}=\frac{V_{t_{\max},SL}}{\sigma^{1/2}} \tag{4-75}$$

Notice that this airspeed is approximately 24 percent slower than that for best range. The remaining maximum-endurance conditions are

$$C_{L,t_{\max}}=\left(\frac{3C_{DO}}{K}\right)^{1/2}=1.732C_{L,E_m} \tag{4-76}$$

$$E_{t_{\max}}=0.866E_m \tag{4-77}$$

$$\gamma_{t_{\max}}=\frac{-1.155}{E_m}=1.155\gamma_{br} \tag{4-78}$$

$$X_{t_{\max}}=0.866E_m(h_1-h_2)=0.866X_{br} \tag{4-79}$$

$$(R/D)_{\min}=\frac{-V_{t_{\max},SL}}{0.866\sigma^{1/2}E_m}=0.88(R/D)_{br} \tag{4-80}$$

$$t_{\max}=\frac{41{,}223E_m}{V_{t_{\max},SL}}(e^{-h_2/47{,}600}-e^{-h_1/47{,}600})=1.14t_{br} \tag{4-81}$$

Let us imagine that the illustrative turbojet transport has again shut down all its engines, but now the pilot is seeking to remain in the air as long as possible. The flight conditions and results are:

$$E=0.866\times16.67=14.44$$

$$X=14.44\times40{,}000=5.76\times10^5\text{ ft}=109\text{ mi}$$

$$\gamma=\frac{-1}{14.44}=-0.069\text{ rad}=-3.97\text{ deg}$$

$$V_{SL}=\left(\frac{2\times100}{\rho_{SL}}\right)^{1/2}\left(\frac{0.06}{3\times0.015}\right)^{1/4}=312\text{ fps}=212\text{ mph}$$

$$V_{40K}=\frac{312}{(0.246)^{1/2}}=629\text{ fps}=429\text{ mph}$$

$$(R/D)_{min,SL} = \frac{-312}{14.4} = -21.7 \text{ fps} = -1{,}300 \text{ fpm}$$

$$(R/D)_{min,40K} = \frac{-629}{14.4} = -43.7 \text{ fps} = -2{,}621 \text{ fpm}$$

$$t_{max} = 1{,}252 \text{ s} = 20.9 \text{ min}$$

The variations in the flight parameters for these two unpowered flight programs are sketched in Fig. 4-7. The differences between these two *maximized* flight programs are not extreme, being of the order of 15 percent. However, flight conditions that differ markedly from those of these two programs can have a dramatic effect upon the unpowered performance. For example, consider a constant-angle glide from 40,000 ft to sea level with an initial true airspeed of 300 mph (440 fps). Our illustrative aircraft would glide at 6.56 deg and have a range of *only* 66 mi, approximately one-half of the maximum range possible.

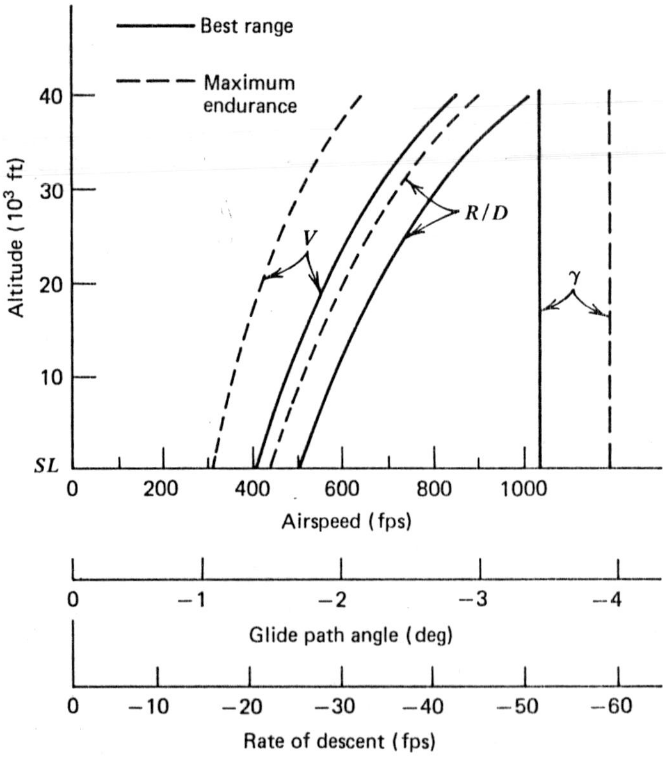

FIGURE 4-7

Flight conditions for maximized unpowered flight.

Let us expand the expression for the minimum rate of descent (Eq. 4-80) in terms of the zero-lift drag coefficient and of the aspect ratio, so that

$$(R/D)_{min} = 2.48 \left(\frac{W/S}{\rho_{SL}\sigma}\right)^{1/2} (K^3 C_{DO})^{1/4} \tag{4-82a}$$

and

$$(R/D)_{min} = 1.05 \left(\frac{W/S}{\rho_{SL}\sigma}\right)^{1/2} \left[\frac{C_{DO}}{(ARe)^3}\right]^{1/4} \tag{4-82b}$$

A sailplane is designed for a minimum rate of descent (the goal is an impossible no-wind zero sinking speed) in accordance with this equation. A 10 percent decrease in the wing loading decreases the rate of descent by 5 percent, a 10 percent decrease in C_{DO} produces a 2.5 percent decrease, and a 10 percent increase in either the aspect ratio or the Oswald span efficiency yields a 7 percent decrease. A well-designed sailplane should be characterized by long narrow wings with a large area, a low gross weight, and a clean, sleek configuration.

One single-sea high-performance sailplane has the following characteristics:

Wing area, S	151 ft^2
Wing span, b	49.7 ft
Aspect ratio, AR	16
Drag polar, C_D	$0.01 + 0.0216 C_L^2$
E_m	34
Gross weight, W	800 lb
Empty weight, OEW	550 lb

At sea level, such a sailplane will have a minimum rate of descent of 125 fpm with a glide angle of 1.95 deg and an airspeed of 42 mph. At 30,000 ft, the glide angle is still 1.95 deg, but the minimum rate of descent has increased to 205 fpm and the corresponding airspeed is now 68 mph. With no thermals (vertical updrafts) the gliding time from 30,000 ft would be 177.6 minutes, almost 3 hours, and the range would be 167 mi. With good thermals, a sailplane like this one can stay up almost indefinitely; the endurance of the pilot becomes the limiting factor.

PROBLEMS

Major characteristics of Aircraft A, Aircraft B, and Aircraft C, used in many of these problems, are listed at the beginning of the problems in Chap. 3.

4-1. a. Using Eq. 4-1, show that maintaining a constant lift coefficient equal to $\mu/2K$ during the ground run prior to rotation minimizes the resistive forces and thus minimizes the ground run itself.

b. For a μ of 0.02, representing a paved, smooth runway, find the values of the ground-run lift and drag coefficients for Aircrafts A, B, and C.

c. Do (b) above for a μ of 0.04, representing the rolling friction coefficient for a hard turf runway.

4-2. Assuming that Aircraft A maintains a constant lift coefficient during the take-off ground run until it reaches an airspeed equal to $1.1 V_s$ (approximately the rotation airspeed), and then using Eq. 4-1:

a. Find the net acceleration at this "rotation" airspeed for a lift coefficient equal to $\mu/2K$ with $\mu = 0.02$. Compare this value with the T/W ratio of the aircraft.

b. Do (a) above, for a μ of 0.04.

c. Do (a) above, for a lift coefficient equal to zero.

4-3. Do Prob. 4-2 for Aircraft B.

4-4. Do Prob. 4-2 for Aircraft C.

4-5. Assuming that the value given for the maximum lift coefficient for Aircraft A is the maximum attainable on take-off:

a. Find the sea-level, standard-day take-off run distance, lift-off airspeed (mph), and time to lift-off (seconds).

b. Do (a) above for a standard-day take-off from an airport that is 5,000 ft above mean sea level.

4-6. Do Prob. 4-5 for Aircraft B.

4-7. Do Prob. 4-5 for Aircraft C.

4-8. A sea-level airport is in a low-pressure area on a hot day. The atmospheric pressure is 1,975 lb/ft^2 and the temperature is 103 deg F.

a. Find the actual density ratio and the equivalent standard-day altitude of the airport.

b. Find the percentage increase in the standard-day, sea-level take-off values for the ground run, the elapsed time, and the lift-off airspeed.

c. What is the additional effect of a 10 percent increase in the gross weight of the aircraft?

4-9. If the T/W ratio of Aircraft B is increased to 0.4 and the maximum lift coefficient is increased to 3.5 by using new flap and boundary layer control technology, what will be the effects on the take-off run, time, and lift-off airspeed?

4-10. Aircraft C is on a sea-level, standard-day take-off run and has reached an airspeed of 125 mph when suddenly one of the four engines fails completely.

a. How far down the runway is the aircraft at the time of engine failure?

b. If the pilot decides to continue the take-off with the remaining three engines, what will be his additional ground run?

c. If, on the other hand, the pilot decides to abort the take-off, how much extra runway will he need to bring the aircraft to a complete stop? It is raining, the runway is wet, and the deceleration factor is 0.2.

d. Which is the safer decision and why?

4-11. a. Plot the sea-level, standard-day take-off run distance in feet as a function of a take-off parameter that is the wing loading divided by the product of the maximum lift coefficient for take-off and the maximum sea-level thrust-to-weight ratio.

b. On the same plot, do (a) above for a density ratio of 0.85 and for a density ratio of 0.75.

c. Calculate the value of the take-off parameter for Aircraft *A* and use it to locate the take-off runs on these plots.

d. Do (c) above for Aircraft *B*.

e. Do (c) above for Aircraft *C*.

4-12. a. Aircraft *A* has just completed a flight and is landing with a gross weight of 19,000 lb at a sea-level airport on a standard day. Find the approach speed ($1.2V_s$) and the landing run required to bring the aircraft to a stop with a deceleration factor of 0.4.

b. Do (a) above, but find the ground run necessary to reduce the airspeed to 20 mph for a rolling turn off the runway onto a taxiway.

c. Do (a) above, at an airfield that is 5,000 ft above mean sea level.

d. If Aircraft *A* has just taken off and is forced to land immediately with a gross weight of 23,500 lb, what will the approach speed and landing run be for the conditions of (a) above?

4-13. Do Prob. 4-12 for Aircraft *B* with a gross weight of 110,000 lb for parts (a) through (c) and a gross weight of 135,000 lb for part (d).

4-14. Do Prob. 4-12 for Aircraft *C* with a gross weight of 450,000 lb for parts (a) through (c) and a gross weight of 580,000 lb for part (d).

4-15. a. With no restrictions whatsoever, find the maximum rate of climb (fpm) for Aircraft *A* on a standard day along with the climb airspeeds (true and calibrated in mph) at sea level, 20,000 ft, 30,000 ft, and 40,000 ft.

b. Using values from (a) above, plot the altitude and true airspeed (mph) as a function of the rate of climb (fpm). Extrapolate to find the absolute ceiling and ceiling airspeed and compare these values with calculated values.

c. If a maximum load factor of 2.5 *g*'s is not to be exceeded, redo (a) above.

4-16. Do Prob. 4-15 for Aircraft *B*.

4-17. Do Prob. 4-15 for Aircraft *C*.

4-18. With no restrictions whatsoever and using the closed-form approximations of this chapter, find the time to climb, fuel used, and distance traveled for Aircraft *A* with C = 0.95 lb/h/lb:

a. From sea level to 20,000 ft

b. From sea level to 30,000 ft

c. From sea level to 40,000 ft

d. From 20,000 to 40,000 ft

4-19. Do Prob. 4-18 for Aircraft *B* with $C = 0.9$ lb/h/lb.

4-20. Do Prob. 4-18 for Aircraft *C* with $C = 0.85$ lb/h/lb.

4-21. With no restrictions whatsoever, find the steepest climb angle and associated airspeed (mph) and rate of climb (fpm) for:

a. Aircraft *A*

b. Aircraft *B*

c. Aircraft *C*

4-22. Do Prob. 4-21 with the restriction of a maximum load factor of 2.5 *g*'s.

4-23. Do Prob. 4-21 at an altitude of 10,000 ft. You have flown up a narrow box canyon and hope to climb out.

4-24. a. Plot the sea-level rate of climb (fpm) of Aircraft *A* as a function of the true airspeed (mph).

b. From this plot, determine the maximum rate of climb and the associated airspeed and compare them with the values obtained from Eqs. 4-35 and 4-32.

4-25. Do Prob. 4-24 for Aircraft *B*.

4-26. Do Prob. 4-24 for Aircraft *C*.

4-27. Find the climb acceleration factor for Aircraft *A* in climbing, with no restrictions, from sea level to 30,000 ft.

4-28. Do Prob. 4-27 for Aircraft *B* in a climb from sea level to 35,000 ft.

4-29. Do Prob. 4-27 for Aircraft *C* in a climb from sea level to 40,000 ft.

4-30. Aircraft *A* has run out of fuel at 30,000 ft.

a. Find the initial and final values of the best-range airspeed, glide angle, rate of descent along with the time and distance to touchdown.

b. The pilot decides to establish a glide angle with an initial true airspeed that is 20 percent higher than the stall speed and then to keep this glide angle constant to touchdown. Do (a) above for this set of conditions.

4-31. Do Prob. 4-30 for Aircraft *B* from an initial altitude of 35,000 ft.

4-32. Do Prob. 4-30 for Aircraft *C* from an initial altitude of 40,000 ft.

4-33. Aircraft *C* is on a straight-in, power-off final approach to a landing with a gross weight of 500,000 lb and an approach airspeed equal to 1.2 times the stall speed. The aircraft is 3 mi from the end of the runway at 1,000 ft.

a. Find the approach airspeed (mph), the rate of descent, and the time and distance to touchdown.

b. Will the aircraft reach the end of the runway? If not, how far short will it be?

c. The pilot raises the nose of the aircraft and slows down to 1.1 times the stall speed. Do (a) above and determine if the pilot has improved or worsened her chances of reaching the runway.

d. Instead of raising the nose, the pilot lowers the nose and allows the airspeed to stabilize at 1.4 times the stall speed. What are her new descent parameters and will she reach the end of the runway?

4-34. A two-place training sailplane has a gross weight of 1,100 lb, a wing area of 140 ft^2, a wing span of 40 ft, and a zero-lift drag coefficient of 0.012. Assume an e of 0.95.

a. Determine the design characteristics of the sailplane, for example, AR, drag polar, maximum lift-to-drag ratio.

b. Find the best-range performance and conditions at sea level and at 30,000 ft.

c. Find the maximum-endurance performance and conditions at sea level and at 30,000 ft.

4-35. Aircraft A is scheduled to fly the following mission profile with a tsfc of 0.9 lb/h/lb:

Phase 1. Taxi for 15 minutes at 20 percent of the maximum thrust

Phase 2. Take-off at sea level on a standard day

Phase 3. Fastest climb to 30,000 ft at 85 percent of the maximum thrust

Phase 4. Cruise 1,000 mi at a constant altitude with a constant airspeed

Phase 5. Descend to sea level (no range, fuel, or time credits)

Phase 6. Loiter at sea level for 1 hour

a. Find the minimum fuel required for each phase and the total amount of fuel required for the mission.

b. Find the time for each phase along with the total time from leaving the blocks on the ramp to the end of the loiter period.

c. Tabulate the airspeeds (mph) for each phase along with the respective values of the wing loading.

d. If Phase 4 is flown as a cruise-climb program at the drag-rise Mach number, what will be the effects on the fuel consumption and on the flight time?

4-36. Aircraft B is scheduled to fly the mission profile of Prob. 4-35 with the changes that the cruise altitude is to be 35,000 ft and the range is to be 2,000 mi. Use a tsfc of 0.85 lb/h/lb. Do Prob. 4-35.

4-37. Aircraft C is scheduled to fly the mission profile of Prob. 4-35 with the changes that the cruise altitude is to be 35,000 mi, the range is to be 3,000 mi, and the tsfc is to be 0.8 lb/h/lb. Do Prob. 4-35.

Chapter Five

Turning Flight in the Horizontal Plane: Turbojets

5-1 COORDINATE SYSTEMS AND GOVERNING EQUATIONS

Even though an aircraft may spend most of a mission in straight flight in the vertical plane, there are times when it must change direction, i.e., turn. For all aircraft there are turns associated with changes in flight headings, collision avoidance, holding patterns, and instrument approaches and landings. In addition, combat aircraft must have a greater degree of maneuverability than transports in order to survive and carry out their assigned operational missions. The maneuvering capability of an aircraft along with the associated design and flight parameters can be determined, to a first approximation, by limiting turning flight to the horizontal plane (with constant altitude), assuming the weight to be constant, and examining the conditions for the maximum bank angle, the maximum turning rate, and the minimum radius of turn.

We shall continue our assumption of quasi-steady-state flight by neglecting any tangential accelerations, but we must consider the acceleration normal to the curved flight path (the centripetal acceleration). We shall also continue to assume coordinated flight (zero sideslip angle) and to neglect the thrust angle of attack. The former assumption means that the velocity vector, lift, and drag all lie in the plane of symmetry (for a symmetrical aircraft), and the latter assumption means that the velocity and thrust vectors are coincident.

Rather than the conventional wind axes used for vertical-plane flight, it is more convenient to use what is known as the principal trihedral, a right-handed cartesian system composed of the three ortogonal unit vectors, **t**, **n**, and **b**. With the origin at the center of gravity of the aircraft as before, the tangent **t** is in the plane of symmetry along the velocity vector (and thus along the x-wind axis), the normal **n** is perpendicular to the plane of symmetry along the radius of curvature and positive toward the center of curvature, and the binormal **b** is perpendicular to both **t** and **n**. For curvilinear flight in the horizontal plane, the flight-path angle is zero, **n** is in the horizontal plane, and **b** is in the vertical plane. The trihedral and the quasi-steady-state forces are shown in Fig. 5-1. Resolving the forces along the trihedral and along the ground axes, X and Y, yields the following dynamic and kinematic

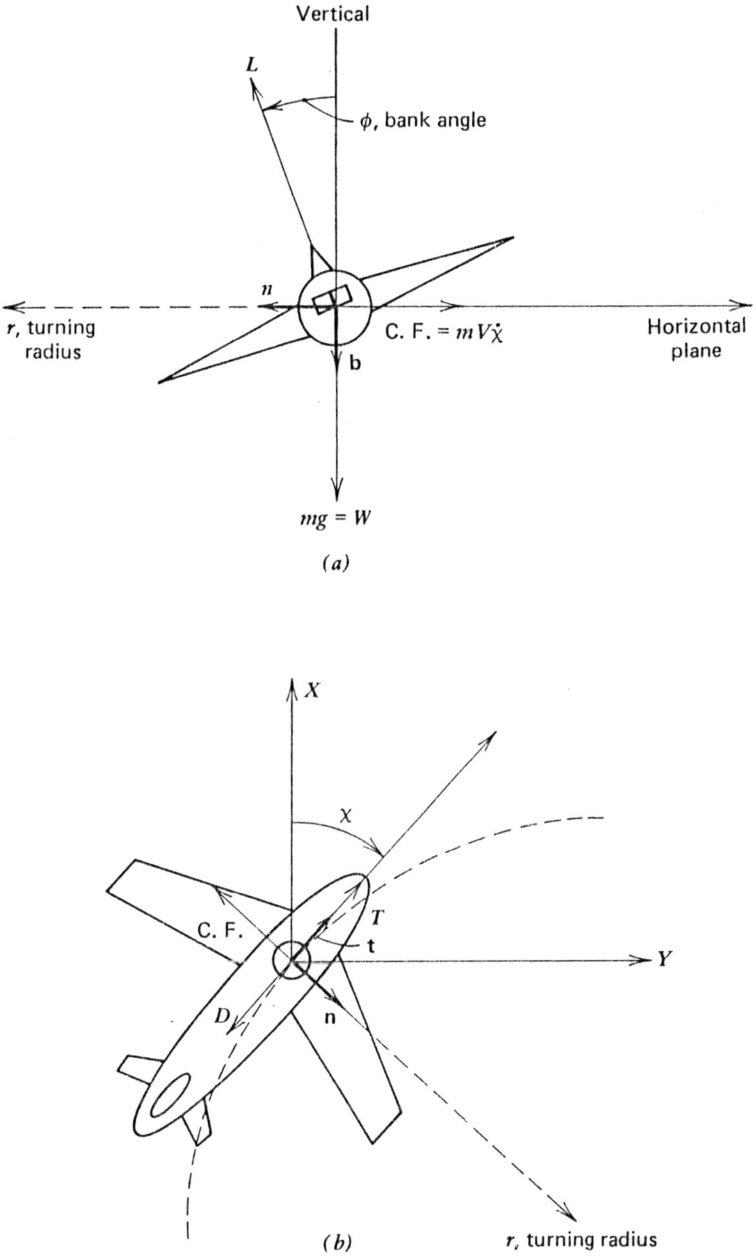

FIGURE 5-1

The principal trihedral coordinate system: (*a*) looking at the front of the aircraft; (*b*) looking down on the aircraft.

equations for turning flight of a turbojet in the horizontal plane:

$$T - D = 0 \tag{5-1}$$

$$L \sin \phi - \frac{W}{g} V\dot{\chi} = 0 \tag{5-2}$$

$$L \cos \phi - W = 0 \tag{5-3}$$

$$\frac{dX}{dt} = V \cos \chi \tag{5-4}$$

$$\frac{dY}{dt} = V \sin \chi \tag{5-5}$$

$$\frac{dW}{dt} = -cT \tag{5-6}$$

where χ is the yaw angle, $\dot{\chi}$ is the turning (yaw) rate, and ϕ is the bank angle; $\dot{\chi}$ is always expressed in radians per second (rad/s). Equation 5-6, the fuel consumption equation for a turbojet, is included for completeness but will not be used in these analyses because the times involved in turning flight will not be sufficiently long to change the weight of the aircraft appreciably. The kinematic equations, Eqs. 5-4 and 5-5, will not be used either as the emphasis is on the maneuverability rather than on the distance traveled.

The three dynamic equations, Eqs. 5-1, 5-2, and 5-3, show that the drag is balanced by the thrust, the centrifugal force is balanced by the horizontal component of the lift, and the weight of the aircraft is balanced by the vertical component of the lift.

5-2 TURNING FLIGHT IN GENERAL

In level (nonturning) flight, the lift is equal to the weight. For horizontal flight, however, Eq. 5-3 shows that

$$L = \frac{W}{\cos \phi} \tag{5-7}$$

so that, as the bank angle ϕ increases, the lift must also be increased if the altitude is to be maintained. The lift-to-weight ratio (L/W) has previously been defined as the *load factor* and given the symbol n. It has the dimensions of g's (as does the thrust-to-weight ratio) and is called the load factor because it is a measure of the forces, or loading, impressed upon the structure. For example, when n is equal to unity, we speak of one-g flight and the lift is equal to the weight. When n is equal to 2, the lift is equal to twice the weight of the aircraft and the wing span, for example, must now accept a load equal to twice the total weight of the aircraft without unacceptable deflections or damage. Similarly, a person in the aircraft will be subjected to an additional force equal to his or her weight. If the load factor exceeds the tolerances of the structures or of the occupants, temporary or permanent damage can take place. The concepts of the maximum possible load factor and the maximum allowable load factor will be discussed in the next section.

With the definition of the load factor, Eq. 5-7 can be written as

$$n=\frac{L}{W}=\frac{1}{\cos\phi}=\left(\frac{T}{W}\right)E \tag{5-8}$$

which shows a direct coupling between the bank angle and the load factor and between the product of the thrust-to-weight and the lift-to-drag ratios. For any value of the load factor, the bank angle can be found from Eq. 5-8, rewritten as

$$\phi=\arccos\frac{1}{n} \tag{5-9}$$

When n is unity, the lift is equal to the weight and the bank angle is zero. When the bank angle is zero, Eq. 5-2 shows that the turning rate must also be zero, indicating that there can be no steady-state coordinated turn without a finite bank angle. When the bank angle is 90 deg, the load factor becomes infinite, which means that an aircraft cannot maintain a steady-state turn with a 90 deg bank angle and still hold a constant altitude. Turns with a bank angle of 90 deg can be executed but only by skidding or by losing or gaining altitude, the latter requiring a T/W ratio greater than one.

An expression for $\dot{\chi}$ can be found by dividing Eq. 5-2 by Eq. 5-3 and rearranging to obtain

$$\dot{\chi}=\frac{g\tan\phi}{V}=\frac{g(n^2-1)^{1/2}}{V} \tag{5-10}$$

where $\dot{\chi}$ is expressed in radians per second, V in feet per second, and g is taken to be equal to 32.2 ft/s^2. The expression to the right of the second equality sign results from the trigonometric relationship that

$$\tan^2\phi-\frac{1}{\cos^2\phi}-1=n^2-1 \tag{5-11}$$

The airspeed V is a tangential velocity and therefore equal to $\dot{\chi}r$. Solving for the turning radius r and using Eq. 5-10 yields the expressions:

$$r=\frac{V}{\dot{\chi}}=\frac{V^2}{g\tan\phi}=\frac{V^2}{g(n^2-1)^{1/2}} \tag{5-12}$$

We can develop relationships among n (and ϕ), the flight conditions, and the aircraft characteristics from Eq. 5-1. With a parabolic drag polar, with the drag written as

$$D=qSC_{D0}+\frac{KL^2}{qS} \tag{5-13}$$

the substitution of L equal to nW results in the horizontal-flight drag function

$$D = qSC_{DO} + \frac{Kn^2W^2}{qS} \tag{5-14}$$

Substituting Eq. 5-14 into Eq. 5-1 and rearranging yields the quadratic equation

$$S^2C_{DO}q^2 - TSq + Kn^2W^2 = 0$$

which can be solved for the dynamic pressure to give

$$q = \frac{T/S}{2C_{DO}}\left\{1 \pm \left[1 - \frac{4KC_{DO}n^2}{(T/W)^2}\right]^{1/2}\right\} \tag{5-15}$$

Consequently, the expression for the airspeed in a turn can be written as

$$V = \left[\frac{T/S}{\rho_{SL}\sigma C_{DO}}\left(1 \pm \left\{1 - \frac{n^2}{[E_m(T/W)]^2}\right\}^{1/2}\right)\right]^{1/2} \tag{5-16}$$

These expressions for the dynamic pressure and airspeed in a turn have several points of interest. When the lift is equal to the weight, the load factor is unity, and the turning values become those for level flight in the vertical plane, as might be expected. When the load factor is greater than unity, as in a turn, the airspeed is less than that for level flight with the same thrust and throttle setting. Consequently, the pilot must increase the thrust when entering a turn if both the airspeed and the altitude are to remain unchanged. It should also be noted that specifying the bank angle (and thus the load factor) and the thrust-to-weight ratio determines the airspeed for a given aircraft at a given altitude and the there may be two possible airspeed values for each bank angle-thrust combination.

Not only does the airspeed drop off in a turn if the thrust is not increased but also the stall speed increases as the square root of the load factor. Using Eq. 5-8,

$$L = nW = \tfrac{1}{2}\rho_{SL}\sigma V^2 SC_L \tag{5-17}$$

Since stall occurs at the maximum lift coefficient, the stall speed in a turn can be written as

$$V_{s,t} = \left[\frac{2n(W/S)}{\rho_{SL}\sigma C_{L\max}}\right]^{1/2} = \left[\frac{2(W/S)}{\rho_{SL}\sigma C_{L\max}\cos\phi}\right]^{1/2} \tag{5-18}$$

or

$$V_{s,t} = n^{1/2}V_{s,l} = \frac{V_{s,l}}{(\cos\phi)^{1/2}} \tag{5-19}$$

where $V_{s,l}$ is the wings-level stall speed. For example, a bank angle of 30 deg ($n = 1.155$) increases the wings-level stall speed by 7.5 percent. By the same token, the calculated lift coefficient for turning flight might very well exceed that available since

$$C_L = \frac{n(W/S)}{q} = \frac{W/S}{q\cos\phi} \tag{5-20}$$

It is wise, therefore, to check turning-flight calculations for airspeeds below stall speeds and for lift coefficients that might exceed the maximum lift coefficient.

Equation 5-18 can also be used to demonstrate the "high-speed stall," which can occur at any airspeed no matter how high the airspeed might be. If an aircraft is in equilibrium at any airspeed and at any load factor (even with $n \leq 1$, as in a climb or level flight) and the pilot makes an abrupt control deflection so as to instantaneously increase the load factor, the stall speed immediately increases and may well exceed the actual airspeed.

Returning to Eq. 5-15, and this time solving for n results in the expressions:

$$n = \frac{1}{W/S}\left[\frac{(T/S)q}{K} - \frac{C_{DO}q^2}{K}\right]^{1/2} = \frac{q}{W/S}\left[\frac{1}{K}\left(\frac{T/S}{q} - C_{DO}\right)\right]^{1/2} \tag{5-21}$$

which can be used to find the bank angle (and load factor) required for a steady turn at a given altitude, thrust-to-weight ratio, and airspeed.

Our illustrative jet transport has a W/S of 100 lb/ft^2 a maximum T/W of 0.25, $C_D = 0.015 + 0.06C_L^2$, $E_m = 16.67$, and a maximum C_L of 1.8, without flaps. Let us suppose that we wish to turn with a bank angle of 30 deg and a T/W of 0.1. From Eq. 5-8, $n = 1/\cos 30\text{ deg} = 1.155$ g's. We shall now use Eq. 5-15 to find the dynamic pressure, using a value of 10 lb/ft^2 (0.1 × 100) for the T/S; therefore,

$$q = \frac{10}{2 \times 0.015}\left\{1 \pm \left[1 - \left(\frac{1.155}{0.1 \times 16.67}\right)^2\right]^{1/2}\right\}$$

$$q_1 = 574 \text{ lb/ft}^2 \qquad q_2 = 92.6 \text{ lb/ft}^2$$

From Eq. 5-20, C_{L1} is equal to 0.2 and C_{L2} is equal to 1.25. Since both values are less than the maximum value, flight is possible at both lift coefficients. At this point, an altitude must be specified. Let us choose sea level first, examine the turning performance there, and then go up to 30,000 ft.

At sea level, the two airspeeds corresponding to the values of the dynamic ratio found above are 695 fps (474 mph) and 279 fps (190 mph). From Eq. 5-10, $\dot{\chi}$ (the turning rate) is found to be 0.0267 rad/s (1.53 deg/s) for the high-speed solution and 0.067 rad/s (3.82 deg/s) for the low-speed solution. The corresponding values for the turning radius, as obtained from Eq. 5-12, are 26,029 and 4,227 ft, respectively. This example shows that the *lower airspeed gives the better turning performance* in terms of a higher turning rate and a smaller turning radius.

At 30,000 ft, the higher airspeed is 1,136 fps, which is in excess of M 1.0 and is not a valid flight condition. The lower airspeed is 456 fps (311 mph), and the corresponding turning rate and radius are 0.041 rad/s (2.33 deg/s) and 11,122 ft. These values, when compared with those for sea level, indicate that turning performance becomes worse with increasing altitude. Conversely, *maneuverability improves as the altitude is lowered.*

Let us consider one more situation in this section, starting with the premise that our illustrative turbojet is in a racetrack holding pattern at 20,000 ft ($\sigma = 0.533$) at an airspeed of 250 mph (367 fps). The altitude and airspeed are to be held constant during the 180 deg turns at each end of the holding pattern. The standard-rate turn for a jet aircraft is 1.5 deg/s or 0.0262 rad/s. Equation 5-10 is used to find that the required bank angle is 16.63 deg and that the associated load factor is 1.044 g's. The turning radius, from Eq. 5-12, is 14,007 ft or 2.65 mi; a 180 deg turn will place the aircraft 5.3 mi from its original track. With q equal to 85.3 lb/ft^2, Eq. 5-16 can be used to obtain the required T/S, which is 8.95 lb/ft^2, so that the T/W required for the turn is 0.0895. In practice, the throttle setting is adjusted to give the desired airspeed.

5-3 MAXIMUM LOAD FACTOR AND MAXIMUM BANK ANGLE

Since the bank angle and the load factor are directly coupled by Eq. 5-8, the flight and design conditions for the maximum possible load factor n_m are those for the maximum bank angle ϕ_m because

$$\phi_m = \arc\cos \frac{1}{n_m} \tag{5-22}$$

To find the airspeed for n_m, take the first derivative of the load factor, as given by Eq. 5-21, with respect to the airspeed and set it equal to zero. Since $dn/dV = \rho V\, dn/dq$, we can set $dn/dq = 0$ and obtain the same results. Differentiating Eq. 5-21 with respect to q, setting the result to zero, and solving for the dynamic pressure produces the expressions:

$$q_{nm} = \frac{T/S}{2C_{DO}} \tag{5-23}$$

and

$$V_{nm} = \left(\frac{T/S}{\rho_{SL}\sigma C_{DO}}\right)^{1/2} = \left[\frac{(T/W)(W/S)}{\rho_{SL}\sigma C_{DO}}\right]^{1/2} \tag{5-24}$$

where the subscript nm denotes the maximum load factor conditions. You may recognize Eq. 5-24 as the ceiling (and minimum-drag) airspeed.

Since the thrust must equal the drag in a steady-state turn, Eq. 5-23 can be rewritten as

$$D_{nm} = 2q_{nm}SC_{DO} \tag{5-25}$$

but, from the definition of the total-drag coefficient,

$$C_{D,nm} = 2C_{DO} = C_{DO} + KC_L^2 \tag{5-26}$$

or

$$C_{L,nm} = \left(\frac{C_{DO}}{K}\right)^{1/2} \tag{5-27}$$

Either recognizing that Eq. 5-27 is the lift coefficient for the maximum lift-to-drag coefficient or by substituting into the definition of E, we can show that *the maximum load factor and maximum bank angle occur when flying at the maximum lift-to-drag ratio.* Substitution of Eq. 5-23 into Eq. 5-21 produces the relationships:

$$n_m = \frac{T/W}{2(KC_{DO})^{1/2}} = \left(\frac{T}{W}\right) E_m \tag{5-28}$$

and

$$\phi_m = \text{arc cos}\left[\frac{1}{(T/W)E_m}\right] \tag{5-29}$$

These expressions represent the maximum possible values of the load factor and of the associated bank angle, for a given T/W ratio, and are directly proportional to the product of the flight thrust-to-weight ratio and the maximum lift-to-drag ratio. Incidentally, we could have easily maximized the load factor by inspection of Eq. 5-8. The far right-hand expression shows that *for a given T/W ratio, the maximum load factor occurs when the lift-to-drag ratio is at a maximum.* We also see that *the largest value of the maximum possible load factor (and bank angle) occurs when both the thrust-to-weight ratio and the lift-to-drag ratio are at their maxima.*

Because of structural limitations or passenger comfort, an aircraft, particularly a noncombat aircraft, is limited to flight conditions that will not exceed a maximum *allowable* load factor. A typical value of the maximum allowable factor for a transport is of the order of 2.5 g's and for a fighter aircraft of the order of 8 g's. Our illustrative transport with a maximum T/W of 0.25 (at sea level) and an E_m of 16.67 has a maximum *possible* load factor of 4.17 g's. If the aircraft is to be limited by the maximum *allowable* value of 2.5 g's, then the T/W ratio must be kept below 0.15 in accordance with the following equation:

$$\frac{T}{W} \leq \frac{n_{m,\ \text{allowable}}}{E_m} \tag{5-30}$$

Returning to Eq. 5-28, we know that at the absolute ceiling for a given throttle setting the T/W ratio is equal to the reciprocal of the maximum lift-to-drag ratio. Therefore, at the absolute ceiling, the maximum load factor is unity and the maximum bank angle is zero degrees, leading to the conclusion that an aircraft cannot make a steady turn at the absolute ceiling without losing altitude.

Without considering the limitations of a maximum allowable load factor, let us look at the maximum possible load factor performance of our illustrative transport, operating at the maximum thrust-to-weight ratio, first at sea level and then at 30,000 ft. Since the lift coefficient for n_m is equal to 0.5 and constant (from Eq. 5-27), it is both independent of the altitude and much less than 1.8, the value of the maximum lift coefficient of the aircraft. Do *not* confuse the lift coefficient for n_m with the lift coefficient for stall in a turn.

At sea level, the maximum load factor is equal to 4.17 g's (0.25 × 16.67), the corresponding bank angle is 76.1 deg, the dynamic pressure is 833 lb/ft^2, and the airspeed is 837 fps (571 mph or M 0.75). With the airspeed and the bank angle known, the turning rate and radius are found to be equal to 0.155 rad/s (8.88 deg/s) and 5,402 ft (1.02 mi), respectively.

At 30,000 ft, the maximum load factor will decrease since the T/W decreases with altitude. Consequently, the maximum load factor at 30,000 ft has decreased to

1.56 g's (4.17 × 0.347), and the bank angle is 50.1 deg. With a constant throttle setting, the dynamic pressure decreases to 311.7 lb/ft^2, but the airspeed remains constant at 837.4 fps, as can be seen from Eq. 5-24. The turning rate and radius at 30,000 ft are equal to 0.046 rad/s (2.6 deg/s) and 18,204 ft (3.45 mi), respectively. As we saw in the preceding section, the turning performance deteriorates with altitude.

Let us introduce a second illustrative aircraft, a fighter with a W/S of 50 lb/ft^2, a maximum sea-level T/W of 0.6 (without afterburner), a subsonic drag polar of $C_D = 0.011 + 0.2\ C_L^2$ (an AR of 2 and an E_m of 10.67), a no-flaps maximum lift coefficient of 1.2 (a bit high), and an allowable maximum load factor of 8 g's. The lift coefficient for the maximum possible load factor is constant and only 0.234. At sea level, the maximum possible load factor is equal to 6.4 g's and the maximum bank angle is 81 deg. The corresponding airspeed is 1,071 fps (730 mph and M 0.96) and the turning rate and radius are 0.19 rad/s (10.9 deg/s) and 5,637 ft (1.07 miles). At 30,000 ft, the maximum load factor is 2.4 g's, the maximum bank angle is 65.3 deg, the airspeed is still 1,071 fps (but the Mach number is 1.08), and the turning rate and radius are 0.0656 rad/s (3.76 deg/s) and 16,327 ft (3.09 mi). It should be noted that the airspeed for the maximum possible load factor is too high for our subsonic drag polar. Ignoring that fact and looking at the turning performance, we find that reducing the wing loading and increasing the thrust-to-weight ratio did not have much of an effect on the turning performance at the maximum load factor, particularly at 30,000 ft. If we limit the maximum load factor for the transport to 2.5 g's, then the sea-level performance would approach that at the higher altitudes.

We conclude this section with the statement that the maximum load factor is not of interest as a desirable or sought after flight condition but rather is an upper limit that imposes structural limitations upon the designer or flight limitations upon the pilot, or both. Equation 5-16 can be used to construct lines of constant load factor as a function of both the airspeed and the altitude for a given aircraft, resulting in what is referred to as a V-n diagram. We shall look at V-n diagrams in more detail in Sec. 8-5.

5-4 MAXIMUM TURNING RATE

Maximum turning rate (*fastest turn*) and minimum turning radius are the true measures of the maneuverability of an aircraft. To find the flight conditions and design characteristics that maximize the turning rate, set the derivative of the turning rate with respect to the airspeed equal to zero. Using Eq. 5-10, the condition for fastest turn can be written as

$$n^2 - 1 - nV\frac{dn}{dV} = 0 \tag{5-31}$$

But dn/dV is equal to $\rho V\, dn/dq$, and dn/dq can be obtained by differentiating Eq. 5-21. Substituting into Eq. 5-31 and solving for the dynamic pressure for the fastest turn produces the expressions:

$$q_{FT} = \left(\frac{W}{S}\right)\left[\frac{K}{C_{DO}}\right]^{1/2} \tag{5-32}$$

and

$$V_{FT} = \left[\frac{2(W/S)}{\rho_{SL}\sigma}\right]^{1/2}\left[\frac{K}{C_{DO}}\right]^{1/4} \tag{5-33}$$

where the subscript FT denotes the fastest-turn conditions. Notice that the fastest-turn airspeed is equal to the level-flight minimum-drag airspeed.

From Eq. 5-32 and the fact that $L = nW$, we find that

$$C_{L,FT} = n_{FT}\left(\frac{C_{DO}}{K}\right)^{1/2} = n_{FT}\,C_{L,Em} \tag{5-34}$$

Using Eq. 5-34, it can be shown that

$$E_{FT} = E_m\left(\frac{2n_{FT}}{1+n_{FT}^2}\right) \tag{5-35}$$

so that the flight lift-to-drag ratio is less than E_m. The remaining conditions for fastest turn are:

$$n_{FT} = \left[2\left(\frac{T}{W}\right)E_m - 1\right]^{1/2} = [2n_m - 1]^{1/2} \tag{5-36}$$

and

$$\dot{\chi}_{FT} = \frac{g\tan\phi}{V_{FT}} = \frac{g(n_{FT}^2 - 1]^{1/2}}{V_{FT}} \tag{5-37}$$

In order to show the parameters affecting the turning rate, Eq. 5-37 can be expanded by appropriate substitutions to obtain

$$\dot{\chi}_{FT} = g\left[\frac{\rho_{SL}\sigma}{W/S}\left(\frac{C_{DO}}{K}\right)^{1/2}(n_m - 1)\right]^{1/2} \tag{5-38}$$

and

$$\dot{\chi}_{FT} = g\left\{\frac{\rho_{SL}\sigma}{W/S}\left[\frac{T/W}{K} - \left(\frac{C_{DO}}{K}\right)^{1/2}\right]\right\}^{1/2} \tag{5-39}$$

These two equations show that a high turning rate calls for a large T/W ratio, a low W/S, a small K (which implies a large AR and Oswald span efficiency), and a low C_{DO}. We see most of these characteristics in a modern fighter designed for maneuvering combat (dog-fighting with guns) except for the large AR. Aspect ratios are kept small on fighter aircraft (of the order of 2) because of the structural, weight, and aerodynamic considerations involved in high g and supersonic flight. Furthermore, as the wing area is increased in order to decrease the wing loading, the wing span will increase, since $b = (\text{AR} \times S)^{1/2}$, thus creating additional structural problems associated with a large wing span if the AR is high. Consequently, modern fighter aircraft rely on lower W/S's and C_{DO}'s and on higher T/W's for good maneuverability and accept the penalties of lower ARs and lower maximum lift-to-drag ratios.

The various equations for the fastest turning rate, particularly Eqs. 5-37 and 5-33, also show that the maximum turning rate is achieved at low airspeed and at sea level. As a consequence, air-to-air combat may start at M 2.0 to M 3.0 and at 50,000 to 60,000 ft but, if continued, will slow down and descend until the low speed-low altitude combat arena is reached or until one of the aircraft is destroyed. The lower limit to the combat altitude is the minimum altitude required for pull-up from an evasive maneuver.

The fastest-turn performance and flight conditions for our two illustrative aircraft are shown in Table 5-1.

TABLE 5-1

Maximum Turning Rate Flight (Fastest Turn)

	h (ft)	C_L	n (g's)	ϕ (deg)	q (lb/ft^2)	V (fps)	$\dot{\chi}_{max}$ (deg/s)	r (ft)
Transport:	SL	1.35	2.71	68.3	200	410	11.3	2,079
	30,000	0.73	1.46	46.6	200	671	2.9	13,215
Fighter:	SL	0.806	3.435	73.1	213	423	14.3	1,693
	30,000	0.457	1.95	59.1	213	692	4.5	8,892

5-5 MINIMUM TURNING RADIUS

The minimum turning radius is found by first expressing the radius as

$$r=\frac{V}{\dot{\chi}}=\frac{V^2}{g(n^2-1)^{1/2}} \tag{5-40}$$

and then setting $dr/dV=0$, which with the fact that dn/dV is equal to $\rho V\,dn/dq$ yields the *tightest turn* condition that

$$n^2-1-qn\frac{dn}{dq}=0 \tag{5-41}$$

With the substitution of dn/dq from Eq. 5-21, the various tightest-turn conditions can be developed and are written below, with the subscript TT denoting tightest turn:

$$q_{\mathrm{TT}}=\frac{2K(W/S)}{T/W} \tag{5-42}$$

$$V_{\mathrm{TT}}=2\left[\frac{K(W/S)}{\rho_{\mathrm{SL}}\sigma(T/W)}\right]^{1/2} \tag{5-43}$$

$$C_{L,\mathrm{TT}}=\frac{n_{\mathrm{TT}}(T/W)}{2K} \tag{5-44}$$

$$n_{\mathrm{TT}} = \left(2 - \frac{1}{n_m^2}\right)^{1/2} \tag{5-45}$$

$$\dot{\chi}_{\mathrm{TT}} = \frac{g \tan \phi_{\mathrm{TT}}}{V_{\mathrm{TT}}} = \frac{g(n_{\mathrm{TT}}^2 - 1)^{1/2}}{V_{\mathrm{TT}}} \tag{5-46}$$

$$r_{\mathrm{TT}} = \frac{V_{\mathrm{TT}}}{\dot{\chi}_{\mathrm{TT}}} = \frac{V_{\mathrm{TT}}^2}{g(n_{\mathrm{TT}}^2 - 1)^{1/2}} \tag{5-47}$$

A more detailed expression for the minimum turning radius that shows the effects of some of the flight and design parameters is

$$r_{\mathrm{TT}} = \frac{4K(W/S)}{\rho_{\mathrm{SL}}\sigma g(T/W)(1 - 1/n_m^2)^{1/2}} \tag{5-48}$$

We see from Eq. 5-48 that a small turning radius calls for a large T/W, a small K (a large eAR), a low C_{D0} (to yield a large E_m), and a low altitude (large σ). Equation 5-47 shows that we also want a low airspeed. These are the same design characteristics and flight conditions called for in the preceding section for high turning rates. Minimum-turning radius (tightest-turn) flight does differ from fastest-turn flight in several ways, however. The first is in the magnitude of the associated load factors. The fastest-turn load factor is proportional to the square root of the maximum possible load factor of the aircraft and has no other theoretical upper limit (see Eq. 5-36). The tightest-turn load factor, on the other hand, can never be larger than the square root of 2 or 1.414 g's (see Eq. 5-45). The second difference is that the tightest-turn airspeeds are even lower than those for fastest turn, so that the theoretical lift coefficients are much higher.

These tightest-turn expressions are very neat and very useful in giving insight into the factors influencing this important aspect of maneuverability. Unfortunately, for most current aircraft, the theoretical airspeed required is almost invariably much less than the corresponding stall speed, which is another way of saying that the required lift coefficients exceed the actual maximum lift coefficients. Equation 5-44 shows that the largest value of the required lift coefficient occurs at sea level, where the thrust is at a maximum.

Without any regard at this time for the size of the lift coefficient required, the minimum turning radius (tightest-turn) performance of our illustrative transport and fighter is shown in Table 5-2. Notice, however, that the lift coefficients at sea level do indeed exceed the respective maximum lift coefficients of 1.8 and 1.2 for both aircraft and that the airspeeds called for are low.

Since flight at airspeeds below the stall speed is not possible, there are four possible solutions. The first of these is to increase the value of the maximum lift coefficient at these low airspeeds by the use of flaps and slats. The second is to reduce the T/W ratio in accordance with Eq. 5-44 until the required lift coefficient is equal to the maximum lift coefficient. The third solution is to maintain maximum thrust and to fly the turn at a lift coefficient equal to the maximum lift coefficient. Although the second and third solutions appear at first glance to be identical, we shall see in subsequent paragraphs that they are not. The fourth solution,

TABLE 5-2

Minimum Turning Radius Flight (Tightest Turn)

	h (ft)	C_L	n (g's)	ϕ (deg)	q (lb/ft^2)	V (fps)	$\dot{\chi}$ (deg/sec)	r_{min} (ft)
Transport:	SL	2.9	1.39	44.1	48	201	8.9	1,292
	30,000	0.89	1.26	37.5	128	537	2.6	11,682
Fighter:	SL	2.11	1.40	44.6	33.3	167	10.9	882
	30,000	0.758	1.35	42.3	89.1	448	3.72	6,850
Alternative 1: Reduce T/W; $C_{L,TT}=C_{L_{max}}$								
Transport: $T/W=0.16$	SL	1.8	1.36	42.7	75.5	252	4.1	2,140
Fighter: $T/W=0.346$	SL	1.2	1.39	76.6	57.8	220	8	1,570
Alternative 2: Maintain T_{max}/W; increase n; $C_L=C_{L_{max}}$								
Transport: $T_{max}/W=0.25$	SL	1.8	2.15	62.3	119	317	11.1	1,640
Fighter: $T_{max}/W=0.6$	SL	1.2	2.4	65.4	100	291	13.8	1,197

which is not a truly responsive solution, is simply to increase the turning altitude until the required lift coefficient becomes equal to the maximum value available.

Let us look at the second solution (reduce the thrust) and call it Alternative 1. Although it keeps the load factor low (below 1.414 g's) and is a straightforward application of the tightest-turn equations, it requires a larger turning radius than does the third solution, which will be discussed in the paragraphs following this one. Using Alternative 1, reducing the sea-level T/W ratio of the transport to 0.159 results in a turning radius of 2,140 ft, compared with the original value of 1,640 ft. For the fighter at sea level, reducing the T/W ratio to 0.346 results in a turning radius of 1,570 ft, compared with 1,197 ft.

We shall call the third solution (maintain the thrust and increase the load factor) Alternative 2. Before we can use this approach, we need expressions for the stall speed in a turn along with the associated load factor. Equation 5-18 can be solved for the load factor at stall in a turn, yielding the expression

$$n_{s,t}=\frac{\rho_{SL}\sigma V_{s,t}^2 C_{L_{max}}}{2(W/S)} \tag{5-49}$$

If the right-hand side of Eq. 5-49 is set equal to the right-hand side of Eq. 5-21, with the appropriate expression for the stall dynamic pressure, we can solve for the corresponding stall speed, i.e.,

$$V_{s,t}=\left[\frac{2(T/S)}{\rho_{SL}\sigma(C_{DO}+KC_{L\max}^2)}\right]^{1/2} \tag{5-50}$$

where $(C_{DO}+KC_{L\max}^2)$ is the drag coefficient at stall. Substitution of Eq. 5-50 back into Eq. 5-49 gives a simpler expression for the stall load factor in a turn,

$$n_{s,t}=\frac{(T/W)C_{L\max}}{C_{DO}+KC_{L\max}^2} \tag{5-51}$$

Equations 5-50 and 5-51 are general equations that can be used to find the load factor and airspeed at stall in any turn. Then Eqs. 5-10 and 5-12 can be used to find the turning rate and the turning radius at the stall point. Applying these equations to our fighter at sea level, maintaining the maximum T/W ratio of 0.6,

$$V_{s,t}=\left\{\frac{2\times 30}{\rho_{SL}[0.011+0.2(1.2)^2]}\right\}^{1/2}=290.6\text{ fps}=198\text{ mph}$$

$$n_{s,t}=\frac{0.6\times 1.2}{0.299}=2.4\ g\text{'s}$$

$$\phi_{s,t}=\arccos\frac{1}{2.4}=65.4\text{ deg}$$

$$\dot{\chi}_{s,t}=\frac{32.2[(2.4)^2-1]^{1/2}}{290.6}=0.243\text{ rad/s}=13.9\text{ deg/s}$$

$$r_{s,t}=\frac{290.6}{0.243}=1{,}197\text{ ft}$$

The procedure for the transport is identical. The results for both aircraft for each of the alternatives are also shown in Table 5-2. Comparison of the results for Alternative 2 with those for sea-level fastest turn, as shown in Table 5-1, shows similar turning performance but at different airspeeds and load factors.

This chapter concludes with the observations that a large value of the maximum lift coefficient is a very important parameter in determining the maneuverability of an aircraft and that the fastest and tightest turns are achieved by using maximum thrust, usually flying on the edge of a stall in the case of fighter aircraft. If you ever have the opportunity to observe an air-to-air combat simulator in action, such as those at NASA-Langley and NASA-Ames, you will quickly realize that the combat arena is on the deck and you will be aware of a horn blowing at frequent intervals. Every time the horn blows, it signifies that one of the aircraft has stalled out.

The qualitative relationships among the various airspeed for turning flight and for level flight are sketched in Fig. 5-2.

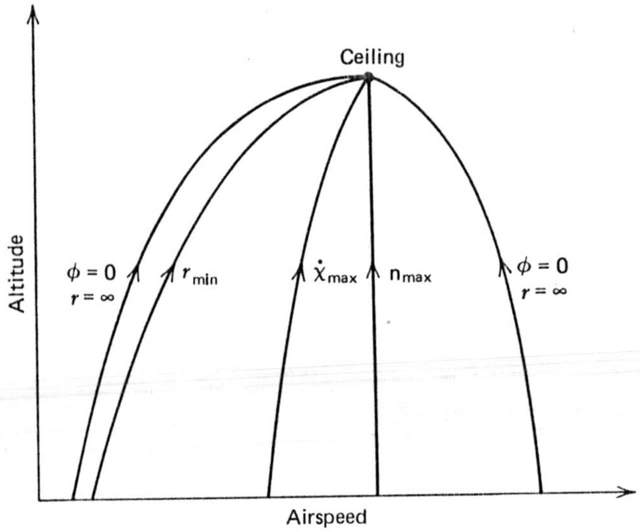

FIGURE 5-2

Relationships among wings-level and turning flight airspeeds.

PROBLEMS

Major characteristics of Aircraft *A*, Aircraft *B*, and Aircraft *C*, used in many of these problems, are listed at the beginning of the problems in Chap. 3.

5-1. Aircraft *A* is in a steady-state turn at sea level with a bank angle of 20 deg and is using maximum thrust.

a. Find the load factor.

b. Find the two possible turning airspeeds (fps and mph) and the associated lift coefficients and lift-to-drag ratios.

c. Are both of these airspeeds physically attainable? Justify your answer.

d. For each physically attainable airspeed, find the turning rate (deg/s) and the turning radius (ft).

e. What would happen to the turning performance if the thrust were reduced by 50 percent?

5-2. Aircraft *A* is in a standard-rate turn of 1.5 deg/s at sea level at a bank angle of 15 deg.

a. Find the load factor, turning airspeed, lift coefficient, and lift-to-drag ratio.

b. Find the stall speed in this turn.

c. Find the turning rate (deg/s) and the turning radius (ft).

d. Find the T/W ratio and the actual thrust required to maintain this turn.

5-3. Do Prob. 5-2 for Aircraft A at 30,000 ft.

5-4. Do Prob. 5-1 for Aircraft B.

5-5. Do Prob. 5-2 for Aircraft B.

5-6. Do Prob. 5-3 for Aircraft B.

5-7. Do Prob. 5-1 for Aircraft C.

5-8. Do Prob. 5-2 for Aircraft C.

5-9. Do Prob. 5-3 for Aircraft C.

5-10. Aircraft C is in a holding pattern at 35,000 ft. The turning rate is 2.5 deg/s and the turning radius is 10,000 ft.

a. Find the true airspeed (fps and mph), the bank angle, and the load factor associated with the maintenance of this steady turn.

b. Find the turning stall speed in fps and in mph. What is the level-flight stall speed?

c. If the load factor is suddenly increased to 1.5 g's what will the instantaneous stall speed be? Will the aircraft stall?

5-11. Aircraft A is turning at 20,000 ft at half-throttle and an airspeed of 300 mph.

a. Find the bank angle and load factor.

b. Find the turning rate (deg/s) and the turning radius (ft).

c. What is the lift coefficient in this turn?

d. What is the stall speed in this turn?

e. Is this turn possible?

5-12 Do Prob. 5-11 for Aircraft B at 30,000 ft with full throttle and an airspeed of 375 mph.

5-13. Do Prob. 5-11 for Aircraft C at 35,000 ft with full throttle and an airspeed of 500 mph.

5-14. For Aircraft A at sea level,

a. Find the maximum steady-state load factor and bank angle.

b. Find the values of the associated airspeed and lift coefficient.

c. Find the associated stall speed and compare it with the actual airspeed.

5-15. Do Prob. 5-14 for Aircraft B at sea level.

5-16. Do Prob. 5-14 for Aircraft C at sea level.

5-17. Do Prob. 5-14 for Aircraft A at 20,000 ft.

5-18. Do Prob. 5-14 for Aircraft B at 30,000 ft.

5-19. Do Prob. 5-14 for Aircraft C at 40,000 ft.

5-20. An interceptor aircraft has a W/S of 94 lb/ft^2, a drag polar of $C_D = 0.013 + 0.18C_L^2$, an unaugmented T/W ratio of 0.8, and a maximum lift coefficient of 0.9. The interceptor launches its missiles at M 2.0 at 50,000 ft and is attacked by an air-to-air fighter.

a. What is the turning performance of the interceptor at 50,000 ft at M 2.0 and at its maximum lift coefficient?

b. Still at 50,000 ft but with the other constraints relaxed, find the fastest-turn performance and conditions.

c. This time, find the tightest-turn performance and conditions at 50,000 ft.

d. The interceptor descends to 5,000 ft and is still engaged in aerial combat. Remembering the constraint of the maximum lift coefficient, find the fastest-turn performance and conditions.

e. Redo (d) above to find the tightest-turn conditions.

5-21. The air-to-air fighter of Prob. 5-20 has a wing loading of 50 lb/ft^2, a drag polar of $C_D = 0.015 + 0.1C_L^2$ an augmented T/W ratio of 1.1, and a maximum lift coefficient of 1.2. Find its turning performance in accordance with Prob. 5-20.

5-22. If you have worked both Prob. 5-20 and Prob. 5-21, which is the better aircraft for air-to-air combat? Why?

5-23. The training sailplane of Prob. 4-34, in the preceding chapter, has a maximum lift coefficient of 3.0.

a. Find the maximum load factor, the maximum bank angle, and the associated airspeed at sea level and at 35,000 ft.

b. Find the fastest-turn performance and conditions at sea level and at 35,000 ft. Be aware of the stall speed and maximum lift coefficient constraints.

c. Find the tightest-turn performance and conditions at sea level and at 35,000 ft, remaining aware of the stall speed and lift coefficient constraints.

Chapter Six

Level Flight in the Vertical Plane: Piston-Props

6-1 INTRODUCTION AND GOVERNING EQUATIONS

We are now ready and prepared to look at the performance of aircraft equipped with reciprocating engines and propellers (piston-props) rather than pure turbojet engines. We shall use the techniques of the preceding chapters and shall find that not only the flight conditions for best performance but also the design parameters are surprisingly different for the piston-prop than for the turbojet.

These differences arise from the fact that, whereas the turbojet produces thrust only, the piston-prop produces power only. (Incidentally, these piston-prop analyses are applicable to any conceivable aircraft propulsion system that delivers power only.) We shall also discover that the evaluation and determination of the best-performance conditions for the piston-prop are generally not as straightforward nor as simple as for the turbojet. Any difficulties that occur in the piston-prop analyses have two root causes. The first is that the fuel consumption rate of a piston-prop engine is proportional to the brake horsepower of the engine. The second, and more troublesome mathematically, is the fact that it is the thrust rather than the power that appears explicitly in the dynamic equations.

Considering only the fuel consumption, the weight balance equation of a piston-prop is the negative of the fuel consumption rate and can be written as

$$\frac{-dW}{dt} = \hat{c}(\mathrm{HP}) \tag{6-1}$$

where HP is the brake horsepower of the engine and $\hat{c}$ is the horsepower specific fuel consumption (hpsfc) expressed as pounds of fuel per hour per horsepower, lb/h/hp. Piston-props burn gasoline, which has a weignt of approximately *6 lb/gal* and thus has a density about 10 percent lower than that of jet fuel.

It is more convenient to replace the engine horsepower in the governing equations by the *thrust power*, which is the power produced by the propeller. It is given the symbol P and is the product of the thrust delivered by the propeller and the true airspeed of the aircraft. The thrust power and the engine horsepower are related by the expressions

$$P = TV = k\eta_p(\text{HP}) \tag{6-2}$$

In Eq. 6-2, η_p is the *propeller efficiency* and k is a conversion factor that has a value of 550 ft-lb/s/hp when V is expressed in fps and a value of 375 mi-lb/h/hp when V is in mph. The thrust power, accordingly, has the units of ft-lb/s or mi-lb/h.

The propeller efficiency varies with the airspeed. However, with a variable-pitch constant-speed propeller, η_p can be assumed to be constant over the design operating speed range. A well-designed propeller will have an efficiency of the order of 80 to possibly 90 percent. In general, we shall assume and use a constant efficiency of 85 percent, unless otherwise stated.

With the substitution of Eq. 6-2, Eq. 6-1 becomes

$$\frac{-dW}{dt} = \frac{\hat{c}P}{k\eta_p} = \frac{\hat{c}TV}{k\eta_p} \tag{6-3}$$

with the fuel consumption rate normally expressed in lb/h, implying that V is in mph and that k has a numerical value of 375; T, of course, is in pounds.

In the preceding chapters, we assumed that the thrust of a properly matched turbojet was essentially independent of the true airspeed. Except for rate-of-climb calculations, thrust power was of no interest but obviously increased linearly with the airspeed. For piston-props, we assume that the horsepower delivered by the engine is independent of the airspeed. The thrust power curve, as sketched in Fig. 6-1*a*, will also be assumed to be flat except in those regions where the propeller efficiency drops off, i.e., at very high and very low airspeeds for a variable pitch propeller. The thrust, on the other hand, will decrease very rapidly, almost exponentially, as shown in Fig. 6-1*b*, as the airspeed increases. Note that the maximum thrust,which is finite and is called the *static thrust*, occurs when the airspeed is zero, such as at the start of the take-off roll. It is this rapid decline in thrust with airspeed, combined with the high engine weight per horsepower of piston engines, that keeps piston-props in the low airspeed region.

The variation of power (either horsepower or thrust power) with altitude for an unsupercharged (aspirated) piston-prop is essentially that for a turbojet. We shall use only the simple relationship of Sec. 2-3 that

$$\frac{P_1}{P_{SL}} = \frac{\rho_1}{\rho_{SL}} = \sigma_1 \tag{6-4}$$

If the engine is turbocharged, the power will remain constant up to the *critical altitude*, which ranges from 12,000 to approximately 20,000 ft, depending primarily on whether the aircraft is pressurized or not.* For altitudes above the critical

---

*Oxygen masks must be worn for flight above 12,000 ft for more than 1 hour in an unpressurized aircraft.

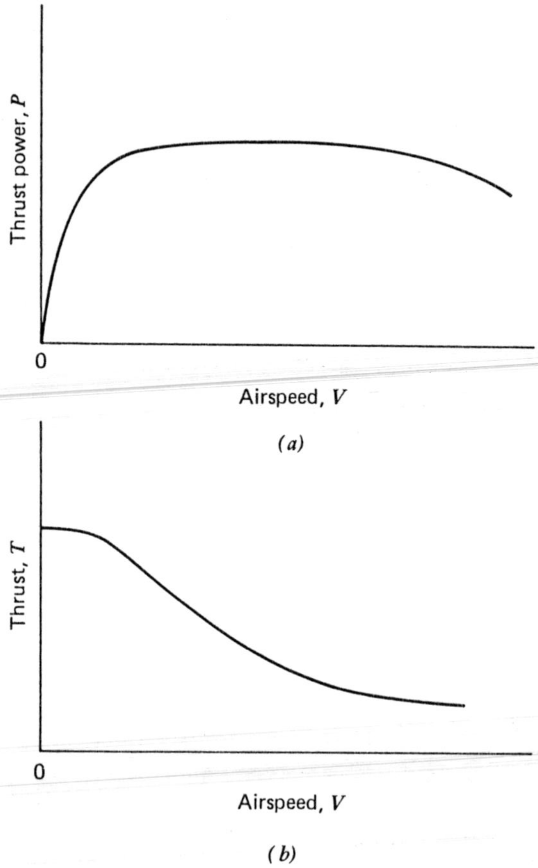

FIGURE 6-1

Available power and thrust of a piston-prop with a constant throttle setting: (a) thrust power; (b) thrust.

altitude, the variation can be represented by

$$\frac{P_1}{P_{SL}} = \frac{\sigma_1}{\sigma_{cr}} \tag{6-5}$$

where σ_{cr} is the density ratio at the critical altitude.

Since only the fuel weight balance equation is changed in going from a turbojet engine to a piston-prop engine, the remaining equations of Sec. 3-1 are still valid. The complete set of governing equations for the quasi-steady flight of a piston-prop aircraft is, therefore,

$$T - D - W \sin \gamma = 0 \tag{6-6}$$

$$L - W \cos \gamma = 0 \tag{6-7}$$

$$\frac{dX}{dt} = V \cos \gamma \tag{6-8}$$

$$\frac{dh}{dt} = V \sin \gamma \tag{6-9}$$

$$\frac{-dW}{dt} = \frac{\hat{c}P}{k\eta_p} \tag{6-10}$$

$$P = TV = k\eta_p(\text{HP}) \tag{6-11}$$

Do not forget that the V in Eqs. 6-6, 6-7, and 6-11 is the true airspeed whereas the V in Eqs. 6-8 and 6-9 is the ground speed. We shall assume, as we did with the turbojet, a no-wind condition so that the ground speed and airspeed are identical. The effects of wind will be discussed in Chap. 10.

6-2 LEVEL FLIGHT AND CEILING CONDITIONS

For level flight in the vertical plane, the flight-path angle is set equal to zero, as was done with the turbojet, and the equations of interest become:

$$T = D \tag{6-12}$$

$$L = W \tag{6-13}$$

$$\frac{dX}{dt} = V \tag{6-14}$$

$$\frac{-dW}{dt} = \frac{\hat{c}P}{k\eta_p} = \frac{\hat{c}TV}{k\eta_p} \tag{6-15}$$

Since now the thrust, as well as the drag, is strongly dependent upon the airspeed, Eq. 6-12, as written, is of little use to us. If both sides are multiplied by the airspeed, this force equation is transformed into the power equation

$$P = TV = DV \tag{6-16}$$

The far right-hand side of Eq. 6-16 is the *drag power*, the power required to overcome the drag. The left-hand terms represent the thrust power, the power that must be available to counteract the drag power if the aircraft is to maintain equilibrium level flight. This *available power* is a function of the engine and propeller characteristics, the altitude, and the throttle setting, and, except at the ceiling, can be larger than the *required (drag) power*. There may be times in our analyses when we shall use the symbol P_A to denote the available power and P_R to denote the required power, but only when necessary to avoid confusion or to remind ourselves that they need not necessarily be equal.

If Eq. 6-16 is divided by Eq. 6-13, we can establish the useful thrust power-to-weight ratio relationship that

$$\frac{P}{W}=\frac{DV}{L}=\frac{V}{E} \tag{6-17}$$

where E is the flight lift-to-drag ratio associated with a particular value of the airspeed. The available thrust power-to-weight ratio (P/W) is analogous to the thrust-to-weight ratio (T/W) of the turbojet and has the units of a velocity (fps or mph). The maximum value of this power-to-weight ratio is obviously related to the horsepower-to-weight ratio of the aircraft by

$$\frac{P_m}{W}=k\eta_p\frac{\text{HP}_m}{W} \tag{6-18}$$

The data provided by the aircraft manufacturers usually includes only the inverse of this ratio (W/HP_m), which is called the *power loading* of the aircraft. At the risk of appearing cynical, perhaps this is done so as to offer a large number because the smaller the value of the HP/W ratio, the larger the manufacturer's power loading.

Returning to Eq. 6-17, substitution of the level-flight drag expression for a parabolic drag polar yields

$$\frac{P}{W}=\frac{DV}{W}=\frac{\rho V^3 C_{DO}}{2(W/S)}+\frac{2K(W/S)}{\rho V} \tag{6-19}$$

This equation establishes the level-flight relationships between the thrust power and the airspeed in terms of the aircraft parameters and the altitude. Examination of the right-hand side of Eq. 6-19 shows that the zero-lift drag power (required power) increases as the cube of the airspeed, whereas the zero-lift drag (required thrust) of the turbojet increases as the square of the airspeed. This cubing effect is an important factor in limiting the maximum airspeed of a piston-prop.

The power required for a typical piston-prop is sketched in Fig. 6-2, along with the maximum available power. If you care to refer back to the typical thrust and drag curves for a typical turbojet (Fig. 3-2), you will see that not only are the shapes of the "required" curves different but also that the relationships between the required and available curves are different. We also notice in Fig. 6-2 that the minimum power (lowest) point appears to be closer to the stall speed and that the region of two level-flight airspeeds for a given throttle setting is quite limited.

The minimum value of the drag power-to-weight ratio is an interesting and important performance parameter. It is given the symbol $P_{\min}/W$ and is found by setting the first derivatives of Eq. 6-19 with respect to the airspeed equal to zero and solving for the minimum drag-power airspeed, $V_{P\min}$. The resulting series of expressions is:

$$V_{P\min}=\left[\frac{2(W/S)}{\rho_{SL}\sigma}\right]^{1/2}\left(\frac{K}{3C_{DO}}\right)^{1/4}=\frac{V_{P\min,SL}}{\sigma^{1/2}} \tag{6-20}$$

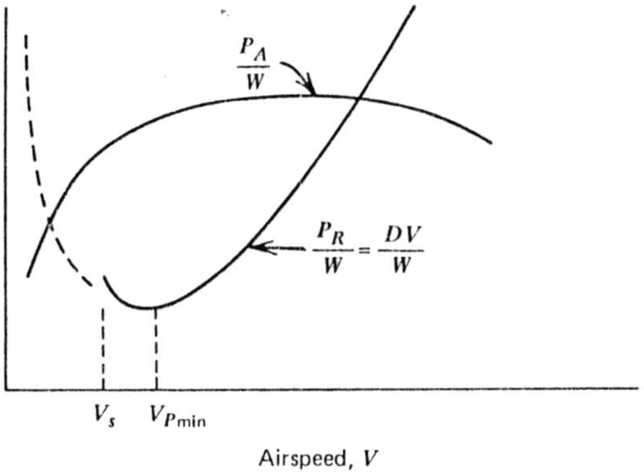

FIGURE 6-2
Available and required power for a given altitude, wing loading, and throttle setting.

$$C_{L,P\min} = \left[\frac{3C_{D0}}{K}\right]^{1/2} = 1.732 C_{L,Em} \tag{6-21}$$

$$E_{P\min} = 0.866 E_m \tag{6-22}$$

$$\frac{P_{\min}}{W} = \frac{V_{P\min}}{0.866 E_m} = \frac{V_{P\min,SL}}{0.866\sigma^{1/2} E_m} \tag{6-23}$$

Notice that, whereas the lift-to-drag ratio remains constant, the airspeed, and thus the minimum required power-to-weight ratio, increases with altitude. Also notice that the required lift coefficient is 73 percent larger than that for E_m (minimum drag). Consequently, it can be shown that the maximum lift coefficient for the aircraft must be at least 2.5 times as large as the lift coefficient for the maximum lift-to-drag ratio (a design characteristic of the aircraft) if the airspeed for minimum power is to be at least 20 percent larger than the stall speed. This may not always be the case so that we must always be aware of the possibility that, for reasons of safety, we may have to fly at airspeeds higher than the minimum-power airspeed even though the latter may be what we want for best performance.

While we are looking at the minimum required power, let us establish the conditions for the absolute ceiling, which occurs when the power available for a given throttle setting is just equal to the minimum required power for that altitude. The *absolute ceiling of a piston-prop* aircraft occurs when the available power is at a maximum, so that

$$\frac{P_{m,c}}{W_c} = \frac{V_{P\min,c}}{0.866 E_m} \tag{6-24}$$

With the assumption that the aircraft weights at sea level and at the ceiling are substantially the same, the absolute ceiling condition for an aspirated engine is given by

$$\left(\frac{P_m}{W}\right)_{SL}\sigma_c = \frac{V_{P_{\min},SL}}{0.866\sigma_c^{1/2}E_m} \tag{6-25}$$

Consequently, the ceiling density ratio for an *aspirated engine* is given by

$$\sigma_c = \left[\frac{1.155V_{P_{\min},SL}}{(P_m/W)_{SL}E_m}\right]^{2/3} \tag{6-26}$$

If the engine is turbocharged, the power-altitude relationship of Eq. 6-5 results in the following relationship for the ceiling density ratio for a *turbocharged engine*:

$$\sigma_c = \left[\frac{1.155V_{P_{\min},SL}\sigma_{cr}}{(P_m/W)_{SL}E_m}\right]^{2/3} \tag{6-27}$$

Equation 6-27 can also be used for an aspirated engine if σ_{cr} is set equal to unity.

The ceiling of a piston-prop is dependent not only on the engine power-to-weight ratio and the maximum lift-to-drag ratio (as was the case with the turbojet) but also on the wing loading. As the wing loading increases (to increase the cruise airspeed), the ceiling lowers whereas the ceiling of a turbojet is independent of the wing loading.

It is time to look at some numbers by introducing an illustrative piston-prop with the following characteristics:

$$\frac{W}{S} = 34 \text{ lb/ft}^2 \qquad HP/W = 0.1 \text{ hp/lb}$$

$$C_D = 0.025 + 0.051C_L^2 \qquad \hat{c} = 0.5 \text{ lb/h/hp}$$

$$AR = 7.34 \qquad e = 0.85$$

$$\eta_p = 0.85 \qquad E_m = 14$$

$$C_{L\max} = 1.8$$

These are typical values for a high-performance twin-engine piston-prop executive aircraft that can carry six to eight passengers.

The first thing to do is to establish the appropriate sea-level values to be used in the ceiling determination.

$$\left(\frac{P_m}{W}\right)_{SL} = 0.85 \times 550 \times 0.1 = 46.75 \text{ fps}$$

$$V_{P_{\min},SL} = \left(\frac{2\times 34}{\rho_{SL}}\right)^{1/2}\left(\frac{0.051}{3\times 0.025}\right)^{1/4} = 153.6 \text{ fps} = 104.7 \text{ mph}$$

$$\frac{P_{\min,SL}}{W} = \frac{153.6}{0.866\times 14} = 12.67 \text{ fps}$$

At this point, let us check the stall speed against the minimum-power airspeed. The former is 86 mph and the latter is 104.7 mph, 22 percent higher than the stall speed, a satisfactory relationship. If the engines are aspirated (not turbocharged), then

$$\sigma_c = \left(\frac{1.155 \times 153.6}{46.75 \times 14}\right)^{2/3} = 0.419$$

The density ratio of 0.419 corresponds to an absolute ceiling of approximately 27,000 ft. Using Eq. 6-20, the ceiling airspeed is 237 fps, or 162 mph.

If the engines are turbocharged with a critical altitude of 20,000 ft, the density ratio at that altitude is 0.533, so that the density ratio at the ceiling is 0.275. The absolute ceiling is approximately 37,500 ft, and the corresponding airspeed is 293 fps, or 200 mph.

Let us return to level flight and Eq. 6-19, which unfortunately does not have a closed-form solution. The simplest approach is to specify the thrust power-to-weight ratio of interest and then introduce iterative values of the airspeed into the right-hand side until the identity is satisfied. In order to find the maximum airspeed with aspirated engines at several altitudes, substitution of the illustrative aircraft characteristics into Eq. 6-19 yields

$$46.75\sigma = 8.74 \times 10^{-7}\sigma V^3 + \frac{1,459}{\sigma V}$$

At sea level, the density ratio is unity, so that

$$46.75 = 8.74 \times 10^{-7} V^3 + \frac{1,459}{V}$$

Since we are looking for the maximum airspeed, the induced drag-power term should be reasonably small. A good starting value might well be one that is slightly lower than the airspeed with no induced drag. Starting with 370 fps, it takes two more interations to obtain a sea-level maximum airspeed of 365 fps, or 249 mph.

Jumping to 20,000 ft, where the density ratio is 0.533, the level-flight equation becomes

$$24.9 = 4.66 \times 10^{-7} V^3 + \frac{2,737}{V}$$

Since the influence of the induced-drag power has increased, we shall start with the lower value of 355 fps for the first trial value. The maximum airspeed at 20,000 ft turns out to be 330 fps, or 225 mph.

It might be interesting and worthwhile to check the value at the ceiling where the maximum airspeed and the minimum-power airspeeds should be identical. At the ceiling, we found the density ratio to be 0.419; therefore,

$$19.58 = 3.66 \times 10^{-7} V^3 + \frac{3,484}{V}$$

Iteration yields a ceiling airspeed of 237 fps, or 162 mph, which is the ceiling airspeed found earlier.

The flight envelope for the illustrative piston-prop with aspirated engines is sketched in Fig. 6-3. Up to approximately 26,000 ft, the maximum airspeed forms the high-speed boundary, and the stall speed the low-speed boundary. In the narrow region between 26,000 ft and the ceiling, the aircraft cannot slow down to the stall speed without losing altitude. In addition, there are two level-flight airspeeds in this region, but the lower one is statically unstable with respect to the airspeed.

With turbocharged engines, our piston-prop will have a constant power output up to the critical altitude of 20,000 ft, so that the airspeed calculations require two equations. The first, for the region from sea level to 20,000 ft, is

$$46.75 = 8.74 \times 10^{-7} \sigma V^3 + \frac{1{,}459}{\sigma V}$$

At sea level, with the density ratio equal to unity, the equation is identical to that for the aspirated engines, so that the sea-level maximum airspeeds are also identical, i.e., 249 mph.

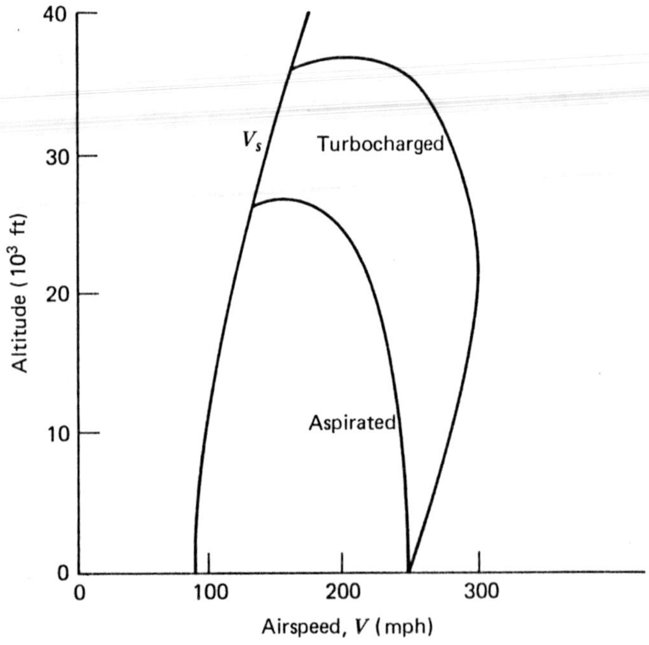

FIGURE 6-3

Level-flight envelope for illustrative piston-prop with aspirated and turbocharged engines.

At 10,000 ft, with the density ratio equal to 0.738, the equation is

$$46.75 = 6.45 \times 10^{-7} V^3 + \frac{1{,}977}{V}$$

and iteration yields a maximum airspeed of approximately 400 fps, or 273 mph, which is greater than the sea-level value.

From 20,000 ft to the absolute ceiling, the equation will be, with a critical altitude density ratio of 0.533,

$$87.7\sigma = 8.74 \times 10^{-7} \sigma V^3 + \frac{1{,}459}{\sigma V}$$

At 20,000 ft, the equation simplifies to

$$46.75 = 4.658 \times 10^{-7} V^3 + \frac{2{,}737}{V}$$

and the maximum airspeed is approximately 440 fps, or 300 mph. At the ceiling, with a density ratio of 0.275, the equation becomes

$$24.1 = 2.4 \times 10^{-7} V^3 + \frac{5{,}300}{V}$$

which is satisfied by the previously calculated ceiling airspeed of 293 fps, or 200 mph.

The flight envelope for the turbocharged piston-prop is also sketched in Fig. 6-3 and clearly displays the improved performance due to turbocharging. A turbocharger will increase the cost, complexity, and weight of an engine somewhat but not significantly. Therefore, the use of turbocharged engines in aircraft used for transportation (rather than for pleasure only) is rapidly increasing.

A maximum sea-level airspeed of 249 mph seems excessively slow in these days of M 0.85 transports and SSTs. Let us double the maximum sea-level airspeed of this aircraft to 498 mph (730 fps) or M 0.65 and use Eq. 6-19 to calculate the new maximum power-to-weight ratio. It is 342 fps rather than 46.75 fps, so that the maximum HP/W ratio is now 0.731 hp/lb rather than 0.1 hp/lb; this is an increase by a factor of 7.31. If the gross weight of the aircraft is 7,500 lb, a maximum HP/W ratio of 0.1 can be satisfied by two engines of 375 hp each for a total of 750 hp for a maximum sea-level airspeed of 249 mph. To double this airspeed to 498 mph, we shall need a total of 5,482 hp, or 15 engines of 375 hp each. If the engine horsepower-to-weight ratio is 0.5 hp/lb, then the engine weight of the low-speed twin-engine version is 1,500 lb, whereas for the high-speed multiengine version, the engine weight would be 10,964 lb, which would drastically increase the gross weight of the aircraft, which in turn would increase the horsepower required, and so on. If we doubled the wing loading, the required increase in power would be cut almost in half. However, the increased wing loading introduces other problems, such as an increase in the take-off run and a lowering of the ceiling.

6-3 BEST RANGE

The expression for the instantaneous (point) range, as obtained by dividing Eq. 6-14 by Eq. 6-15, is

$$\frac{dX}{-dW}=\frac{k\eta_p V}{\hat{c}P} \tag{6-28}$$

which, with the introduction of Eq. 6-17, can be written as

$$\frac{dX}{-dW}=\frac{k\eta_p E}{\hat{c}W} \tag{6-29}$$

We notice that this *piston-prop mileage* is explicitly independent of the airspeed although we must not forget that the value of the lift-to-drag ratio is coupled to the airspeed. For good mileage, the propeller efficiency and lift-to-drag ratio should be as large as possible and the aircraft weight and specific fuel consumption should be minimized.

With the assumptions of constant propeller efficiency and specific fuel consumption, a partial integration of Eq. 6-29 for cruise yields

$$X=\frac{-k\eta_p}{\hat{c}}\int_1^2 \frac{E}{W}\,dW \tag{6-30}$$

If the lift-to-drag ratio is kept constant by flying at a *constant lift coefficient*, Eq. 6-30 can be easily integrated to give the range equation

$$X_{C_L}=\frac{k\eta_p E}{\hat{c}}\ln \mathrm{MR}=\frac{k\eta_p E}{\hat{c}}\ln\left(\frac{1}{1-\zeta}\right) \tag{6-31}$$

Equation 6-31 is the piston-prop version of the *Breguet range equation*, and it is valid for both constant-altitude and cruise-climb flight. If the altitude is held constant, then the airspeed must be reduced along the flight path in order to maintain a constant lift coefficient (to keep E constant) as fuel is consumed. If, on the other hand, the airspeed is held constant, then the aircraft must enter the cruise-climb mode. For the piston-prop, *there is no improvement in range with cruise-climb*, only a reduction in the flight time.

The other flight program of interest is the constant altitude-constant airspeed flight program. Before final integration, the E/W term is replaced by its equivalent $(1/D)$, and Eq. 6-30 becomes

$$X=\frac{-k\eta_p}{\hat{c}}\int_1^2 \frac{dW}{D} \tag{6-32}$$

Introducing the parabolic drag polar and integrating (see Appendix B) produces the following range equation:

$$X_{h,V}=\frac{2k\eta_p E_m}{\hat{c}}\arctan\left[\frac{\zeta E_1}{2E_m(1-KC_{L_1}E_1\zeta)}\right] \tag{6-33}$$

These range equations, Eqs. 6-31 and 6-33, are similar in form to those for the turbojet with the important difference that the airspeed does not appear in either of them. Furthermore, only two equations are required to describe the three flight programs of interest.

The best-range conditions for constant lift coefficient flight can be determined by setting the first derivative of either the instantaneous range or of the range with respect to the airspeed to zero or simply by inspection of Eqs. 6-31 and 6-33. With η_p and $\hat{c}$ assumed constant, the best-range condition is to fly at the maximum lift-to-drag ratio of the aircraft, i.e., at the minimum-drag airspeed. Consequently, the *best-range equation for constant lift coefficient flight* is

$$X_{br;C_L} = \frac{375\eta_p E_m}{\hat{c}} \ln\left(\frac{1}{1-\zeta}\right) \tag{6-34}$$

where k has been replaced by 375 since $\hat{c}$ is expressed in lb/h/hp and X will have the units of statute miles. The *best-range equation for constant altitude-constant airspeed flight* can be closely approximated by starting cruise at the best-range conditions for constant lift coefficient cruise. Then Eq. 6-33 becomes

$$X_{br;h,V} = \frac{750\eta_p E_m}{\hat{c}} \arctan\left(\frac{\zeta}{2-\zeta}\right) \tag{6-35}$$

At this point, we shall examine the differences in best range for the two flight programs. The relative best range is expressed as

$$\frac{X_{br;C_L}}{X_{br;h,V}} = \frac{\ln\,[1/(1-\zeta)]}{2 \arctan\,[\zeta/(2-\zeta)]} \tag{6-36}$$

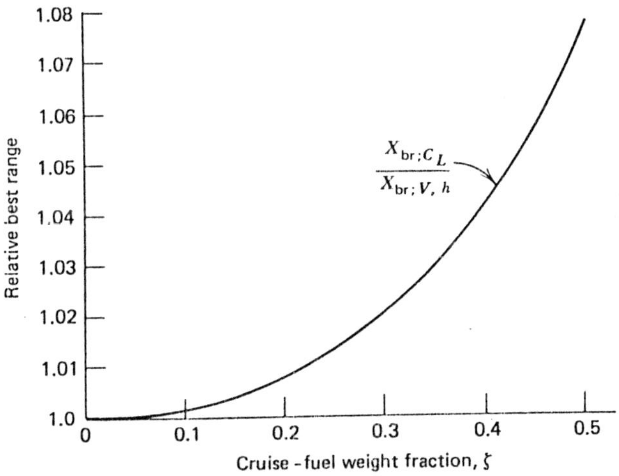

FIGURE 6-4

Piston-prop relative best range as a function of range, i.e., ζ.

and is plotted as a function of the cruise-fuel weight fraction in Fig. 6-4. The increase in range by flying cruise-climb (or constant lift coefficient and altitude) is very small, being less than 1 percent for a cruise-fuel weight fraction of 0.2, which represents a long flight for the average general aviation piston-prop. Therefore, there is no real advantage or need for a piston-prop to fly cruise-climb, even when permitted by air traffic control.

Since there is such a small difference between the performance shown by the two equations, we shall follow the customary practice *of considering only constant altitude-constant airspeed cruise and of using only the Breguet range equation to describe such cruise.*

Returning to the best-range Breguet range equation and conditions, they are:

$$X_{br} = \frac{375\eta_p E_m}{\hat{c}} \ln\left(\frac{1}{1-\zeta}\right) \tag{6-37a}$$

$$V_{br} = V_{md} = \left[\frac{2(W/S)}{\rho_{SL}\sigma}\right]^{1/2} \left(\frac{K}{C_{DO}}\right)^{1/4} \tag{6-37b}$$

$$\frac{P_{br}}{W} = \frac{V_{br}}{E_m} = \frac{V_{md}}{E_m} \tag{6-37c}$$

We see that we cruise at the maximum lift-to-drag ratio and, therefore, at the ceiling for our particular throttle setting. It is important to remember that the best range is independent of both the airspeed and the altitude but that the aircraft must fly at the best-range (minimum-drag) airspeed as specified by Eq. 6-37*b*. Increasing the best-range airspeed by increasing the altitude and/or the wing loading will not increase the range, but it will certainly reduce the flight time and increase the power required.

The illustrative piston-prop of the preceding section has the following characteristics:

$W/S = 34$ lb/ft^2	$HP/W = 0.1$ hp/lb
$C_D = 0.025 + 0.051 C_L^2$	$E_m = 14$
$\eta_p = 0.85$	$P/W = 46.75$ fps
$\hat{c} = 0.5$ lb/h/hp	$C_{L\max} = 1.8$

Let it be turbocharged with a critical altitude of 20,000 ft and a gross weight of 7,500 lb. If the cruise-fuel fraction is 0.2 (which is quite large for an aircraft of this size and type, if it is to have a reasonable payload), the maximum range without reserves will be

$$X_{br} = \frac{375 \times 0.85 \times 14}{0.5} \ln\left(\frac{1}{1-0.2}\right) = 1{,}991 \text{ mi}$$

With our assumptions, this is the maximum range at any altitude. In actuality, however, the range will be slightly larger at altitude since the specific fuel consumption decreases slightly with altitude.

Although the range is unaffected by increasing the altitude, the power required and the airspeed will increase and the flight time will decrease.

$$V_{br} = \left(\frac{2 \times 34}{\rho_{SL}\sigma}\right)^{1/2} \left(\frac{0.055}{0.025}\right)^{1/4} = \frac{202}{\sigma^{1/2}} \text{ fps} = \frac{137.8}{\sigma^{1/2}} \text{ mph}$$

$$\frac{P_{br}}{W} = \frac{202}{140\sigma^{1/2}} = \frac{14.43}{\sigma^{1/2}} \text{ fps}$$

At sea level, the best-range airspeed will be 137.8 mph, and it will take 14.45 hours to fly the 1,991 mi. The required power-to-weight ratio of 14.43 fps represents 30 percent of the maximum power available.

At 20,000 ft, the best-range airspeed will increase to 188.6 mph with an accompanying reduction in the flight time to 10.55 hours, a 27 percent reduction. The power-to-weight ratio increases to 19.76 fps, which represents 42 percent of the maximum power available. Incidentally, if the engines were aspirated (not turbocharged), the range and the airspeed would be the same but the required power would be 79 percent of the maximum power available.

For piston-prop aircraft, such as this one, that are characterized by low gross weights and moderate or low wing loadings, the best-range airspeed is considerably lower than the maximum airspeed. Furthermore, the distances normally flown are generally of the order of 1,000 miles or less, so that the design cruise-fuel weight fraction is low. Consequently, it is customary to cruise at 75 percent of the maximum power available because time of flight is more important than any fuel savings. Operating data issued by the manufacturers include the airspeeds and ranges for both economical (best-range) cruise and for a 75 percent power setting, usually referred to as *max cruise*.

Let us plan a no-wind flight of 1,200 mi at 20,000 ft, first at the best-range conditions and then at 75 percent max power. For best-range flight, the best-range airspeed, which is the minimum-drag airspeed, is 276.6 fps, or 188.6 mph, and the corresponding flight time is 6.36 hours. This airspeed is 63 percent of the maximum airspeed for this altitude and the power required is only 42 percent of the maximum power available. The fuel consumption is found by solving the range equation for the cruise-fuel weight fraction, i.e.,

$$1{,}200 = \frac{375 \times 0.85 \times 14}{0.5} \ln\left(\frac{1}{1-\zeta}\right)$$

so that $\zeta = 0.1258$

and $\Delta W_f = \zeta W = 0.1258 \times 7{,}500 = 943.5 \text{ lb} = 157.2 \text{ gal}$

For 75 percent power (max-cruise) flight at 20,000 ft, the available power must be found first. Since 20,000 ft is the critical altitude, the maximum available power has not yet been affected by the altitude; therefore, the max-cruise power is

$$\frac{P}{W} = 0.75 \times 46.75 = 35.0 \text{ fps}$$

The max-cruise airspeed is found from Eq. 6-19, which with the appropriate substitutions is

$$35.0 = 4.65 \times 10^{-7} V^3 + \frac{2{,}737}{V}$$

The max-cruise airspeed is found, by iteration, to be 392 fps or 267 mph; the corresponding flight time is 4.5 hours. The cruise speed is now 89 percent of the maximum airspeed for this altitude. It is now necessary to find the flight lift-to-drag ratio before solving the general range equation for the cruise-fuel weight fraction. There are two ways of obtaining E. The first and simpler method is to use the relationship of Eq. 6-17 that the thrust power-to-weight ratio is equal to the airspeed divided by the lift-to-drag ratio to find a value of 11.2 for E. The second method is to find the lift coefficient and then use the drag polar to find the drag coefficient, leading to a value for E. Using this method,

$$C_L = \frac{2 \times 34}{\rho_{SL} \times 0.533 \times (392)^2} = 0.3493$$

$$C_D = 0.025 + 0.051(0.3493)^2 = 0.0312$$

$$E = \frac{0.3493}{0.0312} = 11.2$$

Introducing the value of 11.2 for E into Eq. 6-31 results in the range equation:

$$1{,}200 = \frac{375 \times 0.85 \times 11.2}{0.5} \ln\left(\frac{1}{1-\zeta}\right)$$

This equation can be solved to obtain a value of 0.1547 for the cruise-fuel weight fraction, which translates into a fuel consumption of 1,160 lb or 193 gal.

Comparison of the two sets of data shows that flying at 75 percent of the maximum power reduces the flight time by 1.86 hours (29 percent) at the price of an increase in the fuel consumption of 35.8 gal (23 percent). Since the operating (flying) time of an aircraft is the basic unit for most of the costs, such as rental fees, crew salaries, scheduled inspections, and overhauls and since time is of value for the passengers, the savings associated with the reduced flight time will probably outweigh the increased cost of the additional fuel, assuming that the fuel is available and not rationed.

This section, up to this point, can be summarized as follows:

1. The best range of a piston-prop is independent of the airspeed, and thus of the altitude and wing loading. Increasing the best-range airspeed only reduces the flight time.
2. The best range is a function of the propeller efficiency, the maximum lift-to-drag ratio, the specific fuel consumption, and the amount of fuel available for cruise.
3. There is no significant advantage to flying in the cruise-climb mode and, therefore, no significant penalty for flying at a constant altitude and a constant airspeed and for using the Breguet range equation.

The differences between the best-range performance of piston-prop and turbojet aircraft can be seen by comparing the respective Breguet range equations. An interesting and significant observation is that the $375\eta_p$ term in the piston-prop equation is an equivalent best-range airspeed that is independent of the actual airspeed at which the aircraft is flying. This means that if two piston-props have identical maximum lift-to-drag ratios and specific fuel consumptions, the 100 mph aircraft will fly just as far as a 600 mph aircraft if the cruise-fuel weight fractions are also identical. The faster aircraft, obviously, will require considerably less flight time, one-sixth the time in this case. This equivalent best-range airspeed also means that a pure turbojet with its higher specific fuel consumption is not competitive with the piston-prop at the lower airspeeds.

This last statement can be verified by comparing the best range of our illustrative turbojet to that of a hypothetical turbojet that has the same wing loading and drag polar. This comparison is shown in Fig. 6-5 for three altitudes, using two values for the specific fuel consumption of the turbojet because the value of 0.5 is unrealistic. If the altitude were to be further increased without limit, the turbojet and piston-prop curves would eventually intersect.

The best-range equations, and Fig. 6-5, also show that, for comparable aircraft, the best-range airspeed of the turbojet will be 31 percent higher than that of the piston-prop. The best range, however, will always be less than that of the comparable piston-prop (with the same cruise-fuel weight fraction) until the best-

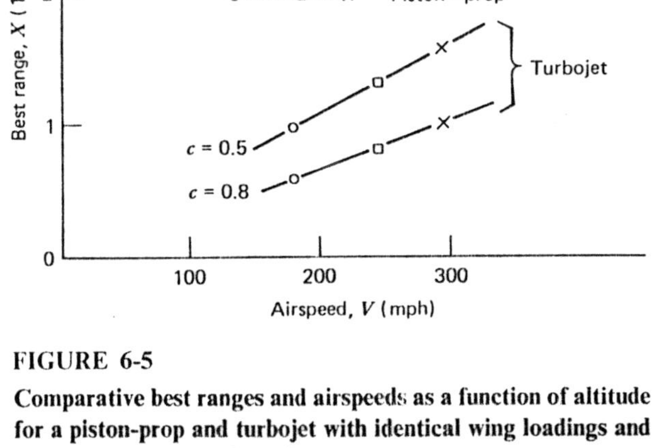

FIGURE 6-5

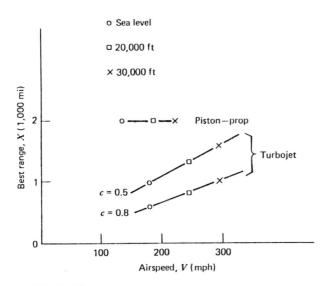

Comparative best ranges and airspeeds as a function of altitude for a piston-prop and turbojet with identical wing loadings and drag polars and with $\zeta = 0.2$.

range airspeed satisfies the condition that

$$V_{br} \geqslant \frac{375\eta_p c}{0.866\hat{c}}$$

If the piston-prop has a propeller efficiency of 85 percent and a specific fuel consumption of 0.5 lb/h/hp and the comparable turbojet has a specific fuel consumption of 0.8 lb/h/lb, then the turbojet must have a wing loading-altitude combination such that the best-range airspeed is at least 588 mph. If the maximum lift-to-drag ratios are not identical (the turbojet usually has the higher value), this airspeed will be changed. For example, with an E_m of 14 for the piston-prop and 18 for the turbojet, the competitive airspeed is reduced from 588 to 457 mph.

6-4 MAXIMUM ENDURANCE

The instantaneous endurance, or flight time per pound of fuel, is easily obtained by inverting Eq. 6-15, so that

$$\frac{dt}{-dW} = \frac{375\eta_p}{\hat{c}P} \tag{6-38}$$

where the thrust power P must be expressed in mph. This equation shows that reducing the power increases the endurance. In fact, with a constant propeller efficiency and a constant specific fuel consumption, the instantaneous endurance will be at a maximum when the power is at a minimum, when $P = P_{min}$. Using the relationship that P/W is equal to V/E, this equation can be written as

$$\frac{dt}{-dW} = \frac{375\eta_p E}{\hat{c}VW} \tag{6-39}$$

which clearly shows the importance of the individual parameters and characteristics. For good endurance, we want a high propeller efficiency, a large lift-to-drag ratio, a small specific fuel consumption, a low airspeed, and a low gross weight.

The simplest flight program would be one in which we maintained constant power throughout the flight. Integration of Eq. 6-38 with constant power yields the deceptively simple endurance equation:

$$t = \frac{375\eta_p \zeta}{\hat{c}(P_1/W_1)} \tag{6-40}$$

where P_1 and W_1 are the values at the start of cruise and t, as in all of the equations to follow, is expressed in hours. The constant thrust power-to-weight ratio (P_1/W_1) represents the power available. As the weight of the aircraft decreases, decreasing the power required, the airspeed must be increased along the flight path in order to maintain equilibrium with a constant power setting. With this program, the instantaneous endurance remains constant along the flight path whereas it increases with the other flight programs as the required power decreases. Consequently, this flight will not be considered further.

If the lift coefficient is to be kept constant, there will be two flight programs to be considered. The first is *cruise-climb*, which, since the airspeed remains constant, can be obtained by merely dividing the range equation by the airspeed to obtain

$$t_{C_L,V}=\frac{375\eta_p E}{\hat{c}V}\ln\left(\frac{1}{1-\zeta}\right) \tag{6-41}$$

The second flight program is at a *constant altitude*. Since the airspeed is continually decreasing as the aircraft weight decreases, the integration is more complicated. A partial integration of Eq. 6-39 results in

$$t_{C_L,h}=\frac{375\eta_p E}{\hat{c}}\int_1^2 \frac{dW}{VW} \tag{6-42a}$$

where

$$V=\left[\frac{2(W/S)}{\rho_{SL}\sigma C_L}\right]^{1/2} \tag{6-42b}$$

A similar, though not identical, integration is shown in Appendix B. Carrying out the integration yields the somewhat awkward expression

$$t_{C_L,h}=\frac{750\eta_p E}{\hat{c}V_1}\left[\frac{1-(1-\zeta)^{1/2}}{(1-\zeta)^{1/2}}\right] \tag{6-43}$$

where V_1 is the airspeed at the start of cruise.

The last, and most realistic, flight program to be considered is one with both a *constant altitude* and a *constant airspeed*. Dividing the range equation, Eq. 6-33, by the airspeed results in

$$t_{h,V}=\frac{750\eta_p E_m}{\hat{c}V}\arctan\left[\frac{\zeta E_1}{2E_m(1-KC_{L_1}\zeta E_1)}\right] \tag{6-44}$$

With these equations, the endurance, or flight time, for a given set of flight conditions can easily be found by substitution and solution. As mentioned earlier in this section, inspection of Eq. 6-38 shows that the maximum instantaneous endurance occurs when the required power is minimized, i.e., when $P=P_{min}$. *If the entire flight is flown so as to maintain this condition, then the aircraft will achieve its maximum endurance.*

Using the previously determined expressions for the conditions associated with the minimum required power, the flight conditions for *maximum endurance* are:

$$E_{t_{max}}=E_{P_{min}}=0.866E_m \tag{6-45a}$$

$$V_{t_{max}}=V_{P_{min}}=\left[\frac{2(W/S)}{\rho_{SL}\sigma}\right]^{1/2}\left(\frac{K}{3C_{DO}}\right)^{1/4} \tag{6-45b}$$

$$\frac{P_{t_{max}}}{W}=\frac{P_{min}}{W}=\frac{V_{t_{max}}}{0.866E_m} \tag{6-45c}$$

The two constant lift coefficient equations for maximum endurance are:

$$t_{max;C_L,V}=\frac{324.75\eta_p E_m}{\hat{c}V_{t_{max}}}\ln\left(\frac{1}{1-\zeta}\right) \tag{6-46}$$

and
$$t_{max;C_L,h}=\frac{649.5\eta_p E_m}{\hat{c}V_{t_{max}}}\left[\frac{1-(1-\zeta)^{1/2}}{(1-\zeta)^{1/2}}\right] \quad (6\text{-}47)$$

The appropriate maximum-endurance equation for constant altitude-constant airspeed flight is

$$t_{max;h,V}=\frac{750\eta_p E_m}{\hat{c}V_{t_{max}}}\arctan\left(\frac{0.433\zeta}{1-0.75\zeta}\right) \quad (6\text{-}48)$$

Let us substitute numbers for the illustrative piston-prop to see what the maximum endurance might be with a cruise-fuel fraction of 0.2. For this aircraft, E_m is 14, $\hat{c}$ is 0.5 lb/h/hp, and η_p is 0.85. All that is needed is the airspeed, which is

$$V_{t_{max}}=\left(\frac{2\times 34}{\rho_{SL}\sigma}\right)^{1/2}\left(\frac{0.051}{3\times 0.025}\right)^{1/4}=\frac{153.6}{\sigma^{1/2}}\text{ fps}=\frac{104.7}{\sigma^{1/2}}\text{ mph}$$

The sea-level maximum-endurance airspeed is 104.7 mph and the respective flight times are:

$$t_{max;C_L,V}=\frac{324.75\times 0.85\times 14}{0.5\times 104.7}\ln\left(\frac{1}{1-0.2}\right)=16.5\text{ hours}$$

$$t_{max;C_L,h}=\frac{649.5\times 0.85\times 14}{0.5\times 104.7}\left[\frac{1-(1-0.2)^{1/2}}{(1-0.2)^{1/2}}\right]=17.4\text{ hours}$$

$$t_{max;h,V}=\frac{750\times 0.85\times 14}{0.5\times 104.7}\arctan\left[\frac{0.433\times 0.2}{1-(0.75\times 0.2)}\right]=17.3\text{ hours}$$

As a matter of interest, the flight time with constant power is only 14.76 hours.

As the altitude increases, so does the maximum-endurance airspeed, so that the flight times are reduced. Remember that with maximum endurance, we are seeking the largest flight times possible. For all the maximum-endurance flight programs, the maximum endurance at any altitude is equal to the sea-level endurance multiplied by the square root of the density ratio. At 20,000 ft, where $\sigma=0.533$, the airspeed increases from 104.7 to 143.4 mph, and the corresponding flight times are all reduced by 27 percent to:

$$t_{max;C_L,V}=12.0\text{ hours}$$

$$t_{max;C_L,h}=12.7\text{ hours}$$

$$t_{max;h,V}=12.6\text{ hours}$$

For maximum endurance, we want to fly as low and as slowly as possible in an aircraft with a low wing loading, a large maximum lift-to-drag ratio, a high propeller efficiency, and a low specific fuel consumption. We also see that there is no advantage to cruise-climb (actually, there is a penalty) and that for moderate values of the cruise-fuel weight fraction, the endurance of constant altitude-constant airspeed flight is comparable to that of constant lift coefficient flight at a constant altitude. Furthermore, it is easier to fly because the airspeed is constant. However, since the power-available curve is tangent to the power-required curve at the minimum-power (maximum-endurance) airspeed (see Fig. 6-2), any inadvertent and small decrease in the actual airspeed will put the aircraft on the back side of the power curve and the airspeed will continue to decrease unless the power is increased. Therefore, the operational maximum-endurance airspeed is normally somewhat higher than the theoretical value so as to establish the cruise airspeed at a stable equilibrium point.

PROBLEMS

The two aircraft, whose major characteristics are given below, will be used in many of the problems of this and subsequent chapters. Each aircraft has the option of being equipped with either aspirated or turbocharged engines. When turbocharged, Aircraft *D* will have a critical altitude of 15,000 ft and Aircraft *E* will have a critical altitude of 20,000 ft. Assume the propeller efficiency to be constant as given unless the problem explicitly states otherwise.

	Aircraft *D* Single Engine	**Aircraft *E* Twin Engine**
Gross weight (lb)	3,000	7,500
Wing area (ft^2)	175	215
Wing span (ft)	36	41
Max HP	235	750
C_{D0}	0.028	0.025
K	0.048	0.045
hpsfc (lb/h/hp)	0.45	0.45
η_p	0.83	0.85
$C_{L_{max}}$	1.8	1.9

6-1. Find the wing loading, the aspect ratio, and the maximum lift-to-drag ratio for:

a. Aircraft *D*

b. Aircraft *E*

6-2. For Aircraft *D*, flying at sea level on a standard day:

a. Find the values of the minimum-drag and minimum-power airpeeds (fps and mph) along with the associated lift and drag coefficients and lift-to-drag ratios.

b. Find the values of the maximum airspeed and of the stall speed along with the associated lift and drag coefficients and lift-to-drag ratios.

c. Find the fuel consumption (lb/h and gal/h).

6-3. Do Prob. 6-2 for Aircraft *D* equipped with an aspirated engine and flying at 5,000 ft.

6-4. Do Prob. 6-2 for Aircraft *D* equipped with an aspirated engine and flying at 15,000 ft.

6-5. Do Prob. 6-2 for Aircraft *D* equipped with a turbocharged engine and flying at 15,000 ft.

6-6. Do Prob. 6-2 for Aircraft *D* equipped with a turbocharged engine and flying at 25,000 ft.

6-7. Do Prob. 6-2 for Aircraft *E*.

6-8. Do Prob. 6-2 for Aircraft *E* equipped with aspirated engines and flying at 20,000 ft.

6-9. Do Prob. 6-2 for Aircraft *E* equipped with turbocharged engines and flying at 20,000 ft.

6-10. Do Prob. 6-2 for Aircraft *E* equipped with turbocharged engines and flying at 30,000 ft.

6-11. Aircraft *D* and Aircraft *E* are equipped with aspirated engines.

a. For Aircraft *D*, find the absolute ceiling density ratio and altitude along with the associated airspeed (fps and mph).

b. Do (a) for Aircraft *E*.

6-12. a. Do Prob. 6-11 with both aircraft equipped with turbocharged engines.

b. If the wing loading of each aircraft is doubled, what will be the effects on the respective ceilings and ceiling airspeeds?

6-13. For Aircraft *D*.

a. Find the best-range airspeed (mph) and best-range mileage (mi/lb and mi/gal) at sea level.

b. Do (a) above at 15,000 ft with an aspirated engine.

c. Do (a) above at 15,000 ft with a turbocharged engine.

d. Do (a) above at 20,000 ft with a turbocharged engine.

6-14. For Aircraft *E*:

a. Find the best-range airspeed (mph) and best-range mileage (mi/lb and mi/gal) at sea level.

b. Do (a) above at 20,000 ft with aspirated engines.

c. Do (a) above at 20,000 ft with turbocharged engines.

d. Do (a) above at 30,000 ft with turbocharged engines.

6-15. Aircraft *D* has 370 lb of fuel available for cruise.

a. Find the maximum range and associated flight time at sea level. What are the corresponding values if the flight is made with 75 percent available power?

b. With an aspirated engine, do (a) above at 15,000 ft.

c. With a turbocharged engine, do (a) above at 15,000 ft.

d. With a turbocharged engine, do (a) above at 20,000 ft.

6-16. Aircraft E has 1,000 lb of fuel available for cruise.

a. Find the maximum range and associated flight time at sea level. What are the corresponding values if the flight is made using 75 percent available power?

b. With aspirated engines, do (a) above at 20,000 ft.

c. With turbocharged engines, do (a) above at 20,000 ft.

d. With turbocharged engines, do (a) above at 30,000 ft.

6-17. Aircraft D is scheduled to fly 1,000 mi at 10,000 ft.

a. What is the minimum amount of fuel required (lb and gal) and what is the flight time?

b. With an aspirated engine, how much additional fuel is required to fly at 75 percent available power and what is the savings in flight time?

c. Do (b) above with a turbocharged engine.

6-18. Aircraft E is scheduled to fly 1,500 mi at 25,000 ft. Do Prob. 6-17.

6-19. Aircraft D and Aircraft E are both loaded with 500 lb of fuel to be used for a sea-level search mission with maximum endurance as the objective. Find and compare the flight times and ranges of the two aircraft.

6-20. Do Prob. 6-19 for a search altitude of 10,000 ft.

6-21. A man-powered aircraft has a gross weight of 207 lb, a wing area of 1,000 ft^2, and a wing span of 95 ft. It is estimated that a human being can generate a maximum of 0.5 hp for a period of 7 minutes and 0.33 hp for a reasonably indefinite period of time. Assume a propeller efficiency of 95 percent.

a. Find the AR, wing loading, and average chord length.

b. Assuming an Oswald span efficiency of 0.95 and a zero-lift drag coefficient of 0.02, write the parabolic drag polar and find the maximum lift-to-drag ratio.

c. Find the sea-level minimum-power and minimum-drag airspeeds (fps and mph).

d. Find the maximum sea-level airspeed (fps and mph) for each of the two available horsepower levels. Do these airspeeds exceed those found in (c) above? In other words, is there enough available power for this aircraft to fly at sea level?

e. Find the density ratio, altitude, and airspeed at the absolute ceiling with the sustained available power. Neglect the increase in the zero-lift drag coefficient as the aircraft moves out of the ground effect region.

f. Find the maximum range using the maximum available power of 0.5 hp for 7 minutes.

g. Find the maximum range that can be flown in 2 hours at the sustained power level of 0.33 hp.

6-22. A long-range, high-altitude reconnaissance aircraft has a gross weight of 16,000 lb, a wing area of 1,600 ft^2, and a wing span of 149.7 ft. It has two turbocharged engines (critical altitude of 20,000 ft) of 1,040 hp each at sea level; the propeller efficiency is 90 percent. The zero-lift drag coefficient is 0.018 and the Oswald span efficiency is 0.98.

a. Find the wing loading, aspect ratio, and average wing chord length.

b. Find the parabolic drag polar and the maximum lift-to-drag ratio.

c. What is the absolute ceiling and what is the ceiling airspeed (fps and mph)?

d. If 4,500 lb of fuel is available for cruise, what is the maximum range of this aircraft with $\hat{c} = 0.4$ lb/h/hp?

e. If the mission is flown at 55,000 ft, what is the flight time?

Chapter Seven

Other Flight: Piston-Props

7-1 TAKE-OFF AND LANDING

If drag and rolling friction are neglected, the take-off ground run approximation for a piston-prop is identical to that for the turbojet, namely,

$$T=\frac{W}{g}\frac{dV}{dt}=\frac{WV}{g}\frac{dV}{dX} \tag{7-1}$$

so that

$$dX=\frac{V}{g(T/W)}\,dV \tag{7-2}$$

Although the thrust is changing throughout the take-off run, it will be assumed constant so that Eq. 7-2 can be integrated to yield the expression

$$d=\frac{V_{\mathrm{LO}}^2}{2g(T/W)} \tag{7-3}$$

where V_{LO}, the lift-off airspeed is in fps and d is in feet.

The problem now is to evaluate the thrust-to-weight ratio. Rather than worry about finding an average value (it is not always easy to obtain data on the thrust variation during take-off), we shall evaluate the thrust-to-weight ratio at lift-off, expressing it as

$$\frac{T}{W}=\frac{550\eta_p(\mathrm{HP}/W)}{V_{\mathrm{LO}}} \tag{7-4}$$

With this substitution, Eq. 7-3 becomes

$$d=\frac{V_{\mathrm{LO}}^3}{1{,}100\eta_p g(\mathrm{HP}/W)} \tag{7-5}$$

If the engines are *aspirated* (not turbocharged), the variation of the horsepower-to-weight ratio is given by the relationship, $\mathrm{HP}/W=(\mathrm{HP}_{\mathrm{SL}}/W)\sigma$, and

$$d=\frac{V_{\mathrm{LO}}^3}{1{,}100\eta_p g\sigma(\mathrm{HP}/W)_{\mathrm{SL}}} \tag{7-6}$$

If the engines are *turbocharged*, then HP/W will be constant and

$$d=\frac{V_{\mathrm{LO}}^3}{1{,}100\eta_p g(\mathrm{HP}/W)_{\mathrm{SL}}} \tag{7-7}$$

For take-offs at sea level *and* on a standard day, the ground run for the two types of engines will be identical. If either of these two conditions is not met, then the appropriate equation must be used.

As with the turbojet, the safe lift-off airspeed will be established as 20 percent higher than the stall speed for the take-off configuration. Since

$$V_s=\left[\frac{2(W/S)}{\rho_{\mathrm{SL}}\sigma C_{L_{\max},\mathrm{TO}}}\right]^{1/2} \tag{7-8}$$

then

$$V_{\mathrm{LO}}=1.2V_s=\left[\frac{2.88(W/S)}{\rho_{\mathrm{SL}}\sigma C_{L_{\max},\mathrm{TO}}}\right]^{1/2} \tag{7-9}$$

These two equations also remind us that the stall and lift-off airspeeds will vary with any changes in the atmospheric density resulting from an airfield elevation other than sea level or from a nonstandard day.

Returning to Eqs. 7-6 and 7-7, the last parameter requiring evaluation is the propeller efficiency. If the variation of η_p with airspeed is known, use the value given for the lift-off airspeed. Otherwise, the selection of an appropriate value requires some judgment based on a knowledge of the lift-off airspeed and the type of aircraft. For a low-performance aircraft with a lift-off airspeed of the order of 100 fps or less, use a value of 0.65. For a medium-performance aircraft lifting off at 100 to 200 fps, a value of 0.75 gives reasonably good results. As the lift-off airspeed increases further, use a value of 0.8 to 0.85. However, for these higher lift-off airspeeds the magnitude of the take-off run, as predicted by Eqs. 7-6 and 7-7 will tend to be on the high side.

Although Eqs. 7-6 and 7-7 would normally be used, since they are reasonably simple and do not require much calculational effort, they do not show the influence of the various parameters on the take-off run. With the substitution of Eq. 7-9 into Eq. 7-6 and with some simplification, the ground-run expression for an *aspirated piston-prop* can be written as

$$d=\frac{2.44}{550\eta_p g\sigma^{2.5}(\mathrm{HP}/W)_{\mathrm{SL}}}\left(\frac{W/S}{\rho_{\mathrm{SL}}C_{L_{\max},\mathrm{TO}}}\right)^{1.5} \tag{7-10}$$

A similar substitution into Eq. 7-7 yields the following ground-run expression for a *turbocharged piston-prop:*

$$d=\frac{2.44}{550\eta_p g\sigma^{1.5}(\mathrm{HP}/W)_{\mathrm{SL}}}\left(\frac{W/S}{\rho_{\mathrm{SL}}C_{L_{\max},\mathrm{TO}}}\right)^{1.5} \tag{7-11}$$

Comparison of these two expressions shows that the take-off run of a turbocharged aircraft is equal to the product of the aspirated distance and the density ratio. Further examination of Eqs. 7-10 and 7-11 shows that in both cases a short take-off run calls for a low wing loading, a high horsepower-to-weight ratio, a high propeller efficiency, and a high atmospheric density. These conclusions are similar to those reached for a turbojet, but the impact of changes in these parameters upon the ground run is different.

A doubling of the HP/W ratio or of the η_p will have the same effect as doubling the T/W ratio: the ground run will be halved. A doubling of the wing loading will double the ground run of a turbojet (a 100 percent increase) but will increase that of the piston-prop by a factor of 2.82 (a 182 percent increase). Doubling the maximum lift coefficient decreases the ground run of a turbojet by 50 percent and that of a piston-prop by 65 percent. It must be remembered that increasing the take-off lift coefficient by mechanical devices will also increase the drag and decrease the acceleration. Finally, a 10 percent decrease in the atmospheric density (an above sea-level airfield and/or a hot day) will increase the ground run of the turbojet by 23 percent, that of the aspirated piston-prop by 30 percent, and that of the turbocharged piston-prop by only 17 percent. Figure 7-1 is a sketch of the take-off ground run as a function of these parameters for two values of the density ratio.

The time to lift-off, in seconds, can be approximated as before by assuming a constant acceleration (and a constant T/W) and by making use of the relationship

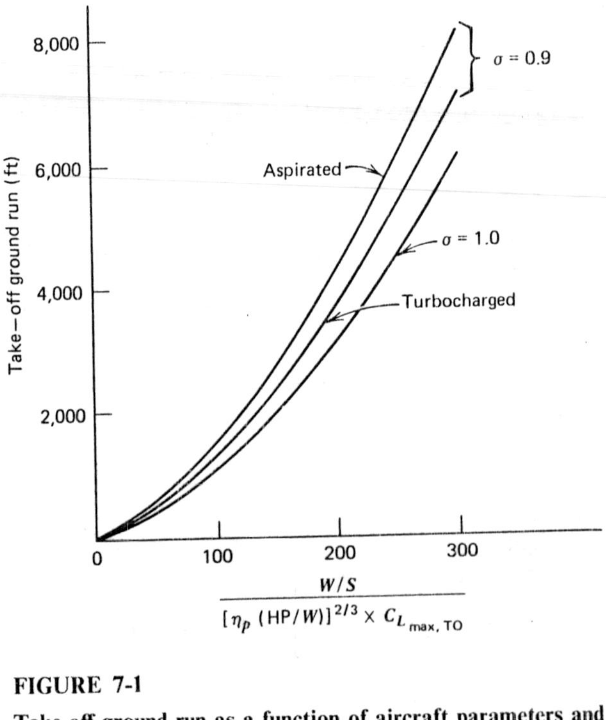

FIGURE 7-1

Take-off ground run as a function of aircraft parameters and density ratio.

that

$$d=\frac{1}{2}at^2=\frac{g(T/W)t^2}{2}=\frac{550\eta_p g(\mathrm{HP}/W)t^2}{2V_{\mathrm{LO}}} \tag{7-12}$$

so that

$$t=\left[\frac{2V_{\mathrm{LO}}d}{550\eta_p g(\mathrm{HP}/W)}\right]^{1/2} \tag{7-13}$$

If the engines are *aspirated*, then

$$t=\left[\frac{2V_{\mathrm{LO}}d}{550\eta_p g\sigma(\mathrm{HP}/W)_{\mathrm{SL}}}\right]^{1/2}=\frac{V_{\mathrm{LO}}^2}{550\eta_p g\sigma(\mathrm{HP}/W)_{\mathrm{SL}}} \tag{7-14}$$

If the engines are *turbocharged*, then

$$t=\left[\frac{2V_{\mathrm{LO}}d}{550\eta_p g(\mathrm{HP}/W)_{\mathrm{SL}}}\right]^{1/2}=\frac{V_{\mathrm{LO}}^2}{550\eta_p g(\mathrm{HP}/W)_{\mathrm{SL}}} \tag{7-15}$$

As with the ground roll, the time to lift-off with turbocharged engines is the product of the aspirated time and the density ratio.

Let us look at some numbers for our illustrative piston-prop, which has the following take-off characteristics at sea level:

$W/S=34$ lb/ft^2	$\mathrm{HP}/W=0.1$ hp/lb
$C_{L_{\max},\mathrm{TO}}=1.8$	Design $\eta_p=0.85$

Assuming a standard day and a sea-level take-off, the first calculation is the lift-off airspeed. From Eq. 7-9,

$$V_{\mathrm{LO}}=\left(\frac{2.88\times 34}{\rho_{\mathrm{SL}}\times 1.8}\right)^{1/2}=151\text{ fps}=103\text{ mph}$$

The next step is to pick a value of 0.75 for the propeller efficiency, based on this lift-off airspeed. Since this is a sea-level take-off on a standard day, the ground run and time to lift-off are independent of the type of engine and we have a choice of equations. Choosing Eqs. 7-6 and 7-14, we find that

$$d=\frac{(151)^3}{1{,}100\times 0.75\times 32.1\times 0.1}=1{,}296\text{ ft}$$

$$t=\frac{(151)^2}{550\times 0.75\times 32.2\times 0.1}=17.2\text{ s}$$

As a check, Eqs. 7-10 and 7-13 yield

$$d=\frac{2.44}{550\times 0.75\times 32.2\times 0.1}\left(\frac{34}{\rho_{\mathrm{SL}}\times 1.8}\right)^{1.5}=1{,}301\text{ ft}$$

$$t=\left(\frac{2\times 151\times 1{,}301}{550\times 0.75\times 32.2\times 0.1}\right)^{1/2}=17.2\text{ s}$$

If, however, for one reason or another, the density ratio is 0.9, then the lift-off airspeed will increase to 159.2 fps (108.5 mph), and a distinction must be made between aspirated and turbocharged engines. With *aspirated engines*, Eqs. 7-6 and 7-14 yield

$$d = \frac{(159.2)^3}{1{,}100 \times 0.75 \times 32.2 \times 0.9 \times 0.1} = 1{,}688 \text{ ft}$$

$$t = \frac{(159.2)^2}{550 \times 0.75 \times 32.2 \times 0.9 \times 0.1} = 21.2 \text{ s}$$

With *turbocharged engines*, Eqs. 7-7 and 7-15 are appropriate, or the aspirated values can simply be multiplied by the density ratio. With either approach, the turbocharged values are 1,519 ft and 19.1 seconds.

With the assumption of a constant deceleration, the landing run of a piston-prop can be estimated by the same crude expression used for the turbojet, namely,

$$d = 0.5 f_d V_s^2 \tag{7-16}$$

The determination of the correct value for f_d is just as complicated as for the turbojet, but the value should be lower because of the larger zero-lift drag (further increased by windmilling propellers) and the greater effectiveness of prop reversal, if such an option is necessary and available.

If the value of f_d is taken to be 0.3 for our piston-prop, which has been on a 0.2 cruise-fuel weight fraction cruise and has a maximum landing lift coefficient of 2.2, the landing wing loading will be 27.2 lb/ft^2, and the sea-level stall speed will be 102 fps (69.5 mph). The ground run from touchdown will be approximately 1,560 ft, which is of the order of the take-off run.

As the wing loading of piston-props is increased in order to attain higher cruising speeds, the landing run will also increase; doubling the wing loading will double the landing run. Decreases in the atmospheric density will also increase the landing roll.

7-2 CLIMBING FLIGHT

With the assumption of constant weight, the governing equations for the quasi-steady-state climb of a piston-prop are those used for the turbojet:

$$T - D - W \sin \gamma = 0 \tag{7-17}$$

$$L - W \cos \gamma = 0 \tag{7-18}$$

$$\frac{dX}{dt} = V \cos \gamma \tag{7-19}$$

$$\text{R/C} = \frac{dh}{dt} = V \sin \gamma \tag{7-20}$$

Solving Eq. 7-17 for the sine of the climb angle and expressing the thrust in terms of the available power yields

$$\sin\gamma = \frac{550\eta_p(\mathrm{HP}/W)}{V} - \frac{D}{W} \tag{7-21}$$

The thrust is expressed in terms of the horsepower instead of the thrust power so that the propeller efficiency will appear explicitly and because the horsepower is a stated aircraft characteristic. With the substitution of Eq. 7-21, Eq. 7-20 becomes

$$\mathrm{R/C} = V\sin\gamma = 550\eta_p(\mathrm{HP}/W) - \frac{DV}{W} \tag{7-22}$$

With the introduction of the parabolic drag polar and the approximation that the climb drag is equal to the level-flight drag, the drag-to-weight ratio and the drag (required) power-to-weight ratio can be written as:

$$\frac{D}{W} = \frac{\rho V^2 C_{D0}}{2(W/S)} + \frac{2K(W/S)}{\rho V^2} \tag{7-23}$$

$$\frac{DV}{W} = \frac{\rho V^3 C_{D0}}{2(W/S)} + \frac{2K(W/S)}{\rho V} \tag{7-24}$$

The rate of climb for a specified airspeed and throttle setting, at any altitude, can be found directly from Eqs. 7-22 and 7-24 by first calculating the sine of the climb angle from Eqs. 7-21 and 7-23 and then multiplying by the airspeed. For example, for our piston-prop starting its climb at sea level at 140 mph (205.3 fps) with a maximum power setting and a propeller efficiency of 85 percent,

$$\mathrm{R/C} = 550 \times 0.85 \times 0.1 - \frac{\rho_{\mathrm{SL}}(205.3)^3 \times 0.025}{2 \times 34} - \frac{2 \times 0.051 \times 34}{\rho_{\mathrm{SL}} \times 205.3}$$

$$\mathrm{R/C} = 32.1 \text{ fps} = 1{,}925 \text{ fpm}$$

$$\sin\gamma = \frac{\mathrm{R/C}}{V} = \frac{32.1}{205.3} = 0.1563$$

$$\gamma = 9.0 \text{ deg}$$

The variations in the climb angle and in the rate of climb as a function of the airspeed (for two values of the propeller efficiency and with the improper extension of the parabolic drag polar down to very low airspeeds) are sketched in Fig. 7-2. Note the shape of the curves and the presence of maximum values.

Steepest climb will obviously occur when the sine of the climb angle is maximized. Setting the first derivative with respect to the airspeed equal to zero leads to the conditions that

$$V^4 + \frac{550\eta_p(\mathrm{HP}/W)(W/S)V}{\rho_{\mathrm{SL}}\sigma C_{D0}} - \frac{4K(W/S)^2}{\rho_{\mathrm{SL}}^2\sigma^2 C_{D0}} = 0 \tag{7-25}$$

which unfortunately does not have an analytic solution. Dropping the fourth-

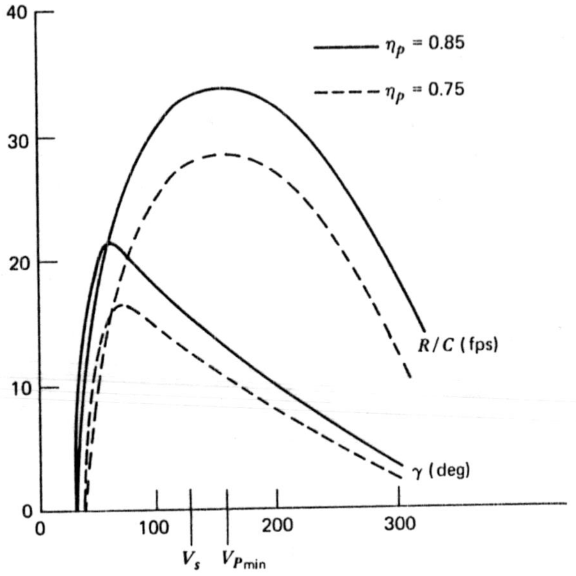

FIGURE 7-2
Sea-level climb values for the illustrative piston-prop as functions of airspeed and η_p with V in fps.

order term, however, does yield the useful approximation that

$$V_{SC} \cong \frac{4K(W/S)}{550\rho_{SL}\sigma\eta_p(HP/W)} \tag{7-26}$$

where the subscript SC denotes steepest-climb conditions. Making use of this expression in Eqs. 7-23 and 7-21 and simplifying leads to the approximation that

$$\sin\gamma_{max} \cong \frac{\rho_{SL}\sigma[550\eta_p(HP/W)]^2}{8K(W/S)} \tag{7-27}$$

For the illustrative piston-prop at sea level,

$$V_{SC} \cong \frac{4\times0.051\times34}{550\times\rho_{SL}\times0.85\times0.1} = 62.4 \text{ fps} = 42.5 \text{ mph}$$

$$\sin\gamma_{max} \cong \frac{\rho_{SL}(550\times0.85\times0.1)^2}{8\times0.051\times34} = 0.374$$

$$\gamma_{max} \cong 22 \text{ deg}$$

These values correlate very well with the exact solution shown in Fig. 7-2 but do lie below the stall speed, which at sea level is 126 fps (86 mph).

This peaking of the climb angle at an airspeed lower than the stall speed is typical. Furthermore, if we increase the steepest-climb performance of a piston-prop by increasing the HP/W ratio, the propeller efficiency, and the aspect ratio and by decreasing the wing loading, the magnitude of the steepest-climb angle will increase but the corresponding airspeed will decrease. Since the theoretical airspeed will be less than the stall speed, we conclude that *the actual steepest-climb airspeed of a piston-prop is in the vicinity of the stall speed and that the larger the calculated value of the steepest-climb angle, the larger the actual value.* For this reasonably high-performance piston-prop, it is of interest to note that the steepest-climb angle is only of the order of 10 to 15 deg. Equations 7-26 and 7-27 are useful in determining the effects of aircraft and flight characteristics upon the steepest-climb performance. When the calculated steepest-climb airspeed falls below the stall speed, as is the usual case, the benefits of increasing the maximum lift coefficient are obvious.

Fastest climb, in which we have a greater interest, is more amenable to analysis than is steepest climb. The rate of climb can be maximized by the inspection of Eq. 7-22, which shows that, for a given power setting, fastest climb occurs when the drag power is at a minimum, i.e., at $P_{\min}$. Accordingly, the expression for fastest climb is

$$(\mathrm{R/C})_{\max} = 550\eta_p(\mathrm{HP}/W) - \frac{V_{P_{\min},\mathrm{SL}}}{0.866\sigma^{1/2}E_m} \tag{7-28}$$

where

$$V_{P_{\min},\mathrm{SL}} = \left[\frac{2(W/S)}{\rho_{\mathrm{SL}}}\right]^{1/2}\left(\frac{K}{3C_{D0}}\right)^{1/4} \tag{7-29}$$

Expressing the true airspeed for fastest climb as

$$V_{\mathrm{FC}} = V_{P_{\min}} = \frac{V_{P_{\min},\mathrm{SL}}}{\sigma^{1/2}} \tag{7-30}$$

shows that even though the true airspeed increases with altitude, the calibrated airspeed remains constant throughout the climb, making climb-out simple for the pilot. Notice in Eq. 7-28 that with the assumption of a constant propeller efficiency the airspeed affects only the drag (required) power and that an increase in the minimum-power airspeed is accompanied by a decrease in both the rate of climb and the climb angle. This effect, which is not present with the turbojet, limits the rate of climb of high wing loading piston-props. The curves of Fig. 7-2 show, however, that for a given wing loading, the magnitude of the rate of climb is not very sensitive to deviations from the minimum-power airspeed. In other words, *climbing at a higher airspeed than the minimum-power airspeed does not significantly penalize the fastest-climb performance.*

The fastest-climb (minimum-power) airspeed (as well as the stall speed) is independent of the type of engine and, for our piston-prop, is $153.6/\sigma^{1/2}$ fps, whereas the stall speed is $126/\sigma^{1/2}$ fps.

With *aspirated engines*, the fastest-climb expression of Eq. 7-28 becomes

$$(\mathrm{R/C})_{\max} = 550\eta_p\sigma(\mathrm{HP}/W)_{\mathrm{SL}} - \frac{V_{P_{\min},\mathrm{SL}}}{0.866\sigma^{1/2}E_m} \tag{7-31}$$

so that, for this aircraft with $\eta_p = 0.85$,

$$(R/C)_{max} = 46.75\sigma - \frac{12.67}{\sigma^{1/2}} \tag{7-32}$$

At sea level, the maximum rate of climb is 34.08 fps (2,045 fpm) and the airspeed is 153.6 fps (104.7 mph). The easiest way to obtain the climb angle is to divide the rate of climb by the airspeed; the corresponding climb angle is 12.8 deg.

At 20,000 ft, where the density ratio is 0.533, the climb airspeed increases to 210.4 fps (143.4 mph), the rate of climb decreases to 7.56 fps (453.6 fpm), and the climb angle decreases to 2 deg. At the ceiling, where the density ratio is 0.419, the airspeed increases to 237 fps (162 mph), and the other values go to zero. Equation 7-32 can obviously be used to determine the various ceilings (absolute, service, and cruise) by using the appropriate value for the rate of climb. The variations of the airspeed, the rate of climb, and the climb angle for fastest climb are sketched as a function of the altitude in Fig. 7-3.

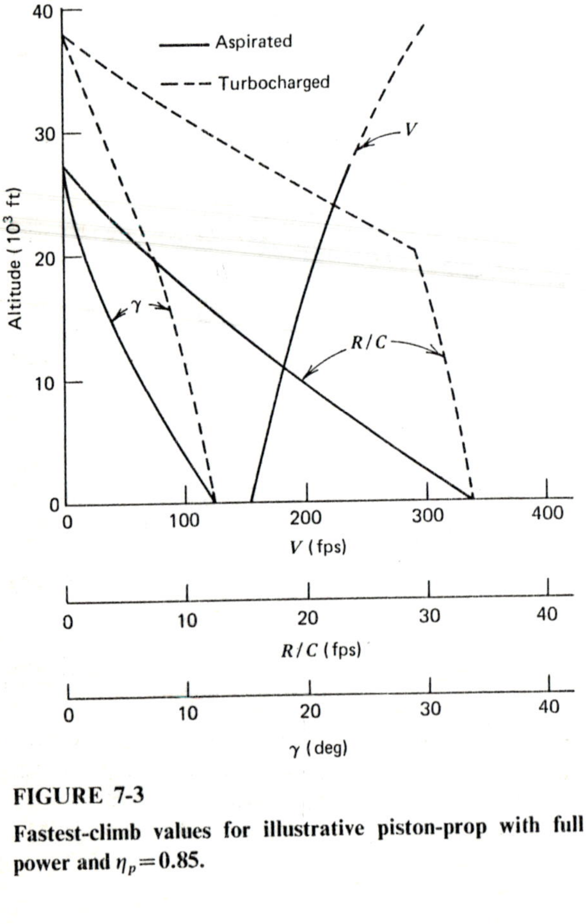

FIGURE 7-3

Fastest-climb values for illustrative piston-prop with full power and $\eta_p = 0.85$.

If the engines are *turbocharged*, two equations are required: one below and one above the critical altitude. *Below the critical altitude,*

$$(R/C)_{max} = 550\eta_p(HP/W)_{SL} - \frac{V_{P_{min},SL}}{0.866\sigma^{1/2}E_m} \tag{7-33}$$

so that, for this aircraft up to and including 20,000 ft,

$$(R/C)_{max} = 46.75 - \frac{12.67}{\sigma^{1/2}}$$

The only difference between this equation and Eq. 7-32 is the absence of the density ratio in the power available term, the airspeed and drag remaining the same as for the aspirated engines. At sea level (on a standard day), the fastest-climb values are identical to the aspirated values but at 20,000 ft, even though the airspeed is still 210.4 fps (143.4 mph), the maximum rate of climb is 29.4 fps (1,764 fpm) and the climb angle is 8 deg; these last values are four times as large as the corresponding aspirated values.

At altitudes *above the critical altitude,*

$$(R/C)_{max} = 550\eta_p\left(\frac{\sigma}{\sigma_{cr}}\right)\left(\frac{HP}{W}\right)_{SL} - \frac{V_{P_{min},SL}}{0.866\sigma^{1/2}E_m} \tag{7-34}$$

so that for this aircraft,

$$(R/C)_{max} = 87.70 - \frac{12.67}{\sigma^{1/2}}$$

At 30,000 ft, with the aspirated piston-prop left below at its absolute ceiling, the turbocharged aircraft is still climbing at 12.1 fps (725 fpm) with a climb angle of 2.75 deg and an airspeed of 251 fps (725 fpm). The fastest-climb values for this turbocharged piston-prop are also sketched in Fig. 7-3.

For tabular solutions for fastest climb to altitude, the following expressions will be used:

$$\Delta t = \frac{\Delta h}{(R/C)_{max,ave}} \tag{7-35}$$

$$\frac{\Delta W_f}{W} = \frac{\hat{c}(HP/W)_{ave}\,\Delta t}{3{,}600} \tag{7-36}$$

$$\Delta X = \frac{V_{ave}(\cos\gamma)_{ave}\,\Delta t}{5{,}280} \tag{7-37}$$

The average and cumulative values for the turbocharged piston-prop are tabulated in Table 7-1 and sketched in Fig. 7-4 for a specific fuel consumption of 0.5 lb/h/hp. To climb to 20,000 ft takes 627.7 seconds (10.5 minutes) with a $\Delta W_f/W$ of 0.00877 (63.5 lb of fuel with an initial gross weight of 7,500 lb) and with 21 mi traveled in

GROUP I. A SAMPLING

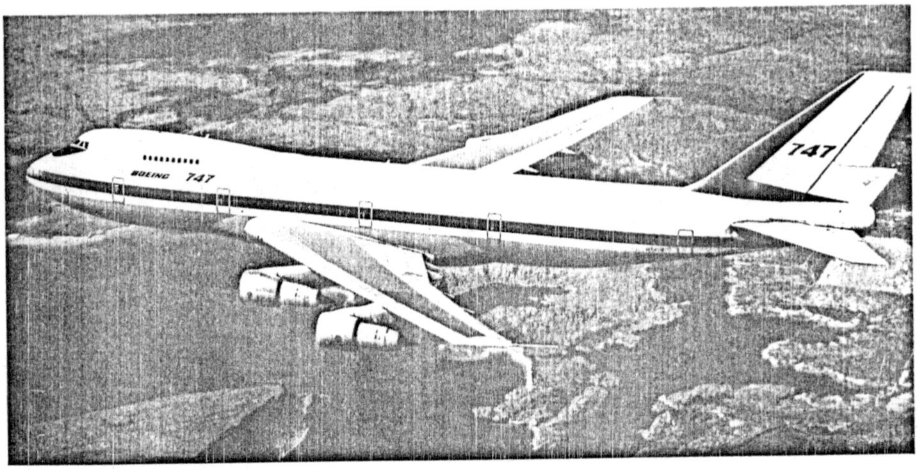

PLATE 1.
The Boeing 747. Manufactured by the Boeing Commercial Airplane Company. A long-range, wide-body turbofan with a maximum gross weight of 800,000 lb, a wing area of 5,500 ft^2, an AR of 7, and four 50,000-lb thrust turbofans.

PLATE 2.
The Beech Baron 58. Manufactured by the Beech Aircraft Corporation. A six-place aspirated piston-prop (two 285 hp engines) with a gross weight of 5,100 lb, a wing area of 200 ft^2, an AR of 7.2, a maximum cruise airspeed of 230 mph, and a maximum range of 1,500 mi.

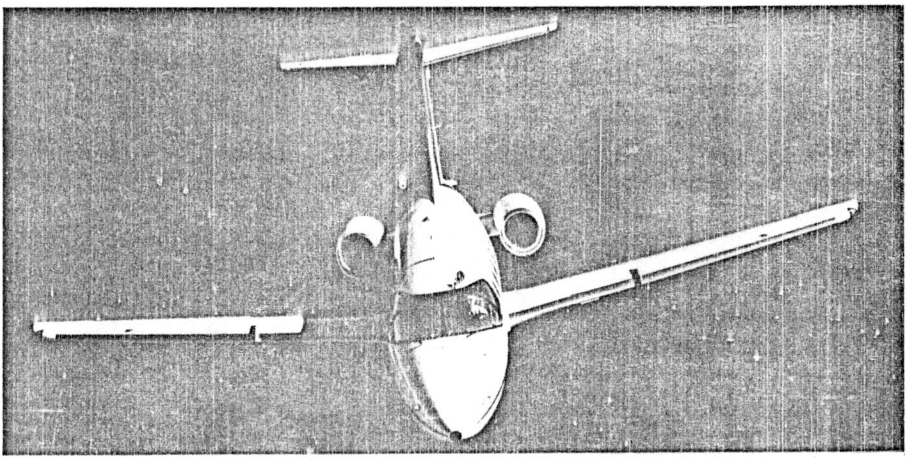

PLATE 3.

The Mitsubishi Diamond I. Manufactured by Mitsubishi Aircraft International, Inc. A nine-place executive transport (two 2,500-lb thrust turbofans) with a gross weight of 14,700 lb, a wing area of 241 ft^2, an AR of 7.5, a cruise airspeed of 460 mph (M 0.7), and a range of 1,400 mi.

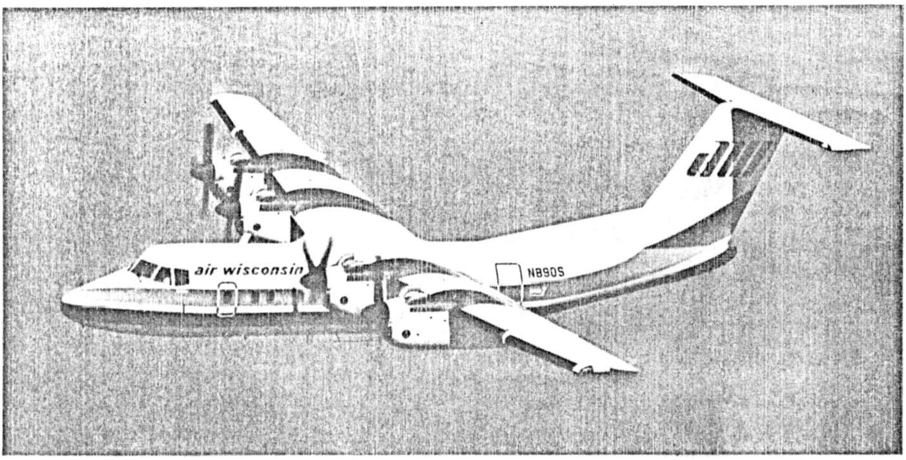

PLATE 4.

The Dash 7 (DHC-7). Manufactured by de Haviland Aircraft of Canada, Ltd. A 50-passenger STOL turboprop (four 1120 eshp engines) with a maximum gross weight of 44,000 lb, a wing area of 860 ft^2, an AR of 10, a maximum cruise airspeed of 265 mph, and a maximum range of 1,400 mi.

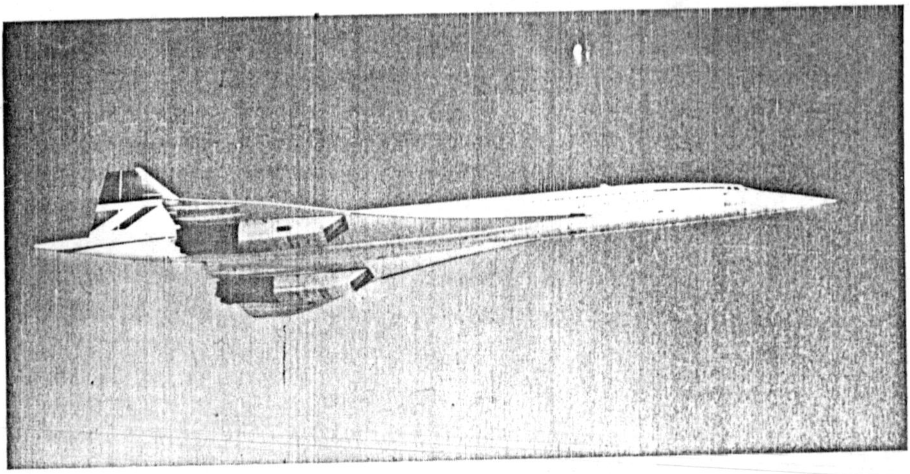

PLATE 5.
The Concorde. Manufactured jointly by British Aerospace and Aerospatiale. A Mach 2 supersonic commercial transport with four 38,000-lb thrust turbojet engines (with 17 percent afterburning), a maximum gross weight of 408,000 lb, a wing area of 3,856 ft^2, an AR of 1.8, and a maximum range of 4,000 mi.

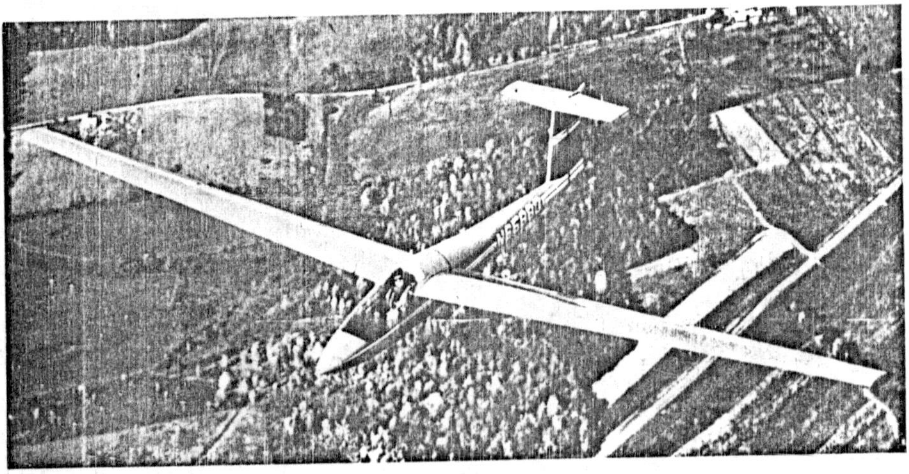

PLATE 6.
The Club-35 Sailplane. Manufactured by the Schweizer Aircraft Corporation. A high performance, one-place sailplane with a maximum gross weight of 660 lb, a wing area of 103.8 ft^2, and an AR of 23.3.

GROUP II. ENGINES

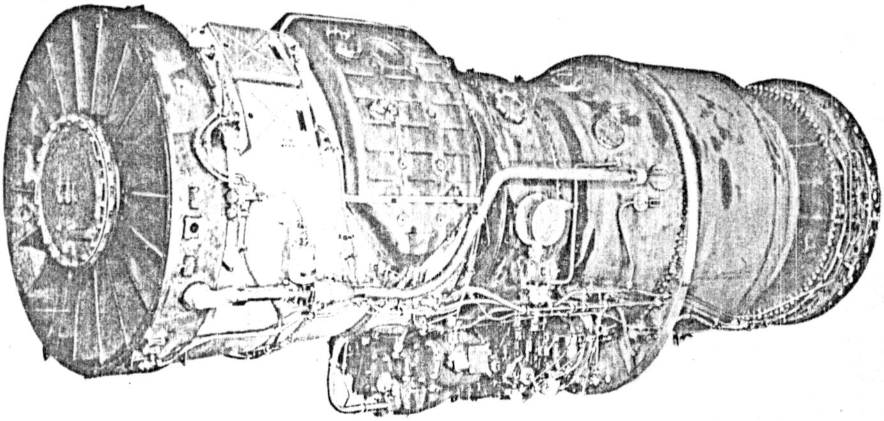

PLATE 7.

The JT3 (J57) Turbojet. Manufactured by Pratt & Whitney Aircraft. A 10,000-lb thrust class engine (18,000 with afterburning) with a dry weight of 3,500 lb, a diameter of 39 in. and a length of 137 in. for a length-to-diameter ratio of 3.5.

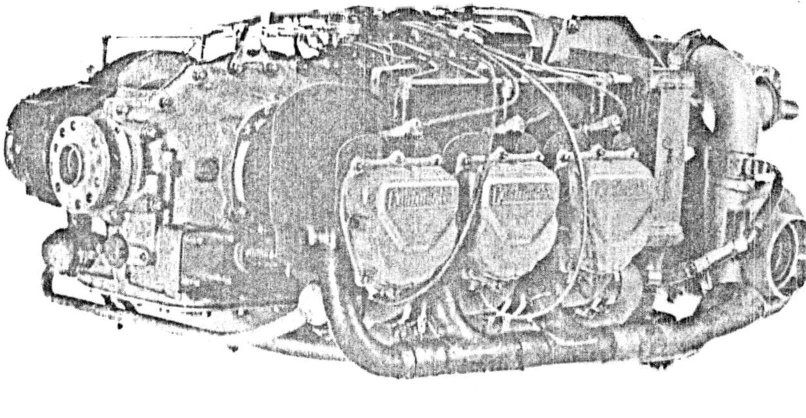

PLATE 8.

The TSIO-520. Manufactured by Teledyne Continental Motors. A 6-cylinder, turbocharged engine with fuel injection, 520 cu in displacement, 325 hp at 2700 rpm, direct propeller drive, and a dry weight of 457 lb.

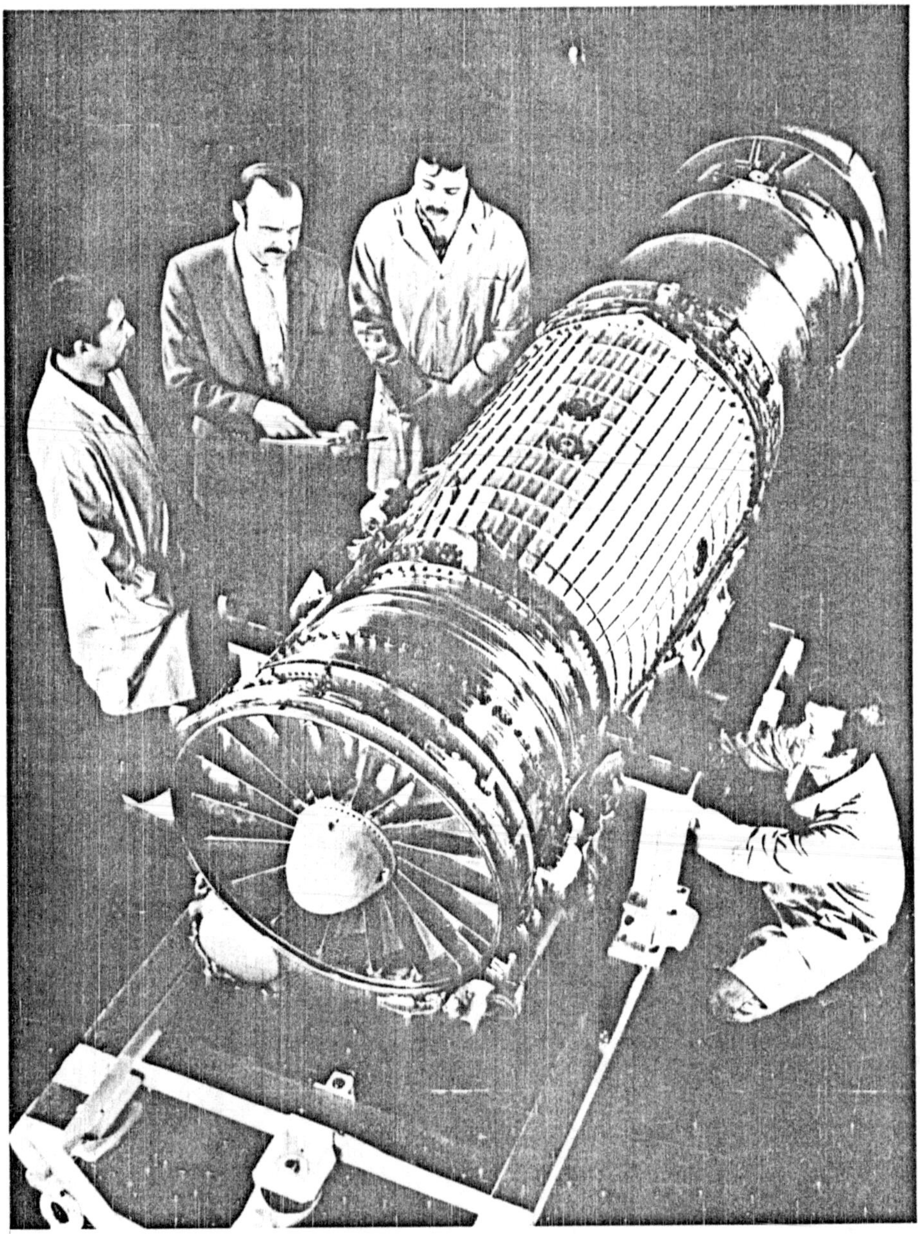

PLATE 9.

The F404-GE-100 Augmented Turbofan. Manufactured by General Electric (USA). A low bypass ratio in the 17,000-lb thrust class with a dry weight of 2,200 lb, a diameter of 35 in. and a length of 158.8 in. for a length-to-diameter ratio of 4.5.

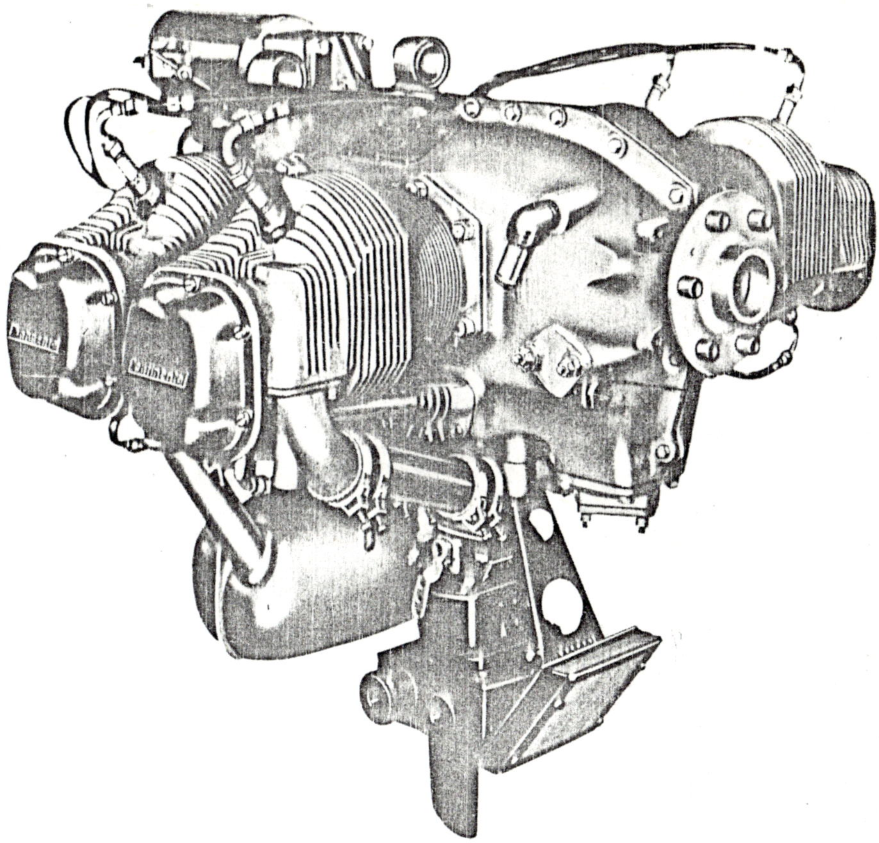

PLATE 10.

The 0-200 Reciprocating Engine. Manufactured by Teledyne Continental Motors. A 4-cylinder aspirated engine with a carburetor, 201 cu in displacement, 100 hp at 2750 rpm, direct propeller drive, and a dry weight of 220 lb.

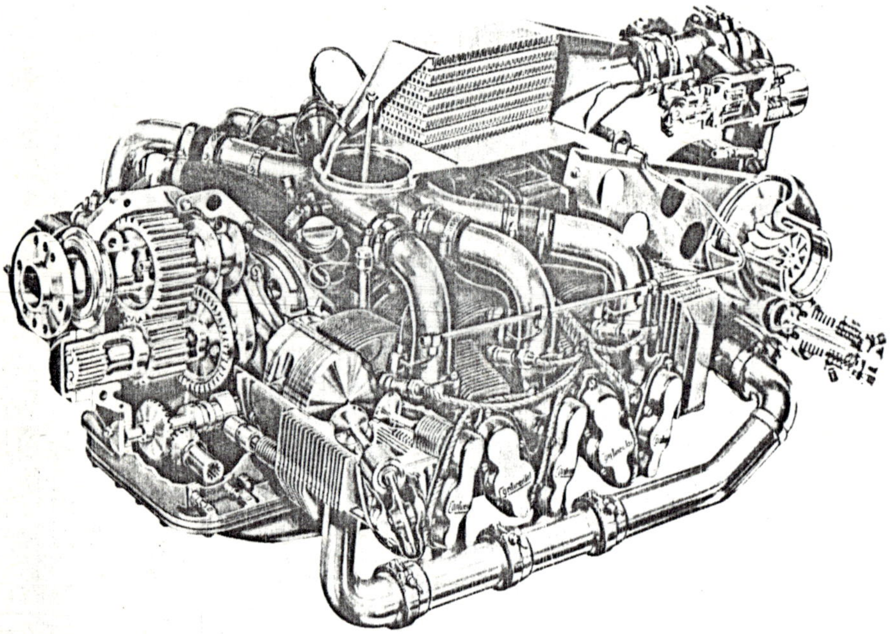

PLATE 11.
The GTSIO-520F Reciprocating Engine. Manufactured by Teledyne Continental Motors. A 6-cylinder turbocharged engine with fuel injection, 520 cu in displacement, 435 hp at 3,400 rpm, and a dry weight of 600 lb.

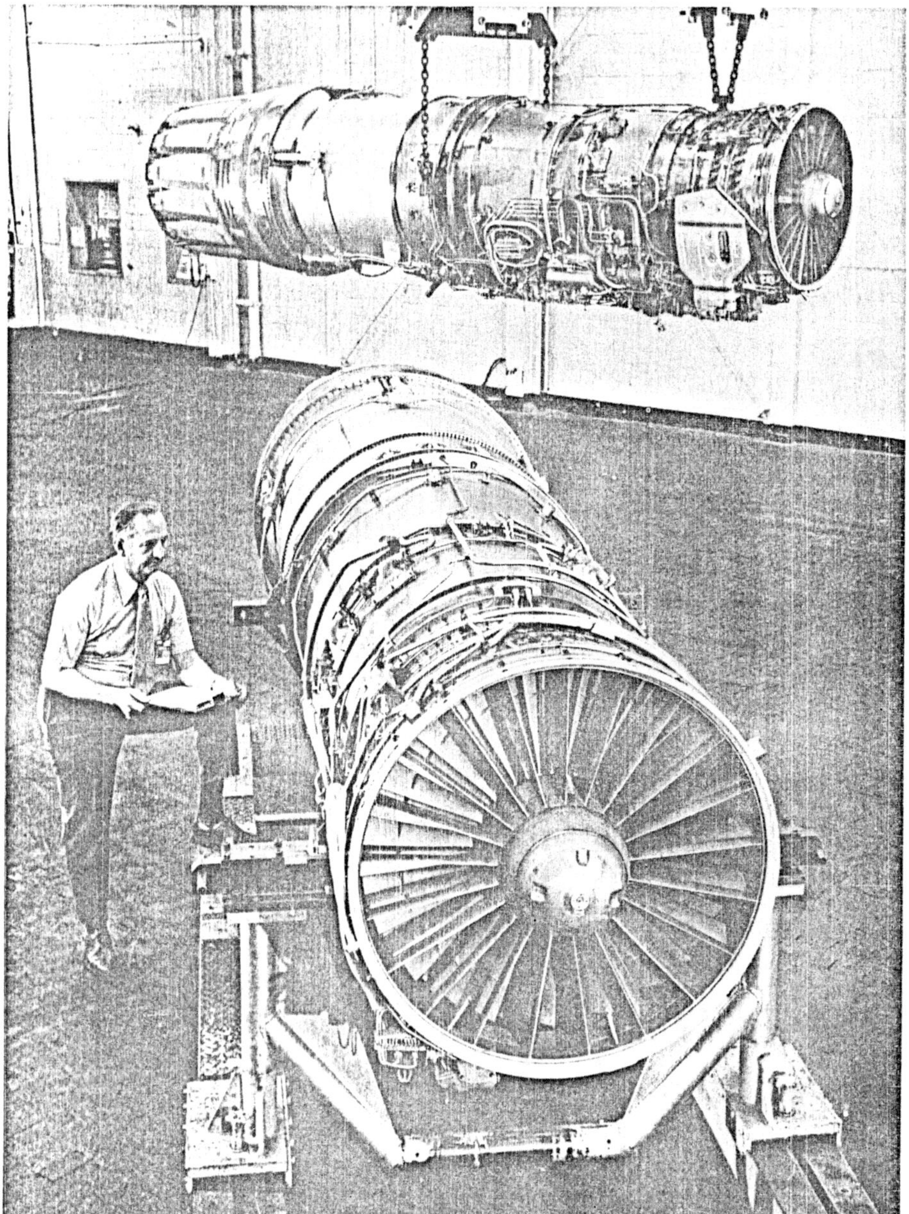

PLATE 12.

The F100 Turbofan. Manufactured by Pratt & Whitney Aircraft. A low bypass ratio (0.63) turbofan of the 14,000-lb thrust class (24,000 lb with afterburning) with a dry weight of 3,000 lb, a diameter of 48 in. and a length of 191 in. for a length-to-diameter ratio of 4.

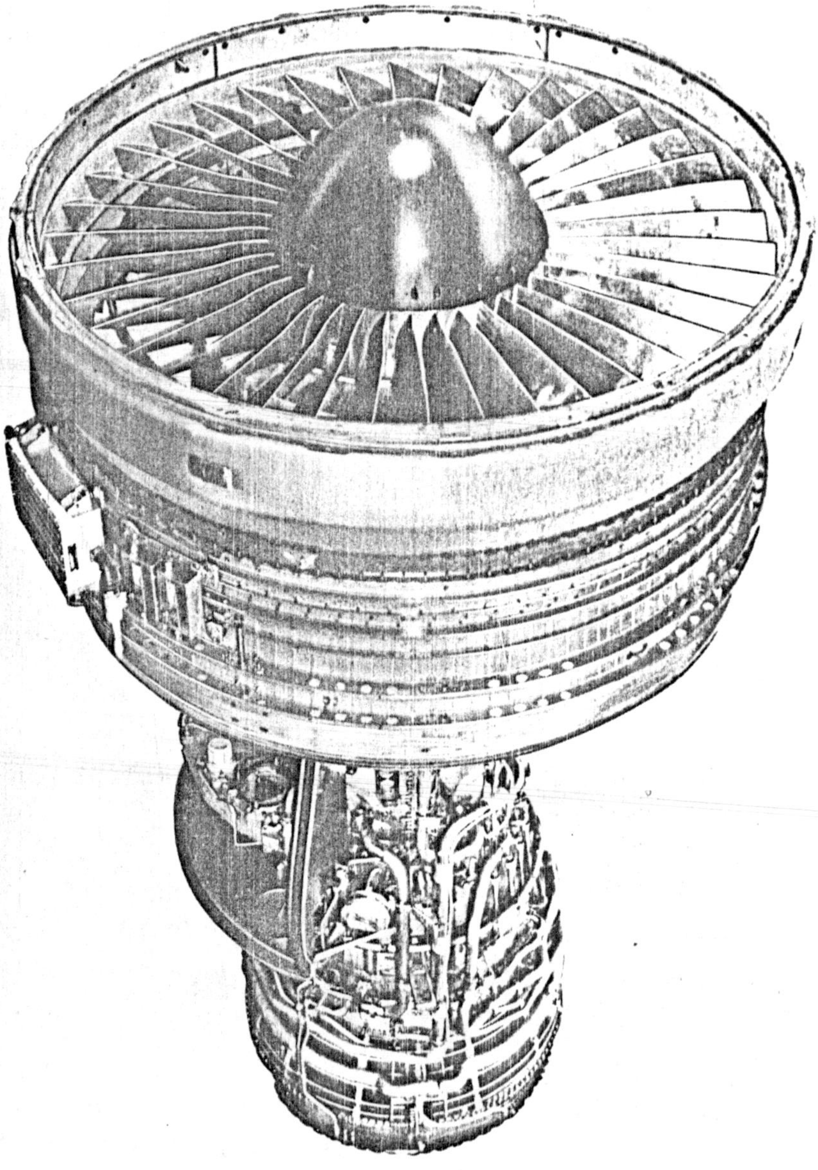

PLATE 13.
The CF6-80A Turbofan. Manufactured by General Electric (USA). A high bypass ratio (4.7) of the 48,000-lb thrust class with a dry weight of 8,400 lb, a diameter of 86.4 in. and a length of 157.4 in. for a length-to-diameter ratio of 1.8.

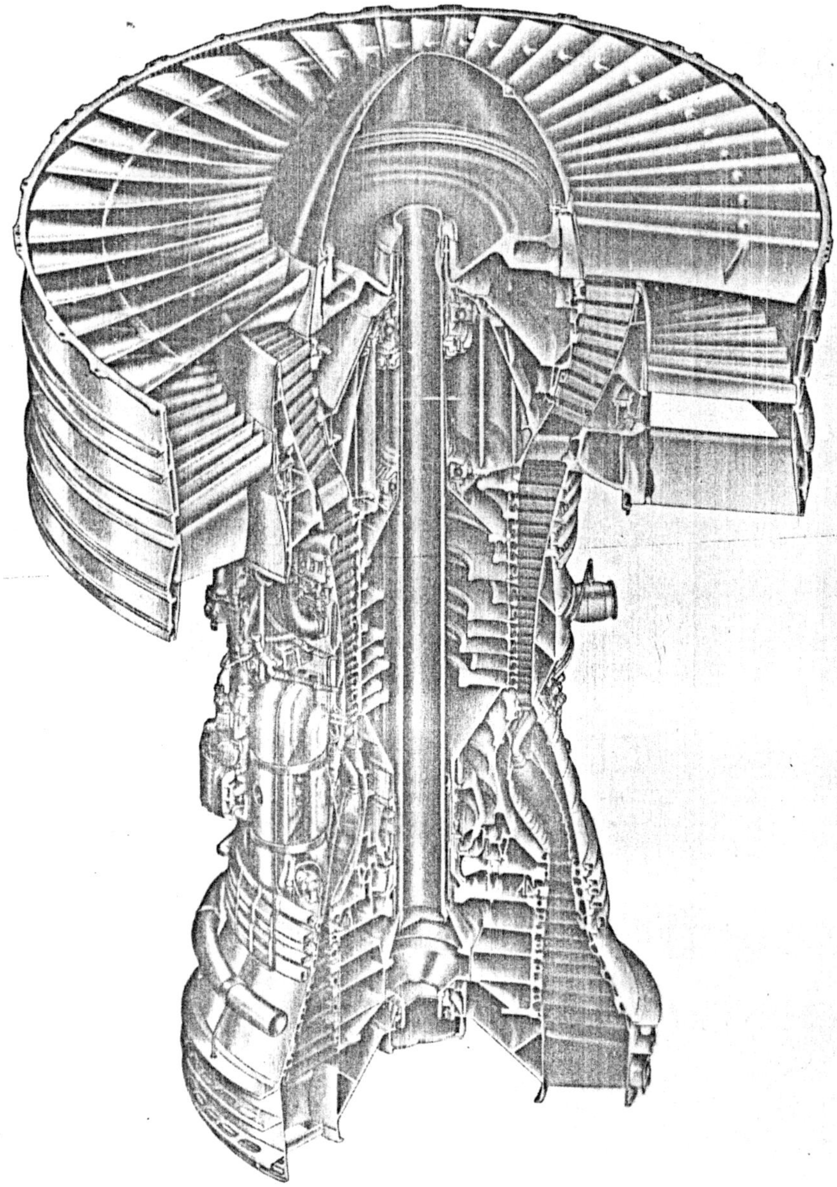

PLATE 14.

The JT9D-7R4 Turbofan. Manufactured by Pratt & Whitney Aircraft. A high bypass ratio (5) turbofan of the 48,000-lb thrust class with a dry weight of 8,900 lb, a diameter of 97 in. and a length of 153.6 in. for a length-to-diameter ratio of 1.6.

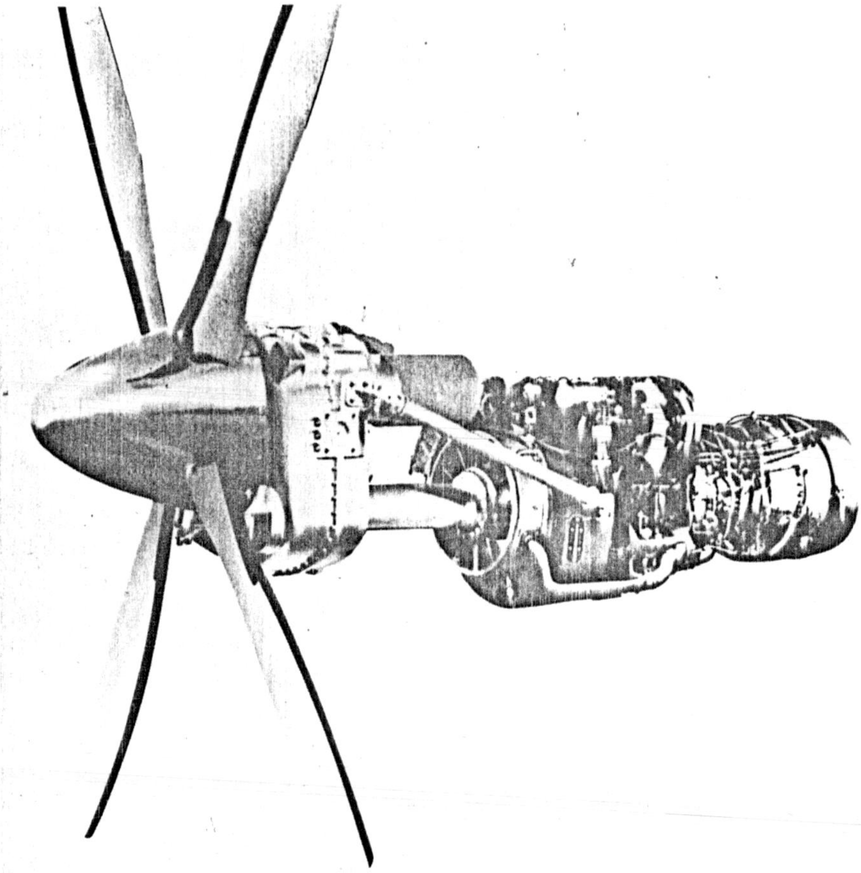

PLATE 15.
The CT7 Turboprop. Manufactured by General Electric (USA). A turboshaft engine used as a turboprop in the 1,700 eshp class with a dry weight of 435 lb without the gearbox and propeller.

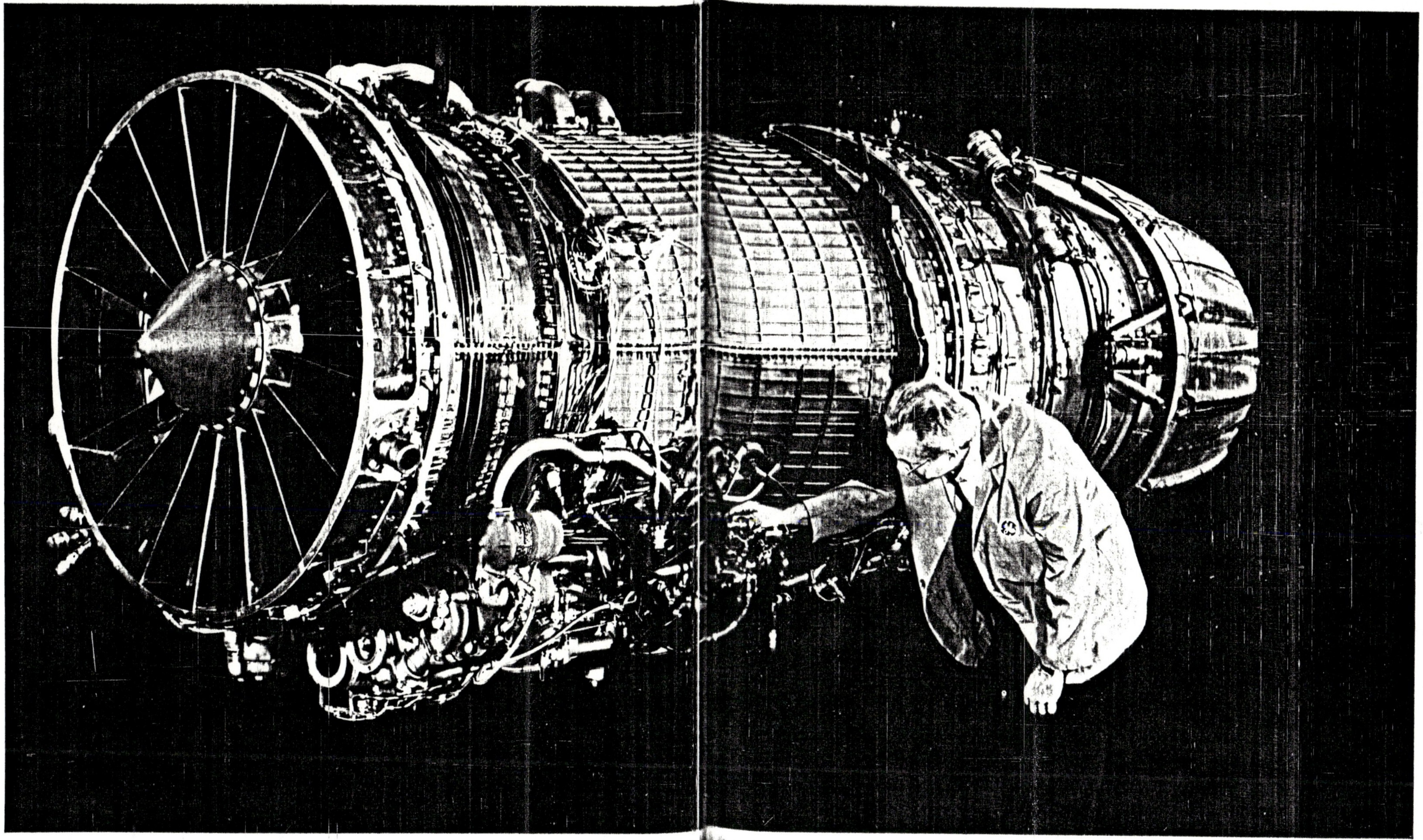

PLATE 16.

The F101 Augmented Turbofan. Manufactured by General Electric (USA). A medium bypass ratio (2) turbofan of the 30,000-lb thrust class, with a dry weight of 4,400 lb, a diameter of 55 in. and a length of 181 in. for a length-to-diameter ratio of 3.3.

PLATE 17.

The JT8D-200 Turbofan. Manufactured by Pratt & Whitney Aircraft. A medium bypass ratio (1.8) turbofan of the 20,000-lb thrust class with a dry weight of 4,400 lb, a diameter of 49 in. and a length of 154 in. for a length-to-diameter ratio of 3.1.

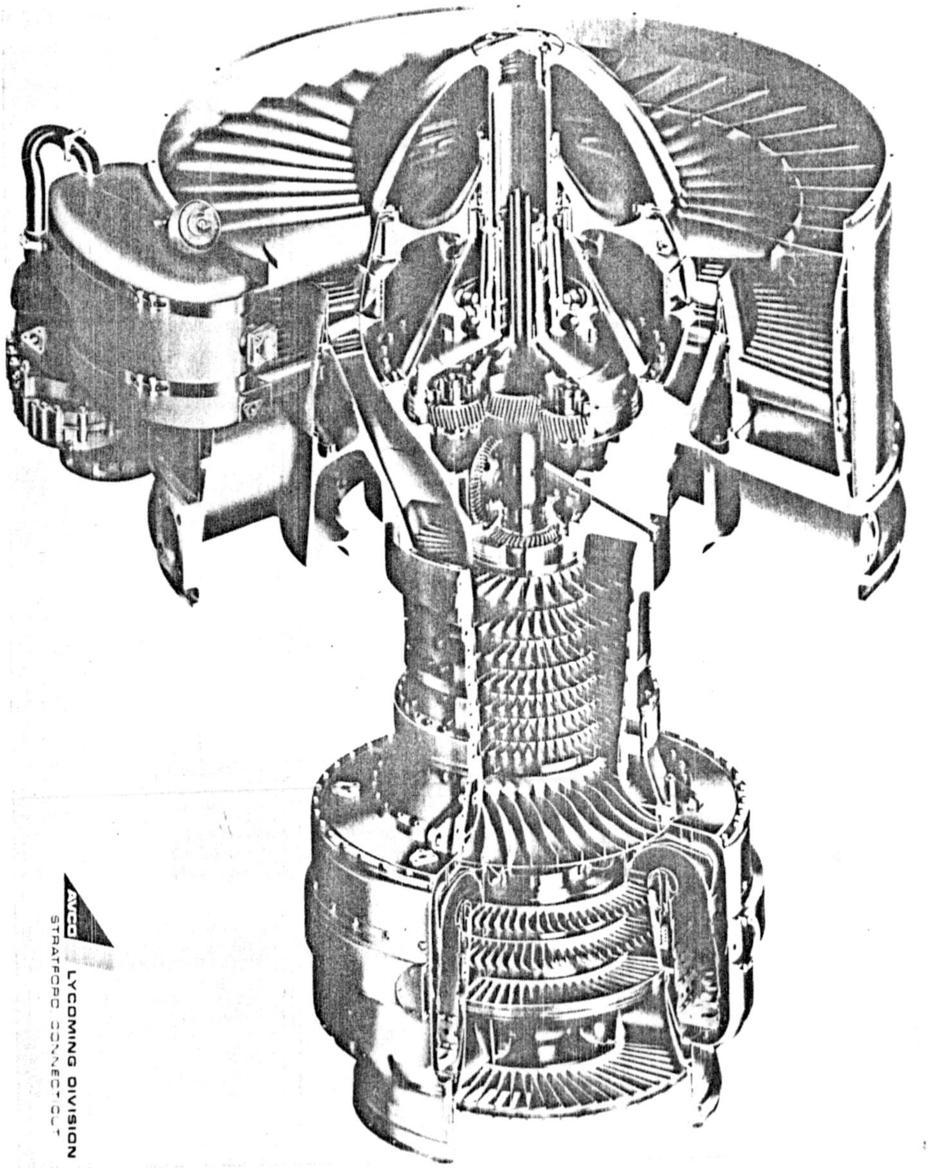

PLATE 18.

The ALF 502 R-3 Turbofan. Manufactured by Avco Lycoming. A high bypass ratio (5.7) turbofan of the 7,000-lb thrust class with a dry weight of 1,245 lb, a diameter of 41.7 in. and a length of 56.8 in. for a length-to-diameter ratio of 1.4.

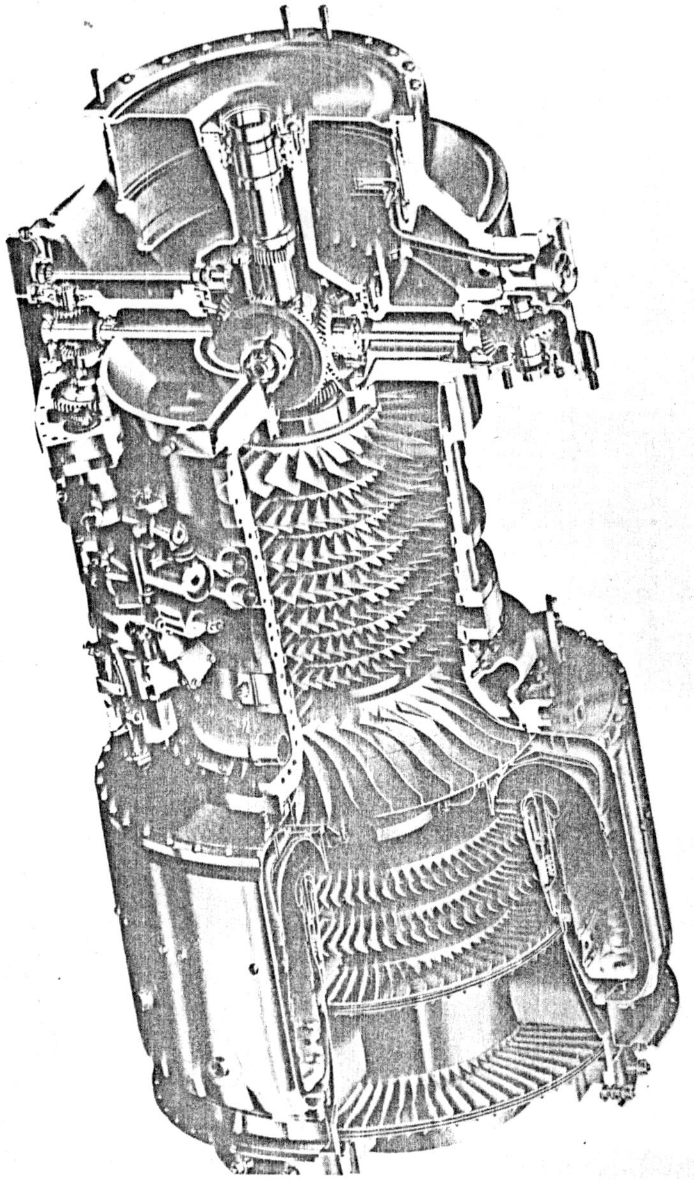

PLATE Plate 19.

The AL 5512 (T55-1-712) Turboshaft. Manufactured by Avco Lycoming. A turboshaft engine of the 4,000 shp class with a dry weight of 725 lb. Used primarily for helicopter propulsion, with a gearbox and propeller could be used as a turboprop.

TABLE 7-1
Average and Cumulative Fastest-Climb Values (Turbocharged Piston-Prop)

Δh (1,000 ft)	$(R/C)_{ave}$ (fps)	Δt (sec)	$\sum t$ (sec)	$(P/W)_{ave}$ (fps)	$\Delta W_f/W$ ($\times 10^3$)	$\sum W_f/W$ ($\times 10^3$)	V_{ave} (fps)	ΔX (mi)	$\sum X$ (mi)
0–5	33.6	148.8	148.8	46.75	2.07	2.07	159.5	4.39	4.39
5–10	32.5	153.6	302.4	46.75	2.13	4.20	172.1	4.92	9.30
10–15	31.4	159.2	461.6	46.75	2.21	6.41	186.2	5.53	14.83
15–20	30.1	166.1	626.7	46.75	2.30	8.71	202.0	6.28	21.11
20–25	24.9	200.8	828.5	43.02	2.57	11.28	219.95	8.31	29.42
25–30	16.25	307.7	1,136.2	36.04	3.29	14.57	240.3	13.97	43.39
30–35	8.25	606.1	1,742.3	30.0	5.40	19.97	263.55	30.24	73.63
35–37.5	2.2	2,273.0	4,015.0	25.7	17.3	37.27	284.45	122.4	196.0

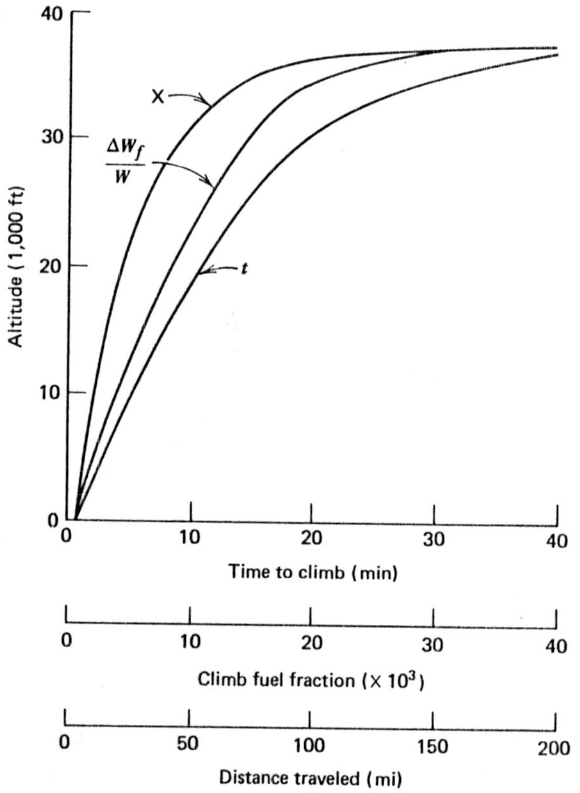

FIGURE 7-4

Fastest-climb values for illustrative piston-prop with turbocharged engines.

the horizontal direction. A climb to 30,000 ft takes 18.9 minutes, 109 lb of fuel, and 43 mi.

As with the turbojet, it would be nice to have closed-form expressions for finding order-of-magnitude values for the climb parameters. Unfortunately, the expressions we have developed do not simplify in a logical manner; however, there are empirical relationships that lead to closed-form expressions that do give reasonable answers.

For *turbocharged aircraft* using maximum power, the appropriate approximations are:

$$(R/C)_{max} = 550\eta_p K_C (HP/W)_{SL} \sigma^{1/2} \tag{7-38}$$

to be used below the critical altitude, and

$$(R/C)_{max} = \frac{550\eta_p K_C (HP/W)_{SL} \sigma^{1.5}}{\sigma_{cr}} \tag{7-39}$$

for altitudes above the critical altitude. K_C has the same value in both expressions and is found by calculating the maximum rate of climb at sea level and then dividing this value by the maximum sea-level value of the thrust power-to-weight ratio, i.e.,

$$K_C = \frac{(R/C)_{max,SL}}{550\eta_p (HP/W)_{SL}} \tag{7-40}$$

From the definition of the rate of climb as dh/dt,

$$t = \int_1^2 \frac{dh}{(R/C)_{max}} \tag{7-41}$$

Substitution of the appropriate expression for the rate of climb and using the atmospheric density ratio approximation,

$$\sigma = e^{-h/30,500} \tag{7-42}$$

and then integrating with a constant power setting, we obtain two expressions for determining the time to climb from one altitude to another. The first expression, *for altitudes below the critical altitude*, is

$$t = \frac{61,000}{550\eta_p K_C (HP/W)_{SL}} \left(e^{h_2/61,000} - e^{h_1/61,000}\right) \tag{7-43}$$

The other, *for altitudes above the critical altitude*, is

$$t = \frac{20,333\sigma_{cr}}{550\eta_p K_C (HP/W)_{SL}} \left(e^{h_2/20,333} - e^{h_{cr}/20,333}\right) \tag{7-44}$$

It so happens that Eq. 7-43 can be used from sea level to altitudes above the critical altitude without any appreciable loss in accuracy and without the need for two integrations.

To determine the fuel consumption during climb, a similar development, for climbs below the critical altitude, produces the expression

$$\ln \mathrm{MR} = \frac{0.03\hat{c}}{\eta_p K_C}\left(e^{h_2/61,000} - e^{h_1 61,000}\right) \tag{7-45}$$

where the MR is still W_1/W_2. Equation 7-45 can be expanded so that the climb-fuel weight fraction can be found directly, but it is simpler to solve for the mass ratio and then use the relationship that

$$\zeta = \left(\frac{\Delta W_f}{W}\right)_{\mathrm{FC}} = \frac{\mathrm{MR}-1}{\mathrm{MR}} \tag{7-46}$$

The range during climb can be approximated by

$$X = \frac{0.01\, V_{P_{\min},\mathrm{SL}}}{\eta_p K_C (\mathrm{HP}/W)_{\mathrm{SL}}}\left(e^{h_2/30,500} - e^{h_1/30,500}\right) \tag{7-47}$$

where V is in fps and X is in miles.

Before deciding whether to develop fuel and distance expressions for climb above the critical altitude, let us use these expressions, along with Eq. 7-43, to calculate the values for a series of climbs from sea level to altitudes that are 5,000 ft apart. It is first necessary to calculate K_C. Since the maximum rate of climb at sea level has already been determined to be 34.08 fps and the value of $550\eta_p(\mathrm{HP}_{\mathrm{SL}}/W)$ is 46.75 fps, K_C must be equal to 0.73. The three expressions for fastest climb, starting at sea level with the illustrative piston-prop with turbocharged engines, a η_p of 0.85, and a specific fuel consumption of 0.5 lb/h/hp, are:

$$t = 1{,}788\left(e^{h/61,000} - 1\right)$$

$$\ln \mathrm{MR} = 0.024\left(e^{h/61,000} - 1\right) = 1.34 \times 10^{-5} t$$

$$\frac{\Delta W_f}{W} = \frac{\mathrm{MR}-1}{\mathrm{MR}}$$

$$X = 25.2\left(e^{h/30,500} - 1\right)$$

The values obtained from these expressions are listed in Table 7-2 with the corresponding values obtained from Table 7-1 shown in parentheses alongside. In general, the agreement is good up to 30,000 ft. Therefore, we will not develop additional relationships for climb above the critical altitude. These expressions can also be used for constant power settings other than maximum power. However, do not use them for final altitudes approaching the absolute ceiling for the specified power setting.

For the piston-prop with *aspirated engines*, one empirical approximation is

$$(\mathrm{R/C})_{\max} = 550\eta_p K_C (\mathrm{HP}/W)_{\mathrm{SL}}\sigma^2 \tag{7-48}$$

where K_C is still defined in accordance with Eq. 7-40. The resulting closed-form expressions for fastest climb are:

$$t_{\min} = \frac{15{,}250}{550\eta_p K_C (\mathrm{HP}/W)_{\mathrm{SL}}}\left(e^{h_2/15,250} - e^{h_1/15,250}\right) \tag{7-49}$$

TABLE 7-2

Closed-Form Climb Values (Turbocharged Piston-Prop)

Δh (ft)	t (sec)	W_f/W ($\times 10^3$)	X (mi)
SL– 5,000	153 (149)	2.05 (2.07)	4.5 (4.4)
SL–10,000	319 (302)	4.27 (4.20)	9.8 (9.3)
SL—15,000	498 (462)	6.68 (6.41)	16.0 (14.8)
SL—20,000	694 (628)	9.23 (8.71)	23.3 (21.1)
SL—25,000	906 (828)	12.1 (11.28)	32.0 (29.4)
SL—30,000	1,136 (1,136)	15.2 (14.57)	42.2 (43.4)
SL—35,000	1,386 (1,742)	18.6 (19.97)	54.2 (73.6)

Note: Numbers in parentheses are the corresponding values from Table 7-1.

$$\ln \mathrm{MR_{FC}} = \frac{7.7 \times 10^{-3}\hat{c}}{\eta_p K_C}\left(e^{h_2/15{,}250} - e^{h_1/15{,}250}\right) \tag{7-50}$$

$$X_{\mathrm{FC}} = \frac{4.2 \times 10^{-3} V_{P_{\mathrm{min,SL}}}}{\eta_p K_C (\mathrm{HP}/W)_{\mathrm{SL}}}\left(e^{h_2/12{,}200} - e^{h_1/12{,}200}\right) \tag{7-51}$$

For the illustrative piston-prop with aspirated engines, K_C will still be equal to 0.73, and the expressions for fastest-climb from sea level with maximum power become:

$$t = 446.8(e^{h/15{,}250} - 1)$$

$$\ln \mathrm{MR} = 6.2 \times 10^{-3}(e^{h_2/15{,}250} - 1) = 1.39 \times 10^{-5} t$$

$$\frac{\Delta W_f}{W} = \frac{\mathrm{MR} - 1}{\mathrm{MR}}$$

$$X = 10.57(e^{h/12{,}200} - 1)$$

The climb values obtained from these expressions are listed in Table 7-3 along with the corresponding values from the tabular method shown in parentheses. Although the correlation could be better, particularly with respect to the fuel consumption to the higher altitudes, it appears reasonable enough to warrant using these expressions rather than attempt to improve the model.

Examination of Eq. 7-28 shows that a high rate of climb for a piston-prop calls not only for a large power-to-weight ratio and a high E_m, as might be expected, but also for a low minimum-power airspeed (a low wing loading), which might not be expected. The most important of all of these parameters is the available power-to-weight ratio. For example, doubling the power of the illustrative piston-prop increases the rate of climb from 34 to 81 fps (a 138 percent increase), and

TABLE 7-3
Closed-Form Climb Values (Aspirated Piston-Prop)

Δh (ft)	t (sec)	W_f/W ($\times 10^3$)	X (mi)
SL— 5,000	173 (165)	2.4 (2.13)	5.3 (5.0)
SL—10,000	414 (380)	5.7 (4.52)	13.2 (12.0)
SL—15,000	748 (681)	10.3 (7.38)	25.2 (22.6)
SL—20,000	1,212 (1,158)	16.7 (11.22)	43.2 (40.9)
SL—25,000	1,856 (2,200)	25.5 (18.32)	70.3 (84.3)

Note: Numbers in parentheses are the corresponding values from Table 7-1.

halving the power-to-weight ratio reduces the rate of climb to 11 fps (a 67 percent reduction) with the same fastest-climb airspeed for all three power levels. On the other hand, doubling the wing loading reduces the airspeed by 30 percent and increases the rate of climb by only 11 percent.

Increasing the fastest-climb airspeed of a piston-prop (by increasing the wing loading) decreases the climb angle but leaves the available power unchanged, thus the decrease in the climb rate. However, increasing the fastest-climb airspeed of a turbojet (also by increasing the wing loading) increases the rate of climb by increasing the available power and leaving the climb angle unchanged.

The analyses in this section have been based on a constant propeller efficiency, and the value used in the calculations has been the design, or best, value. The propeller efficiency not only may have a lower value but also may vary over the airspeed range of interest, particularly for aircraft with low wing loadings. As a consequence, the actual climb rates may be lower than those obtained from the expressions of this section.

7-3 TURNING FLIGHT

With the assumption of a constant weight, the piston-prop equations for a steady-state turn are the same as those for the turbojet, namely,

$$T - D = 0 \tag{7-52}$$

$$L \sin \phi - \frac{W}{g}\dot{\chi}V = 0 \tag{7-53}$$

$$L \cos \phi - W = 0 \tag{7-54}$$

$$n = \frac{L}{W} = \frac{1}{\cos \phi} \tag{7-55}$$

$$\dot{\chi} = \frac{g \tan \phi}{V} = \frac{g(n^2 - 1)^{1/2}}{V} \tag{7-56}$$

$$r = \frac{V}{\dot{\chi}} = \frac{V^2}{g(n^2 - 1)^{1/2}} \tag{7-57}$$

With a parabolic drag polar, the drag-to-weight ratio in a steady turn can be written as

$$\frac{D}{W} = \frac{\rho_{\text{SL}} \sigma V^2 C_{D0}}{2(W/S)} + \frac{2Kn^2(W/S)}{\rho_{\text{SL}} V^2} \tag{7-58}$$

If Eq. 7-52 is rewritten in terms of the available power, divided by the weight, and then combined with Eq. 7-58, the expression relating the turning flight conditions and the aircraft characteristics is

$$\frac{P}{W} = \frac{550\eta_p(\text{HP})}{W} = \frac{\rho_{\text{SL}} \sigma V^3 C_{D0}}{2(W/S)} + \frac{2Kn^2(W/S)}{\rho_{\text{SL}} \sigma V} \tag{7-59}$$

Unfortunately, there is no closed-form solution for the turning airspeed in terms of a specified thrust power-to-weight ratio and load factor (specified bank angle) although iteration is not difficult. It is possible, however, to solve Eq. 7-59 for the load factor in terms of the flight and design parameters to obtain

$$n = \frac{1}{W/S} \left[\frac{550\eta_p \rho_{\text{SL}} \sigma V(\text{HP}/W)}{2K} - \frac{\rho_{\text{SL}}^2 \sigma^2 V^4 C_{D0}}{4K} \right]^{1/2} \tag{7-60}$$

With the load factor known for a given airspeed, Eqs. 7-55 through 7-57 can now be used to determine the turning performance of the piston-prop.

The turning lift coefficient, which is given by

$$C_L = \frac{2n(W/S)}{\rho_{\text{SL}} \sigma V^2} = \frac{2(W/S)}{\rho_{\text{SL}} \sigma V^2 \cos \phi} \tag{7-61}$$

must not exceed the maximum lift coefficient of the aircraft in its turning configuration. Otherwise the calculated airspeed will be less than the stall speed, which is given by

$$V_{s,t} = \left[\frac{2n(W/S)}{\rho_{\text{SL}} \sigma C_{L\text{max}}} \right]^{1/2} = n^{1/2} V_{s,l} \tag{7-62}$$

where $V_{s,l}$ is the wings-level stall speed.

For a given aircraft, the stall speed in a turn can be determined without a prior knowledge of n by rearranging Eq. 7-62 so that

$$n_{s,t} = \frac{\rho_{\text{SL}} \sigma V_{s,t}^2 C_{L\text{max}}}{2(W/S)} \tag{7-63}$$

and then setting the right-hand side equal to the right-hand side of Eq. 7-60 to obtain

$$V_{s,t} = \left[\frac{1.100\eta_p(\text{HP}/W)(W/S)}{\rho_{\text{SL}} \sigma (C_{D0} + KC_{L\text{max}}^2)} \right]^{1/3} \tag{7-64}$$

In Eq. 7-64, the term $C_{D0}+KC_{L_{max}}^2$ is the drag coefficient at stall, with the assumption that the parabolic drag polar is valid down to the stall speed, which it is not. Since substitution of Eq. 7-64 into Eq. 7-63 results in an awkward and unwieldly expression for $n_{s,t}$, we shall instead calculate the stall speed in the turn and substitute the numerical value obtained into Eq. 7-63 to obtain a numerical value for the corresponding load factor.

Let us now look at the sea-level turning performance of our illustrative piston-prop. To refresh our memories, it has the following characteristics: a W/S of 34 lb/ft², a HP/W of 0.1 hp/lb, a $C_D=0.025+0.051\ C_L^2$, and a maximum lift coefficient of 1.8. The level-flight stall speed is $126/\sigma^{1/2}$ fps ($86/\sigma^{1/2}$ mph). With full power and a propeller efficiency of 0.85, the turning values at the stall and at sea level are:

$$V_{s,t}=\left\{\frac{1{,}100\times 0.85\times 0.1\times 34}{\rho_{SL}[0.025+0.51(1.8)^2]}\right\}^{1/3}=191.6\text{ fps}=130.6\text{ mph}$$

$$n_{s,t}=\frac{\rho_{SL}\times(191.6)^2\times 1.8}{2\times 34}=2.31\ g\text{'s}$$

The generalized expression (Eq. 7-60) for the turning load factor at any airspeed can be written, for this aircraft at sea level, as

$$n=0.0294(37V-6.92\times 10^{-7}V^4)^{1/2} \tag{7-65}$$

In order to demonstrate the procedure, let us determine the turning performance at stall and then at another airspeed. At stall,

$$\phi_{s,t}=\text{arc cos}\,\frac{1}{2.31}=64.3\text{ deg}$$

$$\dot{\chi}_{s,t}=\frac{32.2[(2.31)^2-1]^{1/2}}{191.6}=0.35\text{ rad/s}=20\text{ deg/s}$$

$$r_{s,t}=\frac{191.6}{0.35}=547.5\text{ ft}$$

If the other airspeed is to be 250 fps (170 mph), then

$$n=0.0294[37\times 250-6.92\times 10^{-7}(250)^4]^{1/2}=2.38\ g\text{'s}$$

$$\phi=\text{arc cos}\,\frac{1}{2.38}=65.1\text{ deg}$$

$$\dot{\chi}=\frac{g\tan 65.1}{250}=\frac{g[(2.38)^2-1]^{1/2}}{250}=0.278\text{ rad/s}=15.9\text{ deg/s}$$

$$r=\frac{250}{0.278}=899\text{ ft}$$

As a check to be sure that turning flight at 250 fps is possible, we shall calculate the lift coefficient to be sure that it is less than the maximum lift coefficient. Using Eq. 7-61, we find the lift coefficient to be 1.09, which is indeed less than the maximum value of 1.8. As a second, but not independent check, let us calculate the stall speed in this turn to be sure that it is less than the prescribed airspeed of 250 fps. Using the far right-hand term of Eq. 7-62, we find that the stall speed for this turn is 194.4 fps, which is less than the actual airspeed of 250 fps.

The sea-level turning values for maximum power and a propeller efficiency of 0.85 are listed in Table 7-4 and sketched in Fig. 7-5 (with a relative scale only) as a function of the true airspeed. If we take the effort to examine the figure rather carefully and refer to the table for values, we can make several interesting observations. There are two independent conditions that must be met if constant-altitude turning flight is to be possible. The first is that the airspeed be greater than the corresponding stall speed (C_L be less than $C_{L\max}$) and the second is that the flight load factor be greater than unity. In this case, the first condition requires the airspeed to be greater than 191.6 fps, as has already been established, and the second requires that the airspeed be greater than approximately 35 fps. Since the stall speed is the larger of the two, it takes precedence and is the minimum possible airspeed. This may not always be the case, particularly at the higher airspeeds where the requirement for a load factor greater than unity may be the determining limitation.

All three turning-performance parameters have maximum values, but for this aircraft (and for most aircraft) only the maximum load factor occurs at an airspeed higher than the stall speed. The fastest-turn and tightest-turn airspeeds both lie considerably below the stall speed. Consequently, for this aircraft, fastest turn and tightest turn will both occur in the vicinity of the stall speed. An increase in the maximum lift coefficient will obviously raise these airspeeds and increase the maneuverability of the aircraft. It will be shown that increasing the available thrust power will also improve the turning performance and that maneuverability will decrease with an increase in altitude.

Let us now see if we can determine the flight and design conditions for turn at the maximum load factor and for the fastest and the tightest turns. With the assumption of a constant propeller efficiency, setting the first derivative of the load factor, as given by Eq. 7-60, with respect to the airspeed equal to zero yields the maximum load factor condition that

$$V_{nm} = \left[\frac{550\eta_p(\mathrm{HP}/W)(W/S)}{2\rho_{\mathrm{SL}}\sigma C_{D0}}\right]^{1/3} \tag{7-66}$$

Substitution of Eq. 7-66 back into Eq. 7-60 produces, after some reduction, the following expression for *the maximum possible load factor for a piston-prop aircraft*:

$$n_{\max} = 0.687\left\{\frac{\rho_{\mathrm{SL}}\sigma[550\eta_p(\mathrm{HP}/W)]^2 E_m}{K(W/S)}\right\}^{1/3} \tag{7-67}$$

Whereas the maximum possible load factor of a turbojet was a simple function of the T/W ratio and the maximum lift-to-drag ratio (their product), that of the piston-prop is not only more complicated but also involves the wing loading. Obviously at the ceiling, the maximum load factor is unity.

These last two equations will now be used to calculate the sea-level maximum load factor conditions for our piston-prop.

$$V_{nm}=\left(\frac{550\times 0.85\times 0.1\times 34}{2\rho_{SL}\times 0.025}\right)^{1/3}=327\text{ fps}=223\text{ mph}$$

$$n_{max}=0.687\left[\frac{\rho_{SL}\times(550\times 0.85\times 0.1)^2\times 14}{0.051\times 34}\right]^{1/3}=2.387\ g\text{'s}$$

$$\phi_{max}=\arc\cos\frac{1}{2.387}=65.2\text{ deg}$$

$$\dot{\chi}=\frac{g\tan 65.2}{237}=0.294\text{ rad/s}=16.8\text{ deg/s}$$

$$r=\frac{237}{0.294}=805\text{ ft}$$

$$V_{s,t}=126(2.387)^{1/2}=195\text{ fps}<V_{nm}$$

$$C_L=\frac{2\times 2.387\times 34}{\rho_{SL}\times(237)^2}=1.22<C_{L_{max}}$$

These values are consistent with those given in Table 7-4 and with the sketch of Fig. 7-5.

Let us now look at the conditions for *fastest turn*, i.e., maximum turning rate. Setting the first derivative of the turning rate with respect to the airspeed equal to zero produces the general condition (that also holds for the turbojet) that

$$n^2-1-\frac{V}{2}\frac{dn^2}{dV}=0 \tag{7-68}$$

With the introduction of the parabolic drag polar expression of Eq. 7-60, the equation that defines the fastest-turn airspeed is

$$V^4+\frac{550\eta_p(\text{HP}/S)V}{\rho_{SL}\sigma C_{DO}}-\frac{4K(W/S)^2}{\rho_{SL}^2\sigma^2C_{DO}}=0 \tag{7-69}$$

Although this equation does not have an analytic solution, a reasonable and useful approximation can be obtained by dropping the fourth-order term. Therefore,

$$V_{FT}\cong\frac{4K(W/S)}{550\eta_p\rho_{SL}\sigma(\text{HP}/W)} \tag{7-70}$$

where the subscript FT denotes the fastest-turn conditions. With the substitution of Eq. 7-70, Eq. 7-60 becomes

$$n_{FT}=\left\{2-\left[\frac{4K(W/S)}{\rho_{SL}\sigma E_m[550\eta_p(\text{HP}/W)]^2}\right]^2\right\}^{1/2} \tag{7-71}$$

Since the second term in the brackets is several orders of magnitude less than 2, the fastest-turn load factor can be approximated by

$$n_{FT} \cong (2)^{1/2} = 1.414 \tag{7-72}$$

so that

$$\phi_{FT} = \text{arc cos} \frac{1}{(2)^{1/2}} = 45 \text{ deg} \tag{7-73}$$

$$\dot{\chi}_{FT} = \frac{g}{V_{FT}} = \frac{550 \rho_{SL} \sigma g \eta_p (HP/W)}{4K(W/S)} \tag{7-74}$$

$$r_{FT} = \frac{V_{FT}}{\dot{\chi}_{FT}} = \frac{V_{FT}^2}{g} \tag{7-75}$$

For the illustrative piston-prop with full power, the sea-level values are:

$$V_{FT} = \frac{4 \times 0.051 \times 34}{550 \rho_{SL} \times 0.85 \times 0.1} = 62.4 \text{ fps} = 42.6 \text{ mph}$$

$$n_{FT} = \left[2 - \left[\frac{4 \times 0.051 \times 34}{\rho_{SL} \times 14(550 \times 0.85 \times 0.1)^2}\right]^2\right]^{1/2} = (2 - 0.009)^{1/2} = 1.411 \cong 1.414$$

$$\dot{\chi}_{FT} = \frac{g}{62.4} = 0.516 \text{ rad/s} = 29.5 \text{ deg/s}$$

$$r_{FT} = \frac{(62.4)^2}{32.2} = 121 \text{ ft}$$

These values correlate very well with Table 7-4 and Fig. 7-5.

For *tightest turn* (minimum turning radius), the general condition is that

$$n^2 - 1 - \frac{V}{4}\frac{dn^2}{dV} = 0 \tag{7-76}$$

which, for a parabolic drag polar, yields the specific conditions that:

$$V_{TT} = \frac{8K(W/S)}{3 \times 550 \eta_p \rho_{SL} \sigma (HP/W)} \tag{7-77}$$

$$n_{TT} = \left(\frac{4}{3} - \left\{\frac{1.78K(W/S)}{\rho_{SL} \sigma E_m [550 \eta_p (HP/W)]^2}\right\}^2\right)^{1/2} \tag{7-78}$$

$$n_{TT} \cong (\tfrac{4}{3})^{1/2} = 1.154 \tag{7-79}$$

$$\phi_{TT} = \text{arc cos} \frac{1}{1.154} = 30 \text{ deg} \tag{7-80}$$

$$\dot{\chi}_{TT} = \frac{0.577g}{V_{TT}} = \frac{6.97 \times 550 \rho_{SL} \sigma \eta_p (HP/W)}{K(W/S)} \tag{7-81}$$

TABLE 7-4
Sea-Level Turning Flight Values
($\eta_p = 0.85$)

V (fps)	n (g's)	V_s (fps)	$\dot{\chi}$ (rad/s)	r (ft)
10	0.56	94.3	—	—
30	0.98	124.6	—	—
35	1.06	129.6	0.317	110
40	1.13	134.0	0.424	94.3
41.6	1.15	135.3	0.444	93.8
45	1.20	138.0	0.473	95.2
50	1.26	141.6	0.497	100.6
60	1.38	148.0	0.512	117.0
70	1.49	154.0	0.509	137.5
80	1.59	159.0	0.498	160.5
100	1.77	167.0	0.471	212.0
150	2.12	183.4	0.401	373.8
180	2.26	190.0	0.363	495.0
190	2.30	191.1	0.351	540.9
200	2.33	192.0	0.339	589.0
237	2.39	194.6	0.294	805.0
250	2.38	194.3	0.278	899.0
300	2.18	186.0	0.208	1,443.0

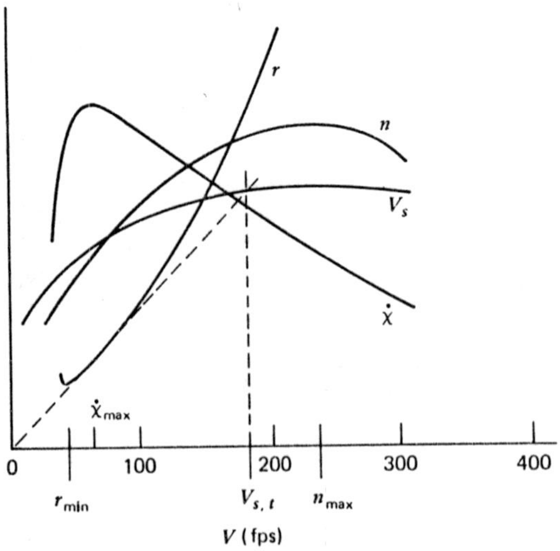

FIGURE 7-5

Sea-level turning rates for illustrative piston-prop with $\eta_p = 0.85$ and maximum power.

$$r_{TT}=\frac{V_{TT}}{\dot{\chi}_{TT}}=\frac{1.73V_{TT}^2}{g}=\frac{12.3}{g}\left[\frac{K(W/S)}{550\rho_{SL}\sigma\eta_p(HP/W)}\right]^2 \tag{7-82}$$

The corresponding sea-level values for our piston-prop are:

$$V_{TT}=\frac{8\times0.051\times34}{3\rho_{SL}\times550\times0.85\times0.1}=41.6\text{ fps}=28.4\text{ mph}$$

$$n_{TT}=1.154$$

$$\dot{\chi}_{TT}=\frac{0.577g}{41.6}=0.446\text{ rad/s}=25.6\text{ deg/s}$$

$$r_{TT}=\frac{1.73(41.6)^2}{32.2}=93.0\text{ ft}$$

These values also correlate well with those in Table 7-4 and Fig. 7-5.

If we examine the expressions for fastest turn (Eq. 7-74) and tightest turn (Eq. 7-82), we see that both are functions of the same set of parameters. For a high degree of maneuverability, we want a high HP/W ratio and propeller efficiency, a large aspect ratio and Oswald span efficiency, a low wing loading, and a low altitude. These are also the features that lead to a high maximum load factor. It is interesting to note that the zero-lift drag coefficient does not appear in any of these expressions and thus does not directly affect the best turning performance of a piston-prop. In fact, a large drag coefficient actually improves the turning performance by reducing the stall speed, as can be seen in Eq. 7-64. Although the turning performance implied in these expressions can seldom be achieved because it calls for airspeeds that usually fall considerably below the turning stall speed for the specified configuration, the higher these theoretical values are, the better the actual turning performance will be. This can be seen by considering Fig. 7-5 and visualizing the turning rate and turning radius curves being appropriately shifted. This figure also emphasizes the importance of a large maximum lift coefficient. As the lift coefficient is increased, the stall speed shifts to the left and the actual turning performance obviously improves.

Since the theoretical fastest-turn and tightest-turn airspeeds for our illustrative piston-prop are both less than the turning stall speed, then the actual fastest turn and the actual tightest turn both occur at the stall speed, as can easily be seen in Fig. 7-5. At sea-level, using Eqs. 7-64 and 7-63,

$$V_{s,t}=\left\{\frac{1{,}100\times0.85\times0.1\times34}{\rho_{SL}[0.025+0.051(1.8)^2]}\right\}^{1/3}=191.6\text{ fps}=130.6\text{ mph}$$

$$n_{s,t}=\frac{\rho_{SL}\times(191.6)^2\times1.8}{2\times34}=2.309\ g\text{'s}$$

$$\dot{\chi}_{s,t}=\frac{g[(2.309)^2-1]^{1/2}}{191.6}=0.348\text{ rad/s}=20.0\text{ deg/s}$$

$$r_{s,t}=\frac{191.6}{0.348}=548\text{ ft}$$

At altitude, we must consider the type of engines being used. At 30,000 ft with turbocharged engines, the HP/W ratio is 0.07 hp/lb [(0.374/0.533) × 0.1] and

$$V_{s,t} = \left[\frac{1{,}100 \times 0.85 \times 0.07 \times 34}{\rho_{SL} \times 0.374 \times [0.025 + 0.051(1.8)^2]}\right]^{1/3} = 236 \text{ fps} = 161 \text{ mph}$$

$$n_{s,t} = \frac{\rho_{SL} \times 0.374(236)^2 \times 1.8}{2 \times 34} = 1.31\ g\text{'s}$$

$$\dot{\chi}_{s,t} = \frac{g[(1.31)^2 - 1]^{1/2}}{236} = 0.116 \text{ rad/s} = 6.6 \text{ deg/s}$$

$$r_{s,t} = \frac{236}{0.116} = 2{,}038 \text{ ft}$$

We see that, as expected, the turning performance has deteriorated with the increase in altitude.

If we had assumed aspirated engines and had forgotten that the ceiling is of the order of 27,000 ft, we would have taken the HP/W ratio to be 0.0374 hp/lb (0.374 × 0.1) at 30,000 ft, so that

$$V_{s,t} = \left\{\frac{1{,}100 \times 0.85 \times 0.0374 \times 34}{\rho_{SL} \times 0.374[0.025 + 0.051(1.8)^2]}\right\}^{1/3} = 191.6 \text{ fps} = 130.6 \text{ mph}$$

$$n_{s,t} = \frac{\rho_{SL} \times 0.374(191.6)^2 \times 1.8}{2 \times 34} = 0.863\ g\text{'s}$$

The fact that the load factor is less than unity should alert us to the fact that the aircraft is above its ceiling and cannot maintain level flight, much less turn. We should also be aware that it is possible, under certain circumstances, to calculate a load factor that is greater than the maximum possible load factor. This obviously is another case of an invalid flight condition.

Let us improve the maneuverability of our piston-prop by doubling the HP/W ratio, halving the wing loading, and increasing the maximum lift coefficient to 2.2. At sea level,

$$V_{nm} = 237 \text{ fps} \quad \text{and} \quad n_{\max} = 4.77\ g\text{'s}$$

$$V_{FT} = 15.6 \text{ fps} \quad \text{and} \quad \dot{\chi}_{\max} = 2.06 \text{ rad/s}$$

$$V_{TT} = 10.4 \text{ fps} \quad \text{and} \quad r_{\min} = 5.8 \text{ ft}$$

$$V_{s,t} = 170 \text{ fps} \quad \text{and} \quad n_{s,t} = 4.45\ g\text{'s}$$

Since only the airspeed for the maximum load factor is greater than the stall speed, fastest turn and tightest turn will occur at the stall speed and have the following values:

$$\dot{\chi}_{s,t} = \frac{g[(4.45)^2 - 1]^{1/2}}{170} = 0.821 \text{ rad/s} = 47 \text{ deg/s}$$

$$r_{s,t} = \frac{170}{0.821} = 207 \text{ ft}$$

Comparing this performance with our basic piston-prop, we see that these changes have increased the turning rate by 135 percent and decreased the turning radius by 62 percent.

We shall conclude this chapter on take-off, climbing, and turning performance of a piston-prop with several observations. The first of these is that the expressions describing this performance were more difficult to obtain, required more approximations and assumptions, and were more unwieldly and awkward to use than those for the turbojet. The theoretical best climbing and turning performance of a piston-prop is characterized by very low airspeeds so that the best actual performance is at the stall speed. Flying at the stall speed, however, is not a good practice except in combat when one's survival is at stake. Consequently, the best airspeeds for climbing and turning are usually of the order of 20 percent higher than the corresponding stall speed. It should be noted that our assumption of a constant propeller efficiency is not valid at the low airspeeds we have been using; in fact, in that speed regime the propeller efficiency is quite sensitive to changes in the airspeed. However, since we cannot fly at the airspeeds we found using a constant propeller efficiency, the extra effort involved in using a variable efficiency does not appear to be warranted.

PROBLEMS

Major characteristics of Aircraft *D* and Aircraft *E*, which are used in these problems, are listed in Chapter 6 at the beginning of the problems. Assume that the value given for the maximum lift coefficient applies to all situations calling for a maximum value.

7-1. For Aircraft *D*:

a. On a sea-level, standard-day take-off, find the lift-off airspeed (fps and mph), the ground run (ft), and the time to lift-off. What value did you use for the propeller efficiency?

b. Do (a) above for a density ratio of 0.88 with an aspirated engine.

c. Do (b) above with a turbocharged engine.

7-2. Do Prob. 7-1 for Aircraft *E*.

7-3. Do Prob. 7-1 with Aircraft *D* 10 percent heavier than its listed gross weight.

7-4. Do Prob. 7-2 with Aircraft *E* 10 percent heavier than its listed gross weight.

7-5. Aircraft *E* is on a sea-level, standard-day take-off and is 500 ft down the runway when one engine fails.

a. What is the airspeed at the time of engine failure?

b. Can the take-off be completed with the remaining engine? If yes, how much remaining runway is required?

c. If the pilot decides to abort the take-off, how much runway will be required to bring the aircraft to a stop with a deceleration factor of 0.3?

7-6. a. Immediately after take-off, Aircraft *D* is required to return and land. With the aircraft at its listed gross weight, what is the approach airspeed and landing roll with a deceleration factor of 0.3?

b. If the aircraft circles the field until 400 lb of fuel is consumed, what will the approach speed and landing roll be?

7-7. Do Prob. 7-6 for Aircraft *E* with the amount of fuel to be consumed in (b) to be 1,000 lb.

7-8. In evaluating the climbing performance of these piston-props, use a propeller efficiency based on the climbing airspeed in conformance with the take-off rule of thumb given in the text of this chapter. For Aircraft *D* at sea level:

a. Find the maximum rate of climb (fpm) and the associated climb angle (deg) and airspeed (fpm). Is this airspeed greater or less than the stall speed? If greater, how much greater is it?

b. Find the steepest climb angle and the associated airspeed. Is this airspeed greater or less than the stall speed? If greater, how much greater is it?

c. Setting the climb speed equal to 1.2 times the stall speed, find the climb angle and rate of climb and compare these values with those obtained in (a) and (b).

7-9. Do Prob. 7-8a for Aircraft *D* at 20,000 ft with an aspirated engine.

7-10. Do Prob. 7-8a for Aircraft *D* at 20,000 ft with a turbocharged engine.

7-11. Do Prob. 7-8 for Aircraft *E* at sea level.

7-12. Do Prob. 7-8a for Aircraft *E* at 25,000 ft with aspirated engines.

7-13. Do Prob. 7-8a for Aircraft *E* at 25,000 ft with turbocharged engines.

7-14. For Aircraft *D*, use the closed-form expressions of this chapter to determine the minimum time to climb (min) from sea level to 20,000 ft along with the fuel required (lb and gal) and the distance traveled (mi) with:

a. An aspirated engine

b. A turbocharged engine

7-15. For Aircraft *D* with a turbocharged engine and a climb airspeed equal to 1.2 times the minimum-drag airspeed:

a. Find the maximum rate of climb at sea level and at 20,000 ft. Use the average rate of climb to determine the time to climb.

b. Use this time to climb in conjunction with an average value for the climb horsepower to determine the fuel used.

c. Use this time to climb in conjunction with an average airspeed and an average climb angle to determine the distance traveled.

d. If you have solved this problem previously, using the closed-form expressions, compare results.

7-16. Do Prob. 7-14 for Aircraft *E* in a climb from sea level to 25,000 ft with turbocharged engines.

7-17. Do Prob. 7-15 for Aircraft *E* in a climb from sea level to 25,000 ft with turbocharged engines.

7-18. For Aircraft *D* in a steady-state turn at sea level:

a. Find the theoretical maximum load factor and the associated flight conditions. Is such a turn possible?

b. Find the theoretical fastest-turn rate (deg/s) and the associated flight conditions. Is such a turn possible?

c. Find the theoretical tightest-turn radius (ft) and the associated flight conditions. Is such a turn possible?

d. For any of the above turns that is not possible, determine the best turning performance.

7-19. Do Prob. 7-18 for Aircraft *E* at sea level.

7-20. For Aircraft *D*, find the stall-speed turning performance and flight conditions:

a. At sea level

b. At 30,000 ft with an aspirated engine

c. At 30,000 ft with a turbocharged engine

7-21. Do Prob. 7-20 for Aircraft *E*.

Chapter Eight

Turboprops, Turbofans, and Other Things

8-1 INTRODUCTION

This chapter has two objectives. The first is to extend the analyses of the preceding chapters to the turboprop and turbofan by relating their performance to that of the piston-prop and the turbojet rather than by separate mathematical developments. The second is to treat several subjects that are not necessarily interrelated but that could either affect the performance and design of a particular aircraft or possibly increase the usefulness and application of the techniques already developed.

Even though there were places in the preceding chapters where these additional subjects might have been introduced, discussion was deferred so as to avoid any possible distraction or confusion. The subjects to be covered are Mach number representation of certain of the performance equations, flight and maneuvering envelopes, and the energy-state approximation.

8-2 THE PISTON-PROP AND TURBOJET REVISITED

Let us return to the Breguet range equations for another look at the differences in the range performance of the piston-prop and of the turbojet. For the piston-prop,

$$X=\frac{375\eta_p E}{\hat{c}}\ln\left(\frac{1}{1-\zeta}\right) \tag{8-1}$$

and for the turbojet,

$$X=\frac{VE}{c}\ln\left(\frac{1}{1-\zeta}\right) \tag{8-2}$$

In each of these equations, the grouping preceding the logarithmic term is the product of the aerodynamic and propulsive efficiencies. This product is sometimes referred to as the *range factor*. Since in both equations the aerodynamic efficiency is simply the lift-to-drag ratio, the propulsive efficiency for the piston-prop is $375\eta_p/\hat{c}$, and for the turbojet V/c.

The aerodynamic efficiency of both types of aircraft is a function of the lift coefficient, and thus for a given configuration and altitude is a function of the airspeed. A typical variation of the lift-to-drag ratio with the lift coefficient and airspeed is shown in Fig. 2-11.

Examining the propulsive efficiency of the piston-prop, we see that it is explicitly independent of the airspeed and is constant if η_p and $\hat{c}$ are constant. Although the propeller efficiency drops off at low and high speeds, it, and the propulsive efficiency, can be considered constant in the design cruising range of the aircraft. The propulsive efficiency of the turbojet, on the other hand, is directly proportional to the airspeed and, although low at low airspeeds, increases linearly and rapidly as the airspeed increases.

In order to establish a frame of reference for comparing the piston-prop and the turbojet and for discussing the turboprop and the turbofan, let us express the level-flight true airspeed of any aircraft, no matter what type of a propulsion system it might have, in terms of the minimum-drag (maximum lift-to-drag ratio) airspeed as

$$V=\left[\frac{2(W/S)}{\rho_{SL}\sigma}\right]^{1/2}\left(\frac{mK}{C_{DO}}\right)^{1/4}=m^{1/4}V_{md} \tag{8-3}$$

where m, the *airspeed parameter*, can be written as

$$m=\left(\frac{V}{V_{md}}\right)^4 \tag{8-3a}$$

Using Eq. 8-3, the lift coefficient can be written as

$$C_L=\left(\frac{C_{DO}}{mK}\right)^{1/2}=\frac{C_{L,Em}}{m^{1/2}} \tag{8-4}$$

and the flight lift-to-drag ratio as

$$E=\frac{2m^{1/2}}{m+1}E_m \tag{8-5}$$

Applying these relationships to the Breguet range equations (application to any of the other range equations gives similar results and conclusions) leads to

$$X=\frac{2\times 375\times m^{1/2}\eta_p E_m}{(m+1)\hat{c}}\ln\left(\frac{1}{1-\zeta}\right) \tag{8-6}$$

for the piston-prop and to

$$X=\frac{2m^{3/4}V_{md}E_m}{(m+1)c}\ln\left(\frac{1}{1-\zeta}\right) \tag{8-7}$$

for the turbojet.

If we either set the derivative of each of these range equations with respect to m equal to zero and solve for m_{br} or recall the conditions for the best-range airspeed for each type of aircraft, we note that for the *piston-prop*

$$V_{br}=V_{md} \qquad \text{(that is, } m_{br}=1\text{)} \tag{8-8}$$

and for the *turbojet*

$$V_{br}=3^{1/4}V_{md} \qquad \text{(that is, } m_{br}=3\text{)} \tag{8-9}$$

Consequently, the value of m_{br} can be used to indicate the type of propulsion system for a particular aircraft. This idea of categorizing aircraft by the magnitude of the best-range airspeed parameter will be developed further in the next section.

The expressions of Eqs. 8-6 and 8-7 are in a convenient form for showing the variations in range as a function of the airspeed parameter (and thus of the airspeed) and for comparing the relative ranges of piston-props and turbojets with the same values of the maximum lift-to-drag ratio and of the cruise-fuel weight fraction. The baseline range is that of a piston-prop with a value of 2 for the ratio $\eta_p/\hat{c}$ (for example, $\eta_p=0.9$ and $\hat{c}=0.45$) and this baseline range is sketched in Fig. 8-1. Since it is unlikely that the ratio of the propeller efficiency to the specific fuel consumption will take on much larger values and since the range of a piston-prop with a constant propeller efficiency is indifferent to the magnitude of the minimum-drag airspeed, this one curve can also serve as the approximate upper limit to the range of a piston-prop.

The range of a turbojet, on the other hand, is strongly dependent on the magnitude of the minimum-drag airspeed. Consequently, the relative range of a turbojet is sketched for several values of V_{md}/c in Fig. 8-1. Since c is of the order of 1.0 lb/hr/lb, the magnitude of the ratio, V_{md}/c, is essentially the minimum-drag airspeed and as

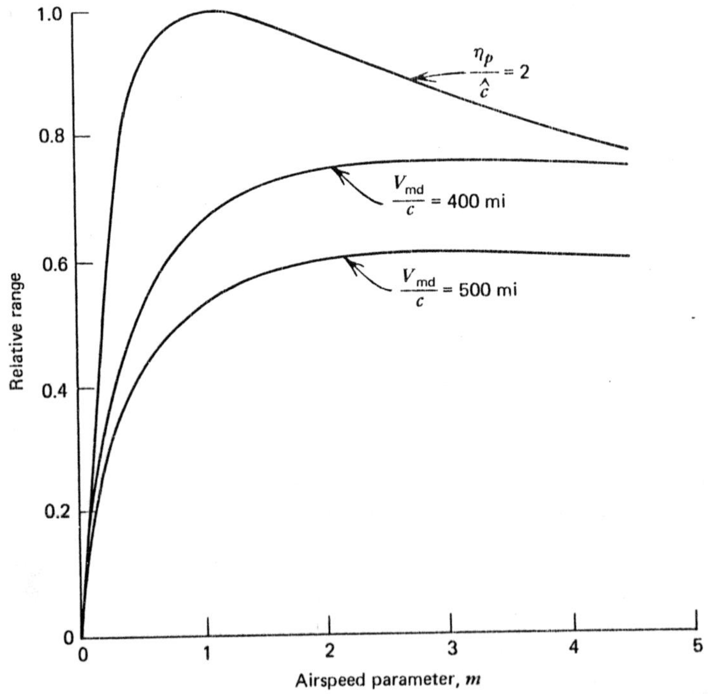

FIGURE 8-1

Relative ranges of a piston-prop and turbojet with the same E_{max} and ζ as a function of the airspeed.

such combines the aerodynamic characteristics of the aircraft with the altitude at which the aircraft is flying.

In Fig. 8-1, we see that the turbojet curves are quite flat in the vicinity of the best-range airspeed because the increase in the propulsive efficiency with increasing airspeed compensates for the decrease in the aerodynamic efficiency, whereas the curve for the piston-prop is pronouncedly peaked, indicating a much greater sensitivity of the range to departures from the best-range airspeed. If the decrease in propeller efficiency with high airspeeds were introduced, the peaking would be even greater. The most striking features of this figure are the built-in range superiority of the piston-prop (this range is independent of both the altitude and the magnitude of the airspeed), the lower best-range airspeed of the piston-prop, and the fact that the turbojet requires high airspeeds (and altitudes) in order to be competitive with respect to range.

Although the piston-prop has a range superiority, it is at a disadvantage with respect to airspeed and flight times since it is inherently slower than the turbojet. For example, with identical wing loadings and drag polars, the best-range airspeed of the piston-prop will be 24 percent lower than that of the turbojet. This relationship (which also holds for many of the other special flight conditions) is implied in Fig. 8-1 but can be better visualized in Fig. 8-2, where the best-range airspeed is sketched as a function of certain aircraft characteristics. If the piston-prop is to fly at the same best-range airspeed as does a specified turbojet, then the wing loading must be significantly larger (particularly since the zero-lift drag coefficient is generally larger than that of a turbojet), which in turn calls for a larger aspect ratio if the maximum lift-to-drag ratio is to be the same. Furthermore, any increase

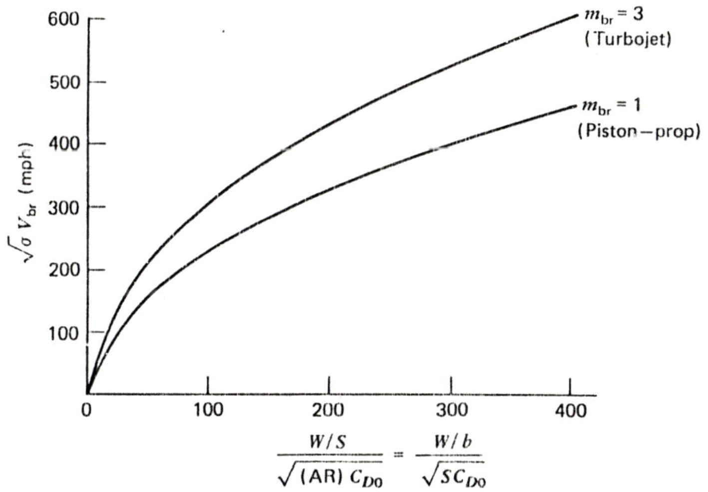

FIGURE 8-2

The best-range airspeeds of piston-props and turbojets as a function of the wing loading and drag polar parameters.

in the maximum airspeed of a piston-prop is accompanied (as was seen in Sec. 2-6) by a disproportionate increase in the required horsepower and engine weight.

In spite of a greater propulsive efficiency, as manifested by the range superiority, the lower airspeeds limit the applicability and desirability of the piston-prop. In addition to the normal desire to go faster and spend less time en route as well as the need to match the competition, higher airspeeds are often mandated by operational requirements or economic considerations. With respect to the latter, turbojets replaced the piston-props in the commercial aviation fleet, not merely through public acceptance of the higher airspeeds but because the reduced flight times increased the aircraft utilization rate, greatly lowering the cost per passenger-mile. Airspeed, therefore, is one of the many nontechnical factors that influence the design and operation of aircraft but that, unfortunately, are beyond the scope of this book.

Comparing performance other than range is much less straightforward and satisfying because it is difficult to establish a true basis of comparability. Perhaps the best that we can do is to specify identical wing loadings and drag polars (although identical C_{D0}'s are very unlikely) and modify or augment our analytical comparisons with observations and comments on the pertinent characteristics of the two types of propulsion systems.

With our definition of comparable aircraft, the one with the larger thrust-to-weight ratio should have the shorter take-off ground run. Evaluating the respective T/W ratios is not simple, particularly since the HP/W ratio and not the T/W ratio is the propulsion characteristic of the piston-prop. Typical values for transport aircraft are 0.25 for the T/W ratio of a turbojet and 0.1 for the HP/W ratio of a piston-prop. If the rule-of-thumb relationship that 1 hp produces approximately 2.5 lb of static thrust is applied to the piston-prop, the two T/W ratios will be the same and so, therefore, should be the take-off ground runs. However, since the thrust of a turbojet falls off at low airspeeds, its actual take-off T/W ratio will be lower. If it is 80 percent of the nominal T/W ratio, then the take-off run of the piston-prop will be of the order of 80 percent of that for a turbojet.

There are further complications to be considered, however. The first of these is the fact that as the wing loading of the piston-prop is increased, in order to obtain higher cruise speeds, the lift-off airspeed will increase, thus reducing the effective T/W ratio of the piston-prop. Another, and probably more important, consideration is that the T/W ratio of the turbojet can be increased with a significantly lower increase in weight than that associated with increasing the HP/W ratio of the piston-prop. After all of this discussion, the only conclusion that can be reached is that it is not possible to make a flat statement as to the take-off superiority of either type of aircraft.

With respect to the maximum rate of climb, the turbojet has a decided superiority at higher wing loadings (of the order of 50 lb/ft^2 or more) and has higher climb airspeeds. At lower wing loadings or at the lower climb speeds imposed by air traffic control, the climb rates of the two aircraft are much closer together. In addition, the turbocharged piston-prop has the definite superiority of minimizing the effects of an increasing altitude on the climb rate.

Whereas the absolute ceiling of a turbojet is independent of the wing loading (as long as the corresponding ceiling airspeed does not exceed the drag-rise Mach number), that of the piston-prop decreases with an increase in the wing loading. For both types of aircraft, increases in the maximum lift-to-drag ratio and in the T/W or HP/W ratios raise the ceiling, with somewhat larger proportionate increases for the piston-prop. In general, the turbojet has a ceiling that is considerably higher than that of the aspirated piston-prop, but the ceiling of the turbocharged piston-prop is much more competitive. In fact, with a low wing loading, the turbocharged ceiling might even be higher than that of a comparable turbojet.

Turning performance is described in terms of the maximum load factor, the fastest turn, and the tightest turn. Since the aircraft characteristics that determine the ceiling also determine the maximum load factor, the maximum load factor of the turbojet will generally be much higher than that of the piston-prop. Turbocharging, however, will not increase the n_{max} of a piston-prop; it will merely retard the decrease with altitude. Looking at fastest and tightest turns for comparable aircraft without regard for whether the aircraft is below the stall speed, the piston-prop seems to have a slight edge over the turbojet that diminishes with increased wing loading. The maximum achievable lift coefficient is always a key factor in determining maneuverability, regardless of the type of propulsion system.

Looking back over this section, we see that it is quite difficult to compare these two classes of aircraft because of the fundamental differences in their characteristics and behavior. The piston-prop has better mileage than the turbojet but is restricted to the low speed regime by the excessive weight of the large engines required for the higher airspeeds. (To see how restrictive the engine weight can be, do a feasibility analysis of a M 0.8 piston-prop transport using the current horsepower-to-engine-weight ratio of 0.5 hp/lb.) Furthermore, the complexity and maintenance problems and costs associated with large internal combustion engines are very high and, finally, the increase in the wing loading required for the higher airspeeds adversely affects other aspects of performance.

Although the piston-prop does not care about such things as airspeed and altitude, the faster (and thus the higher) the turbojet flies, the farther it can fly with a given fuel load. The upper limit to the cruise airspeed is the drag-rise Mach number, at which point the lift-to-drag ratio begins to fall off. The primary disadvantage of the turbojet is its large specific fuel consumption and a secondary disadvantage is low thrust at low airspeeds.

8-3 TURBOPROPS AND TURBOFANS

Turboprops and turbofans are gas turbine engines, as is the turbojet, and are designed to minimize the disadvantages and exploit the advantages inherent in piston-prop and turbojet engines. The fundamental difference among these three turbine engines is in how they produce thrust. The turbojet does it by expansion of hot gases through a nozzle, the turboprop uses a propeller, and the turbofan uses a multibladed fan, which is related in many ways to the propeller. The basic element of a gas turbine engine is the *gas generator* (sometimes called the *core*), which comprises the compressor(s), the burners (the combustion chambers), and the turbine(s) that drive the compressor. The mixture of air and fuel that passes through the gas generator is usually referred to as the *primary flow*. Since the gas generator and the primary flow are common to all gas turbine engines, they can be used as a baseline for a comparative description and evaluation of the characteristics and performance of such engines. In a *turbojet*, the exhaust gases from the gas generator are expanded through a nozzle (the tailpipe), and thrust is the only output. Since the only flow through a turbojet is the primary flow, it is classified as a single-flow engine. The distinguishing characteristics of the turbojet are its light weight, small frontal area, a propulsive efficiency that increases with airspeed, a high specific fuel consumption (the highest of the three), and low thrust at low airspeeds.

In a *turboprop*, the exhaust gases from the gas generator are partially expanded through an additional turbine(s), that drives a propeller through a speed-reduction gearbox, before final expansion through the tailpipe. (In a *turboshaft engine* all of the expansion takes place in the drive turbine(s), which is used to power helicopters, boats, pumps, and generators.) The thrust developed by the nozzle is called the *jet thrust* and is typically of the order of 10 to 15 percent of the total static thrust at sea level on a standard day. The airflow through the propeller is called the *secondary flow*, is considerably larger than the primary flow, and is at ambient temperature. Sometimes it is called the cold gas, and the primary flow is called the hot gas. Since there are two flows, a turboprop can be classified as a multiflow engine.

A turboprop is primarily a power producer and is described in similar terms as the piston-prop, using *equivalent shaft horsepower* (ESHP) instead of brake horsepower, where the equivalent shaft horsepower is the sum of the shaft power delivered to the propeller plus the horsepower equivalent of the jet thrust power ($T_j V$). The following expressions are relevant to the turboprop:

$$\text{ESHP} = \left(\text{SHP} + \frac{T_j V}{375\eta_p}\right) \tag{8-10}$$

$$\frac{dW}{dt} = -\hat{c}^*(\text{ESHP}) \tag{8-11}$$

where V is in mph, and $\hat{c}^*$ is an *equivalent horsepower specific fuel consumption*.

The turboprop is primarily a replacement for the piston-prop since it is capable of higher airspeeds and greater range for a given aircraft weight because of its much lighter engine weight and lower zero-lift drag coefficient. Although heavier than a turbojet or a turbofan, because of the propeller and gearbox, it is of the order of four times lighter than a piston-prop engine of the same horsepower. Furthermore, although the frontal area is somewhat larger than that of a turbojet, it is less than that of a piston-prop, and when the engine is operating, the zero-lift drag coefficient is of the order of that of a turbojet, i.e., less than that of a piston-prop, which means higher lift-to-drag ratios. The presence of the jet thrust, which though relatively small is essentially constant, tends to flatten the thrust curves at the higher airspeeds and to reduce the rate of decrease of the propulsive efficiency. The turboprop has a low specific fuel consumption, of the order of but somewhat higher than that of the piston-prop, with advancements in the technology showing promise of even lower values.

The high-power-to-engine-weight ratio of the turboprop (due to the light weight of the gas generator) is being exploited to make a turboprop behave as though it were turbocharged by a technique known as *derating*. This is done by using a propeller that cannot use all of the shaft power developed by the engine. For example, if a 670 eshp turboprop is derated to 400 eshp, it means that the maximum shp that can be absorbed by the propeller plus the jet-thrust power is 400 eshp even though the engine is capable of producing 670 eshp at sea level. As the altitude increases, the maximum power output of the propeller-jet combination will remain constant at 400 eshp until the maximum engine output, which is decreasing with altitude, drops below the value needed to maintain 400 eshp; this drop will start in the vicinity of 20,000 ft or less. As a consequence, the *derated turboprop* has the characteristics of a turbocharged piston-prop with a critical value of the order of 20,000 ft. One other major advantage of the turboprop over the piston-prop is its much lower maintenance costs. Although its initial cost is higher, it is a simpler engine with a greater reliability, especially with the recent improvements in the gearbox.

A *turbofan* is a multiflow engine similar in many respects to a turboprop except that the additional turbines directly drive a fan that resembles an axial flow compressor. The ratio of the secondary (cold) airflow through the fan to the primary (hot) airflow through the gas generator and the tailpipe is called the *bypass ratio*. The more power that is extracted from the exhaust gases to drive the fan, the higher the bypass ratio is and the smaller the jet thrust. Even though with very high bypass ratios the turbofan may produce more power than thrust and perform more like a turboprop than a turbojet, it is customary to describe the turbofan as though it were a turbojet, with the *equivalent* thrust expressed as

$$T = \left[\frac{375\eta(\text{SHP})}{V} + T_j\right] \tag{8-12}$$

where η represents the conversion efficiency of the fan and SHP is the shaft horsepower delivered to the fan. The weight balance equation for a turbofan can be written as

$$\frac{dW}{dt} = -c^*T \tag{8-13}$$

where c^* is the *equivalent thrust specific fuel consumption*.

The turbofan combines the good propulsive efficiency and high thrust at lower airspeeds of the piston-prop with the constant thrust and increasing efficiency at the higher airspeeds of the turbojet. Since the complexity and weight of the reduction gearbox and of the propeller governor system of the turboprop are eliminated, the turbofan is even simpler and lighter. Furthermore, the airflow through the ducted fan is not greatly affected by the airspeed so that the decrease in propulsive efficiency at high airspeeds is not as significant as the decrease associated with the propeller efficiency of the turboprop. Consequently, the turbofan can be used at airspeeds up to and including low supersonic airspeeds. Although the frontal area is larger than that of the turbojet, the turbofan is considerably shorter and the overall drag is not necessarily any larger. The specific fuel consumption is much less than that of the turbojet and although it is more than that of the turboprop, it is approaching comparable values. The turbofan is also quieter than the turbojet and much quieter than the turboprop, an advantage in these days of increasing concern with and regulation of noise pollution.

It should be noted that, since both the turboprop and the turbofan are multiflow engines, the equivalent specific fuel consumptions are combinations of horsepower and thrust specific fuel consumptions, and thus will vary with airspeed. Any value quoted in the manufacturer's descriptive material is for a specific airspeed, which is not always given. The variation with airspeed is greater for the turbofan than for the turboprop.

Since turboprops and turbofans represent different combinations of the piston-prop and of the turbojet, their performance should fall somewhere between that of the piston-prop and that of the pure turbojet. Returning to Figs. 8-1 and 8-2 and the concept of a best-range airspeed parameter, we might expect the range and best-range airspeed curves to fall between those for the piston-prop and for the turbojet. The value of m_{br} for the turboprop should be of the order of, but somewhat greater than, that for the piston-prop, which is equal to unity. The m_{br} of a turbofan with a low bypass ratio will be of the order of, but less than, that for a turbojet, which is equal to 3. As the bypass ratio is increased, the value of m_{br} will decrease, approaching that of the turboprop in the limit.

It would be nice to have a mathematical model to demonstrate and determine the effects of various combinations of thrust and power upon the range of either a turboprop or a turbofan. With the premise that the total range of a turboprop or turbofan can be represented by the sum of a contribution from the constant-power constituent and one from the constant-thrust constituent, the two Breguet range equations of Eqs. 8-1 and 8-2 can be combined to form a *generalized Breguet range equation* applicable to all four types of propulsion systems, namely,

$$X=\left[\left(\frac{3-k}{2}\right)\frac{375\eta_p E}{\hat{c}}+\left(\frac{k-1}{2}\right)\frac{VE}{c}\right]\ln\left(\frac{1}{1-\zeta}\right) \qquad (8\text{-}14)$$

where k is a *propulsion system designator*. When k is set equal to unity, the turbojet contribution goes to zero and Eq. 8-14 becomes the range equation for a pure piston-prop. Similarly, when k takes on a value of 3, the propeller contribution drops out, leaving the range equation for a pure turbojet. Intermediate values between 1 and 3 for k denote different turboprop and turbofan contributions.

If we now introduce the airspeed parameter of Sec. 8-2, the generalized Breguet equation becomes

$$X=\frac{2m^{1/2}E_m}{m+1}\left[\left(\frac{3-k}{2}\right)\frac{375\eta_p}{\hat{c}}+\left(\frac{k-1}{2}\right)\frac{m^{1/4}V_{md}}{c}\right]\ln\left(\frac{1}{1-\zeta}\right) \qquad (8\text{-}15)$$

The bracketed term represents the overall propulsive efficiency of the aircraft, and the value of k determines the proportionate contribution of the two airflows. For example, if $k = 1.3$, then the overall propulsive efficiency is the sum of 85 percent of the propeller or fan efficiency and 15 percent of the jet efficiency. The grouping in front of the bracketed term is simply the flight lift-to-drag ratio, as given by Eq. 8-5.

The relative range as a function of the airspeed parameter for various values of k is shown in Fig. 8-3, using the values of 2 for $\eta_p/\hat{c}$ and of 400 for V_{md}/c. The curves for k equal to unity (pure propeller) and for k equal to 3 (pure jet) establish the upper and lower limits for our turboprop and turbofan analyses. Starting with the pure turbojet ($k = 3$), we see that increasing the fan contribution by increasing the bypass ratio and decreasing k definitely increases the range of the resulting turbofan. As k is decreased, the jet contribution decreases, going to zero as k goes to unity, reaching the piston-prop as a limit. We also see that as k decreases, the best-range airspeed, expressed in terms of V_{md} also decreases. This means not only that the best-range lift-to-drag ratio and lift coefficient of a turbofan will not be the same as for a turbojet but also that either the wing loading or the altitude of the turbofan best-range airspeed is to be maintained. As k decreases, the shape of the curves changes, becoming less flat in the vicinity of the best-range airspeed parameter.

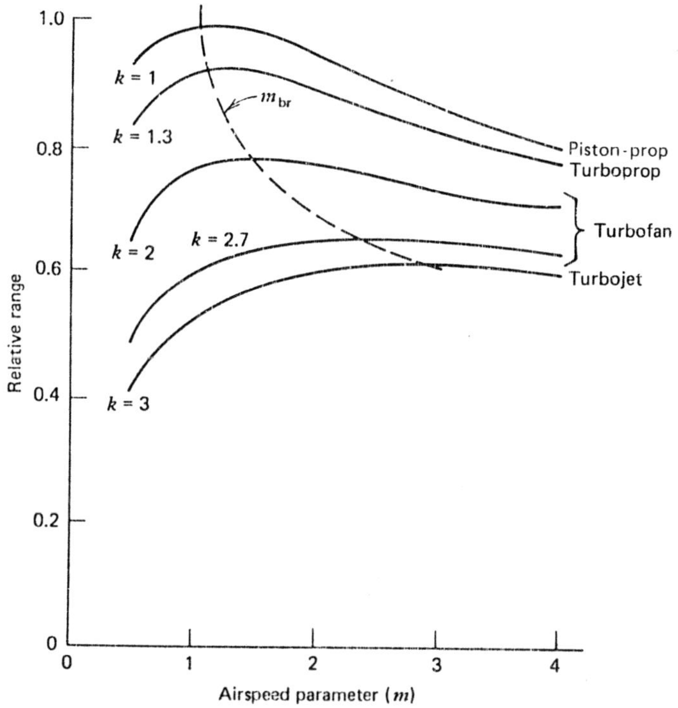

FIGURE 8-3

Relative ranges of a piston-prop, turboprop, turbofan, and turbojet with the same E_{max} and ζ as a function of the airspeed and k.

To examine the effects of increasing the jet thrust contribution on the performance of a turboprop, we merely have to start with k equal to unity and see what happens as k is increased. At first glance, there seems to be no advantage to having any jet thrust in a turboprop. However, with $k=1.3$, there is a slight increase in the best-range airspeed and some flattening of the range curve for the higher airspeeds as a result of the increase in the propulsive efficiency of the jet portion. These trends become more apparent as k is increased to a value of 2.

Returning to Eq. 8-15, it is possible to develop expressions for determining m_{br} if k is known, or k if m_{br} is known, by setting the derivative with respect to m equal to zero. The two expressions are:

$$m_{br} = 1 + \frac{(2V/c)(k-1)}{(750\eta/\hat{c})(3-k)+(V/c)(k-1)} \tag{8-16}$$

$$k = 1 + \frac{4}{2+\left(\dfrac{V/c}{375\eta/\hat{c}}\right)\left(\dfrac{3-m_{br}}{m_{br}-1}\right)} \tag{8-17}$$

where η is the propeller efficiency for a turboprop and the fan efficiency for a turbofan. If V_{br} is not known but V_{md} is, then Eq. 8-16 is still usable with the substitution of $m_{br}{}^{1/4} V_{md}$ for V_{br}. The resulting equation is

$$m_{br} = \left[\left(\frac{750\eta/\hat{c}}{V_{md}/c}\right)\left(\frac{3-k}{k-1}\right)\left(\frac{m_{br}-1}{3-m_{br}}\right)\right]^4 \tag{8-18}$$

which does not have a closed-form solution and must be solved by iteration.

It is not the practice to use a combined Breguet range equation for turboprops and turbofans. Instead, it is customary to use the turbojet equation for turbofans and the piston-prop equation for turboprops. When these simpler expressions are used, the specific fuel consumption is a combined or equivalent specific fuel consumption even though the usual symbols for single-flow engines are normally used. To avoid confusion and to emphasize their combined nature, we are using an asterisk to denote equivalent specific fuel consumptions, that is, $\hat{c}^*$ and c^*. Remember that this is done only in this book and not in practice. With this in mind, the Breguet range equations for turbofans and turboprops become:

$$X = \frac{VE}{c^*} \ln\left(\frac{1}{1-\zeta}\right) \tag{8-19}$$

and

$$X = \frac{375\eta_p E}{\hat{c}^*} \ln\left(\frac{1}{1-\zeta}\right) \tag{8-20}$$

By equating the combined propulsive efficiency from Eq. 8-15 to each of the individual efficiencies from Eqs. 8-19 and 8-20, the following relationships can be obtained:

$$\left(\frac{3-k}{2}\right)\frac{375\eta_p}{\hat{c}} + \left(\frac{k-1}{2}\right)\frac{m^{1/4} V_{md}}{c} = \frac{V}{c^*} = \frac{375\eta_p}{\hat{c}^*} \tag{8-21}$$

where $V = m^{1/4} V_{md}$ and $\hat{c}$ and c are the specific fuel consumptions of the propeller (or fan) power portion and of the jet thrust portion of the engine, respectively. For lack of detailed information on the power plant, which is the reason for all of this, it seems reasonable to assume a best piston-prop value for $\hat{c}$, say 0.4 lb/h/hp, and a best turbojet value for c, say 0.95 lb/h/lb. If we do this, we can now calculate $\hat{c}^*$ and c^* for any given airspeed, including the best-range airspeed, for a specified value of k.

As an example, let us look at a turboprop with a V_{md} at cruising altitude of 400 mph (a bit high perhaps), a propeller efficiency of 80 percent, an E_m of 16, and a k of 1.3, and determine its best range with a cruise-fuel fraction of 0.2. We first go to Eq. 8-18, since V_{br} is not known, which with numerical substitutions is

$$m_{br} = \left[2 \left(\frac{375 \times 0.8/0.4}{400/0.95} \right) \left(\frac{3-1.3}{1.3-1} \right) \left(\frac{m_{br}-1}{3-m_{br}} \right) \right]^4$$

or

$$m_{br} = \left[20.1875 \left(\frac{m_{br}-1}{3-m_{br}} \right) \right]^4$$

Notice that we used our "best" values of 0.95 and 0.4 for c and $\hat{c}$. After several iterations with a first trial value of 1.1, m_{br} is found to be approximately 1.097, so that the best range airspeed, from Eq. 8-3a, is

$$V_{br} = (1.097)^{1/4} \times 400 = 409.4 \text{ mph}$$

In order to find $\hat{c}^*$, Eq. 8-21 becomes

$$\frac{375 \times 0.8}{\hat{c}^*} = \frac{0.85 \times 375 \times 0.8}{0.4} + \frac{0.15 \times 409.4}{0.95}$$

or

$$\hat{c}^* = 0.427 \text{ lb/h/eshp}$$

We now need to find E_{br} from Eq. 8-5;

$$E_{br} = \left[\frac{2 \times (1.097)^{1/2}}{1.097+1} \right] \times 16 = 15.98$$

and finally, using Eq. 8-20,

$$X = \frac{375 \times 0.8 \times 15.98}{0.427} \ln \left(\frac{1}{1-0.2} \right) = 2{,}505 \text{ mi}$$

The corresponding flight time is 6.1 hours.

We could just as well have used Eq. 8-19 with the appropriate value for c^*. We can find c^* by substituting in the complete expression of Eq. 8-21, but it is simpler to use the identity

$$\frac{V}{c^*} = \frac{375\eta_p}{\hat{c}^*}$$

so that

$$c^* = \frac{V\hat{c}^*}{375\eta_p} = \frac{409.4 \times 0.427}{375 \times 0.8} = 0.583 \text{ lb/h/lb}$$

With Eq. 8-19,

$$X = \frac{409.4 \times 15.98}{0.583} \ln\left(\frac{1}{1-0.2}\right) = 2{,}504 \text{ mi}$$

With k and m_{br} known, Eq. 8-15 can also be used to determine the range of this aircraft; i.e.,

$$X = 15.98\left[\frac{0.85 \times 375 \times 0.8}{0.4} + \frac{0.15 \times 409.4}{0.95}\right] \ln\left(\frac{1}{1-0.2}\right) = 2{,}504 \text{ mi}$$

If we had treated this turboprop (with a jet thrust of the order of 15 percent) as though it were a piston-prop and had used the latter's best-range conditions and Breguet range equation, the best-range airspeed and lift-to-drag ratio would have been 400 mph and 16, respectively. With a $\hat{c}$ of 0.45, the range would be 2,380 mi and the flight time would be 5.95 hours, giving errors of the order of 5 and 3 percent, respectively.

Do not forget that as the airspeed changes, say with an altitude change, then $\hat{c}^*$ and c^* must be recalculated. In other words, *when a value is given for the specific fuel consumption of a turboprop or of a turbofan, the airspeed and altitude should also be specified.* Typical variations in $\hat{c}^*$ and c^* as a function of the airspeed parameter for several values of k are sketched in Fig. 8-4.

Unfortunately, the model being used to demonstrate the fundamental differences in performance resulting from the use of turboprops and turbofans cannot be extended to show the variations in thrust as a function of the airspeed. To do so requires internal analyses of the engines that are way beyond the scope of this book.

Comparing turboprop, turbofan, and turbojet engines of comparable power (comparable gas generators), the *turboprop* will deliver the largest amount of thrust at the lower airspeeds, to include the aircraft standing still at the start of the take-off run. The thrust, however, will decrease at the most rapid rate of the three as the airspeed increases and at lift-off will probably be less than that of the other two at the same point. The *turbofan* will produce less thrust than the turboprop at the lower speeds but more than the turbojet, which not only improves the take-off and early climb performance but also allows higher gross weights for take-off. The thrust decreases with increasing airspeed but at a slower rate than does that of the turboprop because of the differences between a fan and a propeller and because of the greater jet thrust component. The *turbojet* has the lowest initial thrust of the three, but the thrust essentially remains constant with airspeed. These differences are shown qualitatively in Fig. 8-5 with the reminder that the shape of the turbofan curve is a function of the bypass ratio. As the bypass ratio increases, the performance of the turbofan approaches that of the turboprop at the lower airspeeds but retains some of the characteristics of the turbojet at the higher airspeeds.

With respect to the other aspects of performance, the turboprop is sufficiently similar to the piston-prop so that it is a reasonable approximation to simply use the piston-prop equations without modification using the appropriate value (if it is given) of the specific fuel consumption wherever needed. The turbofan, however,

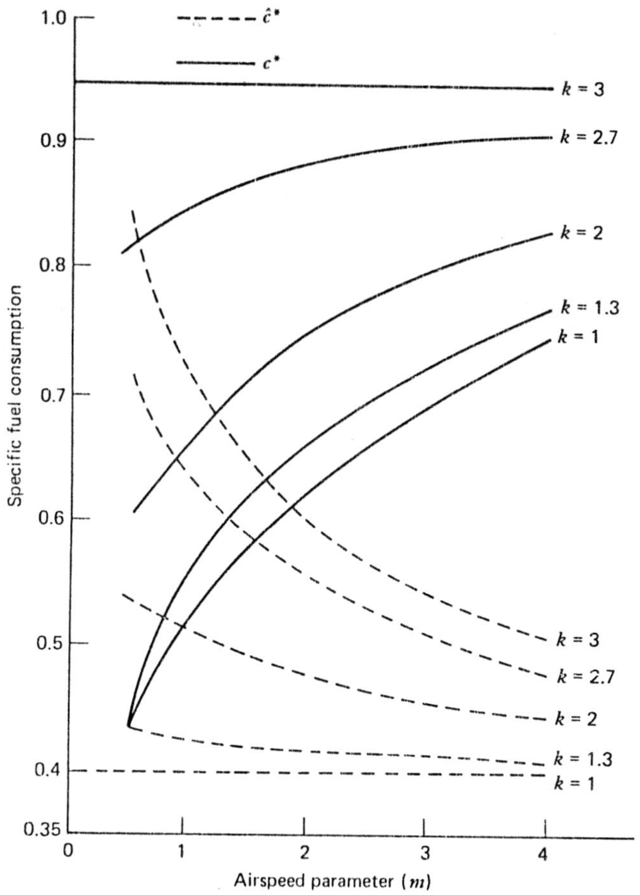

FIGURE 8-4

Equivalent specific fuel consumptions as a function of k and the airspeed.

is not necessarily as simple or as straightforward to handle. If the bypass ratio is low, then the turbojet equations may be used without modification.

As the bypass ratio is increased and the ratio of power to thrust increases, the turbofan takes on more of the characteristics of the turboprop and piston-prop, particularly at the lower airspeeds. It is still possible to use the turbojet equations with the realization that the actual low-speed values might be somewhat different from those obtained from the turbojet equations, e.g., actual take-off runs and unrestricted climb airspeeds and rates will be lower than those for a pure turbojet. At the higher airspeeds, the nature of the fan performance is such that, even with a high bypass ratio, the turbofan will perform more like a turbojet but with a reduced specific fuel consumption. The lower the quoted value of the equivalent thrust specific fuel consumption is, the higher the bypass ratio is apt to be, although it

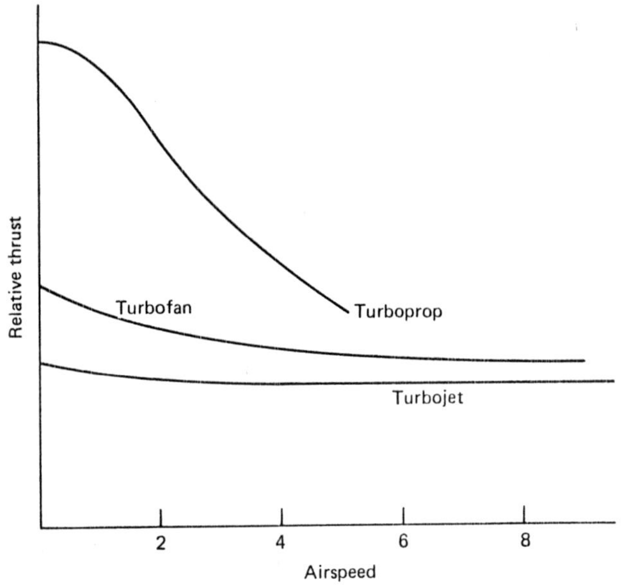

FIGURE 8-5
Relative thrust for gas turbine engines with comparable gas generators as a function of the airspeed.

should be realized that all the improvements in thrust and specific fuel consumption are not necessarily the result of high bypass ratios alone. The internal efficiencies of gas turbine engines are continually improving, not only of turbofans but also of turboprops and turbojets.

If the operating and performance data of the engines and of the aircraft in which they are installed are available, then it is not difficult to determine the type and performance of an aircraft. When detailed data are not available, the primary indicator as to the type of engine is the specific fuel consumption; the lower it is, the larger the power-producing component of the engine is. There may be times when the quoted best-range airspeed may be used to determine the value of m_{br}, particularly when that airspeed is obviously less than the drag-rise Mach number. However, with the current trend toward flying a turbofan at or just below the drag-rise Mach number and then maximizing the range factor by flying as close as possible to the maximum lift-to-drag ratio (by proper choice of altitude), this technique does not always work. It is really not necessary to do this with turboprops, which generally will have an m_{br} of the order of 1.10.

As an example of how m_{br} might be determined for a turbofan, let us look at a turbofan with a manufacturer's quoted best-range airspeed of M 0.8 at 36,000 ft with a corresponding c^* of 0.82 lb/h/lb. The drag polar is $C_D = 0.016 + 0.05C_L^2$ (E_m is 17.67); the wing loading is 90.3 lb/ft^2; and the drag-rise Mach number is 0.85. For this aircraft, the minimum-drag (maximum lift-to-drag ratio) airspeed at 36,000 ft is

$$V_{md} = \left(\frac{2 \times 90.3}{\rho_{SL} \times 0.297}\right)^{1/2} \left(\frac{0.05}{0.016}\right)^{1/4}$$

$$= 672.5 \text{ fps} = 458.5 \text{ mph} = M\,0.695$$

Since M 0.8 at 36,000 ft corresponds to an airspeed of 774 fps (528 mph), from Eq. 8-3a,

$$m_{br} = \left(\frac{V_{br}}{V_{md}}\right)^4 = \left(\frac{774}{672.5}\right)^4 = 1.755$$

Using this value of m_{br}, Fig. 8-3 gives an approximate value of 2.3 for k (which is confirmed by Eq. 8-17) along with a qualitative feel for where the performance of this aircraft fits with respect to a comparable piston-prop, turboprop, or turbojet. Equation 8-5 yields a value of 17 for the best-range lift-to-drag ratio, and now Eq. 8-19 can be used to find the best range once the cruise-fuel fraction is specified. If the latter is 0.2, the corresponding best range will be of the order of 2,400 mi, and for 0.3 it will be of the order of 3,900 mi.

This section can be summarized by several statements. Because of their superiority in their respective speed ranges, the turboprop and turbofan are replacing the piston-prop and turbojet. The turboprop performance can be reasonably approximated by the piston-prop equations whereas judgment has to be exercised in applying the turbojet equations to the turbofan since the latter behaves in certain ways more like a turboprop than a turbojet. For example, the ceiling, maximum possible load factor, and maximum rate of climb will all be lower for a turbofan than for a comparable turbojet. In fact, using the aspirated piston-prop equations with $P/W = (T/W)V$ will give better values for these three variables.

In conclusion, remember that many of the equations in this section were contrived for the purpose of demonstrating and illustrating the performance of turboprops and turbofans, using the piston-prop and the turbojet to establish a frame of reference.

8-4 MACH NUMBER REPRESENTATION

In the preceding chapters, the range has been consistently expressed in statute miles, and the airspeed in either feet per second or miles per hour. Operationally, however, the range is customarily expressed in nautical miles and the airspeed in knots (nautical miles per hour) or as a Mach number. The use of knots and nautical miles does not change the form of any of the equations that we have developed and merely involves the simple conversion relationship that one *nautical mile* (nmi) is equal to approximately 1.15 statute miles (mi), so that one *knot* (kt) is equal to 1.15 miles per hour (mph). The use of the Mach number, on the other hand, does change the form and appearance of our equations and replaces the atmospheric density ratio with the atmospheric pressure ratio.

The *Mach number* is defined as the ratio of the true airspeed to the acoustic velocity (speed of sound), i.e.,

$$M = \frac{V}{a} \tag{8-22}$$

where V is the true airspeed and a is the acoustic velocity.

With the normal and reasonable assumption that air at atmospheric pressure may be treated as an ideal gas, whose equation of state is

$$P = \rho R \Theta \tag{8-23}$$

it can be shown that the *acoustic velocity* is given by the expression that

$$a = (kR\Theta)^{1/2} \tag{8-24}$$

where k is the ratio of specific heats ($k = 1.4$), R is the gas constant for air (see Appendix A), and Θ is the absolute temperature of the air in either degrees Rankine (R) or Kelvin (K). We see from Eq. 8-24 that the acoustic velocity is a function of the temperature, decreasing with altitude in the troposphere and remaining constant in the isothermal stratosphere.

Defining the *sonic ratio* as the ratio of the acoustic velocity at any given altitude to that at sea level, then

$$a^* = \frac{a}{a_{SL}} = \left(\frac{\Theta}{\Theta_{SL}}\right)^{1/2} = \Theta^{*1/2} \tag{8-25}$$

where Θ^* is the temperature ratio. Plotting a^* as a function of altitude, as in Fig. 8-6, shows that the variation in the sonic ratio (and the acoustic velocity) is linear in the troposphere (the slope is proportional to the temperature lapse rate of approximately -2 deg C or -3.6 deg F per 1,000 ft) and that in the troposphere a^* can be expressed in terms of the altitude or the density ratio as

$$a^* = 1 - 3.6 \times 10^{-6} h = 1 + 0.11 \ln \sigma \tag{8-26}$$

where h is in feet. Above the tropopause, in the stratosphere, a^* is constant and equal to 0.867. We also see from Fig. 8-6 that a constant airspeed of 893 fps (609 mph) represents a sea-level Mach number of 0.8 that increases linearly to M 0.92 at the tropopause and remains constant thereafter in the stratosphere.

Turning to the aerodynamic forces, specifically the lift,

$$L = \tfrac{1}{2}\rho V^2 S C_L = \tfrac{1}{2}\rho a^2 M^2 S C_L \tag{8-27}$$

From Eqs. 8-24 and 8-25,

$$\rho a^2 = k\rho R\Theta = kP \tag{8-28}$$

and Eq. 8-27 becomes

$$L = \tfrac{1}{2}kPM^2 S C_L \tag{8-29}$$

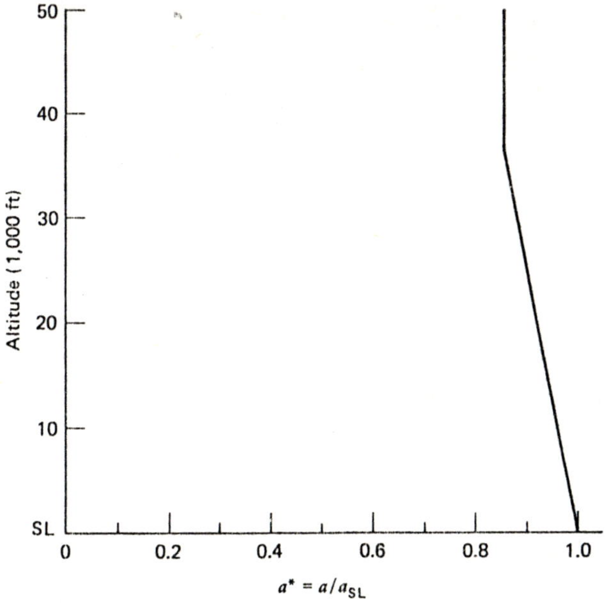

FIGURE 8-6

The sonic ratio as a function of the altitude.

With the pressure ratio defined as

$$\delta = \frac{P}{P_{SL}} = \sigma\Theta^* \tag{8-30}$$

Equation 8-29 becomes, with $k = 1.4$,

$$L = 0.7 P_{SL} \delta M^2 S C_L \tag{8-31}$$

Similarly,

$$D = 0.7 P_{SL} \delta M^2 S C_D \tag{8-32}$$

The dynamic pressure can now be written as

$$q = \tfrac{1}{2}\rho V^2 = 0.7 P_{SL} \delta M^2 \tag{8-33}$$

In level flight, the lift must equal the weight and the thrust must equal the drag; accordingly,

$$\frac{W}{\delta} = 0.7 P_{SL} M^2 S C_L \tag{8-34a}$$

and

$$\frac{T}{\delta} = 0.7 P_{SL} M^2 S C_D \tag{8-34b}$$

In level flight, the Mach number-altitude combination for a specified wing loading and lift coefficient is constant, as can be seen by rewriting Eq. 8-34*a* as

$$\delta M^2 = \frac{W/S}{0.7P_{SL}C_L} \tag{8-35}$$

Consider, for example, an aircraft with a wing loading of 100 lb/ft^2 to be flown at a lift coefficient of 0.5; then

$$\delta M^2 = \frac{100}{0.7 \times 2{,}116 \times 0.5} = 0.135$$

At sea level, the Mach number required would be 0.367 or 243 kt (279 mph); at 20,000 ft, it would be M 0.54 or 332 kt (382 mph); and at 35,000 ft, $M = 0.76$ or 438 kt (504 mph).

Rewriting Eq. 8-34*b* as

$$\frac{T/W}{\delta} = \frac{0.7P_{SL}M^2C_D}{W/S} \tag{8-36}$$

we see that the required thrust-to-weight ratio and altitude combination is also constant for a given wing loading and lift coefficient. Equations 8-35 and 8-36 are valid for all types of propulsion systems.

The Breguet range equation for a turbojet can be written in terms of the Mach number as

$$X = \frac{a_{SL}a^*(\text{ME})}{c} \ln\left(\frac{1}{1-\zeta}\right) \tag{8-37}$$

where the Mach number (from Eq. 8-35) is given by

$$M = \left[\frac{W/S}{0.7P_{SL}\delta C_L}\right]^{1/2} \tag{8-38}$$

In the troposphere, a^*, the sonic ratio, decreases linearly but is constant in the stratosphere (see Eq. 8-26 and Fig. 8-6) whereas δ, the pressure ratio, continually decreases with increasing altitude.

For cruise at a constant Mach number, it is convenient to rewrite Eq. 8-38 as

$$C_L = \frac{W/S}{0.7P_{SL}\delta M^2} \tag{8-39}$$

to show that the lift coefficient required for level flight at the specified Mach number and weight must be increased if the altitude increases (so that δ decreases). As C_L increases, so does the lift-to-drag ratio until the maximum lift-to-drag ratio is reached, at which point an increase in C_L results in a decrease in E (see Fig. 2-11). We also see that as the weight decreases along the flight path, the lift coefficient must also be decreased unless the ratio of the weight to the pressure ratio (W/δ) is kept constant by decreasing δ appropriately; the latter is the cruise-climb condition.

Let us consider a long-range, high-subsonic transport with turbojet engines, a tsfc of 0.8 lb/h/lb, a wing loading of 110 lb/ft^2, and the parabolic drag polar, $C_D = 0.018 + 0.04C_L^2$. Figure 8-7 is a plot of range versus Mach number at various altitudes with a cruise-fuel fraction of 0.3. The overall observation is that flying higher and faster yields greater ranges, which we already know to be true for turbojets.

Looking more closely, we see that *for any given altitude, the maximum range occurs at M_{br} and that increasing the altitude increases both M_{br} and the best range.* The best-range Breguet equation can be written as

$$X_{br} = \frac{0.866 a_{SL} a^* M_{br} E_m}{c} \ln\left(\frac{1}{1-\zeta}\right) \tag{8-40}$$

where

$$M_{br} = \left(\frac{W/S}{0.7 P_{SL}\delta}\right)^{1/2} \left(\frac{3K}{C_{D0}}\right)^{1/4} \tag{8-41}$$

which for our example (with $P_{SL} = 2{,}116$ lb/ft^2) reduces to

$$M_{br} = \frac{0.438}{\delta_{br}^{1/2}} \tag{8-42}$$

Looking at Eq. 8-40, we see that by flying at a constant lift-to-drag ratio (a constant C_L), the only way to increase the range is to increase M_{br}, which Eq. 8-42 shows can be done by decreasing the pressure ratio, i.e., by increasing the cruise altitude. The upper limit to the value of an acceptable cruise Mach number is in the vicinity of the drag-rise Mach number. If we limit M_{br} to a M_{DR} of 0.85, then

$$\delta_{br} = \left(\frac{0.438}{0.85}\right)^2 = 0.2655$$

which corresponds to a best-range altitude of approximately 32,500 ft ($a^* = 0.881$) and a "best" range of

$$X_{br} = \frac{0.866 \times 660.8 \times 0.881 \times 18.63 \times 0.85}{0.8} \ln\left(\frac{1}{1-0.3}\right) = 3{,}564 \text{ nmi}$$

Notice that a value of 660.8 kt is used for a_{SL}, so that the range is in nautical miles. The units of a_{SL} establish the units for the range.

This "best" range should be the maximum range of this aircraft. However, Fig. 8-7 shows that for M 0.85 (and for any other specified Mach number for that matter) the maximum range exceeds the "best" range but at a different altitude, in this case at a higher altitude. That the altitude need not necessarily be higher can be seen by looking at M 0.4 in Fig. 8-7. This apparent contradiction between the best range for a given altitude and the maximum range for a specified Mach number corroborates the conclusion in Sec. 3-7 that *for a specified or restricted airspeed, fly at the altitude where the lift-to-drag ratio is a maximum.*

For a turbojet flying at the maximum lift-to-drag ratio, the Breguet range equation becomes

$$X = \frac{a_{SL} a^* M_{md} E_m}{c} \ln\left(\frac{1}{1-\zeta}\right) \tag{8-43}$$

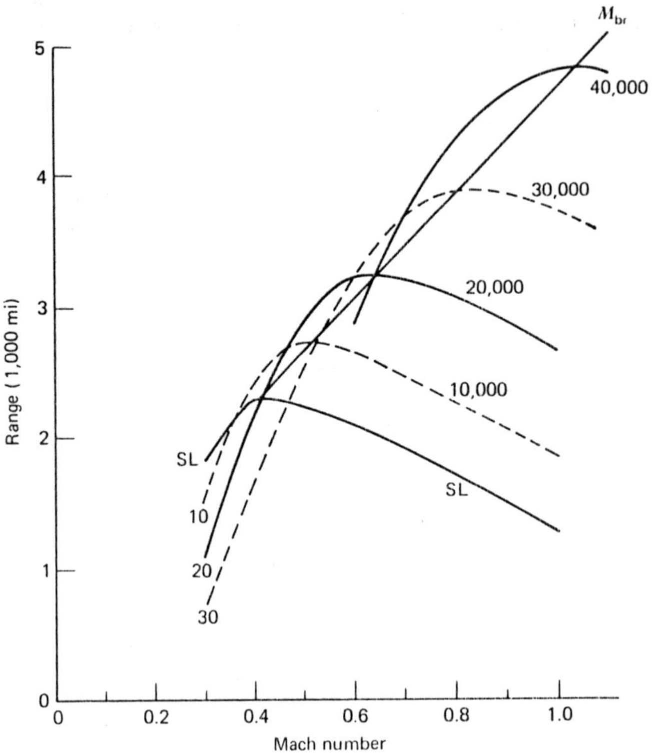

FIGURE 8-7

Range versus Mach number as a function of altitude for a ζ of 0.3.

where
$$M_{md} = \left(\frac{W/S}{0.7P_{SL}\delta}\right)^{1/2}\left(\frac{K}{C_{DO}}\right)^{1/4} = \frac{0.333}{\delta^{1/2}} \tag{8-44}$$

The far right-hand term in Eq. 8-44 is the numerical result for our illustrative turbojet. For cruise at M 0.85 (the assumed drag-rise Mach number), the pressure ratio takes on a value of 0.1535, corresponding to an altitude of approximately 44,000 ft. The associated range is 4,050 nmi (4,658 mi), which is 12 percent larger than that achieved by flying at M 0.85 at the best-range altitude of 32,500 ft. From a practical viewpoint, we must consider the fuel required to climb the additional 12,000 ft as well as whether the thrust required for cruise at the higher altitude is available. With respect to the latter, let us assume a maximum thrust-to-weight ratio at sea level of 0.25. Then at 44,000 ft, where σ is 0.2044, the maximum available thrust-to-weight ratio is approximately 0.25×0.2044 or 0.0511. However, the required cruise T/W ratio is $1/E_m$ or 0.0537, which is in excess of that available so that cruise cannot be maintained at 44,000 ft. From another point of view, the density ratio at the absolute ceiling for a sea-level T/W ratio of 0.25 is 0.215 [$1/(0.25 \times 18.634)$], which corresponds to an altitude of the order of 43,000 ft.

Assuming that we would wish to fly at the absolute ceiling, the corresponding range for this altitude for this aircraft would be somewhat less than the previously determined maximum of 4,050 nmi.

Turning to quasi-steady-state climbing flight,

$$\sin \gamma = \frac{T}{W} - \frac{D}{W} \tag{8-45}$$

and

$$\mathrm{R/C} = \frac{dh}{dt} = V \sin \gamma \tag{8-46}$$

With a parabolic drag polar and using the Mach number in place of the true airspeed, the drag-to-weight ratio is

$$\frac{D}{W} = \frac{0.7 P_{\mathrm{SL}} \delta M^2 C_{DO}}{W/S} + \frac{K(W/S)\cos^2\gamma}{0.7 P_{\mathrm{SL}} \delta M^2} \tag{8-47}$$

With the assumptions of a "small" climb angle and/or a low drag due to lift, Eq. 8-47 can be replaced, as was done before, by the level-flight drag-to-weight ratio, namely,

$$\frac{D}{W} = \frac{0.7 P_{\mathrm{SL}} \delta M^2 C_{DO}}{W/S} + \frac{K(W/S)}{0.7 P_{\mathrm{SL}} \delta M^2} \tag{8-48}$$

In terms of the Mach number, Eq. 8-46 becomes simply

$$\mathrm{R/C} = \frac{dh}{dt} = a_{\mathrm{SL}} a^* M \sin \gamma \tag{8-49}$$

where a^* in the troposphere is given by Eq. 8-26 and a_{SL} and dh/dt are in fps. For our illustrative turbojet,

$$\frac{D}{W} = 0.2424\,\delta M^2 + \frac{0.003}{\delta M^2} \tag{8-50}$$

Consider a constant-Mach number climb at M 0.6 with a sea-level thrust-to-weight ratio of 0.25. (From Eq. 4-32, the sea-level Mach number for fastest climb is 0.596, and at 30,000 ft it is 0.723.) Equation 8-50 becomes

$$\frac{D}{W} = 0.087\,\delta + \frac{0.008}{\delta}$$

Therefore,

$$\sin \gamma = 0.25\sigma - 0.087\,\delta - \frac{0.008}{\delta}$$

and

$$\mathrm{R/C} = 669.6 a^* \sin \gamma$$

At sea level, $\sigma = \delta = a^* = 1$, so that $\gamma = 8.9$ deg, and R/C = 103.8 fps = 6,227 fpm. At 30,000 ft, $\sigma = 0.374$, $\delta = 0.297$, and $a^* = 0.891$. The climb angle has dropped off to 2.33 deg, and the R/C is 24.3 fps or 1,457 fpm. It may be of interest to note that the airspeed has also decreased, from 397 kt to 354 kt. If the airspeed had been kept constant (with an increase in the Mach number to 0.67), then γ would have been 2.22 deg and the R/C would have been 26 fps, or 1,560 fpm, not much difference at all.

Using the relationships and examples of this section, any of the other expressions previously developed can be expressed in terms of the Mach number and used, if so desired.

8-5 FLIGHT AND MANEUVERING ENVELOPES

The flight regime of an aircraft comprises all the possible combinations of airspeed, altitude, and acceleration and is determined by the aerodynamic, propulsion, and structural characteristics of the aircraft. The boundaries of the flight regime are called the *flight limits*; they form the flight and maneuvering envelopes and are defined by the constraints and limitations on the performance of the aircraft. The simplest flight envelope is that of Fig. 3-3, where the level-flight (one-g) regime is defined only by the solution of the level-flight drag equation without consideration of any constraints, such as the stall speed on the low-speed boundary or the drag-rise Mach number on the high-speed boundary. Figure 6-3 is another such simple flight envelope but with the stall speed constraint superimposed. In this section, we shall look more closely at other operational constraints; namely, buffet limits, V-n diagrams, and gust envelopes.

Let us first consider *buffeting*, which is an objectionable shaking of some part of the aircraft (usually part of the wing or horizontal stabilizer) caused by the turbulence arising from the separation of the airflow from the surface of the aircraft. At low speeds, buffeting occurs as the stall is approached and the low-speed buffet limit is essentially established by the value of the maximum lift coefficient; decreasing the airspeed aggravates the buffeting. Since buffeting is caused by the turbulence accompanying airflow separation and since, with the proper set of conditions, separation can occur at any airspeed, buffeting can occur at high speeds as well as at low speeds but at lower lift coefficients and angles of attack. When high-speed buffeting starts, increasing the airspeed aggravates the condition.

The determination of the actual buffet limits is beyond the scope of this book and generally requires intensive flight testing over the complete airspeed range of a particular type of aircraft. In these flight tests, at low speeds the aircraft is accelerated to buffet in a banked turn or pull-up, and at high speeds, if buffet does not occur in level flight, the aircraft is dived until buffeting starts. The *buffet limits* are plotted on a lift coefficient versus Mach number plot, such as the hypothetical and arbitrary one in Fig. 8-8.

The buffet plot is enhanced by superimposing lines of constant $(W/S)/\delta$, which can be calculated from Eq. 8-34a, and which is rewritten here for convenience as

$$\frac{W/S}{\delta}=0.7P_{SL}M^2C_L=1{,}481M^2C_L \tag{8-51}$$

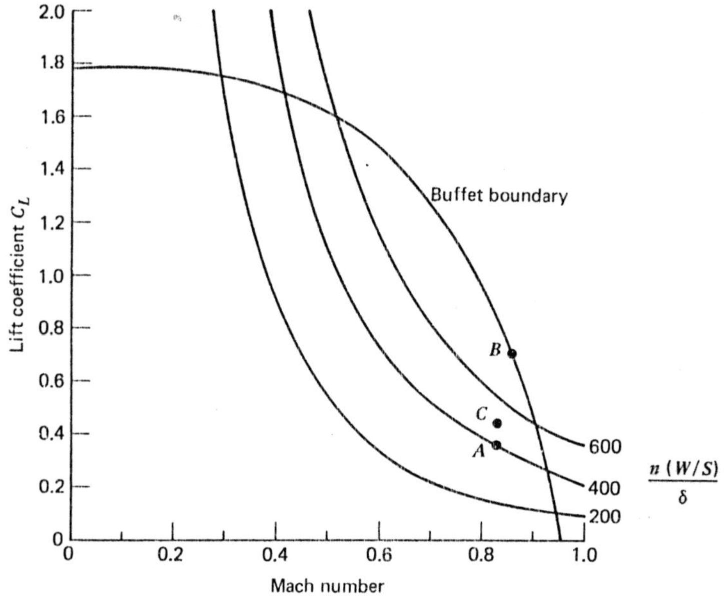

FIGURE 8-8

Buffet limits diagram of C_L versus M as a function of $n(W/S)/\delta$.

One obvious use of such a plot is the determination of whether or not a desired cruise Mach number and altitude combination falls within the acceptable flight regime. Using the illustrative turbojet of Sec. 8-4 (which has a wing loading of 110 lb/ft^2 and the parabolic drag polar $C_D = 0.018 + 0.04C_L^2$), the best-range lift coefficient is found to be 0.387. If the desired best-range Mach number is to be 0.84, then $(W/S)/\delta$ has a value of the order of 400 lb/ft^2, and the cruising condition is shown as point A in Fig. 8-8. Point A is within the acceptable flight regime but is in the general vicinity of the buffet boundary. If the altitude is held constant and the airspeed is increased, the lift coefficient will decrease and the operating point will move along the 400 lb/ft^2 line toward the buffet boundary. Similarly, if the airspeed is held constant and the altitude is decreased, the operating point will move up vertically toward the boundary. The effects of turning flight will be examined in a subsequent paragraph.

Figure 8-8 can also be used to determine and examine other level-flight and cruise conditions. For example, if the illustrative jet is to cruise at E_{max} ($C_L = 0.67$) and M 0.84, then $(W/S)/\delta$ will be 700 lb/ft^2 and the operating point will be located at point B, which happens to be on the buffet boundary. Any increase in speed will induce buffeting. Obviously, point B would not be a desirable flight condition.

As developed in the preceding paragraphs, points A and B represent nonturning, one-g flight conditions. Introducing the load factor ($n = L/W$), Eq. 8-51 can be written as

$$\frac{L/S}{\delta}=\frac{n(W/S)}{\delta}=0.7P_{SL}M^2C_L=1{,}481M^2C_L \tag{8-52}$$

so that the superimposed lines in Fig. 8-8 really represent constant values of $n(W/S)/\delta$ rather than simply $(W/S)/\delta$. Consequently, plots such as Fig. 8-8 can be used to determine the maneuverability of an aircraft with respect to buffeting. For example, assume that an aircraft is cruising at point A in one-g flight and enters a turn with a 30 deg bank, holding airspeed and altitude constant. Since n in the turn is equal to 1.155 g's ($n=1/\cos\phi$), then, from Eq. 8-52, C_L in the turn will be 0.447, and the corresponding flight condition will be at point C, still within the acceptable flight regime but somewhat closer to the buffet boundary.

Again starting with cruise at point A, the maximum bank angle or maximum number of g's that can be sustained without buffeting can be found by locating the point on the boundary for the same Mach number (point B) and then using an expression developed from Eq. 8-52, namely,

$$\frac{n_B}{n_A}=\frac{\delta_B C_{LB}}{\delta_A C_{LA}} \tag{8-53}$$

If the altitude is held constant, then $\delta_A=\delta_B$, and

$$\frac{n_B}{n_A}=\frac{C_{LB}}{C_{LA}} \tag{8-54}$$

But n_A is equal to unity in this case so that $n_B=0.67/0.387=1.73$ g's, which corresponds to a level-turn bank angle of 54.7 deg. Equations 8-52 and 8-53, in these and other forms, can also be used to compare and investigate other flight conditions.

In addition to buffet limits and boundaries, each aircraft type has a structural design speed called the *never-exceed airspeed* V_{NE}, which is often referred to as the *redline airspeed* since the airspeed indicator is marked with a red line at the never-exceed airspeed. Since it is actually the dynamic pressure that determines the structural loading, V_{NE} for high-speed aircraft is often specified as a function of altitude (or as a calibrated rather than a true airspeed) or replaced by a value for the maximum allowable dynamic pressure, q_{max}. High-speed aircraft may also have a designated *design dive speed* V_D, which should not be exceeded in the event the aircraft is inadvertently or intentionally dived. These limiting airspeeds are determined by structural considerations and are prescribed to preclude failure, in other words, to keep the aircraft from breaking apart.

A very useful plot is the *V-n (or V-g) diagram*, sometimes referred to as the *maneuvering envelope*. It is defined by the stall characteristics and by the prescribed maximum allowable load factors, which have been mentioned previously in Secs. 4-4 and 5-3. The *limit load factor* is the published maximum allowable n and is not to be exceeded by the pilot. The *ultimate load factor* is generally 50 percent higher, is not published, and provides a margin of safety for any violation of the limit load factor.

A *V-n* diagram is calculated for a given weight (wing loading) and altitude by calculating the stall speed for various load factors from the expression for the stall speed, i.e.,

$$V_s = \left[\frac{2n(W/S)}{\rho_{SL} \sigma C_{L_{max}}} \right]^{1/2} \tag{8-55}$$

Figure 8-9 is a simple *V-n* diagram for our illustrative turbojet with a wing loading of 110 lb/ft^2, a never-exceed airspeed of 550 mph (807 fps), and limit load factors of +3.5 and −1.5 *g*'s. The maximum lift coefficient for positive angles of attack has been assumed to be 1.8 and for negative angles of attack to be 1.2. If the two maximum values were equal, then the curves would be symmetrical with respect to the zero-*g* line.

The right-hand boundary of the maneuvering envelope is defined by the never-exceed airspeed or by the design dive speed and the left-hand boundary by the one-*g* stall speed. The upper and lower boundaries are defined by the limit load factors. The remaining curved portions of the boundaries represent the stall speeds for various load factors. Flight above the positive curved portions and below the negative curved portions is aerodynamically impossible. Point *A* has a particular significance in that the corresponding airspeed is called the *maneuver speed*, is almost always given the symbol V_A, and is defined as

$$V_A = n_{max}^{1/2} V_S \tag{8-56}$$

where V_s is the wings-level (one-*g*) stall speed and n_{max} is the positive limit factor. In this particular case, V_A, whether determined from Fig. 8-9 or calculated from either Eq. 8-55 or Eq. 8-56, is 289.2 mph. If the aircraft is flying at this airspeed, or slower, and is subjected to a load factor in excess of 3.5 *g*'s, it will stall and the structure will not be stressed beyond the limit load factor. If, however, the aircraft is flying in excess of the maneuvering airspeed, say at 325 mph, then 4.5 *g*'s (point *B*) could be experienced before the aircraft would stall. At 400 mph, the aerodynamically possible load factor exceeds the ultimate load factor with the strong probability of structural failure.

Aircraft normally fly at airspeeds in the vicinity of the *design cruise airspeed* V_c or of the *design normal operating airspeed* V_{NO}. Flight and operating manuals call for slowing the aircraft down to the maneuvering airspeed V_A when encountering or entering areas of severe turbulence. Although flying slower than V_A would reduce the maximum possible load factor, it would affect the control of the aircraft and degrade the ability to maintain a desired altitude. Equations 8-55 and 8-56 show that the maneuvering airspeed increases as the wing loading (weight) is increased and also increases with altitude. The rest of the *V-n* diagram also changes with changes in weight or altitude. With respect to changes in altitude, expressing the airspeed in Fig. 8-9 as the calibrated airspeed rather than as the true airspeed will eliminate the need to redo the *V-n* diagram for different altitudes.

In addition to satisfying the constraints of the *V-n* diagram, the structural design of an aircraft must withstand the load factors within the boundaries of a *gust envelope*, which is based on the assumption that the aircraft penetrates a sharp-

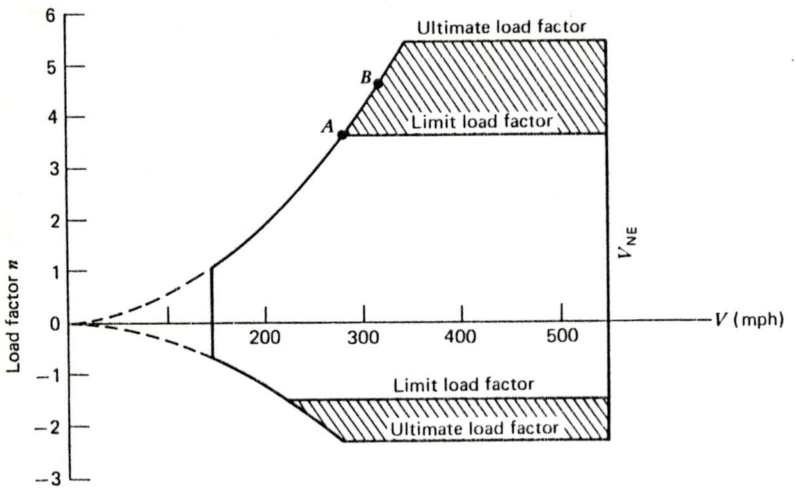

FIGURE 8-9
***V-n* diagram for illustrative turbojet.**

edged vertical gust that has a vertical velocity of U fps. If an aircraft in level one-g flight enters a vertical gust, the angle of attack will suddenly be changed by an increment $\Delta\alpha$ before either the wing begins to move or the airspeed begins to respond to the gust. The change in the angle of attack is equal to

$$\Delta\alpha = \arc\tan\left(\frac{U}{V}\right) \cong \frac{U}{V} \tag{8-57}$$

where V is the true airspeed in fps, so that the resulting change in the lift of the wing can be written as

$$\Delta L = \tfrac{1}{2}\rho V^2 S a_w \left(\frac{U}{V}\right) \tag{8-58}$$

where a_w, which we have encountered before, is the slope of the wing lift curve expressed per radian. The corresponding instantaneous change in the load factor due to this gust is

$$\Delta n = \frac{\Delta L}{W} = \frac{\rho V a_w U}{2(W/S)} \tag{8-59}$$

and since the aircraft is initially in one-g flight, the total load factor becomes

$$n = 1 \pm \frac{\rho V a_w U}{2(W/S)} \tag{8-60}$$

where the minus sign refers to negative or downward gusts.

Since a vertical gust is not truly sharp-edged and since it has been found that the size and wing loading of an aircraft have an effect on the aircraft's response, the gust velocity is multiplied by a modifying factor K_g, referred to as the *gust alleviation factor*. Introducing K_g and conforming to the standard practice of the Federal Air Regulations (FAR) of expressing the airspeed V as the equivalent airspeed in knots while keeping the gust velocity U in fps, Eq. 8-60 becomes

$$n = 1 \pm \frac{K_g V a_w U}{498(W/S)} \tag{8-61}$$

The gust alleviation factor K_g has been empirically defined as

$$K_g = \frac{0.88\mu_g}{5.3 + \mu_g} \tag{8-62}$$

where μ_g, the aircraft mass ratio, is

$$\mu_g = \frac{2(W/S)}{\rho g \bar{c} a_w} \tag{8-63}$$

where $\bar{c}$ is the mean aerodynamic chord of the wing.

These equations lead to two interesting observations. The first is that the faster the aircraft is flying, the larger the gust load factor is and the rougher the ride will be. On the other hand, the larger the wing loading, the smaller the gust factor will be and the smoother the ride.

The gust envelope is constructed in accordance with the criteria established in the FARs, which specify combinations of gust velocities, altitudes, and airspeeds. A typical gust envelope would be of the general shape and form shown in Fig. 8-10, where V_B is called the *design speed for maximum gust intensity* and is defined as the one-g stall speed multiplied by the square root of the limit load

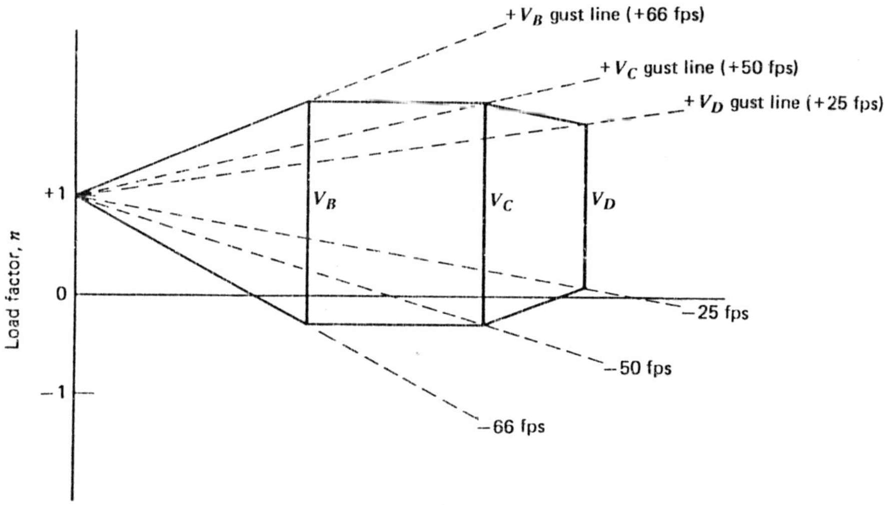

FIGURE 8-10

A typical gust envelope (not to scale).

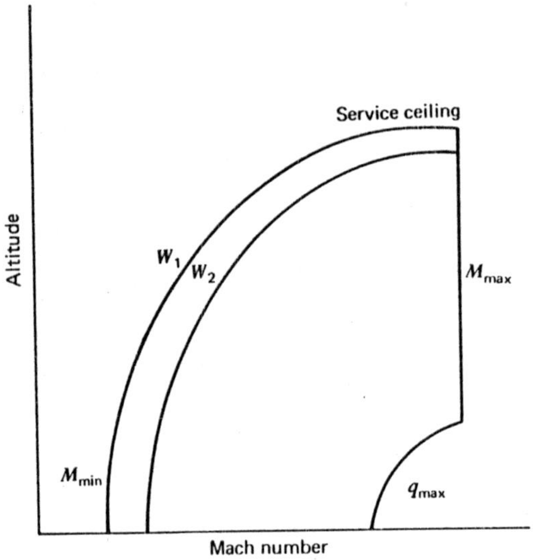

FIGURE 8-11

A typical flight envelope with constraints and with $W_2 > W_1$.

factor at V_c. The combination of the gust envelope and the V-n diagram establishes the acceptable operating regime of an aircraft with respect to its structural integrity. The gust envelope is often superimposed on the V-n diagram to form a single composite maneuvering envelope.

Let us now return to the flight envelope, which is a plot of altitude versus either airspeed or Mach number. If the upper boundary is taken as the more practical value of the service ceiling (100-fpm climb) rather than that of the absolute ceiling, a typical flight envelope would have the general shape shown in Fig. 8-11. Note that increasing the weight (wing loading) reduces the area of the possible flight regime. This is essentially a one-g flight envelope and must be used in conjunction with the composite gust and V-n diagram.

There is an additional flight constraint that must be considered for supersonic and hypersonic flight and that is the temperature or thermal limitations of the aircraft structure and subsystems arising from the aerodynamic heating of the aircraft structures.

8-6 THE ENERGY-STATE APPROXIMATION

The quasi-steady-state approach that we have used in our performance and design analyses is quite effective for most subsonic aircraft. However, for many supersonic and high-performance aircraft in accelerated flight, particularly in climbs and turns, the steady-state approximation is not adequate for determining best or optimal performance. For such aircraft, one approach is to obtain numerical, computer solutions of the complete nonlinear differential equations, applying the calculus of variations to find the optimal trajectories. A simpler and more satisfying, albeit less precise, approach is to describe the state of the aircraft in terms of its total energy and then use a quasi-steady-state approximation that is based on the energy of the aircraft rather than on its airspeed. This is the energy-state approximation that will be discussed in this section, but not in great detail.

The basic precepts of the *energy-state approximation* are to consider the aircraft to be a point mass, introduce the concepts of energy height and specific excess power, and to fly so as to maximize the power in changing from one energy state to another. A few definitions and relationships are in order before applying the energy-state approximation. The total energy of an aircraft comprises its potential energy (due to altitude), its kinetic energy (due to airspeed), and its rotational energy (due to pull-ups, push-overs, and rolls). By considering the aircraft to be a point mass, we are neglecting the rotational energy, so that the energy state of an aircraft is now determined solely by its altitude and airspeed. The "approximate" total energy of the aircraft is given by

$$E = Wh + \frac{WV^2}{2g} \tag{8-64}$$

where E is the total energy (and *not* the lift-to-drag ratio), W is the total weight of the aircraft, h is the true altitude above sea level, and V is the true airspeed. The first term on the right is the potential energy, and the second term is the kinetic energy.

The specific energy (energy per pound of aircraft weight) is defined as the *energy height*, given the symbol h_e, and expressed as

$$h_e = \frac{E}{W} = h + \frac{V^2}{2g} \tag{8-65}$$

The energy height has the units of feet and is determined by specifying the altitude and airspeed of the aircraft.

Figure 8-12 is a typical plot of some constant-energy curves as a function of altitude and airspeed. An aircraft at point A, at an altitude of 16,118 ft and with an airspeed of 500 fps, has an energy height of 20,000 ft, which is theoretically the maximum altitude it could reach by zooming along the energy height curve until it reached an airspeed of zero. Diving along the constant-energy curve would result in an airspeed of 1,135 fps at crash, theoretically, of course.

Let us suppose that point A represents one point on the climb path of the aircraft to a cruise Mach number of 0.8 at 35,000 ft, which is point B, which has an energy height of 44,390 ft. There are obviously many flight paths (trajectories) that could be followed in changing from A to B. One possible trajectory, as shown, entails a constant-airspeed climb to point C and then a constant-altitude acceleration to M 0.8 at point B. No matter which climb schedule is used, there will be a net change in the energy height of 24,390 ft.

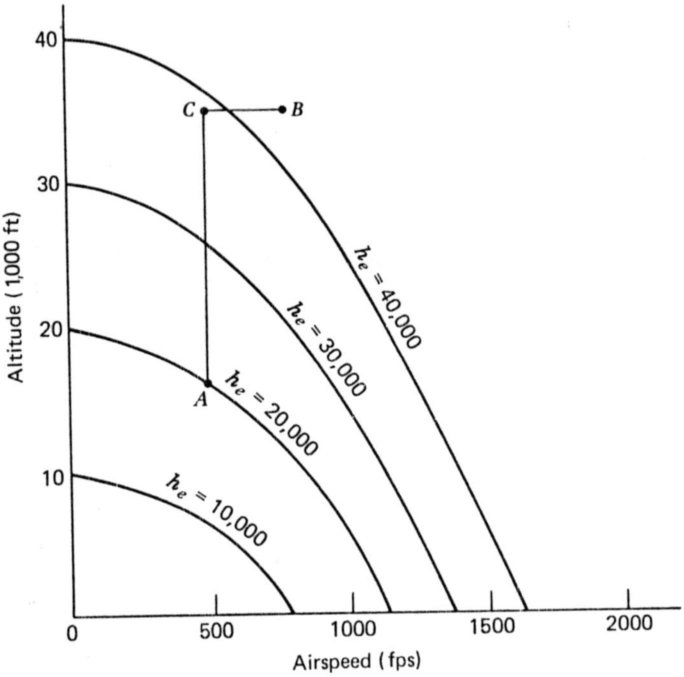

FIGURE 8-12
Constant energy height contours as a function of airspeed and altitude.

It is the rate of change of the energy height that determines the climb path and time to climb. The rate of change of the energy height is called the *specific power*, given the symbol P_s, and is found by differentiating Eq. 8-65 with respect to time to obtain

$$P_s = \frac{dh_e}{dt} = \frac{dh}{dt} + \frac{V}{g}\frac{dV}{dt} \tag{8-66}$$

From Eq. 8-66, we see that the specific power is simply the sum of the rate of climb and the acceleration of the aircraft along the flight path.

The specific power can be related to the aircraft characteristics and the flight parameters through the force equation in the direction of flight, which is

$$T - D - W\sin\gamma = \frac{W}{g}\frac{dV}{dt} \tag{8-67}$$

which assumes that the thrust and velocity vectors are coincident.

Rearranging Eq. 8-67 yields the expressions:

$$\sin\gamma + \frac{1}{g}\frac{dV}{dt} = \frac{T-D}{W} \tag{8-68}$$

$$V \sin \gamma + \frac{V}{g}\frac{dV}{dt} = \frac{V(T-D)}{W} \tag{8-69}$$

However, $V \sin \gamma$ is the rate of climb, so that

$$P_s = \frac{dh_e}{dt} = \frac{dh}{dt} + \frac{V}{g}\frac{dV}{dt} = \frac{V(T-D)}{W} = \frac{TV-DV}{W} \tag{8-70}$$

where the far right-hand term represents the excess specific power of the aircraft that is available for climb and acceleration. The term *specific excess power* (or excess specific power) is often used in describing P_s or dh_e/dt, along with the simpler term, specific power. Note that with the steady-state assumption of a constant airspeed the excess power is used for climbing only and that Eq. 8-70 reduces to Eq. 4-16.

The value of the specific power determines the maneuvering capability of an aircraft at any point in its flight regime. In the event that you may wish to compare the energy-state approximation with the quasi-steady-state approximation, let us use the illustrative turbojet of Chap. 4, which has a maximum sea-level T/W ratio of 0.25, a wing loading of 100 lb/ft^2, and the parabolic drag polar $C_D = 0.015 + 0.06C_L^2$. If it is flying at 30,000 ft at M 0.8 with maximum thrust and with a load factor of unity, then $V = 795.5$ fps, $q = 281.3$ lb/ft^2, $C_L = 0.356$, $C_D = 0.0226$, so that

$$P_s = \frac{dh_e}{dt} = 795.5\left(0.25 \times 0.374 - \frac{281.3 \times 0.0226}{100}\right)$$
$$= 23.8 \text{ fps}$$

This positive value of the excess specific power can be used to climb or to accelerate or for an appropriate combination of the two. If, for example, the pilot chooses to climb at a constant airspeed of M 0.8, his rate of climb leaving 30,000 ft would be 24 fps. If, on the other hand, he levels off at 30,000 ft, with the throttle setting unchanged, the aircraft would have an instantaneous acceleration of

$$\frac{dV}{dt} = \frac{gP_s}{V} = \frac{32.2 \times 24}{777.6} = 0.99 \text{ ft/s}^2$$

If the pilot chooses to cruise at 30,000 ft at M 0.8, he would have to throttle back until P_s becomes equal to zero.

When contours of constant P_s are plotted on an altitude-airspeed diagram, the thrust (or T/W), the weight (or W/S), and the load factor must be specified. Figure 8-13 shows the form and shape of such a plot for a typical subsonic aircraft with a parabolic drag polar and with the assumption that the thrust variation with altitude is directly proportional to the density ratio. The P_s equal to zero contour is the level-flight (one-g) flight envelope discussed in the first paragraph of Sec. 8-5 and sketched in Fig. 3-3. Note that Fig. 3-3 also shows the effects of a more realistic thrust variation on the shape of the flight envelope.

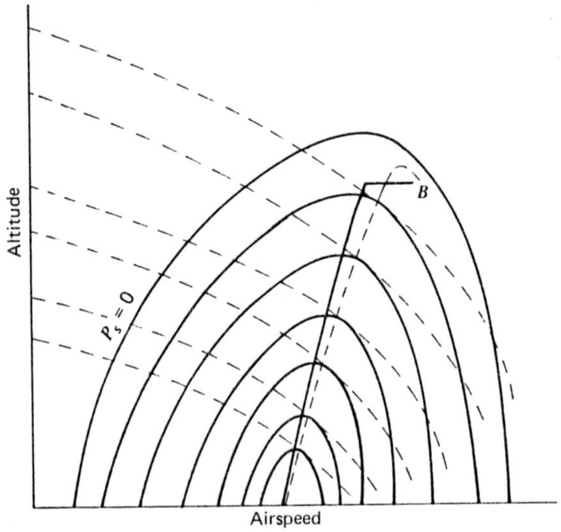

FIGURE 8-13

Maximum rate of climb and maximum-energy climb using excess specific power and energy height contours.

Figure 8-13 can be used to determine a climb schedule (say from sea level to point B) in terms of the airspeed as a function of the actual altitude h. The "maximum rate-of-climb" schedule is determined by maximizing the excess specific power at each altitude, i.e., by flying through the points on the P_s curves that are tangent to the lines of constant altitude. This climb schedule is shown in Fig. 8-12 and is the quasi-steady-state fastest climb schedule of Sec. 4-4.

A somewhat faster "fastest" climb schedule is the *maximum-energy climb*, which is found by maximizing the excess specific power as a function of the energy height; i.e., it is the locus of the points of tangency of the P contours and the curves of constant energy height. It, too, is shown in Fig. 8-13 and does not differ significantly from the steady-state climb schedule because the thrust and drag are reasonably well behaved for such an aircraft.

A supersonic aircraft, however, particularly one whose excess thrust is marginal when passing through the high-drag transonic region, can have significantly different climb schedules. It is for such aircraft that the energy-state methods are useful in determining how to climb to a predetermined cruising altitude and airspeed. A typical, but skeletonized, energy plot for such an aircraft has the form shown in Fig. 8-14. Note the notches in the contours, leading to discontinuities in the contours in the transonic region and to closed contours in the supersonic region. If supersonic cruise is to start at point A, the steady-state solution, as shown in Fig. 8-13, is a steady-state "fastest" climb to the cruising altitude followed by a constant–altitude acceleration to the cruising Mach number. The maximum-energy (minimum-time) climb starts out resembling the steady-state solution

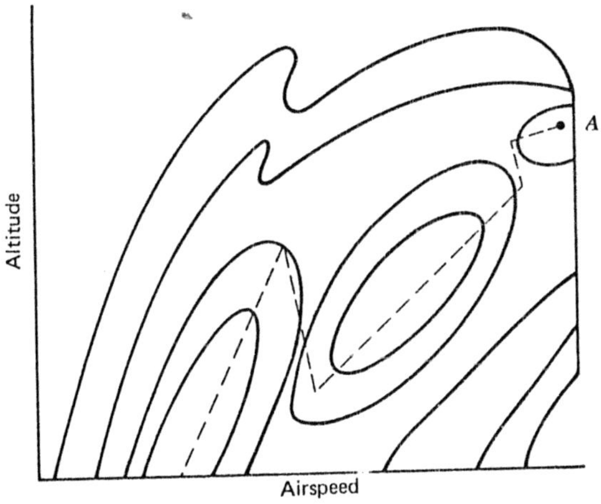

FIGURE 8-14

Maximum-energy climb of a supersonic aircraft with marginal excess transonic thrust.

until the transonic region is approached. At an appropriate point, the aircraft dives along a constant energy height curve to the start of the supersonic climb schedule, which in this example comprises a climb to the next discontinuity, a constant h_e dive, and then another climb. Although this climb schedule is much more difficult to fly, the savings in flight time can be considerable, in some cases as much as 50 percent.

The basic rule for such a minimum-time climb schedule is to fly at all times toward the highest-value P_s contour without decreasing the energy height. When an h_e contour is reached that is tangent to two equal-valued P_s contours (at a discontinuity), the aircraft is put into a constant h_e dive until "normal" climb can be resumed. A zero-g dive yields the most rapid increase in P_s since the drag due to lift is zero and the total drag of the aircraft is at a minimum. Furthermore, in a dive, the available thrust increases as the altitude decreases, leading to an increase in P_s. It should be noted that in an actual climb, pull-outs and nose-overs will decrease the energy height because of the rotational energy required and will modify the flight path somewhat.

The SR-71, which has a high thrust-to-weight ratio, has only one set of closed contours in the supersonic region. Consequently, its climb schedule is a sea-level acceleration to the initial climb Mach number, a subsonic climb at essentially a constant Mach number, a zero-g push-over and dive to the initial supersonic climb Mach number, and then a pull-out followed by an accelerated climb to the cruise altitude and Mach number. The SR-71 pilots call the push-over, dive, and pull-out maneuver the "Dipsy Doodle."

As the thrust-to-weight ratios of supersonic aircraft are further increased, the values of P_s for a given airspeed and altitude will obviously increase and eventually the discontinuities in the transonic region and the closed contours in the supersonic region will disappear. When this happens, the energy diagrams will resemble those of a subsonic aircraft and the maximum-energy climb schedule will also take on the shape of a steady-state climb schedule, as sketched in Fig. 8-15.

The instantaneous rate of climb with acceleration can be found by rearranging Eq. 8-70 into the form

$$\frac{dh}{dt}=\frac{P_s}{1+(V/g)(dV/dh)} \tag{8-71}$$

Compare Eq. 8-71 with the acceleration correction factor of Eq. 4-56.

The time to move from one energy height to another can be found from

$$\Delta t=\int_1^2 \frac{dh_e}{dh_e/dt}=\int_1^2 \frac{dh_e}{P_s} \tag{8-72}$$

By constructing a plot of $1/P_s$ versus h_e for various points along a given climb schedule, a graphical integration will yield the approximate total climb time (the time spent along constant h_e curves is not accounted for).

The energy-state approximation can also be used to determine the minimum-fuel climb schedule by establishing the exchange ratio

$$\frac{dh_e}{dW_f}=\frac{dh_e/dt}{dW_f/dt}=\frac{P_s}{cT} \tag{8-73}$$

where c is the tsfc and T is the thrust. The techniques are the same as those for minimum-time climb with the P_s contours replaced by constant (P_s/cT) contours

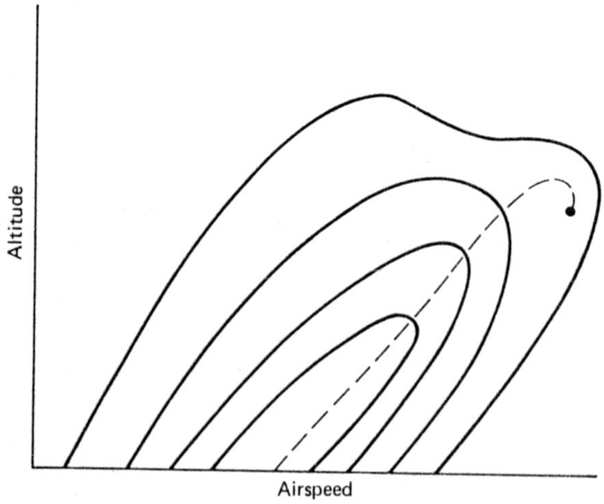

FIGURE 8-15

Maximum-energy climb of a supersonic aircraft with sufficient excess transonic thrust.

and using the expression that

$$\Delta W_f = \int_1^2 \frac{dh_e}{dh_e/dW_f} = \int_1^2 \frac{cT}{P_s}\, dh_e \tag{8-74}$$

The excess specific power is also a function of the load factor, as can be easily seen by assuming a parabolic drag polar and by substituting Eq. 5-14 into Eq. 8-70 to obtain

$$P_s = V\left[\frac{T}{W} - \frac{qC_{DO}}{W/S} - \frac{Kn^2(W/S)}{q}\right] \tag{8-75}$$

which reduces back to Eq. 8-70 when the load factor is unity.

The energy and specific power diagrams of the preceding paragraphs were developed for a 1-g flight. Maneuvering flight can be examined by constructing similar diagrams for other values of n. As the load factor is increased, the corresponding values of P_s will decrease, thus contracting the flight envelope. The zero-P_s contour defines the constant-altitude, constant-airspeed turning performance at that particular load factor and can be related to the steady-state turning analyses of Chap. 5. The P_s equal to zero condition defines the best sustained (steady-state) turning performance at a constant altitude and specified load factor. However, finite values of P_s, can result in better turning performance, either in a dive or a climb, or momentarily if the altitude is held constant.

The energy-power diagrams are very useful in the comparative evaluation of the maneuvering capabilities of two or more different aircraft, with particular reference to air-to-air combat. By overlaying the energy-power diagrams for identical load factors for two aircraft, the regions of superiority and inferiority become readily apparent. In the final determination of the operating flight regime of any aircraft, it is also necessary to include any structural and thermal constraints.

PROBLEMS

In many of the problems in this chapter, reference will be made to various aircraft designated by initials. Aircraft A, B, and C were originally treated as pure turbojets and their major characteristics are listed in Chap. 3 at the beginning of the problems section. Aircraft D and E were originally treated as piston-props, and their major characteristics are listed in Chap. 6 at the beginning of the problems section.

8-1. If a turboprop engine has a shaft horsepower output of 800 shp and a jet thrust of 300 lb, a design airspeed of 300 mph, and a propeller efficiency of 85 percent:

a. What is the equivalent shaft horsepower (eshp)?

b. What would be the equivalent thrust if we wished to express the engine output in such a manner?

c. If the fuel consumption rate at 300 mph is 541 lb/h, what is the equivalent shaft horsepower specific fuel consumption (eshp sfc)?

d. Redo (c) to determine the equivalent thrust specific fuel consumption.

8-2. Do Prob. 8-1 for a design airspeed of 500 mph.

8-3. Do Prob. 8-1 for a design airspeed of 150 mph.

8-4. If a 670 eshp turboprop engine is derated to a sea-level value of 400 eshp, what is the density ratio and altitude at the critical altitude?

8-5. If Aircraft *E* is reequipped with two turboprop engines with a total of 1,200 eshp derated to 660 eshp, the reduction in the engine weight is reflected as a reduction of the gross weight of the aircraft. The wing area (and weight) is reduced to 189 ft^2 so as to maintain the original wing loading. With the gross weight at 6,600 lb:

a. Find the critical altitude of the engines.

b. Find the absolute ceiling and the ceiling airspeed with the new engines and weight.

c. Find the absolute ceiling and airspeed with the original configuration and engines and compare results.

8-6. A turbofan engine has a sea-level jet thrust output of 20,000 lb but is described as having a sea-level thrust of 35,000 lb and a tsfc of 0.65 lb/h/lb at *M* 0.8. Assume a fan efficiency of 85 percent.

a. What is the maximum sea-level shaft power output?

b. Find the equivalent thrust (pounds) and equivalent tsfc at 375 mph.

c. Do (b) above for 150 mph.

8-7. Aircraft *C* is to cruise at 35,000 ft with 150,000 lb of fuel available for cruise. Find the best-range airspeed (mph and *M*), the maximum range, and the equivalent specific fuel consumption for the engine configurations listed below. Do not concern yourself with any changes in the gross weight arising from the engine changes. Assume a fan efficiency of 85 percent.

a. Pure turbojet engines, that is, $k = 3$.

b. Medium bypass ratio turbofans with a k of 2.5.

c. High bypass ratio turbofans with a k of 2.0.

d. Turboprops with a k of 1.3 and a propeller efficiency of 80 percent.

e. Piston-props with a k of 1 and a propeller efficiency of 80 percent.

8-8. Aircraft *C* is retrofitted with high bypass ratio turbofans, and the specifications quote a best-range Mach number of *M* 0.8 at 35,000 ft and a tsfc of 0.65 lb/h/lb.

a. Find the best-range lift-to-drag ratio and the value of k, the engine descriptor.

b. If the available fuel for cruise is 125,000 lb, find the range (miles).

c. As an exercise only, describe these turbofans as though they were turboprops, finding an eshp sfc and the appropriate value for k. Also find the range for two values of the "propeller" efficiency namely, 85 and 80 percent.

8-9. Find three values for the sonic ratio at each of the following altitudes, using Table A-1 and Eq. 8-26:

a. 10,000 ft b. 20,000 ft

c. 30,000 ft d. 35,000 ft

8-10. For Aircraft B, in straight and level flight, use Mach number descriptions to:

a. Find the lift and drag coefficients and lift-to-drag ratio at 10,000 ft and M 0.5, 20,000 ft and M 0.6, and 30,000 ft and M 0.7.

b. Find the best-range Mach number at 10,000 ft, 20,000 ft, and 30,000 ft.

c. Find the best-range T/W ratio at 10,000 ft, 20,000 ft, and 30,000 ft.

8-11. Aircraft C is to cruise-climb at a best-range Mach number of M 0.82.

a. Find the initial cruise altitude and the best-range lift-to-drag ratio.

b. If the fuel available for cruise is 155,000 lb, find the maximum range (nmi and mi) and the final altitude.

c. Find the initial and final values of both the required and available thrust? Is there adequate thrust for this program?

8-12. Do Prob. 8-11 with M 0.82 as a minimum-drag (maximum lift-to-drag ratio) Mach number rather than as a best-range Mach number and compare the range results.

8-13. Aircraft A is flying at 30,000 ft at M 0.6 with a gross weight of 22,000 lb.

a. Find the lift and drag coefficients and the lift-to-drag ratio.

b. Find the required and available T/W ratios at the start and end of cruise.

c. Find the cruise-climb range (nmi), best-range airspeed (kt), and the final altitude with 3,000 lb of available cruise fuel.

8-14. Aircraft B is in a constant-Mach number climb at M 0.55 from sea level to 25,000 ft.

a. Find the rate of climb (fpm) and airspeed (kt and mph) at sea level, 15,000 ft, and 25,000 ft.

b. Using average values, find the time to climb from sea level to 15,000 ft, from 15,000 ft to 25,000 ft, and from sea level to 25,000 ft.

8-15. Aircraft C has a buffet limit boundary which, for tutorial uses only, can be represented by the expression

$$C_{LB} = 2.0 - 2.33M^3$$

where C_{LB} is the buffet limit lift coefficient.

a. For a Mach number of 0.85, what is the buffet limit lift coefficient?

b. If the aircraft is in straight and level flight at 35,000 ft, find the lift coefficient and determine whether this operating point is within the buffet limit and thus an acceptable flight point.

c. Again for straight and level flight at 35,000 ft, at what Mach number will the aircraft begin to buffet?

d. How many g's can this aircraft pull in a steady turn at 35,000 ft at a $M\,0.85$ without buffeting? What is the corresponding bank angle?

e. If the Mach number is held constant at 0.85 and the altitude is increased, at what altitude will the aircraft begin to buffet?

f. If the altitude is held constant at 35,000 ft and the aircraft is slowed down, at what Mach number will buffeting start? Is this buffet speed above or below the stall speed?

8-16. The buffet limits for Aircraft B can be represented by the expression

$$C_{LB} = 1.9 - 2.5M^3$$

Do Prob. 8-15 for an altitude of 30,000 ft.

8-17. Aircraft E has a never-exceed *calibrated* airspeed of 350 mph, i.e., a maximum dynamic pressure of 313 lb/ft^2. The limit load factors are $+3.8$ g's and -1.7 g's, respectively, with a maximum lift coefficient for negative angles of attack of 1.4. Construct a V-n diagram with the ultimate load factors equal to 1.5 times the limit load factors with V as CAS. What is the maneuver speed?

8-18. Aircraft B is restricted to a maximum dynamic pressure of 200 lb/ft^2 and to limit load factors of $+3.5$ g's and -1.5 g's. The maximum lift coefficient for negative angles of attack is 2.0.

a. Construct a simple V-n diagram with the ultimate load factors equal to 1.5 times the limit load factors and with the airspeed expressed as a calibrated airspeed.

b. Find the maneuver speed from the V-n diagram and compare it with the calculated value.

c. What is the maneuver speed if the wing loading is increased by 25 percent?

d. Calculate the gust load factor for a $+66$ fps gust encountered while cruising at a Mach number of $M\,0.75$ at 30,000 ft. The wing sweep angle is 20 deg and $e = 0.8$.

8-19. Aircraft A is cruising at 20,000 ft at a constant best-range airspeed.

a. What is the best-range airspeed (fps and mph)? What are the values of the energy height (ft) and of the specific power (fps)?

b. The available thrust is suddenly increased to its maximum value. What is the instantaneous value of the excess specific power (fps) prior to any change in the airspeed or any action by the pilot.

c. If the pilot decides to use the excess specific power to start climbing without any change in the airspeed, what will the initial rate of climb (fpm) be?

d. If the pilot decides to accelerate to a higher cruise airspeed, what will the initial acceleration (fps^2) be?

e. If the pilot decides to dissipate all of the excess specific power in a steady turn while maintaining the original airspeed, what will the load factor and the bank angle be? Find the associated turning rate (deg/s) and the turning radius (ft).

8-20. Aircraft C is cruising at M 0.82 at 35,000 ft. Do Prob. 8-19.

Chapter Nine

Figures of Merit for Selection and Design

9-1 INTRODUCTION

The objectives of this chapter are to collect in one place the key relationships that define "best" performance and then simplify them to yield "figure of merit" (FOM) expressions that can be used to compare the performance of two or more aircraft (the selection process) or to determine the effects of a change in one or more of the physical characteristics of an aircraft upon its performance (the design process). These FOM expressions will use the basic physical characteristics and dimensions of an aircraft in an explicit and direct manner rather than in such familiar groupings as the lift-to-drag ratio and the best-range airspeed.

There will generally be two FOM expressions written for each performance measure of interest. The first one will use familiar and fundamental relationships, such as the wing loading, aspect ratio, and thrust-to-weight ratio, whereas the second will use individual and specific characteristics, such as the weight, wing area, wing span, and thrust. The differences in the two formats will be obvious when seen. The first format will probably be more useful for selection and the second more useful for design. Let us not forget that the word "design" as used in this book refers only to conceptual or feasibility design.

In comparing or discussing the modification of existing aircraft, the sources of data include *Jane's All the Aircraft of the World*; aviation magazines such as "Flying," "Aviation Week and Space Technology," and "Flight International"; and specifications and operating manuals provided by the manufacturers. Such sources do not normally give values for the zero-lift drag coefficient C_{D0}, the Oswald span efficiency e, the specific fuel consumption c, or $\hat{c}$, or the propeller efficiency, η_p. Since we need these values, we must either back-calculate them from given data or estimate them, taking into consideration the type of aircraft being examined. Since similar type aircraft normally have values of the same order of magnitude for each of these parameters, estimation is usually adequate. If an aircraft under consideration is of a radical design or configuration, back-calculation may be necessary.

The other data that we need are usually available and include:

1. The maximum gross weight W (lb)
2. The wing area S (ft^2)
3. The wing span b (ft)

4. The maximum sea-level thrust T (lb), or horsepower, HP (hp), or equivalent shaft horsepower, ESHP (eshp)
5. The operational empty weight, OEW (lb)
6. The maximum payload weight W_{PL} (lb), or the maximum number of passengers and cargo
7. The maximum fuel weight W_f (lb), or maximum fuel capacity (gal)
8. The stall speed V_s (mph), preferably with full flaps and gear down
9. The service ceiling h_c (ft); not really needed but, when given, serves as a convenient and partial check of certain of the estimated values

The operational empty weight OEW is needed in order to determine the *maximum useful load,* which is the difference between the operational empty weight and the maximum weight of the aircraft. Since the maximum payload weight and the maximum fuel weight represent the weight of each individually but not collectively, their sum usually exceeds the maximum useful load. In other words, it is not possible to carry both the maximum payload and the maximum fuel at the same time without exceeding the maximum allowable gross weight of the aircraft. There is always a trade-off between payload and fuel in planning an operational flight, a trade-off to be determined by the requirements of that particular flight. In the absence of a standardized mission, the generalized selection (comparison) process will, somewhat arbitrarily, use the maximum fuel weight to evaluate the maximum range and the maximum payload to evaluate the maximum-payload range.

The figure of merit expressions will be grouped and developed in the following order:

1. Level flight
   - A. Range
     1. Best mileage (mi/lb)
     2. Maximum range (mi)
     3. Maximum-payload range (lb-mi)
     4. Best-range airspeed (mph)
   - B. Endurance
     1. Minimum fuel-flow rate (lb/h)
     2. Maximum endurance (h)
   - C. Fastest airspeed (mph)
2. Vertical flight
   - A. Minimum take-off run (ft)
   - B. Maximum ceiling (ft)
   - C. Climbing flight

1. Steepest climb angle (deg)
2. Maximum rate of climb (fpm)
3. Minimum time to altitude (min)

3. Turning flight
   A. Maximum load factor (g's)
   B. Fastest turning rate (deg/s)
   C. Tightest turn (ft)

The FOM expressions will be written in final form without development or explanation as the origins of the complete expressions have been covered in preceding chapters. When the density ratio is included in an FOM expression, it will be primarily for use in the design process. In the selection process, the density ratio will normally be set equal to unity; the values obtained will represent the approximate sea-level performance.

Turbojet aircraft will be treated first, and the resultant FOM expressions will be considered to be applicable to turbofans as well. Then the FOM expressions for piston-props will be developed and used for turboprops as well.

In the subsequent paragraphs and sections, we will see two new groupings (aircraft parameters) that have not been seen or used in this book. The first of these is the ratio of the aircraft weight to the wing span (W/b), with the dimensions of lb/ft. This ratio is sometimes referred to as the *span loading*. Reducing its magnitude usually improves many, but not all, aspects of the performance of an aircraft. The second new grouping is the product of the wing area and the zero-lift drag coefficient (SC_{D0}), with the dimensions of ft^2. It sometimes is referred to as the *equivalent flat-plate area* and given the symbol f. For all types of aircraft and for all aspects of performance, we want the lowest possible value for SC_{D0}, since it is a measure of the lowest possible drag.

9-2 TURBOJETS AND TURBOFANS

The *best mileage* (specific range or range factor) is a good measure of the overall efficiency of an aircraft. The customary units of statute miles per pound of fuel (mi/lb) can be converted to the easy-to-visualize units of miles per gallon (mpg) by multiplying the miles per pound value by 6.75, the approximate weight of 1 gal of jet fuel. The complete expression, in miles per pound, is

$$\frac{dX_{\text{br}}}{-dW} = \frac{0.866 V_{\text{br}} E_m}{cW} \tag{9-1a}$$

The corresponding figure of merit (FOM) expressions, also in miles per pound, are

$$\frac{dX_{\text{br}}}{-dW} = \frac{25}{cW}\left(\frac{W/S}{\sigma}\right)^{1/2}\left[\frac{e(\text{AR})}{C_{D0}^3}\right]^{1/4} \tag{9-1b}$$

or
$$\frac{dX_{br}}{-dW}=\frac{25}{c}\left[\frac{1}{(W/b)\sigma}\right]^{1/2}\left[\frac{e}{(SC_{DO})^3}\right]^{1/4} \tag{9-1c}$$

The *maximum range* can be found from the level-flight approximation of the cruise-climb Breguet range equation using the best-range conditions, namely,

$$X_{br}=\frac{0.866V_{br}E_m}{c}\ln \text{MR} \tag{9-2a}$$

where the mass ratio (MR) is given by

$$\text{MR}=\frac{1}{1-(W_f/W)} \tag{9-2b}$$

The FOM expressions are found by first replacing the ln MR by its approximation, the fuel-weight fraction W_f/W. With this substitution, the maximum range is approximately equal to the product of the best mileage and the fuel weight (in pounds), so that

$$X_{br}\cong\left(\frac{dX_{br}}{-dW}\right)\times W_f \tag{9-2c}$$

Equation 9-2*c*, in conjunction with the FOM values from Eq. 9-1*b* or Eq. 9-1*c*, can be used as the FOM expression or we can use the following detailed FOM expressions:

$$X_{br}=\frac{25(W_f/W)}{c}\left(\frac{W/S}{\sigma}\right)^{1/2}\left[\frac{e(\text{AR})}{C_{DO}^3}\right]^{1/4} \tag{9-2d}$$

or
$$X_{br}=\frac{25W_f}{c}\left[\frac{1}{(W/b)\sigma}\right]^{1/2}\left[\frac{e}{(SC_{DO})^3}\right]^{1/4} \tag{9-2e}$$

where the range is in miles. For the maximum range W_f should be the weight of the maximum useable fuel that the aircraft can carry, remembering that the corresponding maximum payload that can be carried at the same time is the difference between the useful load and the maximum fuel weight. This maximum range obviously does not provide any allowances for taxi, take-off, climb, descent, landing, reserves, wind, etc. and is often referred to as the *maximum still-air range*.

The *maximum-payload range*, with the units of pound-miles (lb-mi), is a measure of the payload-carrying capability of an aircraft based on the manufacturer's division of the maximum useful load between fuel and payload. Since the fuel weight with the maximum payload will not normally be the maximum, the maximum-payload range will be defined as the product of the best mileage, the maximum payload weight, and the corresponding fuel weight, i.e.,

$$X_{PL,max}=\left(\frac{dX_{br}}{-dW}\right)\times W_{PL,max}\times W_f \tag{9-3a}$$

where
$$W_f=(W-\text{OEW})-W_{PL,max}=W_{useful}-W_{PL,max} \tag{9-3b}$$

To illustrate the differences between maximum range, maximum-payload range, and ordinary payload range, consider an aircraft with a best mileage of 0.4 mi/lb, a maximum useful load of 10,000 lb, a maximum usable fuel weight of 9,000 lb, and a maximum payload capability of 3,500 lb. Obviously, the sum of the maximum fuel and payload weights exceeds the maximum useful load by 2,500 lb.

For this aircraft, the *maximum range* will be of the order of 3,600 mi (0.4 × 9,000). Since the allowable payload weight in this case is 1,000 lb, the corresponding payload range will be 3.6×10^6 lb-mi (3,600 × 1,000).

With the maximum payload of 3,500 lb, the corresponding fuel weight is 6,500 lb. Consequently, the best range with this amount of fuel will be 2,600 mi (0.4 × 6,500) and the *maximum-payload range* will be 9.1×10^6 mi.

For any other combination of payloads and fuel weights, whose sum is equal to the maximum useful load, such as 2,000 lb of payload and 8,000 lb, the procedure is similar. The best range will be 3,200 mi (0.4 × 8,000), and the payload range will be 6.4×10^6 (3,200 × 2,000). Incidentally, if the sum of the payload and fuel weights is less than the maximum useful load, the gross weight of the aircraft will decrease, thus increasing the best mileage.

The familiar expression for the *best-range airspeed*, expressed here in mph rather than fps, is

$$V_{br} = 0.68 \left[\frac{2(W/S)}{\rho_{SL}\sigma}\right]^{1/2} \left(\frac{3K}{C_{DO}}\right)^{1/4} \tag{9-4a}$$

where the 0.68 is the conversion factor from fps to mph. The FOM expressions, still in mph, are:

$$V_{br} = 20 \left(\frac{W/S}{\sigma}\right)^{1/2} \left[\frac{1}{e(\text{AR})C_{DO}}\right]^{1/4} \tag{9-4b}$$

or

$$V_{br} = 20 \left(\frac{W/b}{\sigma}\right)^{1/2} \left[\frac{1}{e(SC_{DO})}\right]^{1/4} \tag{9-4c}$$

Although the values obtained from these expressions are obviously not valid when the calculated V exceeds the actual drag-rise number, they can still be used as figures of merit for comparison and for design modification.

The *minimum fuel-flow rate* will use the units of pounds of fuel per hour (lb/h), which can be converted to gallons per hour (gph) by dividing by 6.75. The complete expression is

$$\left(\frac{dW_f}{dt}\right)_{\min} = \frac{cW}{E_m} \tag{9-5a}$$

and the FOM expressions are:

$$\left(\frac{dW_f}{dt}\right)_{\min} = 1.13cW \left[\frac{C_{DO}}{e(\text{AR})}\right]^{1/2} \tag{9-5b}$$

or

$$\left(\frac{dW_f}{dt}\right)_{\min} = 1.13c \left(\frac{W}{b}\right) \left(\frac{SC_{DO}}{e}\right)^{1/2} \tag{9-5c}$$

Two observations are in order. The first is that in this case the lower the value of the FOM, the better the performance is. The second is that the fuel-flow rate is completely independent of the airspeed, and thus of the altitude also.

Maximum endurance is associated with the minimum fuel-flow rate. The units are hours and the complete expression is

$$t_{\max}=\frac{E_m}{c}\ln \mathrm{MR} \tag{9-6a}$$

With the approximation of W_f/W for the ln MR, as before, the FOM expressions can be written as

$$t_{\max}=\frac{E_m W_f}{cW}=\frac{W_f}{(dW_f/dt)_{\min}} \tag{9-6b}$$

or

$$t_{\max}=\frac{0.886(W_f/W)}{c}\left[\frac{e(\mathrm{AR})}{C_{DO}}\right]^{1/2} \tag{9-6c}$$

or

$$t_{\max}=\frac{0.886W_f}{(W/b)c}\left(\frac{e}{SC_{DO}}\right)^{1/2} \tag{9-6d}$$

Remember that the maximum endurance of the aircraft is with the maximum usable fuel weight. With respect to the maximum-endurance airspeed, since it is approximately 76 percent of the best-range airspeed, the aircraft with the highest best-range airspeed also has the highest maximum-endurance airspeed.

The *fastest airspeed* (in level flight and in mph) can be found from the rather complicated relationship that

$$V_{\max}=0.68\left[\frac{T/S}{\rho_{SL}\sigma C_{DO}}\left(1+\left\{1-\frac{1}{[E_m(T/W)]^2}\right\}^{1/2}\right)\right]^{1/2} \tag{9-7a}$$

The FOM expressions, still in mph, are much simpler:

$$V_{\max}=20\left[\frac{(T/W)(W/S)}{C_{DO}}\right]^{1/2} \tag{9-7b}$$

or

$$V_{\max}=20\left(\frac{T}{SC_{DO}}\right)^{1/2} \tag{9-7c}$$

Even though these values will often exceed the drag-rise Mach number, and may even be supersonic, they are still valid figures of merit.

The *minimum take-off ground run* in feet is approximated by

$$d=\frac{1.44(W/S)}{\rho_{SL}\sigma g(T/W)C_{L_{\max,\mathrm{TO}}}} \tag{9-8a}$$

The FOM expressions are

$$d=\frac{20(W/S)}{\sigma(T/W)C_{L_{\max,\mathrm{TO}}}} \tag{9-8b}$$

or
$$d=\frac{20W^2}{\sigma TSC_{L\max,\mathrm{TO}}} \tag{9-8c}$$

If neither the maximum lift coefficient for take-off nor the lift-off airspeed is available (the usual situation), use 120 percent of the full-flaps stall speed, if it is available, to calculate the value to be used for the lift coefficient. If only the no-flaps stall speed is given, use it to calculate the value to be used. As with the fuel-flow rate, the smaller the value of the FOM, the better the performance is.

One expression for the *maximum ceiling* in feet is

$$h_c=30{,}500\ \ln\left[\left(\frac{T}{W}\right)E_m\right] \tag{9-9a}$$

The FOM expressions, still in feet, are:

$$h_c=30{,}500\ \ln\left\{0.886\left(\frac{T}{W}\right)\left[\frac{e(\mathrm{AR})}{C_{DO}}\right]^{1/2}\right\} \tag{9-9b}$$

or
$$h_c=30{,}500\ \ln\left[\frac{0.886T}{W/b}\left(\frac{e}{SC_{DO}}\right)^{1/2}\right] \tag{9-9c}$$

When the service ceiling is given, as is often the case, these expressions can be used to verify the given value and to determine how to increase the ceiling, if so desired. The given value of the ceiling can also be used to back-calculate E_m or to verify the previously calculated value, using

$$\sigma_c=\frac{1}{E_m(T/W)} \tag{9-9d}$$

The *steepest climb angle* occurs at sea level and, for climb angles of the order of 30 deg or less, can be found (in degrees) from

$$\gamma_{\max}=\arcsin\left(\frac{T}{W}-\frac{1}{E_m}\right) \tag{9-10a}$$

where the arc sin is in degrees. The FOM expressions, still in degrees, are:

$$\gamma_{\max}=60\left\{\left(\frac{T}{W}\right)-1.13\left[\frac{C_{DO}}{e(\mathrm{AR})}\right]^{1/2}\right\} \tag{9-10b}$$

or
$$\gamma_{\max}=\frac{60}{W}\left[T-1.13\left(\frac{W}{b}\right)\left(\frac{SC_{DO}}{e}\right)^{1/2}\right] \tag{9-10c}$$

The *fastest climb* (maximum rate of climb) also occurs at sea level and is probably most easily determined, in fpm, from

$$(\mathrm{R/C})_{\max}=60V\sin\gamma \tag{9-11a}$$

where
$$V=\left[\frac{(T/S)\Gamma}{3\rho_{SL}\sigma C_{DO}}\right]^{1/2}$$

$$\sin\gamma = \frac{T}{W}\left(1-\frac{\Gamma}{6}\right) - \frac{3}{2\Gamma E_m^{\,2}(T/W)}$$

$$\Gamma = 1 + \left[1 + \frac{3}{[E_m(T/W)]^2}\right]^{1/2}$$

The FOM expressions, in fpm, are

$$(\mathrm{R/C})_{\max} = 600\left[\frac{(T/W)^3(W/S)}{C_{DO}}\right]^{1/2} \qquad (9\text{-}11b)$$

or

$$(\mathrm{R/C})_{\max} = 600\left[\frac{T^3}{W^2(SC_{DO})}\right]^{1/2} \qquad (9\text{-}11c)$$

Remember that the T/W ratio actually used may be less than the maximum available in order to lengthen engine life, to avoid the possibility of exceeding a maximum allowable load factor, or to stay within FAR airspeed limits.

The *minimum time to altitude* describes the fastest climb with a constant throttle setting from sea level to a specified altitude. The simplest FOM expression, in minutes, is

$$t_{\min} = \frac{1.8h}{(\mathrm{R/C})_{\max}} \qquad (9\text{-}12)$$

where h is the final altitude in feet and the value to be used for the R/C is obtained from either Eq. 9-11b or Eq. 9-11c.

The *maximum load factor* is one measure of the maneuverability of an aircraft, occurs at sea level, and is given in g's by

$$n_{\max} = \left(\frac{T}{W}\right)E_m \qquad (9\text{-}13a)$$

The FOM expressions are:

$$n_{\max} = 0.9\left(\frac{T}{W}\right)\left[\frac{e(\mathrm{AR})}{C_{DO}}\right]^{1/2} \qquad (9\text{-}13b)$$

or

$$n_{\max} = \frac{0.9T}{W/b}\left(\frac{e}{SC_{DO}}\right)^{1/2} \qquad (9\text{-}13c)$$

The *fastest turning rate* at constant altitude can be expressed in degrees per second as

$$\dot{\chi}_{\max} = 57.3g\left[\frac{\rho_{\mathrm{SL}}\sigma}{2K(W/S)}\left(\frac{T}{W} - \frac{1}{E_m}\right)\right]^{1/2} \qquad (9\text{-}14a)$$

The FOM expressions, in deg/s, are:

$$\dot{\chi}_{\max} = 100\left[\frac{e(\mathrm{AR})(T/W)}{W/S}\right]^{1/2} \qquad (9\text{-}14b)$$

or

$$\dot{\chi}_{\max} = \frac{100(eT)^{1/2}}{W/b} \qquad (9\text{-}14c)$$

The *tightest turn* (minimum-radius turn), expressed in feet, is given by

$$r_{\min} = \frac{4\,K(W/S)}{\rho g(T/W)\left\{1 - \dfrac{1}{[E_m(T/W)]^2}\right]^{1/2}} \qquad (9\text{-}15a)$$

The FOM expressions, also in feet, are:

$$r_{\min} = \frac{20(W/S)}{e(\mathrm{AR})(T/W)} \qquad (9\text{-}15b)$$

or

$$r_{\min} = \frac{20(W/b)^2}{eT} \qquad (9\text{-}15c)$$

Remember that, for both fastest turns and for tightest turns, the theoretical airspeeds are generally lower than the corresponding stall speeds. Therefore, the maximum lift coefficient, which does not appear explicitly, and the actual stall speeds should be considered in a comparative evaluation.

9-3 A TURBOJET-TURBOFAN COMPARISON

In this section, two existing and successful twin-engine executive aircraft will be evaluated and compared using basic data taken from *Jane's All the World's Aircraft*. Both aircraft were originally equipped with turbojet engines, which were subsequently replaced by turbofans; in fact, both aircraft are now using the same engine. One aircraft will be designated Aircraft *A* and the other will be Aircraft *B*.

From the data given in *Jane's*, only the following items were used in calculating the figures of merit:

	Aircraft *A*	**Aircraft *B***
Weight W (lb)	23,500	24,000
Wing area S (ft^2)	308.26	380.0
Wing span b (ft)	44.79	50.43
Thrust T (lb)	7,400	7,400
OEW (lb)	12,700	14,154
Maximum payload:		
No. of passengers	10 (2,000 lb)	8 (1,600 lb)
Baggage (lb)	1,050	545
Maximum fuel W_f (lb)	9,464	8,708
Stall speed V_s (mph)		
(full flaps and gear down)	114	104
Service ceiling h_c (ft)	45,000	45,000

On the basis of these data, it appears that the aircraft are almost identical and that there should be no significant differences in their performance. We shall, however, continue with our comparative evaluation.

The three pieces of missing data are, as expected, the zero-lift drag coefficient, the Oswald span efficiency, and the thrust specific fuel consumption. Since the aircraft are so similar, let us estimate values and use identical values for both aircraft. A value of 0.016 for C_{D0} seems reasonable (it may be a bit on the low side) since the aircraft are of moderate size and range with wings of moderate thickness. With aspect ratios of the order of 7, an Oswald span efficiency of 0.85 also seems reasonable. The thrust specific fuel consumption is taken to be 0.85 lb/h/lb, lower than normal for a pure turbojet and a bit high for a high bypass turbofan.

The data we have acquired and estimated can be combined or used to calculate some familiar and two not so familiar parameters and groupings, namely:

	Aircraft *A*	Aircraft *B*
W/S (lb/ft^2)	76.2	63.2
T/W(g's)	0.315	0.308
AR	6.5	6.7
K	0.0576	0.0559
E_m	16.5	16.7
Useful load (lb)	10,800	9,846
W/b (lb/ft)	524.7	475.9
SC_{D0} (ft)	4.93	6.08
$C_{L_{max},TO}$	1.60	1.59

At this point, let us use the given service ceiling to check the reasonableness of the estimated values for e and for C_{D0}; they are reflected in the calculated values of E_m. Using the approximation for the density ratio at the ceiling, i.e.,

$$\sigma_{cA}=\frac{1}{0.315\times 16.5}=0.192$$

$$\sigma_{cB}=\frac{1}{0.308\times 16.7}=0.194$$

the density ratio for Aircraft A is 0.192 and for Aircraft B is 0.194, which closely correspond to ceilings of 45,000 ft, the value given in *Jane's* for each aircraft.

The values shown for $C_{L_{max}}$ were determined by using 120 percent of the full-flaps stall speed in fps and the relationship that

$$C_L=\frac{2(W/S)}{\rho_{SL}\sigma V^2}$$

For Aircraft A, 120 percent of the stall speed is 200 fps, and the corresponding lift coefficient is 1.6. For Aircraft B, 120 percent of the stall speed is 183 fps, and the corresponding lift coefficient is 1.59.

The figures of merit can now be calculated for each aircraft, using the expressions of the preceding section; they are shown in Table 9-1. Although the primary purpose of these figures of merit is either for comparing one or more aircraft with each other or for evaluating the effects of modifying the characteristics of an individual aircraft, the values themselves are representative of the sea-level performance of the aircraft.

The FOM calculations are quite straightforward, with the possible exception of the maximum range and the maximum-payload range. In calculating the FOM for maximum range, the maximum fuel weights of 9,464 and 8,708 lb were used for Aircraft *A* and Aircraft *B*, respectively. As a matter of interest only, the corresponding payload weights are 1,336 and 1,138 lb.

TABLE 9-1

Comparative Evaluation of Two Turbojet Aircraft

Figure of Merit	Aircraft A	Aircraft B	Relative Value[b]
Best mileage (mi/lb)	0.37	0.33	1.11
(mpg)	2.51	2.26	1.11
Maximum range (mi)	3,520	2,908	1.21
Maximum-payload range (lb-mi)	8.8×10^6	5.5×10^6	1.59
Best-range airspeed (mph)	320	289	1.11
Minimum fuel flow (lb/hr)[a]	1,214	1,221	1.00
(gph)[a]	180	181	1.00
Maximum endurance (h)	8.8	8.0	1.09
Fastest airspeed (mph)	775	698	1.11
Shortest TO run (ft)[a]	3,025	2,580	0.85
Ceiling (ft)	49,500	49,980	1.00
Steepest climb angle (deg)	15.2	14.9	1.02
Fastest climb rate (fpm)	7,320	6,454	1.13
Fastest climb to 30,000 ft (min)[a]	7.4	8.4	1.14
Maximum load factor (*g*'s)	5.3	5.2	1.02
Fastest turn (deg/s)	15.1	16.7	0.90
Tightest turn (ft)[a]	876	720	0.82

[a]The smaller the value, the better the performance.

[b]The relative performance of Aircraft *A* with respect to Aircraft *B*; Aircraft *B* is the baseline aircraft.

In calculating the maximum-payload range, Aircraft *A* uses the given maximum payload weight of 3,050 lb with a corresponding fuel weight of 7,750 lb. Aircraft *B* uses the maximum payload weight of 2,145 lb and a corresponding fuel weight of 7,701 lb.

Looking down the second and third columns of Table 9-1, we see that Aircraft *A* appears to be better than Aircraft *B* in most areas of performance. The last column, entitled "Relative Value," is a quantitative measure of the relative performance of Aircraft *A* with respect to Aircraft *B*, using the performance of *B* as the baseline.

In level and climbing flight, the superiority of Aircraft A is quite apparent, particularly with respect to the maximum range and the maximum-payload range. Aircraft B, on the other hand, has an 85 percent shorter take-off run, by virtue of its lower wing loading, and a 10 to 18 percent better turning performance (without consideration of stall speeds) by virtue of its lower span loading.

Aircraft B's performance is penalized by its 23 percent larger wing area, leading directly to a 23 percent larger equivalent flat plate area (SC_{DO}), if we assume identical C_{DO}'s as we have, and by its 11% larger operational empty weight fraction (0.59 versus 0.54). If the wing span of Aircraft A were to be increased to 49.38 ft, then the span loading would decrease and become equal to that of Aircraft B, as would the turning performance. Decreasing W/b would also improve the range performance of Aircraft A even more, with the exception of the best-range airspeed, which would decrease by about 5 percent. The best way to decrease the take-off run of A would be to increase the lift coefficient rather than increase either the wing area or the thrust, which would increase SC_{DO} in the first case and the engine weight (and thus the gross weight) in the second. The performance of Aircraft B, with the exception of the take-off run, can be improved by simply reducing its wing area.

If we could have only one *gross figure of merit* to use in the evaluation and comparison of turbojet and turbofan aircraft, it might well be the product of the span loading (W/b), the equivalent flat-plate area (SC_{DO}), and the operational empty weight fraction (OEW/W). This product reduces to the product of the equivalent flat-plate area and the *operational empty weight span loading* (OEW/b), a new parameter. The expressions for this single GFOM are

$$\text{GFOM} = \left(\frac{W}{b}\right)(SC_{DO})\left(\frac{\text{OEW}}{W}\right) \tag{9-16a}$$

or

$$\text{GFOM} = \left(\frac{\text{OEW}}{b}\right)(SC_{DO}) \tag{9-16b}$$

When using the GFOM, the smaller the value is, the better the performance. This is the opposite of the majority of the FOM's.

For our two aircraft, A and B, the relevant parameters and the GFOM are:

	Aircraft A	Aircraft B	Relative Value[a]
W/b (lb/ft)	524.7	475.9	0.91
SC_{DO} (ft^2)	4.93	6.08	1.23
OEW/W	0.54	0.59	1.09
OEW/b (lb/ft)	283.5	280.7	0.99
GFOM	1,397	1,707	1.22

[a] With Aircraft B as the baseline, the larger the relative value, the better the relative performance of A.

Looking at these figures above, we see that the GFOM for Aircraft A is of the order of 20 percent lower than that for Aircraft B, which is reflected in the value of 1.22 for the relative value. If we look at the numbers making up the GFOM, we see that the larger span loading of Aircraft A is compensated for by the smaller OEW/W fraction, so that the OEW/b ratios of both aircraft are essentially equal. Therefore, it is the 23 percent larger SC_{DO} (in this case, due to the larger wing area alone) that leads to the larger GFOM for Aircraft B.

Comparing the GFOM with the Relative Values of Table 9-1, we see that the GFOM is definitely a "gross" (a rough or order of magnitude) figure of merit and that it is primarily a measure of the relative performance with respect to level flight, particularly the range. Although it provides no details on comparative performance, the GFOM approach is quick and easy to use. Furthermore, it emphasizes the importance of the new design parameters that have appeared in this chapter, namely, the *span loading* (W/b), the *equivalent flat-plate area* (SC_{DO}), the *operational empty weight fraction* (OEW/W), and the *operational empty weight span loading* (OEW/b).

9-4 PISTON-PROPS AND TURBOPROPS

Although the figure of merit (FOM) expressions to be developed in this section for piston-prop aircraft will be applied to all propeller-driven aircraft, it may be necessary at times to consider the type of power plant being used. The four types of propeller power plants in current use are the aspirated piston-prop, the turbocharged piston-prop, the straight turboprop, and the derated (or flat-rated) turboprop. The power delivered by the aspirated piston-prop and the straight turboprop immediately starts to decrease with an increase in altitude (or a decrease in the atmospheric density), whereas that delivered by the other two propulsion systems remains constant up to the critical altitude, usually of the order of 20,000 ft or less, before starting to decrease. When using FOMs, the distinction among power plants is normally required only when comparing aircraft with different types of power plants or in considering changes in the design characteristics of an aircraft.

The development of this section will parallel that of the turbojet section. The propeller efficiency η_p is an additional parameter to be estimated or back-calculated, along with C_{DO}, e, and $\hat{c}$ (the horsepower specific fuel consumption). As each of the propeller FOM expressions is developed, you may wish to compare it with the corresponding turbojet expression. It is interesting to see how the various aircraft characteristics affect the performance of the different classes of aircraft.

The units for the *best mileage* are statute miles per pound of fuel (mi/lb) and may be converted to miles per gallon by multiplying by 6 lb/gal for piston-props using gasoline and by 6.75 lb/gal for turboprops using jet fuel. The complete expression is

$$\frac{dX_{br}}{-dW} = \frac{375\eta_p E_m}{\hat{c}W} \qquad (9\text{-}17a)$$

The FOM expressions, in mi/lb, are:

$$\frac{dX_{br}}{-dW}=\frac{330\eta_p}{\hat{c}W}\left[\frac{e(AR)}{C_{DO}}\right]^{1/2} \tag{9-17b}$$

or

$$\frac{dX_{br}}{-dW}=\frac{330\eta_p}{\hat{c}(W/b)}\left(\frac{e}{SC_{DO}}\right)^{1/2} \tag{9-17c}$$

The absence of the density ratio in any of these expressions indicates that the mileage is independent of the altitude. Remember, however, that the specific fuel consumption does decrease slightly with altitude within the troposphere.

The *maximum range*, expressed in statute miles (mi), is given by the Breguet range equation, namely,

$$X_{br}=\frac{375\eta_p E_m}{\hat{c}}\ln \text{MR} \tag{9-18a}$$

Approximating the ln MR by the fuel weight fraction (W_f/W), the FOM expressions, in miles, are

$$X_{br}=\left(\frac{dX_{br}}{-dW}\right)W_f \tag{9-18b}$$

or

$$X_{br}=\frac{330\eta_p(W_f/W)}{\hat{c}}\left[\frac{e(AR)}{C_{DO}}\right]^{1/2} \tag{9-18c}$$

or

$$X_{br}=\frac{330\eta_p W_f}{\hat{c}(W/b)}\left(\frac{e}{SC_{DO}}\right)^{1/2} \tag{9-18d}$$

where, for maximum range, the fuel weight is the maximum usable fuel that can be loaded aboard the aircraft.

The *maximum-payload range* is simply the best mileage multiplied by the maximum payload weight and by the corresponding fuel weight, which, in lb-mi, is

$$X_{PL,max}=\left(\frac{dX_{br}}{-dW}\right)W_{PL,max}\times W_f \tag{9-19}$$

where the best-mileage value is obtained from one of the Eq. 9-17 relationships and where

$$W_f=(W-\text{OEW})-W_{PL,max}=W_{useful}-W_{PL,max}$$

The *best-range airspeed*, in miles per hour, is

$$V_{br}=0.68\left[\frac{2(W/S)}{\rho_{SL}\sigma}\right]^{1/2}\left[\frac{K}{C_{DO}}\right]^{1/4} \tag{9-20a}$$

The FOM expressions, in mph, are

$$V_{br}=15\left(\frac{W/S}{\sigma}\right)^{1/2}\left[\frac{1}{e(AR)C_{DO}}\right]^{1/4} \tag{9-20b}$$

or $$V_{br} = 15\left[\frac{(W/b)}{\sigma}\right]^{1/2}\left[\frac{1}{e(SC_{DO})}\right]^{1/4} \tag{9-20c}$$

These expressions are identical to those for a turbojet with the exception of the smaller multiplier, which reflects the fact that the propeller-aircraft airspeeds are approximately 76 percent of the corresponding turbojet airspeeds. Remember that most of the propeller aircraft, particularly the slower ones, cruise at airspeeds higher than the best-range airspeeds in order to reduce the flight time.

The *minimum fuel-flow rate*, in pounds per hour, is given by

$$\left(\frac{dW_f}{dt}\right)_{min} = \frac{\hat{c}VW}{0.866 \times 375 \times \eta_p E_m} \tag{9-21a}$$

where $$V = V_{P_{min}} = 0.68\left[\frac{2(W/S)}{\rho_{SL}\sigma}\right]^{1/2}\left(\frac{K}{3C_{DO}}\right)^{1/4} \tag{9-21b}$$

The units of V are mph.

The FOM expressions, in lb/h, are

$$\left(\frac{dW_f}{dt}\right)_{min} = \frac{0.04\hat{c}W}{\eta_p}\left[\frac{(W/S)}{\sigma}\right]^{1/2}\left\{\frac{C_{DO}}{[e(AR)]^3}\right\}^{1/4} \tag{9-21c}$$

or $$\left(\frac{dW_f}{dt}\right)_{min} = \frac{0.04\hat{c}}{\eta_p}\left[\frac{(W/b)^3}{\sigma}\right]^{1/2}\left(\frac{SC_{DO}}{e^3}\right)^{1/4} \tag{9-21d}$$

The *maximum endurance*, in hours, is simply

$$t_{max} = \frac{W}{(dW_f/dt)_{min}} \ln \text{MR} \tag{9-22a}$$

With the ln MR replaced by its approximation, W_f/W, the corresponding FOM expression is

$$t_{max} = \frac{W_f}{(dW_f/dt)_{min}} \tag{9-22b}$$

where W_f is the maximum fuel weight and the value of $(dW_f/dt)_{min}$ is obtained from one of the preceding equations.

There is no closed-form solution for the *fastest airspeed* of a piston-prop. (It can be found by an iterative solution of Eq. 6-19.) Consequently, there is no analytic expression that can be used as the complete expression. However, with the not unreasonable assumption that the drag due to lift at the fastest airspeed is of the order of 10 percent of the total drag, that is, $C_D = 1.1C_{DO}$, the approximate fastest airspeed can be found from the solution of

$$TV = DV$$

and $$550\eta_p(\text{HP}) = \tfrac{1}{2}\rho_{SL}\sigma V^3 SC_D$$

so that, in mph,

$$V_{max} = 0.68\left[\frac{2 \times 550\eta_p(\text{HP}/S)}{1.1\rho_{SL}\sigma C_{DO}}\right]^{1/3} \tag{9-23a}$$

The corresponding FOM expressions, in mph. are

$$V_{max}=50\left[\frac{\eta_p(\mathrm{HP}/W)(W/S)}{\sigma C_{DO}}\right]^{1/3} \tag{9-23b}$$

or

$$V_{max}=50\left[\frac{\eta_p(\mathrm{HP})}{\sigma(SC_{DO})}\right]^{1/3} \tag{9-23c}$$

With turbocharged piston-props and derated turboprops, the fastest airspeed will increase with altitude, reaching a maximum at the critical altitude.

A complete expression, in feet, for the *minimum take-off ground run* is

$$d=\frac{2.44}{550\eta_p g\sigma^{1.5}(\mathrm{HP}/W)}\left[\frac{W/S}{\rho_{SL}C_{L\max,\mathrm{TO}}}\right]^{1.5} \tag{9-24a}$$

The FOM expressions, which are awkward, are

$$d=\frac{1.2}{\eta_p\sigma^{1.5}(\mathrm{HP}/W)}\left(\frac{W/S}{C_{L\max,\mathrm{TO}}}\right)^{1.5} \tag{9-24b}$$

or

$$d=\frac{1.2W^{2.5}}{\eta_p\sigma^{1.5}\mathrm{HP}(SC_{L\max,\mathrm{TO}})^{1.5}} \tag{9-24c}$$

One complete expression for the approximate *maximum ceiling*, in feet, is

$$h_c=\frac{2\times 30{,}500}{3}\ln\left[\frac{325\eta_p(\mathrm{HP}/W)E_m}{\sigma_{cr}V}\right] \tag{9-25a}$$

where HP/W and V are sea-level values and V, in mph, is found from Eq. 9-21*b* with σ equal to unity, and where σ_{cr} is the critical altitude density ratio for turbocharged engines or derated turboprops. For aspirated engines and straight turboprops, set σ_{cr} equal to unity.

The FOM expressions, in feet, are

$$h_c=20{,}000\ \ln\left[\frac{32\eta_p(\mathrm{HP}/W)}{\sigma_{cr}(W/S)^{1/2}}\left\{\frac{[e(\mathrm{AR})]^3}{C_{DO}}\right\}^{1/4}\right] \tag{9-25b}$$

or

$$h_c=20{,}000\ \ln\left[\frac{32\eta_p(\mathrm{HP})}{\sigma_{cr}(W/b)^{1.5}}\left(\frac{e^3}{SC_{DO}}\right)^{1/4}\right] \tag{9-25c}$$

These are not simple expressions, unfortunately, but they do give insight into the effects of physical characteristics upon the ceiling. If a given service ceiling is to be used to back-calculate or to verify any of the parameters, use Eq. 6-27, which is repeated here for convenience.

$$\sigma_c=\left[\frac{1.155\sigma_{cr}V}{(P/W)E_m}\right]^{2/3} \tag{9-26}$$

where V is given by Eq. 9-21*b*.

The *steepest climb angle* occurs at sea level and its approximate value, in degrees, is given by

$$\gamma_{max} = \text{arc sin}\left\{\frac{\rho_{SL}[550\eta_p(\text{HP}/W)]^2}{8\ K(W/S)}\right\} \tag{9-27a}$$

where the arc sin is in degrees.

The FOM expressions, in degrees, are

$$\gamma_{max} = 1{,}600\left\{\frac{e(\text{AR})}{W/S}[\eta_p(\text{HP}/W)]^2\right\} \tag{9-27b}$$

or

$$\gamma_{max} = 1{,}600\left\{\frac{e[\eta_p(\text{HP})]^2}{W(W/b)^2}\right\} \tag{9-27c}$$

Do not forget that the corresponding airspeed may be less than the stall speed.

The *fastest climb* (maximum rate of climb) also occurs at sea level, and the complete expression, in fpm, is

$$(\text{R/C})_{max} = 60\left[550\eta_p(\text{HP}/W) - \frac{1.47V}{0.866E_m}\right] \tag{9-28a}$$

where V, in mph, is given by Eq. 9-21*b*.

The FOM expressions, in fpm, are:

$$(\text{R/C})_{max} = 60\left[550\eta_p(\text{HP}/W) - 22(W/S)^{1/2}\left\{\frac{C_{DO}}{[e(\text{AR})]^3}\right\}^{1/4}\right] \tag{9-28b}$$

or

$$(\text{R/C})_{max} = \frac{33{,}000}{W}\left[\eta_p(\text{HP}) - 0.04(W/b)^{1.5}\left(\frac{SC_{DO}}{e^3}\right)^{1/4}\right] \tag{9-28c}$$

These are not as simple as we would like them to be, but the second term can be significant.

The *minimum time to altitude* is obviously related to the fastest climb. An empirical relationship that uses the fastest climb values or expressions of the preceding paragraph is, in minutes,

$$t_{min} = \frac{H}{(\text{R/C})_{max}}\left[\exp\left(\frac{h}{H}\right) - 1\right] \tag{9-29}$$

where h is the specified altitude in feet and H is an empirical constant that has two values, 15,250 for aspirated engines and straight turboprops and 61,000 for turbocharged engines and derated turboprops. This expression is valid for constant-throttle climbs only and should not be used for climbs to altitudes approaching the ceiling.

The *maximum load factor* attainable in steady-state level flight occurs at sea level and is given (in g's) by

$$n_{max} = 0.687\left\{\frac{\rho_{SL}[550\eta_p(\text{HP}/W)]^2 E_m}{K(W/S)}\right\}^{1/3} \tag{9-30a}$$

The FOM expressions are:

$$n_{\max} = 8.5\left(\frac{[\eta_p(\mathrm{HP}/W)]^2}{W/S}\left\{\frac{[e(\mathrm{AR})]^3}{C_{DO}}\right\}^{1/2}\right)^{1/3} \tag{9-30b}$$

or

$$n_{\max} = \frac{8.5}{W/b}\left\{[\eta_p(\mathrm{HP})]^2\left(\frac{e^3}{SC_{DO}}\right)^{1/2}\right\}^{1/3} \tag{9-30c}$$

The *fastest turning rate* in steady-state level flight occurs at sea level and can be expressed in degrees per second as

$$\dot{\chi}_{\max} = \frac{57.3 \times 550 \times \rho_{SL}\sigma g(\mathrm{HP}/W)}{4K(W/S)} \tag{9-31a}$$

The FOM expressions, in deg/s, are

$$\dot{\chi}_{\max} = \frac{1{,}900\eta_p(\mathrm{HP}/W)e(\mathrm{AR})}{W/S} \tag{9-31b}$$

or

$$\dot{\chi}_{\max} = \frac{1{,}900\eta_p e(\mathrm{HP})}{(W/b)^2} \tag{9-31c}$$

The *tightest turn* (minimum-radius turn) for steady-state level flight also occurs at sea level and can be expressed, in feet, as

$$r_{\min} = \frac{12.3}{g}\left[\frac{K(W/S)}{550\rho_{SL}\eta_p(\mathrm{HP}/W)}\right]^2 \tag{9-32a}$$

The FOM expressions, in feet, are

$$r_{\min} = 0.023\left[\frac{W/S}{e(\mathrm{AR})\eta_p(\mathrm{HP}/W)}\right]^2 \tag{9-32b}$$

or

$$r_{\min} = 0.023\left[\frac{(W/b)^2}{\eta_p e(\mathrm{HP})}\right]^2 \tag{9-32c}$$

A with the turbojet, do not forget to check the actual stall speeds against the theoretical airspeeds for both fastest turn and tightest turn.

9-5 A PISTON-PROP COMPARISON

In this section, we shall use the FOM's to compare two single-engine turbocharged piston-prop aircraft. One will be designated as Aircraft *C* and the other as Aircraft *D*. The relevant data from Jane's are:

	Aircraft C	Aircraft D
Weight W (lb)	4,000	3,600
Wing area S (ft^2)	175	174.5
Wing span b (ft)	36.75	32.75
Horsepower, HP	310	300
OEW (lb)	2,287	2,065
Maximum payload:		
No. of passengers	6 (1,200 lb)	6 (1,200 lb)
Baggage (lb)	240	200
Maximum fuel W_f (lb)	540	564
Stall speed (mph)	67	69.5
Critical altitude (ft)	12,000	12,000
Service ceiling (ft)	27,000	N/A

Notice that the critical altitude for the engines is only 12,000 ft, indicating light turbocharging. However, these aircraft are not pressurized; consequently, all occupants would have to wear oxygen masks whenever the cruising altitude remained above 12,000 ft for more than an hour. If the aircraft were pressurized, we might expect the critical altitude to be higher, as would be the ceiling.

Since the aircraft are similar, the missing data will be estimated and the same values used for each aircraft. The following values will be given to the missing data: 0.025 for the zero-lift drag coefficient, 0.85 for the Oswald span efficiency, 0.8 for the propeller efficiency, and 0.5 lb/h/hp for the horsepower specific fuel consumption.

The additional groupings and parameters, to include a gross figure of merit (GFOM), are:

	Aircraft C	Aircraft D
W/S (lb/ft^2)	22.9	20.6
HP/W (hp/lb)	0.078	0.083
AR	7.72	6.15
E_m	14.36	12.81
Useful load (lb)	1,713	1,535
W/b (lb/ft)	108.8	109.9
SC_{DO} (ft^2)	4.38	4.36
OEW/W	0.57	0.57
OEW/b (lb/ft)	62.2	63.0
GFOM	272.1	274.7
$C_{Lmax,TO}$	1.38	1.19

According to the GFOMs, the range performance of these two aircraft should be very close to each other. The individual FOMs are shown in Table 9-2, along with a column showing the relative performance of Aircraft *C* with respect to Aircraft *D*. As we look down the columns, we see that there are no truly significant differences in performance with the exception of the maximum-payload range, which indicates that Aircraft *C* has twice the payload range of *D*. If we look more closely at the original data, we see that the manufacturer of Aircraft *D* is showing a maximum payload weight (6 passengers and 200 lb of baggage) that is only 40 lb lighter than that of Aircraft *C*, and yet *D* has a 178 lb smaller maximum useful load, in other words, 138 lb less fuel. With a maximum payload, Aircraft *C* has 273 lb of fuel and a range of 587 mi, whereas Aircraft *D* has only 135 lb of fuel and a range of only 286 mi, half that of Aircraft *C*.

TABLE 9-2

Comparative Evaluation of Two Single-Engine Turbocharged Piston-Prop Aircraft

Figure of Merit	Aircraft *C*	Aircraft *D*	Relative Value[b]
Best mileage (mi/lb)	2.15	2.12	1.12
(mpg)	12.9	12.7	1.12
Maximum range (mi)	1,160	1,200	0.97
Maximum-payload range (lb-mi)	8.8×10^5	4.0×10^5	2.11
Best-range airspeed (mph)	113	113	1.00
Minimum fuel flow (lb/h)[a]	46.4	47.0	1.01
(gph)[a]	7.7	7.8	1.01
Maximum endurance (h)	11.6	12.0	0.97
Fastest airspeed (mph)	192	190	1.01
Shortest TO run (ft)[a]	1,295	1,350	1.04
Ceiling (ft)	24,250	23,200	1.05
Steepest climb angle (deg)	17.9	18.0	1.00
Fastest climb rate (fpm)	1,460	1,510	0.96
Fastest climb to 25,000 ft (min)[a]	21.3	25.7	0.96
Maximum load factor (*g*'s)	2.2	2.2	1.00
Fastest turn (deg/s)	34.0	32.0	1.06
Tightest turn (ft)[a]	71.7	80.1	1.12

[a]The smaller the value, the better the performance.

[b]The relative performance of Aircraft *C* with rspect to *D*; Aircraft *D* is the baseline aircraft.

9-6 A STRAIGHT TURBOPROP COMPARISON

In this section, we shall compare two twin-engine aircraft, similar in size and mission and equipped with straight (not derated or flat-rated) turboprop engines. One will be designated as Aircraft *E* and the other as Aircraft *F*.

The relevant data from *Jane's* are:

	Aircraft *E*	Aircraft *F*
Weight W (lb)	9,000	10,325
Wing area S (ft^2)	229	266
Wing span b (ft)	42.7	46.7
Total ESHP (eshp)	1,240	1,400
OEW (lb)	4,976	6,733
Maximum payload:		
No. of passengers	7 (1,400 lb)	7 (1,400 lb)
Baggage	400	600
Maximum fuel (lb)	2,578	2,592
Stall speed V_s (mph)	96	94
Service ceiling (ft)	31,600	32,800

The estimated values for the missing data are 0.22 for the zero-lift drag coefficient, 0.85 for the Oswald span efficiency, 0.85 for the propeller efficiency, and 0.55 for the equivalent shaft horsepower specific fuel consumption. The remaining groupings and parameters of interest are:

	Aircraft *E*	Aircraft *F*
W/S (lb/ft^2)	39.3	38.8
HP/W (eshp/lb)	0.14	0.14
AR	7.96	8.19
E_m	15.55	15.72
Useful load (lb)	4,024	3,559
W/b (lb/ft)	210.8	221.2
SC_{D0} (ft^2)	5.04	5.85
OEW/W	0.55	0.65
OEW/b	116.4	144.2
GFOM	587.2	843.6
$C_{L\max,\mathrm{TO}}$	1.67	1.72

There are differences in some of these parameters, and these differences are reflected in the GFOMs. The GFOM of Aircraft *E* is 40 percent lower than that of Aircraft *F*, which should be an indication that the range performance of *E* might be of the order of 40 percent better.

The individual FOMs are tabulated in Table 9-3 along with the relative values of Aircraft *E*'s performance with respect to that of *F*. We see that the performances are generally comparable, except that *E* does have better mileage and ranges.

TABLE 9-3
Comparative Evaluation of Two Straight Turboprops

Figure of Merit	Aircraft *E*	Aircraft *F*	Relative Value[b]
Best mileage (mi/lb)	0.99	0.88	1.13
(mpg)	6.7	5.9	1.13
Maximum range (mi)	2,560	2,280	1.12
Maximum-payload range (lb-mi)	3.7×10^6	2.7×10^6	1.35
Best-range airspeed (mph)	151	149	1.01
Minimum fuel flow (lb/h)[a]	134	150	1.12
(gph)[a]	19.8	22.2	1.12
Maximum endurance (h)	19.2	17.3	1.10
Fastest airspeed (mph)	297	294	1.01
Shortest TO run (ft)[a]	1,170	1,110	0.96
Ceiling (ft)	32,100	32,800	0.99
Steepest climb angle (deg)	38.0	38.0	1.00
Fastest climb rate (fpm)	3,100	3,075	1.00
Fastest climb to 25,000 ft (min)[a]	20.0	20.6	1.02
Maximum load factor (g's)	2.9	3.0	0.97
Fastest turn (deg/s)	38.0	39.0	0.97
Tightest turn (ft)[a]	56	54	0.96

[a]The smaller the value, the better the performance.

[b]The relative performance of Aircraft *E* with respect to Aircraft *F*; Aircraft *F* is the baseline aircraft.

PROBLEMS

Rather than solve constructed problems, it is suggested that the person who is interested in applying the techniques of this chapter to the comparative evaluation of two or more competitive aircraft or to the examination of possible ways to improve specific aspects of a particular aircraft actually do so for aircraft that are of interest to him or to her. As mentioned in the text, sources of needed data include *Jane's All the World's Aircraft*, which can be found in many libraries, "Aviation Week & Space Technology" (The Annual Inventory Issue is very good for comparative data), "Flying" and other such aviation magazines, and manufacturers' specifications and operating data.

A good exercise is to examine the FOMs developed for the aircraft in this chapter with an eye to improving certain aspects of their performance without degrading the good performance features. You may wish to develop FOMs for the illustrative aircraft in the text as well as for Aircraft *A* through *E* that have been used in the problems at the end of certain chapters.

Chapter Ten

Effects of Wind on Performance

10-1 INTRODUCTION

In all our performance analyses we have assumed a no-wind condition, which means that we have assumed that the ground speed and the airspeed are identically equal. This is a reasonable assumption in the design of an aircraft and in the comparative evaluation of the performance of competing designs. Operationally, however, the no-wind assumption is unrealistic. Not only is there usually a wind, but its presence can strongly affect the performance of an aircraft. It affects take-off and landing distances, climb speeds and climb rates, and cruising performance.

With respect to the latter, wind does not affect the *endurance* of either turbojets or piston-props (and accordingly of turbofans and turboprops) because neither the wind speed nor the ground speed appears, either explicitly or implicitly, in any of the endurance equations. This means that the no-wind and with-wind maximum endurance airspeeds are identical. It does not mean that the ground speeds and airspeeds are identical, as can be seen from Eq. 2-7, which is repeated here, rearranged, and renumbered for convenience.

$$V_g = V \pm V_w = V\left(1 \pm \frac{V_w}{V}\right) \tag{10-1}$$

In Eq. 10-1, V is the true airspeed, V_w is the x component of the wind velocity, and the plus and minus signs denote a tailwind and a headwind, respectively. Since the ground speed changes, the actual range will be affected but there is nothing that can be done operationally to minimize the effects of wind other than by a judicious choice of cruise altitudes. There will be no further discussion in this chapter of maximum-endurance cruise.

Best-range cruise performance in the presence of a wind is another story, however, and most of this chapter will be devoted to its examination with special emphasis on fuel consumption. We will develop analytic expressions that are relatively simple and are based on the no-wind best-range airspeed of a particular aircraft or class of aircraft. In the interests of simplicity, a parabolic drag polar is assumed, cross-wind components are neglected, and only cruise-climb flight is considered. Expressions for the best-range airspeed and relative range in the presence of a constant headwind or tailwind component are derived individually, first for turbojets and then for piston-props. These expressions are then used to obtain common relationships for the relative flight time, the relative fuel consumption, and the actual fuel savings. There will be a certain amount of repetition in the development of the equations in this chapter to avoid the need for frequent back-referencing and to make the chapter relatively self-contained.

The chapter will close with a discussion of the effects of wind on other flight, namely, take-off, landing, and climb.

10-2 CRUISE PERFORMANCE

10-2-1 Best-Range Conditions

For a *turbojet aircraft*, the governing equations for quasi-steady flight with a small or zero climb angle are

$$T=D, \qquad L=W, \qquad \frac{dX}{dt}=V_g \tag{10-2}$$

Since the fuel flow rate of a turbojet engine is proportional to the thrust, the weight balance equation of the aircraft can be written as

$$\frac{-dW}{dt}=cT \tag{10-3}$$

where c, the thrust specific fuel consumption, is assumed to be constant. If the expression for the instantaneous range is obtained from Eqs. 10-2 and 10-3, and then integrated from start to end of cruise with both the lift coefficient and the airspeed held constant, the resulting Breguet range equation is

$$X=\frac{V_g E}{c}\ln\left(\frac{1}{1-\zeta}\right) \tag{10-4}$$

where E is the flight lift-to-drag ratio and ζ, the cruise-fuel weight fraction, is the ratio of the fuel used during cruise to the gross weight of the aircraft at the start of cruise.

At this point we will define a *relative airspeed parameter* n as

$$n=\frac{V}{V_{\text{br}}} \tag{10-5}$$

where V_{br}, the no-wind best-range airspeed, is given by

$$V_{\text{br}}=\left[\frac{2(W/S)}{\rho_{\text{SL}}\sigma}\right]^{1/2}\left(\frac{3K}{C_{DO}}\right)^{1/4}=3^{1/4}V_{\text{md}} \tag{10-6}$$

where V_{md} is the minimum-drag airspeed of the aircraft.

Using the relative airspeed parameter n, the lift-to-drag ratio can be expressed as

$$E=2\sqrt{3}\left(\frac{n^2}{3n^4+1}\right)E_m \tag{10-7}$$

and the ground speed as

$$V_g = V_{br}\left(n \pm \frac{V_w}{V_{br}}\right) = V_{br}(n \pm \text{WF}) \tag{10-8}$$

where WF, the *wind fraction*, is the ratio of the wind speed to the no-wind best-range airspeed.

Consequently, a generalized version of the turbojet Breguet range equation that acknowledges the presence of a wind can be written as

$$X = \frac{2\sqrt{3}\,V_{br}E_m(n \pm \text{WF})}{c} \times \left(\frac{n^2}{3n^4+1}\right)\ln\left(\frac{1}{1-\zeta}\right) \tag{10-9}$$

Maximizing the range with respect to the true airspeed, by setting the first derivative of Eq. 10-9 with respect to n equal to zero, yields the best-range condition that

$$n_{br} = \left(\frac{n_{br} \pm \frac{2}{3}\text{WF}}{n_{br} \pm 2\text{WF}}\right)^{1/4} \tag{10-10}$$

where the subscript br denotes the best-range conditions. When the wind fraction is set equal to zero (the no-wind condition), n_{br} becomes unity and the no-wind best-range airspeed is that defined in Eq. 10-6, and the true airspeed and the ground speed are equal. When the WF is not equal to zero, n_{br} *is the ratio of the best-range airspeed in the presence of a wind to the no-wind best-range airspeed.* This ratio, or *relative best-range airspeed,* is shown in Fig. 10-1 for various values of the wind fraction. It can be seen that the increase in airspeed required for a headwind is not only disproportionately larger than the decrease for a tailwind but also increases at a rapid rate, whereas the airspeed for a tailwind levels off, approaching in the limit the value of the minimum-drag airspeed.

If n is set equal to unity in Eq. 10-9, we can define X_{brwu} as the uncorrected best range with a wind, where

$$X_{brwu} = \frac{\sqrt{3}\,V_{br}E_m}{2c}(1 \pm \text{WF})\ln\left(\frac{1}{1-\zeta}\right) \tag{10-11}$$

When n is other than unity, Eq. 10-9 can be thought of as the range corrected for wind and given the symbol X_{wc}. Obviously, when n takes on the appropriate value of n_{br} for a given WF, X_{wc} becomes X_{brwc}. Defining the *relative range* R as the ratio of X_{wc} to X_{brwu} for a given fuel weight, R can be expressed as

$$R = \frac{X_{wc}}{X_{brwu}} = 4\left(\frac{n^2}{3n^2+1}\right)\left(\frac{n \pm \text{WF}}{1 \pm \text{WF}}\right) \tag{10-12}$$

When $n = n_{br}$, R represents the best relative range *for a given fuel weight*, that is, the maximum improvement in range that can be obtained by correcting the no-wind best-range airspeed. Figure 10-1 shows that the improvement in range with a tailwind is slight, being less than 2 percent for a WF of 0.5 and calls for a 10 percent

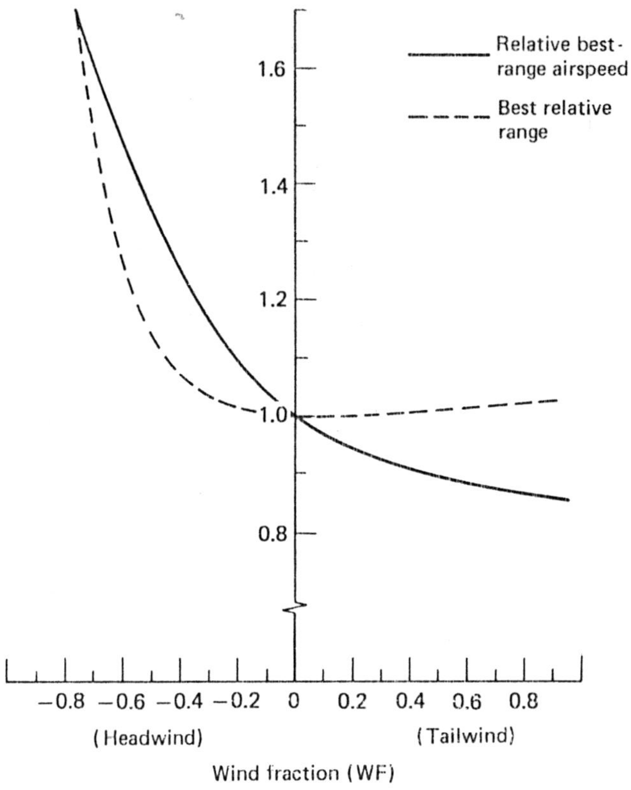

FIGURE 10-1

Relative best-range airspeed and best relative range for a turbojet

reduction in airspeed with an accompanying increase in the flight time. The improvement in range with a headwind is appreciably larger but does not become significant until the wind fraction becomes of the order of −0.3 or larger. For example, with a WF of −0.4, a 6.4 percent increase in range can be obtained by increasing the no-wind best-range airspeed by 22 percent.

It may not be possible or desirable to fly at the appropriate value of n_{br}. Figure 10-2 is a plot of the relative range as a function of the relative airspeed for various values of the wind fraction, with emphasis on headwinds. Examination of the headwind curves shows that (1) the relative range is relatively insensitive to the value of n in the vicinity of n_{br}; (2) for larger values of the headwind fraction, significant improvements in range can be attained for values of n less than n_{br}; and (3) flying at values of less than unity introduces range penalties that increase rapidly as the wind fraction increases. Looking at the single tailwind curve, we see that decreasing the airspeed beyond the best-range point, say, to maintain a block time, results in a rapidly increasing range penalty.

It must not be forgotten that the baseline for the relative range is the uncorrected best range in the presence of a wind and not the no-wind best range. In order to

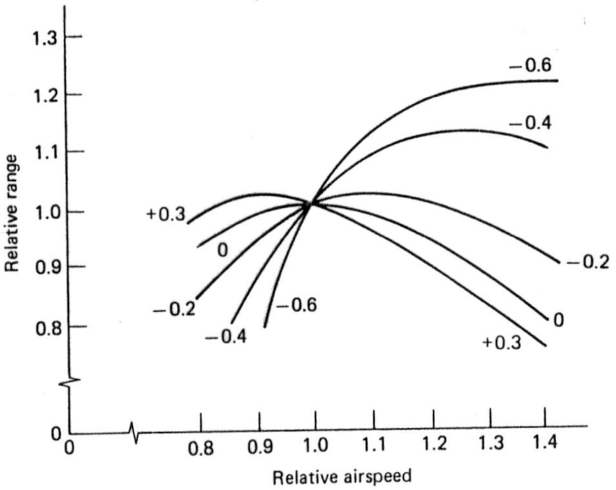

FIGURE 10-2

Relative range as a function of the relative airspeed and wind fraction for a turbojet.

compare the effect of the wind on the no-wind best range, multiply the relative range by $1 \pm \mathrm{WF}$. Thus, a relative range of 1.25 for a headwind fraction of -0.6 indicates a corrected best range of 0.5 of the no-wind best range. If there were no airspeed correction for the headwind, the relative range would be unity and the actual range would be 0.4 of the no-wind best range.

Before examining the effects of an airspeed correction upon the flight time and fuel consumption for a given range, a more practical operational situation, the best-range conditions for *piston-props* will be developed. The relationships of Eq. 10-2 are still valid, but the weight balance equation is

$$\frac{-dW}{dt} = \frac{\hat{c}P}{k\eta_p} \tag{10-13}$$

where P, the thrust power, is the product of the thrust and the true airspeed, η_p is the propeller efficiency, k is a horsepower conversion factor, and $\hat{c}$ is the horsepower specific fuel consumption, assumed to be constant. Combining equations and integrating yields the Breguet range equation in the form

$$X = \frac{k\eta_p E(1 \pm \mathrm{WF})}{\hat{c}} \ln\left(\frac{1}{1-\zeta}\right) \tag{10-14}$$

The definition of the relative airspeed parameter remains unchanged; however, the no-wind best-range airspeed for a piston-prop is

$$V_{\mathrm{br}} = \left[\frac{2(W/S)}{\rho_{\mathrm{SL}}\sigma}\right]^{1/2} \left(\frac{K}{C_{DO}}\right)^{1/4} = V_{\mathrm{md}} \tag{10-15}$$

The lift-to-drag ratio can be written as

$$E=\left(\frac{2n^2}{n^4+1}\right)E_m \tag{10-16}$$

and the range equation in the presence of a wind becomes

$$X=\frac{2k\eta_p E_m(n\pm \mathrm{WF})}{\hat{c}}\left(\frac{n}{n^4+1}\right)\ln\left(\frac{1}{1-\zeta}\right) \tag{10-17}$$

Maximizing the range with respect to n results in the best-range condition that

$$n_{\mathrm{br}}=\left(\frac{2n_{\mathrm{br}}\pm \mathrm{WF}}{2n_{\mathrm{br}}\pm 3\mathrm{WF}}\right)^{1/4} \tag{10-18}$$

Again using the uncorrected best range for a given fuel weight as the baseline, the relative range is

$$R=2\left(\frac{n}{n^4+1}\right)\left(\frac{n\pm \mathrm{WF}}{1\pm \mathrm{WF}}\right) \tag{10-19}$$

Figure 10-3 shows both the relative best-range airspeed and the best relative range as a function of the wind fraction, and Fig. 10-4 shows the relative range as a function of the relative airspeed for several values of the wind fraction. Comparison with Figs. 10-1 and 10-2 shows similar relationships with corresponding values for

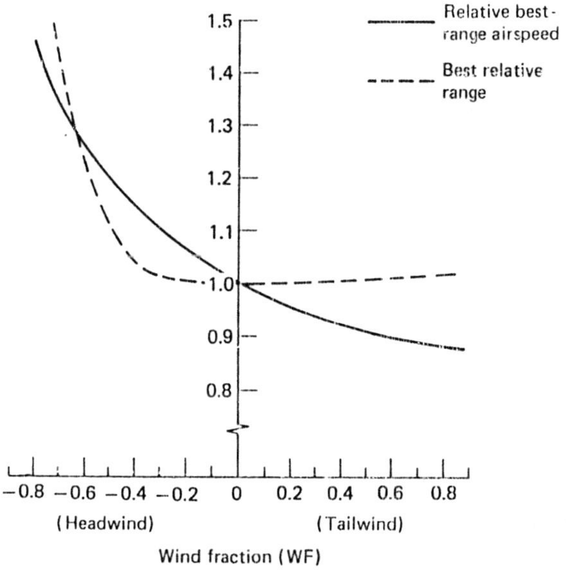

FIGURE 10-3

Relative best-range airspeed and best relative range for a piston-prop.

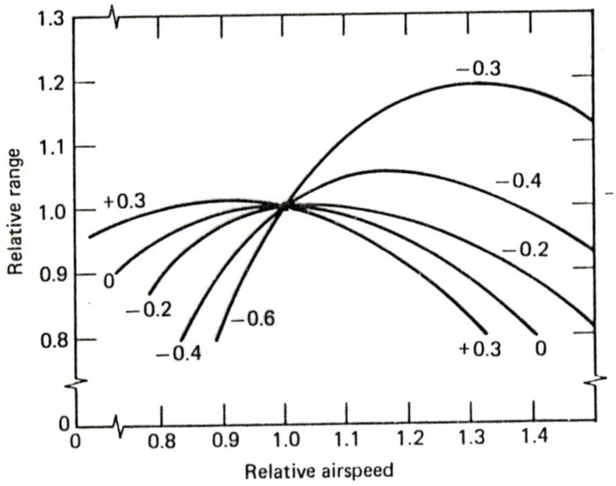

FIGURE 10-4

Relative range as a function of the relative airspeed and wind fraction for a piston-prop.

the piston-prop generally lower than those for the turbojet. The airspeed limit for an increasing tailwind is the minimum-power airspeed.

10-2-2 Flight Time and Fuel Consumption

The results of the preceding section will now be applied to a fixed range, such as an airline route segment, in order to determine the effects of wind on both the flight time (and thus the block time) and the fuel consumption. The *relative flight time* (RFT) is defined as the ratio of the flight time with an airspeed correction to the flight time using the no-wind best-range airspeed. An expression for the RFT that is applicable to both turbojets and piston-props is

$$\mathrm{RFT}=\frac{1\pm \mathrm{WF}}{n\pm \mathrm{WF}} \tag{10-20}$$

Figure 10-5 shows the RFT as a function of the relative airspeed for various wind fractions, again with emphasis on headwinds.

The *relative fuel consumption* (RFC) is defined as the ratio of the cruise-fuel weight fraction with a corrected airspeed (ζ_c) to the cruise-fuel weight fraction with the no-wind best-range airspeed (ζ_u). The expression for the RFC, again applicable to both turbojets and piston-props, is

$$\mathrm{RFC}=\frac{\zeta_c}{\zeta_u}=\frac{1-(1-\zeta_u)^{1/R}}{\zeta_u} \tag{10-21}$$

In Eq. 10-21, the value of R to be used in the exponent is obtained from Eq. 10-12 (or Fig. 10-2) for a turbojet and from Eq. 10-19 (or Fig. 10-4) for a piston-prop.

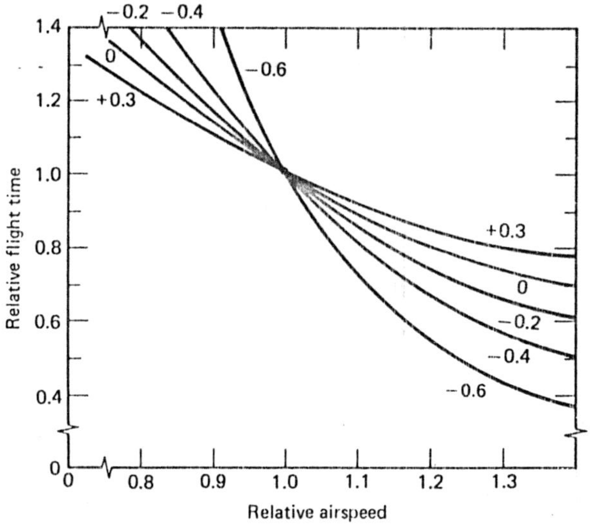

FIGURE 10-5
Relative flight time as a function of the relative airspeed and wind fraction.

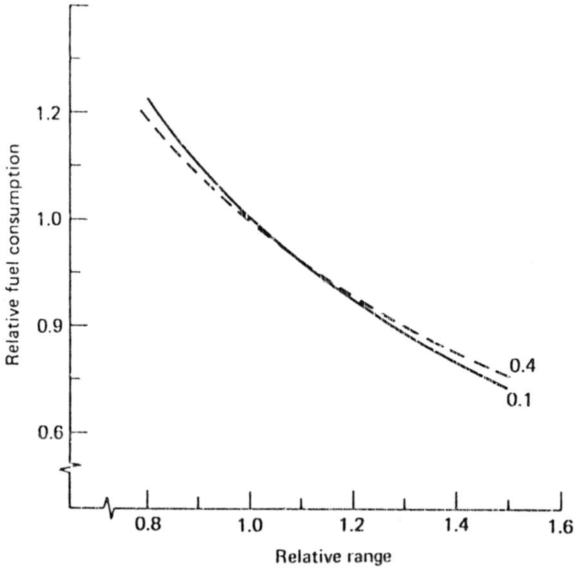

FIGURE 10-6
Relative fuel consumption as a function of the relative range and the uncorrected cruise-fuel weight fraction.

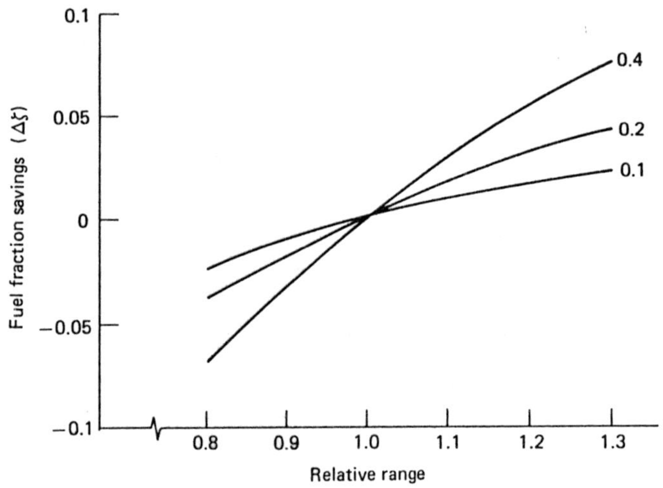

FIGURE 10-7

Relative fuel savings as a function of the relative range and the uncorrected cruise-fuel weight fraction.

Figure 10-6 shows the RFC as a function of the relative range for two values of the uncorrected cruise-fuel weight fraction; the larger the weight fraction, the greater the actual range.

The *fuel fraction savings* $\Delta\zeta$ and the *actual fuel savings* W_{fs} are of greater interest and significance than the RFC and can be found from the relationships

$$\Delta\zeta = \frac{W_{fs}}{W_1} = \zeta_u(1 - \text{RFC}) \tag{10-22}$$

where W_1 is the gross weight of the aircraft at the start of cruise. A negative $\Delta\zeta$ represents an increase in the fuel consumption rather than a savings. Figure 10-7 shows that the greater the range to be flown (the larger the ζ_u), the larger the fuel fraction savings are. In addition, Eq. 10-22 also shows that the heavier the aircraft is at start of cruise, the greater the actual fuel savings.

In view of the similarity of the relative airspeed and relative range curves for turbojet and piston-prop aircraft, no attempt has been made to develop specific relationships for turbofans and turboprops. Instead, the turbojet and piston-prop solutions will serve as upper and lower limits, respectively, using wind fractions based on the manufacturer's posted no-wind best-range airspeed. For turboprops, the piston-prop solutions can be used with very little loss in accuracy, but turbofans will require some judgment, based on the bypass ratio, as to the values to be used.

10-2-3 Conclusions and Examples

Since the endurance of an aircraft is unaffected by the wind, there is no need to correct the no-wind airspeed for endurance. The range, however, is obviously affected by the wind, beneficially by a tailwind and adversely by a headwind. The range can be improved by decreasing the airspeed for a tailwind and increasing it for a headwind. An improvement in range translates into a fuel savings for a fixed range, or into a payload increase.

With a knowledge of the magnitude of the tailwind or headwind component and of the no-wind best-range airspeed, it is possible to determine the effects of a wind upon the range performance along with the airspeed corrections to be made. For a given wind fraction, a turbojet requires larger airspeed changes but achieves larger improvements in range than do the piston-prop and turboprop. The turbofan falls somewhere in between, depending upon the bypass ratio.

In the presence of a tailwind, the improvement in range (and fuel savings) is small, being less than 2 percent for a wind fraction of +0.5 and is accompanied by a 10 percent reduction in the airspeed. If the airspeed is to be reduced in order to meet scheduled block times, the fuel consumption will also decrease until the best-range airspeed for that particular tailwind is reached. Further airspeed reductions, however, will progressively increase the fuel consumption.

In the presence of a headwind, relatively large increases in airspeed are required, and the improvement in range is appreciably larger than that for a tailwind but does not become significant until the wind fraction becomes of the order of −0.3 or larger. The increased airspeed also reduces the flight time and thus the block time. For a given headwind and two different aircraft, the aircraft with the lower no-wind best-range airspeed will experience a higher wind fraction, leading to larger increases in both airspeed and relative range. In addition, the larger the specified range is, the greater the fuel savings will be. For each wind fraction, there is a best-range relative airspeed which yields maximum benefits but which may not be operationally attainable. In such a situation, fly as fast as possible.

Let us now look at the range performance of a series of aircraft, ranging from a large M 0.8 transport down to a single-engine piston-prop, in the presence of a headwind. In the interests of uniformity, a headwind component of 106 mph and an uncorrected cruise-fuel fraction of 0.3 will be used for all of the examples, whether completely realistic or not. We shall use the appropriate equations, rather than the figures, to obtain any values needed for our solutions. Remember that any fuel savings obtained by correcting the airspeed for the wind effect can be exchanged for additional payload.

Let us first consider a wide-body turbofan with a no-wind best-range Mach number of 0.8 (528 mph in the stratosphere) and a drag-rise Mach number of 0.85. Long-range aircraft such as this one are characterized by high wing loadings and high cruising altitudes, resulting in theoretical no-wind best-range airspeeds that approach and even exceed the drag-rise Mach number. These high airspeeds not only make it impossible to increase the airspeed to the best relative airspeed but also tend to keep the value of the wind fraction low, even for the strongest winds. For this aircraft, for example, the wind fraction is −0.2 and the maximum relative airspeed is 1.06. A wind fraction of −0.2 calls for an n_{br} of 1.085 for a turbojet and of 1.06 for a piston-prop. With an n of 1.06, the relative range for a turbojet is 1.009, and for a piston-prop 1.007. Making allowance for the bypass ratio, let us use a value of 1.008 for R. For a fixed range, the flight time will be reduced by approximately 7 percent and, with our fuel-cruise fraction of 0.3, the reduction will be 1.98×10^{-3}. If the initial gross weight is 600,000 lb, the savings in fuel will be 1,190 lb (176 gal), or 0.6 percent of the uncorrected fuel consumption.

General aviation aircraft are characterized by lower wing loadings and lower no-wind best-range airspeeds, shorter ranges, and much lower gross weights than commercial aircraft. For example, a current corporate turbojet with a gross weight of the order of 20,000 lb has a no-wind best-range airspeed of 463 mph and a maximum cruise airspeed of 528 mph. Now the wind fraction is -0.23, and the uncorrected best-range ground speed is 357 mph. With the appropriate value of n_{br}, which is 1.10, the corrected best-range airspeed and ground speed are 509 and 403 mph, respectively, and the relative range is 1.014. For a fixed range, the flight time is reduced 11 percent, and the fuel savings will be of the order of 69 lb, or 1.2 percent of the uncorrected fuel consumption.

Now consider a twin-engine turboprop with a gross weight of 12,000 lb and a no-wind best-range airspeed of 250 mph. The wind fraction is -0.42, and the corresponding best relative airspeed is 1.16. Consequently, the corrected airspeed and ground speed are 290 and 184 mph, respectively, and the relative range is 1.053. The flight time for a fixed range will be reduced by 22 percent, and the fuel savings will be approximately 152 lb, or 4.2 percent of the uncorrected fuel consumption.

The final example is a single-engine piston-prop with a gross weight of 3,400 lb, a no-wind best-range airspeed of 156 mph, and a maximum cruising airspeed of 200 mph. The wind fraction increases to -0.68 so that n_{br} is of the order of 1.33 and the corrected airspeed should be 207 mph. Limiting the airspeed to the maximum cruising value of 200 mph (a ground speed of 94 mph) reduces n to 1.28, and the corresponding relative range is 1.3. The flight time will be reduced by 47 percent, and the fuel savings will be 206 lb, or 22 percent, with an uncorrected cruise-fuel fraction of 0.3.

Let us examine this last example in more detail in order to see what these savings, corrections, and parameters really mean in a practical sense. First, a cruise-fuel fraction of 0.3 is reasonable for a long-range aircraft but not for the low-performance aircraft of this example. A more realistic value is 0.1, which for this aircraft corresponds to a fuel load of 340 lb. Let us also assume that the aircraft has a maximum lift-to-drag ratio of 16, a propeller efficiency of 85 percent, and a specific fuel consumption of 0.5 lb/h/hp. With no wind whatsoever and flying at 156 mph, the range would be 1,075 mi with a flight time of 6.9 hours. With a headwind of 106 mph, maintaining the airspeed at 156 mph results in a ground speed of only 50 mph; the flight time does not change but the range is drastically reduced to 345 mi. If we fly this 345 mi at 200 mph, the flight time is cut almost in half, from 6.9 to 3.7 hours, and there will be 75 lb of fuel in the tanks at the destination. If all the fuel is used, the range will be increased by 30 percent (the R of 1.3) to 448 mi.

We will close this section with a brief discussion of the assumptions leading to this simplified and analytic approach. Neglecting the wind and crab angle effects introduces maximum errors of the order of only 1.5 percent for a direct crosswind and a wind fraction of 0.5. A computer solution is required to properly account for the effects of the other assumptions. The assumption of a constant headwind or tailwind component ignores the variations of the wind velocity with range and altitude. The variations with range can be handled by updating the wind fraction and the resulting effects; treatment of the altitude dependence leads to trajectory optimization. The assumption of a constant specific fuel consumption implies operation at the engine design point with a flat curve and ignores the off-design point variation with altitude, Mach number, and thrust. The parabolic drag polar assumption can only be considered valid up to the region of the drag-rise Mach number and even then should be limited to aircraft of conventional design for which the parabolic drag polar is applicable.

10-3 OTHER FLIGHT

We are all aware of the operational practice of taking off and landing as directly into the wind as possible so as to maximize the headwind, but we may not be aware of how effective this practice is in shortening the take-off and landing rolls. For take-off the headwind has the effect of increasing the airspeed, increasing the lift and drag, affecting the thrust, and reducing the time (and distance) required to achieve the lift-off airspeed, which is determined by aerodynamic considerations. On landing, the headwind reduces the touchdown ground speed for a prescribed touchdown airspeed.

Returning to Sec. 4-1, the simplified take-off equation can be rewritten as

$$T = \frac{W}{g} V_g \frac{dV_g}{dX} \tag{10-23}$$

which, when integrated from zero ground speed to lift-off, becomes

$$d = \frac{V_{gLO}^2}{2g(T/W)} \tag{10-24}$$

Using Eq. 10-1, the lift-off ground speed can be expressed as

$$V_{gLO} = V_{LO}\left(1 \pm \frac{V_w}{V_{LO}}\right) \tag{10-25}$$

where V_{LO} is the lift-off airspeed, determined by the techniques of Sec. 4-1, and we have a wind fraction based on the lift-off airspeed. Expanding Eq. 10-25 gives

$$V_{gLO}^2 = V_{LO}^2\left[1 \pm 2\frac{V_w}{V_{LO}} + \left(\frac{V_w}{V_{LO}}\right)^2\right] \tag{10-26}$$

Since take-off wind fractions are of the order of 0.3 or less, the square of the wind fraction term can be neglected, leaving the approximation

$$V_{gLO}^2 \cong V_{LO}^2\left(1 \pm 2\frac{V_w}{V_{LO}}\right) \tag{10-27}$$

Substituting Eq. 10-27 into Eq. 10-24 yields the expression

$$d=\frac{V_{LO}^2}{2g(T/W)}\left(1\pm 2\frac{V_w}{V_{LO}}\right) \tag{10-28}$$

where the minus sign is associated with a headwind and the plus sign with a tailwind. Comparing Eq. 10-28 with Eq. 4-3, we see that they are identical with the exception of the wind factor $1\pm 2(V_w/V_{LO})$, which shortens the take-off roll for an upwind take-off and lengthens it for a downwind take-off.

Similarly, we can show that the time to lift off in the presence of a wind can be found from the expression

$$t=\frac{V_{LO}[1\pm(V_w/V_{LO})]}{g(T/W)} \tag{10-29}$$

In Sec. 4-1, we found that a long-range subsonic transport with a T/W ratio of 0.25 and a lift-off airspeed of 170 mph (260 fps) had a minimum take-off roll of 4,200 ft and an elapsed time of 32 seconds. If the component of the wind down the runway has a magnitude of 30 mph (44 fps), the wind fraction is equal to 0.176. Consequently, for a take-off into the wind, the take-off roll will be shortened to 2,722 ft, a 35 percent reduction, and the time to lift off is shortened to 26 seconds, an 18 percent reduction. If, however, the pilot for some unknown reason decides to take off downwind, the take-off distance increases by 35 percent to 5,678 ft and the elapsed time increases 18 percent to 37.6 seconds.

The wind has a similar effect on the landing roll and elapsed time. In the example in Sec. 4-1, the aircraft touched down at a stall speed of 108.4 mph (159 fps) and had a deceleration factor of 0.4 g's for a landing roll of the order of 5,000 ft with an elapsed time of 12 seconds. With the same runway wind component of 44 fps, the wind fraction becomes equal to 0.277. Landing into this wind reduces the landing roll to 2,230 ft, a reduction of over 50 percent, and shortens the time of roll to approximately 9 seconds. Landing downwind, on the other hand, increases the landing roll and time to 7,770 ft and 15 seconds.

Before leaving the effects of wind on take-off and landing, we must remember that all our analyses are simplified and approximate and have not considered the impact of rotation phases, nonstandard-day temperatures and pressures, nor of the requirements of the Federal Air Regulations (FARs) with respect to balanced field lengths, etc. For example, in operational take-off and landing calculations, the FARs state that only 50 percent of the reported headwind values can be used and that 150 percent of the reported tailwind values must be used.

With respect to climbing flight, a detailed analysis of all the effects of wind is not simple and not relevant to this book. We can say a few words about the subject, however. First of all, a horizontal wind with a constant velocity will not affect the rate of climb but will affect the climb angle and the horizontal distance traveled. A headwind component will increase the climb angle, whereas a tailwind component will decrease the angle. If the magnitude of the headwind or tailwind component changes with altitude, which is normal, we speak of a wind gradient, which does affect the climb performance. An increasing headwind during climb is called a positive gradient. A positive gradient will increase both the climb angle and the rate of climb. Conversely, a negative gradient will decrease the climb angle and rate of climb. Changes in the wind direction will change the magnitude of the headwind or tailwind component and have the same effect as a gradient.

Sudden or rapid changes in the magnitude of the horizontal wind component, either with time or with a change in altitude, are called *wind shears.* They usually occur close to the ground, often on a final approach to a landing, are hard to handle, and have been the cause of some major accidents. Vertical wind components are transitory and can contribute to wind shear. The vertical components often appear as *gusts,* which are oscillatory, usually in a random manner, and manifest themselves as turbulence, which stresses both the aircraft and its occupants.

PROBLEMS

10-1. Aircraft *D* is scheduled for a 500-mi flight at the best-range airspeed at 15,000 ft.

a. Find the no-wind cruise fuel required (lb and gal) and the associated flight time (h).

b. Do (a) at the same best-range airspeed but in the presence of a 30-mph headwind component.

c. With the proper headwind correction, what will the new cruise airspeed be? Find the fuel required and flight time and compare with the results from (b).

10-2. Do Prob. 10-1 for Aircraft *D*, only this time the baseline airspeed will be the no-wind 75 percent maximum power airspeed. Is this an improvement over the best-range airspeed?

10-3. Do Prob. 10-1 for Aircraft *D* but with a 30-mph tailwind.

10-4. Do Prob. 10-2 for Aircraft *D* but with a 30-mph tailwind.

10-5. Do Prob. 10-1 for Aircraft *E* at 20,000 ft for a range of 750 mph and with a 45-mph headwind.

10-6. Do Prob. 10-2 for Aircraft *E* with a 45-mph headwind.

10-7. Do Prob. 10-1 for Aircraft *E* with a 45-mph tailwind.

10-8. Do Prob. 10-2 for Aircraft *E* with a 45-mph tailwind.

10-9. Aircraft *A* is flying at 25,000 ft at its best-range airspeed with 4,000 lb of fuel available for cruise. Use a sfc of 0.85 lb/h/lb.

a. What is the best-range airspeed and how far can the aircraft fly (mi) and how long will it take (h) with a no-wind condition?

b. With a headwind component of 75 mph, do (a) without changing the airspeed.

c. Correct the airspeed for the headwind. What is the new airspeed? Find the range and time of flight using this airspeed.

10-10. Do Prob. 10-9 for a tailwind component of 75 mph.

10-11. Do Prob. 10-9 for Aircraft *B* flying at 30,000 ft at its best-range airspeed and with 24,000 lb of fuel available for cruise. Use an sfc of 0.8 lb/h/lb and do not exceed the drag-rise Mach number.

10-12. Do Prob. 10-9 for Aircraft *B* with a tailwind component of 75 mph.

10-13. Do Prob. 10-9 for Aircraft *C* flying at 35,000 ft at its best-range airspeed or drag-rise Mach number, whichever is lower. The fuel available for cruise is 150,000 lb. Use a sfc of 0.70 lb/h/lb and a headwind component of 150 mph.

10-14. Do Prob. 10-9 for Aircraft *C* with a tailwind component of 150 mph.

10-15. What effect will a 20-mph headwind have on the sea-level, standard-day take-off performance of

a. Aircraft *A*?

b. Aircraft *B*?

c. Aircraft *C*?

d. Aircraft *D*?

e. Aircraft *E*?

10-16. What effect will a 40-mph headwind and a density ratio of 0.8 have on the take-off performance of

a. Aircraft *A*?

b. Aircraft *B*?

c. Aircraft *C*?

d. Aircraft *D*?

e. Aircraft *E*?

10-17. Do Prob. 10-15 with a tailwind.

10-18. Do Prob. 10-16 with a tailwind.

Chapter Eleven

Stability and Control Considerations

11-1 INTRODUCTION

In our performance analyses and preliminary designs the aircraft has been treated as a point mass, and no consideration has been given to whether or not it is flyable. Obviously an aircraft that cannot be flown at all or cannot be flown in an acceptable manner is of little value, no matter how excellent its theoretical performance might be. In other words, in addition to meeting performance specifications, an aircraft must meet flying (handling) quality specifications as well.

The term *stability and control* is customarily used in the examination and description of the flying qualities of an aircraft. With an aircraft initially in equilibrium, *stability* can be thought of as the response of the basic, or bare, aircraft to a disturbance or to a specified input. *Control* can be thought of as the techniques and hardware involved in shaping or modifying this response so that the aircraft flies and maneuvers in an acceptable manner, acceptable being defined by the operational requirements and specifications.

Since the study of stability and control can be quite complex and is definitely beyond the scope of this book, the treatment in this chapter will of necessity be sketchy and brief. The objectives are to impart an awareness of the significance and possible implications of stability and control upon the design and performance of an aircraft and to introduce the reader to some of the vocabulary. The emphasis is primarily on longitudinal static stability, which deals with the sizing of the horizontal stabilizer, the location of both the stabilizer and the center of gravity of the aircraft, and the effects of these on performance.

Returning to stability, we differentiate between static stability and dynamic stability. The test for static stability is to consider an instantaneous and small displacement of the aircraft from a static (steady-state) equilibrium condition. If the first motion of the aircraft is back toward the original equilibrium condition, the aircraft is *statically stable*. If the first motion is away from the equilibrium condition, in the direction of the displacement, the aircraft is *statically unstable*. If the aircraft remains in equilibrium in the displaced attitude, i.e., does not move, the aircraft is *neutrally stable*. At this point you may wish to return to Sec. 3-2 and review the static stability of points 1 and 2 in Fig. 3-2.

To be dynamically stable, an aircraft must be statically stable; however, static stability does not guarantee dynamic stability. Consider, for example, a statically stable aircraft in level steady-state flight with its nose on the horizon. The nose is suddenly displaced a few degrees above the horizon by a force that is just as suddenly removed. Since the aircraft is statically stable, the nose will initially move toward the horizon. If the nose comes to rest the first time it reaches the horizon, the aircraft is *dynamically stable* and its transient response is said to be overdamped. If the nose oscillates about the horizon with an ever-decreasing amplitude so that it eventually comes to rest on the horizon, then the aircraft is still *dynamically stable* but has an underdamped transient response. If, however, the nose drops below the horizon and does not come back up or if the nose oscillates about the horizon with an ever-increasing amplitude, then the aircraft is *dynamically unstable*. The basic criteria for dynamic stability are that, in response to a disturbance, the resulting transient response disappears with time and that the aircraft returns to its original equilibrium attitude.

11-2 EQUATIONS AND COORDINATE SYSTEMS

Since we are now interested in the attitude of the aircraft, i.e., its rotational mode, we must consider the moment equations as well as the force equations. In their most general form, Newton's laws of motion can be written in vector form as

$$\mathbf{F} = \int \mathbf{a}\, dm \tag{11-1}$$

and

$$\mathbf{M} = \int \mathbf{r} \times \mathbf{a}\, dm \tag{112}$$

where $\mathbf{r}$ is the vector moment arm from the point about which the moments are taken to a particle mass. This point is generally the center of mass of the aircraft, customarily referred to as the center of gravity or cg.

With the assumption of rigidity, an aircraft has 6 degrees of freedom; 3 are translational and the other 3 are rotational. Stability and control analyses use *body* or *Euler* axes. They are similar to the wind axes that are used in performance analyses except that they are fixed to and rotate with the aircraft. The attitude of an aircraft with respect to the surface of the earth is described in terms of the three angles that relate the body axes to the local horizon axes. These three angles are called the *Euler angles*, have the symbols Θ, ψ, ϕ, and are respectively referred to as the *pitch angle*, the *yaw angle*, and the *roll angle*.

Moments about the center of gravity, the cg, follow the right-hand rule. Therefore, moments about the y axis, which are called the *pitching moments*, are positive when the nose comes up. The moments about the z axis are the *yawing moments* and are positive when the nose goes to the right. The moments about the x axis are the *rolling moments* and are positive with a roll to the right.

In examining the stability and control of a rigid symmetrical aircraft, it is customary to separate the 6 degrees of freedom into two groups of three each. The first group is called the symmetric degrees of freedom, describes the longitudinal motion of the aircraft, and comprises the x and z force equations and the pitching moment equation. The second group is called the asymmetric degrees of freedom, describes the lateral motion of the aircraft, and comprises the remaining three equations, namely, the y force equation and the yawing and rolling moment equations.

11-3 STATIC STABILITY

For an aircraft to be in static equilibrium, the sum of all the forces and moments must be equal to zero. Consequently, the longitudinal static-equilibrium equations are

$$F_x = 0 \tag{11-3}$$

$$F_z = 0 \tag{11-4}$$

$$M_y \equiv M = 0 \tag{11-5}$$

and the lateral static-equilibrium equations are

$$F_y = 0 \tag{11-6}$$

$$M_x \equiv L = 0 \tag{11-7}$$

$$M_z \equiv N = 0 \tag{11-8}$$

The conventional notation for the rolling, pitching, and yawing moments calls for script L's, M's, and N's rather than the style used above, primarily to avoid possible confusion with the symbols for lift and Mach number.

The vertical stabilizer on a conventional aircraft gives it inherent static stability in yaw. If the aircraft is given a slight change in yaw so as to produce a sideslip angle, the vertical stabilizer (part of the tail or empennage) will provide a restoring moment that will swing the nose back toward its equilibrium position in the same manner as the feathers of an arrow. An aircraft may or may not have inherent static stability in roll. Actually, most aircraft are slightly unstable in roll, both statically and dynamically (this instability is known as *spiral divergence*), so as to improve the Dutch Roll characteristics. Wing dihedral is helpful in increasing roll stability if such is desired.

Longitudinal static stability is of the greatest concern of the three since it is very sensitive to the location of the center of gravity of the aircraft (and thus to the cargo and passenger loading) and to the sizing and location of the horizontal stabilizer, both of which affect the structure (and thus the aircraft weight) and the performance of the aircraft. Consequently, we shall spend most of this chapter looking at longitudinal static stability and even then not in great detail.

11-4 LONGITUDINAL STATIC STABILITY

The governing equation for longitudinal static stability is the pitching moment equation, Eq. 11-5. The sum of the pitching moments can be written as

$$M = qS\bar{c}C_m \tag{11-9}$$

where C_m, the *pitching moment coefficient*, is a function of the angle of attack (and thus the lift coefficient) and is defined by the expression

$$C_m = \frac{M}{qS\bar{c}} \tag{11-10}$$

In Eqs. 11-9 and 11-10, S is the wing area, q is the dynamic pressure ($\frac{1}{2}\rho V^2$), and $\bar{c}$ is the mean aerodynamic chord (mac), which is not necessarily equal to the mean geometric chord that we used in our performance analyses. It is close enough in value, however, for us to use the value of S/b for the mac in our examples and problems.

The pitching moment coefficient is plotted as a function of the lift coefficient in Fig. 11-1 for two different aircraft. When C_m is equal to zero, Eq. 11-5 is satisfied and the aircraft is in equilibrium. We see that both aircraft have the same equilibrium lift coefficient, C_L. If Aircraft 1 is pitched up by some disturbance, such as a vertical gust, the immediate effect will be to increase the angle of attack and thus C_L, say, to point B. For this new C_L the corresponding pitching moment is positive (corresponding to point C), further increasing the angle of attack and C_L, which in turn increases C_m so that the nose comes up even farther, and so on. If no action is taken by the pilot or by an autopilot, the aircraft will continue to pitch up until it stalls. Aircraft 1 is statically (and dynamically) unstable.

If Aircraft 2 is disturbed in a similar manner to the same point B, the resulting pitching moment at D is negative, producing a pitchdown of the nose and moving the aircraft back toward C_{L_A}. Aircraft 2 is statically stable, but not necessarily dynamically stable. A disturbance that decreases C_m to point E by nosing the aircraft down results in a negative C_m and a further pitching down for Aircraft 1, an unstable reaction, and a positive C_m and pitchup for Aircraft 2, a stable reaction.

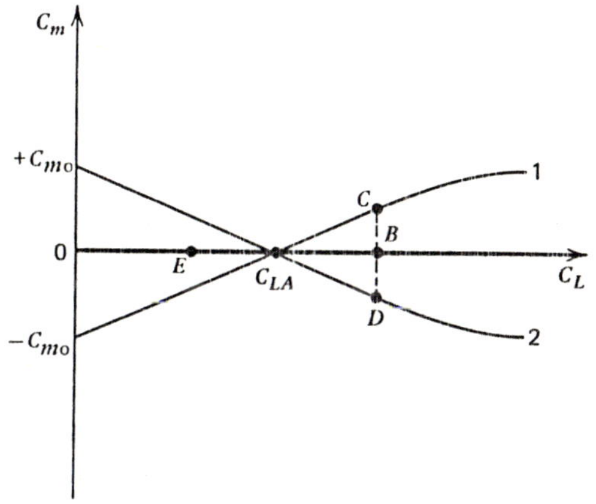

FIGURE 11-1

The pitching moment coefficient C_m as a function of the lift coefficient C_L.

Figure 11-1 can be used to establish a criterion *for static stability that the slope of the pitching moment curve be negative* ($dC_m/dC_L<0$). If the slope is positive ($dC_m/dC_L>0$), then the aircraft is *statically unstable.* Finally, if the slope is zero ($dC_m/dC_L=0$), then the aircraft is *neutrally stable.* For an aircraft to be useful, it must also have an equilibrium condition, which simply means that there must be a C_L at which C_m is equal to zero. The curves of Fig. 11-2 show that the requirement for equilibrium adds the condition that C_{mo} *be greater than zero,* where C_{mo} is the pitching moment coefficient when C_L is zero. (C_{mo} is the zero-lift pitching moment coefficient.) When the pitching moment curve can be approximated by a straight line, it can be represented by the equation for a straight line, written as

$$C_m=C_{mo}+\frac{dC_m}{dC_L}C_L \tag{11-11}$$

The two principal contributors to the pitching moment of a conventional aircraft are the lift of the wing and the lift of the horizontal stabilizer, which is usually a tail. For aircraft without a horizontal stabilizer ("tailless" aircraft), the pitching moment of the wing about its aerodynamic center (ac) can be of significance. The drag of the wing and of the tail can be neglected to a first approximation as can the pitching moment of the tail about its aerodynamic center and any pitching moments generated by other elements of the aircraft, such as the fuselage, engine nacelles, and engine thrust.

Figure 11-3 shows the major pitching moments about the cg of the aircraft, to include the wing's pitching moment about its aerodynamic center, M_{ac}. The angle of attack of the wing is assumed to be sufficiently small so that its sine can be set equal to zero and its cosine equal to unity. The distances to the aerodynamic center of the wing and to the cg of the aircraft are measured from the leading edge of the

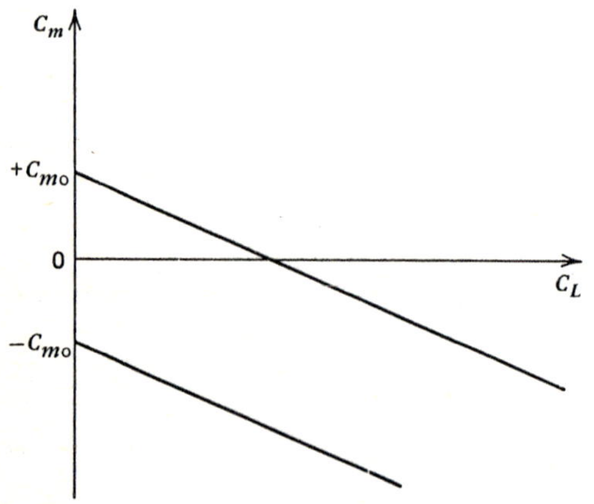

FIGURE 11-2

Equilibrium for a statically stable aircraft as a function of C_{mo}.

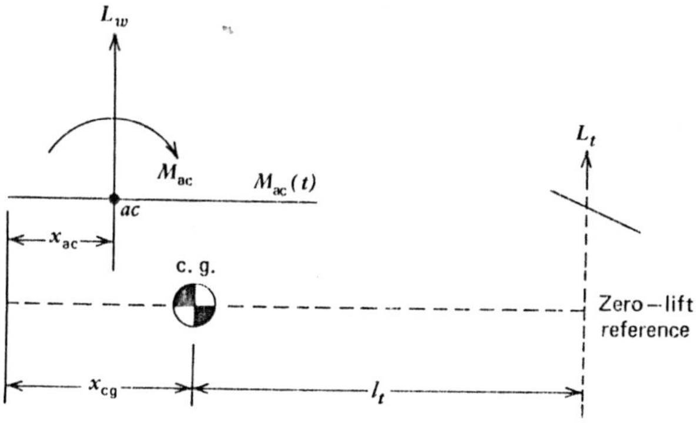

FIGURE 11-3

Major pitching moments from wing and tail.

mac (which is located and marked on the fuselage reference line) and are positive to the rear. This figure shows the horizontal stabilizer to the rear of the cg and called a *tail*; this is the conventional location. The horizontal stabilizer can be mounted ahead of the cg, in which case it is called a *canard.* Orville and Wilbur Wright used a canard, and canards are reappearing on certain modern aircraft. Tail, and canard, lengths are measured from the cg to the ac of the horizontal stabilizer.

Summing moments about the cg yields the following equation:

$$M = q_w S_w \bar{c} C_m = M_{ac} + L(x_{cg} - x_{ac}) - L_t l_t \tag{11-12}$$

Dividing through by $q_w S_w \bar{c}$ leaves the expression for the pitching moment coefficient as

$$C_m = C_{mac} + \left(\frac{x_{cg}}{\bar{c}} - \frac{x_{ac}}{\bar{c}}\right) C_{L_w} - \left(\frac{q_t}{q_w}\right)\left(\frac{S_t l_t}{S_w \bar{c}}\right) C_{L_t} \tag{11-13}$$

The C_{mac} of the wing is independent of the angle of the attack, and thus of the lift coefficient. It is zero for a symmetrical-airfoil wing, negative for a wing with positive camber, and positive for a wing with negative camber. The dynamic pressure of the tail can be less than that of the wing due to wing wake, slipstream, etc. The ratio of q_t to q_w is called the *tail efficiency* and is given the symbol η_t. Every attempt is made to keep it as close to unity as possible. The ratio of $S_t l_t$ to $S_w \bar{c}$ is called the *tail volume coefficient*, given the symbol V_H, and is of the order of 0.6 for a conventional subsonic transport. For a canard, this ratio would obviously be called the *canard volume coefficient* and would have a smaller value.

The slope of the pitching moment curve with respect to the lift coefficient of the wing can be written as

$$\frac{dC_m}{dC_L} = \left(\frac{x_{cg}}{\bar{c}} - \frac{x_{ac}}{\bar{c}}\right) - \left(\frac{a_t}{a_w}\right) \eta_t V_H \frac{d\alpha_t}{d\alpha_w} \tag{11-14}$$

where a_w is the slope of the wing lift curve and a_t is the slope of the tail lift curve.

Since C_{mac} is independent of C_{Lw}, it does not appear in Eq. 11-14 and therefore has no influence on the stability. The first term on the right-hand side of Eq. 11-14 reveals the importance of the cg location with respect to the ac of the wing. When the cg is behind the ac, $x_{cg} > x_{ac}$, the term is positive and thus destabilizing. If the cg is ahead of the ac, $x_{cg} < x_{ac}$, the term is negative and stabilizing. The slope of the pitching moment curve and the cg location, expressed as a percentage of the mac, have a one-to-one relationship. If $x_{cg}/\bar{c}$ is increased a tenth (the cg is shifted $0.1\bar{c}$ to the rear), the slope becomes a tenth more positive and correspondingly less stable.

The second term on the right-hand side represents the contribution of the horizontal tail to the static stability of the aircraft. The first factor is the ratio of the slope of the tail lift curve to the slope of the wing lift curve. If the tail is aerodynamically similar to the wing, this factor is unity. [For large aircraft the aspect ratio (AR) of the tail is usually less than that of the wing so that this factor is a little less than unity.] The next two factors are the tail efficiency and the tail volume coefficient; they have already been mentioned. The remaining factor in this term is the derivative of the tail angle of attack with respect to the wing angle of attack, namely, $d\alpha_t/d\alpha_w$, which will be shown in the next paragraph to be always positive. Consequently, since all the factors are positive and the sign of the term is negative, the contribution of the horizontal stabilizer is always negative and stabilizing if it is located behind the cg (a tail) and positive and destabilizing if ahead of the cg (a canard).

The *incidence angles* of the wing and tail, i_w and i_t, are the angles that their mean chord lines make with the fuselage reference line. The *downwash angle* ε is the change in the relative wind direction as it passes over the wing. From the relationships shown in Fig. 11-4, the angle of attack of the tail can be written as

$$\alpha_t = \alpha_w - \varepsilon + i_t - i_w \tag{11-15}$$

Therefore,

$$\frac{d\alpha_t}{d\alpha_w} = 1 - \frac{d\varepsilon}{d\alpha_w} \tag{11-16}$$

and Eq. 11-14 can now be written as

$$\frac{dC_m}{dC_{Lw}} = \left(\frac{x_{cg}}{\bar{c}} - \frac{x_{ac}}{\bar{c}}\right) - \left(1 - \frac{d\varepsilon}{d\alpha_w}\right)\left(\frac{a_t}{a_w}\right)\eta_t V_H \tag{11-17}$$

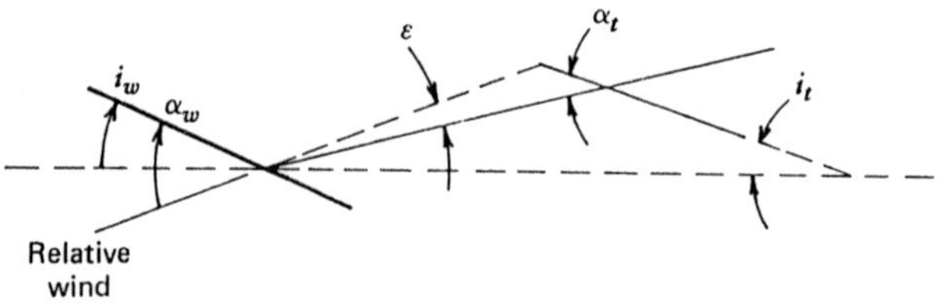

FIGURE 11-4

The tail angle of attack relationships.

The downwash angle can be expressed as

$$\varepsilon = \varepsilon_0 + \frac{d\varepsilon}{d\alpha_w}\alpha_w \tag{11-18}$$

where ε_0 is the zero-lift downwash angle; it is small and is equal to zero for a symmetrical-airfoil wing. The slope, $d\varepsilon/d\alpha_w$, is a complicated relationship among the wing characteristics and the size and location of the tail. It is always positive and of the order of 0.3 to 0.5.

If we express the tail lift coefficient as the product of its lift-curve slope and its angle of attack, Eq. 11-13 can be rewritten as

$$C_m = C_{mac} + \left(\frac{x_{cg}}{\bar{c}} - \frac{x_{ac}}{\bar{c}}\right) C_{L_w} - \eta_t V_H a_t \alpha_t \tag{11-19}$$

where α_t is defined by Eq. 11-15.

In addition to a negative pitching moment slope, an aircraft must have a positive C_{mo} in order to have an equilibrium condition. Setting C_{L_w} equal to zero in Eq. 11-19 and making use of Eq. 11-15 yields

$$C_{mo} = C_{mac} - \eta_t V_H a_t(\alpha_{wo} - \varepsilon_0 + i_t - i_w) \tag{11-20}$$

Without a horizontal stabilizer C_{mo} has a fixed value, namely, the value of C_{mac}, which is established by the wing camber. If the wing airfoil section is symmetrical, then $C_{mac} = \alpha_{wo} = \varepsilon_0 = 0$ and Eq. 11-20 reduces to

$$C_{mo} = \eta_t V_H a_t(i_w - i_t) \tag{11-21}$$

for an aircraft with a tail. We see that the tail may be needed to provide a suitable C_{mo} and that its value may be controlled by varying the incidence angle of the tail. Changing C_{mo} by changing i_t does not affect the slope of the pitching moment curve, and thus the stability, but does shift the equilibrium C_L, as can be seen in Fig. 11-5.

Returning to Eq. 11-11, the general form of a straight-line pitching moment curve, C_{mo} can be found from Eq. 11-21 (or 11-20), and dC_m/dC_L from Eq. 11-19.

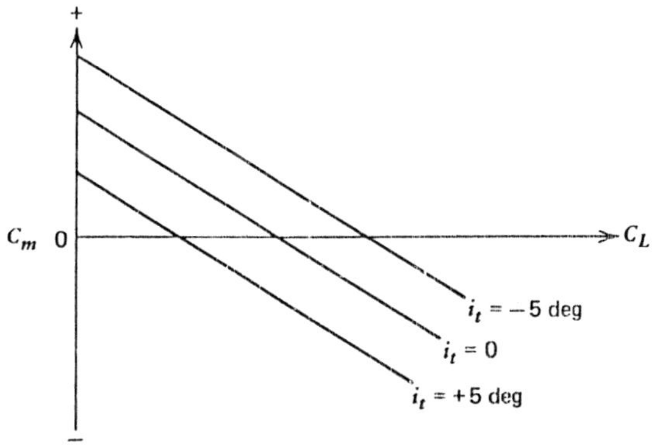

FIGURE 11-5

The static equilibrium condition as a function of the incidence angle of the tail.

The wing characteristics are determined by the wing design and the cg location by the operational requirements. The horizontal stabilizer is then used to obtain the desired stability or the desired equilibrium condition, or both. Table 11-1 shows some possible wing-stabilizer combinations along with one possible tailless configuration.

Horizontal stabilizers normally have symmetrical airfoil sections, as they have to generate negative as well as positive lift, and lower aspect ratios than wings, of the order of 4, primarily for structural and aeroelastic reasons. For most subsonic aircraft the ac of the wing is located in the vicinity of the quarter-chord point, that is, $x_{ac}=0.25\bar{c}$. As an aircraft becomes supersonic the ac moves toward the trailing edge.

Let us use the equations that we have developed in a numerical example, using a conventional subsonic aircraft with a wing loading of 100 lb/ft^2, a symmetrical wing with a sweep angle of 35 deg, and an aspect ratio of 8. The horizontal stabilizer has a symmetrical airfoil section, an aspect ratio of 4, and the same sweep angle as the wing, a customary design practice. The tail volume coefficient is 0.6, a typical value. The design cg is located $0.47\bar{c}$ behind the leading edge of the mac, that is, $x_{cg}/\bar{c}=0.47$.

Let us first find the slope of the pitching moment curve and see if the aircraft is statically stable. We will assume the tail efficiency to be unity and $d\varepsilon/d\alpha_w$ to be 0.33, a typical value for cruising flight. Using Eq. 2-9, we find the lift-curve slopes of the wing and tail to be 0.07 per degree and 0.06 per degree, respectively. Substituting values into Eq. 11-17, we get, with $x_{ac}/\bar{c}=0.25$,

$$\frac{dC_m}{dC_L}=(0.47-0.25)-(1-0.33)\times\left(\frac{0.06}{0.07}\right)\times 1\times 0.6$$

$$\frac{dC_m}{dC_L}\doteq -0.12$$

TABLE 11-1

Wing-Horizontal Stabilizer Combinations

CG Location	Wing Characteristics	Stabilizer Purpose
1. Behind ac, i.e., $x_{cg}-x_{ac}>0$ and wing alone is unstable.	a. Positive camber $C_{mac}<0$	a. Provide $+C_{mo}$ and negative slope.
	b. Negative camber $C_{mac}>0$	b. Provide negative slope only.
	c. Zero camber $C_{mac}=0$	c. Same as (a).
2. Ahead of ac, i.e., $x_{cg}-x_{ac}<0$ and wing alone is stable.	a. Positive camber	a. Provide $+C_{mo}$.
	b. Negative camber	b. None. (tailless)
	c. Zero camber	c. Same as 1(a).

Since the slope is negative, the aircraft is statically (but not necessarily dynamically) stable.

If the aircraft is designed to cruise at M 0.8 at 36,000 ft, let us find the C_{mo} required for equilibrium. The equilibrium C_L is

$$C_L = \frac{2(W/S)}{\rho_{SL}\sigma V^2} = \frac{2\times 100}{\rho_{SL}\times 0.297(774)^2} = 0.473$$

Setting $C_m = 0$, the equilibrium condition, in Eq. 11-1, we find

$$0 = C_{mo} - 0.12 \times 0.473$$

so that $C_{mo} = 0.057$. But from Eq. 11-21,

$$0.057 = 1 \times 0.6 \times 0.06(i_w - i_t)$$

and

$$i_w - i_t = 1.58 \text{ deg}$$

If the fuselage reference line (FRL) is horizontal and we would like the aircraft to be level in cruise so that the flight attendants will not have to push the beverage carts uphill, then the wing incidence angle should be set equal to the cruise angle of attack. With the cruise C_L equal to 0.473 and $a_w = 0.07$, the cruise angle of attack is 6.76 deg. Thus, i_w is set equal to 6.76 deg, and the tail incidence angle should be set equal to

$$i_t = 6.76 - 1.58 = 5.18 \text{ deg}$$

The angle of attack of the tail can be found from Eq. 11-15, using Eq. 11-18 to find ε, to be

$$\alpha_t = 6.76 - (0.33 \times 6.76) - 1.58) = 2.95 \text{ deg}$$

The tail lift coefficient is $0.06 \times 2.95 = 0.177$.

We shall develop this example further after we examine other aspects of longitudinal static stability.

11-5 STATIC MARGIN AND TRIM

An aircraft is designed with a design cg location, that is, a specific $x_{cg}/\bar{c}$ for a given set of design conditions. As the cg is moved toward the rear, the slope of the pitching moment curve becomes more positive (the aircraft becomes less stable statically), and the slope will eventually become equal to zero. That particular cg location, where $dC_m/dC_L = 0$, is the *neutral point* and is denoted by N_0. At that point the aircraft is neutrally stable, as you might suspect. If the cg is moved to the rear of the neutral point, the aircraft will become statically unstable with the degree of instability increasing with increasing distance behind the neutral point. The location of the neutral point, with respect to the leading edge of the mac, can be found from Eq. 11-17 by setting dC_m/dC_L equal to zero and replacing x_{cg} by N_0.

So doing and solving for $N_0/\bar{c}$ yields

$$\frac{N_0}{\bar{c}} = \frac{x_{ac}}{\bar{c}} + \left(1 - \frac{d\varepsilon}{d\alpha_w}\right)\left(\frac{at}{aw}\right)\eta_t V_H \tag{11-22}$$

The neutral point locates and specifies the most rearward position of the cg of a statically stable aircraft. The *static margin* (*SM*) of an aircraft indicates how far the cg can be moved behind the design cg location before the aircraft becomes neutrally stable, and then unstable. The static margin, by definition, is

$$\mathrm{SM} = \frac{N_0}{\bar{c}} - \frac{x_{cgd}}{\bar{c}} \tag{11-23}$$

where x_{cgd} is the design cg position. The static margin is numerically equal to the magnitude of dC_m/dC_L when the cg is actually at its design location. For the example of the previous section, the static margin is 0.12 and the neutral point is located $0.12\bar{c}$ ft behind the design cg location and $0.59\bar{c}$ ft behind the leading edge of the mac of the wing.

The static margin is of significance in loading all aircraft, especially transport aircraft. A typical value for the SM of a large cargo transport is of the order of 0.2 or less. This means that a C-5 transport with a mac length of approximately 30 ft and an SM of 0.2 must have its 200-ft cargo compartment so loaded that the actual cg position is less than 6 ft behind the design cg. If the mac were 10 ft, the maximum allowable cg shift to the rear would be 2 ft. If our example aircraft weighs 150,000 lb, the wing area will be 1,500 ft^2, the wing span 109.5 ft, and the mac will be of the order of 13.7 ft. The design cg will be 6.4 ft behind the leading edge of the mac. Since the static margin is 0.12, the neutral point will be only 1.6 ft behind the nominal cg and 8 ft behind the leading edge of the mac.

An aircraft with a given C_{m0} and a given slope of the pitching moment curve has only one static equilibrium condition, such as at point A in Fig. 11-6*a*. In level flight, the lift is equal to the weight and $C_L = 2(W/S)/\rho V^2$, so that the value of C_L at point A is unique for a specified combination of W, ρ, V, and, S. If we wish to cruise at a lower airspeed or with a larger weight without changing any of the other flight parameters, we must increase the lift coefficient, say, to point B, and find some way to set the pitching moment to zero so as to establish a new equilibrium point.

One way to increase the equilibrium lift coefficient the desired amount would be to shift the location of the cg an appropriate distance to the rear. C_{mo} will not change but the slope will become more positive, and the static stability will be reduced. The effect of a cg shift to the rear is shown in Fig. 11-6*b*. Shifting the cg forward decreases the equilibrium lift coefficient (also shown in Fig. 11-6*b*) and increases the static stability, still without changing C_{mo}. In addition to affecting the static stability, shifting the cg is not very practical even though the required cg movement is not necessarily large. For example, decreasing the airspeed of the aircraft of the preceding section from M 0.8 to M 0.75 increases the lift coefficient from 0.473 to 0.538. Using Eq. 11-11, the slope required for equilibrium decreases from -0.12 to -0.106, requiring a rearward shift of the cg of $0.014\bar{c}$, or 0.2 ft.

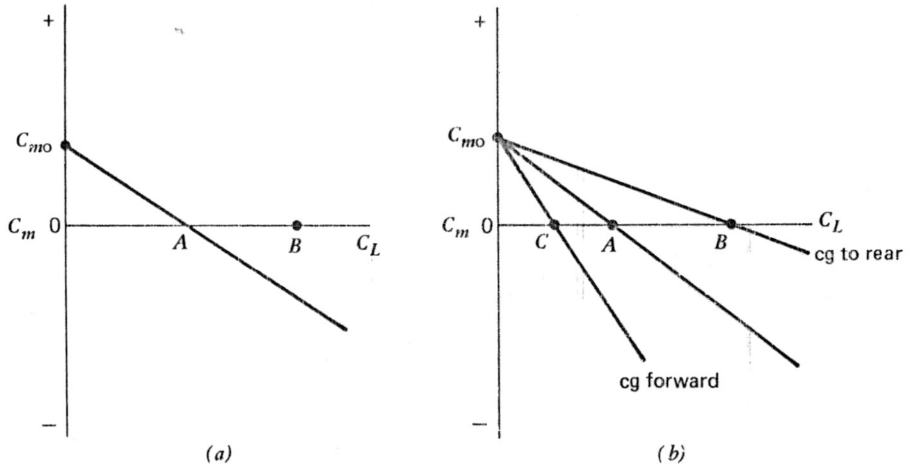

FIGURE 11-6

(*a*) Pitching moment coefficient curve for nominal cg location; (*b*) effect of moving cg.

Incidentally, the Wright Brothers put the pilot on a sliding pallet so that he could shift the cg somewhat and trim the aircraft.

A second way to change the equilibrium, or *trim*, condition is to add a movable section, called an *elevator*, to the trailing edge of the horizontal stabilizer. Deflecting the elevator generates a pitching moment that will change the equilibrium position without affecting the slope, and thus the static stability. If the elevator deflection is defined as in Fig. 11-7*a*, a positive deflection will result in a positive (pitchup) moment and will shift the trim point, as shown in Fig. 11-7*b*. The generalized pitch-

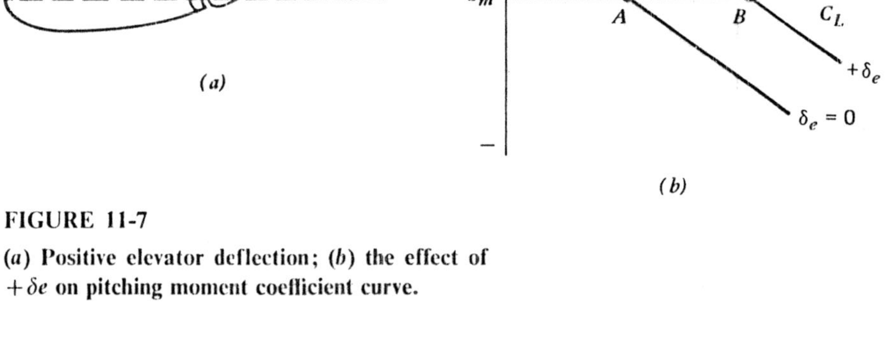

FIGURE 11-7

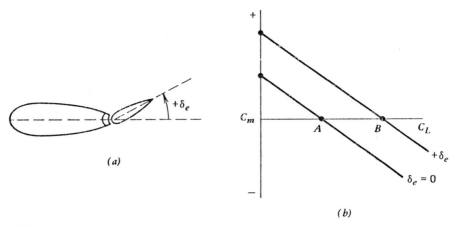

(*a*) Positive elevator deflection; (*b*) the effect of $+\delta e$ on pitching moment coefficient curve.

ing moment equation, Eq. 11-11, can now be expanded to

$$C_m = C_{mo} + \left(\frac{dC_m}{dC_{L_W}}\right) C_{Lw} + \left(\frac{dC_m}{d\delta_e}\right) \delta e \tag{11-24}$$

where the *elevator effectiveness* is

$$\frac{dC_m}{d\delta_e} = -\eta_t V_H a_t \frac{d\alpha_t}{d\delta_e} \tag{11-25}$$

In Eq. 11-24, the term $d\alpha_t/d\delta_e$ (the change in the tail angle of attack due to an elevator deflection) is an empirical function of the ratio of the elevator area to the total area of the stabilizer. It ranges from zero, for no elevator, to $-$one for all elevator and is usually of the order of -0.5. The product of the elevator effectiveness and the maximum elevator deflection (of the order of 25 deg) represents the maximum incremental pitching moment that can be generated by the elevator and is sometimes referred to as the *elevator strength.*

The neutral point defines the most rearward location of the cg in terms of static stability. The most forward cg location is determined by the elevator strength and the most demanding pitchup requirement, usually the landing flare. A good landing flare puts the aircraft in equilibrium at an airspeed close to the stall speed and thus at a lift coefficient close to the maximum lift coefficient. This is illustrated in Fig. 11-8, showing the steepening of the curve as the cg moves forward as well

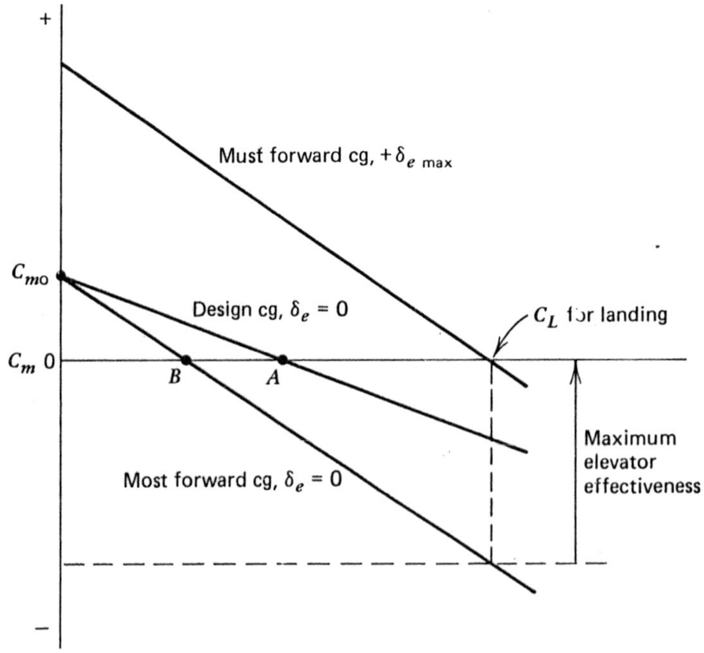

FIGURE 11-8

Most forward cg location as function of the maximum C_L required and maximum elevator effectiveness.

as the positive elevator strength. If the cg moves ahead of the most forward point, the elevator cannot produce enough positive pitching moment to make the total pitching moment equal to zero. In addition to trimming the aircraft, the elevator is also used for maneuvering.

As mentioned in Sec. 11-3 changing the tail incidence angle changes C_{mo} and thus the trim condition in the same manner as does an elevator deflection. Let $i_t = i_{to} + \delta i_t$, where i_{to} is the design incidence angle. Then Eq. 11-15 can be written, with the effect of an elevator deflection included, as

$$\alpha_t = (\alpha_w - \varepsilon + i_{to} - i_w) + \delta i_t - \left(\frac{d\alpha_t}{d\delta_e}\right)\delta e \tag{11-26}$$

and Eq. 11-24 can be further expanded to

$$C_m = C_{mo} + \left(\frac{dC_m}{dC_{L_w}}\right) C_{L_w} + \left(\frac{dC_m}{d\delta_e}\right)\delta e + \left(\frac{dC_m}{d\delta i_t}\right)\delta i_t \tag{11-27}$$

where the *tail effectiveness* is

$$\frac{dC_m}{d\delta i_t} = -\eta_t V_H a_t \tag{11-28}$$

Most large transport aircraft have variable incidence horizontal stabilizers with a limited range of δi_t to aid in trimming the aircraft as well as elevators for maneuvering. Many modern fighter aircraft, on the other hand, have no elevators per se and use the entire horizontal stabilizer as the sole longitudinal control surface. Such configurations are referred to as all-movable horizontal stabilizers or simply as *flying tails*. Comparing Eqs. 11-28 and 11-25 shows the greater effectiveness of the flying tail since $d\alpha_t/d\delta_e$ is always less than unity and $d\alpha_t/d\delta i_t$, which does not appear explicitly, is unity.

The pitching moments about the cg, and thus the static stability and trim conditions, are affected by such things as the fuselage and engine nacelle characteristics, vertical location of the cg with respect to the wing ac, thrust and slipstream, dive brakes, spoilers, flaps, and extended landing gear. The design of the elevator and its associated hardware is not a trivial task, especially if the elevator is connected directly to the pilot's control column. Among the factors that must be considered are stick forces per g, variations of the pitching moment curve and stick forces with airspeed and altitude, reversibility of controls, etc. Most modern aircraft, other than light-weight low-performance aircraft, use power-boost controls, analogous to the power steering in an automobile, to move the control surfaces.

11-6 STABILIZER SIZING, LIFT, AND DRAG

The tail volume coefficient of our example aircraft was arbitrarily chosen. Let us now actually determine the tail volume coefficient required to satisfy a set of operational requirements. Let us leave the cruise condition (M 0.8 and 36,000 ft) and the design SM (0.12) unchanged. The value of the SM is determined by dynamic stability considerations which are beyond the scope of this discussion. The most rearward cg position allowed will be the neutral point, although in practice it would be somewhat forward of this point. We would like the maximum allowable cg shift to be $0.3\bar{c}$, or 4.11 ft, for this particular aircraft. This cg shift is the distance between the most forward and most aft locations and is generally given the symbol $\Delta x_{\mathrm{cg,max}}$. The most forward cg position is to be determined by the landing flare and touchdown at a wing lift coefficient of 2.0. The maximum elevator deflections are +25 deg up and −15 deg down, typical values.

Since the slope of the pitching moment curve with the cg at its design location is equal to −SM, Eq. 11-17 can be used to find an expression for x_{cgd} in terms of V_H, as follows

$$-0.12=\left(\frac{x_{\mathrm{cgd}}}{\bar{c}}-0.25\right)-(1-0.33)\left(\frac{0.06}{0.07}\right)\times 1\times V_H$$

$$\frac{x_{\mathrm{cgd}}}{\bar{c}}=0.574V_H+0.13 \tag{a}$$

A similar expression for the neutral point can be found by setting the slope equal to zero in Eq. 11-17 or by remembering that the neutral point is $\mathrm{SM}\bar{c}$ units behind the design cg location.

$$\frac{N_0}{\bar{c}}=0.574V_H+0.25=\frac{x_{\mathrm{cgd}}}{\bar{c}}+0.12 \tag{b}$$

The next step is to find C_{mo} from Eq. 11-24 using the design equilibrium conditions with C_m and δ_e both set equal to zero.

$$0=C_{mo}-(0.12)\times 0.473$$

so that

$$C_{mo}=0.057$$

The elevator effectiveness can be found from Eq. 11-25, with $d\alpha_t/d\delta_e=-0.5$, to be

$$\frac{dC_m}{d\delta_e}=-1\times V_H\times 0.06\times -0.5=0.03V_H$$

Equation 11-17 is used again to find the slope at the most forward cg location, to wit

$$\frac{dC_m}{dC_L}=\left(\frac{x_{\mathrm{cgf}}}{\bar{c}}-0.25\right)-0.574V_H$$

This expression is substituted into Eq. 11-24, with $C_m=0$ and $\delta_{e_{\max}}=+25$ deg, to give an expression for the most forward cg location.

$$0=0.057+\left(\frac{x_{\mathrm{cgf}}}{\bar{c}}-0.25-0.574V_H\right)\times 2+0.03V_H\times 25$$

$$\frac{x_{\mathrm{cgf}}}{\bar{c}}=0.199V_H+0.222 \tag{c}$$

At this point a solution can be found either mathematically or graphically. The mathematical solution is obtained by subtracting Eq. (*c*) from Eq. (*b*) and setting the difference equal to 0.3, the maximum allowable cg shift. Doing so yields a value for the tail volume coefficient, which can be then used to solve for the other values of interest, namely,

$$V_H = 0.725$$

$$x_{xgd} = 0.546\bar{c}$$

$$N_0 = 0.666\bar{c}$$

$$x_{cgf} = 0.366c$$

$$i_w - i_t = 1.31 \text{ deg}$$

The graphical solution shown in Fig. 11-9, obtained by ploting Eqs. (a), (b), and (c), shows the effects of varying V_H. For example, since the value of 0.725 is a bit on the high side, we might wish to reduce it to 0.6. The cg shift is reduced to $0.25\bar{c}$, or from 4.11 to 3.42 ft, which may be acceptable, and the design cg moves forward, which may be desirable.

The requirement for a horizontal stabilizer introduces problems for the aircraft designer. First, it adds weight and introduces structural complexities that call for a

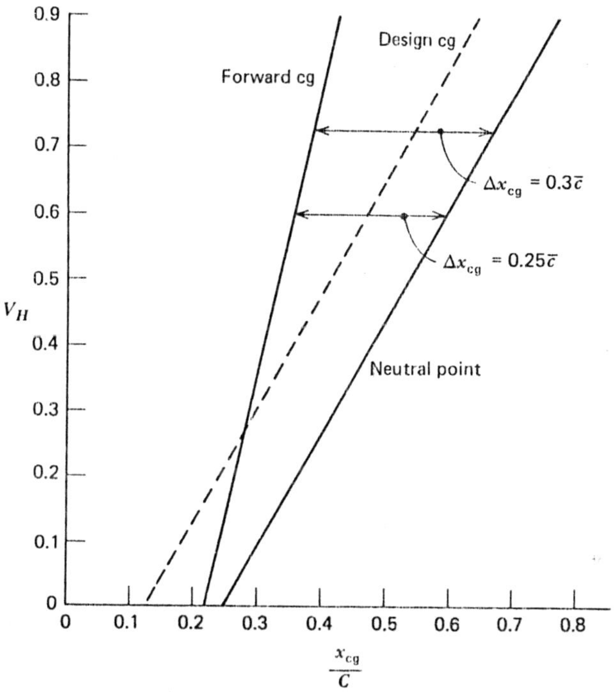

FIGURE 11-9

Graphical solution for sizing a horizontal stabilizer.

judicious trade-off between the tail length and the stabilizer area. If, for example, the former is increased in order to reduce the area and weight of the stabilizer, the weight of the connecting structure should be minimized, leading to the possibility of aeroelastic problems. If the tail length is kept short, the increased stabilizer area is accompanied by an increase in weight and by increases in the lift and drag generated by the stabilizer. These increases are directly proportional to the stabilizer area.

Let us look first at the lift. Figure 11-10*a* shows that with the cg behind the ac of the wing the tail lift is positive, thus increasing the net lift of the aircraft. Note that with this configuration the aircraft without a tail is statically unstable. If, however, the cg is ahead of the ac, as in Fig. 11-10*b*, the tailless aircraft is statically stable but the tail lift is negative, reducing the net lift of the aircraft. Accordingly, the wing area must be increased as well as the final approach and landing airspeeds. To give an idea as to the size of the tail lift, for our illustrative aircraft with a tail length three times that of the mac (41.1 ft), it is approximately one-tenth of the wing lift for design cruise with no elevator deflection required.

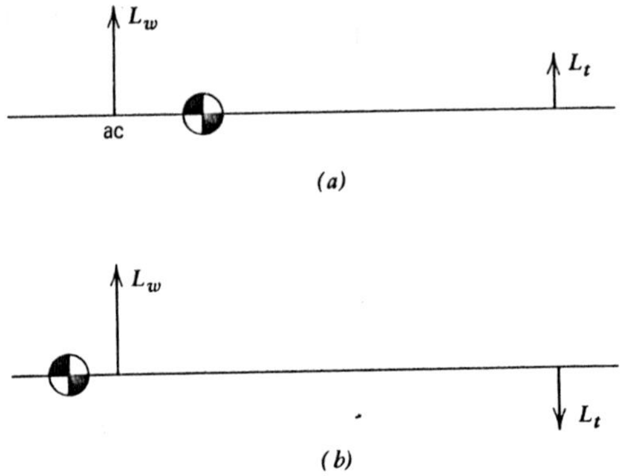

FIGURE 11-10

Relationships among tail lift, cg location, and net lift: (*a*) cg behind ac; (*b*) cg ahead of ac.

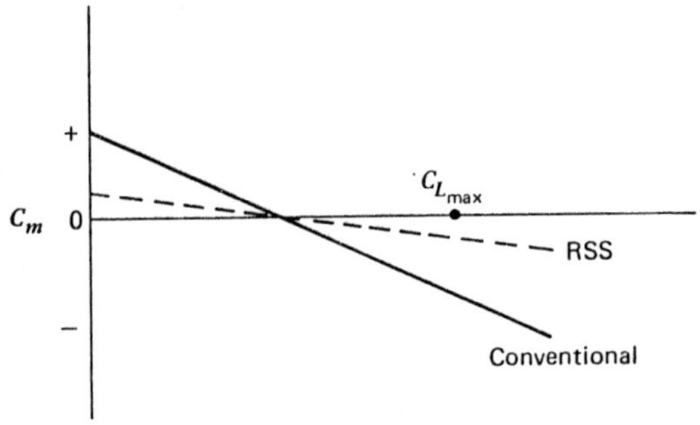

FIGURE 11-11

Pitching moment coefficient curves with a conventional static margin and with relaxed static stability (RSS).

The tail drag poses an even greater problem that is independent of the relationship between the cg and the ac. With our aircraft still in design cruise, the tail drag is of the order of 15 percent of the aircraft drag. This translates into a considerable loss in performance and increase in fuel consumption. If the static margin, and thus the stability, can be reduced, then the elevator strength required for equilibrium at high angles of attack is correspondingly reduced, as can be seen in Fig. 11-11. Consequently, a much smaller tail volume coefficient will suffice. There is, therefore, much interest in the concept of *relaxed static stability*, whereby a bare aircraft that is not flyable because of its lack of inherent stability is given the necessary stability, both static and dynamic, by an automatic control system.

11-7 DYNAMIC STABILITY AND RESPONSE

Static stability, as mentioned earlier, does not guarantee dynamic stability; it merely means that the first motion of an aircraft disturbed from an equilibrium condition is back toward the equilibrium position. Dynamic stability, on the other hand, is concerned with the behavior of the aircraft over a "long" period of time, where "long" may mean only a few seconds of time. Although static stability is a necessary condition for dynamic stability, it is not a sufficient condition; in fact, excessive static stability may result in dynamic instability.

Whereas static stability considers only the effects of small disturbances, dynamic stability deals with both disturbances and commands. Commands are control actions to change the flight path of the aircraft, e.g., change altitude, direction, airspeed, etc., in order to maneuver the aircraft. The dynamic response to a command or a disturbance input is the set of time responses (the time history) of the flight variables over a significant period of time. The time response can be separated into a transient response and a steady-state (final) response. If we are examining the pitch angle Θ, for example, its time response can be written as

$$\Theta(t)=\Theta_{tr}(t)+\Theta_{ss}(t) \tag{11-29}$$

The aircraft is *dynamically stable if the transient response* $\Theta_{tr}(t)$ *goes to zero* in the limit as time increases, leaving only the steady-state response $\Theta_{ss}(t)$ as an equilibrium condition. For disturbance inputs this equilibrium condition will normally be the original one, whereas for command inputs it will be a function of the type and magnitude of the command. The transient response can be exponentials, damped oscillations, or various combinations thereof. Typical time responses of a dynamically stable aircraft to a disturbance and to a command are shown in Fig. 11-12.

The transient response of an unstable aircraft will theoretically increase without limit as time increases, and there will be no steady-state response as such. In actuality, the transient response will be terminated by a stall or some other physical limitation, such as a structural failure or a crash. Typical unstable responses to any input are shown in Fig. 11-13. If the transient response is a bounded oscillation, as

GROUP III. TURBOJETS AND TURBOFANS

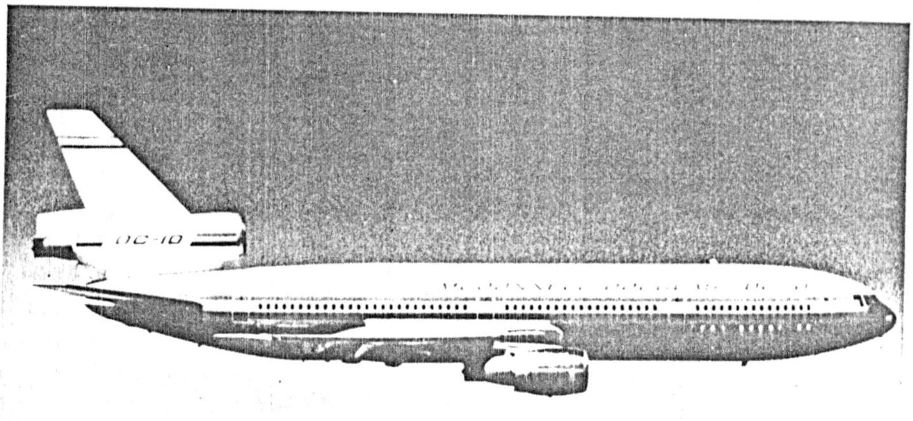

PLATE 20.

The Douglas DC-10. Manufactured by McDonnell Douglas Corporation. A wide-body long-range turbofan (three high bypass ratio turbofans of the 52,000-lb thrust class) with a maximum gross weight of 572,000 lb, a wing area of 3,958 ft^2, and an AR of 6.9.

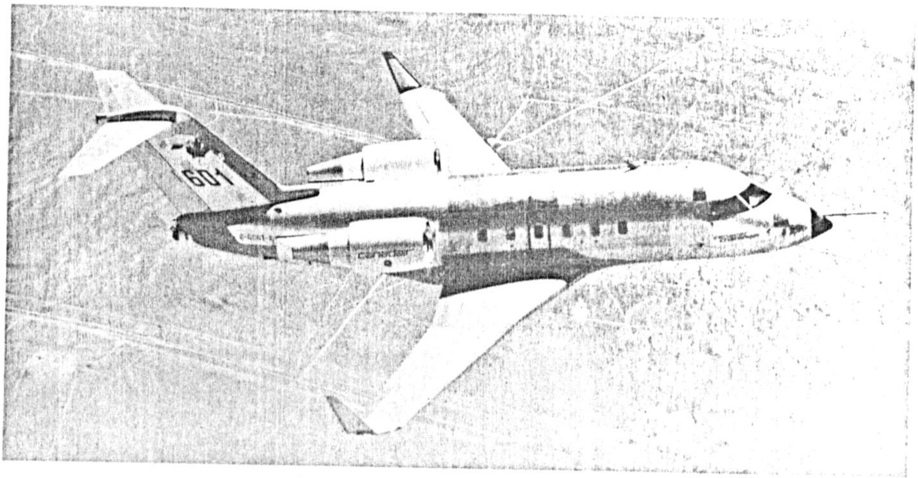

PLATE 21.

The Canadair Challenger 601. Manufactured by Canadair Limited. A 24-passenger wide-body business transport (two 8,650-lb thrust turbofans) with a maximum gross weight of 36,000 lb, a wing area of 450 ft^2, an AR of 8.5, a cruise airspeed of 495 mph (M 0.75), and a maximum range of 4,000 mi.

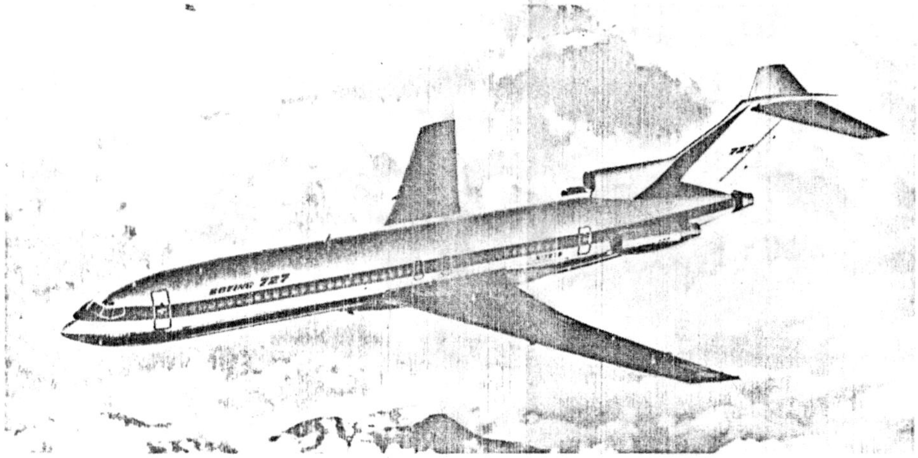

PLATE 22.

The Boeing 727-200. Manufactured by the Boeing Commercial Airplane Company. The world's best-selling airliner, a medium-range narrow-body transport (three low bypass ratio turbofans of the 14,000-lb thrust class) with a maximum gross weight of 210,000 lb, a wing area of 1,700 ft^2, and an AR of 6.9.

PLATE 23.

The BAe 146. Manufactured by British Aerospace Limited. A short-range, 70-90 passenger feederliner (four high bypass ratio turbofans of the 7,000-lb thrust class) with a maximum gross weight of 80,000 lb, a wing area of 832 ft^2, an AR of 9, a cruise airspeed of 436 mph (M 0.66), and a range of the order of 1,000 mi.

PLATE 24.

The Lockheed L-1011 TriStar. Manufactured by Lockheed-California Company. A long-range wide-body transport (three high bypass ratio turbofans of the 50,000-lb thrust class) with a maximum gross weight of 496,000 lb, a wing area of 3,456 ft^2, and an AR of 7.

PLATE 25.

The A300B2 Airbus. Manufactured by Airbus Industrie. A medium-range wide-body transport (two high bypass ratio turbofans of the 52,000-lb thrust class) with a maximum gross weight of 360,000 lb, a wing area of 2,800 ft^2, and an AR of 7.7.

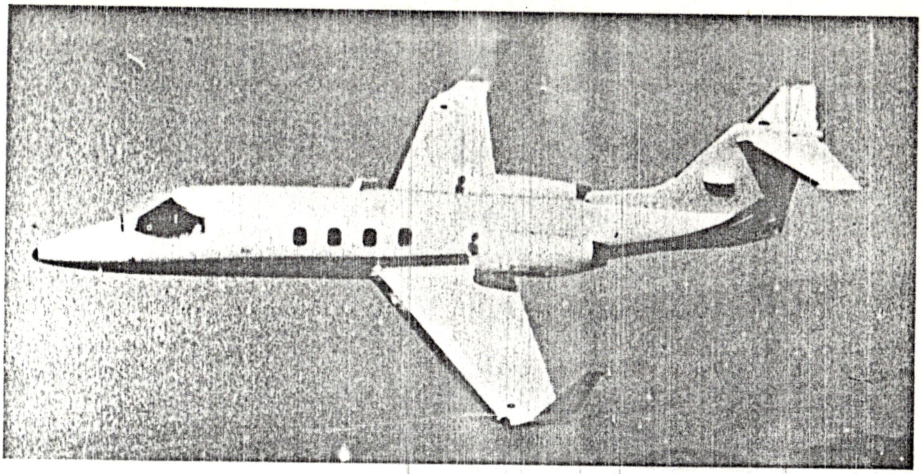

PLATE 26.

The Gates Learjet 56 Longhorn. Manufactured by Gates Learjet Aircraft Corporation. A 13-place executive transport (two 3,700-lb thrust turbofans) with a maximum gross weight of 20,500 lb, a wing area of 265 ft², an AR of 7.2, a cruise airspeed of 508 mph (M 0.77), and a range of 3,000 mi.

PLATE 27.

The Boeing 767. Manufactured by Boeing Commercial Airplane Company. A medium-range wide-body advanced technology transport (two high bypass ratio turbofans of the 48,000-lb thrust class) with a maximum gross weight of 310,000 lb, a wing area of 3,050 ft², and an AR of 8.

GROUP IV. PISTON-PROPS AND TURBOPROPS

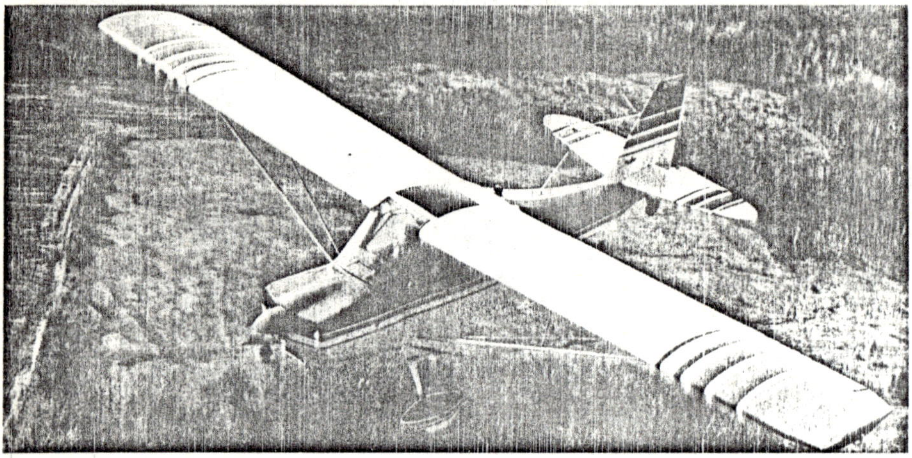

PLATE 28.

The Bellanca Scout. Manufactured by Bellanca Aircraft Corporation. A two-place piston-prop (one aspirated 180 hp engine) with a maximum gross weight of 2,150 lb, a wing area of 165 ft^2, an AR of 7.9, a maximum cruise airspeed of 125 mph, and a range of 385 mi.

PLATE 29.

The Cessna 210 Pressurized Centurion. Manufactured by the Cessna Aircraft Company. A six-place turbocharged piston-prop (one 300 hp engine) with cabin pressurization to an equivalent 12,125 ft at an altitude of 23,000 ft. Has a maximum gross weight of 3,800 lb, a wing area of 179 ft^2, an AR of 7.7, a maximum cruise airspeed of 197 mph, and a range of 1,200 mi.

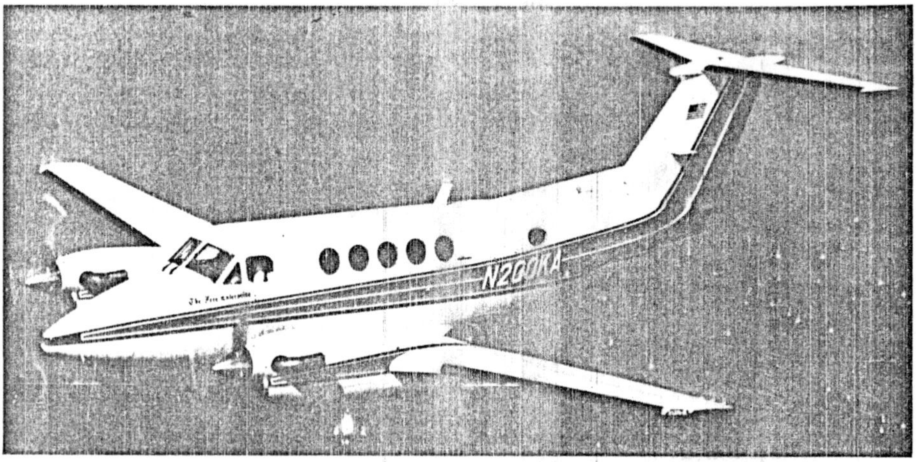

PLATE 30.

The Beech Super King Air 200. Manufactured by the Beech Aircraft Corporation. A ten-place business turboprop transport (two 750 eshp turboprop engines) with a maximum gross weight of 11,000 lb, a wing area of 280 ft^2, an AR of 7.5, a maximum cruise airspeed of 307 mph, and a range of 1,800 mi.

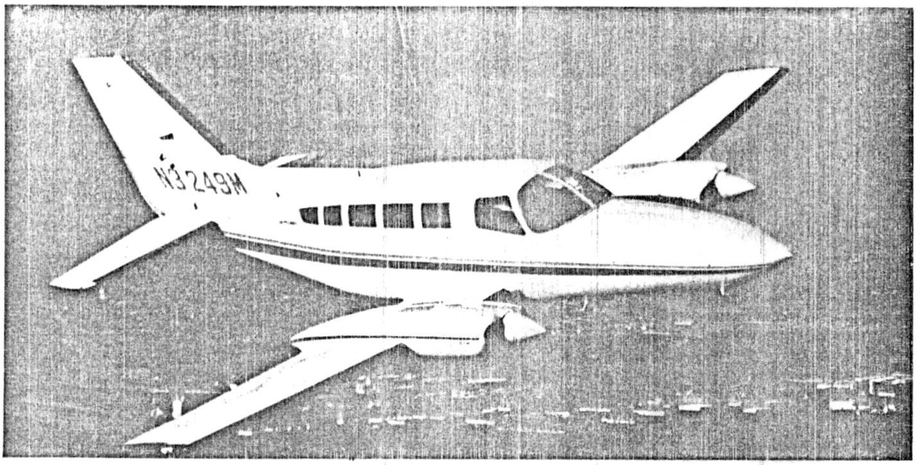

PLATE 31.

The Cessna 402. Manufactured by Cessna Aircraft Company. A ten-place turbocharged piston-prop (two 325 hp engines) with a maximum gross weight of 6,850 lb, a wing area of 226 ft^2, an AR of 8.6, a maximum cruise airspeed of 245 mph, and a range of 1,100 mi.

PLATE 32.
The de Haviland DHC-2 Beaver. Manufactured by de Haviland Aircraft of Canada Limited. An eight-place single-engine (450 hp) aspirated piston-prop with a maximum gross weight of 5,100 lb, a wing area of 250 ft^2, an AR of 9.2, a cruise airspeed of 143 mph, and a range of 500 mi.

PLATE 33.
The EMBRAER EMB-121 Xingu. Manufactured by Empresa Brasileira de Aeronautica, SA. A nine-passenger pressurized turboprop (two 680 eshp engines) with a maximum gross weight of 12,300 lb, a wing area of 296 ft^2, an AR of 7.3, a cruise airspeed of 250 mph, and a range of 1,600 mi.

PLATE 34.
The Mitsubishi Solitaire. Manufactured by Mitsubishi Aircraft International, Inc. A nine-place pressurized turboprop (two 727 eshp engines) with a maximum gross weight of 10,470 lb, a wing area of 178 ft^2, an AR of 8.6, a maximum cruise airspeed of 370 mph, and a range of 1,650 mi.

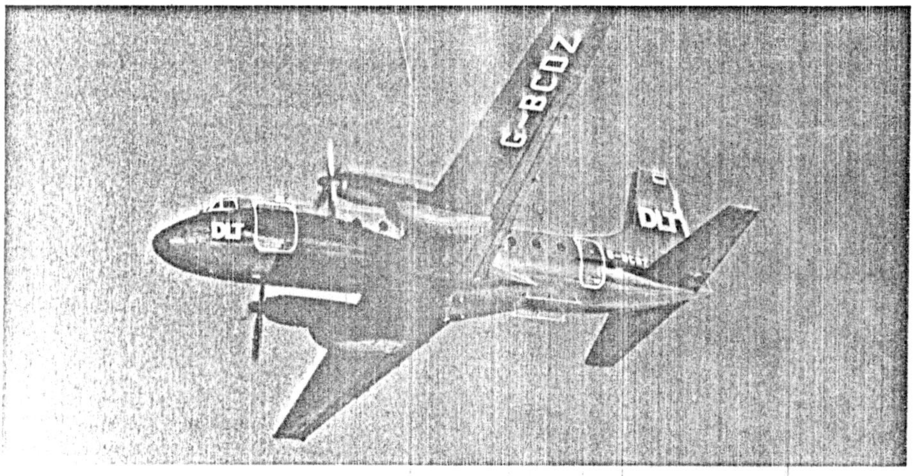

PLATE 35.
The BAe 748. Manufactured by British Aerospace Limited. A short/medium-range turboprop transport (two 2,280 eshp engines) with a maximum gross weight of 51,000 lb, a wing area of 829 ft^2, an AR of 12.7, a cruise airspeed of 281 mph, and a maximum range of 1,500 mi.

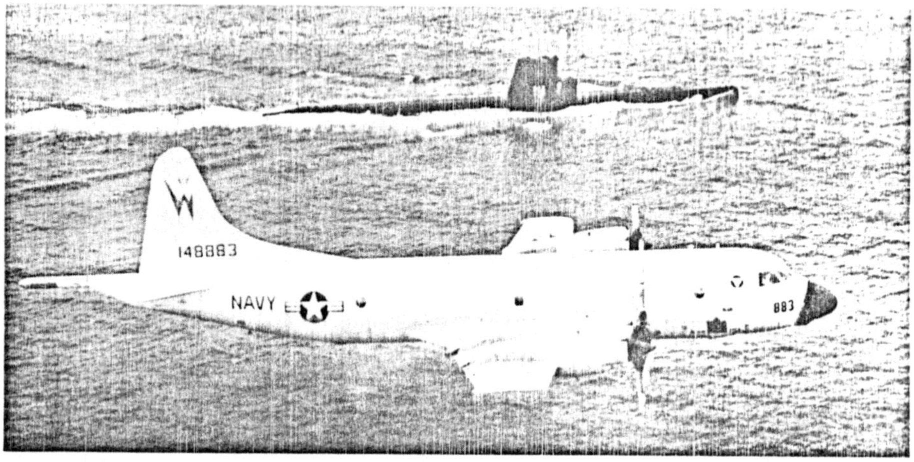

PLATE 36.

The Lockheed P-3C Orion. Manufactured by Lockheed-California Company. An anti-submarine warfare (ASW) patrol aircraft (four 4,910 eshp turboprop engines) with a gross weight of 142,000 lb, a wing area of 1,300 ft^2, an AR of 7.5, and a patrol airspeed of 237 mph. This aircraft is a derivative of the Lockheed Electra, which was used in the 1950s as a commercial airliner.

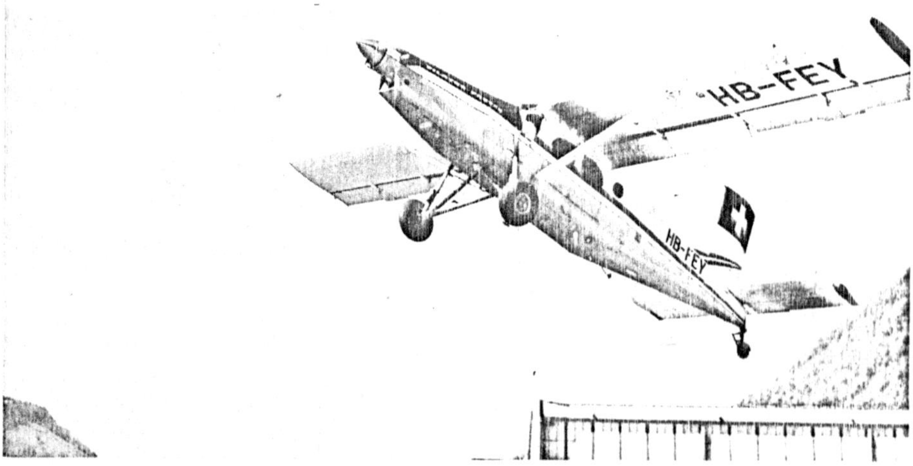

PLATE 37.

The Pilatus PC-6 Turbo Porter. Manufactured by Pilatus Aircraft Limited. An 8-to 11-place STOL utility single-engine turboprop (680 eshp derated to 550 eshp) with a gross weight of 4,840 lb, a wing area of 310 ft^2, an AR of 8, a cruise airspeed of 150 mph, and a maximum range of 600 mi.

PLATE 39.

The de Haviland DHC-6 Twin Otter. Manufactured by de Haviland Aircraft of Canada Limited. A STOL turboprop transport (two 650 eshp engines) with a gross weight of 12,500 lb, a wing area of 420 ft^2, an AR of 10, a maximum cruise airspeed of 210 mph, and a maximum range of 850 mi.

GROUP V. MILITARY AIRCRAFT AND THE SST

PLATE 40.

The Hawk. Manufactured by British Aerospace Limited. A two-place advanced jet trainer with an operational role as an attack aircraft. Has one low bypass ratio (0.75) non-afterburning turbofan of the 5,200-lb thrust class with a maximum gross weight of 17,000 lb, a wing area of 180 ft^2, an AR of 5.3, and a maximum level airspeed of M 0.88.

PLATE 41.

The McDonnell F-15A Eagle. Manufactured by McDonnell Douglas Corporation. A one-place air superiority fighter with two low bypass ratio (0.63) turbofans (with afterburning) of the 24,000-lb thrust class and with a maximum gross weight of 56,000 lb, a wing area of 608 ft^2, an AR of 3, and a maximum airspeed greater than M 2.5.

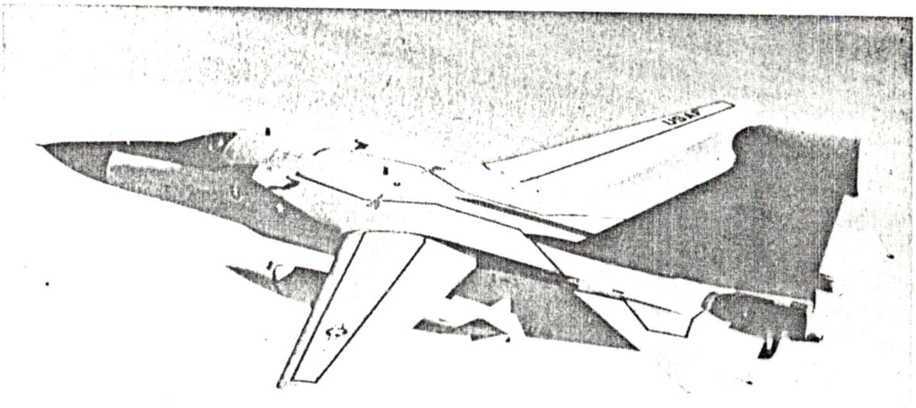

PLATE 42.

The Grumman F-14 Tomcat. Manufactured by Grumman Aerospace Corporation. A two-place multimission aircraft with two low bypass ratio (0.9) turbofans of the 21,000-lb thrust class with afterburning and a maximum gross weight of 74,000 lb. Has a variable sweep wing with a maximum wing area of 565 ft^2 and an associated AR of 7.3. Has a maximum airspeed of M 2.4.

PLATE 43.
The McDonnell F-18A Hornet. Manufactured by McDonnell Douglas Corporation. A one-place multimission fighter aircraft with two low bypass turbofans of the 16,000-lb thrust class with afterburning, a maximum gross weight of 48,000 lb, a wing area of 400 ft^2, an AR of 3.5, and a maximum airspeed in excess of M 1.8.

PLATE 44.
The Av-8A Harrier. Manufactured by British Aerospace Limited. A true VTOL fighter aircraft with a single low bypass ratio turbofan of the 21,500-lb thrust class and with a maximum gross weight of 25,000 lb, a wing area of 210 ft^2, and an AR of 3.2.

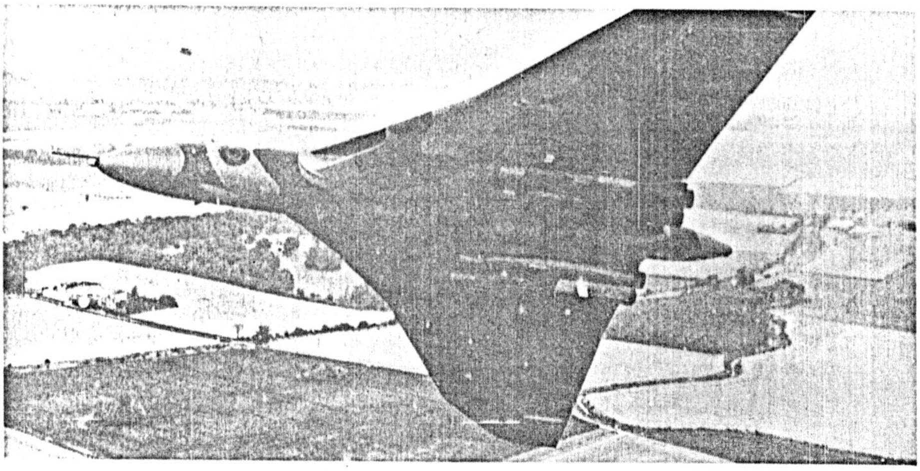

PLATE 45.

The Vulcan Medium Bomber. Manufactured by British Aerospace Limited. The first delta wing bomber, it became operational in the 1950s equipped with four turbojet engines of the 20,000-lb thrust class. It is still operation as a low-lvel bomber and has on occasions been used as a tanker. It has a a maximum airspeed of M 0.94, in the transonic region.

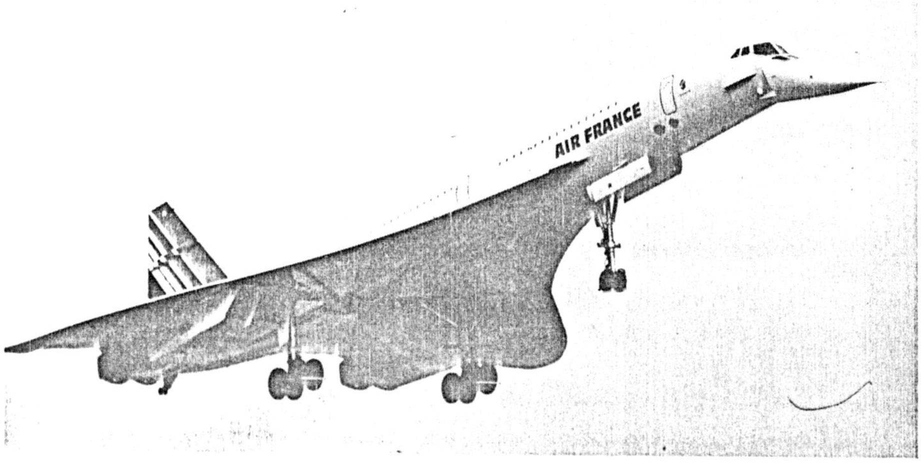

PLATE 46.

The Concorde SST on final approach to a landing. Note the large angle of attack and the drooped nose to provide the pilot with a view of the runway.

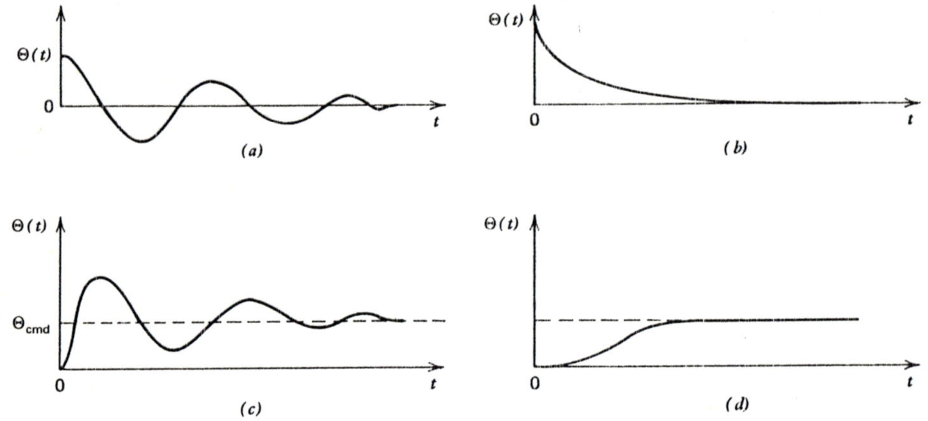

FIGURE 11-12

Typical responses of dynamically stable aircraft: (*a*) and (*b*): disturbance inputs; (*c*) and (*d*): command inputs.

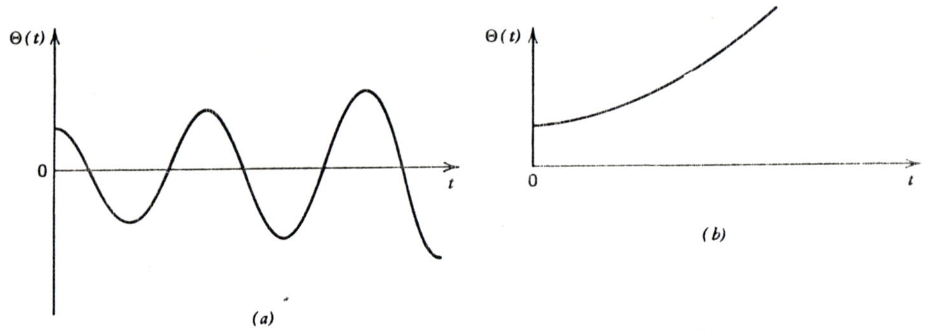

FIGURE 11-13

Transient response of dynamically unstable aircraft to a disturbance input.

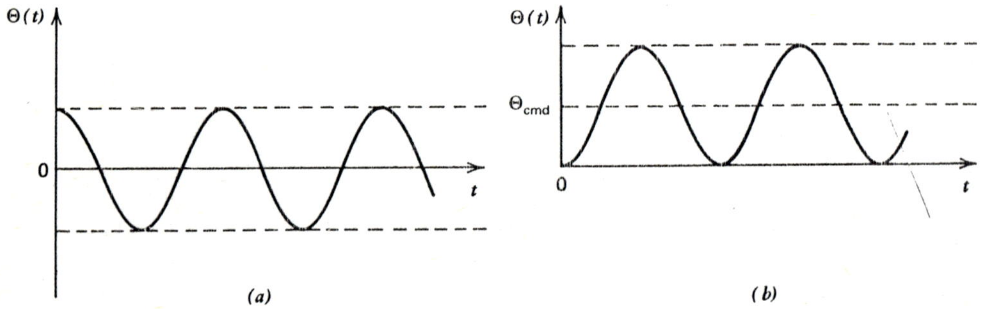

FIGURE 11-14

Transient responses of a marginally stable aircraft: (*a*) disturbance input; (*b*) command input.

shown in Fig. 11-14, the aircraft is described as being *marginally stable*. It should be noted that a bare aircraft that is unstable may still be useful and flyable if the degree of instability does not exceed the ability of a pilot or a flight control system to control the aircraft. In effect, the combination of an unstable bare aircraft and a suitable pilot or flight control system can form a stable system.

The dynamic analysis of a rigid aircraft with a vertical plane of symmetry customarily uses a specific coordinate system known as *disturbed* or *stability axes*. This right-handed orthogonal system is fixed to the aircraft with the origin at the cg, x in the plane of symmetry and aligned with the relative wind at the start of the analysis, z in the plane of symmetry, and y perpendicular to x and z.

Since the duration of the dynamic response is relatively brief, we can assume that the weight of the aircraft remains constant and that the earth is flat and non-rotating. Writing the Newtonian equations of motion in scalar form yields six ordinary differential equations. Unfortunately, these equations are nonlinear. For example, the three longitudinal equations of motion are:

$$F_x = m(\dot{U} - RV + QW) \qquad (11\text{-}30)$$

$$F_z = m(\dot{W} + PV - QV) \qquad (11\text{-}31)$$

$$M = I_{yy}\dot{Q} + (I_{xx} - I_{zz})PR + I_{xz}(P^2 - R^2) \qquad (11\text{-}32)$$

where U, V, and W are the scalar components of the velocity of the aircraft with respect to inertial space and P, Q, and R are the angular velocities about each of the aircraft axes.

Since nonlinear equations generally do not have analytic solutions and require numerical methods which yield numerical solutions, it is customary to linearize these equations about an equilibrium condition. There are many techniques available for analyzing linear systems, and the closed-form analytical solutions provide information as to the physical significance of the system parameters. The right-hand sides of Eqs. 11-30 to 11-32 are the inertial terms and can be linearized by direct application of small perturbation (disturbance) theory. The left-hand sides of these equations represent the external forces and moments, which are nonlinear functions of the flight and aircraft variables. They are linearized by a Taylor's series expansion about the reference equilibrium condition with rejection of the higher-order terms using the small perturbation assumption.

After linearization, the equations of motion will form a set of linear ordinary differential equations with coefficients that can be taken to be constant. These coefficients are usually put into a special nondimensionalized form and called the *NACA stability derivatives*. These stability derivatives are functions of the aircraft characteristics and of the flight conditions. They strongly influence the dynamic characteristics and behavior of the aircraft. Evaluating the stability derivatives is an important, and in many cases a difficult task. The following techniques are all used: analytical evaluation, wind tunnel testing, ground and airborne simulation, and flight tests of the actual aircraft itself.

The dynamic analysis of the linearized aircraft equations requires an understanding of the transfer function concept of classical control as a minimum and may often get into the techniques of modern control theory. We shall limit ourselves to a general discussion of the transient modes of the longitudinal dynamics to be followed by a discussion of the lateral transient modes. The characteristics of these transient modes are determined by the roots (eigenvalues) of the characteristic equation of the aircraft, which is the determinant of the coefficient matrix of the set of linearized equations.

With respect to the longitudinal dynamics, the three flight variables of interest are the changes in the forward airspeed, in the angle of attack, and in the pitch angle. By changes, we mean the changes from the values at the initial equilibrium condition, which is taken to be straight, unaccelerated, coordinated flight with the wings level. The primary aerodynamic control is the elevator or a flying tail. Some aircraft may have a canard for either primary or auxiliary control, and spoilers are used by some aircraft as a fast-response longitudinal control, usually to complement one of the other types of controls.

Nearly all aircraft have two longitudinal transient modes, each second-order and representing a damped oscillation composed of a sinusoid multiplied by a decaying exponential. One oscillation is characterized by a low frequency (long period) and very light damping and is called the *phugoid* mode. It consists primarily of variations in both the airspeed and the pitch angle with the angle of attack remaining essentially constant. The phugoid is a gentle mode and is easily controlled by the pilot, even if it is slightly unstable. In fact, most pilots are unaware of the presence of the phugoid and of their control of it. The other oscillation is known as the *short-period* mode. It is characterized by a high frequency (short period), much higher damping, and by variations in the angle of attack and the pitch angle with very little change in the airspeed. It tends to be a brief but somewhat violent mode. It is the short-period mode that throws passengers and cabin crew around in clear air turbulence.

The periods and damping of the longitudinal modes vary with the aircraft characteristics and with the flight conditions. If our aircraft is cruising at the design conditions with a lift-to-drag ratio of the order of 17, the phugoid mode will have a natural frequency of the order of 0.07 rad/s (a period of 86 seconds) and a damping ratio of 0.03, which is virtually undamped. One way to indicate the amount of damping of an oscillation is by the time required for the amplitude to decay to 5 percent of its maximum value; this time is called the *settling time*. For this phugoid, the settling time is approximately 1,280 seconds, or 21.3 min. The short-period mode, on the other hand, has a natural frequency of the order of 1.15 rad/s (a period of 5.5 seconds) and a damping ratio of the order of 0.35. The corresponding settling time is only 7.4 seconds.

Figure 11-15 is a qualitative sketch of the response of the aircraft to a sharp pulse deflection of the elevator whereby the elevator is deflected +6 deg for approximately 1 second and then returned to zero. Note the brief but comparatively violent nature of the short-period mode, particularly with respect to the change in the angle of attack. The initial two oscillations represent substantial vertical velocity changes that correspond to pulling approximately 1.4 g's. There is no sign of the phugoid in the angle of attack response. The airspeed response,

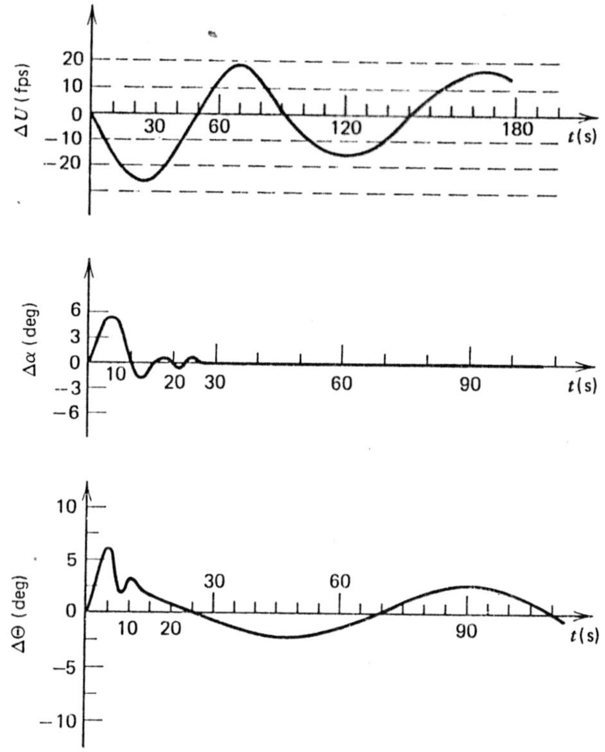

FIGURE 11-15

Transient responses for a pulse elevator deflection of +6 deg for 1 second.

however, is all phugoid; the short-period mode is too brief to have any apparent effect on the airspeed. The pitch angle response shows both the phugoid mode and the short-period mode.

With respect to the phugoid mode, it can be shown that the *damping ratio* (*DR*) can be approximated by the relationship

$$\mathrm{DR} = \frac{C_D}{\sqrt{2}C_L} = \frac{1}{\sqrt{2}E} \tag{11-33}$$

and an order of magnitude value for the *undamped natural frequency* (ω_n) can be obtained from

$$\omega_n = \frac{g\sqrt{2}}{U} \quad \text{rad/s} \tag{11-34}$$

where g is the acceleration of gravity in ft/s^2 and U is the equilibrium airspeed in fps.

Similar approximate relationships for the short-period characteristics are not as straightforward. It can be shown, however, that the damping ratio is directly proportional to the square root of the atmospheric density and that the undamped natural frequency is directly proportional to the product of the equilibrium airspeed and the square root of the atmospheric density. For both modes the settling time is approximately equal to 3 divided by the product of the damping ratio and the undamped natural frequency.

Turning to the lateral dynamics, the relevant equations are the y force equation and the rolling and yawing moment equations. The three flight variables of interest are the bank (or roll) angle ϕ, the change in the yaw (or heading) angle ψ, and the sideslip angle β. If the initial steady-state equilibrium condition is straight, unaccelerated, coordinated flight with wings level, the equilibrium values of the roll and yaw angles are zero.

The distinction between the yaw angle and the sideslip angle can be confusing. It may help to think of the yaw angle as a heading angle that tells where the nose of the aircraft is pointing with respect to some reference on the surface of the earth, such as the north pole. The sideslip angle, on the other hand, only appears when the velocity vector is not in the plane of symmetry and is approximately equal to the y (or sideways) component of the velocity divided by the equilibrium airspeed. When the sideslip angle is zero, we speak of coordinated flight but the yaw angle and yaw rate can be either zero or finite. When the sideslip angle is not zero (uncoordinated flight), there must be a yaw angle. In the one special case where the flight path is straight and the aircraft is slipping with its wings level, the sideslip angle is equal to the negative of the yaw angle. Normally, the sideslip angle and the yaw angle are neither equal nor directly related.

The lateral characteristic equation, which determines the number and nature of the transient modes, is fifth-order. One root of this equation is zero with the result that an aircraft is insensitive to changes in the yaw angle. This means that if the aircraft is disturbed, it will not return to its original heading; it cannot remember which way it was heading.

There are three other transient modes: two first-order modes and one second-order mode. A first-order mode is exponential in shape. If the mode is stable, the exponent is negative and the mode decays to approximately 5 percent of its original value when the magnitude of the exponent becomes equal to 3. If the mode is unstable, the exponent is positive and the mode increases without limit unless checked by some external control.

One of the first-order transients is slightly unstable and is called the *spiral divergence*. If an aircraft is disturbed so that one wing drops, such as to the right, there is no restoring moment. If the pilot or autopilot does not apply left aileron to level the wings, the right wing will slowly continue to drop and a yawing moment will be generated, causing the aircraft to start a turn to the right. The vertical component of the lift will no longer be large enough to balance the weight and the aircraft will lose altitude. The aircraft will start an ever-tightening spiral, losing altitude, and will eventually crash. In very early aircraft, this spiral divergence mode was sufficiently fast to be called "the death spiral." In modern aircraft the wing drop is so slow that a pilot or autopilot has no trouble in controlling it. In fact, the average pilot is not even aware of this instability. This mode could easily be made stable but doing so would decrease the damping of the second-order transient mode.

The other first-order mode is known as the *roll subsidence* mode. It is a stable mode and primarily affects the roll rate and the roll angle; the aileron provides the primary control force. This mode determines the rolling response of an aircraft to an aileron deflection.

The second-order mode is called the *Dutch roll* mode. It is a damped oscillation, characterized for modern aircraft by a high undamped natural frequency (a short period) and light damping. It is primarily actuated by rudder deflection and yawing disturbances. Although there is some change in the roll angle, the Dutch roll can be approximated by considering only the sideslip and yaw angles. For our jet aircraft flying at sea level at 300 mph, the natural frequency is of the order of 1.3 rad/s (a period of about 5 seconds) and the damping ratio is of the order of 0.14 so that the settling time is approximately 16 seconds. This mode, when excited during a landing approach, can be too fast for a pilot to control and too violent to ignore. Consequently, most modern jet aircraft use yaw-rate feedback to increase the damping artificially in order to improve the transient response.

In order to make a coordinated turn (with the sideslip angle equal to zero), it is necessary to use both aileron and rudder in the proper proportions so as to match the roll angle with the turning (yawing) rate.

PROBLEMS

11-1. For each of the aircraft, whose pitching moment coefficient equation is given below, describe the static stability and state whether or not there is a static equilibrium flight condition.

a. $C_m = -0.3C_L$ b. $C_m = 0.06 + 0.15C_L$

c. $C_m = 0.06 - 0.15C_L$ d. $C_m = -0.03$

e. $C_m = +0.03$ f. $C_m = 0.020C_L$

11-2. For the aircraft whose pitching moment coefficient equation is given below

$$C_m = 0.06 - 0.15C_L + 0.02\delta e$$

Note: δe is expressed in degrees.

a. Determine the static stability.

b. Find the static margin.

c. Locate the neutral point, N_0, with respect to the design location of the cg.

d. What is the value of the equilibrium lift coefficient?

e. What is the value of the elevator effectiveness?

f. If the maximum elevator deflections are +25 deg and −15 deg, respectively, and if the landing flare lift coefficient is 2.2, locate the most forward cg position with respect to the design cg location.

g. What is the maximum allowable cg shift (from N_0 to the most forward position) in terms of the mac?

11-3. Do Prob. 11-2 for

a. $C_m = 0.03 - 0.1C_L + 0.05\delta e$

b. $C_m = 0.09 - 0.25C_L + 0.025\delta e$

11-4. An aircraft has the pitching moment coefficient equation,

$$C_m = 0.06 - 0.15C_L + 0.02\delta e$$

a. Plot the pitching moment coefficient as a function of the lift coefficient for elevator deflections of −3 deg, 0 deg, and +3 deg, respectively. Find the value of the equilibrium lift coefficient in each case.

b. Plot the elevator deflection (deg) required to establish equilibrium as function of the lift coefficient from 0 to 2.0.

c. Shift the cg forward a distance of 0.1 of the mac and do (b) on the same plot.

d. Shift the cg aft a distance of 0.1 of the mac and do (b) on the same plot.

11-5. Do Prob. 11-4 for the pitching moment coefficient equations of Prob. 11-3.

11-6. Aircraft *B* (whose major characteristics are listed in the problems section of Chap. 3) has a sweep angle of 20 deg for both the wing and horizontal stabilizer (a tail) although the tail aspect ratio is 4. The design conditions are best-range cruise at 30,000 ft, a SM of 0.10, and a maximum allowable cg shift of 0.25 of the mac. Assume that the most rearward cg position is the neutral point. The slope of the downwash angle curve at landing is 0.35, and the maximum elevator deflections are +25 deg and −15 deg.

a. Find the minimum value of the tail volume coefficient, assuming a tail efficiency of unity.

b. If the fuselage is to be kept horizontal during cruise, to ease the burdens on the cabin crew, what should the wing incidence angle be? What will the tail incidence angle be?

c. If the ratio of the tail distance to the mac length is selected to be equal to 3, what is the tail area? Assume that the values of the zero-lift drag coefficient and of the Oswald span efficiency of the tail are those of the wing. Find the cruise-equilibrium tail drag and tail lift in pounds and as a percentage of the wing drag and lift.

11-7. Aircraft *E* (whose major characteristics are listed in the problems section of Chap. 6) has a straight wing and tail, the latter having an aspect ratio of 4.2. The design conditions are cruise at 75 percent max power at 20,000 ft, a SM of 0.18, and a maximum allowable cg shift of 0.3 of the mac. Do Prob. 11-6.

11-8. An aircraft has a tail volume coefficient of 0.65, a SM of 0.15, and the slopes of the wing and tail lift curves are 0.07 per deg and 0.06 per deg, respectively. The aircraft has a flying tail with maximum deflections of ± 12 deg with respect to the design (fixed) incidence angle. The wing incidence angle is 3 deg, the equilibrium value of the lift coefficient is 0.33, and the maximum value of the lift coefficient is 2.5.

a. Write the pitching moment coefficient equation.

b. What is the design (fixed) tail incidence angle (deg)?

c. What is the slope of the pitching moment coefficient curve at the most forward cg position?

d. What is the maximum allowable cg shift?

11-9. The following expressions represent the short-period time response of the change in pitch angle to a constant elevator deflection. Which of these represents the response of a dynamically stable aircraft? For the stable aircraft, what will be the final value of the pitch angle when the transient mode has died out?

a. $\Theta(t)=5-5e^{-0.3t}\sin(1.5t+1.57)$

b. $\Theta(t)=-5+5e^{-0.3t}\sin(1.5t+1.57)$

c. $\Theta(t)=4-6e^{+0.3t}\sin(1.07t+0.73)$

d. $\Theta(t)=-4+6e^{+0.3t}\sin(1.07t+0.73)$

11-10. a. Find the damping ratio and the undamped natural frequency of the phugoid mode of Aircraft C flying at M 0.8 at 35,000 ft.

b. Do (a) for Aircraft B at 30,000 ft at M 0.75.

c. Do (a) for Aircraft A at 20,000 ft at M 0.6.

11-11. a. Find the damping ratio and undamped natural frequency of the phugoid mode of Aircraft D at 10,000 ft and 175 mph.

b. Do (a) for Aircraft E at 20,000 ft and 275 mph.

Chapter Twelve

Some Design Examples

12-1 INTRODUCTION

In this concluding chapter, we shall apply the knowledge and techniques that we have acquired to perform a feasibility design of three transport aircraft. The first two aircraft will have the same set of operational requirements: One will be a turbojet and the other will be a turbofan. The third aircraft will have a different set of operational requirements and will be a piston-prop.

There are many design techniques in use today, and the ones presented here are not unique nor are they necessarily the best. They are, however, relatively simple, they are compatible with a hand-held calculator, and they do give a feeling for some of the elements of the design process. Aircraft design is essentially an iterative process and, as such, is eminently suited for the high-speed and powerful computers that are currently available and that are continually being improved. Their large memories (that allow the storage of vast amounts of aerodynamic, propulsion, and structural data), their graphics capability, and their ability to interact directly with the designer are responsible for the great interest in and popularity of computer-aided design and computer-aided manufacturing (CAD/CAM) in the aerospace industry.

In our design examples, we will not perform any iterations but rather will stick with our first configuration. We shall, however, take our first aircraft through a typical mission profile and calculate the fuel consumed during each phase. Our typical mission profile will comprise the following phases: taxi from the ramp to the end of the runway, the take-off run, fastest climb to the cruise altitude, and cruise. Our profile will end with the completion of cruise and the determination of the amount of fuel remaining. We will not go through the descent and landing phase nor the diversion to an alternate destination if the weather does not permit landing at the original destination.

As mentioned previously, our designs will not be complete. There will be no iterations or sensitivity analyses, nor will there necessarily be a complete set of operational constraints, such as a minimum approach airspeed. Furthermore, we will not have any technical groups, such as an aerodynamic section or a propulsion group or a structures division, to provide us with the technical data we need, nor will we have an extensive data base as is provided in many computer design programs. Instead, we will make estimates, educated and reasonable we hope, as to the values of the data we need for our feasibility designs.

With these disclaimers out of the way, let us proceed with our design examples.

12-2 A TURBOJET EXAMPLE

The operational requirements for this design example are for a turbojet that will carry 200 passengers (at 200 lb per passenger) plus 5,000 lb of cargo for 2,400 mi at a cruise Mach number of M 0.8 or better. Cruise is to start at 33,000 ft, where the density ratio is 0.336 and the sonic ratio is 0.879. The overall runway length for take-off is to be 5,000 to 6,000 ft or less, which means that our calculated take-off ground run should be of the order of 2,500 to 3,000 ft or less.

Using our experience, judgment, and knowledge of the type of aircraft with which we are dealing, we shall choose a zero-lift drag coefficient of 0.016, an aspect ratio of 8, and an Oswald span efficiency of 0.85. These values will give us the parabolic drag polar $C_D=0.016+0.0468C_L^2$ and a maximum lift-to-drag ratio of 18.3. In addition, we will assume a thrust specific fuel consumption of 0.95 lb/h/lb, a thrust-to-engine weight ratio of 5, and a maximum lift coefficient for take-off of 2.2.

We are now ready to start our design analysis. There are many ways to start and many approaches to use. Rather than assume a gross weight, we shall go immediately to the cruise portion of the mission profile and find a cruise-fuel weight fraction. For our first try, we will cruise at the best-range conditions with a best-range airspeed of M 0.8 (784.8 fps or 535 mph at 33,000 ft) and with a best-range lift-to-drag ratio of 15.8 (0.866 × 18.2). Substituting into the Breguet range equation and solving,

$$2{,}400=\frac{535\times 15.6}{0.95}\ln \mathrm{MR}$$

$$\mathrm{MR}=1.3096 \qquad \zeta=\frac{1.3096-1}{1.3096}=0.236$$

yields a value of 1.3096 for the mass ratio, which translates into a cruise-fuel weight fraction of 0.236.

Since this will be a reasonably large aircraft, let us now assume that 10 percent of the total fuel loaded aboard the aircraft will be used for taxi, take-off, and climb, and that another 10 percent will be kept for the descent, landing, and reserve. Consequently, the cruise-fuel weight fraction will be equal to 0.8 of the total fuel load divided by the aircraft weight at the start of cruise. If we further assume, in the interests of simplicity and conservatism, that the initial cruise weight and the ramp gross weight are identical, then the fuel-weight fraction of the aircraft will be 0.295 (0.236/0.8).

Although we do not really need it yet, let us find the initial cruise wing loading from the level-flight equation,

$$C_L=\left(\frac{0.016}{3\times 0.0468}\right)^{1/2}=0.338$$

$$\frac{W}{S}=\tfrac{1}{2}\rho_{\mathrm{SL}}\times 0.336\times(784.8)^2\times 0.338=82.9\ \mathrm{lb/ft^2}$$

where the value of 0.338 for the best-range lift coefficient was found from $(C_{D0}/3K)^{1/2}$. Making a small allowance for the decrease in weight prior to cruise, let us set the *maximum wing loading equal to* 85 *lb/ft*^{2}.

In order to determine the maximum thrust-to-weight ratio, let us assume a ceiling of 45,000 ft, where the density ratio is equal to 0.194, so that

$$\frac{T}{W} = \frac{1}{0.194 \times 18.2} = 0.28$$

We will round off this value and set the *T/W ratio equal to* 0.3. With a thrust-to-engine weight ratio of 5, the engine weight fraction will be 0.06 (0.3/5).

With the assumption of a value of 0.45 for the structural weight fraction, the payload weight fraction can be found from

$$1 = 0.45 + 0.06 + \frac{W_{PL}}{W} + 0.254$$

to be equal to 0.1945. With this value and a total payload weight of 45,000 lb, the corresponding gross weight is 231,362 lb. Let us round off this number and set the initial *gross weight of the aircraft equal to* 230,000 *lb*.

We will now pick the size and number of the engines. The maximum total thrust will be 69,000 lb (0.3 × 230,000). We need at least two engines, for reasons of safety, but otherwise want the minimum number of engines. Let us settle for *two engines of* 35,000 *lb of thrust each for a total maximum thrust of* 70,000 *lb*.

Now we can lay out some of the specifications for this first trial configuration, as follows:

Powerplant:

Two turbojet engines, each with 35,000 lb of thrust at sea-level.

Weights:

Gross weight, lb	230,000
Operational empty weight, lb	117,500
Maximum useful load, lb	112,500
Payload, lb	45,000
Fuel, lb	67,500
Maximum wing loading, lb/ft^2	85

Dimensions:

Wing area, ft^2	2,705.9
Wing span, ft	147.1
Average wing chord, ft	18.4

We now need to evaluate the performance of this configuration to see if it will meet the operational requirements. In this particular case, let us take our aircraft through a mission profile up to the end of cruise.

We shall allow 15 min for taxiing at a thrust level of 20 percent of the maximum thrust. The fuel consumed can be found from

$$\Delta W_f = cT\,\Delta t = 0.95 \times 0.2 \times 70{,}000 \times \frac{15}{60}$$

$$= 3{,}325 \text{ lb}$$

The amount of fuel consumed during taxiing and waiting for take-off clearance (jet engines idle at thrust levels of the order of 20 percent of maximum) is not trivial. At some large airports aircraft are often towed to the maintenance areas, and serious consideration has been given to towing aircraft to the end of the runway and to not starting the engines until the clearance has been received.

At the start of take-off, the aircraft weight is 226,675 lb and the wing loading is 83.8 lb/ft^2. With a maximum take-off lift coefficient of 2.2, the sea-level stall speed is 179 fps (122.2 mph) and the lift-off airspeed is 215 fps (146.7 mph). Using Eq. 4-3, the take-off ground run is 2,394 ft, which satisfies the operational constraint, and the elapsed time is 22.2 seconds. Using the same fuel-consumption equation as for taxi, the fuel consumed at maximum thrust during take-off is 410 lb.

At the start of climb, the aircraft weighs 226,265 lb and has a wing loading of 83.6 lb/ft^2. The calculations for the maximum rate of climb at sea level are

$$\Gamma = 1 + \left[1 + \frac{3}{(18.2 \times 0.3)^2}\right]^{1/2} = 2.0$$

$$V_{FC} = \left(\frac{83.7 \times 0.3 \times 2.0}{3\rho_{SL} \times 0.016}\right)^{1/2} = 663.4 \text{ fps} = 452 \text{ mph} = M\,0.59$$

$$\sin \gamma_{FC} = 0.3\left(1 - \frac{2}{6}\right) - \frac{3}{2 \times 2 \times (18.2)^2 \times 0.3} = 0.1924$$

$$\gamma_{FC} = 11.1 \text{ deg}$$

$$(R/C)_{max} = 663.4 \times 0.1924 = 127.6 \text{ fps} = 7{,}656 \text{ fpm}$$

For fastest climb to 33,000 ft, with maximum thrust and no restrictions or limitations, the closed-form expressions of Chap. 4 are used to find the following values: a climb time of 8.97 minutes, a climb-fuel weight fraction of 0.195 and a fuel consumption of 4,414 lb, and a distance traveled of 67.6 mi. Incidentally, the amount of fuel used from start of engines to start of cruise is 8,149 lb, which is 12.1 percent of the total fuel loaded aboard the aircraft.

The aircraft weight at the start of cruise is 221,851 lb and the wing loading is 82 lb/ft^2. With this wing loading, the best-range airspeed is 780 fps (532 mph) or $M\,0.795$. Using the Breguet range equation, the cruise mass ratio for 2,400 mi is found to be 1.312. The corresponding cruise-fuel weight fraction is 0.2376, so that the cruise fuel consumption is 52,706 lb. The fuel remaining at the end of cruise is 6,645 lb, or 9.8 percent of the total fuel loaded aboard the aircraft.

Let us now return to the climb phase and see what the climb performance would be if we limited the thrust so that the maximum allowable load factor would be 2.5 g's. The appropriate T/W ratio is 0.11, so that Γ is 2.2, the sea-level climb airspeed is 423 fps (288 mph) or M 0.38, which by coincidence is the air traffic control maximum airspeed of 250 knots, and the rate of climb is dramatically reduced to 1,293 fpm.

If during the climb, the throttle setting is increased so as to keep the T/W ratio equal to 0.11 as long as possible, the maximum T/W ratio at 33,000 ft will be 0.1 and the rate of climb will have increased to 1,779 fpm, primarily because the airspeed has increased to 489 mph or M 0.73. The average rate of climb becomes 1,536 fpm and the time to climb is 21.5 minutes. With an average T/W ratio of 0.105, the climb fuel will be 5,831 lb, which, though 32 percent higher than that for fastest climb without restrictions, does not significantly affect our analysis.

We shall end our preliminary design of this turbojet at this point, although in practice this would be just the beginning of many iterations and sensitivity analyses. For example, knowing the order of magnitude of the gross weight, the wing area, and the wing span, a more detailed breakdown of the weights can be made and our technical estimates can be refined. Also, it may have crossed your mind as to why it was decided to cruise at the best-range conditions inasmuch as the cruise airspeed was specified. It seems more logical to cruise at M 0.8 and at the maximum lift-to-drag ratio, and it is. This flight program will be examined in the turbofan example of the next section, using the same set of operational requirements.

12-3 A TURBOFAN EXAMPLE

The operational requirements for our turbofan will be those of the preceding example, namely, a payload weight of 45,000 lb, a M 0.8 cruise for 2,400 mi at 33,000 ft, and a calculated take-off run of the order of 2,500 to 3,000 ft.

Since the frontal area of a turbofan is larger than that of a turbojet, it seems reasonable to increase the zero-lift drag coefficient to 0.018. If we leave the other wing characteristics unchanged, the parabolic drag polar for the turbofan will be $C_D = 0.018 + 0.0468C_L^2$, which leads to a maximum lift-to-drag ratio of 17.23. Although this is lower than the typical value of 18+ that we would expect to find in a modern high subsonic transport, we will not attempt to raise it, at least not at this time. Perhaps the most important change from the turbojet values is the decrease in the thrust specific fuel consumption to a value of 0.65 at a Mach number of 0.8 and at an altitude of 35,000 ft. The thrust-to-engine weight will remain at 5, and the maximum lift coefficient for take-off will remain at 2.2. We shall also use a structural weight fraction of 0.45 and a thrust-to-weight ratio of 0.3, as before.

Moving to the cruise phase, our first cruise will be at M 0.8 (784.8 fps or 535 mph) and at the maximum lift-to-drag ratio. From the Breguet range equation, we find that the mass ratio is 1.184 and that the cruise-fuel weight fraction is 0.1557. If we assume that 80 percent of the total fuel loaded is available for cruise, the fuel weight fraction can be taken as 0.1946, leading to a payload weight fraction of

0.2954 and a gross weight of 152,336 lb. This should be the minimum-weight configuration for this set of conditions. Since the cruise lift coefficient is 0.620 (the square root of the ratio of the zero-lift drag coefficient to K), the initial cruise wing loading will be 152.5 lb/ft^2, which seems high. Let us check the take-off ground run, which with this wing loading and a maximum lift coefficient will be 4,347 ft with a lift-off airspeed of 197.6 mph.

Since the take-off ground run is too large with a wing loading this high, we shall reduce the wing loading to 130 lb/ft^2, keeping the airspeed at M 0.8. Knowing the wing loading, airspeed, and altitude, the lift coefficient can be found from

$$C_L = \frac{2 \times 130}{\rho_{SL} \times 0.336 \times (784.8)^2} = 0.5286$$

to be 0.5286. Using the drag polar to find the corresponding drag coefficient, the new flight lift-to-drag ratio is 17. From the Breguet range equation, we can obtain values of 1.187 and 0.1576 for the mass ratio and cruise-fuel weight fraction, respectively, leading to a payload weight fraction of 0.293 and a gross weight of 153,600 lb. The corresponding take-off ground run is 3,706 ft, which is still too long.

Continuing with a similar set of calculations for various values of the wing loading, the results are shown in Fig. 12-1. One curve shows the gross weight as a function of the wing loading, and the other the take-off ground run. If we use the operational constraint that the take-off ground run be not more than 3,000 ft, we

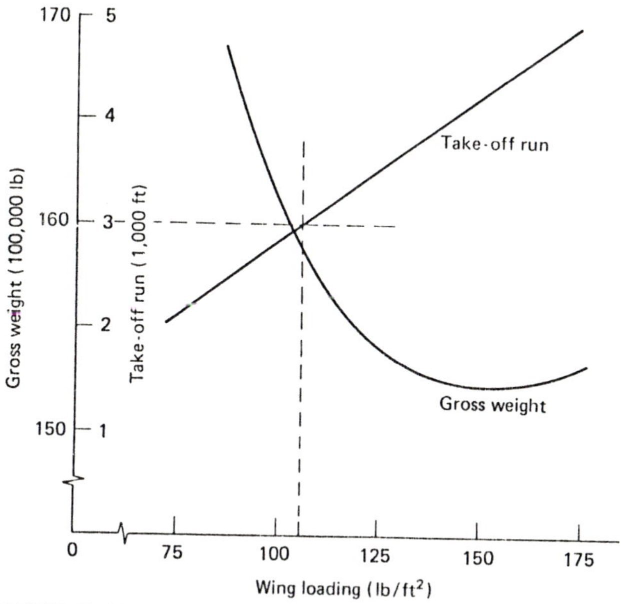

FIGURE 12-1

Trade-offs between gross weight and take-off run versus wing loading for the turbofan example.

find that we can satisfy this constraint with this configuration and set of characteristics with a *wing loading of* 105 *lb/ft*^{2} *and a gross weight of* 159,000 *lb*. We could have changed one or more of the aircraft characteristics and plotted the corresponding curves in Fig. 12-1. For example, an increase in the T/W ratio would decrease the ground run but increase the gross weight, whereas an increase in the maximum lift coefficient would affect only the ground run. We could also have superimposed other operational constraints, such as the ceiling with one engine and a minimum approach airspeed, so that the figure would have been covered by many curves. Such a figure is commonly referred to as a *carpet plot*, and the acceptable configuration is the one that satisfies all the operational constraints.

The specifications for this preliminary configuration are:

Powerplant:

Two turbofan engines, each with 24,000 lb of thrust at sea level and a thrust specific fuel consumption of 0.65 lb/h/lb at M 0.8 and 35,000 ft.

Weights:

Gross weight, lb	159,000
Operational empty weight, lb	81,100
Maximum useful load, lb	77,900
Payload, lb	45,000
Fuel, lb	32,900
Maximum wing loading, lb/ft^2	105

Dimensions:

Wing area, ft^2	1,514.3
Wing span, ft	110.1
Average wing chord, ft	13.8

Looking now at the maximum sea-level rate of climb, with maximum thrust and with no restrictions or limitations, the climb speed is 485.7 mph, or M 0.64, and the rate of climb is 8,137 fpm. If the aircraft is limited to 2.5 g's, the maximum T/W ratio will be 0.145 and the corresponding sea-level rate of climb is 2,590 fpm at an airspeed of 350 mph, or M 0.46. With the air traffic control restriction of 250 knots (288 mph) and keeping the thrust-to-weight ratio at 0.145, the sea-level rate of climb is 2,164 fpm.

12-4 A PISTON-PROP EXAMPLE

The operational requirements for the piston-prop are a payload capability of 20 passengers and 1,000 lb of cargo (a total payload weight of 5,000), a cruise range of 1,000 mi at 25,000 ft at the highest airspeed compatible with a maximum calculated take-off ground run of the order of 2,000 ft (a balanced field length of the order of 4,000 ft).

We will use turbocharged engines with a total HP/W ratio of 0.1, a critical altitude of 20,000 ft, a horsepower specific fuel consumption of 0.45 lb/h/hp, a horsepower-to-engine weight ratio of 0.5 hp/lb, and a design propeller efficiency of 0.85. We will keep the aspect ratio of 8 and the Oswald span efficiency of 0.85 but will increase the zero-lift drag coefficient to 0.025. The corresponding parabolic drag polar will be $C_D = 0.025 + 0.0468C_L^2$, and the maximum lift-to-drag ratio has dropped to 14.6.

Moving on to the cruise phase and flying at the best-range conditions of the maximum lift-to-drag ratio, the Breguet range equation is

$$1{,}000 = \frac{0.85 \times 375 \times 14.6}{0.45} \ln \text{MR}$$

and the mass ratio is 1.1015, leading to a cruise-fuel weight fraction of 0.092. Since this should be a much smaller aircraft than the two preceding aircraft, we will allow 20 percent of our total fuel for taxi, take-off, and climb and the usual 10 percent for letdown, landing, and reserve, leaving 70 percent for cruise. With these assumptions, the total fuel weight fraction for this aircraft becomes 0.13. Leaving the structural weight fraction at 0.45 (which may be a bit on the low side),

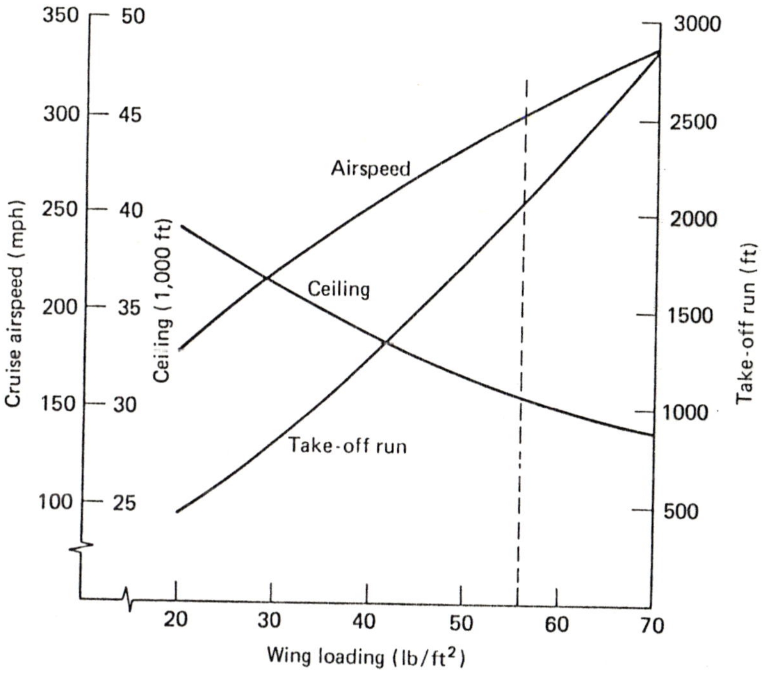

FIGURE 12-2

Trade-offs among airspeed, take-off run, and ceiling versus wing loading for the piston-prop example.

the weight fraction equation becomes

$$1 = 0.45 + 0.2 + \frac{W_{PL}}{W} + 0.13$$

so that the payload weight fraction is 0.219. With a total payload of 5,000 lb, the corresponding gross weight will be 22,831 lb, which we will round off to 23,000 lb.

With a horsepower-to-weight ratio of 0.1, the required horsepower is 2,300 hp. Rather than two large engines, we shall use four smaller engines, each with 575 hp at sea level, for a total power output of the 2,300 hp that we need. This is a standard-size engine with a good production and maintenance history.

Since the range and, to a first approximation, the gross weight, are independent of the airspeed, we can select the wing loading that will give us a cruise airspeed that is acceptable. We must not forget, however, that the wing loading adversely affects both the take-off run and the ceiling. Consequently, we have three factors to consider in selecting the wing loading, namely, the cruise airspeed, the take-off ground run, and the ceiling. These three are plotted as a function of the wing loading in Fig. 12-2.

If we choose 300 mph as the cruise airspeed (this is a very good cruise airspeed for a piston-prop and higher than typical), the corresponding take-off run is 2,050 ft, the ceiling is 30,800 ft, and the wing loading is 56 lb/ft^2. The specifications for this configuration are:

Powerplant:

Four turbocharged piston-prop engines, each with 575 hp at sea level, a critical altitude of 20,000 ft, and a specific fuel consumption of 0.45 lb/h/hp.

Weights:

Gross weight, lb	23,000
Operational empty weight, lb	14,950
Maximum useful load, lb	8,050
Payload, lb	5,000
Fuel, lb	3,050
Maximum wing loading, lb/ft^2	56

Dimensions:

Wing area, ft^2	410.7
Wing span, ft	57.3
Average wing chord, ft	7.2

In addition to a ceiling of 30,800 ft, the maximum rate of climb at sea level is 1,900 fpm at an airspeed of 131.5 mph (192.9 fps).

It might be interesting to run through a mission profile for this aircraft. Taxiing for 15 minutes at 20 percent max power would use approximately 52 lb of fuel.

The take-off run would be 2,032 ft with a lift-off airspeed of 120 mph and a fuel consumption of 6.7 lb. Using the closed-form expressions of Chap. 7, the climb to 25,000 ft would take 16.4 minutes with a fuel consumption of 269 lb and a distance traveled of 43 mi. The amount of fuel used to the start of cruise is 327.4 lb, which is 10.7 percent of the total fuel. The cruise fuel consumption is 2,089.7 lb and the fuel remaining is 633 lb, or 20.7 percent of the total fuel loaded. These fuel figures indicate that the assumption of 20 percent of the fuel needed prior to cruise was overly conservative and that the fuel load and gross weight can probably be reduced somewhat.

We shall conclude this section and chapter with the realization that these design examples have been greatly simplified and are primarily tutorial. However, as simple and superficial as they might be, they are very useful for roughly sizing a propsed aircraft quickly and easily and for giving insight as to how a design can be improved and as to the importance of the different aircraft and subsystem characteristics. It might be interesting to go back over these examples and see just where improvements could be made, always keeping in mind the penalties associated with each improvement. Aircraft performance and design are classic examples of the many trade-offs that are always being made in engineering and in life itself; in other words, there is a price of some kind associated with each change and improvement. The secret of success is to have the value of the change exceed the cost or associated penalty.

PROBLEMS

These sets of problems are designed to show that there are operating regions in which a particular type of aircraft has an edge as well as to exercise you in the techniques of feasibility designs that are somewhat rough and approximate but that do give quick and easy answers.

This first set of operational requirements and constraints will apply, either as is or as later modified, to all the problems to follow. You are to look at an aircraft to carry 40,000 lb of payload (175 passengers and 5,000 lb of cargo) 4,000 mi with a no-wind condition, using *all* the fuel loaded aboard, at a cruise Mach number of 0.8 with cruise starting at 30,000 ft. The aircraft regardless of its type or power plant, is to have a wing loading of 100 lb/ft^2, a maximum lift-to-drag ratio of 18, an Oswald span efficiency of 0.8, and a structural weight fraction of 0.44.

12-1. The first look is at a jet aircraft with a zero-lift drag coefficient of 0.018, a thrust-to-aircraft weight ratio of 0.25, a sfc of 0.8 lb/h/lb, and a thrust-to-engine weight of 4. Find the minimum gross weight of the aircraft, along with the fuel weight, the aspect ratio, and the wing span and area.

12-2. Now consider a piston-prop aircraft with a zero-lift drag coefficient of 0.025, an HP-to-aircraft weight ratio of 0.1 hp/lb, a sfc of 0.45 lb/h/hp, a propeller efficiency of 80 percent, and an HP-to-engine weight of 0.5 hp/lb.

Find the minimum gross weight of the aircraft, the total fuel weight, the aspect ratio, and the wing span and area.

12-3. Now consider a turboprop with a zero-lift drag coefficient of 0.023, an ESHP-to-aircraft weight ratio of 0.1 eshp/lb, a sfc of 0.5 lb/h/eshp, a propeller efficiency of 80 percent, and an ESHP-to-engine weight of 2 eshp/lb. Find the minimum gross weight, the total fuel weight, the aspect ratio, and the wing span and area.

12-4. Which of these three types of aircraft is best suited for this high-altitude, high-airspeed, long-range mission?

The original operational requirements are now modified to reduce the cruise airspeed to 250 mph and the wing loading to 40 lb/ft^2. All other requirements and constraints as well as the characteristics of the various types of aircraft remain unchanged.

12-5. Do Prob. 12-1.

12-6. Do Prob. 12-2.

12-7. Do Prob. 12-3.

12-8. Do Prob. 12-4.

This marks the end of the "design" problems. You are encouraged to construct your own set of operational requirements and constraints in sufficient detail so that you can define your configuration. Do not be reluctant to estimate missing technical characteristics on your first try. Once you have an idea as to size of your aircraft, you are in a much better position to refine these estimates. Finally, let your imagination loose with regards to your operational requirements: you might be surprised at the results. For example, an interesting requirement is to lay out an aircraft to take 4,000 to 10,000 passengers on a nonstop trip half-way around the world to see a soccer match.

Some Selected References

Anderson, J. D., Jr., *Introduction to Flight*, McGraw-Hill, New York, 1978.

Babister, A. W., *Aircraft Dynamic Stability and Response*, Pergamon Press, Oxford, 1980.

Blakelock, J. H., *Automatic Control of Aircraft and Missiles*, John Wiley, New York, 1965.

Clancy, L. J., *Aerodynamics*, John Wiley, New York, 1975.

Dommasch, D. O., Shelby, S. S., and Connolly, T. F., *Aeroplane Aerodynamics*, Third Edition, Isaac Pitman, London, 1961.

Etkin, B., *Dynamics of Flight—Stability and Control*, Second Edition, John Wiley, New York, 1982.

Hill, P. G., and Peterson, C. R., *Mechanics and Thermodynamics of Propulsion* Addison-Wesley, Reading, Mass., 1965.

Houghton, E. L., and Brock, A. E., *Aerodynamics for Engineering Students*, Edward Arnold, London, 1972.

Kerrebrock, J. L., *Aircraft Engines and Gas Turbines*, MIT Press, Cambridge, Mass., 1977.

Keuthe, A. M., and Chow, C-Y., *Foundations of Aerodynamics*, John Wiley, New York, 1976.

Kuchemann, F. R. S., *The Aerodynamic Design of Aircraft*, Pergamon Press, Oxford, 1978.

McCormick, B. W., *Aerodynamics, Aeronautics, and Flight Mechanics*, John Wiley, New York, 1979.

Miele, A., *Flight Mechanics, Vol. 1, Theory of Flight Paths*, Addison-Wesley, Reading, Mass., 1962.

Nicolai, L. M., *Fundamentals of Aircraft Design*, University of Dayton, Dayton, Ohio, 1975.

Perkins, C. D., and Hage, R. E., *Airplane Performance Stability and Control*, John Wiley, New York, 1949.

Appendix A

Properties of the Standard Atmosphere

TABLE A-1

Standard Atmosphere Property Ratios (English Units)

Altitude	Pressure	Temperature	Density	Speed of Sound
10^3 ft (10^3 m)	$\delta = P/P_{SL}$	$\Theta^* = \Theta/\Theta_{SL}$	$\sigma = \rho/\rho_{SL}$	$a^* = a/a_{SL}$
0 (0)	1.000	1.000	1.000	1.000
5 (1.52)	0.832	0.966	0.862	0.983
10 (3.05)	0.688	0.931	0.738	0.965
15 (4.57)	0.564	0.897	0.629	0.947
20 (6.10)	0.460	0.862	0.533	0.929
25 (7.62)	0.371	0.828	0.448	0.910
30 (9.14)	0.297	0.794	0.374	0.891
35 (10.67)	0.235	0.759	0.310	0.871
36 (10.97)	0.223	0.752	0.297	0.867
40 (12.19)	0.185	0.752	0.246	0.867
45 (13.72)	0.145	0.752	0.194	0.867
50 (15.24)	0.114	0.752	0.152	0.867
55 (16.76)	0.090	0.752	0.120	0.867
60 (18.29)	0.0708	0.752	0.094	0.867
65 (19.81)	0.0557	0.752	0.074	0.867
70 (21.34)	0.0438	0.752	0.058	0.867
75 (22.86)	0.0344	0.752	0.046	0.867
80 (24.38)	0.0271	0.752	0.036	0.867
82 (25.0)	0.0246	0.752	0.033	0.867
85 (25.91)	0.0213	0.761	0.028	0.872

Sea-level values:

$P = 2{,}116$ lb/ft^2 $\qquad \rho = 23.769 \times 10^{-4}$ lb-s^2/ft^4

$\Theta = 519$ deg $R = 59$ deg F $\qquad a = 1{,}116$ ft/s

$k = 1.4$ $\qquad R = 1.7165 \times 10^3$ ft^2/s^2-deg R

TABLE A-2
Standard Atmosphere Property Ratios (SI Units)

Altitude	Pressure	Temperature	Density	Speed of Sound
10^3 m	$\delta = P/P_{SL}$	$\Theta^* = \Theta/\Theta_{SL}$	$\sigma = \rho/\rho_{SL}$	$a^* = a/a_{SL}$
0	1.000	1.000	1.000	1.000
1	0.887	0.977	0.907	0.988
2	0.784	0.955	0.822	0.977
3	0.692	0.932	0.742	0.965
4	0.608	0.901	0.668	0.949
5	0.533	0.887	0.601	0.942
6	0.465	0.865	0.538	0.930
7	0.405	0.842	0.481	0.918
8	0.351	0.819	0.428	0.905
9	0.303	0.797	0.380	0.893
10	0.261	0.774	0.337	0.880
11	0.223	0.752	0.297	0.867
12	0.191	0.752	0.254	0.867
13	0.163	0.752	0.217	0.867
14	0.128	0.752	0.185	0.867
15	0.119	0.752	0.158	0.867
16	0.101	0.752	0.135	0.867
17	0.087	0.752	0.115	0.867
18	0.074	0.752	0.098	0.867
19	0.063	0.752	0.084	0.867
20	0.054	0.752	0.072	0.867

Sea-level values:
$P = 1.013 \times 10^5$ N/m^2 $\rho = 1.226$ kg/m^3
$\Theta = 288$ K $= 15$ deg C $a = 340$ m/s
$k = 1.4$ $R = 287$ m^2/s^2-K

Appendix B

Turbojet Range Equation Integrations

The general integral range equation, with c constant, is

$$X=\frac{-1}{c}\int_1^2 \frac{V}{D}\,dW \tag{B-1}$$

B-1 THE CONSTANT ALTITUDE-CONSTANT LIFT COEFFICIENT FLIGHT PROGRAM

$$D=W\left(\frac{D}{W}\right)=W\left(\frac{D}{W}\right)=\frac{W}{E} \tag{B-2}$$

$E=C_L/C_D$ but C_L is constant and $C_D=C_{D0}+KC_L^2$ is also constant. Therefore, E is constant along the flight path and

$$X=-\frac{E}{c}\int_1^2 V\frac{dW}{W} \tag{B-3}$$

$$V=\sqrt{\frac{2W}{\rho SC_L}}=\sqrt{\frac{2}{\rho SC_L}}\,W^{1/2} \tag{B-3a}$$

so that

$$X=-\frac{E}{c}\sqrt{\frac{2}{\rho SC_L}}\int_1^2 W^{-1/2}\,dW \tag{B-4}$$

Performing the indicated integration progressively yields:

$$X=\frac{E}{c}\sqrt{\frac{2}{\rho SC_L}}\left[2(W_1{}^{1/2}-W_2{}^{1/2})\right]$$

$$=\frac{2E}{c}\sqrt{\frac{2W_1}{\rho SC_L}}\left(1-\sqrt{\frac{W_2}{W_1}}\right)$$

$$=\frac{2EV_1}{c}\left(1-\frac{1}{\sqrt{\mathrm{MR}}}\right)$$

so that

$$X_{h,CL}=\frac{2EV_1}{c}\left(1-\sqrt{1-\zeta}\right) \tag{B-5}$$

B-2 THE CONSTANT AIRSPEED-CONSTANT LIFT COEFFICIENT (CRUISE-CLIMB) PROGRAM

With $D = W/E$ from Eq. B-2 and E constant by virtue of C_L and thus C_D being constant, Eq. B-1 can be written as

$$X = -\frac{EV}{c}\int_1^2 \frac{dW}{W} = \frac{EV}{c}[\ln W_1 - \ln W_2]$$

$$= \frac{EV}{c}\ln\left(\frac{W_1}{W_2}\right) = \frac{EV}{c}\ln \mathrm{MR} \tag{B-6}$$

With the $\mathrm{MR} = 1/(1-\zeta)$, the range equation can be written as

$$X_{V,C_L} = \frac{EV}{c}\ln\left(\frac{1}{1-\zeta}\right) \tag{B-7}$$

B-3 THE CONSTANT ALTITUDE-CONSTANT AIRSPEED FLIGHT PROGRAM

$$X = -\frac{V}{c}\int_1^2 \frac{dW}{D} = \frac{V}{cqSC_{DO}}\int_1^2 \frac{-dW}{1+aW^2} \tag{B-8}$$

with $$D = qSC_{DO} + \frac{KW^2}{qS}$$

and where $$a = \frac{K}{q^2S^2C_{DO}} \tag{B-9}$$

Performing the integration yields

$$X = \frac{V}{qSC_{DO}c\sqrt{a}}[\tan^{-1}\sqrt{a}W_1 - \tan^{-1}\sqrt{a}W_2] \tag{B-10}$$

But $$W_2 = W_1 - \Delta W_f = W_1\left(1 - \frac{\Delta W_f}{W_1}\right) = W_1(1-\zeta) \tag{B-11}$$

so that with Eq. B-9, Eq. B-10 becomes

$$X = \frac{V}{c\sqrt{KC_{DO}}}[\tan^{-1}\sqrt{a}W_1 - \tan^{-1}\sqrt{a}W_1(1-\zeta)] \tag{B-12}$$

The bracketed term represents the difference between two angles and can be written as

$$\Theta_1 - \Theta_2 = \tan^{-1}[\tan(\Theta_1 - \Theta_2)] \tag{B-13}$$

where
$$\tan \Theta_1 = \sqrt{a}W_1$$
$$\tan \Theta_2 = \sqrt{a}W_1(1-\zeta) \quad \text{(B-14)}$$

Substituting for $\tan(\Theta_1 - \Theta_2)$ in Eq. B-13 yields

$$\Theta_1 - \Theta_2 = \tan^{-1}\left[\frac{\tan \Theta_1 - \tan \Theta_2}{1 + \tan \Theta_1 \tan \Theta_2}\right]$$

which with the expressions of Eq. B-14 and B-9 becomes

$$\Theta_1 - \Theta_2 = \tan^{-1}\left(\frac{\sqrt{\dfrac{K}{C_{DO}}\dfrac{W_1}{qS}}\,\zeta}{1 + \dfrac{K}{C_{DO}}\dfrac{W_1{}^2}{q^2S^2}(1-\zeta)}\right)$$

which can be rearranged to become

$$\Theta_1 - \Theta_2 = \tan^{-1}\left(\frac{\sqrt{KC_{DO}}\,W_1\zeta}{qSC_{DO} + \dfrac{KW_1{}^2}{qS} - \dfrac{KW_1{}^2\zeta}{qS}}\right)$$

But $qSC_{DO} + \dfrac{KW_1{}^2}{qS} = D_1$, the total initial drag,

and $\dfrac{W_1}{qS} = C_{L_1}$ so that

$$\Theta_1 - \Theta_2 = \tan^{-1}\left[\frac{\sqrt{KC_{DO}}\,W_1\zeta}{D_1\left(1 - \dfrac{KC_{L_1}W_1\zeta}{D_1}\right)}\right] \quad \text{(B-15)}$$

Since $W_1/D_1 = E_1$, the initial lift-to-drag ratio, and $\sqrt{KC_{Do}} = 1/(2E_{max})$ for a parabolic drag polar, we can rewrite Eq. B-15 and substitute in Eq. B-12 to obtain the range equation in the form

$$X_{h,V} = \frac{2E_{max}V}{c} \arctan\left[\frac{E_1\zeta}{2E_{max}(1 - KC_{L_1}E_1\zeta)}\right] \quad \text{(B-16)}$$

INDEX

FLIGHT STABILITY AND AUTOMATIC CONTROL

FLIGHT STABILITY AND AUTOMATIC CONTROL

Dr. Robert C. Nelson

Aerospace and Mechanical Engineering Department
University of Notre Dame

McGraw-Hill Book Company

New York St. Louis San Francisco Auckland Bogotá Caracas Colorado Springs Hamburg
Lisbon London Madrid Mexico Milan Montreal New Delhi Oklahoma City
Panama Paris San Juan São Paulo Singapore Sydney Tokyo Toronto

This book was set in Times Roman.
The editors were Anne T. Brown and John M. Morriss;
the designer was Amy Becker
the production supervisor was Leroy A. Young.
Project supervision was done by The Universities Press.
R. R. Donnelley & Sons Company was printer and binder.

FLIGHT STABILITY AND AUTOMATIC CONTROL

234567890DOCDOC89321098

ISBN 0-07-046218-6

Nelson, Robert C., (date).
Flight stability and automatic control.

Includes bibliographies and index.
1. Stability of airplanes. 2. Airplanes—Control systems. I. Title.
TL574.S7N45 1989 629.132'36 87-32481
ISBN 0-07-046218-6

ABOUT THE AUTHOR

Dr. Nelson received his B.S. and M.S. degrees in Aerospace Engineering from the University of Notre Dame and his Ph.D. in Aerospace Engineering from the Pennsylvania State University. Prior to joining Notre Dame, Dr. Nelson was an instructor of Aerospace Engineering at the Pennsylvania State University and an engineer for the Air Force Flight Dynamics Laboratory at Wright Patterson Air Force Base, Ohio. While employed by AFFDL, he worked on an advanced development program to develop the technology for an air to air short range bomber defense missile. For his contribution to this effort he received a Technical Achievement award from the Air Force Systems Command.

In 1975, Dr. Nelson joined the faculty at Notre Dame and has been active in research dealing with the aerodynamics and flight dynamics of both aircraft and missiles. His present research interests include the aerodynamics of slender bodies at large angles of attack, flow visualization techniques and delta wing aerodynamics. He has published over forty articles and papers on his research. Dr. Nelson has also contributed a chapter to the AIAA Progress Series book *Missile Aerodynamics* entitled, "The Role of Flow Visualization in the Study of High Angle of Attack Aerodynamics," which was published by the AIAA in 1986.

Dr. Nelson has also been active as a consultant to government and industrial organizations. He is a Registered Professional Engineer and Associate Fellow of the American Institute of Aeronautics and Astronautics (AIAA). He served as the General Chairman of the AIAA Atmospheric Flight Mechanics Conference in 1982 and was the Chairman of the AIAA Atmospheric Flight Mechanics Technical Committee from May 1983 to 1985.

CONTENTS

PREFACE

The goal of this book is to present an integrated treatment of the basic elements of aircraft stability, flight control, and autopilot design. An understanding of flight stability and control played an important role in the ultimate success of the earliest aircraft designs. In later years the design of automatic controls ushered in the rapid development of the commercial and military aircraft. Today, both military and civilian aircraft rely heavily on automatic control systems to provide artifical stabilization and autopilots to aid pilots in navigating and landing their aircraft in adverse weather conditions.

This book is intended as a textbook for a course in aircraft flight dynamics for senior undergraduate or first year graduate students. The material presented includes static stability, aircraft equations of motion, dynamic stability, flying or handling qualities, and automatic control. Chapter 1 reviews some basic concepts of aerodynamics, properties of the atmosphere, several of the primary flight instruments and nomenclature. In Chapter 2 the concepts of airplane static stability and control are presented. The design features that can be incorporated into an aircraft design to provide static stability and sufficient control power are discussed. The rigid body aircraft equations of motion are developed along with techniques to model the aerodynamic forces and moments acting on the airplane in Chapter 3. The dynamic characteristics of an airplane for free and forced response are presented in Chapters 4 and 5. Chapter 4 discusses the longitudinal dynamics while Chapter 5 presents the lateral dynamics. In both chapters the relationship between the rigid body motions and the pilot's opinion of the ease or difficulty of flying the airplane is explained. Handling or flying qualities are those control and dynamic characteristics that govern how well a pilot can fly a particular control task. Chapter 6 discusses the solution of the equations of motion for either arbitrary control input or atmospheric disturbances. The last two chapters, 7 and 8, deal with the application of control theory to airplane stabilization and control. Autopilot design concepts are presented for controlling aircraft attitude, flight speed as well as conceptual designs for an automatic landing systems. Both classical root locus and modern matrix control analysis techniques are used to design simple autopilots.

To help in understanding the concepts presented in the text I have included a number of worked-out example problems throughout the book, and at the end of each chapter one will find a problem set. A major feature of the textbook is that the material is introduced by way of simple exercises. For example, dynamic stability is presented first by restricted single degree of freedom motions. This approach permits the reader to gain some experience in the mathematical representation and physical understanding or aircraft response before the more complicated multiple degree of freedom motions are analysized. Several appendices have also been included to provide additional data on airplane aerodynamic, mass, and geometric characteristics as well as review material of some of the mathematical and analysis techniques used in the text.

I am indebted to all the students who used the early drafts of this book. Their many suggestions and patience as the book evolved is greatly appreciated. In addition, I would like to acknowledge the support of Dr. Albin A. Szewczyk, Chairman of the Department of Aerospace and Mechanical at the University of Notre Dame, for making department resources available to me. As in any large undertaking one needs encouragement from time to time and I am greatful for the support from my wife, Julie, and my colleagues Stephen M. Batill and Thomas J. Mueller. I would like to express my thanks for the many useful comments and suggestions provided by colleagues who reviewed this text during the course of its development, especially to John Anderson, University of Maryland; Richard Duprey, United States Air Force Academy; Paul J. Hermann, Iowa State University; Ira D. Jacobson, University of Virginia; James F. Marchman III, Virginia Polytechnic Institute and N. X. Vinh, University of Michigan.

Finally, I would like to express my appreciation to Marilyn Walker for her patience in typing the many versions of this manuscript and to Cecilia Brendel for providing quality figures and artwork from my rough sketches.

Robert C. Nelson

CHAPTER 1

INTRODUCTION

"For some years I have been afflicted with the belief that flight is possible to man."
Wilbur Wright, May 13, 1900

1.1 ATMOSPHERIC FLIGHT MECHANICS

Atmospheric flight mechanics is a broad heading that encompasses three major disciplines, namely, performance, flight dynamics and aeroelasticity. In the past, each of these subjects was treated independently of the other. However, because of the structural flexibility of modern airplanes, the interplay between the disciplines can no longer be ignored. For example, if the flight loads cause significant structural deformation of the aircraft, one can expect changes in the airplane's aerodynamic and stability characteristics which in turn will influence its performance and dynamic behavior.

Airplane performance deals with the determination of performance indices such as range, endurance, rate of climb, and take off and landing distance as well as flight path optimization. To evaluate these indices, one normally treats the airplane as a point mass that is acted on by gravity, lift, drag and thrust. The accuracy of the performance calculations depends upon how accurately the lift, drag and thrust can be determined.

Flight dynamics is concerned with the motion of an airplane due to internally or externally generated disturbances. We are particularly interested in the vehicle's stability and control capabilities. To describe adequately the rigid-body motion of an airplane one needs to consider the complete equations of motion with six degrees of freedom. Again, this will require accurate estimates of the aerodynamic forces and moments acting on the airplane.

The final subject included under the heading of atmospheric flight mechanics is aeroelasticity. Aeroelasticity deals with both static and dynamic aeroelastic phenomena. Basically, aeroelasticity is concerned with phenomena associated with interactions between inertial, elastic and aerodynamic forces. Problems that arise for a flexible aircraft include control reversal, wing divergence, and control surface flutter, to name just a few.

This book is divided into three sections; the first section deals with the properties of the atmosphere, static stability and control concepts, development of aircraft equations of motion and aerodynamic modeling of the airplane; the second part examines aircraft motions due to control inputs or atmospheric disturbances; and the third portion is devoted to aircraft autopilots. Although no specific chapters are devoted entirely to performance or aeroelasticity, an effort is made to show the reader, at least in a qualitative way, how performance specifications and aeroelastic phenomena influence aircraft stability and control characteristics.

The interplay between the three disciplines that make up atmospheric flight mechanics is best illustrated by the experimental high-performance airplane shown in Fig. 1.1. The X-29A aircraft incorporates the latest in advanced technologies in controls, structures, and aerodynamics. These technologies will provide substantial performance improvements over more conventional fighter designs. Such a design could not be developed without paying close attention to the interplay between performance, aeroelasticity, stability and control. In fact, the evolution of this radical design was developed using tradeoff studies between the various disciplines to justify the expected performance improvements.

The forces and moments acting on an airplane depend upon the properties of the atmosphere through which it is flying. In the following sections we will review some basic concepts of fluid mechanics that will help us develop an appreciation of the atmospheric properties essential to our understanding of airplane flight mechanics. In addition we will discuss some of the important aircraft instruments that provide flight information to the pilot.

1.2 BASIC DEFINITIONS

The aerodynamic forces and moments generated on an airplane are due to its geometric shape, attitude to the flow, airspeed, and to the properties of the ambient air mass through which it is flying. Air is a fluid and, as such, possesses certain fluid properties. The properties that we are interested in are the pressure, temperature, density, viscosity and speed of sound of air at the flight altitude.

FLUID. A fluid can be thought of as any substance that flows. To have such a property, the fluid must deform continuously when acted on by a shearing

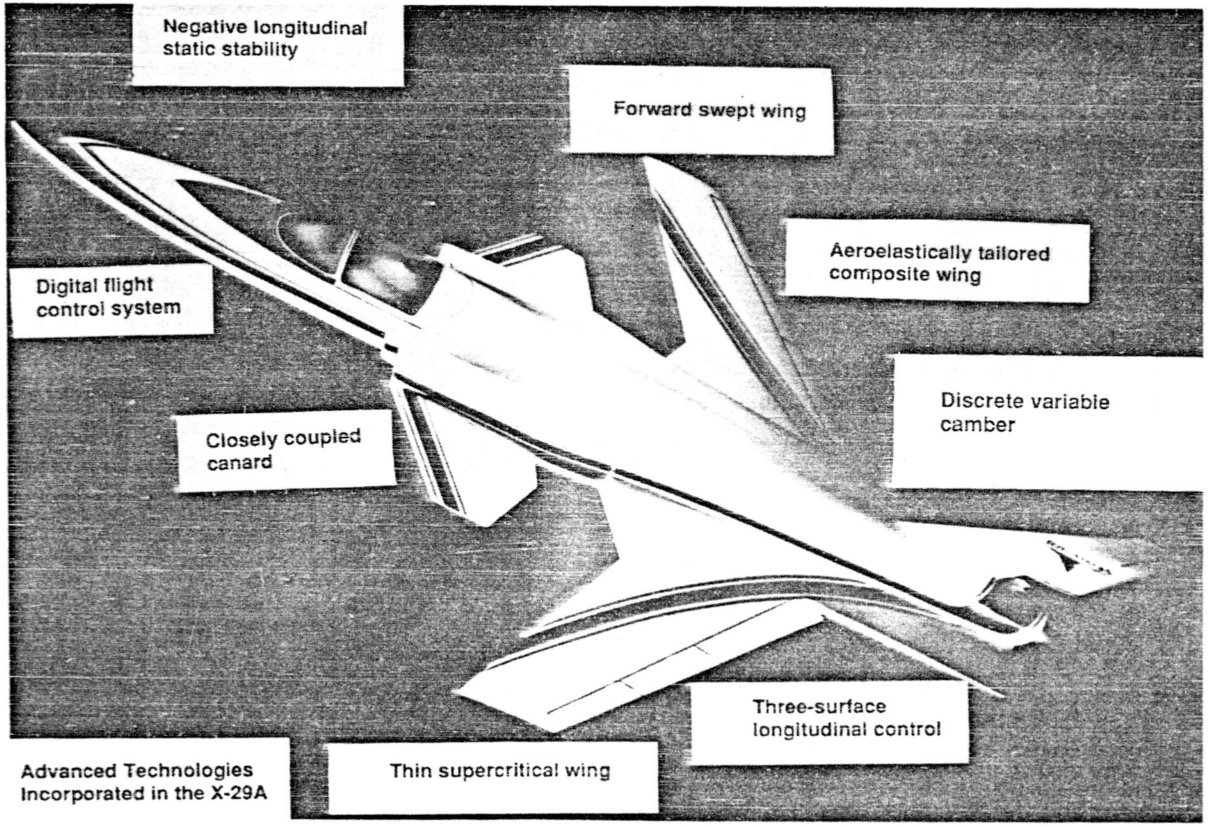

FIGURE 1.1
Advanced technologies incorporated on the X-29A aircraft.

force. A shear force is a force tangent to the surface of the fluid element. There are no shear stresses present in the fluid when it is at rest. A fluid can transmit forces normal to any chosen direction. The normal force and the normal stress are the pressure force and pressure, respectively.

Both liquids and gases can be considered to be fluids. Liquids under most conditions do not change their weight per unit volume appreciably and can be considered to be incompressible for most engineering applications. Gases, on the other hand, do change their weight or mass per unit volume appreciably under the influences of pressure or temperature and, therefore, must be considered to be compressible.

PRESSURE. Pressure is the normal force per unit area acting on the fluid. The average pressure is calculated by dividing the normal force to the surface by the surface area:

$$P = \frac{F}{A} \tag{1.1}$$

The static pressure in the atmosphere is nothing more than the weight per unit area of the air above the elevation being considered. The ratio of the pressure P at altitude to sea-level standard pressure P_0 is given the symbol δ:

$$\delta = \frac{P}{P_0} \tag{1.2}$$

The relationship between pressure, density ρ, and temperature T is given by the equation of state,

$$P = \rho RT \tag{1.3}$$

where R is a constant, the magnitude depending on the gas being considered. For air, R has a value 287 J/(kg K) or 1716 ft^2/(sec^2 °R). Atmospheric air follows the equation of state provided that the temperature is not too high and that air can be treated as a continuum.

TEMPERATURE. In aeronautics the temperature of air is an extremely important parameter, in that it affects the properties of air such as density and viscosity. Temperature is an abstract concept but can be thought of as a measure of the motion of molecular particles within a substance. The concept of temperature also serves as a means of determining the direction in which heat energy will flow when two objects of different temperatures come into contact. Heat energy will flow from the higher temperature object to that at lower temperature.

As we will show later, the temperature of the atmosphere varies significantly with altitude. The ratio of the ambient temperature at altitude, T, to a sea-level standard value, T_0 is denoted by the symbol θ:

$$\theta = \frac{T}{T_0} \tag{1.4}$$

where the temperatures are measured using the absolute Kelvin or Rankine scales.

DENSITY. The density of a substance is defined as the mass per unit volume:

$$\rho = \frac{\text{Mass}}{\text{Unit volume}} \tag{1.5}$$

From the equation of state, it can be seen that the density of a gas is directly proportional to the pressure and inversely proportional to the absolute temperature. The ratio of ambient air density ρ to standard sea level air density ρ_0 occurs in many aeronautical formulas and is given the designation σ.

$$\sigma = \rho/\rho_0 \tag{1.6}$$

VISCOSITY. Viscosity can be thought of as the internal friction of a fluid. Both liquids and gases possess the property of viscosity, with liquids being much more viscous than gases. As an aid in visualizing the concept of viscosity, consider the following simple experiment. Consider the motion of the fluid between two parallel plates separated by the distance h. If one plate is held fixed while the other plate is being pulled with a constant velocity u, then the velocity distribution of the fluid between the plates will be linear as shown in Fig. 1.2.

To produce the constant velocity motion of the upper plate, a tangential force must be applied to the plate. The magnitude of the force must be equal to the friction forces in the fluid. It has been established from experiments that the force per unit area of the plate is proportional to the velocity of the moving plate and inversely proportional to the distance between the plates. Expressed mathematically we have

$$\tau \propto \frac{u}{h} \tag{1.7}$$

where τ is the force per unit area, which is called the shear stress.

A more general form of Eq. (1.7) can be written by replacing u/h with the derivative du/dy. The proportionality factor is denoted by μ, the coefficient of absolute viscosity, which is obtained experimentally.

$$\tau = \mu \frac{du}{dy} \tag{1.8}$$

Equation (1.8) is known as Newton's law of friction.

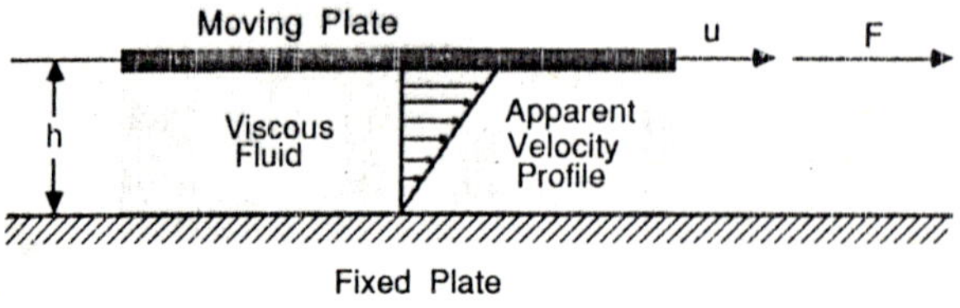

FIGURE 1.2
Shear stress between two plates.

For gases, the absolute viscosity depends only on temperature, with increasing temperature causing an increase in viscosity. To estimate the change in viscosity with temperature, several empirical formulations are commonly used. The simplest formula is Rayleigh's, which is

$$\frac{\mu_1}{\mu_0} = \left(\frac{T_1}{T_0}\right)^{3/4} \tag{1.9}$$

where the temperatures are on the absolute scale and the subscript zero denotes the reference condition.

An alternate expression for calculating the variation of absolute viscosity with temperature was developed by Sutherland. The empirical formula developed by Sutherland is valid provided the pressure is greater than 0.1 atmosphere, and is

$$\frac{\mu_1}{\mu_0} = \left(\frac{T_1}{T_0}\right)^{3/2} \frac{T_1 + S_1}{T_0 + S_1} \tag{1.10}$$

where S_1 is a constant. When the temperatures are expressed in the Rankine scale, $S_1 = 198°\text{R}$; when the temperatures are expressed in the Kelvin scale, $S_1 = 110\text{ K}$.

The ratio of the absolute viscosity to the density of the fluid is a parameter that appears frequently and has been identified with the symbol ν; it is called the kinematic viscosity:

$$\nu = \frac{\mu}{\rho} \tag{1.11}$$

An important dimensionless quantity known as the Reynolds number is defined as

$$R = \frac{\rho V l}{\mu} = \frac{V l}{\nu} \tag{1.12}$$

and can be thought of as the ratio of the inertial to viscous forces of the fluid.

MACH NUMBER AND THE SPEED OF SOUND. The ratio of an airplane's speed V to the local speed of sound a is an extremely important parameter, called Mach number after the Austrian physicist Ernst Mach. The mathematical definition of Mach number is

$$M = \frac{V}{a} \tag{1.13}$$

As an airplane moves through the air, it creates pressure disturbances that propagate away from the airplane in all directions with the speed of sound. If the airplane is flying at a Mach number less than 1, the pressure disturbances travel faster than the airplane and influence the air ahead of the airplane. An example of this phenomenon is the upwash field created in front of a wing.

However, for flight at Mach numbers greater than 1 the pressure disturbances are moving more slowly than the airplane and, therefore, the flow ahead of the airplane has no warning of the oncoming aircraft.

The aerodynamic characteristics of an airplane depend upon the flow regime existing around the airplane. As the flight Mach number is increased, the flow around the airplane can be completely subsonic, a mixture of subsonic and supersonic flow, or completely supersonic. The flight Mach number is used to classify the various flow regimes. An approximate classification of the flow regimes is given below.

Incompressible subsonic flow	$0 < M < 0.5$
Compressible subsonic flow	$0.5 < M < 0.8$
Transonic flow	$0.8 < M < 1.2$
Supersonic flow	$1.2 < M < 5$
Hypersonic flow	$5 < M$

In order to have accurate aerodynamic predictions at $M > 0.5$ compressibility effects must be included.

The local speed of sound must be known to determine the Mach number. The speed of sound can be shown to be related to the absolute ambient temperature by the following expression:

$$a = (\gamma RT)^{1/2} \tag{1.14}$$

where γ is the ratio of specific heats and R is the gas constant. The ambient temperature will be shown in a later section to be a function of altitude.

1.3 AEROSTATICS

Aerostatics deals with the state of a gas at rest. It follows from the definition given for a fluid that all forces acting on the fluid must be normal to any cross-section within the fluid. Unlike a solid, a fluid at rest cannot support a shearing force. A consequence of this is that the pressure in a fluid at rest is independent of direction. That is to say that, at any point, the pressure is the same in all directions. This fundamental concept owes its origin to Pascal, a French scientist (1623–1662).

VARIATION OF PRESSURE IN A STATIC FLUID. Consider the small vertical column of fluid shown in Fig. 1.3. Because the fluid is at rest, the forces in both the vertical and horizontal directions must sum to zero. The forces in the vertical direction are due to the pressure forces and the weight of the fluid column. The force balance in the vertical direction is given by

$$PA = (P + dP)A + \rho gA\,dh \tag{1.15}$$

or

$$dP = -\rho g\,dh \tag{1.16}$$

Equation (1.16) tells us how the pressure varies with elevation above some

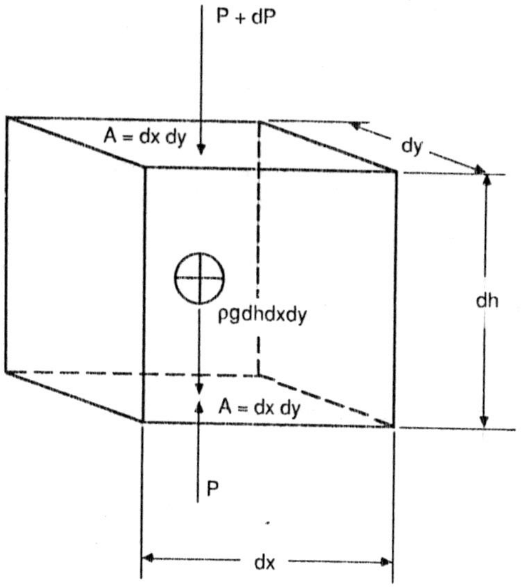

FIGURE 1.3
Element of fluid at rest.

reference level in a fluid. As the elevation is increased, the pressure will decrease. Therefore, the pressure in a static fluid is equal to the weight of the column of fluid above the point of interest.

One of the simplest means of measuring pressure is by way of a fluid manometer. Figure 1.4 shows two types of manometers. The first manometer consists of a U-shaped tube containing a liquid. When pressures of different magnitudes are applied across the manometer the fluid will rise on the side of the lower pressure and fall on the side of the higher pressure. By writing a force balance for each side, one can show that

$$P_1A + \rho gxA = P_2A + \rho g(x+h)A \tag{1.17}$$

which yields a relationship for the pressure difference in terms of the change in

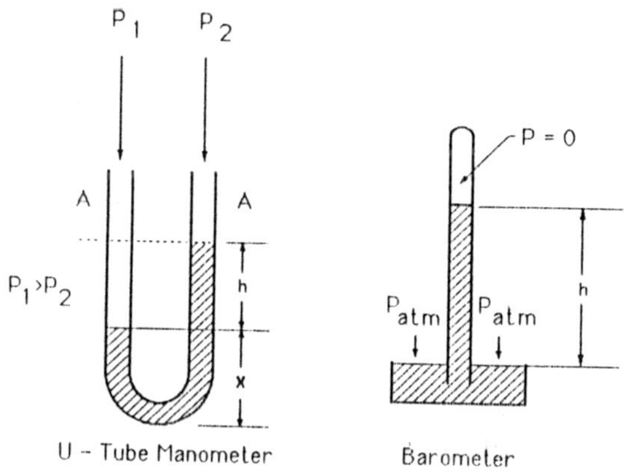

FIGURE 1.4
Sketch of U-tube manometer and barometer.

height of the liquid column:

$$P_1 - P_2 = \rho g h \tag{1.18}$$

The second sketch shows a simple mercury barometer. The barometer can be thought of as a modified U-tube manometer. One leg of the tube is closed off and evacuated. The pressure at the top of this leg is zero and atmospheric pressure acts on the open leg. The atmospheric pressure is therefore equal to the height of the mercury column, i.e.

$$P_{\text{atm}} = \rho g h \tag{1.19}$$

In practice the atmospheric pressure is commonly expressed as so many inches or millimeters of mercury. Remember, however, that neither inches nor millimeters of mercury are units of pressure.

1.4 DEVELOPMENT OF BERNOULLI'S EQUATION

Bernoulli's equation establishes the relationship between pressure, elevation, and velocity of the flow along a stream tube. For this analysis, the fluid is assumed to be a perfect fluid, i.e. we will ignore viscous effects. Consider the element of fluid in the stream tube shown in Fig. 1.5. The forces acting on the differential element of fluid are due to pressure and gravitational forces. The pressure force acting in the direction of the motion is given by

$$F_{\text{pressure}} = P\,\mathrm{d}A - \left(P + \frac{\partial P}{\partial s}\,\mathrm{d}s\right)\mathrm{d}A \tag{1.20}$$

or

$$= \mathrm{d}P\,\mathrm{d}A \tag{1.21}$$

The gravitational force can be expressed as

$$F_{\text{gravitational}} = -g\,\mathrm{d}m \sin\alpha \tag{1.22}$$

$$= g\,\mathrm{d}m\,\frac{\mathrm{d}z}{\mathrm{d}s} \tag{1.23}$$

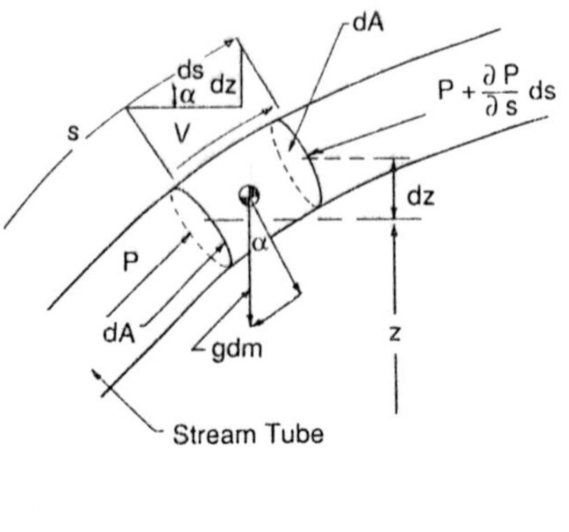

FIGURE 1.5
Forces acting on an element of flow in a stream tube.

Applying Newton's second law yields

$$-dP\,dA - g\,dm\frac{dz}{ds} = dm\frac{dV}{dt} \tag{1.24}$$

The differential mass dm can be expressed in terms of the mass density of the fluid element times its respective volume, i.e.

$$dm = \rho\,dA\,ds \tag{1.25}$$

Inserting the expression for the differential mass, the acceleration of the fluid can be expressed as

$$\frac{dV}{dt} = -\frac{1}{\rho}\frac{dP}{ds} - g\frac{dz}{ds} \tag{1.26}$$

The acceleration can be expressed as

$$\frac{dV}{dt} = \frac{\partial V}{\partial t} + \frac{\partial V}{\partial s}\frac{ds}{dt} \tag{1.27}$$

The first term on the right-hand side, $\partial V/\partial t$, denotes the change in velocity as a function of time for the entire flow field. The second term denotes the acceleration due to a change in location. If the flow field is steady, the term $\partial V/\partial t = 0$ and Eq. (1.27) reduces to

$$\frac{\partial V}{\partial s}\frac{ds}{dt} = -\frac{1}{\rho}\frac{dP}{ds} - g\frac{dz}{ds} \tag{1.28}$$

The changes of pressure as a function of time cannot accelerate a fluid particle. This is because the same pressure would be acting at every instant on all sides of the fluid particles. Therefore, the partial differential can be replaced by the total derivative in Eq. (1.28):

$$V\frac{dV}{ds} = -\frac{1}{\rho}\frac{dP}{ds} - g\frac{dz}{ds} \tag{1.29}$$

Integrating Eq. (1.29) along a streamline yields

$$\int_1^2 V\,dV = -\int_1^2 \frac{dP}{\rho} - g\int_1^2 dz \tag{1.30}$$

which is known as Bernoulli's equation. Bernoulli's equation establishes the relationship between pressure, elevation, and velocity along a stream tube.

INCOMPRESSIBLE BERNOULLI EQUATION. If the fluid is considered to be incompressible, Eq. (1.29) readily can be integrated to yield the incompressible Bernoulli equation:

$$P_1 + \tfrac{1}{2}\rho V_1^2 + \rho g z_1 = P_2 + \tfrac{1}{2}\rho V_2^2 + \rho g z_2 \tag{1.31}$$

The differences in elevation can usually be ignored when dealing with the flow of gases such as air. An important application of Bernoulli's equation is the determination of the so-called stagnation pressure of a moving body or a body exposed to a flow. The stagnation point is defined as that point on the body at which the flow comes to rest. At that point the pressure is

$$P_0 = P_\infty + \tfrac{1}{2}\rho V_\infty^2 \tag{1.32}$$

where P_∞ and V_∞ are the static pressure and velocity far away from the body, that is the pressures and velocities that would exist if the body were not present. In the case of a moving body, V_∞ is equal to the velocity of the body itself and P_∞ is the static pressure of the medium through which the body is moving.

BERNOULLI'S EQUATION FOR A COMPRESSIBLE FLUID. At higher speeds (of the order of 100 m/s), the assumption that the fluid density of gases is constant, becomes invalid. As speed is increased, the air undergoes a compression and, therefore, the density cannot be treated as a constant. If the flow can be assumed to be isentropic, the relationship between pressure and density can be expressed as

$$P = c\rho^\gamma \tag{1.33}$$

where γ is the ratio of specific heats for the gas. For air, γ is approximately 1.4.

Substituting Eq. (1.33) into Eq. (1.30) and performing the indicated integrations yields the compressible form of Bernoulli's equation:

$$\frac{\gamma}{\gamma-1}\frac{P}{\rho} + \tfrac{1}{2}V^2 + gz = \text{constant.} \tag{1.34}$$

As noted earlier, the elevation term is usually quite small for most aeronautical applications and, therefore, can be ignored. The stagnation pressure can be found by letting $V = 0$, in Eq. (1.34).

$$\frac{\gamma}{\gamma-1}\frac{P}{\rho} + \tfrac{1}{2}V^2 = \frac{\gamma}{\gamma-1}\frac{P_0}{\rho_0} \tag{1.35}$$

If we divide Eq. (1.35) by the term P/ρ, we obtain

$$1 + \frac{\gamma-1}{2}\frac{1}{\gamma P/\rho}V^2 = \frac{P_0/P}{\rho_0/\rho} \tag{1.36}$$

Equation (1.36) can be solved for the velocity by substituting the following expressions,

$$a^2 = \gamma RT = \gamma P/\rho \tag{1.37}$$

and

$$\frac{P_0}{P} = \left(\frac{\rho_0}{\rho}\right)^\gamma \tag{1.38}$$

into Eq. (1.36) and rearranging to yield a relationship for the velocity and Mach number as follows.

$$V = \frac{2a^2}{\gamma - 1}\left[\left(\frac{P_0}{P}\right)^{(\gamma-1)/\gamma} - 1\right]^{1/2} \tag{1.39}$$

$$M = \frac{2}{\gamma - 1}\left[\left(\frac{P_0}{P}\right)^{(\gamma-1)/\gamma} - 1\right]^{1/2} \tag{1.40}$$

Equations (1.39) and (1.40) can be used to find the velocity and Mach number provided the flow regime is below $M = 1$.

1.5 THE ATMOSPHERE

The performance characteristics of an airplane depend on the properties of the atmosphere through which it flies. Because the atmosphere is continuously changing with time, it is impossible to determine airplane performance parameters precisely without first defining the state of the atmosphere.

The earth's atmosphere is a gaseous envelope surrounding the planet. The gas which we call air is actually a composition of numerous gases. The composition of dry air at sea level is shown in Table 1.1. The relative percentages of the constituents remains essentially the same up to an altitude of 90 km or 300 000 ft owing primarily to atmospheric mixing caused by winds and turbulence. At altitudes above 90 km the gases begin to settle or separate. The variability of water vapor in the atmosphere must be taken into account by the performance analyst. Water vapor can comprise up to 4 percent by volume of atmospheric air. When the relative humidity is high, the air density is lower than that for dry air for the same conditions of pressure and temperature. Under these conditions the density may be reduced by as much as 3 percent. A change in air density will cause a change in the aerodynamic forces acting on the airplane and therefore influence its performance capabilities. Furthermore, changes in air density created by water vapor will affect engine performance, which again influences the performance of the airplane.

TABLE 1.1
Composition of atmospheric air

	Density		Percentage by volume	Percentage by weight
	kg/m³	slugs/ft³		
Air	1.2250	2.3769×10^{-3}	100	100
Nitrogen			78.03	75.48
Oxygen			20.99	23.18
Argon			0.94	1.29

The remaining small portion of the composition of air is made up of neon, helium, krypton, xenon, CO_2 and water vapor

The atmosphere can be thought of as being composed of various layers, with each layer of the atmosphere having its own distinct characteristics. For the purpose of this discussion we will divide the atmosphere into four regions. In ascending order the layers are the troposphere, stratosphere, ionosphere and exosphere. The four layers are illustrated in Fig. 1.6. The troposphere and stratosphere are extremely important to aerospace engineers since most aircraft fly in these regions. The troposphere extends from the earth's surface to an altitude of approximately 6–13 miles or 10–20 km. The air masses in the troposphere are in constant motion and the region is characterized by unsteady or gusting winds and turbulence. The influence of turbulence and wind shear on aircraft structural integrity and flight behavior continues to be an important area of research for the aeronautical community. The structural loads imposed on an aircraft during an encounter with turbulent air can reduce the structural life of the airframe or, in an encounter with severe turbulence, can cause structural damage to the airframe.

Wind shear is an important atmospheric phenomenon that can be hazardous to aircraft during take-off or landing. Wind shear is the variation of the wind vector in both magnitude and direction. In vertical wind shear, the wind speed and direction change with altitude. An airplane landing in such a

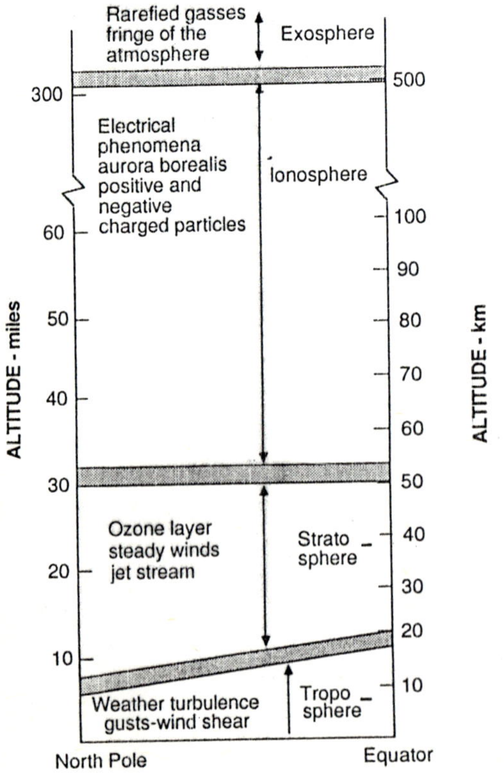

FIGURE 1.6
Layers of earth's atmosphere.

wind shear may be difficult to control; this can cause deviations from the intended touchdown point. Wind shears are created by the movement of air masses relative to one another or to the earth's surface. Thunderstorms, frontal systems, and the earth's boundary layer all produce wind shear profiles, that at times are severe enough to be hazardous to aircraft flying at low altitude.

The next layer above the troposphere is called the stratosphere. The stratosphere extends up to 50–70 miles, or 80–113 km, above the earth's surface. Unlike the troposphere, the stratosphere is a relatively tranquil region, free of gusts and turbulence, but it is characterized by high, steady winds. Wind speeds of the order of 37 m/s or 120 ft/s have been measured in the stratosphere.

The ionosphere extends from the upper edge of the stratosphere to an altitude of 200–300 miles, or 124–186 km (The name is derived from the word "ion", which describes a particle which has either a positive or negative electric charge.) This is the region where the air molecules undergo dissociation and many electrical phenomena occur. The aurora borealis is a visible electrical display that occurs in the ionosphere.

The last layer of the atmosphere is called the exosphere. The exosphere is the outermost region of the atmosphere and is made up of rarefied gas. In effect this is a transition zone between the earth's atmosphere and interplanetary space. For many applications we can consider air resistance to cease in the exosphere.

As stated previously, the properties of the atmosphere change with time and location on the earth. In order to compare the flight performance characteristics of airplanes and flight instruments, a standard atmosphere was needed. The modern standard atmosphere was first developed in the 1920s, independently in the United States and in Europe. The National Advisory Committee for Aeronautics (NACA) generated the American Standard Atmosphere. The European standard was developed by the International Commission for Aerial Navigation (ICAN). The two standard atmospheres were essentially the same except for some slight differences. These differences were resolved by an international committee and an international standard atmosphere was adopted by the International Civil Aviation Organization (ICAO) in 1952.

The standard atmosphere assumes a unique temperature profile which was determined by an extensive observation program. The temperature profile consists of regions of linear variations of temperature with altitude, and regions of constant temperature (isothermal regions). Figure 1.7 shows the temperature profile through the standard atmosphere. The standard sea-level properties of air are listed in Table 1.2.

The properties of the atmosphere can be expressed analytically as a function of altitude. However, before proceeding with the development of the analytical model of the atmosphere, we must define what we mean by altitude. For the present we will be concerned with three different definitions of

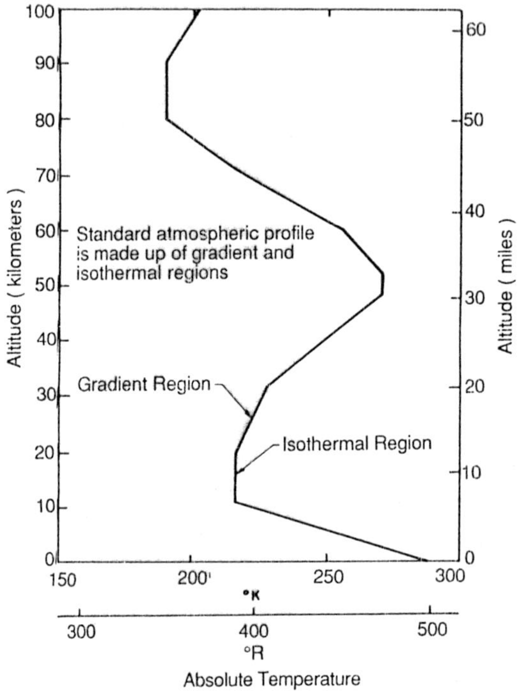

FIGURE 1.7
Temperature profile in the standard atmosphere.

altitude; absolute, geometric, and geopotential. Figure 1.8 shows the relationship between absolute and geometric altitude. Absolute altitude is the distance from the center of the Earth to the point in question, whereas the geometric altitude is the height of the point above sea level. The absolute and geometric altitudes are related to each other in the following manner:

$$h_a = h_G + R_0 \tag{1.41}$$

where h_a, h_G, and R_0 are the absolute altitude, geometric altitude, and radius of the Earth, respectively.

TABLE 1.2
Properties of air at sea level in the standard atmosphere

	English units	SI units
Gas constant, R	1716 ft·lb/(slug·°R)	287 $m^2/(K{\cdot}s^2)$
Pressure, P	2116.2 lb/ft^2	1.012×10^5 N/m^2
	29.92 in Hg	760 mm Hg
Density, ρ	(2.377×10^{-3}) $slug/ft^3$	1.225 kg/m^3
Temperature	518.69°R	288.16 K
Absolute viscosity, μ	3.737×10^{-7} $lb{\cdot}s/ft^2$	1.789×10^{-3} $N{\cdot}s/m^2$
Kinematic viscosity, ν	1.572×10^{-4} ft^{-2}/s	1.460×10^{-3} m^{-2}/s
Speed of sound, a	1116.4 ft/s	340.3 m/s

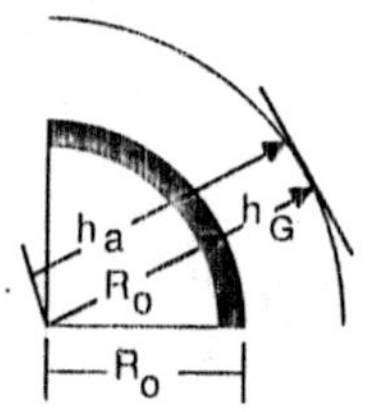

FIGURE 1.8
Definition of geometric and absolute altitude.

Historically, measurements of atmospheric properties have been based upon the assumption that the acceleration due to gravity is constant. This assumption leads to a fictitious altitude called the geopotential altitude. The relationship between the geometric and geopotential altitudes can be determined from an examination of the hydrostatic equation (Eq. (1.16)). Rewriting the hydrostatic equation,

$$dP = -\rho g \, dh \tag{1.42}$$

we see that the change in pressure is a function of the fluid density, and if we employ the acceleration due to gravity at sea level then h is the geopotential altitude. Therefore, we have

$$dP = -\rho g_0 \, dh \tag{1.43}$$

when h is the geopotential height, and

$$dP = -\rho g \, dh_G \tag{1.44}$$

when h_G is the geometric height.

Equations (1.43) and (1.44) can be used to establish the relationship between the geometric and geopotential altitude. Upon comparing these equations we see that

$$dh = \frac{g}{g_0} \, dh_G \tag{1.45}$$

Further, it can be shown that

$$g = g_0 \left(\frac{R_0}{R_0 + h_G} \right)^2 \tag{1.46}$$

which, when substituted into Eq. (1.45) yields

$$dh = \frac{R_0^2 \, dh_G}{(R_0 + h_G)^2} \tag{1.47}$$

Equation (1.47) can be integrated to give an expression relating the two altitudes:

$$h = \frac{R_0}{R_0 + h_G} h_G \tag{1.48}$$

or

$$h_G = \frac{R_0}{R_0 - h} h \tag{1.49}$$

In practice, the difference between the geometric and geopotential altitudes is quite small for altitudes below 15.2 km or 50 000 ft. However, for the higher altitudes the difference must be taken into account for accurate performance calculations.

Starting with the relationship for the change in pressure with altitude and the equations of state

$$\mathrm{d}P = -\rho g_0 \,\mathrm{d}h \tag{1.50}$$

and

$$P = \rho R T \tag{1.51}$$

we can obtain the following expression by dividing Eq. (1.50) by (1.51):

$$\frac{\mathrm{d}P}{P} = -\frac{g_0}{R}\frac{\mathrm{d}h}{T} \tag{1.52}$$

If the temperature varies with altitude in a linear manner, Eq. (1.52) yields

$$\int_{P_1}^{P} \frac{\mathrm{d}P}{P} = -\frac{g_0}{R}\int_{h_1}^{h} \frac{\mathrm{d}h}{T_1 + \lambda(h - h_1)} \tag{1.53}$$

which, upon integration, gives

$$\ln \frac{P}{P_1} = -\frac{g_0}{R\lambda} \ln \frac{T_1 + \lambda(h - h_1)}{T_1} \tag{1.54}$$

where P_1, T_1, and h_1, are the pressure, temperature and altitude at the start of the linear region and λ is the rate of temperature change with altitude, which is called the lapse rate. Equation (1.54) can be rewritten in a more convenient form as

$$\frac{P}{P_1} = \left(\frac{T}{T_1}\right)^{-g_0/(R\lambda)} \tag{1.55}$$

Equation (1.55) can be used to calculate the pressure at various altitudes in any one of the linear temperature profile regions, provided the appropriate constants P_1, T_1, h_1 and λ are used.

The density variation can be easily determined as follows:

$$\frac{P}{P_1} = \frac{\rho T}{\rho_1 T_1} \tag{1.56}$$

and therefore

$$\frac{\rho}{\rho_1} = \left(\frac{T}{T_1}\right)^{-(1+g_0/(R\lambda))} \tag{1.57}$$

In the isothermal regions, the temperature remains constant as altitude varies. Starting again with equation (1.52), we obtain

$$\ln \frac{P}{P_1} = -\frac{g_0}{RT_1}(h - h_1) \tag{1.58}$$

or

$$\frac{P}{P_1} = e^{-g_0(h-h_1)/(RT_1)} \tag{1.59}$$

where P_1, T_1, and h_1 are the values of pressure, temperature and altitude at the start of the isothermal region. The density variation in the isothermal regions can be obtained as

$$\frac{\rho}{\rho_1} = e^{-g_0(h-h_1)/(RT_1)} \tag{1.60}$$

Equations (1.55), (1.57), (1.59) and (1.60) can be used to predict accurately the pressure and density variation in the standard atmosphere up to approximately 57 miles, or 91 km. Table 1.3 gives the values of temperature pressure and density at the boundaries between the various temperature segments. The properties of the standard atmosphere as a function of altitude are presented in tabular form in the Appendix.

Example Problem 1.1. The temperature from sea level to 30 000 ft is found to decrease in a linear manner. The temperature and pressure at sea level are measured to be 40°F and 2050 lb/ft², respectively. If the temperature at 30 000 ft is −60°F, find the pressure and density at 20 000 ft.

Solution. The temperature can be represented by the linear equation

$$T = T_1 + \lambda h \quad \text{where} \quad T_1 = 499.6°\text{R} \quad \text{and} \quad \lambda = \frac{T - T_1}{h} = -0.00333°\text{R/ft}$$

TABLE 1.3
Properties of the atmosphere at the isothermal gradient boundaries

Geopotential altitude, H, km	Geometric altitude, Z, km	T_0, K	P, N/m²	ρ, kg/m³	dT/dH, K/km
0	0	288.15	1.01325×10^5	1.225	−6.5
11	11.019	216.65	2.2636×10^4	3.639×10^{-1}	0
20	20.063	216.65	5.474×10^3	8.803×10^{-2}	1
32	32.162	228.65	8.6805×10^2	1.332×10^{-2}	1
47	47.350	270.65	1.1095×10^2	1.427×10^{-3}	2.8
52	52.429	270.65	5.9002×10^1	7.594×10^{-4}	0
61	61.591	252.65	1.8208×10^1	2.511×10^{-4}	−2
79	79.994	180.65	1.03757	2.001×10^{-5}	−4
88.74	90.0	180.65	0.16435	3.170×10^{-5}	0

The temperature at 20 000 ft can be obtained as

$$T = 499.6 - (0.00333°\text{R/ft})h$$

When $h = 20\,000$ ft, $T = 432.9°$R. The pressure can be calculated from Eq. (1.54), i.e.

$$\frac{P}{P_1} = \left(\frac{T}{T_1}\right)^{-g_0/R\lambda} \qquad P = P_1\left(\frac{T}{T_1}\right)^{-g_0/R\lambda} = (2050\ \text{lb/ft}^2)\left(\frac{432.9°\text{R}}{499.6°\text{R}}\right)^{5.63} = 915\ \text{lb/ft}^2$$

The density can be found from either Eq. (1.3) or (1.56). Using the equation of state,

$$P = \rho RT \qquad \rho = \frac{P}{RT}$$

$$\rho = \frac{915\ \text{lb/ft}^2}{(1718\ \text{ft}^2/(\text{s}^2\cdot°\text{R}))(432.9°\text{R})}\ 0.00123\ \text{slug/ft}^3$$

1.6 AERODYNAMIC NOMENCLATURE

To describe the motion of an airplane it is necessary to define a suitable coordinate system for formulation of the equations of motion. For most problems dealing with aircraft motion, two coordinate systems are used. One coordinate system is fixed to the earth and may be considered for the purpose of aircraft motion analysis to be an inertial coordinate system. The other coordinate system is fixed to the airplane and is referred to as a body coordinate system. Figure 1.9 shows the two right-handed coordinate systems.

The forces acting on an airplane in flight consist of aerodynamic, thrust and gravitational forces. These forces can be resolved along an axis system fixed to the airplane's center of gravity, as illustrated in Fig. 1.10. The force components are denoted X, Y and Z, T_x, T_y and T_z and W_x, W_y and W_z for the

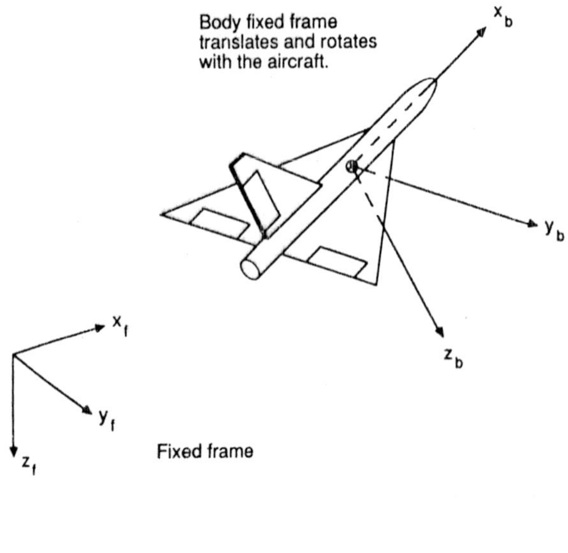

FIGURE 1.9
Body axes coordinate system.

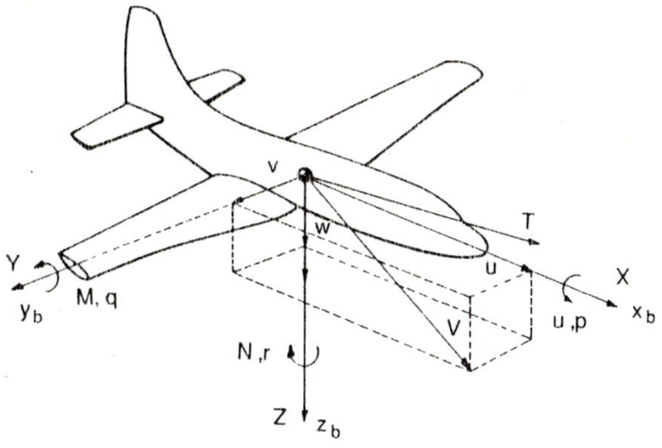

	Roll Axis x_b	Pitch Axis y_b	Yaw Axis z_b
Angular rates	p	q	r
Velocity components	u	v	w
Aerodynamic force components	X	Y	Z
Aerodynamic moment components	L	M	N
Moment of inertia about each axis	I_x	I_y	I_z
Products of inertia	I_{yz}	I_{xz}	I_{xy}

FIGURE 1.10
Definition of forces, moments and velocity components in a body fixed coordinate frame.

aerodynamic, thrust and gravitational force components along the x, y and z axes, respectively. The aerodynamic forces are defined in terms of dimensionless coefficients, the flight dynamic pressure and a reference area as follows:

$$X = C_x QS \qquad \text{axial force} \tag{1.61}$$

$$Y = C_y QS \qquad \text{side force} \tag{1.62}$$

$$Z = C_z QS \qquad \text{normal force} \tag{1.63}$$

In a similar manner, the moments on the airplane can be divided into moments created by the aerodynamic load distribution and the thrust force not acting through the center of gravity. The components of the aerodynamic moment are also expressed in terms of dimensionless coefficients, flight dynamc pressure, reference area and a characteristic length as follows:

$$L = C_l QSl \qquad \text{rolling moment} \tag{1.64}$$

$$M = C_m QSl \qquad \text{pitching moment} \tag{1.65}$$

$$N = C_n QSl \qquad \text{yawing moment} \tag{1.66}$$

For airplanes, the reference area S is taken as the wing platform area and the characteristic length l is taken as the wing span for the rolling and yawing moment and the mean chord for the pitching moment. For rockets and missiles, the reference area is usually taken as the maximum cross-sectional area, and the characteristic length is taken as the maximum diameter.

The aerodynamic coefficients C_x, C_y, C_z, C_l, C_m, and C_n are primarily a function of Mach number, Reynolds number, angle of attack and sideslip angle, and are secondary functions of the time rate of change of angle of attack and sideslip, and the angular velocity of the airplane.

The aerodynamic force and moment acting on the airplane and its angular and translational velocity are illustrated in Fig. 1.10. The x and z axes are in the plane of symmetry with the x axis pointing along the fuselage and the positive y axis along the right wing. The resultant force and moment, as well as the airplane's velocity, can be resolved along these axes.

The angle of attack and sideslip can be defined in terms of the velocity components as illustrated in Fig. 1.11. The equations for α and β are given below.

$$\alpha = \tan^{-1}\frac{w}{u} \tag{1.67}$$

and

$$\beta = \sin^{-1}\frac{v}{V} \tag{1.68}$$

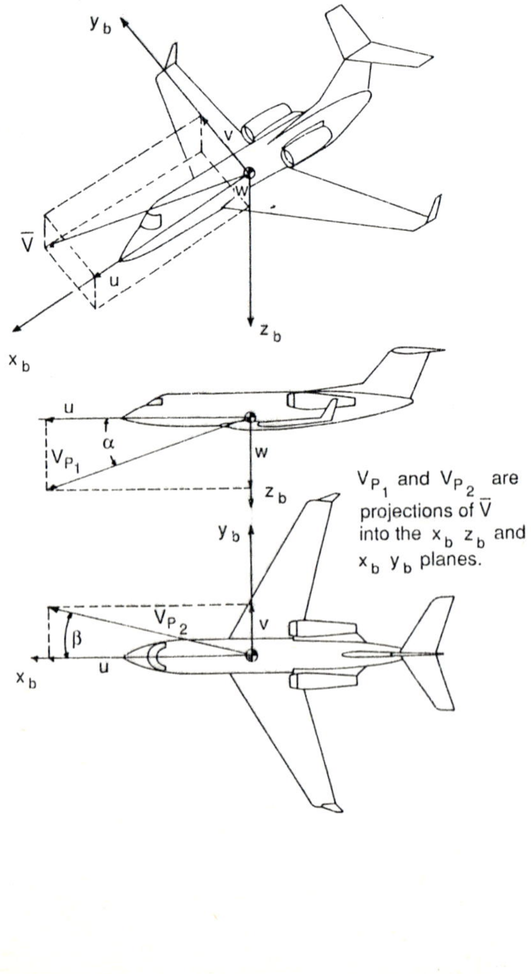

FIGURE 1.11
Definition of angle of attack and sideslip.

where

$$V = (u^2 + v^2 + w^2)^{1/2} \tag{1.69}$$

If the angle of attack and sideslip are small, i.e. <15°, then Eq. (1.67) and (1.68) can be approximated by

$$\alpha = \frac{w}{u} \tag{1.70}$$

and

$$\beta = \frac{v}{u} \tag{1.71}$$

where α and β are in radians.

1.7 AIRCRAFT INSTRUMENTS

The earliest successful airplanes were generally flown without the aid of aircraft instruments.[1] The pilots of these early vehicles were primarily preoccupied with maneuvering and controlling their sometimes temperamental aircraft. However, as new designs were developed, the performance, stability, and control steadily improved to the point where the pilot needed more information about the airplane's flight conditions to fly the airplane safely. One of the major changes in aircraft design that lead to improved performance was the evolution of the open air cockpit. Prior to this development, pilots flew their airplanes in either a crouched or inclined position, exposed to the oncoming airstream. Besides providing the pilot with shelter from the airstream, the cockpit also provided a convenient place for the location of aircraft instruments. The early open-cockpit pilots were hesitant to fly from a closed cockpit because this eliminated their ability to judge sideslip (or skid) by the wind blowing on one side of their face. They also used the sound of the slipstream to provide an indication of the airspeed.

The actual chronological development of aircraft instruments is not readily available; however, one can safely guess that some of the earliest instruments to appear on the cockpit instrument panel were a magnetic compass for navigation, airspeed and altitude indicators for flight information, and engine instruments such as rpm and fuel gauges. The flight decks of modern airplanes are equipped with a multitude of instruments that provide the flight crew with information they need to fly their aircraft. The instruments can be categorized according to their primary use as flight, navigation, power plant, environmental, and electrical systems instruments.

Several of the instruments that compose the flight instrument group will be discussed in the following sections. The instruments include the airspeed

[1] The Wright brothers used several instruments on their historic flight. They had a tachometer to measure engine rpm, an anemometer to measure air speed, and a stop watch.

indicator, altimeter, rate of climb indicator, and Mach meter. These four instruments, along with angle of attack and sideslip indicators, are extremely important for flight test measurement of performance and stability data.

AIR DATA SYSTEMS. The Pitot static system of an airplane is used to measure the total pressure created by the forward motion of the airplane and the static pressure of the ambient atmosphere. The difference between total and static pressures is used to measure airspeed and Mach number and the static pressure is used for measurement of altitude and rate of climb. The Pitot static system is illustrated in Fig. 1.12. The Pitot static probe normally consists of two concentric tubes; the inner tube is used to determine the total pressure and the outer tube is used to determine the static pressure of the surrounding air.

AIRSPEED INDICATOR. The pressures measured by the Pitot static probe can be used to determine the airspeed of the airplane. For low flight speeds, when compressibility effects can be safely ignored, we can use the incompressible form of Bernoulli's equation to show that the difference between the total and the static pressure is the dynamic pressure:

$$P_0 = P + \tfrac{1}{2}V_\infty^2 \tag{1.72}$$

$$\tfrac{1}{2}\rho V_\infty^2 = P_0 - P \tag{1.73}$$

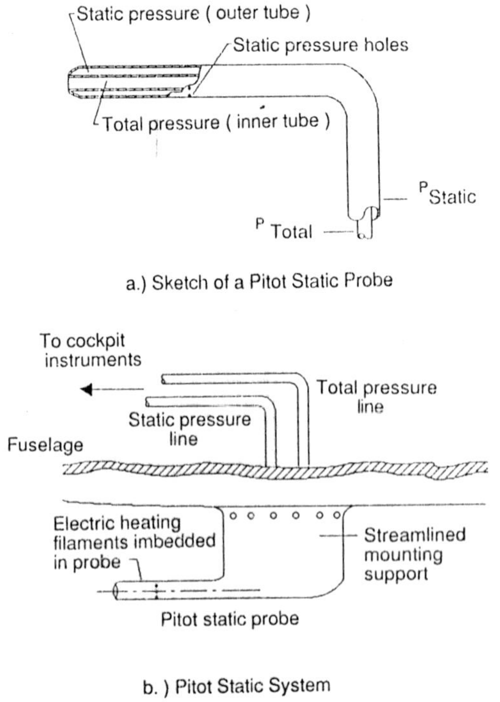

FIGURE 1.12
Pitot static system.

or

$$V_\infty = \left(\frac{2(P_0 - P)}{\rho}\right)^{1/2} \tag{1.74}$$

The airspeed indicator in the cockpit consists of a differential pressure gauge that measures the dynamic pressure and deflects an indicator hand proportionally to the pressure difference. As indicated by Eq. (1.74), the airspeed is both a function of the measured pressure difference and the air density ρ. As was shown earlier, air density is a function of altitude and atmospheric conditions. To obtain the true airspeed, the airspeed indicator would be required to measure the change in both pressure and air density. This is not feasible for a simple instrument and therefore the scale on the airspeed indicator is calibrated using standard sea-level air. The speed measured by the indicator is called the indicated airspeed (IAS).

The speed measured by an airspeed indicator can be used to determine the true flight speed provided that the indicated airspeed is corrected for instrument error, position error, compressibility effects, and density corrections for altitude variations. Instrument error includes those errors inherent to the instrument itself, for example, pressure losses or mechanical inaccuracies in the system. Position error has to do with the location of the Pitot static probe on the airplane. Ideally, the probe should be located so that it is in the undisturbed freestream; in general this is not possible and thus the probe is affected by flow distortion due to the fuselage and/or wing. The total pressure measured by a Pitot static probe is relatively insensitive to flow inclination. Unfortunately, this is not the case for the static measurement and care must be used to position the probe so as to minimize the error in the static measurement. If one knows the instrument and position errors, one can correct the indicated airspeed to give what is referred to as the calibrated airspeed (CAS).

At high speeds, the Pitot static probe must be corrected for compressibility effects. This can be demonstrated by examining the compressible form of the Bernoulli equation:

$$\frac{V^2}{2} + \frac{\gamma}{\gamma - 1}\frac{P}{\rho} = \frac{\gamma}{\gamma - 1}\frac{P_0}{\rho_0} \tag{1.75}$$

Equation (1.75) can be expressed in terms of Mach number as follows:

$$P_0 = P\left(1 + \frac{\gamma - 1}{2}M^2\right)^{\gamma/(\gamma-1)} \tag{1.76}$$

Recall that the airspeed indicator measures the difference between the total and static pressure. Equation (1.76) can be rewritten as

$$q_c = P_0 - P = P\left[\left(1 + \frac{\gamma - 1}{2}M^2\right)^{\gamma/(\gamma-1)} - 1\right] \tag{1.77}$$

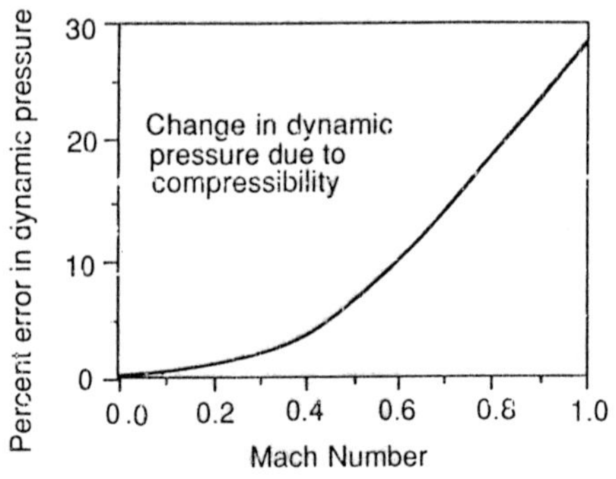

FIGURE 1.13
Percent error in dynamic pressure if compressibility is neglected.

where q_c is the compressible equivalent to the dynamic pressure. Figure 1.13 shows the percentage error in dynamic pressure if compressibility is ignored.

The equivalent airspeed (EAS) can be thought of as the flight speed in the standard sea-level air mass which produces the same dynamic pressure as the actual flight speed. To obtain the actual, or true, airspeed, (TAS) the equivalent airspeed must be corrected for density variations. Using the fact that the dynamic pressures are the same, one can develop a relationship between the true and equivalent airspeeds as follows:

$$\tfrac{1}{2}\rho_0 V_{\mathrm{EAS}}^2 = \tfrac{1}{2}\rho V_{\mathrm{TAS}}^2 \tag{1.78}$$

$$V_{\mathrm{TAS}} = \frac{V_{\mathrm{EAS}}}{\sqrt{\sigma}} \tag{1.79}$$

where $\sigma = \rho/\rho_0$.

The definitions for the various airspeed designations are summarized in Table 1.4.

TABLE 1.4
Airspeed designations

Airspeed*	**Definition**
V_{IAS} Indicated airspeed	Airspeed indicated by the airspeed instrument. The indicated airspeed is affected by altitude, compressibility, instrument and position error.
V_{CAS} Calibrated airspeed	Indicated airspeed corrected for instrument and position errors.
V_{EAS} Equivalent airspeed	Calibrated airspeed corrected for compressibility.
V_{TAS} True airspeed	Equivalent airspeed corrected for density altitude.

* When the prefix K is used, the airspeed is in knots.

ALTIMETER. An altimeter is a device used to measure the altitude of an airplane. The control of an airplane's altitude is very important for safe operation of an aircraft. Pilots use an altimeter to maintain adequate vertical spacing between their aircraft and other airplanes operating in the same area and to establish sufficient distance between the airplane and the ground.

Earlier in this chapter we briefly discussed the mercury barometer. A barometer can be used to measure the atmospheric pressure. As we have shown, the static pressure in the atmosphere varies with altitude, so that if we use a device similar to a barometer we can measure the static pressure outside the airplane, and then relate that pressure to a corresponding altitude in the standard atmosphere. This is the basic idea behind a pressure altimeter.

The mercury barometer would of course be impractical for application in aircraft, because it is both fragile and sensitive to the motion of the airplane. To avoid this difficulty, the pressure altimeter uses the same principle as an aneroid[2] barometer. This type of barometer measures the pressure by magnifying small deflections of an elastic element that deforms as pressure acts upon it.

The altimeter is a sensitive pressure transducer that measures the ambient static pressure and displays an altitude value on the instrument dial. The altimeter is calibrated using the standard atmosphere and the altitude indicated by the instrument is referred to as the pressure altitude. The *pressure altitude* is the altitude in the standard atmosphere corresponding to the measured pressure. The pressure altitude and actual or geometric altitude will be the same only when the atmosphere through which the airplane is flying is identical to the standard atmosphere.

In addition to pressure altitude, there are two other altitudes that are important for performance analysis. They are the density and temperature altitudes. The *density altitude* is the altitude in the standard atmosphere corresponding to the ambient density. In general, the ambient density is not measured, but rather is calculated from the pressure altitude given by the altimeter and the ambient temperature measured by a temperature probe. The *temperature altitude* is, as you might guess, the altitude in the standard atmosphere corresponding to the measured ambient temperature.

As noted earlier, the atmosphere is continuously changing; therefore, to compare performance data for an airplane from one test to another, or for comparison of different airplanes, the data must be referred to a common atmospheric reference. The density altitude is used for airplane performance data comparisons.

An altimeter is an extremely sophisticated instrument, as illustrated by the drawing in Fig. 1.14. This particular altimeter uses two aneroid capsules to increase the sensitivity of the instrument. The deflections of the capsules are magnified and represented by the movement of the pointer with respect to a scale on the surface plate of the meter and a counter. This altimeter is

[2] Aneroid is derived from the Greek word aneros which means "not wet."

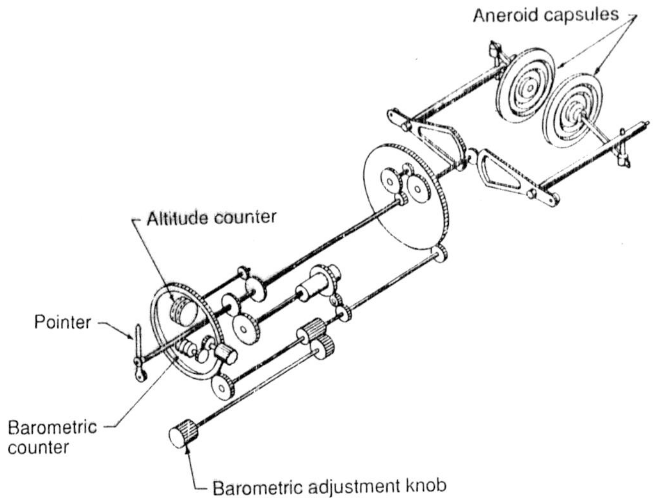

FIGURE 1.14
Cutaway drawing of an altimeter.

equipped with a barometric pressure setting mechanism. The adjusting mechanism allows the pilot manually to correct the altimeter for variations in sea-level barometric pressure. With such adjustments, the altimeter will indicate an altitude that closely approaches the true altitude above sea level.

RATE OF CLIMB INDICATOR. One of the earliest instruments used to measure rate of climb was called a statoscope. This instrument was used by balloonists to detect variation from a desired altitude. The instrument consisted of a closed atmospheric chamber connected by a tube containing a small quantity of liquid to an outer chamber vented to the atmosphere. As the altitude changed, air would flow from one chamber to the other to equalize the pressure. Air passing through the liquid would create bubbles and the direction of the flow of bubbles indicated whether the balloon was ascending or descending. A crude indication of the rate of climb was obtained by observing the frequency of the bubbles passing through the liquid.

Although the statoscope provided the balloonist with a means of detecting departure from a constant altitude, it was difficult to use as a rate of climb indicator. A new instrument called the balloon variometer was developed for rate of climb measurements. The variometer was similar to the statoscope; however, the flow into the chamber took place through a capillary leak. The pressure difference across the leak was measured with a sensitive liquid manometer that was calibrated to indicate the rate of climb.

Present-day rate of climb indicators are similar to the variometer. An example of a leak type rate of climb indicator is shown in Fig. 1.15. This instrument consists of an insulated chamber, a diaphragm, a calibrated leak,

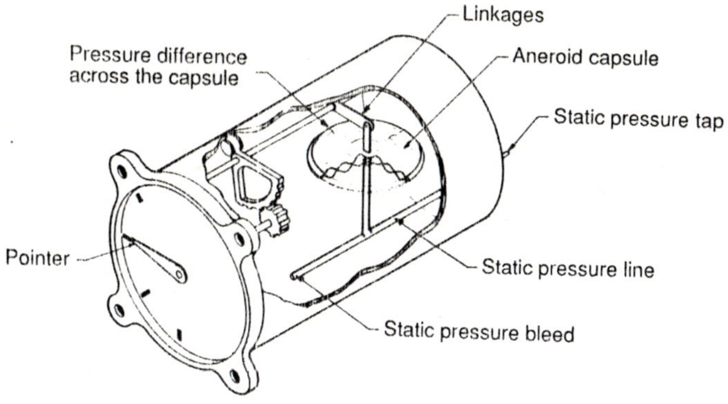

FIGURE 1.15
Sketch of the basic components of a rate of climb indicator.

and an appropriate mechanical linkage to measure the deflection of the diaphragm. The static pressure is applied to the interior of the diaphragm and is also allowed to leak into the chamber by way of a capillary or orifice opening. The diaphragm measures the differential pressure across the leak and the deflection of the diaphragm is transmitted to the indicator dial by a mechanical linkage, as illustrated in the sketch in Fig. 1.15.

MACHMETER. The Pitot static tube can be used to determine the Mach number of an airplane from the measured stagnation and static pressure. If the Mach number is less than 1, Eq. (1.40) can be used to find the Mach number of the airplane:

$$\frac{P_0}{P} = \left(1 + \frac{\gamma - 1}{2} M^2\right)^{\gamma/(\gamma-1)} \tag{1.80}$$

However, when the Mach number is greater than unity, a bow wave forms ahead of the Pitot probe, as illustrated in Fig. 1.16. The bow wave is a curved

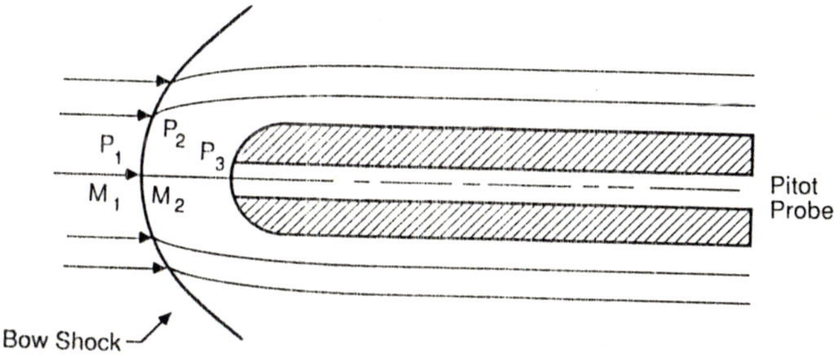

FIGURE 1.16
Detached shock wave ahead of a Pitot static probe.

detached shock wave. In the immediate vicinity of the Pitot orifice, the shock wave can be approximated as a normal shock wave. Using the normal shock relationships, the pressure ratio across the shock can be written as

$$\frac{P_2}{P_1} = \left(\frac{2\gamma}{\gamma - 1}\right) M_1^2 - \left(\frac{\gamma - 1}{\gamma + 1}\right) \tag{1.81}$$

where M_1 is the Mach number ahead of the shock wave. The relationship between the Mach number M_1 ahead of the normal shock and the Mach number M_2 behind the shock is given by Eq. (1.82):

$$M_2^2 = \frac{\frac{1}{2}(\gamma - 1)M_1^2 + 1}{\gamma M_1^2 - \frac{1}{2}(\gamma - 1)} \tag{1.82}$$

After passing through the shock wave, the air is slowed down adiabatically to zero velocity at the total pressure orifice of the Pitot probe. The pressure ratio behind the shock can be expressed as

$$\frac{P_3}{P_2} = \left(1 + \frac{\gamma - 1}{2} M_2^2\right)^{\gamma/(\gamma - 1)} \tag{1.83}$$

Upon combining the previous equations, the ratio of stagnation pressure to static pressure in terms of the flight Mach number can be written.

$$\frac{P_3}{P_1} = \left[\left(\frac{2\gamma}{\gamma + 1}\right) M_1^2 - \left(\frac{\gamma - 1}{\gamma + 1}\right)\right]\left[1 + \frac{\gamma - 1}{2}\left[\left(\frac{\frac{1}{2}(\gamma - 1)M_1^2 + 1}{\gamma M_1^2 - \frac{1}{2}(\gamma - 1)}\right)\right]\right]^{\gamma/(\gamma - 1)} \tag{1.84}$$

The above expression is known as the Rayleigh Pitot tube formula, named after Lord Rayleigh who first developed this equation in 1910. If we assume that the ratio γ of specific heats for air is 1.4, the above expression can be rewritten as

$$\frac{P_3}{P_1} = \frac{7M_1^2 - 1}{6}\left[1 + 0.2\left(\frac{M_1^2 + 5}{7M_1^2 - 1}\right)\right]^{3.5}. \tag{1.85}$$

The preceding equations can be used to design a Mach meter.

The use of Rayleigh's formula is invalid for very high Mach numbers or altitudes. When the Mach number is high, there will be appreciable heat exchange, which violates the assumption of adiabatic flow used in the development of the equation. At very high altitude, air cannot be considered as a continuous medium and again the analysis breaks down.

ANGLE OF ATTACK INDICATORS. The measurement of angle of attack is important for cruise control and stall warning. There are several devices that can be used to measure the angle of attack of an airplane, two of which are the vane and pressure-sensor type indicator. The pivot vane sensor is a mass-balanced wind vane that is free to align itself with the oncoming flow. The vane type angle of attack sensor has been used extensively in airplane flight test programs. For flight test applications the sensor is usually mounted on a nose

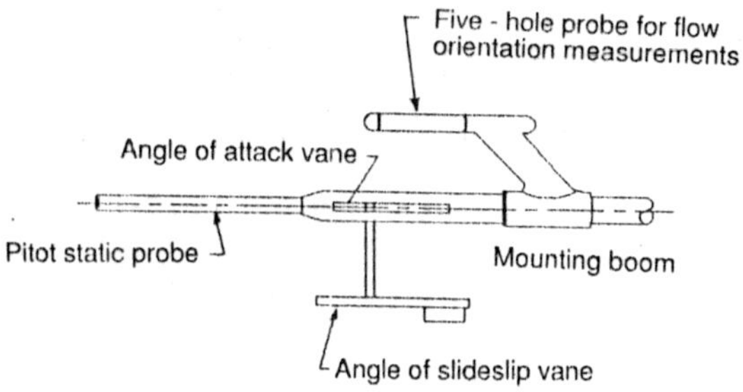

FIGURE 1.17
Flight test instrumentation, Pitot static probe, angle of attack and sideslip varies, five-hole probe mounted on a nose or wing boom.

boom or a boom mounted to the wing tips along with a Pitot static probe, as illustrated in Fig. 1.17. Note that a second vane system is mounted on the boom to measure the side-slip angle.

The angle measured by the vane is influenced by the distortion of the flow field created by the airplane. Actually, the sensor only measures the local angle of attack. The difference between the measured and actual angle of attack is called the position error. Position error can be minimized by mounting the sensor on the fuselage, where the flow distortion is small. The deflection of the vane is recorded by means of a potentiometer.

A null-seeking pressure sensor can also be used to measure the angle of attack. Figure 1.18 is a schematic of a null-seeking pressure sensor. The sensor consists of the following components; a rotatable tube containing two orifices spaced at equal angles to the tube axis; a pressure transducer to detect the difference in pressure between the two orifices; a mechanism for rotating the probe until the pressure differential is zero; and a device for measuring the rotation or angle of attack. The device shown in Fig. 1.18 consists of a rotatable probe that protrudes through the fuselage and an air chamber that is

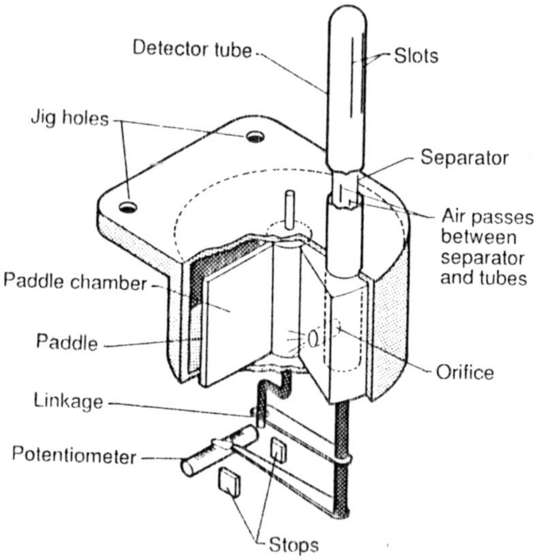

FIGURE 1.18
Null sensing pressure probe for measuring angle of attack.

mounted inside the fuselage. The pressures from the two slits are vented to air chambers by a swivel paddle. If a pressure difference exists at the two slots, the swivel paddle rotates. The paddle is connected by way of linkages so that as the paddle moves the pressure tube is rotated until the pressures are equalized. The angular position of the probe is recorded by a potentiometer.

Example Problem 1.2. An aircraft altimeter calibrated to the standard atmosphere reads 10 000 ft. The airspeed indicator has been calibrated for both instrument and position errors and reads a velocity of 120 knots. If the outside air temperature is 20°F, determine the true airspeed.

Solution. The altimeter is a pressure gauge calibrated to the standard atmosphere. If the altimeter reads 10 000 ft, the static pressure it senses must correspond to the static pressure at 10 000 ft in the standard atmosphere. Using the standard atmospheric table in the Appendix, the static pressure at 10 000 ft is given as

$$P = 1455.6 \text{ lb/ft}^2$$

The ambient density can be calculated using the equation of state:

$$\rho = \frac{P}{RT}$$

$$\rho = \frac{1455.6 \text{ lb/ft}^2}{(1716 \text{ ft}^2/(\text{s}^2\cdot{}^\circ\text{R}))(479.7^\circ\text{R})}$$

$$\rho = 0.001768 \text{ slug/ft}^3$$

A low-speed airspeed indicator corrected for instrument and position error reads the equivalent airspeed. The true speed and equivalent airspeed are related by

$$V_{\text{TAS}} = \frac{V_{\text{EAS}}}{\sqrt{\sigma}}$$

where σ is the ratio of the density at altitude to the standard sea level value of density:

$$\sigma = \rho/\rho_0 = (0.001768/0.002739) = 0.7432$$

Now, solving for the true airspeed:

$$V_{\text{KTAS}} = \frac{V_{\text{KEAS}}}{\sqrt{\sigma}} = \frac{120 \text{ knots}}{\sqrt{0.7432}}$$

$$= 139 \text{ knots}$$

1.8 SUMMARY

In this chapter we have examined the properties of air and how those properties vary with altitude. For the comparison of flight test data and for calibrating aircraft instruments, a standard atmosphere is a necessity: the 1962 U.S. Standard Atmosphere provides the needed reference for the aerospace community. The standard atmosphere was shown to be made up of gradient and isothermal regions.

Finally, we discussed the basic concepts behind several basic flight instruments that play an important role in flight test measurements of aircraft performance, stability and control. In principle these instruments seem to be quite simple: they are, in fact, extremely complicated mechanical devices. Although we have discussed several mechanical instruments, most of the information presented to the flight crew on the newest aircraft designs comes from multifunctional electronic displays. Color cathode ray tubes are used to display air data such as attitude, speed and altitude. Additional displays include navigation, weather, and engine performance information, to name just a few items. The improvements offered by this new technology can be used to reduce the workload of the flight crew and improve the flight safety of the next generation of airplane designs.

1.9 PROBLEMS

1.1. An altimeter set for sea-level standard pressure indicates an altitude of 20 000 ft. If the outside ambient temperature is −15°F, find the air density and the density altitude.

1.2. An airplane is flying at an altitude of 5000 m as indicated by the altimeter, and the outside air temperature is 120°C. If the airplane is flying at a true airspeed of 300 m/s, determine the indicated airspeed.

1.3. A high-altitude remotely piloted communications platform is flying at a pressure altitude of 60 000 ft and an indicated airspeed of 160 ft/s. The outside ambient temperature is −75°F. Estimate the Reynolds number of the wing based on a mean chord of 3.5 ft.

1.4. An airplane is flying at a pressure altitude of 10 000 ft and the airspeed indicator reads 100 knots. If there is no instrument error and the position error is given by Fig. P1.4, find the true airspeed of the airplane.

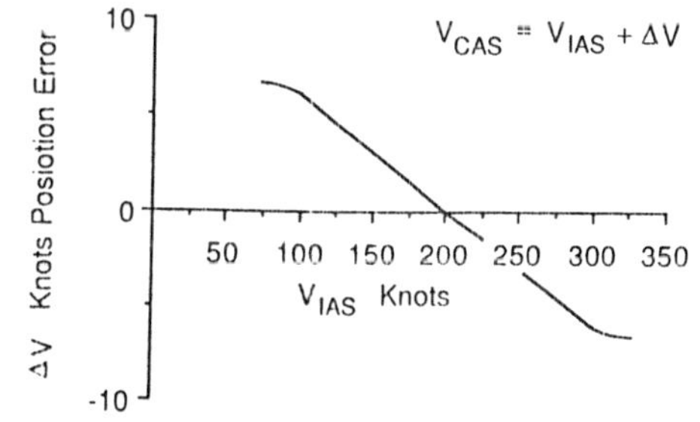

FIGURE P1.4
Position error versus indicated airspeed.

1.5. Under what conditions are the following relationships valid?

$$V_{\text{CAS}} = V_{\text{EAS}} = V_{\text{TAS}}$$

$$V_{\text{CAS}} = V_{\text{EAS}} \neq V_{\text{TAS}}$$

$$V_{\text{CAS}} \neq V_{\text{EAS}} = V_{\text{TAS}}$$

1.6. A small right circular cylinder is used to measure the angle of attack of an airplane by measuring the difference in pressure at two port locations which are located at $\theta = 20°$. Assuming that the flow on the forward face of the cylinder can be accurately modeled as an inviscid flow, the velocity along the cylinder surface can be expressed as

$$V_\theta = 2V_\infty \sin\theta$$

If, while flying at 200 ft/s under sea-level standard conditions, the pressure difference is 32.5 lb/ft^2, what is the angle of the airplane?

REFERENCES

1.1. Anderson, J. D.: *Introduction to Flight,* McGraw-Hill, New York, 1978.
1.2. Domnasch, D. O., S. S. Sherby, and T. F. Connolly: *Airplane Aerodynamics,* Pitman, New York, 1967.
1.3. Pallett, E. H. J.: *Aircraft Instruments* Pitman, London, England, 1982.
1.4. *U.S. Standard Atmosphere, 1962,* prepared under sponsorship of the National Aeronautics and Space Administration, United States Air Force, and United States Weather Bureau, Washington, D.C., December 1962.
1.5. Putnam, T. W.: "The X-29 Flight-Research Program," *AIAA Student Journal,* Fall, 1984.

CHAPTER 2

STATIC STABILITY AND CONTROL

"Isn't it astonishing that all these secrets have been preserved for so many years just so that we could discover them!"
Orville Wright/June 7, 1903

2.1 HISTORICAL PERSPECTIVE

By the start of the twentieth century, the aeronautical community had solved many of the technical problems necessary for achieving powered flight of a heavier-than-air aircraft. One of the problems still beyond the grasp of these early investigators was a lack of understanding of the relationship between stability and control, as well as the influence of the pilot on the pilot–machine system. Most of the ideas regarding stability and control came from experiments with uncontrolled hand-launched gliders. Through such experiments, it was quickly discovered that for a successful flight, the glider had to be inherently stable. Earlier aviation pioneers such as Albert Zahm in the United States, Alphonse Penaud in France, and Frederick Lanchester in England, all contributed to the notion of stability. Zahm, however, was the first to correctly outline the requirements for static stability in a paper he presented in 1893. In his paper, he analyzed the conditions necessary for obtaining a stable equilibrium for an airplane descending at a constant speed. Figure 2.1 shows a sketch of a glider from Zahm's paper. Zahm concluded that the center of gravity had to be in front of the aerodynamic force and the vehicle would require what he referred to as "longitudinal dihedral" to have a stable equilibrium point. In the terminology of today, he showed that if the

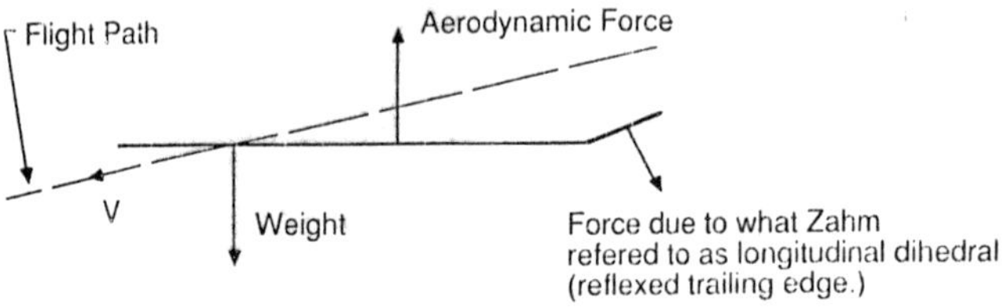

FIGURE 2.1
Zahm's description of longitudinal stability.

center of gravity was ahead of the wing aerodynamic center, then one would need a reflexed airfoil in order to be stable at a positive angle of attack.

In the twenty years prior to the Wright brothers' successful flight, there were many individuals in the United States and Europe working with gliders and unmanned powered models. These investigators were constantly trying to improve their vehicles, with the ultimate goal of achieving powered flight of a airplane under human control. Three men who would leave lasting impressions on the Wright brothers were Otto Lilienthal of Germany and Octave Chanute and Samuel Pierpont Langley of the United States.

Lilienthal made a significant contribution to aeronautics by his work with model and man-carrying gliders. His experiments included the determination of the properties of curved or cambered wings. He carefully recorded the details of over 2000 glider flights. The information in his journal includes data on materials, construction techniques, handling characteristics of his gliders, and aerodynamics. His successful flights and recorded data inspired and aided many other aviation pioneers. Lilienthal's glider designs were statically stable but had very little control capability. For control, Lilienthal would shift his weight to maintain equilibrium flight, much as hang-glider pilots do today. The lack of suitable control proved to be a fatal flaw for Lilienthal. In 1896, he lost control of his glider; the glider stalled and plunged to earth from an altitude of 50 ft. Lilienthal died a day later from the injuries incurred in the accident.

In the United States, Octave Chanute became interested in gliding flight in the mid 1890s. Initially, he built gliders patterned after Lilienthal's designs. After experimenting with modified versions of Lilienthal's gliders, he developed his own designs. His gliders incorporated biplane and multiplane wings, controls to adjust the wings to maintain equilibrium, and a vertical tail for steering. These design changes represented substantial improvements over Lilienthal's monoplane gliders. Many of Chanute's innovations would be incorporated in the Wright brothers' designs. In addition to corresponding with the Wright brothers, Chanute visited their camp at Kitty Hawk to lend his experience and advice to their efforts.

Another individual who helped the Wright brothers was Samuel Pierpont Langley, secretary of the Smithsonian Institution. The Wright brothers knew of Langley's work and wrote to the Smithsonian asking for the available aeronautical literature. The Smithsonian informed the Wright brothers of the activities of many of the leading aviation pioneers and this information was, no doubt, very helpful to them.

Around 1890, Langley became interested in problems of flight. Initially, his work consisted of collecting and examining all the available aerodynamic data. From the study of these data and his own experiments, he concluded that heavier-than-air powered flight was possible. Langley then turned his attention to designing and perfecting unmanned powered models. On May 6, 1896, his powered model flew for one and one-half minutes and covered a distance of three-quarters of a mile. Langley's success with powered models pioneered the practicality of mechanical flight.

After his successful model flights, Langley was engaged by the War Department to develop a man-carrying airplane. Congress appropriated $50 000 for the project. Langley and his engineering assistant, Charles Manley, started work on their own design in 1899. For the next four years, they were busy designing, fabricating and testing the full-size airplane which was to be launched by a catapult fixed to the top of a houseboat. The first trial was conducted on September 7, 1903, in the middle of the Potomac River near Tidewater, Virginia. The first attempt ended in failure as the airplane pitched down into the river at the end of the launch rails. A second attempt was made on December 8, 1903; this time, the airplane pitched up and fell back into the river. In both trails, the launching system prevented the possibility of a successful flight. For Langley, it was a bitter disappointment, and the criticism he received from the press deeply troubled him. He was, however, one of the pioneering geniuses of early aviation and it is a shame that he went to his grave still smarting from the ridicule. Some twenty years later his airplane was modified, a new engine was installed, and the airplane flew successfully.

The time had come for someone to design a powered airplane capable of carrying a man aloft. As we all know, the Wright brothers made their historic first flight on a powered airplane at Kitty Hawk, North Carolina, on December 17, 1903. Orville Wright made the initial flight which lasted only 12 seconds and covered approximately 125 feet. Taking turns operating the aircraft, Orville and Wilbur made three more flights that day. The final flight lasted 59 seconds and covered a distance of 852 feet while flying into a 20 mph headwind. The airplane tended to fly in a porpoising fashion, with each flight ending abruptly as the vehicle's landing skids struck the ground. The Wright brothers found their powered airplane to be much more responsive than their earlier gliders and, as a result, had difficulty controlling their airplane.

Figure 2.2 shows two photographs of the Kitty Hawk Flyer. The first photograph shows Orville Wright making the historical initial flight and the second shows the airplane after the fourth and last flight of the day. Notice the damaged horizontal rudder (the term used by the Wright brothers). Today we use the term canard to describe a forward control surface. The word canard comes to us from the French word that means "duck." The French used the term canard to describe an early French airplane that had its horizontal tail located far forward of the wing. They thought this airplane looked like a duck with its neck stretched out in flight.

The first flight of a powered
heavier - than - air airplane

On the fourth flight the horizontal rudder
(canard) was broken during the landing

FIGURE 2.2
Photographs of the Wright brothers' airplane, December 17, 1903, Kitty Hawk, North Carolina.

From this very primitive beginning, we have witnessed a remarkable revolution in aircraft development. In a matter of decades, airplanes have evolved into an essential part of our national defense and commercial transportation system. The success of the Wright brothers can be attributed to their step-by-step experimental approach. After reviewing the experimental data of their contemporaries, the Wright brothers were convinced that additional information was necessary before a successful airplane could be designed. They embarked upon an experimental program which included wind-tunnel and flight-test experiments. The Wright brothers designed and constructed a small wind tunnel and made thousands of model tests to determine the aerodynamic characteristics of curved airfoils. They also conducted thousands of glider experiments in developing their airplane. Through their study of the works of others and their own experimental investigations, the Wright brothers were convinced that the major obstacle to achieving powered flight was the lack of sufficient control. Therefore, much of their work was directed towards improving the control capabilities of their gliders. They felt strongly that powerful controls were essential for the pilot to maintain equilibrium and to prevent accidents such as the ones that caused the tragic deaths of Lilienthal and other glider enthusiasts.

This approach represented a radical break with the design philosophy of the day. The gliders and airplanes designed by Lilenthal, Chanute, Langley and other aviation pioneers were designed to be inherently stable. In these designs, the pilot's only function was to steer the vehicle. Although such vehicles were statically stable, they lacked maneuverability and were susceptible to upset by atmospheric disturbances. The Wright brothers' airplane was statically unstable but quite maneuverable. The lack of stability made their work as pilots very difficult. However, through their glider experiments they were able to teach themselves to fly their unstable airplane.

The Wright brothers succeeded where others failed because of their dedicated scientific and engineering efforts. Their accomplishments were the foundation on which others could build. Some of the major accomplishments are listed below.

1. They designed and built a wind tunnel and balance system to conduct aerodynamic tests. With their tunnel they developed a systematic airfoil aerodynamic data base.
2. They developed a complete flight control system with adequate control capability.
3. They designed a lightweight engine and an efficient propeller.
4. Finally, they designed an airplane with a sufficient strength-to-weight ratio, capable of sustaining powered flight.

These early pioneers provided much of the understanding that we have today regarding static stability, maneuverability, and control. However, it is not clear whether any of these men truly comprehended the relationship between these topics.

2.2 INTRODUCTION

How well an airplane flies and how easily it can be controlled are subjects studied in aircraft stability and control. By stability we mean the tendency of the airplane to return to its equilibrium position after it has been disturbed. The disturbance may be generated by the pilot's control actions or by atmospheric phenomena. The atmospheric disturbances can be wind gusts, wind gradients, or turbulent air. An airplane must have sufficient stability that the pilot does not become fatigued by constantly having to control the airplane owing to external disturbances. Although airplanes with little or no inherent aerodynamic stability can be flown, they are unsafe to fly, unless they are provided artificial stability by way of an electromechanical device called a stability augmentation system (SAS).

Two flight conditions are necessary for an airplane to fly its mission successfully. The airplane must be able to achieve equilibrium flight and it must have the capability to maneuver for a wide range of flight velocities and altitudes. To achieve equilibrium or to perform maneuvers, the airplane must be equipped with aerodynamic and propulsive controls. The design and performance of control systems is an integral part of airplane stability and control.

The stability and control characteristics of an airplane are referred to as the vehicle's handling or flying qualities. It is important to the pilot that the airplane possesses satisfactory handling qualities. Airplanes with poor handling qualities will be difficult to fly and could be potentially dangerous. Pilots form their opinions of the airplane on the basis of its handling characteristics. An airplane will be considered to be of poor design if it lacks adequate handling qualities, regardless of how outstanding the airplane's performance might be. In the study of airplane stability and control, we are interested in what makes an airplane stable, how to design the control systems, and what conditions are necessary for good handling qualities. In the following chapters we will discuss each of these topics from the point of view of how they influence the design of the airplane.

STATIC STABILITY. Stability is a property of an equilibrium state. To discuss stability we must first define what is meant by equilibrium. If an airplane is to remain in steady uniform flight, the resultant force as well as the resultant moment about the center of gravity must both be equal to zero. An airplane satisfying this requirement is said to be in a state of equilibrium or flying at a trim condition. On the other hand, if the forces and moments do not sum to zero, the airplane will be subjected to translational and rotational accelerations.

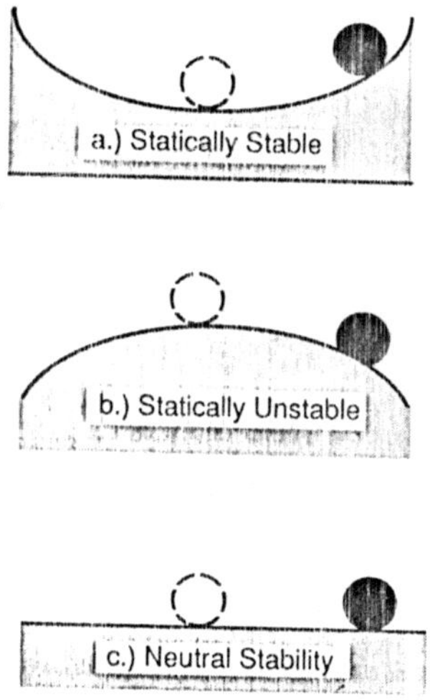

FIGURE 2.3
Sketches illustrating various conditions of static stability.

The subject of airplane stability is generally divided into static and dynamic stability. Static stability is the initial tendency of the vehicle to return to its equilibrium state after a disturbance. An example of the various types of static stability that can exist are illustrated in Fig. 2.3. If the ball were to be displaced from the bottom of the bowl (Fig. 2.3(a)), the ball would, by virtue of the gravitational attraction, roll back to the bottom of the bowl (i.e. the force and moment would tend to restore the ball to its equilibrium point). Such a situation would be referred to as a stable equilibrium point. On the other hand, if we were able to balance a ball on the bowl shown in Fig. 2.3(b), then any displacement from the equilibrium point would cause the ball to roll off the bowl. In this case, the equilibrium point would be classified as an unstable equilibrium point. The last example is shown in Fig. 2.3(c), where the ball is placed on a flat surface. Now, if the fall were to be displaced from its initial equilibrium point to another position, the ball would remain at the new position. This example would be classified as a neutrally stable equilibrium point and represents the limiting (or boundary) between static stability and static instability. The important point in this simple example is that if we are to have a stable equilibrium point, the vehicle must develop a restoring force and/or moment which tends to bring the vehicle back to the equilibrium condition.

DYNAMIC STABILITY. In the study of dynamic stability we are concerned with the time history of the motion of the vehicle after it is disturbed from its equilibrium point. Figure 2.4 shows several possible airplane motions that could occur if the airplane were disturbed from its equilibrium conditions. Note that the vehicle can be statically stable but dynamically unstable. Static stability, therefore, does not guarantee the existence of dynamic stability. However, if the vehicle is dynamically stable it must be statically stable.

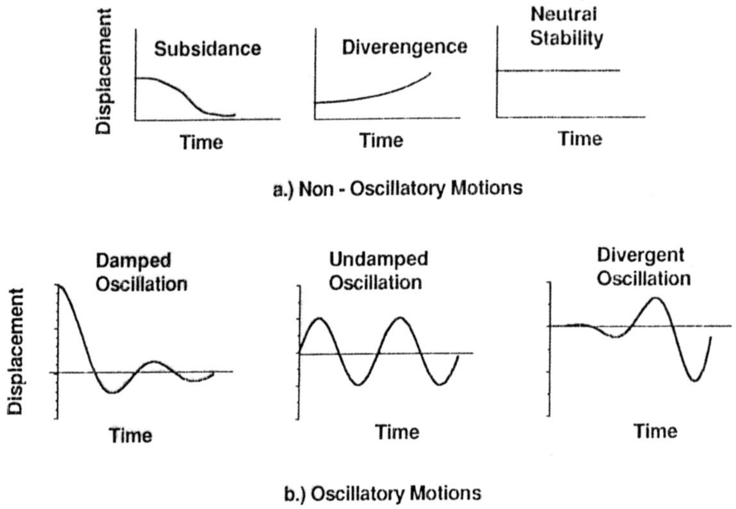

FIGURE 2.4
Examples of stable and unstable dynamic motions.

The reduction of the disturbance with time indicates that there is resistance to the motion and, therefore, that energy is being dissipated. The dissipation of energy is called positive damping. If energy is being added to the system, then we have a negative damping. Positive damping for an airplane is provided by forces and moments which arise owing to the airplanes motion. In the case of positive damping, these forces and moments will oppose the motion of the airplane and cause the disturbance to damp out with time. An airplane that has negative aerodynamic damping will be dynamically unstable; to fly such an airplane, artificial damping must be designed into the vehicle. The artificial damping is provided by a Stability Augmentation System (SAS). Basically, a Stability Augmentation System is an electromechanical device which senses the undesirable motion and then moves the appropriate controls to damp out the motion. This is usually accomplished with small control movements and, therefore, the pilot's control actions are not influenced by the system.

Of particular interest to the pilot and designer is the degree of dynamic stability. Dynamic stability is usually specified by the time it takes a disturbance to be damped to half of its initial amplitude or, in the case of an unstable motion, the time it takes for the initial amplitude of the disturbance to double. In the case of an oscillatory motion, the frequency and period of the motion are extremely important.

So far, we have been discussing the response of an airplane to external disturbances while the controls are held fixed. When we add the pilot to the system, additional complications can arise. For example, an airplane that is dynamically stable to external disturbances with the controls fixed can become unstable by the pilot's control actions. If the pilot attempts to correct for a disturbance and his or her control input is out of phase with the oscillatory motion of the airplane, the control actions would increase the motion rather than correct it. This type of pilot/vehicle response is called Pilot-Induced-Oscillation (PIO). There are many factors which contribute to the PIO tendency of an airplane. A few of the major contributions are insufficient aerodynamic damping, insufficient control system damping, and pilot reaction time.

2.3 STATIC STABILITY AND CONTROL

DEFINITION OF LONGITUDINAL STATIC STABILITY. In the first example we showed that, to have static stability, we need to develop a restoring moment on the ball when it is displaced from its equilibrium point. The same requirement exists for an airplane. Let us consider the two airplanes and their respective pitching moment curves shown in Fig. 2.5. The pitching moment curves have been assumed to be linear until the wing is close to stalling.

In Fig. 2.5, both airplanes are flying at the trim point denoted by B, i.e. $C_{m_{cg}} = 0$. Suppose the airplanes suddenly encounter an upward gust such that the angle of attack is increased to point C. At the angle of attack denoted by C, airplane 1 develops a negative (nose-down) pitching moment which tends to rotate the airplane back towards it equilibrium point. However, for the same disturbance, airplane 2 develops a positive (nose-up) pitching moment which tends to rotate the aircraft away from the equilibrium point. If we were to encounter a disturbance which reduced the angle of attack, e.g to point A, we would find that the airplane 1 develops a nose-up moment which rotates the aircraft back toward the equilibrium point. On the other hand, airplane 2 is found to develop a nose-down moment which rotates the aircraft away from the equilibrium point. On the basis of this simple analysis, we can conclude that to have static longitudinal stability the aircraft pitching moment curve

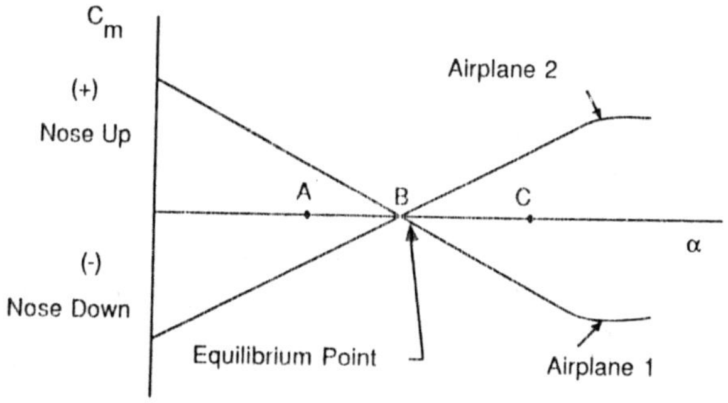

FIGURE 2.5
Pitching moment coefficient versus angle of attack.

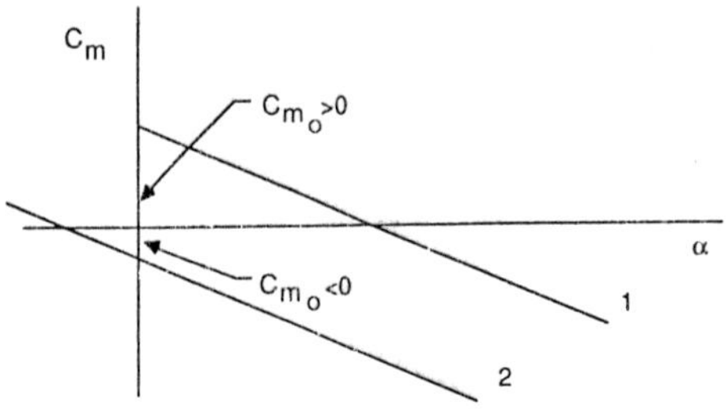

FIGURE 2.6
Pitching moment coefficient versus angle of attack for a stable airplane.

must have a negative slope. i.e.

$$\frac{dC_m}{d\alpha} < 0 \tag{2.1}$$

through the equilibrium point.

Another point that we must make is illustrated in Fig. 2.6. Here we see two pitching moment curves which both satisfy the condition for static stability. However, only curve 1 can be trimmed at a positive angle of attack. Therefore, in addition to having static stability, we must also have a positive intercept, i.e. $C_{m_0} > 0$ in order to trim at positive angles of attack. Although we developed the criterion for static stability from the C_m versus α curve, we could have just as easily accomplished the same result by working with a C_m versus C_L curve. In this case, the requirement for static stability would be as follows:

$$\frac{dC_m}{dC_L} < 0 \tag{2.2}$$

The two conditions are related by the following expression:

$$C_{m_\alpha} = \frac{dC_m}{d\alpha} = \frac{dC_m}{dC_L}\frac{dC_L}{d\alpha} \tag{2.3}$$

which shows that the derivatives differ only by the slope of the lift curve.

CONTRIBUTION OF AIRCRAFT COMPONENTS. In discussing the requirements for static stability, we have so far only considered the total airplane pitching moment curve. However, it is of interest (particularly to airplane designers) to know the contribution of the wing, fuselage, tail, propulsion system, etc., to the pitching moment and static stability characteristics of the airplane. In the following sections, each of the components will be considered separately. We will start by breaking the airplane down into its basic components such as the wing, fuselage, horizontal tail, and propulsion unit. Detailed methods for estimating the aerodynamic stability coefficients can be found in the United States Air Force Stability and Control Datcom [2.7]. The

Datcom, short for Data Compendium, is a collection of methods for estimating the basic stability and control coefficients for flight regimes of subsonic, transonic, supersonic, and hypersonic speeds. Methods are presented in a systematic body build-up fashion, e.g. wing alone, body alone, wing–body and wing–body–tail techniques. The methods range from techniques that are based upon simple expressions developed from theory to correlations obtained from experimental data. In the following sections, as well as in later chapters, we shall develop simple methods for computing the aerodynamic stability and control coefficients. Our emphasis will be for the most part on methods that can be derived from simple theoretical considerations. These methods are in general accurate for preliminary design purposes and show the relationship between the stability coefficients and the geometric and aerodynamic characteristics of the airplane. Furthermore, the methods are generally only valid for the subsonic flight regime. A complete discussion of how to extend these methods to higher speed flight regimes is beyond the scope of this book and the reader is referred to Ref. 2.7 for the high-speed methods.

WING CONTRIBUTION. The contribution of the wing to an airplane's static stability can be examined with aid of Fig. 2.7. In this sketch we have replaced the wing by its mean aerodynamic chord $\bar{c}$. The distances from the wing leading edge to the aerodynamic center and the center of gravity are denoted by X_{ac} and X_{cg}, respectively. The vertical displacement of the center of gravity is denoted by Z_{cg}. The angle the wing chord line makes with the fuselage reference line is denoted as i_w. This is the angle at which the wing is mounted onto the fuselage.

If we sum the moments about the center of gravity, the following equation is obtained:

$$\sum \text{Moments} = M_{cg_w}$$

$$\begin{aligned} M_{cg_w} = {} & L_w \cos(\alpha_w - i_w)[X_{cg} - X_{ac}] + D_w \sin(\alpha_w - i_w)[X_{cg} - X_{ac}] \\ & + L_w \sin(\alpha_w - i_w)[Z_{cg}] - D_w \cos(\alpha_w - i_w)[Z_{cg}] + M_{ac_w} \end{aligned} \tag{2.4}$$

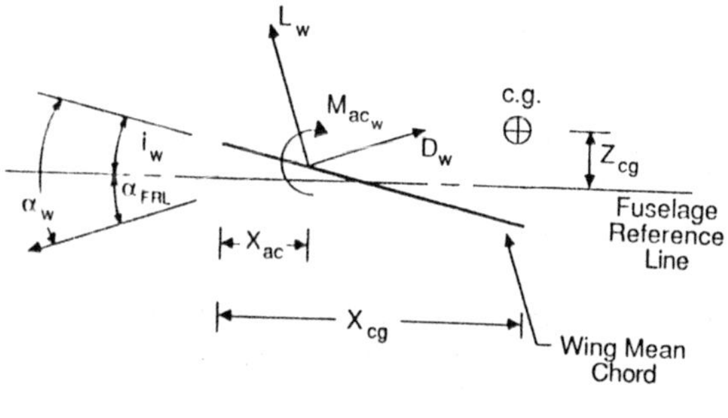

FIGURE 2.7
Wing contribution to the pitching moment.

Dividing by $\frac{1}{2}\rho V^2 S\bar{c}$ yields

$$C_{m_{cg_w}} = C_{L_w}\left(\frac{X_{cg}}{\bar{c}} - \frac{X_{ac}}{\bar{c}}\right)\cos(\alpha_w - i_w) + C_{D_w}\left(\frac{X_{cg}}{\bar{c}} - \frac{X_{ac}}{\bar{c}}\right)\sin(\alpha_w - i_w)$$

$$+ C_{L_w}\frac{(Z_{cg})}{\bar{c}}\sin(\alpha_w - i_w) - C_{D_w}\frac{(Z_{cg})}{\bar{c}}\cos(\alpha_w - i_w) + C_{m_{ac_w}} \quad (2.5)$$

Equation (2.5) can be simplified by assuming that the angle of attack is small. With this assumption the following approximations can be made.

$$\cos(\alpha_w - i_w) = 1, \qquad \sin(\alpha_w - i_w) = \alpha_w - i_w, \qquad C_L \gg C_D$$

If we further assume that the vertical contribution is negligible, then Eq. (2.5) reduces to:

$$C_{m_{cg_w}} = C_{m_{ac_w}} + C_{L_w}\left(\frac{X_{cg}}{\bar{c}} - \frac{X_{ac}}{\bar{c}}\right) \quad (2.6)$$

or

$$C_{m_{cg_w}} = C_{m_{ac_w}} + (C_{L_{0_w}} + C_{L_{\alpha_w}}\alpha_w)\left(\frac{X_{cg}}{\bar{c}} - \frac{X_{ac}}{\bar{c}}\right) \quad (2.7)$$

where $C_{L_w} = C_{L_{0_w}} + C_{L_{\alpha_w}}\alpha_w$. Applying the condition for static stability yields:

$$C_{m_{0_w}} = C_{m_{ac_w}} + C_{L_{0_w}}\left(\frac{X_{cg}}{\bar{c}} - \frac{X_{ac}}{\bar{c}}\right) \quad (2.9)$$

$$C_{m_{\alpha_w}} = C_{L_{\alpha_w}}\left(\frac{X_{cg}}{\bar{c}} - \frac{X_{ac}}{\bar{c}}\right) \quad (2.10)$$

For a wing-alone design to be statically stable, Eq. (2.10) tells us that the aerodynamic center must lie aft of the center of gravity to make $C_{m_\alpha} < 0$. Since we also want to be able to trim the aircraft at a positive angle of attack, the pitching moment coefficient at zero angle of attack, C_{m_0}, must be greater than zero. A positive pitching moment about the aerodynamic center can be achieved by using a negative-cambered airfoil section or an airfoil section that has a reflexed trailing edge. For many airplanes, the center of gravity position is located slightly aft of the aerodynamic center (see data in the Appendix). Also, the wing is normally constructed of airfoil profiles having positive camber. Therefore, the wing contribution to static longitudinal stability is usually destabilizing for most conventional airplanes.

TAIL CONTRIBUTION—AFT TAIL. The horizontal tail surface can be located either forward or aft of the wing. When the surface is located forward of the wing, the surface is called a canard. Both surfaces are influenced by the flow field created by the wing. The canard surface is affected by the upwash flow from the wing, whereas the aft tail is subjected to the downwash flow. Figure 2.8 is a sketch of the flow field surrounding a lifting wing. The wing flow field

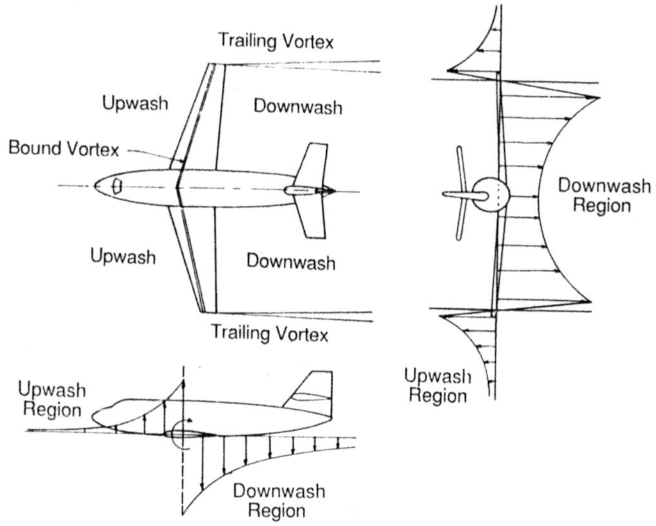

FIGURE 2.8
Flow field around an airplane created by the wing.

is due, primarily, to the bound and trailing vortices. The magnitude of the upwash or downwash depends upon the location of the tail surface with respect to the wing.

The contribution that a tail surface located aft of the wing makes to the airplane's lift and pitching moment can be developed with the aid of Fig. 2.9. In this sketch, the tail surface has been replaced by its mean aerodynamic chord. The angle of attack at the tail can be expressed as

$$\alpha_t = \alpha_w - i_w - \varepsilon + i_t \tag{2.11}$$

where ε and i_t are the downwash and tail incidence angles, respectively. If we assume small angles and neglect the drag contribution of the tail, the total lift

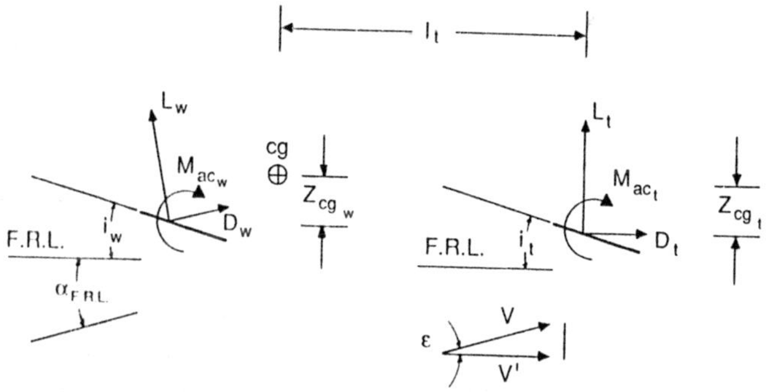

FIGURE 2.9
Aft tail contribution to the pitching moment.

of the wing and tail can be expressed as

$$L = L_w + L_t \tag{2.12}$$

or

$$C_L = C_{L_w} + \eta \frac{S_t}{S} C_{L_t} \tag{2.13}$$

where

$$\eta = \frac{\frac{1}{2}\rho V_t^2}{\frac{1}{2}\rho V_w^2} \tag{2.14}$$

The ratio of the dynamic pressures is called the tail efficiency and can have values in the range 0.8–1.2. The magnitude of η depends upon the location of the tail surface. If the tail is located in the wake region of the wing or fuselage, η will be less than unity because $Q_t < Q_w$ due to the momentum loss in the wake. On the other hand, if the tail is located in a region where $Q_t > Q_w$, then η will be greater than unity. Such a situation could exist if the tail were located in either the slip stream of the propeller or in the exhaust wake of a jet engine.

The pitching moment due to the tail can be obtained by summing the moments about the center of gravity:

$$\begin{aligned} M_t = &-l_t[L_t \cos(\alpha_{\mathrm{FRL}} - \varepsilon) + D_t \sin(\alpha_{\mathrm{FRL}} - \varepsilon)] \\ &- Z_{\mathrm{cg}_t}[D_t \cos(\alpha_{\mathrm{FRL}} - \varepsilon) - L_t \sin(\alpha_{\mathrm{FRL}} - \varepsilon)] + M_{\mathrm{ac}_t} \end{aligned} \tag{2.15}$$

Usually only the first term of the preceding equation is retained; the other terms are generally small in comparison to the first term. If we again use the small-angle assumption and that $C_{L_t} \gg C_{D_t}$, then Eq. (2.15) reduces to

$$M_t = -l_t L_t = -l_t C_{L_t} \tfrac{1}{2}\rho V_t^2 S_t \tag{2.16}$$

$$C_{m_t} = \frac{M_t}{\frac{1}{2}\rho V^2 S\bar{c}} = -\frac{l_t S_t}{S\bar{c}} \eta C_{L_t} \tag{2.17}$$

or

$$C_{m_t} = -V_H \eta C_{L_t} \tag{2.18}$$

where $V_H = l_t S_t / S\bar{c}$ and is called the horizontal tail volume ratio.

From Fig. 2.9, the angle of attack of the tail is seen to be

$$\alpha_t = \alpha_w - i_w - \varepsilon + i_t \tag{2.19}$$

The coefficient C_{L_t} can be written as

$$C_{L_t} = C_{L_{\alpha_t}} \alpha_t = C_{L_{\alpha_t}}(\alpha_w - i_w - \varepsilon + i_t) \tag{2.20}$$

where $C_{L_{\alpha_t}}$ is the slope of the tail lift curve. The downwash angle ε can be expressed as

$$\varepsilon = \varepsilon_0 + \frac{d\varepsilon}{d\alpha}\alpha_w \tag{2.21}$$

where ε_0 is the downwash at zero angle of attack.

The downwash behind a wing with an elliptic lift distribution can be derived from finite-wing theory and can be shown to be related to the wing lift coefficient and aspect ratio:

$$\varepsilon = \frac{2C_{L_w}}{\pi AR_w} \tag{2.22}$$

where the downwash angle is in radians. The rate of change of downwash angle with angle of attack is determined by taking the derivative of Eq. (2.22).

$$\frac{d\varepsilon}{d\alpha} = \frac{2C_{L_{\alpha_w}}}{\pi AR_w} \tag{2.23}$$

where $C_{L_{\alpha_w}}$ is per radian. The above expressions do not take into account the position of the tail plane relative to the wing, i.e. its vertical and longitudinal spacing. More accurate methods for estimating the downwash at the tailplane can be found in Ref. 2.7. An experimental technique for determining the downwash using wind tunnel force and moment measurements will be presented by way of a problem assignment at the end of this chapter.

Rewriting the tail contribution to the pitching moment yields

$$C_{m_{cg_t}} = -V_H \eta C_{L_t} \tag{2.24}$$

$$C_{m_{cg_t}} = \eta V_H C_{L_{\alpha_t}}(\varepsilon_0 + i_w - i_t) - \eta V_H C_{L_{\alpha_t}} \alpha \left(1 - \frac{d\varepsilon}{d\alpha}\right) \tag{2.25}$$

Comparing Eq. (2.25) with the linear expression for the pitching moment given as

$$C_{m_{cg_t}} = C_{m_0} + C_{m_\alpha} \alpha \tag{2.26}$$

yields expressions for the intercept and slope:

$$C_{m_{0_t}} = \eta V_H C_{L_{\alpha_t}}(\varepsilon_0 + i_w - i_t) \tag{2.27}$$

$$C_{m_{\alpha_t}} = -\eta V_H C_{L_{\alpha_t}} \left(1 - \frac{d\varepsilon}{d\alpha}\right) \tag{2.28}$$

Recall that earlier we showed that the wing contribution to C_{m_0} was negative for an airfoil having positive camber. The tail contribution to C_{m_0} can be used to ensure that C_{m_0} for the complete airplane is positive. This can be accomplished by adjusting the tail incidence angle i_t. Note that we would want to mount the tail plane at a negative angle of incidence to the fuselage reference line to increase C_{m_0} due to the tail.

The tail contribution to the static stability of the airplane ($C_{m_{\alpha_t}} < 0$) can be controlled by proper selection of V_H and $C_{L_{\alpha_t}}$. The contribution of $C_{m_{\alpha_t}}$ will become more negative by increasing the tail moment air l_t or tail surface area S_t and by increasing $C_{L_{\alpha_t}}$. The tail lift curve slope $C_{L_{\alpha_t}}$ can be increased most easily by increasing the aspect ratio of the tail planform. The designer can adjust anyone of these parameters to achieve the desired slope. As noted here, a tail surface located aft of the wing can be used to ensure that the airplane has a positive C_{m_0} and a negative C_{m_α}.

CANARD—FORWARD TAIL SURFACE. A canard is a tail surface located ahead of the wing. There are several attractive features of the canard surface. The canard, if properly positioned, can be relatively free from wing or propulsive flow interference. Canard control is more attractive for trimming the large nose-down moment produced by high-lift devices. To counteract the nose-down pitching moment, the canard must produce lift which will add to the lift being produced by the wing. An aft tail must produce a down load to counteract the pitching moment and thus reduce the airplane's overall lift force. The major disadvantage of the canard is that it produces a destabilizing contribution to the aircraft's static stability. However, this is not a severe limitation. By proper location of the center of gravity, one can ensure the airplane is statically stable.

FUSELAGE CONTRIBUTION. The primary function of the fuselage is to provide room for the flight crew and payload such as passengers and cargo. The optimum shape for the internal volume at minimum drag is a body for which the length is larger that the width or height. For most fuselage shapes used in airplane designs, the width and height are of the same order of magnitude and for many designs a circular cross-section is used.

The aerodynamic characteristics of long, slender bodies were studied by Max Munk [2.8] in the earlier 1920s. Munk was interested in the pitching moment characteristics of airship hulls. In his analysis, he neglected viscosity and treated the flow around the body as an ideal fluid. Using momentum and energy relationships, he showed that the rate of change of the pitching moment with angle of attack (per radian) for a body of revolution is proportional to the body volume and dynamic pressure:

$$\frac{dM}{d\alpha} = \text{fn}(\text{volume}, \tfrac{1}{2}\rho V^2) \tag{2.29}$$

Multhopp [2.9] extended this analysis to account for the induced flow along the fuselage due to the wings for bodies of arbitrary cross-section. A summary of Multhopp's method for C_{m_0} and C_{m_α} due to the fuselage is presented as follows:

$$C_{m_{0_f}} = \frac{k_2 - k_1}{36.5S\bar{c}} \int_0^{l_f} w_f^2(\alpha_{0_w} + i_f)\, dx \tag{2.30}$$

which can be approximated as

$$C_{m_{0_f}} = \frac{k_2 - k_1}{36.5S\bar{c}} \sum_{x=0}^{x=l_f} w_f^2(\alpha_{0w} + i_f)\,\Delta x \qquad (2.31)$$

where $k_2 - k_1$ = correction factor for body fineness ratio
S = the wing reference area
$\bar{c}$ = the wing mean aerodynamic cord
w_f = the average width of the fuselage sections
α_{0w} = the wing zero-lift angle relative to the fuselage reference line
i_f = the incidence of the fuselage camber line relative to the fuselage reference line at the center of each fuselage increment. The incidence angle is defined as negative for nose droop and aft upsweep.
Δx = the length of the fuselage increments

Figure 2.10 illustrates how the fuselage can be divided into segments for the calculation of C_{m_0} and also defines the body width w_f for various body cross-sectional shapes. The correction factor $(k_2 - k_1)$ is given in Fig. 2.11.

The local angle of attack along the fuselage is greatly affected by the flow field created by the wing, as was illustrated in Fig. 2.8. The portion of the fuselage ahead of the wing is in the wing upwash, while the aft portion is in the wing downwash flow. The change in pitching moment with angle of attack is given by

$$C_{m_{\alpha_f}} = \frac{1}{36.5S\bar{c}} \int_0^{l_f} w_f^2 \frac{\partial \varepsilon_u}{\partial \alpha}\, dx \qquad (\text{deg}^{-1}) \qquad (2.32)$$

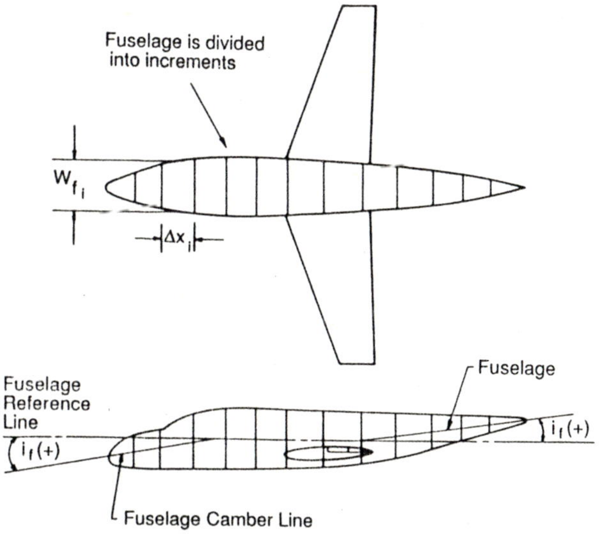

FIGURE 2.10
Procedure for calculating C_{m0} due to the fuselage.

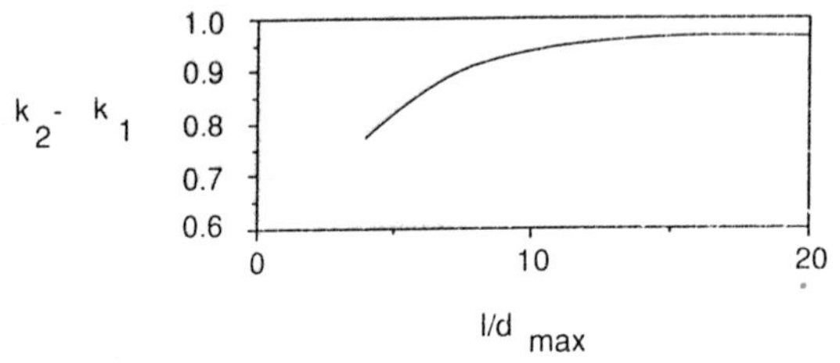

FIGURE 2.11
$k_2 - k_1$ vs l/d.

which can be approximated by

$$C_{m_{\alpha_f}} = \frac{1}{36.5 S\bar{c}} \sum_{x=0}^{x=l_f} w_f^2 \frac{\partial \varepsilon_u}{\partial \alpha} \, dx \tag{2.33}$$

where S = the wing reference area, and $\bar{c}$ = the wing mean aerodynamic chord.

The fuselage can again be divided into segments and the local angle of attack of each section, which is composed of the geometric angle of attack of the section plus the local induced angle due to the wing upwash or downwash for each segment, can be estimated. The change in local flow angle with angle of attack, $\partial\varepsilon_u/\partial\alpha$, varies along the fuselage and can be estimated from Fig. 2.12. For locations ahead of the wing, the upwash field creates large local angles of attack, therefore $\partial\varepsilon_u/\partial\alpha > 1$. On the other hand, a station behind the wing is in the downwash region of the wing vortex system and the local angle of attack is reduced. For the region behind the wing, $\partial\varepsilon_u/\partial\alpha$ is assumed to vary linearly from zero to $(1 - \partial\varepsilon/\partial\alpha)$ at the tail. The region between the wing leading edge and trailing edge is assumed to be unaffected by the wing flow field, $\partial\varepsilon_u/\partial\alpha = 0$. Figure 2.13 is a sketch showing the application of Eq. (2.33)

POWER EFFECTS. The propulsion unit can have a significant effect on both the longitudinal trim and static stability of the airplane. If the thrust line is offset from the center of gravity, the propulsive force will create a pitching moment that must be counteracted by the aerodynamic control surface.

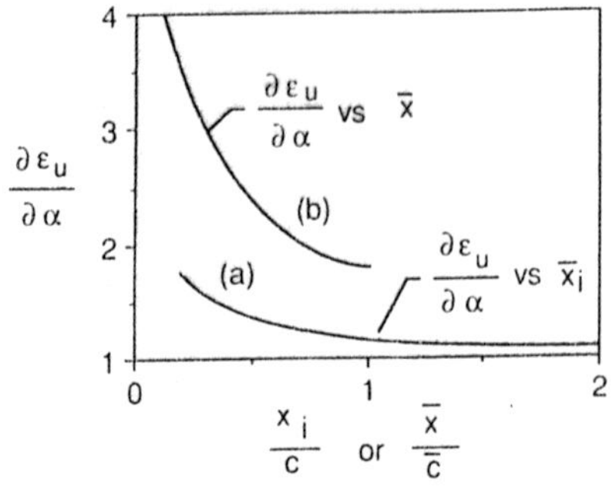

FIGURE 2.12
Variation of local flow angle along the fuselage.

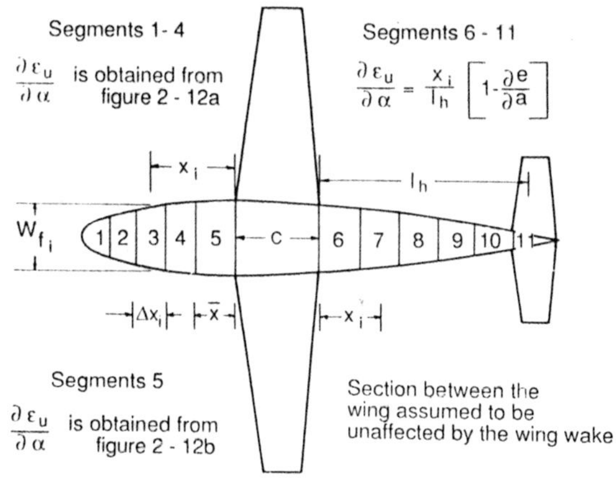

FIGURE 2.13
Procedure for calculating C_{m_α} due to the fuselage.

The static stability of the airplane is also influenced by the propulsion system. For a propeller driven airplane the propeller will develop a normal force in its plane of rotation when the propeller is at an angle of attack. The propeller normal force will create a pitching moment about the center of gravity thus producing a propulsion contribution to C_{m_α}. Although one can derive a simple expression for C_{m_α} due to the propeller, the actual contribution of the propulsion system to the static stability is much more difficult to estimate. This is due to the indirect effects that the propulsion system has on the airplanes characteristics. For example, the propeller slipstream can have an effect on the tail efficiency η and the downwash field. Because of these complicated interactions the propulsive effects on airplane stability are commonly estimated from powered wind tunnel models.

A normal force will be created on the inlet of a jet engine when it is at an angle of attack. As in the case of the propeller powered airplane, the normal force will produce a contribution to C_{m_α}.

STICK FIXED NEUTRAL POINT. The total pitching moment for the airplane can now be obtained by summing the wing, fuselage, and tail contributions.

$$C_{m_{cg}} = C_{m_0} + C_{m_\alpha}\alpha \tag{2.34}$$

where

$$C_{m_0} = C_{m_{0_w}} + C_{m_{0_f}} + \eta V_H C_{L_{\alpha_t}}(\varepsilon_0 + i_w - i_t) \tag{2.35}$$

$$C_{m_\alpha} = C_{L_{\alpha_w}}\left(\frac{X_{cg}}{\bar{c}} - \frac{X_{ac}}{\bar{c}}\right) + C_{m_{\alpha_f}} - \eta V_H C_{L_{\alpha_t}}\left(1 - \frac{d\varepsilon}{d\alpha}\right) \tag{2.36}$$

Notice that the expression for C_{m_α} depends upon the center of gravity position as well as the aerodynamic characteristics of the airplane. The center of gravity

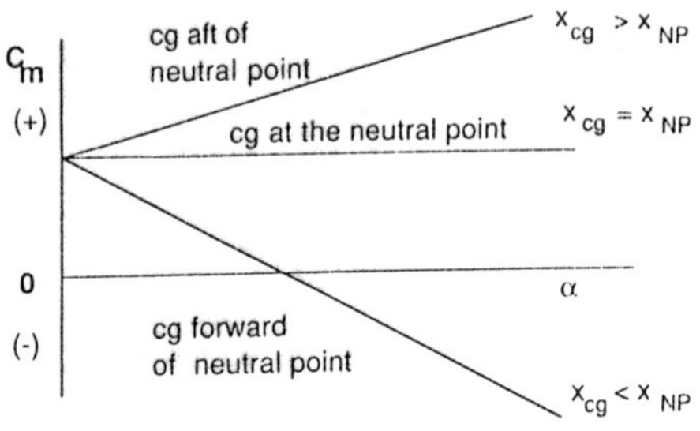

FIGURE 2.14
The influence of center of gravity position on longitudinal static stability.

of an airplane varies during the course of its operation and, therefore, it is important to know if there are any limits to the center of gravity travel. To ensure that the airplane possesses static longitudinal stability, we would like to know at what point $C_{m_\alpha} = 0$. Setting C_{m_α} equal to zero and solving for the center of gravity position yields

$$\frac{X_{\mathrm{NP}}}{\bar{c}} = \frac{X_{\mathrm{ac}}}{\bar{c}} - \frac{C_{m_{\alpha_f}}}{C_{L_{\alpha_w}}} + \eta V_{\mathrm{H}} \frac{C_{L_{\alpha_t}}}{C_{L_{\alpha_w}}} \left(1 - \frac{\mathrm{d}\varepsilon}{\mathrm{d}\alpha}\right) \tag{2.37}$$

We call this location the stick fixed neutral point. If the airplane's center of gravity ever reaches this point, the airplane will be neutrally stable. Movement of the center of gravity beyond the neutral point causes the airplane to be statically unstable. The influence of center of gravity position on static stabililty is shown in Fig. 2.14.

Example Problem 2.1. Given the general aviation airplane shown in Fig. 2.15, determine the contribution of the wing, tail and fuselage to the C_m versus α curve. Also determine the stick fixed neutral point. For this problem, assume standard sea-level atmospheric conditions.

Solution. The lift curve slopes for the two-dimensional sections making up the wing and tail must be corrected for a finite aspect ratio. This is accomplished by using the formula

$$C_{L_\alpha} = \frac{C_{l_\alpha}}{1 + C_{l_\alpha}/(\pi AR)}$$

Substituting the two-dimensional lift curve slope and the appropriate aspect ratio yields

$$C_{L_{\alpha_w}} = \frac{C_{l_{\alpha_w}}}{1 + C_{l_{\alpha_w}}/(\pi AR_w)}$$

$$= \frac{(0.097/\mathrm{deg})(57.3\ \mathrm{deg/rad})}{1 + (0.097/\mathrm{deg})(57.3\ \mathrm{deg/rad})/(\pi(6.06))}$$

$$= 4.3\ \mathrm{rad}^{-1}$$

In a similar manner the lift curve slope for the tail can be found:

$$C_{L_{\alpha_t}} = 3.91\ \mathrm{rad}^{-1}$$

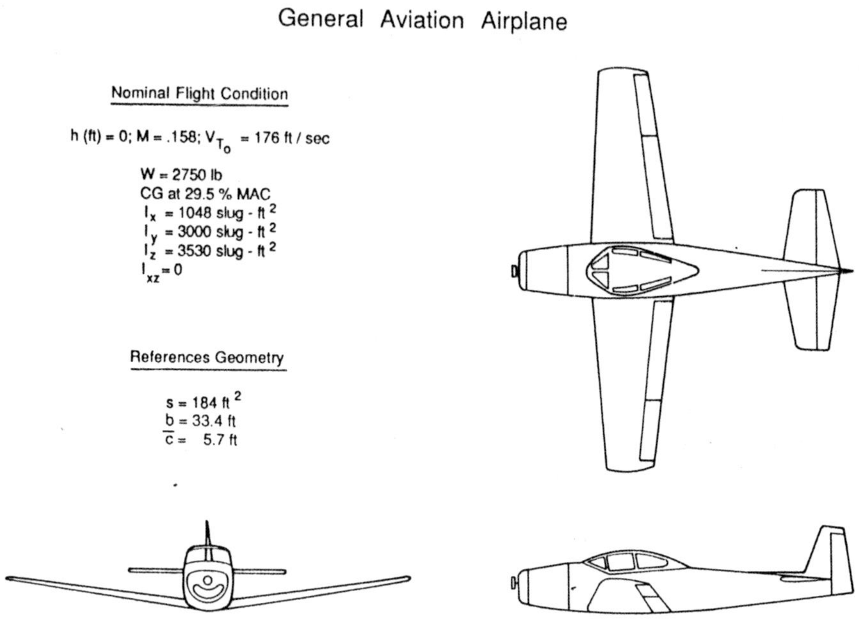

General Aviation Airplane
Nominal Flight Condition
h (ft) = 0; M = .158; V_{T_0} = 176 ft / sec
W = 2750 lb
CG at 29.5 % MAC
I_x = 1048 slug - ft^2
I_y = 3000 slug - ft^2
I_z = 3530 slug - ft^2
I_{xz} = 0
References Geometry
s = 184 ft^2
b = 33.4 ft
$\bar{c}$ = 5.7 ft

FIGURE 2.15
General Aviation airplane.

The wing contribution to C_{m_0} and C_{m_α} is found from Eq. (2.9) and (2.10):

$$C_{m_{0_w}} = C_{m_{ac_w}} + C_{L_{0_w}}\left(\frac{X_{cg}}{\bar{c}} - \frac{X_{ac}}{\bar{c}}\right)$$

and

$$C_{m_{\alpha_w}} = C_{L_{\alpha_w}}\left(\frac{X_{cg}}{\bar{c}} - \frac{X_{ac}}{\bar{c}}\right)$$

The lift coefficient at zero angle of attack is obtained by multiplying the absolute value of the zero lift angle of attack by the lift curve slope:

$$\begin{aligned} C_{L_{0_w}} &= C_{L_{\alpha_w}}|\alpha_0| \\ &= (4.3\ \text{rad}^{-1})(5\ \text{deg})/(57.3\ \text{deg/rad}) \\ &= 0.375 \end{aligned}$$

Substituting the approximate information into the equations for $C_{m_{0_w}}$ and $C_{m_{\alpha_w}}$ yields

$$\begin{aligned} C_{m_{0_w}} &= C_{m_{ac_w}} + C_{L_{0_w}}\left(\frac{X_{cg}}{\bar{c}} - \frac{X_{ac}}{\bar{c}}\right) \\ &= -0.116 + (0.375)(0.295 - 0.250) \\ &= -0.099 \end{aligned}$$

$$\begin{aligned} C_{m_{\alpha_w}} &= C_{L_{\alpha_w}}\left(\frac{X_{cg}}{\bar{c}} - \frac{X_{ac}}{\bar{c}}\right) \\ &= (4.3\ \text{rad}^{-1})(0.295 - 0.250) \\ &= 0.1935\ \text{rad}^{-1} \end{aligned}$$

For this particular airplane, the wing contribution to C_{m_α} is destabilizing.

The tail contribution to the intercept and slope can be estimated from Eqs (2.27) and (2.28):

$$C_{m0_t} = \eta V_H C_{L_{\alpha_t}}(\varepsilon_0 + i_w - i_t)$$

$$C_{m_{\alpha_t}} = -\eta V_H C_{L_{\alpha_t}}\left(1 - \frac{d\varepsilon}{d\alpha}\right)$$

The tail volume ratio V_H is given by

$$V_H = \frac{l_t S_t}{S\bar{c}}$$

or

$$V_H = \frac{(16\text{ ft})(43\text{ ft}^2)}{(184\text{ ft}^2)(5.7\text{ ft})} = 0.66$$

The downwash term is estimated using the expression

$$\varepsilon = \frac{2C_{L_w}}{\pi AR}$$

where ε is the downwash angle in radians:

$$\varepsilon_0 = \frac{2C_{L_{w0}}}{\pi AR}$$

$$= \frac{2(0.375)(57.3\text{ deg/rad})}{\pi(6.06)} = 0.04\text{ rad} = 2.3°$$

and

$$\frac{d\varepsilon}{d\alpha} = \frac{2C_{L_{\alpha_w}}}{\pi AR}$$

where $C_{L_{\alpha_w}}$ is per radians:

$$\frac{d\varepsilon}{d\alpha} = \frac{2(4.3)}{\pi(6.06)} = 0.45$$

Substituting the preceding information into the formulas for the intercept and slope yields

$$C_{m0_t} = \eta V_H C_{L_{\alpha_t}}(\varepsilon_0 + i_w - i_t)$$

$$= (0.66)(3.91)[2.3° + 1.0° - (-1.0°)]/57.2\text{ deg/rad}$$

$$= 0.194$$

and

$$C_{m_{\alpha_t}} = -\eta V_H C_{L_{\alpha_t}}\left(1 - \frac{d\varepsilon}{d\alpha}\right)$$

$$= -(0.66)(3.91)(1 - 0.45)$$

$$= -1.42\text{ rad}^{-1}$$

In this example, the ratio η of tail to wing dynamic pressure was assumed to be unity.

The fuselage contribution to C_{m_0} and C_{m_α} can be estimated from Eqs (2.29) and (2.31) respectively. To use these equations, we must divide the fuselage into segments, as indicated in Fig. 2.16. The summation in Eq. (2.31) can easily be

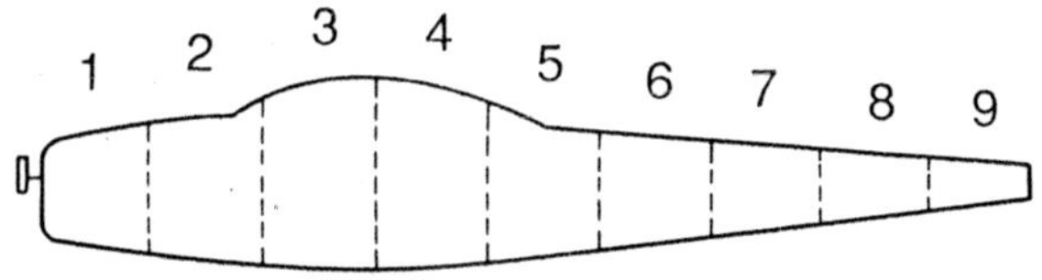

Station	Δx ft	w_f ft	$\alpha_{0_w} + i_f$	$w_f^2 [\alpha_{0_w} + i_f] \Delta x$
1	3.0	3.6	-1.5	-58.3
2	3.0	4.6	-1.5	-95.2
3	3.0	4.6	-1.5	-95.2
4	3.0	4.6	-1.5	-95.2
5	3.0	4.1	-1.5	-75.6
6	3.0	3.1	-1.5	-43.2
7	3.0	2.3	-1.5	-23.8
8	3.0	1.5	-1.5	-10.1
9	3.0	0.8	-1.5	-2.9
				Sum = -499.5

FIGURE 2.16
Sketch of segmented fuselage for calculating $C_{m\alpha}$ for the example problem.

estimated from the geometry and is found by summing the individual contributions as illustrated by the table in Fig. 2.16.

$$\sum_{x=0}^{l_B} w_f^2(\alpha_{0_w} + i_f)\,\Delta x = 500$$

The body fineness ratio is estimated from the geometrical data given in Fig. 2.15:

$$\frac{l_f}{d_{\max}} = 6.2$$

and the correction factor $k_2 - k_1$ is found from Fig. 2.11, $k_2 - k_1 = 0.86$. Substituting these values into Eq. (2.31) yields

$$C_{m_{0_f}} = -0.010$$

In a similar manner C_{m_α} can be estimated. A table is included in Fig. 2.16 that shows the estimate of the summation. $C_{m_{\alpha_f}}$ was estimated to be

$$C_{m_{\alpha_f}} = 0.12\ \text{rad}^{-1}$$

The individual contributions and the total pitching moment curve are shown in Fig. 2.17.

The stick fixed neutral point can be estimated from Eq. (2.37):

$$\frac{X_{\text{NP}}}{\bar{c}} = \frac{X_{\text{ac}}}{\bar{c}} - \frac{C_{m_{\alpha_f}}}{C_{L_{\alpha_w}}} + \eta V_{\text{H}} \frac{C_{L_{\alpha_t}}}{C_{L_{\alpha_w}}}\left(1 - \frac{d\varepsilon}{d\alpha}\right)$$
$$= 0.37$$

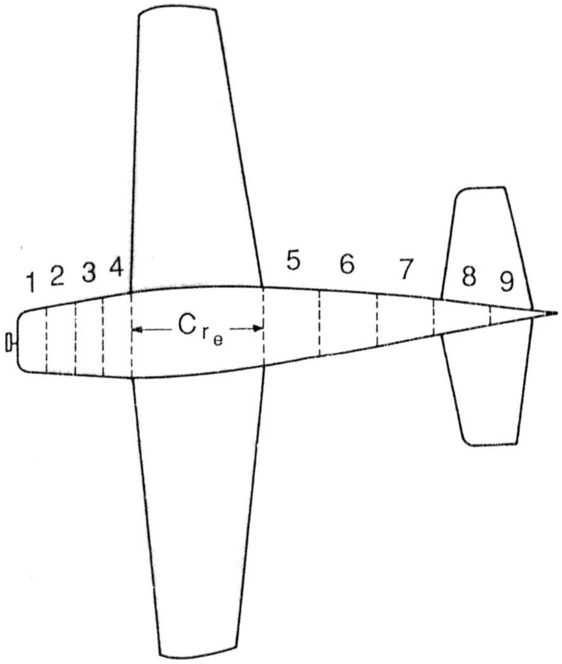

Station	Δx ft	w_f ft	x	$\frac{\partial \varepsilon_u}{\partial \alpha}$	$w_f^2 \frac{\partial \varepsilon_u}{\partial \alpha} \Delta x$
1	1.5	3.0	5.25	1.2	16.2
2	1.5	3.4	4.5	1.3	22.5
3	1.5	3.8	3.75	1.4	30.3
4	1.5	4.2	1.5	3.2	84.7
5	2.9	3.8	1.45	0.06	2.5
6	2.9	3.1	4.35	0.18	5.0
7	2.9	2.3	7.25	0.31	4.8
8	2.9	1.5	10.15	0.43	2.8
9	2.9	0.8	13.05	0.55	1.0

C_{r_e} = 6.5 ft $\quad$ l_n = 13 ft $\quad$ Sum = 85.1

FIGURE 2.16 *(contd.)*

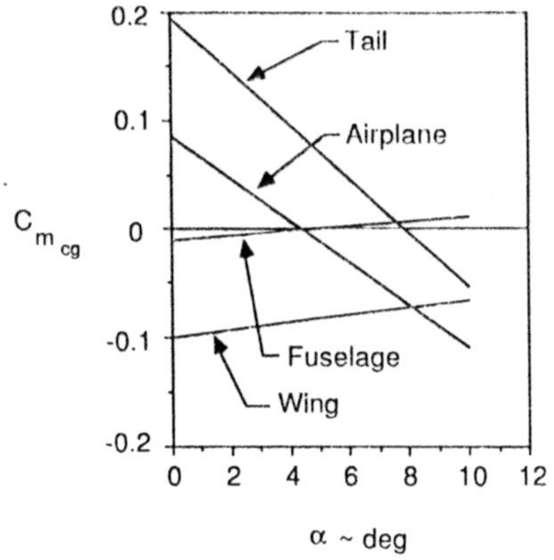

FIGURE 2.17
Component contributions to pitching moment for example problem.

2.4 LONGITUDINAL CONTROL

Control of an airplane can be achieved by providing an incremental lift force on one or more of the airplane's lifting surfaces. The incremental lift force can be produced by deflecting the entire lifting surface or by deflecting a flap incorporated in the lifting surface. Owing to the fact that the control flaps or movable lifting surfaces are located at some distance from the center of gravity, the incremental lift force creates a moment about the airplane's center of gravity. Figure 2.18 shows the three primary aerodynamic controls. Pitch control can be achieved by changing the lift on either a forward or aft control surface. If a flap is used, the flapped portion of the tail surface is called an elevator. Yaw control is achieved by deflecting a flap on the vertical tail called the rudder and roll control can be achieved by deflecting small flaps located outboard toward the wing tips in a differential manner. These flaps are called ailerons. A roll moment can also be produced by deflecting a wing spoiler. As the name implies a spoiler disrupts the lift. This is accomplished by deflecting a section of the upper wing surface so that the flow separates behind the spoiler

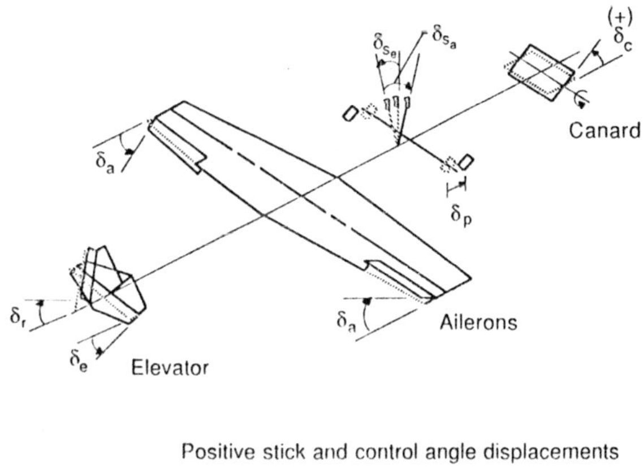

FIGURE 2.18
Primary aerodynamic controls.

which causes a reduction in the lifting force. To achieve a roll moment, only one spoiler needs to be deflected.

In this section we shall be concerned with longitudinal control. Control of the pitch attitude of an airplane can be achieved by deflecting all or a portion of either a forward or an aft tail surface. Factors affecting the design of a control surface are control effectiveness, hinge moments, and aerodynamic and mass balancing. Control effectiveness is a measure of how effective the control deflection is in producing the desired control moment. As we shall show shortly, control effectiveness is a function of the size of the flap and tail volume ratio. Hinge moments are also important because they are the aerodynamic moments that must be overcome to rotate the control surface. The hinge moment governs the magnitude of force required of the pilot to move the control surface. Therefore, great care must be used in designing a control surface so that the control forces are within acceptable limits for the pilots. Finally, aerodynamic and mass balancing deal with techniques to vary the hinge moments so that the control stick forces stay within an acceptable range.

ELEVATOR EFFECTIVENESS. We need some form of longitudinal control to fly at various trim conditions. As shown earlier, the pitch attitude can be controlled by either an aft tail or forward tail (canard). We shall examine how an elevator on an aft tail provides the required control moments. Although we are restricting our discussion to an elevator on an aft tail, the same arguments could be made with regard to a canard surface. In Fig. 2.19 we see a sketch of the horizontal tail as well as the influence of the elevator on the pitching moment curve. Notice that the elevator does not change the slope of the pitching moment curves, but only shifts them so that different trim angles can be achieved.

When the elevator is deflected, it changes the lift and pitching moment of the airplane. The change in lift for the airplane can be expressed as follows:

$$\Delta C_L = C_{L_{\delta_e}} \delta_e \qquad \text{where} \qquad C_{L_{\delta_e}} = \frac{\mathrm{d}C_L}{\mathrm{d}\delta_e} \tag{2.38}$$

$$C_L = C_{L_\alpha} \alpha + C_{L_{\delta_e}} \delta_e \tag{2.39}$$

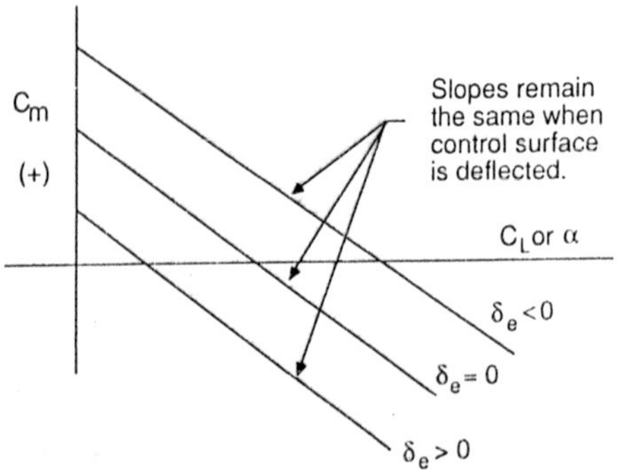

FIGURE 2.19
The influence of the elevator on the C_m vs α curve.

On the other hand, the change in pitching moment acting on the airplane can be written as

$$\Delta C_m = C_{m_{\delta_e}} \delta_e \qquad \text{where} \qquad C_{m_{\delta_e}} = \frac{dC_m}{d\delta_e} \tag{2.40}$$

The stability derivative $C_{m_{\delta_e}}$ is called the elevator control power. The larger the value of $C_{m_{\delta_e}}$ the more effective the control is in creating the control moment.

Adding ΔC_m to the pitching moment equation yields

$$C_m = C_{m_0} + C_{m_\alpha}\alpha + C_{m_{\delta_e}}\delta_e \tag{2.41}$$

The derivatives $C_{L_{\delta_e}}$ and $C_{m_{\delta_e}}$ can be related to the aerodynamic and geometric characteristics of the horizontal tail in the following manner. The change in lift of the airplane due to deflecting the elevator is equal to the change in lift force acting on the tail:

$$\Delta L = \Delta L_t \tag{2.42}$$

$$\Delta C_L = \frac{S_t}{S}\eta\, \Delta C_{L_t} = \frac{S_t}{S}\eta \frac{dC_{L_t}}{d\delta_e}\delta_e \tag{2.43}$$

where $dC_{L_t}/d\delta_e$ is the elevator effectiveness. The elevator effectiveness is proportional to the size of the flap being used as an elevator and can be estimated from the equation

$$\frac{dC_{L_t}}{d\delta_e} = \frac{dC_{L_t}}{d\alpha_t}\frac{d\alpha_t}{d\delta_e} = C_{L_{\alpha_t}}\tau \tag{2.44}$$

The parameter τ can be determined from Fig. 2.20:

$$C_{L_{\delta_e}} = \frac{S_t}{S}\eta \frac{dC_{L_t}}{d\delta_e} \tag{2.45}$$

The increment in airplane pitching moment is

$$\Delta C_m = -V_H\eta\, \Delta C_{L_t} = -V_H\eta \frac{dC_{L_t}}{d\delta_e}\delta_e \tag{2.46}$$

or

$$C_{m_{\delta_e}} = -V_H\eta \frac{dC_{L_t}}{d\delta_e} \tag{2.47}$$

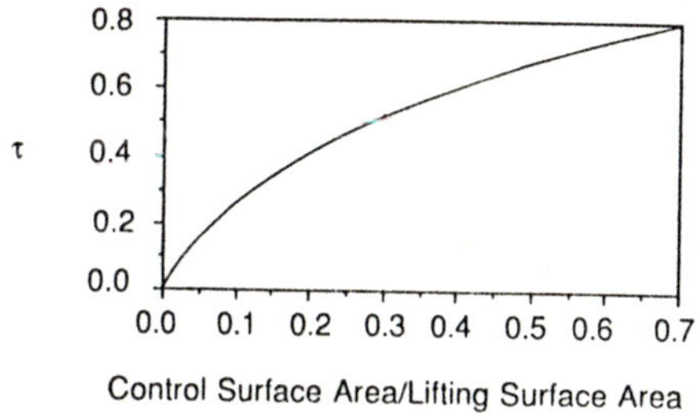

FIGURE 2.20
Flap effectiveness parameter.

The designer can control the magnitude of the elevator control effectiveness by proper selection of the volume ratio and flap size.

ELEVATOR ANGLE TO TRIM. Now let us consider the trim requirements. An airplane is said to be trimmed if the forces and moments acting on the airplane are in equilibrium. Setting the pitching moment of equation equal to zero (definition of trim) we can solve for the elevator angle required to trim the airplane:

$$C_m = 0 = C_{m_0} + C_{m_\alpha}\alpha + C_{m_{\delta_e}}\delta_e \tag{2.48}$$

or

$$\delta_{\text{trim}} = -\frac{C_{m_0} + C_{m_\alpha}\alpha_{\text{trim}}}{C_{m_\delta}} \tag{2.49}$$

The lift coefficient to trim is

$$C_{L\text{trim}} = C_{L_\alpha}\alpha_{\text{trim}} + C_{L_{\delta_e}}\delta_{\text{trim}} \tag{2.50}$$

We can use the above equation to obtain the trim angle of attack.

$$\alpha_{\text{trim}} = \frac{C_{L\text{trim}} - C_{L_{\delta_e}}\delta_{\text{trim}}}{C_{L_\alpha}} \tag{2.51}$$

If we substitute this equation back into Eq. (2.49) we get the following equation for the elevator angle to trim:

$$\delta_{\text{trim}} = -\frac{C_{m_0}C_{L_\alpha} + C_{m_\alpha}C_{L_{\text{trim}}}}{C_{m_{\delta_e}}C_{L_\alpha} - C_{m_\alpha}C_{L_{\delta_e}}} \tag{2.52}$$

The elevator angle to trim can also be obtained directly from the pitching moment curves shown in Fig. 2.19.

FLIGHT MEASUREMENT OF X_{NP}. The equation developed for estimating the elevator angle to trim the airplane can be used to determine the stick fixed neutral point from flight test data. Suppose we conducted a flight test experiment in which we measured the elevator angle of trim at various air speeds for different positions of the center of gravity. If we did this, we could develop curves as shown in Fig. 2.21.

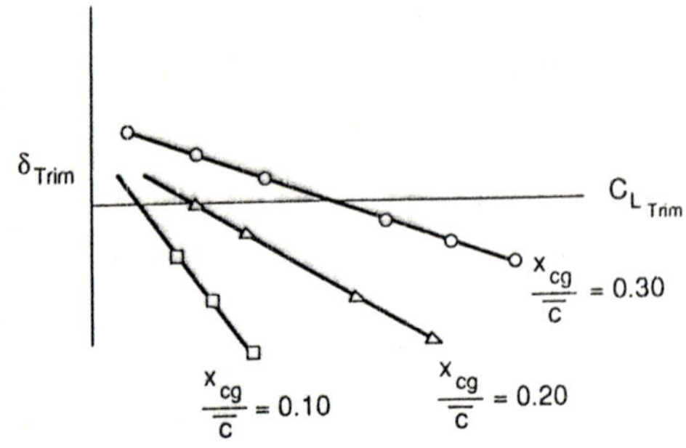

FIGURE 2.21
$\delta_{e\text{trim}}$ vs $C_{L\text{trim}}$.

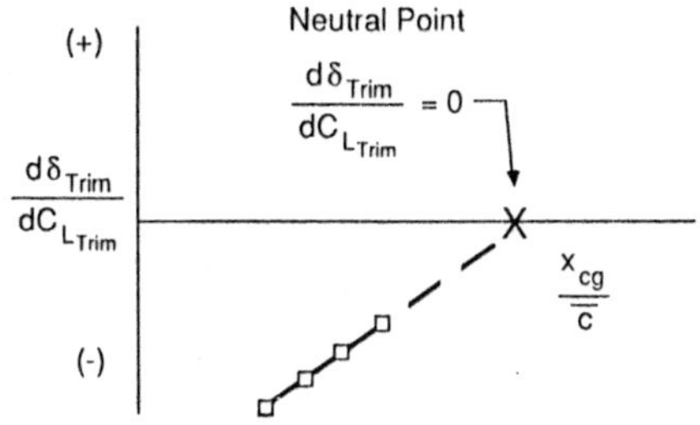

FIGURE 2.22
$d\delta_{etrim}/dC_{Ltrim}$ vs C.G.

Now, differentiating Eq. (2.52) with respect to $C_{L_{\text{trim}}}$ yields

$$\frac{d\delta_{\text{trim}}}{dC_{L_{\text{trim}}}} = -\frac{C_{m_\alpha}}{C_{m_{\delta_e}} C_{L_\alpha} - C_{m_\alpha} C_{L_{\delta_e}}} \tag{2.53}$$

Note that when $C_{m_\alpha} = 0$ (i.e. the center of gravity is at the neutral point) Eq. (2.53) equals zero. Therefore, if we measure the slopes of the curves in Fig. 2.21 and plot them as a function of center of gravity location, we can estimate the stock fixed neutral point as illustrated in Fig. 2.22 by extrapolating to find the center of gravity position that makes $d\delta_{\text{trim}}/dC_{L_{\text{trim}}}$ equal to zero.

ELEVATOR HINGE MOMENT. It is important to know the moment acting at the hinge line of the elevator (or other type of control surface). The hinge moment is, of course, the moment the pilot must overcome by exerting a force on the control stick. Therefore, to design the control system properly we must know the hinge moment characteristics. The hinge moment is defined as shown below in Fig. 2.23. If we assume that the hinge moment can be expressed as the addition of the effects of angle of attack, elevator deflection angle, and tab angle taken separately, then we can express the hinge moment coefficient in the following manner:

$$C_h = C_{h_0} + C_{h_{\alpha_t}} \alpha_t + C_{h_{\delta_e}} \delta_e + C_{h_{\delta_t}} \delta_t \tag{2.54}$$

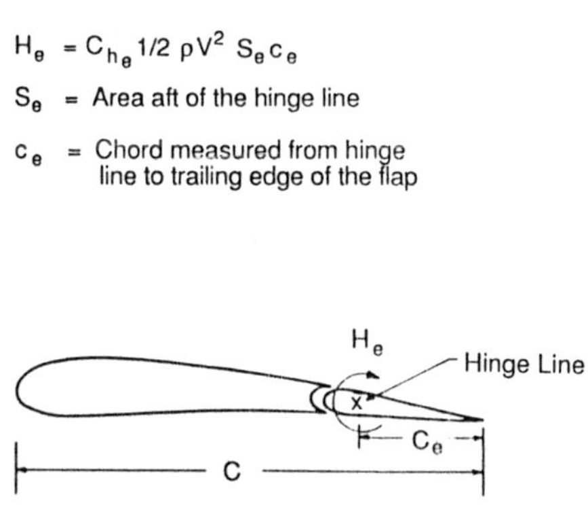

FIGURE 2.23
Definition of hinge moments.

where C_{h_0} is the residual moment, and

$$C_{h_{\alpha_t}} = \frac{dC_h}{d\alpha_t} \qquad C_{h_{\delta_e}} = \frac{dC_h}{d\delta_e} \qquad C_{h_{\delta_t}} = \frac{dC_h}{d\delta_t} \tag{2.55}$$

The hinge moment parameters defined above are very difficult to predict analytically with great precision. Wind tunnel tests are usually required to provide the control system designer with the information needed to design the control system properly.

When the elevator is set free, i.e. the control stick is released the stability and control characteristics of the airplane are affected. For simplicity, we shall assume that both δ_t and C_{h_0} are equal to zero. Then, for the case when the elevator is allowed to be free,

$$C_{h_e} = 0 = C_{h_{\alpha_t}}\alpha_t + C_{h_{\delta_e}}\delta_e \tag{2.56}$$

Solving for δ_e yields

$$(\delta_e)_{\text{free}} = -\frac{C_{h_{\alpha_t}}}{C_{h_{\delta_e}}}\alpha_t \tag{2.57}$$

Usually, the coefficients $C_{h_{\alpha_t}}$ and $C_{h_{\delta_e}}$ are negative. If this is indeed the case, then Eq. (2.55) tells us that the elevator will float upwards as the angle of attack is increased. The lift coefficient for a tail with a free elevator is given by

$$C_{L_t} = C_{L_{\alpha_t}}\alpha_t + C_{L_{\delta_e}}\delta_{e\text{free}} \tag{2.58}$$

$$C_{L_t} = C_{L_{\alpha_t}}\alpha_t - C_{L_{\delta_e}}\frac{C_{h_{\alpha_t}}}{C_{h_{\delta_e}}}\alpha_t \tag{2.59}$$

which simplifies to

$$C_{L_t} = C_{L_{\alpha_t}}\alpha_t\left(1 - \frac{C_{L_{\delta_e}}C_{h_{\alpha_t}}}{C_{L_{\alpha_t}}C_{h_{\delta_e}}}\right) = C'_{L_{\alpha_t}}\alpha_t \tag{2.60}$$

where

$$C'_{L_{\alpha_t}} = C_{L_{\alpha_t}}\left(1 - \frac{C_{L_{\delta_e}}C_{h_{\alpha_t}}}{C_{L_{\alpha_t}}C_{h_{\delta_e}}}\right) = C_{L_{\alpha_t}}f \tag{2.61}$$

The slope of the tail lift curve is modified by the term in the parentheses. The factor f can be greater than or less than unity depending on the sign of the hinge parameters $C_{h_{\alpha_t}}$ and $C_{h_{\delta_e}}$. Now, if we were to develop the equations for the total pitching moment for the free elevator case, we would obtain an equation similar to Eqs (2.35)–(2.36). The only difference would be that the term $C_{L_{\alpha_t}}$ would be replaced by $C'_{L_{\alpha_t}}$. Substituting $C'_{L_{\alpha_t}}$ into Eqs (2.35) and

(2.36) yields

$$C'_{m0} = C_{m0_w} + C_{m0_f} + C'_{L_{\alpha_t}}\eta V_H(\varepsilon_0 + i_w - i_t) \tag{2.62}$$

$$C'_{m_\alpha} = C_{L_{\alpha_w}}\left(\frac{X_{cg}}{\bar{c}} - \frac{X_{ac}}{\bar{c}}\right) + C_{m_{\alpha_f}} - C'_{L_{\alpha_t}}\eta V_H\left(1 - \frac{\partial \varepsilon}{\partial \alpha}\right) \tag{2.63}$$

where the prime indicates elevator free values. To determine the influence of a free elevator on the static longitudinal stability, we again examine the condition in which $C_{m_\alpha} = 0$. Setting C'_{m_α} equal to zero in Eq. (2.63) and solving for $x/\bar{c}$ yields the stick free neutral point:

$$\frac{X'_{NP}}{\bar{c}} = \frac{X_{ac}}{\bar{c}} + V_H\eta\frac{C'_{L_{\alpha_t}}}{C_{L_{\alpha_w}}}\left(1 - \frac{d\varepsilon}{d\alpha}\right) - \frac{C_{m_{\alpha_f}}}{C_{L_{\alpha_w}}} \tag{2.64}$$

The difference between the stick fixed neutral point and the stick free neutral point can be expressed as follows:

$$\frac{X_{NP}}{\bar{c}} - \frac{X'_{NP}}{\bar{c}} = (1 - f)V_H\eta\frac{C_{L_{\alpha_t}}}{C_{L_{\alpha_w}}}\left(1 - \frac{d\varepsilon}{d\alpha}\right) \tag{2.65}$$

The factor f determines whether the stick free neutral point lies forward or aft of the stick fixed neutral point.

A term which appears frequently in the literature is static margin. The static margin is simply the distance between the neutral point and the actual center of gravity position.

$$\text{Stick fixed static margin} = \frac{X_{NP}}{\bar{c}} - \frac{X_{cg}}{\bar{c}} \tag{2.66}$$

$$\text{Stick free static margin} = \frac{X'_{NP}}{\bar{c}} - \frac{X_{cg}}{\bar{c}} \tag{2.67}$$

For most aircraft designs, it is desirable to have a stick fixed static margin of approximately 5 percent of the mean chord. The stick fixed or stick free static neutral points represent an aft limit on the center of gravity travel for the airplane.

2.5 STICK FORCES

In order to deflect a control surface, the pilot must move the control stick or rudder pedals. The forces exerted by the pilot to move the control surface is called the stick force or pedal force, depending upon which control is being used. The stick force is proportional to the hinge moment acting on the control surface:

$$F = \text{fn}(H_e) \tag{2.68}$$

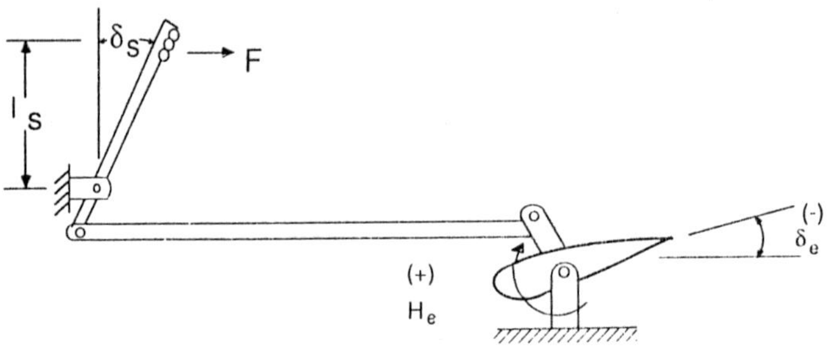

FIGURE 2.24
Relationship between stick force and hinge moment.

Figure 2.24 is a sketch of a simple mechanical system used for deflecting the elevator. The work of displacing the control stick is equal to the work done in moving the control surface to the desired deflection angle. From Fig. 2.24 we can write the expressions for the work performed at the stick and elevator:

$$F l_s \delta_s = H_e \delta_e \tag{2.69}$$

or

$$F = \frac{\delta_e}{l_s \delta_s} \tag{2.70}$$

or

$$F = G H_e \tag{2.71}$$

where $G = \delta_e / l_s \delta_s$ is called the gearing ratio and is a measure of the mechanical advantage provided by the control system.

Substituting the expression for the hinge moment defined earlier into the stick force equation yields

$$F = G C_{h_e} \tfrac{1}{2} \rho V^2 S_e \bar{c}_e \tag{2.72}$$

From this expression we see that the magnitude of the stick force increases with the size of the airplane and with the square of the airplane's speed. Similar expressions can be obtained for the rudder pedal force and aileron stick force.

The control system is designed to convert the stick and pedal movements into control surface deflections. Although this may seem to be a relatively easy task, it is in fact quite complicated. The control system must be designed so that the control forces are within acceptable limits. On the other hand, the control forces required in normal maneuvers must not be too small, otherwise it might be possible to overstress the airplane. Proper control system design will provide the pilot with stick force magnitudes that give the pilot a feel for the commanded maneuver. The magnitude of the stick force provides the pilot with an indication of the severity of the motion that will result from the stick movement.

The convention for longitudinal control is that a pull force should always rotate the nose upward which causes the airplane to slow down. A push force will have the opposite effect, i.e. the nose will rotate downward and the airplane will speed up. The control system designer must also be sure that the airplane does not experience control reversals due to aerodynamic or aeroelastic phenomena.

TRIM TABS. In addition to making sure that the stick and rudder pedal forces required to maneuver or trim the airplane are within acceptable limits, it is important that some means be provided to zero out the stick force at the trimmed flight speed. If such a provision is not made, the pilot will become fatigued by trying to maintain the necessary stick force. The stick force at trim can be made zero by incorporating a tab on either the elevator or the rudder. The tab is a small flap located at the trailing edge of the control surface. The trim tab can be used to zero out the hinge moment and thereby eliminate the stick or pedal forces. Figure 2.25 is a sketch illustrating the concept of a trim tab. Although the trim tab has a great influence over the hinge moment, it has only a slight effect on the lift produced by the control surface.

STICK FORCE GRADIENTS. Another important parameter in the design of a control system is called the stick force gradient. Figure 2.26 shows the variation of the stick force with speed. The stick force gradient is a measure of the change in stick force needed to change the speed of the airplane. To provide the airplane with speed stability, the stick force gradient must be

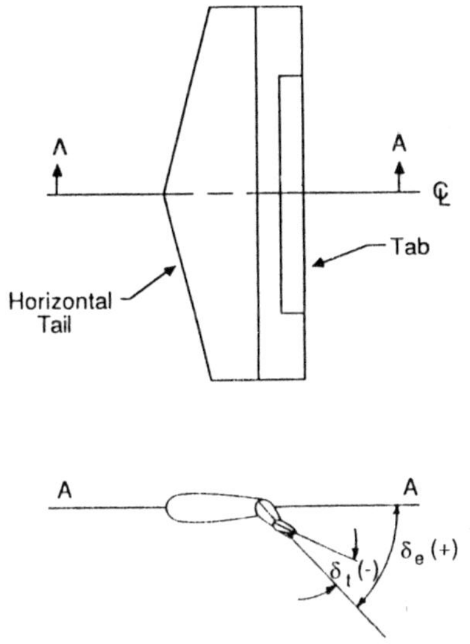

FIGURE 2.25
Trim tabs.

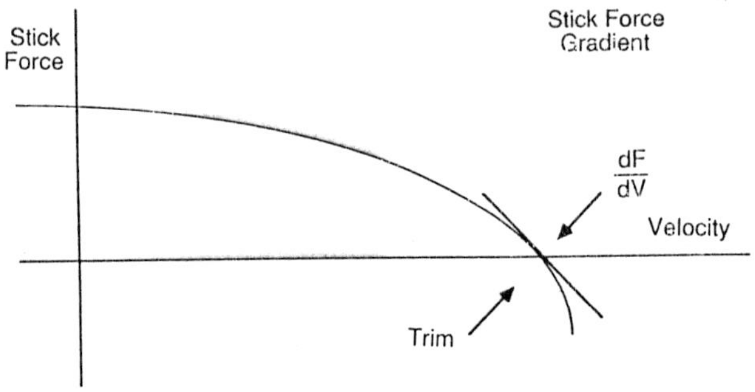

FIGURE 2.26
Stick force versus velocity.

negative, i.e.

$$\frac{dF}{dV} < 0 \tag{2.73}$$

The need for a negative stick force gradient can be appreciated by examining Fig. 2.26. If the airplane slows down, a positive stick force occurs which rotates the nose of the airplane downwards, which causes the airplane to increase its speed back towards the trim velocity. For the case in which the airplane exceeds the trim velocity, a negative (pull) stick force causes the airplane's nose to pitch up, which causes the airplane to slow down. The negative stick force gradient provides the pilot and airplane with speed stability. The larger the gradient, the more resistant the airplane will be to disturbances in the flight speed. If an airplane did not have speed stability the pilot would have to continuously monitor and control the airplane's speed. This would be highly undesirable from the pilot's point of view.

2.6 DEFINITION OF DIRECTIONAL STABILITY

Directional, or weathercock, stability is concerned with the static stability of the airplane about the z axis. Just as in the case of longitudinal static stability, it is desirable that the airplane should tend to return to an equilibrium condition when subjected to some form of yawing disturbance. Figure 2.27 shows the yawing moment coefficient versus sideslip angle β for two airplane configurations. To have static directional stability, the airplane must develop a yawing moment which will restore the airplane to its equilibrium state. Assume that both airplanes are disturbed from their equilibrium conditions, so that the airplanes are flying with a positive sideslip angle β. Airplane 1 will develop a restoring moment which will tend to rotate the airplane back to its equilibrium condition, i.e. zero sideslip angle. Airplane 2 will develop a yawing moment that will tend to increase the sideslip angle. Examining these curves, we see that to have static directional stability the slope of the yawing moment curve must be positive ($C_{n_\beta} > 0$). Note that an airplane possessing static directional

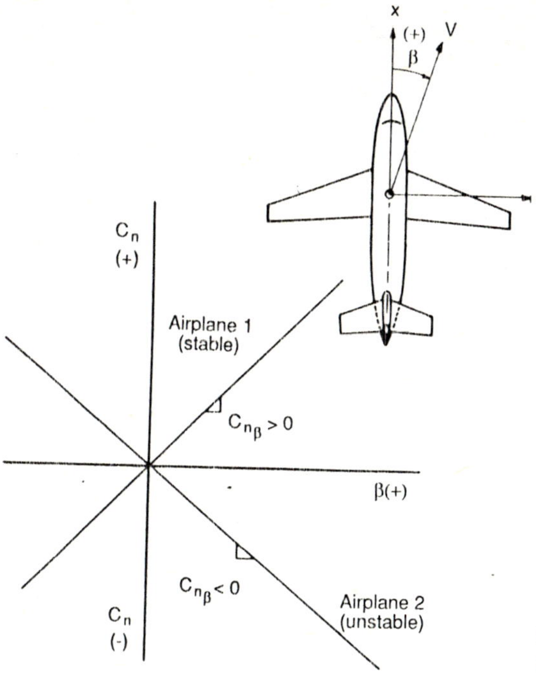

FIGURE 2.27
Static directional stability.

stability will always point into the relative wind, thus the name "weathercock stability."

CONTRIBUTION OF AIRCRAFT COMPONENTS. The contribution of the wing to directional stability is usually quite small in comparison to the fuselage, provided the angle of attack is not large. The fuselage and engine nacelles, in general, create a destabilizing contribution to directional stability. The wing fuselage contribution can be calculated from the following empirical expression taken from Ref. 2.7.

$$C_{n_{\beta_{wf}}} = -k_n k_{Rl} \frac{S_{fs}}{S_w} \frac{l_f}{b} \quad \text{(per deg)} \tag{2.74}$$

where k_n = an empirical wing–body interference factor that is a function of the fuselage geometry

k_{Rl} = an empirical correction factor that is a function of the fuselage Reynolds number

S_{fs} = the projected side area of the fuselage

l_f = the length of the fuselage

The empirical factors k_n and k_{Rl} are determined from Figs 2.28 and 2.29, respectively.

Since the wing–fuselage contribution to directional stability is destabilizing, the vertical tail must be properly sized to ensure that the airplane has

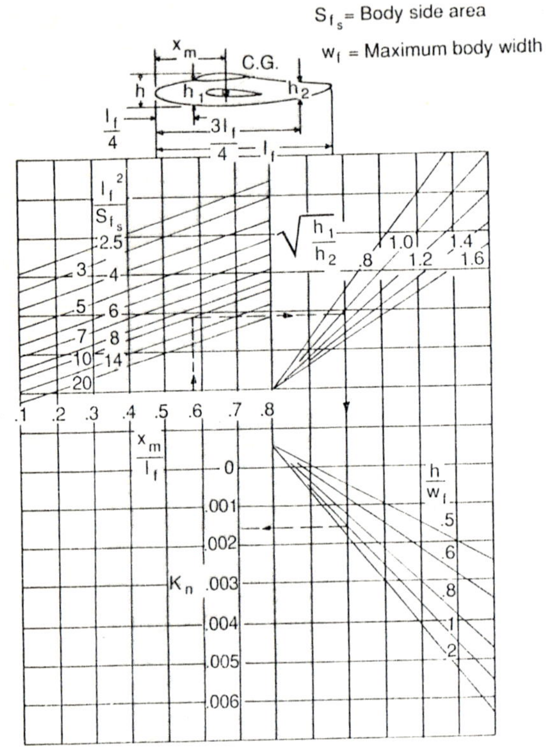

FIGURE 2.28
Wing body interference factor.

directional stability. The mechanism by which the vertical tail produces directional stability is shown in Fig. 2.30. If we consider the vertical tail surface in Fig. 2.30, we see that when the airplane is flying at a positive sideslip angle, the vertical tail produces a side force (lift force in *xy* plane) which tends to rotate the airplane about its center of gravity. The moment produced is a restoring moment. The side force acting on the vertical tail can be expressed as

$$Y_v = -C_{L_{\alpha_v}} \alpha_v Q_v S_v \tag{2.75}$$

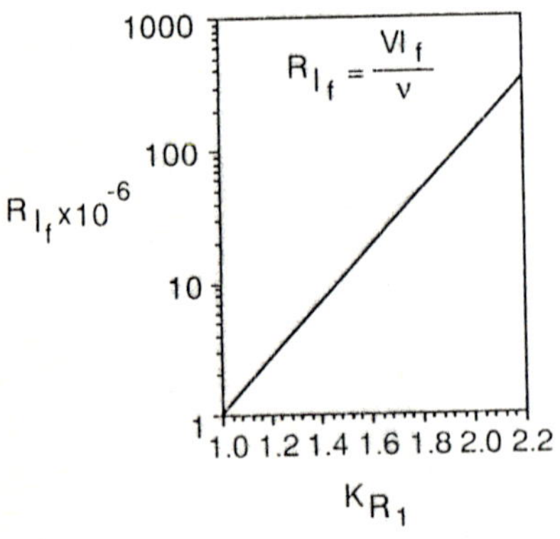

FIGURE 2.29
Reynolds number correction factor.

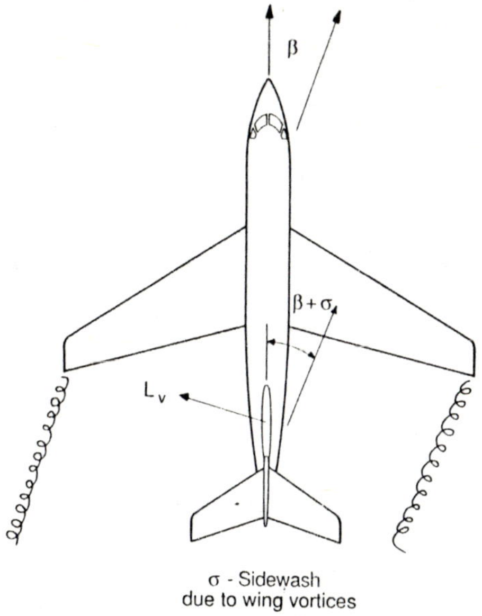

FIGURE 2.30
Vertical tail contribution to directional stability.

where the subscript v refers to properties of the vertical tail. The angle of attack α_v that the vertical tail plane will experience can be written as

$$\alpha_v = \beta + \sigma \tag{2.76}$$

where σ is the sidewash angle. The sidewash angle is analogous to the downwash angle ε for the horizontal tail plane. The sidewash is caused by the flow field distortion due to the wings and fuselage. The moment produced by the vertical tail can be written as a function of the side force acting on it:

$$N_v = l_v Y_v = l_v C_{L_{\alpha_v}}(\beta + \sigma)Q_v S_v \tag{2.77}$$

or in coefficient form:

$$C_n = \frac{N_v}{Q_w Sb} = \frac{l_v S_v}{Sb}\frac{Q_v}{Q_w}(C_{L_{\alpha_v}}(\beta + \sigma) \tag{2.78}$$

$$= V_v \eta_v C_{L_{\alpha_v}}(\beta + \sigma) \tag{2.79}$$

where V_v is the vertical tail volume ratio and η_v is the ratio of the dynamic pressure at the vertical tail to the dynamic pressure at the wing.

The contribution of the vertical tail to directional stability can now be obtained by taking the derivative of Eq. (2.77) with respect to β:

$$C_{n_{\beta_v}} = V_v \eta_v C_{L_{\alpha_v}}\left(1 + \frac{d\sigma}{d\beta}\right) \tag{2.80}$$

A simple algebraic equation for estimating the combined sidewash and tail efficiency factor η_v is presented in Ref. 2.7 and is reproduced below.

$$\eta_v\left(1+\frac{d\sigma}{d\beta}\right)=0.724+306\frac{S_v/S}{1+\cos\Lambda_{c/4w}}+0.4\frac{z_w}{d}+0.009AR_w \qquad (2.81)$$

where S = the wing area

S_v = the vertical tail area, including the submerged area to the fuselage centerline

z_w = is the distance, parallel to the z axis, from wing root quarter chord point to fuselage centerline

d = is the maximum fuselage depth

AR_w = is the aspect ratio of the wing

2.7 DIRECTIONAL CONTROL

Directional control is achieved by a control surface called a rudder which is located on the vertical tail, as shown in Fig. 2.31. The rudder is a hinged flap which forms the aft portion of the vertical tail. By rotating the flap, the lift force (side force) on the fixed vertical surface can be varied to create a yawing moment about the center of gravity. The size of the rudder is determined by the directional control requirements. The rudder control power must be sufficient to accomplish the requirements listed in Table 2.1.

The yawing moment produced by the rudder depends upon the change in lift on the vertical tail due to the deflection of the rudder times its distance from the center of gravity. For a positive rudder deflection, a positive side force is created on the vertical tail. A positive side force will produce a negative yawing moment:

$$N=-l_vY_v \qquad (2.82)$$

where the side force is given by

$$Y_v=C_{L_v}Q_vS_v$$

Rewriting this equation in terms of a yawing moment coefficient yields

$$C_n=\frac{N}{Q_wSb}=-\frac{Q_v}{Q_w}\frac{l_vS_v}{Sb}\frac{dC_{L_v}}{d\delta_r}\delta_r \qquad (2.83)$$

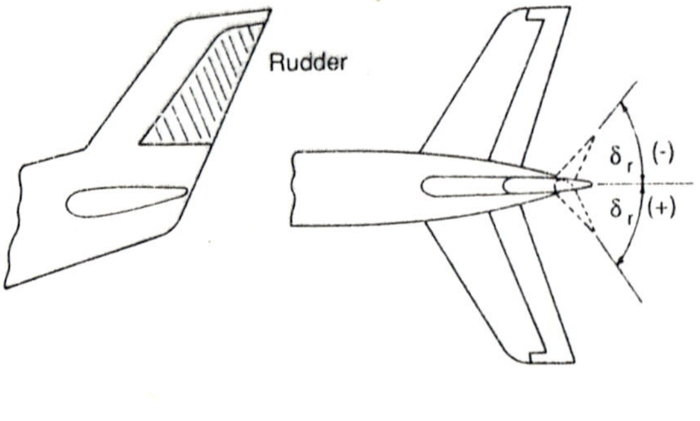

FIGURE 2.31
Directional control by means of the rudder.

TABLE 2.1
Requirements for directional control

Rudder requirements	Implication for rudder design
Adverse yaw	When an airplane is banked, in order to execute a turning maneuver, the ailerons may create a yawing moment that opposes the turn (i.e. adverse yaw). The rudder must be able to overcome the adverse yaw so that a coordinated turn can be achieved. The critical condition for adverse yaw occurs when the airplane is flying slow (i.e. high C_L).
Cross-wind landings	To maintain alignment with the runway during a cross-wind landing requires the pilot to fly the airplane at a sideslip angle. The rudder must be powerful enough to permit the pilot to trim the airplane for the specified cross-winds. For transport airplanes, landing may be carried out for cross-winds up to 15.5 m/s or 51 ft/s.
Asymmetric power condition	The critical asymmetric power condition occurs for a multi-engine airplane when one engine fails at low flight speeds. The rudder must be able to overcome the yawing moment produced by the asymmetric thrust arrangement.
Spin recovery	The primary control for spin recovery in many airplanes is a powerful rudder. The rudder must be powerful enough to oppose the spin rotation.

or

$$C_n = -\eta_v V_v \frac{dC_{L_v}}{d\delta_r} \delta_r \tag{2.84}$$

The rudder control effectiveness is the rate of change of yawing moment with rudder deflection angle:

$$C_n = C_{n_{\delta_r}} \delta_r = -\eta_v V_v \frac{dC_{L_v}}{d\delta_r} \delta_r \tag{2.85}$$

or

$$C_{n_{\delta_r}} = -\eta V_v \frac{dC_{L_v}}{d\delta_r} \tag{2.86}$$

where

$$\frac{dC_{L_v}}{d\delta_r} = \frac{dC_{L_v}}{d\alpha_v}\frac{d\alpha_v}{d\delta_r} = C_{L_{\alpha_v}} \tau \tag{2.87}$$

and the factor τ can be estimated from Fig. 2.20.

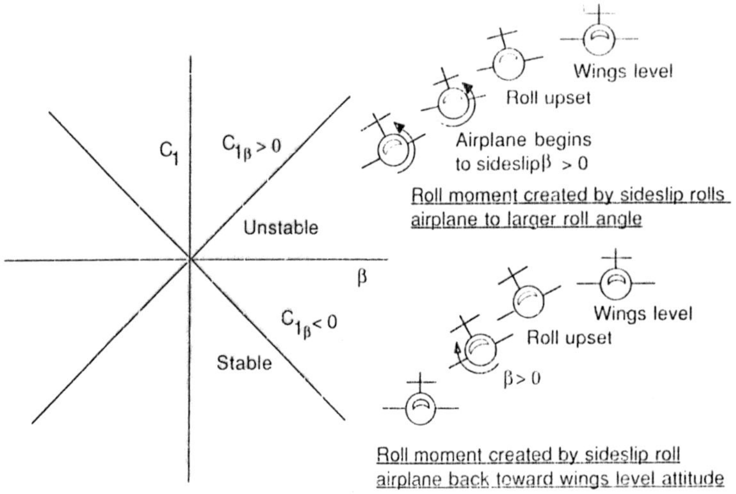

FIGURE 2.32
Static roll stability.

2.8 ROLL STABILITY

An airplane possesses static roll stability if, when it is disturbed from a wings-level attitude, a restoring moment is developed. The restoring rolling moment can be shown to be a function of the sideslip angle β as illustrated in Fig. 2.32. The requirement for stability is that $C_{l_\beta} < 0$. The roll moment created on an airplane when it starts to sideslip depends upon wing dihedral, wing sweep, position of the wing on the fuselage, and the vertical tail. Each of these contributions will be discussed qualitatively in the following paragraphs.

The major contributor to C_{l_β} is the wing dihedral angle Γ. The dihedral angle is defined as the spanwise inclination of the wing with respect to the horizontal. If the wing tip is higher than the root section, then the dihedral angle is positive; if the wing tip is lower than the root section, then the dihedral angle is negative. A negative dihedral angle is commonly called anhedral.

When an airplane is disturbed from a wings-level attitude, it will begin to sideslip as shown in Fig. 2.32. Once the airplane starts to sideslip a component of the relative wind is directed toward the side of the airplane. The leading wing experiences an increased angle of attack and consequently an increase in lift. The trailing wing experiences the opposite effect. The net result is a rolling moment that tries to bring the wing back to a wings-level attitude. This restoring moment is often referred to as the "dihedral effect".

The additional lift created on the downward-moving wing is created by the change in angle of attack produced by the sideslipping motion. If we resolve the sideward velocity component into components along and normal to the wing span the local change in angle of attack can be estimated as

$$\Delta\alpha = \frac{v_n}{u} \tag{2.88}$$

where

$$v_n = V \sin \Gamma \tag{2.89}$$

By approximating the sideslip angle as

$$\beta = \frac{v}{u} \tag{2.90}$$

and assuming that Γ is a small angle, then the change in angle of attack can be written as

$$\Delta\alpha \cong \beta\Gamma \tag{2.91}$$

The angle of attack on the upward-moving wing will be decreased by the same amount. Methods for estimating the wing contribution to C_{l_β} can again be found in Ref. 2.7.

Wing sweep also contributes to the dihedral effect. In the case of a sweptback wing, the windward wing has an effective decrease in sweep angle while the trailing wing experiences an effective increase in sweep angle. For a given angle of attack, a decrease in sweepback angle will result in a higher lift coefficient. Therefore, the windward wing (less effective sweep) will experience more lift than the trailing wing. It can be concluded that sweepback adds to the dihedral effect. On the other hand, sweepforward will decrease the effective dihedral effect.

The fuselage contribution to dihedral effect is illustrated in Fig. 2.33. The

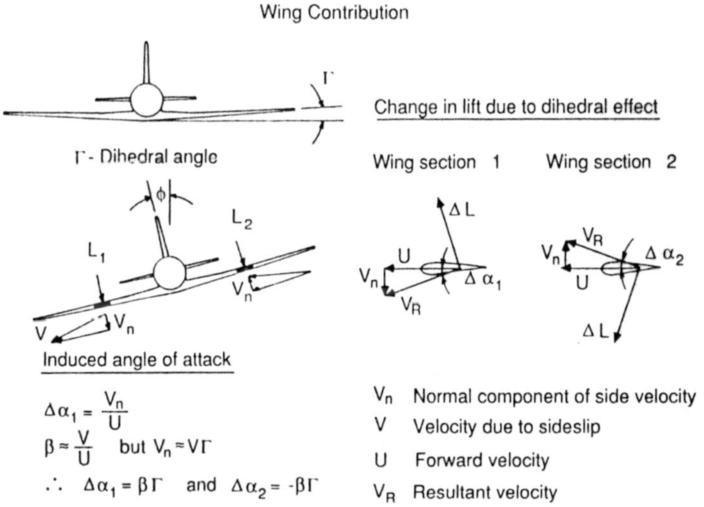

FIGURE 2.33
Wing and fuselage contribution to the dihedral effect.

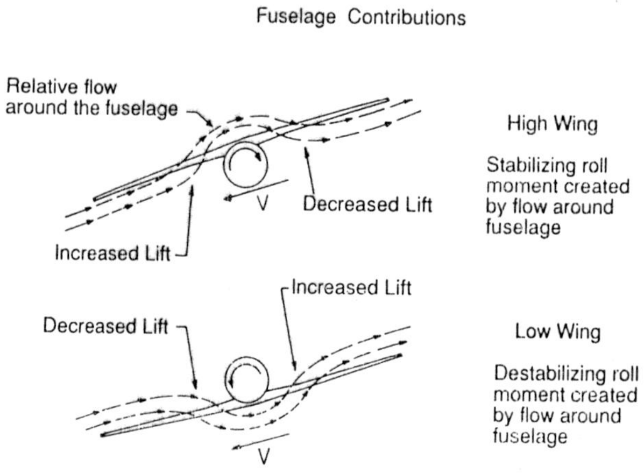

FIGURE 2.33 *(contd.)*

sideward flow turns in the vicinity of the fuselage and creates a local change in wing angle of attack at the inboard wing stations. For a low wing position, the fuselage contributes a negative dihedral effect; the high wing produces a positive dihedral effect. In order to maintain the same C_{l_β}, a low-wing aircraft will require a considerably greater wing dihedral angle than a high-wing configuration.

The horizontal tail can also contribute to the dihedral effect in a manner similar to the wing. However, owing to the size of the horizontal tail with respect to the wing, its contribution is usually small. The contribution to dihedral effect from the vertical tail is produced by the side force on the tail due to sideslip. The side force on the vertical tail produces both a yawing and a rolling moment. The rolling moment occurs because the center of pressure for the vertical tail is located above the aircraft's center of gravity. The rolling moment produced by the vertical tail tends to bring the aircraft back to a wings-level attitude.

2.9 ROLL CONTROL

Roll control is achieved by the differential deflection of small flaps called ailerons which are located outboard on the wings, or by the use of spoilers. Figure 2.34 is a sketch showing both types of roll control devices. The basic principle behind these devices is to modify the spanwise lift distribution so that a moment is created about the x axis. An estimate of the roll control power for an aileron can be obtained by a simple strip integration method as illustrated in Fig. 2.35 and the equations that follow. The incremental change in roll moment due to a change in aileron angle can be expressed as

$$\Delta L = (\Delta\,\text{Lift})y \tag{2.92}$$

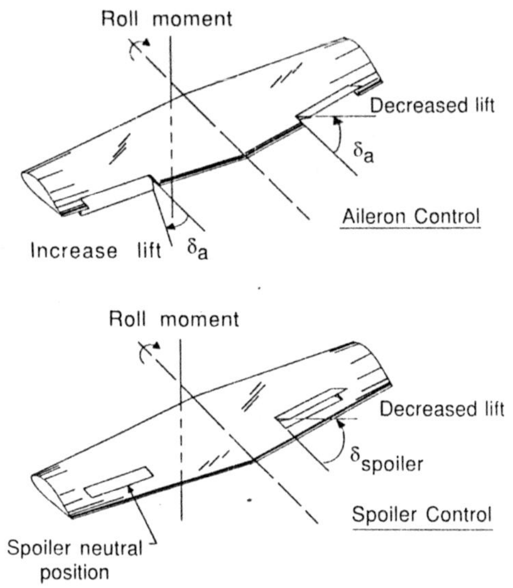

FIGURE 2.34
Aileron and spoilers for roll control.

which can be written in coefficient form as

$$\Delta C_l = \frac{\Delta L}{QSb} = \frac{C_L Qcy\,\mathrm{d}y}{QSb} \tag{2.93}$$

$$= \frac{C_L cy\,\mathrm{d}y}{Sb} \tag{2.94}$$

The lift coefficient on the stations containing the aileron can be written as

$$C_l = C_{L_\alpha}\frac{\mathrm{d}\alpha}{\mathrm{d}\delta_a}\delta_a = C_{L_\alpha}\tau\delta_a \tag{2.95}$$

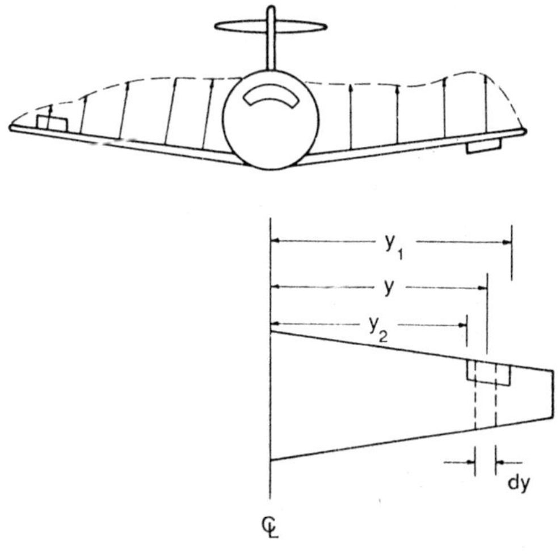

FIGURE 2.35
Strip theory approximation of roll control effectiveness.

which is similar to the technique used to estimate the control effectiveness of an elevator and rudder. Substituting Eq. (2.94) into Eq. (2.95) and integrating over the region containing the aileron yields

$$C_l = \frac{2C_{L_{\alpha_w}}\tau\delta_a}{Sb}\int_{y_1}^{y_2} cy\,dy \tag{2.96}$$

where $C_{L_{\alpha_w}}$ and τ have been corrected for three-dimensional flow and the factor of 2 has been introduced to account for the other aileron. The control power $C_{l_{\delta_a}}$ can be obtained by taking the derivative with respect to δ_a:

$$C_{l_{\delta_a}} = \frac{2C_{L_{\alpha_w}}\tau}{Sb}\int_{y_1}^{y_2} cy\,dy \tag{2.97}$$

2.10 SUMMARY

The requirements for static stability were developed for longitudinal, lateral directional and rolling motions. It is easy to see why a pilot would require the airplane that he or she is flying to possess some degree of static stability. Without static stability the pilot would have to continuously control the airplane to maintain a desired flight path. This would be quite fatiguing for the pilot. The degree of static stability desired by the pilot has been determined through flying quality studies and will be discussed in a later chapter. The important point at this time is to recognize that the airplane must be made statically stable, either because of the inherent aerodynamic characteristics or by artificial means through the use of an automatic control system, for the pilot to find the airplane acceptable for flying.

The inherent static stability tendencies of the airplane were shown to be a function of the geometric and aerodynamic properties of the airplane. The designer can control the degree of longitudinal and lateral directional stability by proper sizing of the horizontal and vertical tail surfaces, whereas roll stability was shown to be a consequence of dihedral effect, which is controlled by wing placement and or wing dihedral angle.

In addition to static stability, the pilot also wants sufficient control authority to keep the airplane in equilibrium (i.e. trim) or to maneuver. Aircraft response to control input and control force requirements are important flying quality characteristics which are determined by the control surface size. The stick force and stick force gradient are important parameters that influence how the pilot feels about the flying characteristics of the airplane. Stick forces must provide the pilot with a feel for the maneuver initiated. In addition, we show that the stick force gradient provides the airplane with speed stability. If the longitudinal stick force gradient is negative at the trim flight speed, then the airplane will resist disturbances in speed and will fly at a constant speed.

Finally, the relationship between static stability and control was examined. An airplane that is very stable statically will not be very maneuverable; if the airplane has very little static stability, it will be very maneuverable. The degree of maneuverability or static stability is determined by the designer on the basis of the mission requirements of the airplane being designed.

2.11 PROBLEMS

2.1. If the slope of the C_m versus C_L curve is -0.15 and the pitching moment at zero lift is equal to 0.08, determined the trim lift coefficient. If the center of gravity of the airplane is located at $X_{cg}/\bar{c} = 0.3$, determine the stick fixed neutral point.

2.2. For the data shown in Fig. P2.2, determine the following.

(a) The stick fixed neutral point.

(b) If we wish to fly the airplane at a velocity of 125 ft/s at sea level, what would be the trim lift coefficient and what would be the elevator angle for trim?

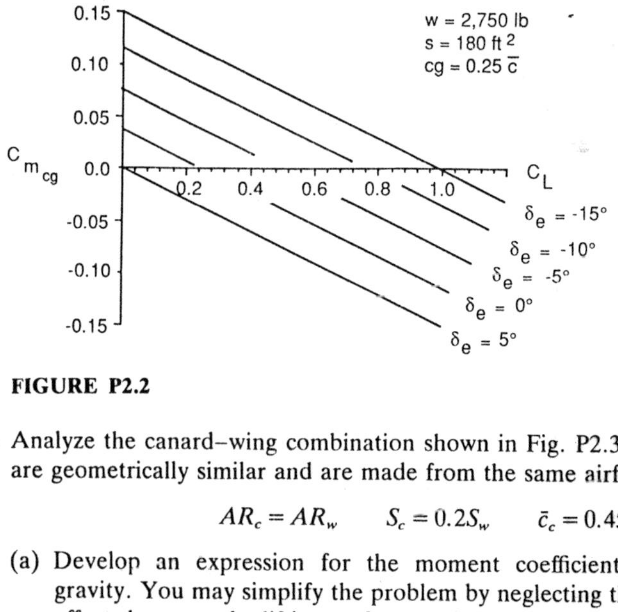

FIGURE P2.2

2.3. Analyze the canard–wing combination shown in Fig. P2.3. The canard and wing are geometrically similar and are made from the same airfoil section.

$$AR_c = AR_w \qquad S_c = 0.2S_w \qquad \bar{c}_c = 0.45\bar{c}_w$$

(a) Develop an expression for the moment coefficient about the center of gravity. You may simplify the problem by neglecting the upwash (downwash) affects between the lifting surfaces and the drag contribution to the moment. Also assume small angle approximations.

(b) Find the neutral point for this airplane.

2.4. Using the data for the business jet aircraft included in the Appendix, determine the following longitudinal stability information at subsonic speeds:

(a) wing contribution to pitching moment

(b) tail contribution to pitching moment

(c) fuselage contribution to pitching moment

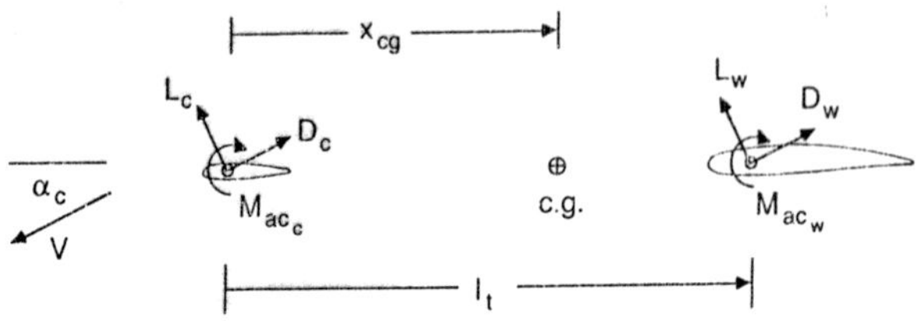

FIGURE P2.3

(d) total pitching moment
(e) plot the various contributions
(f) estimate the stick fixed neutral point

2.5. The downwash angle at zero angle of attack and the rate of change of downwash with angle of attack can be determined experimentally by several techniques. The downwash angle can be measured directly by using a five- or seven-hole pressure probe to determine the flow direction at the position of the tail surface, or it can be obtained indirectly from pitching moment data measured from wind tunnel models. This latter technique will be demonstrated by way of this problem. Suppose that a wind tunnel test was conducted to measure the pitching moment as a function of the angle of attack for various tail incidence settings as well as for the case when the tail surface is removed. Figure P2.5 is a plot of such information. Notice that the tail-off data intersect the complete configuration data at several points. At the points of intersection, the contribution of the tail surface to the pitching moment curve must be zero. For this to be the case, the lift on the tail surface is zero, which implies that the tail angle of attack is zero at these points. From the definition of the tail angle of attack,

$$\alpha_t = \alpha_w - i_w - \varepsilon + i_t$$

we obtain

$$\varepsilon = \alpha_w - i_w + i_t$$

at the interception points. Using the data of Fig. P2.5, determine the downwash angle versus the angle of attack of the wing. From this information estimate ε_0 and $d\varepsilon/d\alpha$.

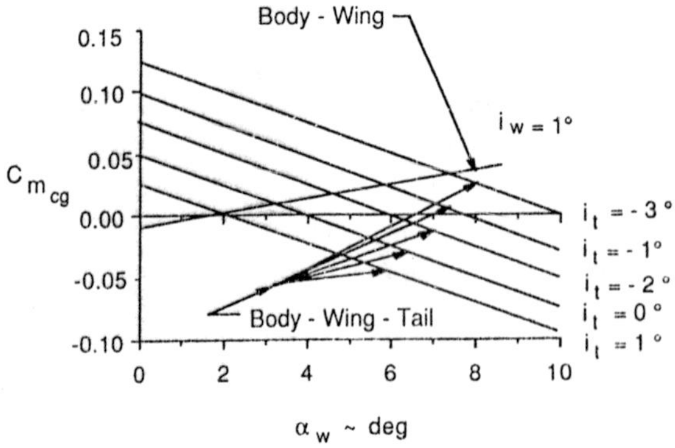

FIGURE P2.5

2.6. If the airplane in Example Problem 2.1 has the following hinge moment characteristics,

$$C_{h_\alpha} = -0.003 \qquad C_{h_\delta} = -0.005$$
$$C_{h_0} = 0.0 \qquad S_e/S_t = 0.35$$

What would be the stick fixed neutral point location?

2.7. As an airplane nears the ground its aerodynamic characteristics are changed by the presence of the ground plane. This change is called ground effect. A simple model for determining the influence of the ground on the lift drag and pitching moment can be obtained by representing the airplane by a horseshoe vortex system and its image as shown in Fig. P2.7. Using this sketch shown qualitatively the changes that one might expect, i.e. whether the forces and moment increase or decrease?

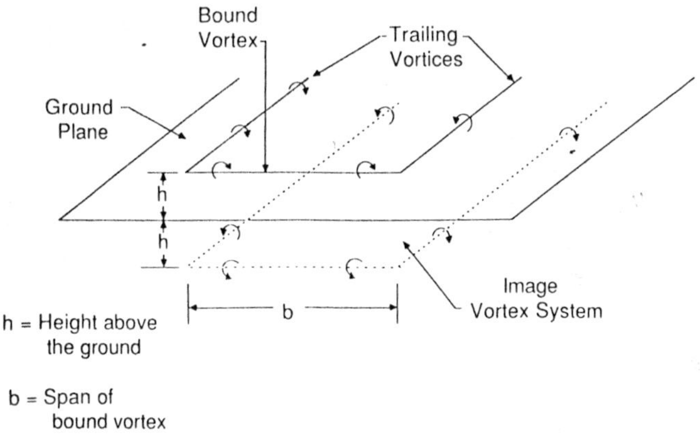

FIGURE P2.7

2.8. If the control characteristics of the elevator used in Example Problem 2.1 are as given below, determine the forward-most limit on the center of gravity travel so that the airplane can be controlled during landing, i.e. at $C_{L_{\max}}$. Neglect ground effects on the airplane's aerodynamic characteristics.

$$C_{m_{\delta e}} = -1.03 \qquad \delta_{e\max} = \begin{cases} +10° \\ -20° \end{cases} \qquad C_{L\max} = 1.4$$

2.9. Size the vertical tail for the airplane configuration shown in Fig. P2.9 so that its weathercock stability has a value of $C_{n_\beta} = +0.1\ \text{rad}^{-1}$. Clearly state your assumptions.

2.10. Figure P2.10 is a sketch of a wing planform for a business aviation airplane.

(a) Use strip theory to determine the roll control power.

(b) Comment on the accuracy of the strip in integration technique.

2.11. Suppose the wing planform in Problem 2.10 is incorporated into a low-wing aircraft design. Find the wing dihedral angle necessary to produce a dihedral effect of $C_{l_\beta} = -0.1$. Neglect the fuselage interference on the wing dihedral contribution.

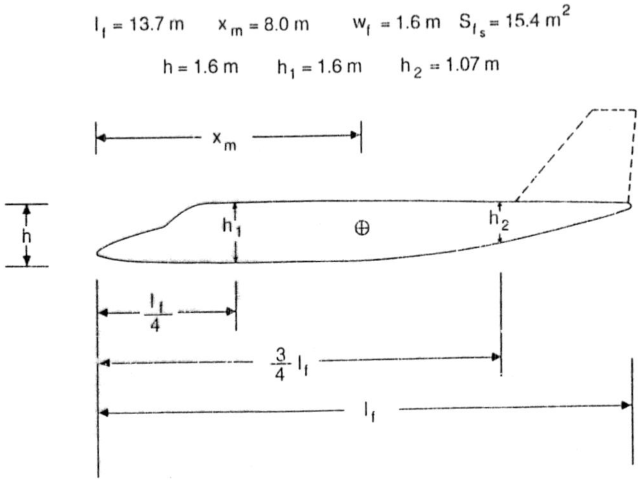

FIGURE P2.9

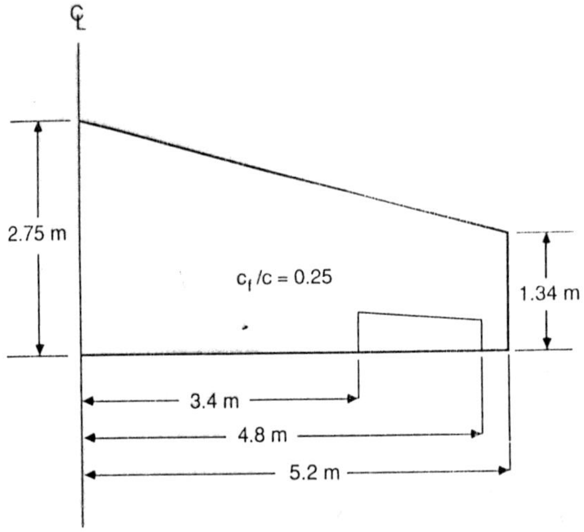

FIGURE P2.10

REFERENCES

2.1. Dickman, E. W.: *This Aviation Business*, Brentano's, New York, 1926.

2.2. Gibbs-Smith, C. H., and T. D. Crouch: "Wilbur and Orville Wright—A 75th Anniversary Commemoration," *Astronautics and Aeronautics, pp.* 8–15, *December*, 1978.

2.3. Perkins, C. D.: "Development of Airplane Stability and Control Technology," *AIAA Journal of Aircraft*, vol. 7, no. 4, 1970.

2.4. Etkin, B: *Dynamics of Flight*, Wiley, New York, 1959.

2.5. Perkins, C. D., and R. E. Hage,: *Airplane Performance Stability and Control*, Wiley, New York, 1949.

2.6. McRuer, D., I. Ashkenas, and D. Graham,: *Aircraft Dynamics and Automatic Control,* Princeton University Press, Princeton, NJ, 1973.
2.7. *USAF Stability and Control Datcom,* Flight Control Division, Air Force Flight Dynamics Laboratory, Wright Patterson Air Force Base, OH.
2.8. Munk, M. M.: *The Aerodynamic Forces on Airship Hulls,* NACA TR 184, 1924.
2.9. Multhopp, H.: *Aerodynamics of Fuselage,* NACA TM-1036, 1942.

CHAPTER 3

AIRCRAFT EQUATIONS OF MOTION

Success four flights Thursday morning all against twenty-one mile wind—started from level with engine power alone average speed through air thirty-miles—longest 57 seconds inform press home Christmas

Telegram message sent by Orville Wright/December 17, 1903

3.1 INTRODUCTION

In Chapter 2, the requirements for static stability were examined. It was shown that static stability is a tendency of the aircraft to return to its equilibrium position. In addition to static stability, the aircraft must also be dynamically stable. An airplane can be considered to be dynamically stable if, after being disturbed from its equilibrium flight condition, the ensuing motion diminishes with time. Of particular interest to the pilot and designer is the degree of dynamic stability. The required degree of dynamic stability is usually specified by the time it takes the motion to damp to half of its initial amplitude or, in the case of an unstable motion, the time it takes for the initial amplitude or disturbance to double. Also of interest is the frequency or period of the oscillation.

An understanding of the dynamic characteristics of an airplane is important in assessing the handling or flying qualities of an airplane as well as for designing autopilots. Flying qualities of an airplane are dependent upon pilot opinion, that is, the pilot's likes or dislikes with regard to the various vehicle motions. It is possible to design an airplane that has excellent performance but is considered to be an unsatisfactory airplane by the pilot. From the early 1960s to the present, there has been a considerable amount of

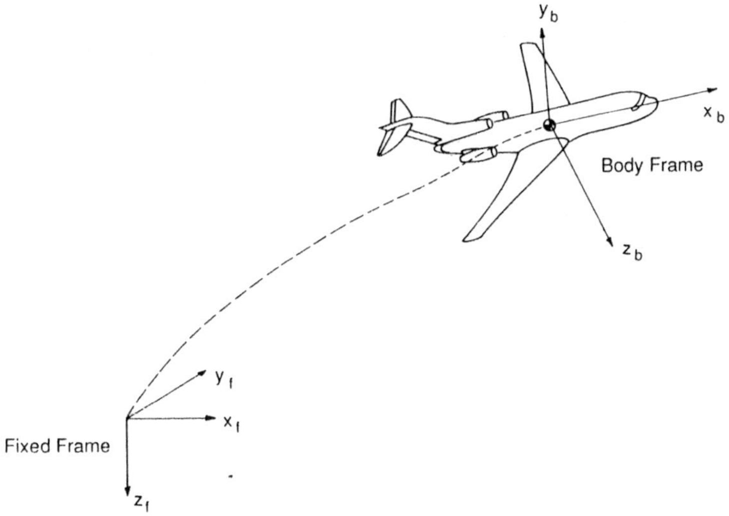

FIGURE 3.1
Body and inertial axes systems.

research directed toward quantifying pilot opinion in terms of aircraft motion characteristics, such as frequency and damping ratio of the aircraft's various modes of motion. Thus, it is important to understand the dynamic characteristics of an airplane and the relationship of the motion to the vehicle's aerodynamic characteristics and to pilot opinion.

Before developing the equations of motion, it is important to review the axis system that was specified earlier. Figure 3.1 shows the body axes system that is fixed to the aircraft and the inertial axes system that is fixed to the Earth.

3.2 DERIVATION OF RIGID BODY EQUATIONS OF MOTION

The rigid body equations of motion are obtained from Newton's second law, which states that the summation of all external forces acting on a body is equal to the time rate of change of the momentum of the body, and the summation of the external moments acting on the body is equal to the time rate of change of the moment of momentum (angular momentum). The time rates of change of linear and angular momentum are referred to an absolute or inertial reference frame. For many problems in airplane dynamics, an axis system fixed to the Earth can be used as an inertial reference frame. Newton's second law can be expressed in the following vector equations:

$$\sum \mathbf{F} = \frac{\mathrm{d}}{\mathrm{d}t}(m\mathbf{v}) \tag{3.1}$$

$$\sum \mathbf{M} = \frac{\mathrm{d}}{\mathrm{d}t}\mathbf{H} \tag{3.2}$$

The vector equations can be rewritten in scalar form and then consist of three force equations and three moment equations. The force equations can be expressed as follows:

$$F_x = \frac{d}{dt}(mu) \qquad F_y = \frac{d}{dt}(mv) \qquad F_z = \frac{d}{dt}(mw) \tag{3.3}$$

where F_x, F_y, F_z and u, v, w are the components of the force and velocity along the x, y and z axes, respectively. The force components are composed of contributions due to the aerodynamic, propulsive, and gravitational forces acting on the airplane. The moment equations can be expressed in a similar manner:

$$L = \frac{d}{dt}H_x \qquad M = \frac{d}{dt}H_y \qquad N = \frac{d}{dt}H_z \tag{3.4}$$

where L, M, N and H_x, H_y, H_z are the components of the moment and moment of momentum along the x, y and z axes, respectively.

Consider the airplane shown in Fig. 3.2. If we let δm be an element of mass of the airplane, $\mathbf{v}$ be the velocity of the elemental mass relative to an absolute or inertial frame, $\delta\mathbf{F}$ be the resulting force acting on the elemental mass; then Newton's second law yields

$$\delta\mathbf{F} = \delta m \frac{d\mathbf{v}}{dt} \tag{3.5}$$

and the total external force acting on the airplane is found by summing all the elements of the airplane:

$$\sum \delta\mathbf{F} = \mathbf{F} \tag{3.6}$$

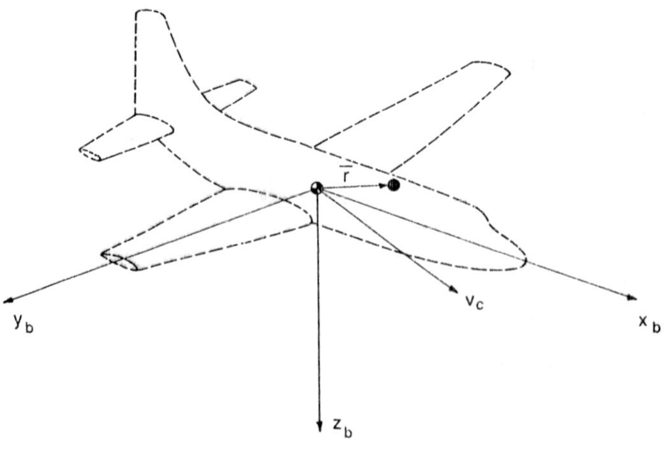

FIGURE 3.2
An element of mass on an airplane.

The velocity of the differential mass dm is

$$\mathbf{v} = \mathbf{v}_c + \frac{d\mathbf{r}}{dt} \tag{3.7}$$

where $\mathbf{v}_c$ is the velocity of the center of mass of the airplane and d$\mathbf{r}$/dt is the velocity of the element relative to the center of mass. Substituting this expression for the velocity into Newton's second law yields

$$\sum \delta\mathbf{F} = \mathbf{F} = \frac{d}{dt}\sum\left(\mathbf{v}_c + \frac{d\mathbf{r}}{dt}\right)\delta m \tag{3.8}$$

If we assume that the mass of the vehicle is constant, Eq. (3.8), can be rewritten as

$$\mathbf{F} = m\frac{d\mathbf{v}_c}{dt} + \frac{d}{dt}\sum \frac{d\mathbf{r}}{dt}\,\delta m \tag{3.9}$$

or

$$\mathbf{F} = m\frac{d\mathbf{v}_c}{dt} + \frac{d^2}{dt^2}\sum \mathbf{r}\,\delta m \tag{3.10}$$

Because $\mathbf{r}$ is measured from the center of mass, the summation $\sum \mathbf{r}\,\delta m$ is equal to zero. The force equation then becomes

$$\mathbf{F} = m\frac{d\mathbf{v}_c}{dt} \tag{3.11}$$

which relates the external force on the airplane to the motion of the vehicle's center of mass.

In a similar manner, we can develop the moment equation referred to a moving center of mass. For the differential element of mass, δm, the moment equation can be written as

$$\delta\mathbf{M} = \frac{d}{dt}\,\delta\mathbf{H} = \frac{d}{dt}(\mathbf{r}\times\mathbf{v})\,\delta m \tag{3.12}$$

The velocity of the mass element can be expressed in terms of the velocity of the center of mass and the relative velocity of the mass element to the center of mass:

$$\mathbf{v} = \mathbf{v}_c + \frac{d\mathbf{r}}{dt} = \mathbf{v}_c + \boldsymbol{\omega}\times\mathbf{r} \tag{3.13}$$

where $\boldsymbol{\omega}$ is the angular velocity of the vehicle and $\mathbf{r}$ is the position of the mass element measured from the center of mass. The total moment of momentum can be written as

$$\mathbf{H} = \sum \delta\mathbf{H} = \sum(\mathbf{r}\times\mathbf{v}_c)\,\delta m + \sum[\mathbf{r}\times(\boldsymbol{\omega}\times\mathbf{r})]\,\delta m \tag{3.14}$$

The velocity $\mathbf{v}_c$ is a constant with respect to the summation and can be taken outside the summation sign

$$\mathbf{H} = \sum \mathbf{r}\,\delta m \times \mathbf{v}_c + \sum [\mathbf{r} \times (\boldsymbol{\omega} \times \mathbf{r})]\,\delta m \tag{3.15}$$

The first term in Eq. (3.15) is zero because the term $\sum \mathbf{r}\,\delta m = 0$, as explained previously. If we express the angular velocity and position vector as

$$\boldsymbol{\omega} = p\hat{\mathbf{i}} + q\hat{\mathbf{j}} + r\hat{\mathbf{k}} \tag{3.16}$$

and

$$\mathbf{r} = x\hat{\mathbf{i}} + y\hat{\mathbf{j}} + z\hat{\mathbf{k}} \tag{3.17}$$

then after expanding Eq. (3.15), **H** can be written as

$$\mathbf{H} = (p\hat{\mathbf{i}} + q\hat{\mathbf{i}} + r\hat{k}) \sum (x^2 + y^2 + z^2)\,\delta m - \sum (x\hat{\mathbf{i}} + y\hat{\mathbf{i}} + z\hat{\mathbf{k}})(px + qy + rz)\,\delta m \tag{3.18}$$

The scalar components of **H** are

$$\begin{aligned}
H_x &= p \sum (y^2 + z^2)\,\delta m - q \sum xy\,\delta m - r \sum xz\,\delta m \\
H_y &= -p \sum xy\,\delta m + q \sum (x^2 + x^2)\,\delta m - r \sum yz\,\delta m \\
H_z &= -p \sum xz\,\delta m - q \sum yz\,\delta m + r \sum (x^2 + y^2)\,\delta m
\end{aligned} \tag{3.19}$$

The summations in the above equations are the mass moment and products of inertia of the airplane and are defined as follows:

$$\begin{aligned}
I_x &= \iiint_v (y^2 + z^2)\,\delta m & I_{xy} &= \iiint_v xy\,\delta m \\
I_y &= \iiint_v (x^2 + z^2)\,\delta m & I_{xz} &= \iiint_v xz\,\delta m \\
I_x &= \iiint_v (x^2 + y^2)\,\delta m & I_{yz} &= \iiint_v yz\,\delta m
\end{aligned} \tag{3.20}$$

The terms I_x, I_y and I_z are the mass moments of inertia of the body about the x, y and z axes, respectively. The terms with the mixed indices are called the products of inertia. Both the moments and products of inertia depend on the shape of the body and the manner in which its mass is distributed. The larger the moments of inertia the greater the resistance the body will have to rotation. The scaler equations for the moment of momentum are given below.

$$\begin{aligned}
H_x &= pI_x - qI_{xy} - rI_{xz} \\
H_y &= -pI_{xy} + qI_y - rI_{yz} \\
H_z &= -pI_{xz} - qI_{yz} + rI_z
\end{aligned} \tag{3.21}$$

If the reference frame is not rotating then, as the airplane rotates, the moments and products of inertia will vary with time. To avoid this difficulty we will fix the axis system to the aircraft (body axis system). Now we must determine the derivatives of the vectors **v** and **H** referred to the rotating body frame of reference.

It can be shown that the derivative of an arbitrary vector **A** referred to a rotating body frame having an angular velocity **ω** can be represented by the following vector identity:

$$\left.\frac{d\mathbf{A}}{dt}\right|_I = \left.\frac{d\mathbf{A}}{dt}\right|_B + \boldsymbol{\omega} \times \mathbf{A} \tag{3.22}$$

where the subscript I and B refer to the inertial and body fixed frames of reference. Applying this identity to the equations derived earlier yields

$$\mathbf{F} = m\left.\frac{d\mathbf{v}_c}{dt}\right|_B + m(\boldsymbol{\omega} \times \mathbf{v}_c) \tag{3.23}$$

$$\mathbf{M} = \left.\frac{d\mathbf{H}}{dt}\right|_B + \boldsymbol{\omega} \times \mathbf{H} \tag{3.24}$$

The scalar equations are

$$\begin{aligned} F_x &= m(\dot{u} + qw - rv) \qquad & F_y &= m(\dot{v} + ru - pw) \qquad & F_z &= m(\dot{w} + pv - qu) \\ L &= \dot{H}_x + qH_z - rH_y \qquad & M &= \dot{H}_y + rH_x - pH_z \qquad & N &= \dot{H}_z + pH_y - qH_x \end{aligned} \tag{3.25}$$

The components of the force and moment acting on the airplane are composed of aerodynamic, gravitational, and propulsive contributions.

By proper positioning of the body axis system, one can make the products of inertia $I_{yz} = I_{xy} = 0$. To do this we are assuming that the xz plane is a plane of symmetry of the airplane. With this assumption, the moment equations can be written as

$$\begin{aligned} L &= I_x\dot{p} - I_{xz}\dot{r} + qr(I_z - I_y) - I_{xz}pq \\ M &= I_y\dot{q} + rp(I_x - I_z) + I_{xz}(p^2 - r^2) \\ N &= -I_{xz}\dot{p} + I_z\dot{r} + pq(I_y - I_x) + I_{xz}qr \end{aligned} \tag{3.26}$$

3.3 ORIENTATION AND POSITION OF THE AIRPLANE

The equations of motion have been derived for an axis system fixed to the airplane. Unfortunately, the position and orientation of the airplane cannot be described relative to the moving body axis frame. The orientation and position of the airplane can be defined in terms of a fixed frame of reference as shown in Fig. 3.3. At time $t = 0$, the two reference frames coincide.

The orientation of the airplane can be described by three consecutive rotations, whose order is important. The angular rotations are called the Euler angles. The orientation of the body frame with respect to the fixed frame can

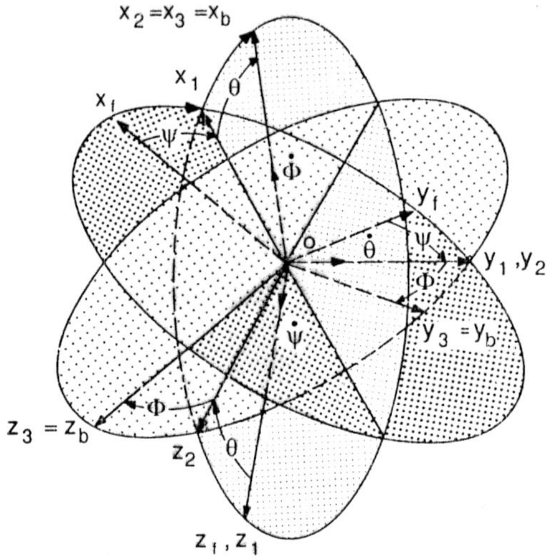

FIGURE 3.3
Relationship between body and inertial axes systems.

be determined in the following manner. Imagine the airplane to be positioned so that the body axis system is parallel to the fixed frame and then apply the following rotations:

1. Rotate the x_f, y_f, z_f frame about $0z_f$ through the yaw angle ψ to the frame to x_1, y_1, z_1.
2. Rotate the x_1, y_1, z_1 frame about $0y_1$ through the pitch angle θ bringing the frame to x_2, y_2, z_2.
3. Rotate the x_2, y_2, z_2 frame about $0x_2$ through the roll angle Φ to bring the frame to x_3, y_3, z_3, the actual orientation of the body frame relative to the fixed frame.

Remember that the order of rotation is extremely important.

Having defined the Euler angles, one can determine the flight velocities components relative to the fixed reference frame. To accomplish this, let the velocity components along the x_f, y_f, z_f frame be dx/dt, dy/dt, dz/dt and, similarly, let the subscripts 1 and 2 denote the components along x_1, y_1, z_1 and x_2, y_2, z_2, respectively. Examining Fig. 3.3, we can show that

$$\frac{dx}{dt} = u_1 \cos\psi - v_1 \sin\psi \qquad \frac{dy}{dt} = u_1 \sin\psi + v_1 \cos\psi \qquad \frac{dz}{dt} = w_1 \tag{3.27}$$

Before proceeding further, let us use the shorthand notation $S_\psi \equiv \sin\psi$, $C_\psi \equiv \cos\psi$, $S_\theta \equiv \sin\theta$, etc. In a similar manner to Eq. (3.27), u_1, v_1, and w_1 can be expressed in terms of u_2, v_2 and w_2:

$$u_1 = u_2 C_\theta + w_2 S_\theta \qquad v_1 = v_2 \qquad w_1 = -u_2 S_\theta + w_2 C_\theta \tag{3.28}$$

and

$$u_2 = u \qquad u_2 = v\mathrm{C}_\Phi - w\mathrm{S}_\Phi \qquad w_2 = v\mathrm{S}_\Phi + w\mathrm{C}_\Phi \tag{3.29}$$

where u, v and w are the velocity components along the body axes x_b, y_b, z_b.

If we back-substitute the above equations, we can determine the absolute velocity in terms of the Euler angles and velocity components in the body frame:

$$\begin{bmatrix} \dfrac{dx}{dt} \\ \dfrac{dy}{dt} \\ \dfrac{dz}{dt} \end{bmatrix} = \begin{bmatrix} \mathrm{C}_\theta \mathrm{C}_\psi & \mathrm{S}_\Phi \mathrm{S}_\theta \mathrm{C}_\psi - \mathrm{C}_\Phi \mathrm{S}_\psi & \mathrm{C}_\Phi \mathrm{S}_\theta \mathrm{C}_\psi + \mathrm{S}_\Phi \mathrm{S}_\psi \\ \mathrm{C}_\theta \mathrm{S}_\psi & \mathrm{S}_\Phi \mathrm{S}_\theta \mathrm{S}_\psi + \mathrm{C}_\Phi \mathrm{C}_\psi & \mathrm{C}_\Phi \mathrm{S}_\theta \mathrm{S}_\psi - \mathrm{S}_\Phi \mathrm{C}_\psi \\ -\mathrm{S}_\theta & \mathrm{S}_\Phi \mathrm{C}_\theta & \mathrm{C}_\Phi \mathrm{C}_\theta \end{bmatrix} \begin{bmatrix} u \\ v \\ w \end{bmatrix} \tag{3.30}$$

Integration of these equations yields the airplane's position relative to the fixed frame of reference.

The relationship between the angular velocities in the body frame (p, q and r) and the Euler rates ($\dot{\psi}$, $\dot{\theta}$, and $\dot{\Phi}$) can also be determined from Fig. 3.3:

$$\begin{bmatrix} p \\ q \\ r \end{bmatrix} = \begin{bmatrix} 1 & 0 & -\mathrm{S}_\theta \\ 0 & \mathrm{C}_\Phi & \mathrm{C}_\theta \mathrm{S}_\Phi \\ 0 & -\mathrm{S}_\Phi & \mathrm{C}_\theta \mathrm{C}_\Phi \end{bmatrix} \begin{bmatrix} \dot{\Phi} \\ \dot{\theta} \\ \dot{\psi} \end{bmatrix} \tag{3.31}$$

Equation (3.31) can be solved for the Euler rates in terms of the body angular velocities and is given by Eq. (3.32)

$$\begin{bmatrix} \dot{\Phi} \\ \dot{\theta} \\ \dot{\psi} \end{bmatrix} = \begin{bmatrix} 1 & \mathrm{S}_\Phi \tan\theta & \mathrm{C}_\Phi \tan\theta \\ 0 & \mathrm{C}_\Phi & -\mathrm{S}_\Phi \\ 0 & \mathrm{S}_\Phi \sec\theta & \mathrm{C}_\Phi \sec\theta \end{bmatrix} \begin{bmatrix} p \\ q \\ r \end{bmatrix} \tag{3.32}$$

By integrating the above equations, one can determine the Euler angles ψ, θ and Φ.

3.4 GRAVITATIONAL AND THRUST FORCES

The gravitational force acting on the airplane acts through the center of gravity of the airplane. Because the body axis system is fixed to the center of gravity, the gravitational force will not produce any moments. It will, however, contribute to the external force acting on the airplane and will have components along the respective body axes. Figure 3.4 shows that the gravitational force components acting along the body axis are a function of the airplane's orientation in space. The gravitational force components along the

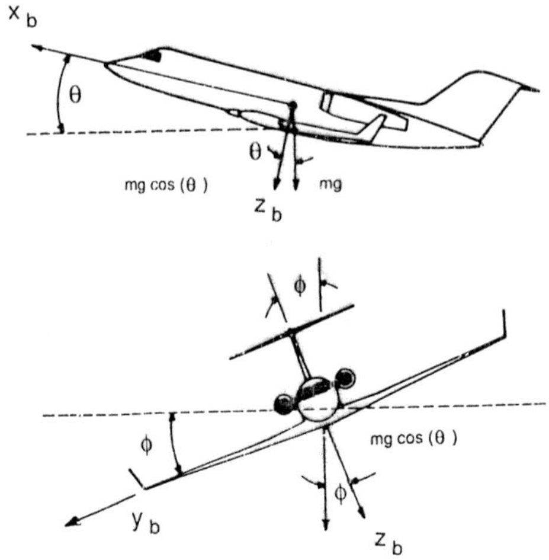

FIGURE 3.4
Components of gravitational force acting along the body axis.

x, y and z axes can be easily shown to be

$$(F_x)_{\text{gravity}} = -mg \sin \theta$$
$$(F_y)_{\text{gravity}} = mg \cos \theta \sin \Phi \qquad (3.33)$$
$$(F_z)_{\text{gravity}} = mg \cos \theta \cos \phi$$

The thrust force due to the propulsion system can have components that act along each of the body axis directions. In addition, the propulsive forces can also create moments if the thrust does not act through the center of gravity. Figure 3.5 shows some examples of moments created by the propulsive system.

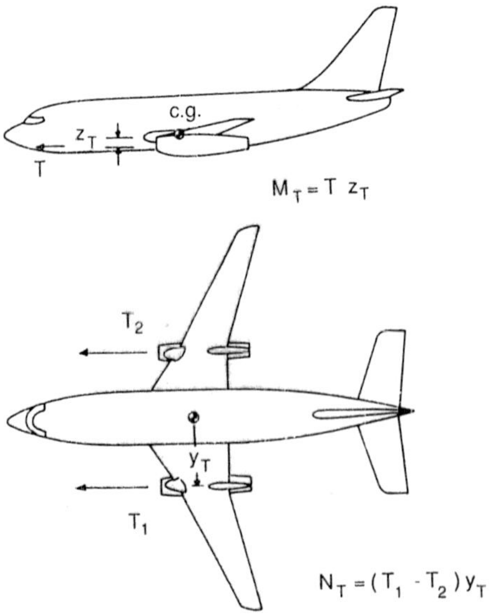

FIGURE 3.5
Force and moments due to propulsion system.

TABLE 3.1
Summary of kinematic and dynamic equations

$X - mg\,S_\theta = m(\dot{u} + qw - rv)$ $Y + mg\,C_\theta S_\Phi = m(\dot{v} + ru - pw)$ $Z + mg\,C_\theta\,C_\Phi = m(\dot{w} + pv - qu)$	Force equations
$L = I_{xz}\dot{p} - I_{xz}\dot{r} + qr(I_z - I_y) - I_{xz}pq$ $M = I_y\dot{q} + rq(I_x - I_z) + I_{xz}(p^2 - r^2)$ $N = -I_{xz}\dot{p} + I_z\dot{r} + pq(I_y - I_x) + I_{xz}qr$	Moment equations
$p = \dot{\Phi} - \dot{\psi}\,S_\theta$ $q = \dot{\theta}C_\Phi + \dot{\psi}\,C_\theta\,S_\Phi$ $r = \dot{\psi}\,C_\theta\,C_\Phi - \dot{\theta}\,S_\Phi$	Body angular velocities in terms of Euler angles and Euler rates
$\dot{\theta} = q\,C_\Phi - r\,S_\Phi$ $\dot{\Phi} = p + q\,S_\Phi\,T_\theta + r\,C_\Phi\,T_\theta$ $\dot{\psi} = (qS_\Phi + rC_\Phi)\sec\theta$	Euler rates in terms of Euler angles and body angular velocities

Velocity of aircraft in the fixed frame in terms of Euler angles and body velocity components

$$\begin{bmatrix} \dfrac{dX}{dt} \\ \dfrac{dY}{dt} \\ \dfrac{dZ}{dt} \end{bmatrix} = \begin{bmatrix} C_\theta C_\psi & S_\Phi S_\theta C_\psi - C_\Phi S_\psi & C_\Phi S_\theta C_\psi + S_\Phi S_\psi \\ C_\theta S_\psi & S_\Phi S_\theta S_\psi + C_\Phi C_\psi & C_\Phi S_\theta S_\psi - S_\Phi S_\psi \\ -S_\theta & S_\Phi C_\theta & C_\Phi C_\theta \end{bmatrix} \begin{bmatrix} u \\ v \\ w \end{bmatrix}$$

The propulsive forces and moments acting along the body axis system are denoted as follows:

$$(F_x)_{\text{propulsive}} = X_T \qquad (F_y)_{\text{propulsive}} = Y_T \qquad (F_z)_{\text{propulsive}} = Z_T \tag{3.34}$$

and

$$(L)_{\text{propulsive}} = L_T \qquad (M)_{\text{propulsive}} = M_T \qquad (N)_{\text{propulsive}} = N_T \tag{3.35}$$

Table 3.1 gives a summary of the rigid body equations of motion.

3.5 SMALL-DISTURBANCE THEORY

The equations developed in the previous section can be linearized by using small-disturbance theory. In applying small-disturbance theory we are assuming that the motion of the airplane consists of small deviations about a steady flight condition. Obviously, this theory cannot be applied to problems in which large-amplitude motions are to be expected (e.g. spinning or stalled flight). However, in many cases small-disturbance theory yields sufficient accuracy for practical engineering purposes.

All the variables in the equations of motion are replaced by a reference value plus a perturbation or disturbance:

$$
\begin{aligned}
u &= u_0 + \Delta u & v &= v_0 + \Delta v & w &= w_0 + \Delta w \\
p &= p_0 + \Delta p & q &= q_0 + \Delta q & r &= r_0 + \Delta r \\
X &= X_0 + \Delta X & Y &= Y_0 + \Delta Y & Z &= Z_0 + \Delta Z \\
M &= M_0 + \Delta M & N &= N_0 + \Delta N & L &= L_0 + \Delta L \\
\delta &= \delta_0 + \Delta\delta
\end{aligned}
\tag{3.36}
$$

For convenience, the reference flight condition is assumed to be symmetric and the propulsive forces are assumed to remain constant. This implies that

$$v_0 = p_0 = q_0 = r_0 = \phi_0 = \psi_0 = 0 \tag{3.37}$$

Furthermore, if we initially align the x axis so that it is along the direction of the airplane's velocity vector, then $w_0 = 0$.

Now, if we introduce the small-disturbance notation into the equations of motion, we can simplify the equations of motion. As an example, consider the X force equation:

$$X - mg \sin\theta = m(\dot{u} + qw - rv) \tag{3.38}$$

Substituting the small-disturbance variables into the above equations yields

$$
\begin{aligned}
&X_0 + \Delta X - mg \sin(\theta_0 + \Delta\theta) \\
&\quad = m\left[\frac{\mathrm{d}}{\mathrm{d}t}(u_0 + \Delta u) + (q_0 + \Delta q)(w_0 + \Delta w) - (r_0 + \Delta r)(v_0 + \Delta v)\right]
\end{aligned}
\tag{3.39}
$$

If we neglect products of the disturbance and assume that

$$w_0 = v_0 = p_0 = q_0 = r_0 = \phi_0 = \psi_0 = 0 \tag{3.40}$$

then the X equation becomes

$$X_0 + \Delta X - mg \sin(\theta_0 + \Delta\theta) = m\,\Delta\dot{u} \tag{3.41}$$

This equation can be reduced further by applying the following trigonometric identity:

$$\sin(\theta_0 + \Delta\theta) = \sin\theta_0 \cos\Delta\theta + \cos\theta_0 \sin\Delta\theta = \sin\theta_0 + \Delta\theta \cos\theta_0$$

Therefore,

$$X_0 + \Delta X - mg(\sin\theta_0 + \Delta\theta\cos\theta_0) = m\,\Delta\dot{u} \tag{3.42}$$

If all the disturbance quantities are set equal to zero in the above equation, we have the reference flight condition

$$X_0 - mg \sin\theta_0 = 0 \tag{3.43}$$

This reduces the X-force equation to

$$\Delta X - mg\,\Delta\theta\cos\theta_0 = m\,\Delta\dot{u} \tag{3.44}$$

The force ΔX is the change in aerodynamic and propulsive force in the x direction and can be expressed by means of a Taylor series in terms of the perturbation variables. If we assume that ΔX is a function only of u, w, δ_e and δ_T, then ΔX can be expressed as

$$\Delta X = \frac{\partial X}{\partial u}\Delta u + \frac{\partial X}{\partial w}\Delta w + \frac{\partial X}{\partial \delta_e}\Delta\delta_e + \frac{\partial X}{\partial \delta_T}\delta_T \tag{3.45}$$

where $\partial X/\partial u$, $\partial X/\partial w$, $\partial X/\partial \delta_e$ and $\partial X/\partial \delta_T$ are called stability derivatives and are evaluated at the reference flight condition. The variables δ_e and δ_T are the change in elevator angle and throttle setting, respectively. If a canard or all-moveable stabilator is used for longitudinal control, then the control term would be replaced by

$$\frac{\partial X}{\partial \delta_H}\Delta\delta_H \qquad \text{or} \qquad \frac{\partial X}{\partial \delta_c}\Delta\delta_c$$

Substituting this expression into the force equation yields:

$$\frac{\partial X}{\partial u}\Delta u + \frac{\partial X}{\partial w}\Delta w + \frac{\partial X}{\partial \delta_e}\Delta\delta_e + \frac{\partial X}{\partial \delta_T}\Delta\delta_T - mg\,\Delta\theta\cos\theta_0 = m\,\Delta\dot{u} \tag{3.46}$$

or, upon rearranging,

$$\left(m\frac{\mathrm{d}}{\mathrm{d}t} - \frac{\partial X}{\partial u}\right)\Delta u - \left(\frac{\partial X}{\partial w}\right)\Delta w + (mg\cos\theta_0)\,\Delta\theta = \frac{\partial X}{\partial \delta_e}\Delta\delta_e + \frac{\partial X}{\partial \delta_T}\Delta\delta_T$$

The equation can be rewritten in a more convenient form by dividing through by the mass m:

$$\left(\frac{\mathrm{d}}{\mathrm{d}t} - X_u\right)\Delta u - X_w\,\Delta w + (g\cos\theta_0)\,\Delta\theta = X_{\delta e}\,\Delta\delta_e + \frac{\partial X}{\partial \delta_T}\Delta\delta_T \tag{3.47}$$

where $X_u = \partial X/\partial u/m$, $X_w = \partial X/\partial w/m$, etc., are aerodynamic derivatives divided by the airplane mass.

The change in aerodynamic forces and moments are functions of the motion variables Δu, Δw, etc. The aerodynamic derivatives that are usually the most important for conventional airplane motion analysis are given below.

$$\left.\begin{aligned}
\Delta X &= \frac{\partial X}{\partial u}\Delta u + \frac{\partial X}{\partial w}\Delta w + \frac{\partial X}{\partial \delta_e}\Delta\delta_e + \frac{\partial X}{\partial \delta}\Delta\delta_T \\
\Delta Y &= \frac{\partial Y}{\partial v}\Delta v + \frac{\partial Y}{\partial p}\Delta p + \frac{\partial Y}{\partial r}\Delta r + \frac{\partial Y}{\partial \delta_r}\Delta\delta_r \\
\Delta Z &= \frac{\partial Z}{\partial u}\Delta u + \frac{\partial Z}{\partial w}\Delta w + \frac{\partial Z}{\partial \dot{w}}\Delta\dot{w} + \frac{\partial Z}{\partial q}\Delta q + \frac{\partial Z}{\partial \delta_e}\Delta\delta_e + \frac{\partial Z}{\partial \delta_T}\Delta\delta_T
\end{aligned}\right\} \tag{3.48}$$

TABLE 3.2
The linearized small-disturbance longitudinal and lateral rigid body equations of motion

Longitudinal equations

$$\left(\frac{d}{dt} - X_u\right)\Delta u - X_w\,\Delta w + (g\cos\theta_0)\,\Delta\theta = X_{\delta e}\,\Delta\delta_e + X_{\delta_T}\,\Delta\delta_T$$

$$-Z_u\,\Delta u + \left((1 - Z_{\dot{w}})\frac{d}{dt} - Z_w\right)\Delta w - \left((u_0 + Z_q)\frac{d}{dt} - g\sin\theta_0\right)\Delta\theta = Z_{\delta e}\,\Delta\delta_e\; Z_{\delta_T}\,\Delta\delta_T$$

$$-M_u\,\Delta u - \left(M_{\dot{w}}\frac{d}{dt} + M_w\right)\Delta w + \left(\frac{d^2}{dt^2} - M_q\frac{d}{dt}\right)\Delta\theta = M_{\delta e}\,\Delta\delta_e + M_{\delta_T}\,\Delta\delta_T$$

Lateral equations

$$\left(\frac{d}{dt} - Y_v\right)\Delta v + (u_0 - Y_r)\,\Delta r - (g\cos\theta_0)\,\Delta\phi = Y_{\delta r}\,\Delta\delta_r$$

$$-L_v\,\Delta v + \left(\frac{d}{dt} - L_p\right)\Delta p - \left(\frac{I_{xz}}{I_x}\frac{d}{dt} + L_r\right)\Delta r = L_{\delta a}\,\Delta\delta_a + L_{\delta r}\,\Delta\delta_r$$

$$-N_v\,\Delta v - \left(\frac{I_{xz}}{I_z}\frac{d}{dt} + N_p\right)\Delta p + \left(\frac{d}{dt} - N_r\right)\Delta r = N_{\delta a}\,\Delta\delta_a + N_{\delta a}\,\Delta\delta_r$$

$$\left.\begin{aligned}
\Delta L &= \frac{\partial L}{\partial v}\Delta v + \frac{\partial L}{\partial p}\Delta p + \frac{\partial L}{\partial r}\Delta r + \frac{\partial L}{\partial \delta r}\Delta\delta r + \frac{\partial L}{\partial \delta_a}\partial\delta_a \\
\Delta M &= \frac{\partial M}{\partial u}\Delta u + \frac{\partial M}{\partial w}\Delta w + \frac{\partial M}{\partial \dot{w}}\Delta\dot{w} + \frac{\partial M}{\partial q}\Delta q + \frac{\partial M}{\partial \delta_e}\Delta\delta_e + \frac{\partial M}{\partial \delta_T}\Delta\delta_T \\
\Delta N &= \frac{\partial N}{\partial v}\Delta v + \frac{\partial N}{\partial p}\Delta p + \frac{\partial N}{\partial r}\Delta r + \frac{\partial N}{\partial \delta r}\Delta\delta r + \frac{\partial N}{\partial \delta_a}\Delta\delta_a
\end{aligned}\right\} \tag{3.49}$$

The aerodynamic forces and moments can be expressed as a function of all the motion variables however, in the above equations only the terms that are usually significant have been retained. Note also that the longitudinal aerodynamic control surface was assumed to be an elevator. For aircraft that use either a canard or combination of longitudinal controls the elevator terms in the preceding equations can be replaced by the appropriate control derivatives and angular deflections.

The complete set of linearized equations of motion is presented in Table 3.2.

3.6 AIRCRAFT TRANSFER FUNCTIONS

The longitudinal and lateral equations of motion were described by a set of linear differential equations. A very useful concept in analysis and design of control systems is the transfer function. The transfer function gives the relationship between the output of and input to a system. In the case of dynamics it specifies the relationship between the motion variables and the control input. The transfer function is defined as the ratio of Laplace transform of the output to the Laplace transform of the input, with all the initial conditions set to zero (i.e. the system is assumed to be initially at rest). For the reader who is not familiar with theory of Laplace transformations, a brief review of the basic concepts of Laplace transformation theory is included in one of the appendices at the end of this book.

LONGITUDINAL TRANSFER FUNCTION. The longitudinal transfer functions can be obtained from the equations of motion in the following manner. First, we simplify the equations of motion by assuming that

$$\theta_0 = 0 \rightarrow \cos\theta_0 = 1 \qquad \text{and} \qquad \sin\theta_0 = 0 \tag{3.50}$$

and

$$Z_q = Z_{\dot{w}} = 0 \tag{3.51}$$

Incorporating these assumptions yields the following set of longitudinal differential equations:

$$\begin{aligned}
\left(\frac{\mathrm{d}}{\mathrm{d}t} - X_u\right)\Delta u - X_w\,\Delta w + g\,\Delta\theta &= X_\delta\,\Delta\delta \\
-Z_u\,\Delta u + \left(\frac{\mathrm{d}}{\mathrm{d}t} - Z_w\right)\Delta w - u_0\frac{\mathrm{d}}{\mathrm{d}t}\Delta\theta &= Z_\delta\,\Delta\delta \\
-M_u\,\Delta u - \left(M_{\dot{w}}\frac{\mathrm{d}}{\mathrm{d}t} + M_w\right)\Delta w + \frac{\mathrm{d}}{\mathrm{d}t}\left(\frac{\mathrm{d}}{\mathrm{d}t} - M_q\right)\Delta\theta &= M_\delta\,\Delta\delta
\end{aligned} \tag{3.52}$$

If we take the Laplace transformation of the equations and then divide by the control deflection we can find the transfer functions $\overline{\Delta u}/\overline{\Delta\delta_e}$, $\overline{\Delta\theta}/\overline{\Delta\delta}$, and $\overline{\Delta w}/\overline{\Delta\delta_0}$.

$$\begin{aligned}
(s - X_u)\frac{\overline{\Delta u}}{\overline{\Delta\delta}} - X_w\frac{\overline{\Delta w}}{\overline{\Delta\delta}} + g\frac{\overline{\Delta\theta}}{\overline{\Delta\delta}} &= X_\delta \\
-Z_u\frac{\overline{\Delta u}}{\overline{\Delta\delta}} + (s - Z_w)\frac{\overline{\Delta w}}{\overline{\Delta\delta}} - u_0 s\frac{\overline{\Delta\theta}}{\overline{\Delta\delta}} &= Z_\delta \\
-M_u\frac{\overline{\Delta u}}{\overline{\Delta\delta}} = (M_{\dot{w}}s + M_w)\frac{\overline{\Delta w}}{\overline{\Delta\delta}} + s(s - M_q)\frac{\overline{\Delta\theta}}{\overline{\Delta\delta}} &= M_\delta
\end{aligned} \tag{3.53}$$

These equations can be solved by means of Cramer's rule to find $\overline{\Delta u}/\overline{\Delta\delta}$, $\overline{\Delta\theta}/\overline{\Delta\delta}$ and $\overline{\Delta w}/\overline{\Delta\delta}$, as follows:

$$\frac{\overline{\Delta\theta}}{\overline{\Delta\delta}} = \frac{\begin{vmatrix} s - X_u & -X_w & X_\delta \\ -Z & s - Z_w & Z_\delta \\ -M_u & -(M_{\dot{w}}s + M_w) & M_\delta \end{vmatrix}}{\begin{vmatrix} s - X_u & -X_w & g \\ -Z_u & s - Z_w & -u_0 s \\ -M_u & -(M_{\dot{w}}s + M_w) & s(s - M_q) \end{vmatrix}} \tag{3.54}$$

Expressions for $\overline{\Delta u}/\overline{\Delta \delta}$ and $\overline{\Delta w}/\overline{\Delta \delta}$ can be found in a similar fashion. By expanding the determinants, the transfer functions can be expressed as a ratio of two polynomials. For example,

$$\frac{\overline{\Delta u}}{\overline{\Delta \delta}} = \frac{N_\delta^u(s)}{\Delta_{\text{long}}} = \frac{A_u s^3 + B_u s^2 + C_u s + D_u}{\Delta_{\text{long}}} \tag{3.55}$$

$$\frac{\overline{\Delta w}}{\overline{\Delta \delta}} = \frac{N_\delta^w(s)}{\Delta_{\text{long}}} = \frac{A_w s^3 + B_w s^2 + C_w s + D_w}{\Delta_{\text{long}}} \tag{3.56}$$

and

$$\frac{\overline{\Delta \theta}}{\overline{\Delta \delta}} = \frac{N_\delta^\theta(s)}{\Delta_{\text{long}}} = \frac{A_\theta s^2 + B_\theta s + C_\theta}{\Delta_{\text{long}}} \tag{3.57}$$

where

$$\Delta_{\text{long}} = As^4 + Bs^3 + Cs^2 + Ds + E$$

The longitudinal functions are presented in Table 3.3.

LATERAL TRANSFER FUNCTIONS. The lateral transfer functions can be derived in a manner identical to that presented for the longitudinal transfer functions. If we take the Laplace transform of the lateral equations given in Table 3.2 and divide through by the control input, we obtain the following equations:

$$\begin{aligned}
(s - Y_v)\frac{\overline{\Delta v}}{\overline{\Delta \delta}} + g\cos\theta_0 \frac{\overline{\Delta \phi}}{\overline{\Delta \delta}} + (u_0 - Y_r)s\frac{\overline{\Delta \psi}}{\overline{\Delta \delta}} &= Y_\delta \\
-L_v \frac{\overline{\Delta v}}{\overline{\Delta \delta}} + (s^2 - L_p s)\frac{\overline{\Delta \phi}}{\overline{\Delta \delta}} - \left(\frac{I_{xz}}{I_x}s^2 - L_r s\right)\frac{\overline{\Delta \psi}}{\overline{\Delta \delta}} &= L_\delta \\
-N_v \frac{\overline{\Delta v}}{\overline{\Delta \delta}} - \left(\frac{I_{xz}}{I_z}s^2 + N_p s\right)\frac{\overline{\Delta \phi}}{\overline{\Delta \delta}} + (s^2 - N_r s)\frac{\overline{\Delta \psi}}{\overline{\Delta \delta}} &= N_\delta
\end{aligned} \tag{3.58}$$

where Y_δ, L_δ, etc., and $\overline{\Delta \delta}$ represent either the aileron or rudder derivatives and the corresponding aileron or rudder control derivatives. The lateral transfer functions $\overline{\Delta v}/\overline{\Delta \delta}$, $\overline{\Delta \phi}/\overline{\Delta \delta}$ and $\overline{\Delta \psi}/\overline{\Delta \delta}$ can be determined by solving the above equations. The transfer functions will have the following form:

$$\frac{\overline{\Delta v}}{\overline{\Delta \delta}} = \frac{A_v s^3 + B_v s^2 + C_v s + D_v}{(As^4 + Bs^3 + Cs^2 + Ds + E)} \tag{3.59}$$

where the coefficients A_v, B_v, . . . , A, B, . . . and E are functions of the lateral stability derivatives. The transfer function $\overline{\Delta v}/\overline{\Delta \delta_a}$ is obtained from Table 3.4 by replacing δ with δ_a and the control derivatives by Y_{δ_a}, L_{δ_a} etc. In a similar manner $\overline{\Delta v}/\overline{\Delta \delta_r}$ can be obtained.

TABLE 3.3
Longitudinal control transfer functions

	A	B	C	D	E
Δ_{long}	1	$-M_q - u_0M_w - Z_w - X_u$	$-Z_wM_q - u_0M_w - X_wZ_u$ $+ X_u(M_q + u_0M_w + Z_w)$	$-X_u(Z_wM_q - u_0M_w)$ $+ Z_u(X_wM_q + gM_w)$ $- M_u(u_0X_w - g)$	$g(Z_uM_w - M_uZ_w)$
N_δ^θ	$M_\delta + Z_\delta M_w$	$X_\delta(Z_uM_w + M_u)$ $+ Z_\delta(M_w - X_uM_w)$ $- M_\delta(X_u + Z_w)$	$X_\delta(Z_uM_w - Z_wM_u)$ $+ Z_\delta(M_uX_w - M_wX_u)$ $+ M_\delta(Z_wX_u - X_wZ_u)$		
N_δ^w	Z_δ	$X_\delta Z_u - Z_\delta(X_u + M_q) + M_\delta u_0$	$X_\delta(u_0M_u - Z_uM_q)$ $+ Z_\delta X_uM_q - u_0M_\delta X_u$	$g(Z_\delta M_u - M_\delta Z_u)$	
N_δ^u	X_δ	$-X_\delta(Z_w + M_q + u_0M_w) + Z_\delta X_w$	$X_\delta(Z_wM_q - u_0M_w)$ $- Z_\delta(X_wM_q + gM_w)$ $+ M_\delta(u_0X_w - g)$	$g(M_\delta Z_w - Z_\delta M_w)$	

TABLE 3.4
Lateral control transfer functions

	A	B	C	D	E
Δ	$1-\frac{I_{xz}^2}{I_x I_z}$	$-Y_v\left(1-\frac{I_{xz}^2}{I_x I_z}\right)-L_p-N_r$ $-\frac{I_{xz}}{I_x}N_p-\frac{I_{xz}}{I_z}L_r$	$u_{\theta}N_v+L_p(Y+N_r)+N_p\left(\frac{I_{xz}}{I_x}Y_v-L_r\right)$ $+Y_v\left(\frac{I_{xz}}{I_z}L_r+N_r\right)+u_0\frac{I_{xz}}{I_z}L_v$	$-u_0N_vL_p+Y_v(N_pL_r-L_pN_r)$ $+u_0N_pL_v-g\left(L_v+\frac{I_{xz}}{I_x}N_v\right)$	$g(L_vN_r-L_vL_r)$
N_δ^v	$Y_\delta\left(1-\frac{I_{xz}^2}{I_x I_z}\right)$	$-Y_\delta\left[L_p+N_r+\frac{I_{xz}}{I_x}N_p+\frac{I_{xz}}{I_z}L_r\right]$ $\times u_0\left(\frac{I_{xz}}{I_z}L_\delta+N_\delta\right)$	$Y_\delta(L_pN_r-N_pL_r)+u_0(N_\delta L_p-L_\delta N_p)$ $+g\left(L_\delta+\frac{I_{xz}}{I_x}N_\beta\right)$	$g(N_\delta L_r-L_\delta N_r)$	
N_δ^ϕ	$L_\delta+\frac{I_{xz}}{I_x}N_\delta$	$Y_\delta\left(L_v-\frac{I_{xz}^2}{I_x I_z}N_v\right)-L_\delta(N_r+Y_v)$ $+N_\delta\left(L_r-\frac{I_{xz}}{I_x}Y_v\right)$	$Y_\delta(L_rN_v-L_vN_r)$ $+L_\delta(Y_vN_r+u_0N_v)$ $-N_\delta(u_0L_v+Y_vL_r)$		
N_δ^r	$N_\delta+\frac{I_{xz}}{I_z}L_\delta$	$Y_\delta\left(N_v+\frac{I_{xz}}{I_z}L_v\right)$ $+L_\delta\left(N_p-\frac{I_{xz}}{I_z}Y_v\right)$ $-N_\delta(Y_v+L_p)$	$Y_\delta(L_vN_p-N_vL_p)$ $-L_\delta Y_vN_p+N_\delta Y_vL_p$	$g(L_\delta N_v-N_\delta L_v)$	

3.7 AERODYNAMIC FORCE AND MOMENT REPRESENTATION

In the previous sections we represented the aerodynamic force and moment contributions by means of the aerodynamic stability coefficients. We did this without explaining the rationale behind the approach.

The method of representing the aerodynamic forces and moments by stability coefficients was first introduced by Bryan over a half century ago [3.1, 3.3]. The technique proposed by Bryan assumes that the aerodynamic forces and moments can be expressed as a function of the instantaneous values of the perturbation variables. The perturbation variables are the instantaneous changes from the reference conditions of the translational velocities, angular velocities, control deflection, and their derivatives. With this assumption, we can express the aerodynamic forces and moments by means of a Taylor series expansion of the perturbation variables about the reference equilibrium condition. For example, the change in the force in the x direction can be expressed as follows:

$$\Delta X(u, \dot{u}, w, \dot{w}, \ldots, \delta_e, \dot{\delta}_e) = \frac{\partial X}{\partial u}\Delta u + \frac{\partial X}{\partial \dot{u}}\Delta \dot{u} + \cdots \frac{\partial X}{\partial \delta_e}\Delta \delta_e + \text{H.O.T.} \tag{3.60}$$

The term $\partial X/\partial u$ is called the stability derivative and is evaluated at the reference flight condition.

The contribution of the change in the velocity u to the change ΔX in the X force is just $(\partial X/\partial u)\,\Delta u$. We can also express $\partial X/\partial u$ in terms of the stability coefficient C_{x_u} as follows:

$$\frac{\partial X}{\partial u} = C_{x_u}\frac{1}{u_0}QS \tag{3.61}$$

where

$$C_{x_u} = \frac{\partial C_x}{\partial (u/u_0)} \tag{3.62}$$

Note that the stability derivative has dimensions, whereas the stability coefficient is defined so that it is nondimensional.

The preceding discussion may seem as though we are making the aerodynamic force and moment representation extremely complicated. However, by making use of the assumption that the perturbations are small, we only need to retain the linear terms in Eq. (3.60). Even though we have retained only the linear terms, the expressions may still include numerous first order terms. Fortunately, many of these terms can also be neglected because their contribution to a particular force or moment is negligible. For example, we have examined the pitching moment in detail in Chapter 2. If we express the pitching moment in terms of the perturbation variables, as indicated below,

$$M(u, v, w, \dot{u}, \dot{v}, \dot{w}, p, q, r, \delta_a, \delta_e, \delta_r) = \frac{\partial M}{\partial u}\Delta u + \frac{\partial M}{\partial v}\Delta v + \frac{\partial M}{\partial w}\Delta w + \cdots + \frac{\partial M}{\partial p}\Delta p + \cdots \tag{3.63}$$

it should be quite obvious that terms such as $(\partial M/\partial v)\,\Delta v$ and $(\partial M/\partial p)\,\Delta p$ are not going to be significant for an airplane. Therefore, we can neglect these terms in our analysis.

In the following sections, we shall use the stability derivative approach to represent the aerodynamic forces and moments acting on the airplane. The expressions developed for each of the forces and moments will only include the terms that are usually important in studying the airplane's motion. The remaining portion of this chapter is devoted to presentation of methods for predicting the longitudinal and lateral stability coefficients. We will confine our discussion to methods which are applicable to subsonic flight speeds. Note that many of the stability coefficients vary significantly with Mach number. This can be seen by examining the data on the A-4D airplane in the Appendix or by examining Fig. 3.6.

We have developed a number of relationships for estimating the various stability coefficients; for example, expressions for some of the static stability coefficients such as C_{m_α}, C_{n_β} and C_{l_β} were formulated in Chapter 2. Developing prediction methods for all of the stability derivatives necessary for performing vehicle motion analysis would be beyond the scope of this book. Therefore, we shall confine our attention to the development of several important dynamic derivatives and simply refer the reader to Ref. 3.4, the *US Air Force Stability and Control DATCOM*. This report is a comprehensive collection of aerodynamic stability and control prediction techniques which is widely used through the aviation industry.

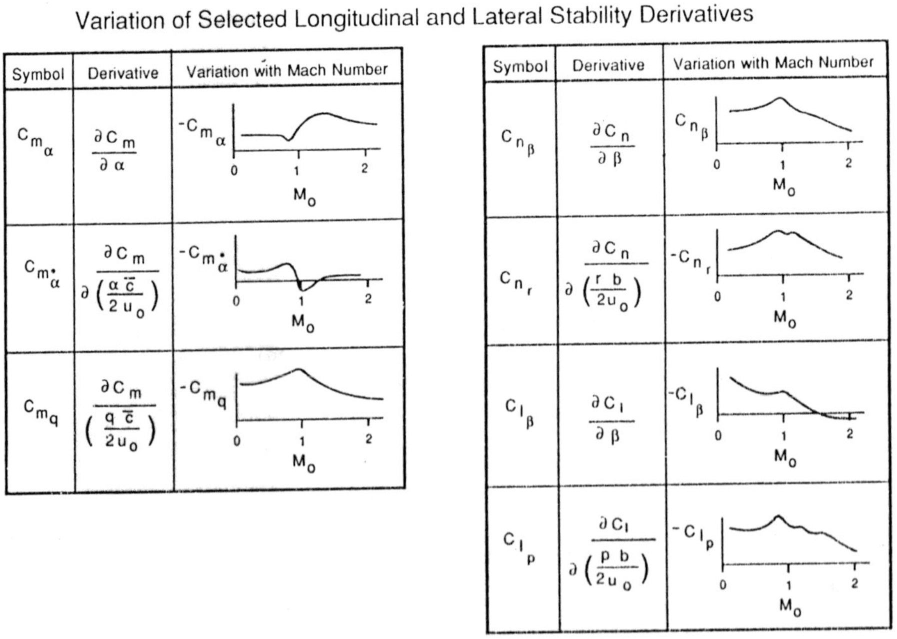

FIGURE 3.6
Variation of selected longitudinal and lateral derivatives with Mach number.

DERIVATIVES DUE TO THE CHANGE IN FORWARD SPEED. The drag, lift and pitching moment vary with changes in the airplane's forward speed. In addition the thrust of the airplane is also a function of the forward speed. The aerodynamic and propulsive forces acting on the airplane along the X body axes are the drag force and the thrust. The change in the X force i.e. ΔX due to a change in forward speed can be expressed as

$$\Delta X = \frac{\partial X}{\partial u}\Delta u = -\frac{\partial D}{\partial u}\Delta u + \frac{\partial T}{\partial u}\Delta u \tag{3.64}$$

or

$$\frac{\partial X}{\partial u} = -\frac{\partial D}{\partial u} + \frac{\partial T}{\partial u} \tag{3.65}$$

The derivative $\partial X/\partial u$ is called the speed damping derivative. Equation (3.65) can be rewritten as

$$\frac{\partial X}{\partial u} = -\frac{\rho S}{2}\left(u_0^2\frac{\partial C_D}{\partial u} + 2u_0 C_{D_0}\right) + \frac{\partial T}{\partial u} \tag{3.66}$$

where the subscript 0 indicates the reference condition. Expressing $\partial X/\partial u$ in coefficient form yields.

$$C_{X_u} = -[C_{D_u} + 2C_{D_0}] + C_{T_u} \tag{3.67}$$

where

$$C_{D_u} = \frac{\partial C_D}{\partial(u/u_0)} \quad \text{and} \quad C_{T_u} = \frac{\partial C_T}{\partial(u/u_0)} \tag{3.68}$$

are the changes in drag and thrust coefficient with forward speed. These coefficients have been made nondimensional by differentiating with respect to (u/u_0). The coefficient C_{D_u} can be estimated from a plot of the drag coefficient versus Mach number.

$$C_{D_u} = \mathbf{M}\frac{\partial C_D}{\partial \mathbf{M}} \tag{3.69}$$

where $\mathbf{M}$ is the Mach number of interest. The thrust term C_{T_u} is zero for gliding flight and is also a good approximation for jet powered aircraft. For a variable pitch propeller and piston engine power plant, C_{T_u} can be approximated by assuming it to be equal to the negative of the reference drag coefficient (i.e. $C_{T_u} = -C_D$).

The change in the Z force with respect to forward speed can be shown to be

$$\frac{\partial Z}{\partial u} = -\rho S u_0[C_{L_u} + C_{L_0}] \tag{3.70}$$

or in coefficient form

$$C_{Z_u} = -[C_{L_u} + 2C_{L_0}] \tag{3.71}$$

The coefficient C_{L_u} arises from the change in lift coefficient with Mach number. C_{L_u} can be estimated from the Prandtl–Glauent formula which corrects the incompressible lift coefficient for Mach number effects.

$$C_L = \frac{C_L|_{M=0}}{\sqrt{1-\mathbf{M}^2}} \tag{3.72}$$

Differentiating the lift coefficient with respect to Mach number yields

$$\frac{\partial C_L}{\partial \mathbf{M}} = \frac{\mathbf{M}}{1-\mathbf{M}^2} C_L \tag{3.73}$$

but

$$C_{L_u} = \frac{\partial C_L}{\partial (u/u_0)} = \frac{u_0}{a} \frac{\partial C_L}{\partial\left(\dfrac{u}{a}\right)}. \tag{3.74}$$

$$= \mathbf{M} \frac{\partial C_L}{\partial \mathbf{M}} \tag{3.75}$$

where a is the speed of sound.

C_{L_u} can therefore be expressed as

$$C_{L_u} = \frac{\mathbf{M}^2}{\sqrt{1-\mathbf{M}^2}} C_{L_0} \tag{3.76}$$

This coefficient can be neglected at low flight speeds but can become quite large near the critical Mach number for the airplane.

The change in the pitching moment due to variations in the forward speed can be expressed as

$$\Delta M = \frac{\partial M}{\partial u} \Delta u \tag{3.77}$$

or

$$\frac{\partial M}{\partial u} = C_{m_u} \rho S \bar{c} u_0 \tag{3.78}$$

The coefficient C_{m_u} can be estimated as follows.

$$C_{m_u} = \frac{\partial C_m}{\partial \mathbf{M}} \mathbf{M} \tag{3.79}$$

The coefficient C_{m_u} depends upon Mach number but is also affected by the elastic properties of the airframe. At high speeds aeroelastic bending of the airplane can cause large changes in the magnitude of C_{m_u}.

DERIVATIVES DUE TO PITCHING VELOCITY q. The stability coefficients C_{z_q} and C_{m_q} represent the change in the Z force and pitching moment coefficients with respect to the pitching velocity q. The aerodynamic characteristics of both the wing and horizontal tail are affected by the pitching motion

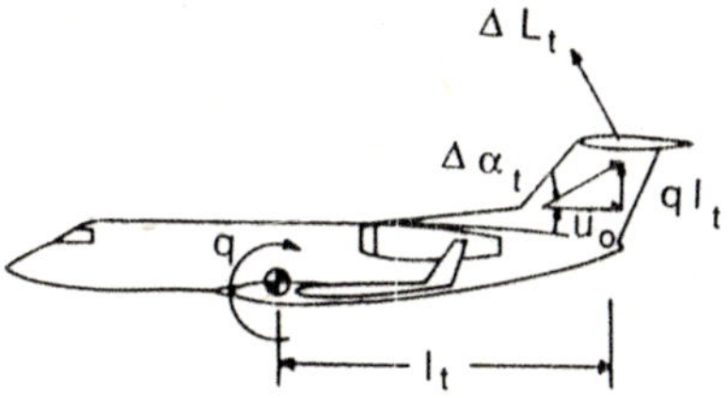

FIGURE 3.7
Mechanism for aerodynamic force due to pitch rate.

of the airplane. The wing contribution is usually quite small in comparison to that produced by the tail. A common practice is to compute the tail contribution and then increase it by 10 percent to account for the wing. Figure 3.7 is a sketch of an airplane undergoing a pitching motion.

As illustrated in Fig. 3.7, the pitching rate q causes a change in the angle of attack at the tail, which results in a change in the lift force acting on the tail:

$$\Delta L_t = C_{L_{\alpha_t}} \Delta\alpha_t Q_t S_t \tag{3.80}$$

or

$$\Delta Z = -\Delta L_t = -C_{L_{\alpha_t}} \frac{ql_t}{u_0} Q_t S_t \tag{3.81}$$

$$C_Z = \frac{Z}{Q_w S} \tag{3.82}$$

$$\Delta C_Z = -C_{L_{\alpha_t}} \frac{ql_t}{u_0} \eta \frac{S_t}{S} \tag{3.83}$$

$$C_{Z_q} \equiv \frac{\partial C_z}{\partial(q\bar{c}/2u_0)} = \frac{2u_0}{\bar{c}} \frac{\partial C_z}{\partial q} \tag{3.84}$$

$$C_{Z_q} = -2C_{L_{\alpha_t}} \eta V_H \tag{3.85}$$

The pitching moment due to the change in lift on the tail can be calculated as follows:

$$\Delta M_{cg} = -l_t \Delta L_t \tag{3.86}$$

$$\Delta C_{m_{cg}} = -V_H \eta C_{L_{\alpha_t}} \frac{ql_t}{u_0} \tag{3.87}$$

$$C_{m_q} \equiv \frac{\partial C_m}{\partial(q\bar{c}/2u_0)} = \frac{2u_0}{\bar{c}} \frac{\partial C_m}{\partial q} \tag{3.88}$$

$$C_{m_q} = -2C_{L_{\alpha_t}} \eta V_H \frac{l_t}{\bar{c}} \tag{3.89}$$

Equations (3.85) and (3.89) represent the tail contribution to C_{z_q} and C_{m_q} respectively. The coefficients for the complete airplane are obtained by

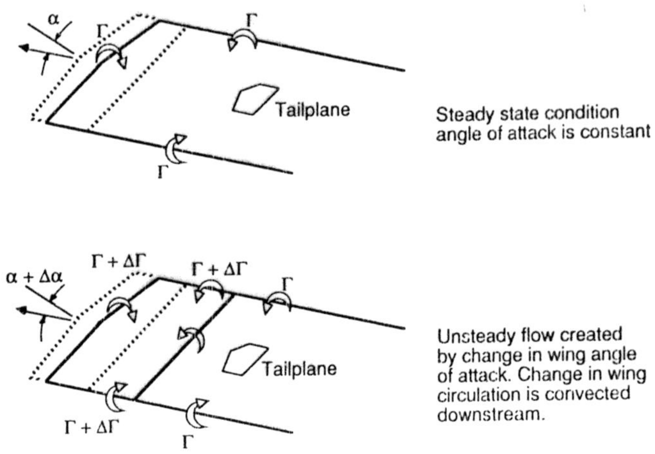

FIGURE 3.8
Mechanism for aerodynamic force due to the lag in flow field development.

increasing the tail values by 10 percent to account for the wind and fuselage contributions.

DERIVATIVES DUE TO THE TIME RATE OF CHANGE OF THE ANGLE OF ATTACK. The stability coefficients $C_{z\dot{\alpha}}$ and $C_{m\dot{\alpha}}$ arise because of the lag in the wind downwash getting to the tail. As the wing angle of attack changes, the circulation around the wing will be altered. The change in circulation alters the downwash at the tail; however, it takes a finite time for the alteration to occur. Figure 3.8 illustrates the lag in flow field development. If the airplane is traveling with a forward velocity u_0, then a change in circulation imparted to the trailing vortex wake will take the increment in time $\Delta t = l_t/u_0$ to reach the tail surface.

The lag in angle of attack at the tail can be expressed as:

$$\Delta \alpha_t = \frac{d\varepsilon}{dt} \Delta t \tag{3.90}$$

where

$$\Delta t = l_t/u_0 \tag{3.91}$$

or

$$\Delta \alpha_t = \frac{d\varepsilon}{dt}\frac{l_t}{u_0} = \frac{d\varepsilon}{d\alpha}\frac{d\alpha}{dt}\frac{l_t}{u_0} \tag{3.92}$$

$$\Delta \alpha_t = \frac{d\varepsilon}{d\alpha} \dot{\alpha} \frac{l_t}{u_0} \tag{3.93}$$

$$\Delta L_t = C_{L_{\alpha_t}} \Delta \alpha_t Q_t S_t \tag{3.94}$$

$$\Delta C_z = -\frac{\Delta L_t}{Q_w S} = -C_{L_{\alpha_t}} \Delta\alpha_t \eta \frac{S_t}{S} \tag{3.95}$$

$$= -C_{L_{\alpha_t}} \frac{d\varepsilon}{d\alpha} \dot{\alpha} \frac{l_t}{u_0} \eta \frac{S_t}{S} \tag{3.96}$$

$$C_{Z_{\dot{\alpha}}} = \frac{\partial C_z}{\partial(\dot{\alpha}\bar{c}/2u_0)} = \frac{2u_0}{\bar{c}} \frac{\partial C_z}{\partial \dot{\alpha}} \tag{3.97}$$

$$C_{Z_{\dot{\alpha}}} = -2V_H \eta C_{L_{\alpha_t}} \frac{d\varepsilon}{d\alpha} \tag{3.98}$$

The pitching moment due to the lag in the downwash field can be calculated as follows:

$$\Delta M_{cg} = -l_t \, \Delta L_t = -l_t C_{L_{\alpha_t}} \Delta\alpha_t \, Q_t S_t \tag{3.99}$$

$$\Delta C_{m_{cg}} = -V_H \eta C_{L_{\alpha_t}} \frac{d\varepsilon}{d\alpha} \dot{\alpha} \frac{l_t}{u_0} \tag{3.100}$$

$$C_{m_{\dot{\alpha}}} = \frac{\partial C_m}{\partial(\dot{\alpha}\bar{c}/2u_0)} = \frac{2u_0}{\bar{c}} \frac{\partial C_m}{\partial \dot{\alpha}} \tag{3.101}$$

$$C_{m_{\dot{\alpha}}} = -2C_{L_{\alpha_t}} \eta V_H \frac{l_t}{\bar{c}} \frac{d\varepsilon}{d\alpha} \tag{3.102}$$

Equations (3.99) and (3.102) yield only the tail contribution to these stability coefficients. To obtain an estimate for the complete airplane these coefficients are increased by 10 percent.

The stability coefficients C_{l_p}, C_{n_r}, C_{z_q}, C_{m_q}, $C_{z_{\dot{\alpha}}}$, and $C_{m_{\dot{\alpha}}}$ all oppose the motion of the vehicle and thus can be considered as damping terms. This will become more apparent as we analyze the motion of an airplane in Chapters 4 and 5.

As noted earlier, there are many more derivatives for which we could develop prediction methods. The few simple examples presented here should give the reader an appreciation of how one would go about determining estimates of the aerodynamic stability coefficients. A summary of some of the theoretical prediction methods for some of the more important longitudinal and lateral stability coefficients is presented in Tables 3.5 and 3.6.

3.8 SUMMARY

The nonlinear differential equations of motion of a rigid airplane were developed from Newton's second law of motion. Linearization of these equations was accomplished using small-disturbance theory. In following chapters we shall solve the linearized equations of motion. These solutions will yield valuable information on the dynamic characteristics of airplane motions.

TABLE 3.5
Equations for estimating the longitudinal stability coefficients

	X-force derivatives	Z-force derivatives	Pitching moment derivatives
u	$C_{X_u} = -[C_{D_u} + 2C_{D0}] + C_{T_u}$	$C_{Z_u} = -\frac{M^2}{1-M^2} C_{L0} - 2C_{L0}$	$C_{m_u} = \frac{\partial C_m}{\partial M} M_0$
α	$C_{X_\alpha} = C_{L0} - \frac{2C_{L0}}{\pi e}$	$C_{Z_\alpha} = -C_{L\alpha} + C_{D0}$	$C_{m_\alpha} = C_{L\alpha w}\left(\frac{X_{cg}}{\bar{c}} - \frac{X_{ac}}{\bar{c}}\right) + C_{m\alpha\,\mathrm{fus}} - \eta V_H C_{L\alpha_t}\left(1 - \frac{d\varepsilon}{d\alpha}\right)$
$\dot{\alpha}$	0	$C_{Z_{\dot{\alpha}}} = -2\eta C_{L\alpha_t} V_H \frac{d\varepsilon}{d\alpha}$	$C_{m_{\dot{\alpha}}} = -2\eta C_{L\alpha_t} V_H \frac{l_t}{2} \frac{d\varepsilon}{d\alpha}$
q	0	$C_{Z_q} = -2\eta C_{L\alpha_t} V_H$	$C_{m_q} = -2\eta C_{L\alpha_t} V_H \frac{l_t}{\bar{c}}$
δ_e	0	$C_{Z_{\delta_e}} = -C_{L\delta_e} = -\frac{S_t}{S} \eta \frac{dC_{Lt}}{d\delta_e}$	$C_{m_{\delta\varepsilon_e}} = -\eta V_H \frac{dC_{Lt}}{d\delta_e}$

* Subscript 0 indicates reference values and M is the Mach number.

AR	Aspect ratio
b	Wing span
C_{D0}	Reference drag coefficient
C_{L0}	Reference lift coefficient
$C_{L\alpha}$	Airplane lift curve slope
$C_{L\alpha_w}$	Wing lift curve slope
$C_{L\alpha_t}$	Tail lift curve slope
$\bar{c}$	Mean aerodynamic chord
e	Oswald's span efficiency factor
l_t	Distance from c.g. to tail quarter chord
l_v	Distance from c.g. to vertical tail aerodynamic center
V_H	Horizontal tail volume ratio
V_v	Vertical tail volume ratio
M	Flight mach number
S	Wing area
S_t	Horizontal tail area
S_v	Vertical tail area
z_v	Distance from center of pressure of vertical tail to fuselage centerline
Γ	Wing dihedral angle
Λ	Wing sweep angle
$\frac{d\varepsilon}{d\alpha}$	Change in downwash due to a change in angle of attack
η	Efficiency factor of the horizontal tail
η_v	Efficiency factor of the vertical tail
λ	Taper ratio (tip chord/root chord)
$\frac{d\sigma}{d\beta}$	Change in sidewash angle with a change in side slip angle

TABLE 3.6
Equations for estimating the lateral stability coefficients

	Y-force derivatives	Yawing moment derivatives	Rolling moment derivatives
β	$C_{y\beta} = \eta \frac{S_v}{S} C_{L\alpha_v}\left(1 + \frac{\partial\sigma}{d\beta}\right)$	$C_{n\beta} = C_{n\beta_{fus}} + \eta_v V_v C_{L\alpha_v}\left(1 + \frac{d\sigma}{d\beta}\right)$	$C_{l\beta} = \left(\frac{C_{l\beta}}{\Gamma}\right)\Gamma + \Delta C_{l\beta}$ (see Fig. 3.9)
p	$C_{y_p} = C_L \frac{AR+\cos\Lambda}{AR+4\cos\Lambda}\tan\Lambda$	$C_{n_p} = -\frac{C_L}{8}$	$C_{l_p} = -\frac{C_{L\alpha}}{12}\frac{1+3\lambda}{1+\lambda}$
r	$C_{y_r} = -2\left(\frac{l_v}{b}\right)(C_{Y\beta})_{tail}$	$C_{n_r} = -2\eta_v V_v\left(\frac{l_v}{b}\right)C_{L\alpha_v}$	$C_{l_r} = \frac{C_L}{4} - 2\frac{l_v}{b}\frac{z_v}{b}C_{y\beta_{tail}}$
δ_a	0	$C_{n\delta_a} = 2KC_{L0}C_{l\delta_a}$ (see Fig. 3.10)	$C_{l\delta_a} = \frac{2C_{L\delta_a}}{Sb}\int_{y_1}^{y_2} cy\, dy$
δ_r	$C_{Y\delta_r} = \frac{S_v}{S}\tau C_{L\alpha_v}$	$C_{n\delta_r} = -V_v\eta_v\tau C_{L\alpha_v}$	$C_{l\delta_r} = \frac{S_v}{S}\left(\frac{Z_v}{b}\right)\tau C_{L\alpha_v}$

AR	Aspect ratio
b	Wing span
C_{D0}	Reference drag coefficient
C_{L0}	Reference lift coefficient
$C_{L\alpha}$	Airplane lift curve slope
$C_{L\alpha_w}$	Wing lift curve slope
$C_{L\alpha_t}$	Tail lift curve slope
$\bar{c}$	Mean aerodynamic chord
e	Oswald's span efficiency factor
l_t	Distance from c.g. to tail quarter chord
l_v	Distance from c.g. to vertical tail aerodynamic center
V_H	Horizontal tail volume ratio
V_v	Vertical tail volume ratio
M	Flight mach number
S	Wing area
S_t	Horizontal tail area
S_v	Vertical tail area
z_v	Distance from center of pressure of vertical tail to fuselage centerline
Γ	Wing dihedral angle
Λ	Wing sweep angle
$\frac{d\varepsilon}{d\alpha}$	Change in downwash due to a change in angle of attack
η	Efficiency factor of the horizontal tail
η_v	Efficiency factor of the vertical tail
λ	Taper ratio (tip chord/root chord)
$\frac{d\sigma}{d\beta}$	Change in sidewash angle with a change in side slip angle

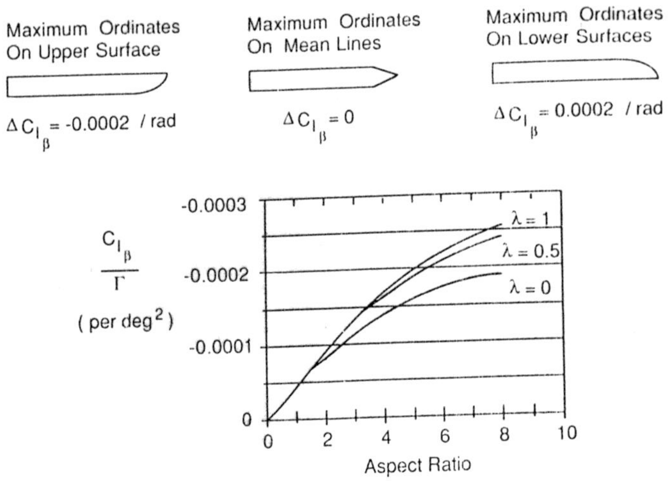

FIGURE 3.9
Tip shape and aspect ratio effect on C_{l_B}.

It is important to keep in mind that the linearization of the equations of motion and the aerodynamic representation by stability derivatives is only valid provided that the motions of the aircraft are small and that the angles of attack and sideslip are in a range in which the aerodynamic forces and moments are linear.

3.9 PROBLEMS

3.1. Starting with the Y force equation, use small-disturbance theory to determine the linearized force equation. Assume steady level flight for the reference flight conditions.

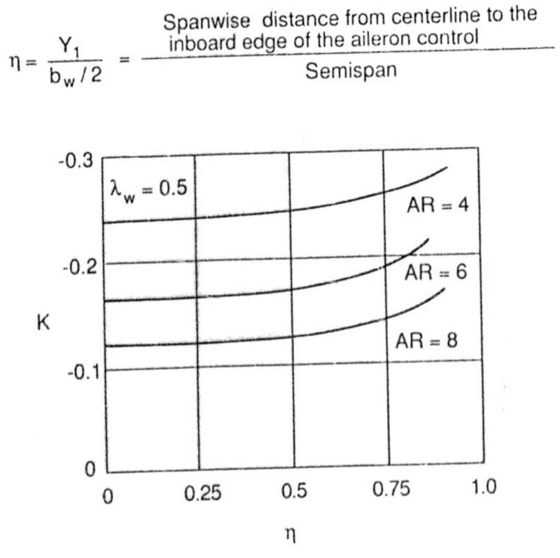

FIGURE 3.10
Empirical factor for $C_{n_{\delta_a}}$ estimate.

3.2. Starting with the Z force equation, use small-disturbance theory to determine the linearized force equations. Assume steady level flight for the reference flight conditions.

3.3. Repeat Problem 3.2 assuming the airplane is experiencing a steady pull up maneuver, i.e. $q_0 = \text{constant}$.

3.4. Using the geometric data given below and in Figure P3.4, estimate C_{m_α}, $C_{m_{\dot{\alpha}}}$, C_{m_q} and $C_{m_{\delta_e}}$.

Geometric Data

$S = 232\ \text{ft}^2$ $\quad b = 36$

$S_H = 54\ \text{ft}^2$ $\quad l_t = 21\ \text{ft}$

$S_v = 37.4\ \text{ft}^2$

$\Gamma = 2.5°$

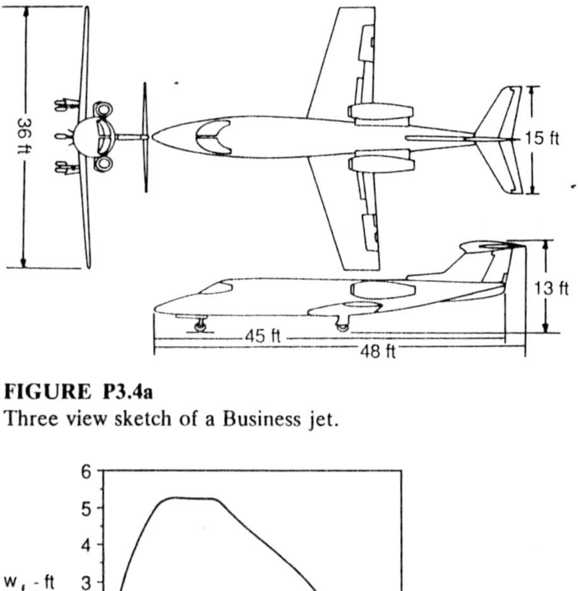

FIGURE P3.4a
Three view sketch of a Business jet.

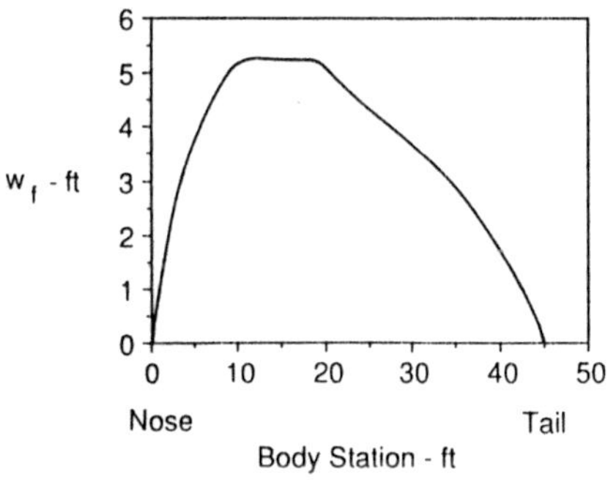

FIGURE P3.4b
Aircraft fuselage width as a function of body station.

3.5. Estimate C_{l_p} and C_{n_r} for the airplane described in Problem 3.4.

3.6. Develop an expression for the yaw moment due to yaw rate, C_{n_r}. Consider only the vertical tail in your analysis.

3.7. Develop an expression for C_{m_q} due to a canard surface.

3.8. Estimate C_{y_β}, C_{n_β} and C_{n_r} for the Boeing 747 at subsonic speeds. Compare your predictions with the data in the Appendix.

3.9. Explain why deflecting the ailerons produce a yawing moment.

REFERENCES

3.1. McRuer, D., I. Ashkenas, and D. Graham, *Aircraft Dynamics and Automatic Control,* Princeton University Press, Princeton, NJ, 1973.
3.2. Bryan, G. H., and W. E. Williams: "The Longitudinal Stability of Aerial Gliders," *Proceedings of the Royal Society of London, Series A,* Vol. 73, pp. 110–116, 1904.
3.3. Bryan, G. H.: *Stability in Aviation,* Macmillan, London, England, 1911.
3.4. *USAF Stability and Control DATCOM,* Flight Control Division, Air Force Flight Dynamics Laboratory, Wright-Patterson Air Force Base, Oh.
3.5. Smetana, F. O., D. C. Summey, and W. D. Johnson: *Riding and Handling Qualities of Light Aircraft—A Review and Analysis,* NASA CR-1975, March 1972.

CHAPTER 4

LONGITUDINAL MOTION (STICK FIXED)

"*The equilibrium and stability of a bird in flight, or an aerodome or flying machine, has in the past been the subject of considerable speculation, and no adequate explanation of the principles involved has hitherto been given*".
From Frederick W. Lanchester's book *Aerodonetics* published in 1908 in which he develops an elementary theory of longitudinal dynamic stability. [4.1]

4.1 HISTORICAL PERSPECTIVE

The theoretical basis for the analysis of flight vehicle motions developed almost concurrently with the successful demonstration of a powered flight of a man-carrying airplane. As early as 1897, Frederick Lanchester was studying the motion of gliders. He conducted experiments with hand-launched gliders and found that his gliders would fly along a straight path if they were launched at what he called the glider's "natural speed". Launching the glider at a higher or lower speed would result in an oscillatory motion. He also noticed that, if the glider were launched at its "natural speed" and then disturbed from its flight path, the glider would start oscillating along its flight trajectory. What Lanchester had discovered was that all flight vehicles possess certain natural frequencies or motions when disturbed from their equilibrium flight.

Lanchester called the oscillatory motion the phugoid motion. He wanted to use the Greek word meaning "to fly" to describe his newly discovered motion; actually, phugoid, literally means "to flee". Today, we still use the term "phugoid" to describe the long-period slowly damped oscillation associated with the longitudinal motion of an airplane.

The mathematical treatment of flight vehicle motions was first developed by G. H. Bryan. He was aware of Lanchester's experimental observations and set out to develop the mathematical equations for dynamic stability analysis. His stability work was published in 1911. Bryan made significant contributions to the analysis of vehicle flight motions. He laid the mathematical foundation for airplane dynamic stability analysis, developed the concept of the aerodynamic stability derivative, and recognized that the equations of motion could be separated into a symmetric longitudinal motion and an unsymmetric lateral motion. Although the mathematical treatment of airplane dynamic stability was formulated shortly after the first successful human-controlled flight, the theory was not used by the inventors because its mathematical complexity and the lack of information on the stability derivatives.

Experimental studies were initiated by L. Bairstow and B. M. Jones of the National Physical Laboratory (NPL) in England, and by Jerome Hunsaker of the Massachusetts Institute of Technology (MIT) to determine estimates of the aerodynamic stability derivatives used in Bryan's theory. In addition to determining stability derivatives from wind-tunnel tests of scale models, Bairstow and Jones nondimensionalized the equations of motion and showed that, with certain assumptions, there were two independent solutions, i.e. one longitudinal and one lateral. During the same period, Hunsaker and his group at MIT were conducting wind-tunnel studies of scale models of several flying airplanes. The results from these early studies were extremely valuable in establishing relationships between aerodynamics, geometric and mass characteristics of the airplane and its dynamic stability.[1]

Although these early investigators could predict the stability of the longitudinal and lateral motions, they were unsure how to interpret their findings. They were perplexed by the fact that when their analysis predicted an airplane would be unstable, the airplane was flown successfully. They wondered how the stability analysis could be used to assess whether an airplane was of good or bad design. The missing factor in analyzing airplane stability in these early studies was the failure to consider the pilot as an essential part of the airplane system.

In the late 1930s the National Advisory Committee of Aeronautics (NACA) conducted an extensive flight test program. Many airplanes were tested with the goal of quantitatively relating the measured dynamic characteristics of the airplane with the pilot's opinion of its handling characteristics. These experiments laid the foundation for modern flying qualities research. In 1943, R. Gilruth published the results of the NACA research program in the form of flying qualities' specifications. For the first time, the designer had a list of specifications which could be used in designing the airplane. If the design complied with the specifications, one could be reasonably sure that the airplane would have good flying qualities [4.1–4.4].

In this chapter we shall examine the longitudinal motion of an airplane

[1] The first technical report by the National Advisory Committee of Aeronautics, NACA (forerunner of the National Aeronautics and Space Administration, (NASA)), summarizes the MIT research in dynamic stability.

disturbed from its equilibrium state. Several different analytical techniques will be presented for solving the longitudinal differential equations. Our objectives are for the student to understand the various analytical techniques employed in airplane motion analysis and to develop an appreciation of the importance of aerodynamic or configuration changes on the airplane's dynamic stability characteristics. Later, we shall discuss what constitutes good flying qualities in terms of the dynamic characteristics presented here. Before attempting to solve the longitudinal equations of motion, we will examine the solution of a simplified aircraft motion. By studying the simpler motions with a single degree of freedom, we shall gain some insight into the more complicated longitudinal motions we shall study later in this chapter.

4.2 SECOND-ORDER DIFFERENTIAL EQUATIONS

Many physical systems can be modeled by second-order differential equations. For example, control servomotors, special cases of aircraft dynamics, as well as many electrical and mechanical systems, are governed by second-order differential equations. Because the second-order differential equation plays such an important role in aircraft dynamics we shall examine its characteristics before proceeding with our discussion of aircraft motions.

To illustrate the properties of a second-order differential equation, we shall examine the motion of a mechanical system composed of a mass, a spring, and a damping device. The forces acting on the system are shown in Fig. 4.1. The spring provides a linear restoring force that is proportional to the extension of the spring, and the damping device provides a damping force that is proportional to the velocity of the mass. The differential equation for the system can be written as

$$m\frac{d^2x}{dt^2} + c\frac{dx}{dt} + kx = F(t) \tag{4.1}$$

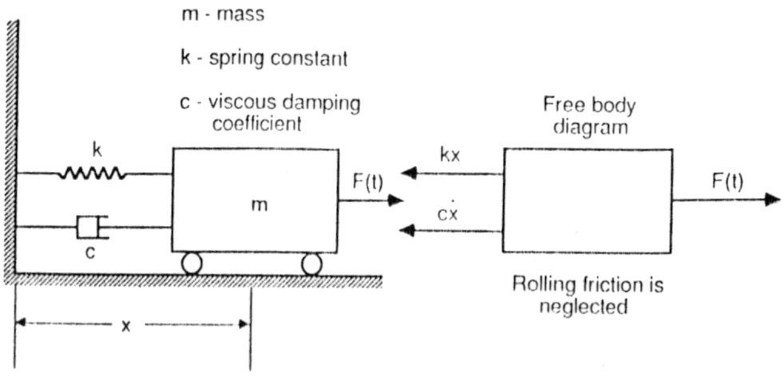

FIGURE 4.1
Sketch of a spring mass damper system.

or

$$\frac{d^2x}{dt^2} + \frac{c}{m}\frac{dx}{dt} + \frac{k}{m}x = \frac{1}{m}F(t) \tag{4.2}$$

This is a nonhomogeneous, second-order differential equation with constant coefficients. The coefficients in the equation are determined from the physical characteristics of the mechanical system being modeled, i.e. its mass, damping coefficient, and spring constant. The function $F(t)$ is called the forcing function. If the forcing function is zero, the response of the system is referred to as the free response. When the system is driven by a forcing function $F(t)$ the response is referred to as the forced response. The general solution of the nonhomogeneous differential equation is the sum of the homogeneous and particular solutions. The homogeneous solution is the solution of the differential equation when the right-hand side of the equation is zero. This corresponds to the free response of the system. The particular solution is a solution which when substituted into the left-hand side of the differential equation yields the nonhomogeneous or right-hand side of the differential equation. In the following section we will restrict our discussion to the solution of the free response or homogeneous equation.

The solution of the differential equation with constant coefficients is found by letting

$$x = Ae^{\lambda t} \tag{4.3}$$

and substituting into the differential equation yields.

$$\lambda^2 Ae^{\lambda t} + \frac{c}{m}\lambda Ae^{\lambda t} + \frac{k}{m}Ae^{\lambda t} = 0 \tag{4.4}$$

Clearing the equation of $Ae^{\lambda t}$ yields

$$\lambda^2 + \frac{c}{m}\lambda + \frac{k}{m} = 0 \tag{4.5}$$

which is called the characteristic equation. The roots of the characteristic equation are called the characteristic roots or eigenvalues of the system.

The roots of Eq. (4.5) are

$$\lambda_{1,2} = -\frac{c}{2m} \pm \sqrt{\left(\frac{c}{2m}\right)^2 - \frac{k}{m}} \tag{4.6}$$

The solution of the differential equation can now be written as

$$x(t) = C_1 e^{\lambda_1 t} + C_2 e^{\lambda_2 t} \tag{4.7}$$

where C_1 and C_2 are arbitrary constants that are determined from the initial conditions of the problem. The type of motion that occurs if the system is displaced from its equilibrium position and released depends upon the value of λ. But λ depends upon the physical constants of the problem, namely m, c, and k. We shall consider three possible cases for λ.

When $(c/2m) > \sqrt{k/m}$, the roots are negative and real, which means that the motion will die out exponentially with time. This type of motion is referred to as an overdamped motion. The equation of motion is given by

$$x(t) = C_1 \exp\left[-\frac{c}{2m} + \sqrt{\left(\frac{c}{2m}\right)^2 - \frac{k}{m}}\right]t + C_2 \exp\left[-\frac{c}{2m} - \sqrt{\left(\frac{c}{2m}\right)^2 - \frac{k}{m}}\right]t \tag{4.8}$$

For the case where $(c/2m) < \sqrt{k/m}$, the roots are complex:

$$\lambda = -\frac{c}{2m} \pm i\sqrt{\frac{k}{m} - \left(\frac{c}{2m}\right)^2} \tag{4.9}$$

The equation of motion is as follows:

$$x(t) = \exp\left(-\frac{c}{2m}t\right)\left[C_1 \exp\left(i\sqrt{\frac{k}{m} - \left(\frac{c}{2m}\right)^2}\,t\right) + C_2 \exp\left(-i\sqrt{\frac{k}{m} - \left(\frac{c}{2m}\right)^2}\,t\right)\right] \tag{4.10}$$

which can be rewritten as

$$x(t) = \exp\left(-\frac{c}{2m}t\right)\left[A \cos\left(\sqrt{\frac{k}{m} - \left(\frac{c}{2m}\right)^2}\,t\right) + B \sin\left(\sqrt{\frac{k}{m} - \left(\frac{c}{2m}\right)^2}\,t\right)\right] \tag{4.11}$$

The solution given by Eq. (4.11) is a damped sinusoid having a natural frequency given by

$$\omega = \sqrt{\frac{k}{m} - \left(\frac{c}{2m}\right)^2} \tag{4.12}$$

The last case we consider is when $(c/2m) = \sqrt{k/m}$. This represents the boundary between the overdamped exponential motion and the damped sinusoidal motion. This particular motion is referred to as the critically damped motion. The roots of the characteristic equation are identical, i.e.

$$\lambda_{1,2} = -\frac{c}{2m} \tag{4.13}$$

The general solution for repeated roots has the form

$$x(t) = (C_1 + C_2 t)e^{-\lambda t} \tag{4.14}$$

If λ is a positive constant, then $e^{-\lambda t}$ will go to zero faster than $C_2 t$ goes to infinity as time increases. Figure 4.2 shows the motion for the three cases analyzed here.

The constant for the critically damped case is called the critical damping constant and is defined as

$$c_{cr} = 2\sqrt{k/m} \tag{4.15}$$

For oscillatory motion, the damping can be specified in terms of the critical

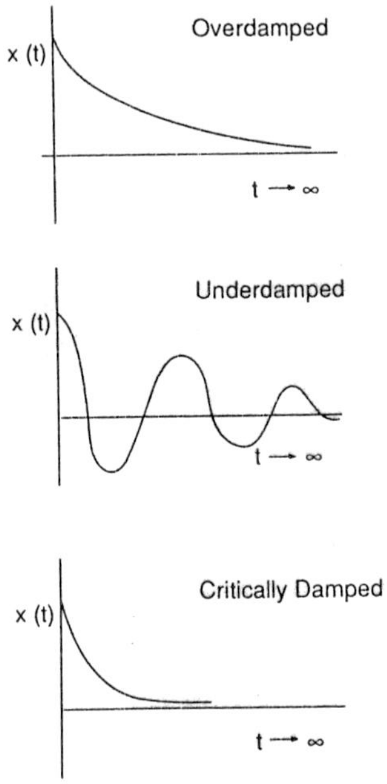

FIGURE 4.2
Typical motions of a dynamic system.

damping:

$$c = \zeta c_{cr} \tag{4.16}$$

where ζ is called the damping ratio,

$$\zeta = \frac{c}{c_{cr}} \tag{4.17}$$

For a system that has no damping, i.e. $c = 0$, which implies that the $\zeta = 0$ the motion is an undamped oscillation. The natural frequency is called the undamped natural frequency and can be obtained from Eq. (4.12) by setting $c = 0$:

$$\omega_n = \sqrt{\frac{k}{m}} \tag{4.18}$$

Since both the damping ratio and undamped natural frequency are specified as functions of the system physical constants, we can rewrite the differential equation in terms of the damping ratio and undamped natural frequency as follows:

$$\frac{d^2x}{dt^2} + 2\zeta\omega_n \frac{dx}{dt} + \omega_n^2 x = f(t) \tag{4.19}$$

Equation (4.19) is referred to as the standard form of a second-order differential equation with constant coefficients. Although we developed the standard form of a second-order differential equation from a mechanical mass–spring–damper system, the equation could have been developed using any one of an almost limitless number of physical systems. For example, a torsional spring–mass–damper equation of motion is given by

$$\frac{d^2\theta}{dt^2}+\frac{c}{I}\frac{d\theta}{dt}+\frac{k}{I}\theta=f(t) \tag{4.20}$$

where c, k and I are the torsional damping coefficient, torsional spring constant and moment of inertia, respectively.

The characteristic equation for the standard form of the second-order differential equation with constant coefficients can be shown to be

$$\lambda^2+2\zeta\omega_n\lambda+\omega_n^2=0 \tag{4.21}$$

The roots of the characteristic equation are

$$\lambda_{1,2}=-\zeta\omega_n\pm i\omega_n\sqrt{1-\zeta^2} \tag{4.22}$$

or

$$\lambda_{1,2}=\eta\pm i\omega \tag{4.23}$$

where

$$\eta=-\zeta\omega_n \tag{4.24}$$

$$\omega=\omega_n\sqrt{1-\zeta^2} \tag{4.25}$$

The real η of the root governs the damping of the response and ω is the damped natural frequency.

Figure 4.3 shows the relationship between the roots of the characteristic

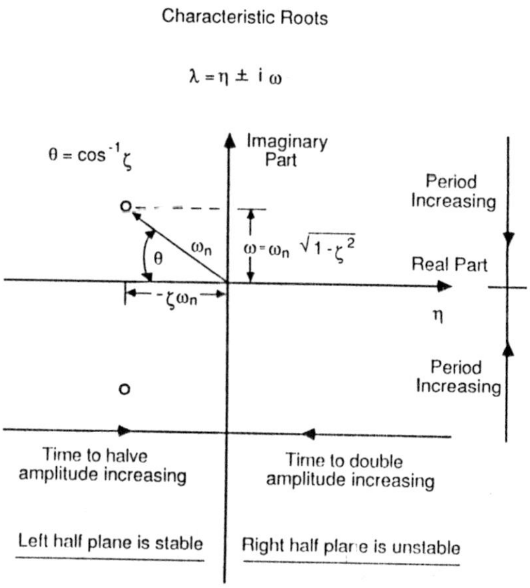

FIGURE 4.3
Relationship between η, ω, ζ and ω_n.

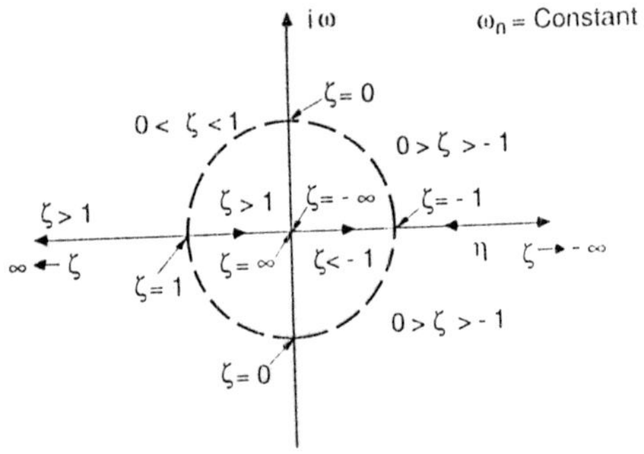

FIGURE 4.4
Variation of roots with damping ratio.

equation and η, ω, ζ and ω_n. When the roots are complex, the radial distance from the origin to the root is the undamped natural frequency. The system damping η is the real part of the complex root and the damped natural frequency is the imaginary part of the root. The damping ratio ζ is equal to the cosine of the angle between the negative real axis and the radial line from the origin to the root:

$$\cos(\pi - \theta) = -\cos\theta = \frac{-\zeta\omega_n}{\omega_n} \tag{4.26}$$

or

$$\zeta = \cos\theta \tag{4.27}$$

The influence of the damping ratio on the roots of the characteristic equation can be examined by holding the undamped natural frequency constant and varying ζ from $-\infty$ to ∞ as shown in Fig. 4.4. The response of the homogeneous equation to a displacement from its equilibrium condition can take on many forms depending upon the magnitude of the damping ratio. The classification of the response is given in Table 4.1.

TABLE 4.1
Variation of response with damping ratio

Magnitude of damping ratio	Type of root	Time response
$\zeta < -1$	Two real distinct roots	Exponentially growing motion
$0 > \zeta > -1$	Complex roots with positive real part	Exponentially growing sinusoidal motion
$\zeta = 0$	Complex roots with real part zero	Undamped sinusoidal motion Pure harmonic motion
$0 < \zeta < 1$	Complex roots with real part negative	Underdamped exponentially decaying sinusoidal motion
$\zeta = 1$	Two negative equal real roots	Critically damped exponentially decaying motion
$\zeta > 1$	Two negative distinct real roots	Overdamped exponentially decaying motion

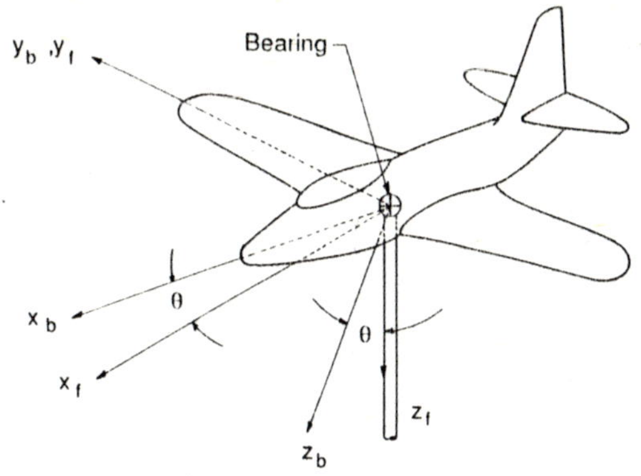

FIGURE 4.5
A model constrained to a pure pitching motion.

4.3 PURE PITCHING MOTION

Consider the case in which the airplane's center of gravity is constrained to move in a straight line at constant speed but the aircraft is free to pitch about its center of gravity. Figure 4.5 is the sketch of a wind-tunnel model constrained so that it can perform only in a pitching motion.

The equation of motion can be developed from the rigid body equations developed in Chapter 3 by making the appropriate assumptions. However, to aid our understanding of this simple motion, we shall rederive the governing equation from first principles. The equation governing this motion is obtained from Newton's second law:

$$\sum \text{Pitching moments} = \sum M_{cg} = I_y \ddot{\theta} \tag{4.28}$$

The pitching moment M and pitch angle θ can be expressed in terms of an initial reference value indicated by a subscript (0) and the perturbation by the Δ symbol.

$$M = M_0 + \Delta M \tag{4.29}$$

$$\theta = \theta_0 + \Delta\theta \tag{4.30}$$

Equation (4.28) then reduces to

$$\Delta M = I_y \, \Delta\ddot{\theta} \tag{4.31}$$

For the restricted motion that we are examining, the variables are the angle of attack, pitch angle, the time rate of change of these variables, and the elevator angle. The pitching moment is not a function of the pitch angle but is a function of the other variables and can be expressed in functional form as follows:

$$\Delta M = \text{fn}(\Delta\alpha, \Delta\dot{\alpha}, \Delta q, \Delta\delta_e) \tag{4.32}$$

Equation (4.32) can be expanded in terms of the perturbation variables by means of a Taylor series:

$$\Delta M = \frac{\partial M}{\partial \alpha}\Delta\alpha + \frac{\partial M}{\partial \dot{\alpha}}\Delta\dot{\alpha} + \frac{\partial M}{\partial q}\Delta q + \frac{\partial M}{\partial \delta_e}\Delta\delta_e \tag{4.33}$$

If we align the body and fixed frames so they coincide at $t = 0$, the change in angle of attack and pitch angles are identical, i.e.

$$\Delta\alpha = \Delta\theta \qquad \text{and} \qquad \Delta\dot{\theta} = \Delta q = \Delta\dot{\alpha} \tag{4.34}$$

This is true only for the special cases where the center of gravity is constrained. Substituting this information into Eq. (4.31) yields

$$\Delta\ddot{\alpha} - (M_q + M_{\dot{\alpha}})\,\Delta\dot{\alpha} - M_\alpha\,\Delta\alpha = M_{\delta e}\,\Delta\delta_e \tag{4.35}$$

where

$$M_q = \frac{\partial M}{\partial q}\Big/ I_y, \qquad M_{\dot{\alpha}} = \frac{\partial M}{\partial \dot{\alpha}}\Big/ I_y \quad \text{etc.}$$

Equation (4.35) is a nonhomogeneous second-order differential equation, having constant coefficients. This equation is similar to a torsional spring–mass–damper system with a forcing function, which was mentioned briefly in the previous section. The static stability of the airplane can be thought of as the equivalent of an aerodynamic spring, while the aerodynamic damping terms are similar to a torsional damping device. The characteristic equation for Eq. (4.35) is

$$\lambda^2 - (M_q + M_{\dot{\alpha}})\lambda - M_\alpha = 0 \tag{4.36}$$

This equation can be compared with the standard equation of a second-order system:

$$\lambda^2 + 2\zeta\omega_n\lambda + \omega_n^2 = 0 \tag{4.37}$$

where ζ is the damping ratio and ω_n is the undamped natural frequency. By inspection, we see that

$$\omega_n = \sqrt{-M_\alpha} \tag{4.38}$$

and

$$\zeta = \frac{-(M_q + M_{\dot{\alpha}})}{2\sqrt{-M_\alpha}} \tag{4.39}$$

Note that the frequency is related to the airplane's static stability and that the damping ratio is a function of the aerodynamic damping and static stability.

If we solve the characteristic Eq. (4.37), we obtain the following roots:

$$\lambda_{1,2} = \frac{-2\zeta\omega_n \pm \sqrt{4\zeta^2\omega_n^2 - 4\omega_n^2}}{2} \tag{4.40}$$

or

$$\lambda_{1,2} = -\zeta\omega_n \pm i\omega_n\sqrt{1-\zeta^2} \tag{4.41}$$

Expressing the characteristic roots as

$$\lambda_{1,2} = \eta \pm i\omega \tag{4.42}$$

and comparing Eq. (4.42) with Eq. (4.41), yields

$$\eta = -\zeta\omega_n \tag{4.43}$$

and

$$\omega = \omega_n\sqrt{1-\zeta^2} \tag{4.44}$$

which are the real and imaginary parts of the characteristic roots. The angular frequency ω is called the damped natural frequency of the system.

The general solution to Eq. (4.35) for a step change $\Delta\delta_e$ in the elevator angle can be expressed as

$$\Delta\alpha(t) = \Delta\alpha_{\text{trim}}\left[\left(1 + \frac{e^{-\zeta\omega_n t}}{\sqrt{1-\zeta^2}}\sin(\sqrt{1-\zeta^2}\,\omega_n t + \phi)\right)\right] \tag{4.45}$$

where $\Delta\alpha_{\text{trim}}$ = change in trim angle of attack $= -(M_{\delta e}\,\Delta\delta_e)/M_\alpha$

ζ = damping ratio $= -\dfrac{(M_q + M_{\dot\alpha})}{2\sqrt{-M_\alpha}}$

ω_n = undamped natural frequency $= \sqrt{-M_\alpha}$

ϕ = phase angle $= \tan^{-1}(-\sqrt{I-\zeta^2}/-\zeta)$

The solution is a damped sinusoidal motion with the frequency a function of C_{m_α} and the damping rate a function of $C_{m_q} + C_{m_{\dot\alpha}}$ and C_{m_α}. Figure 4.6 illustrates the angle of attack time history for various values of the damping

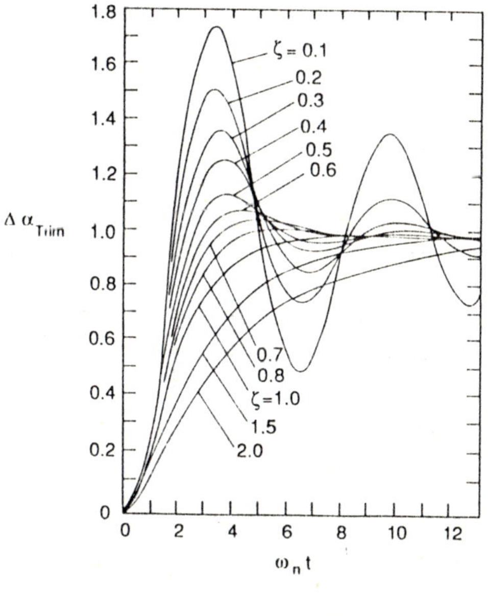

FIGURE 4.6
Angle of attack time history of a pitching model for various damping ratios.

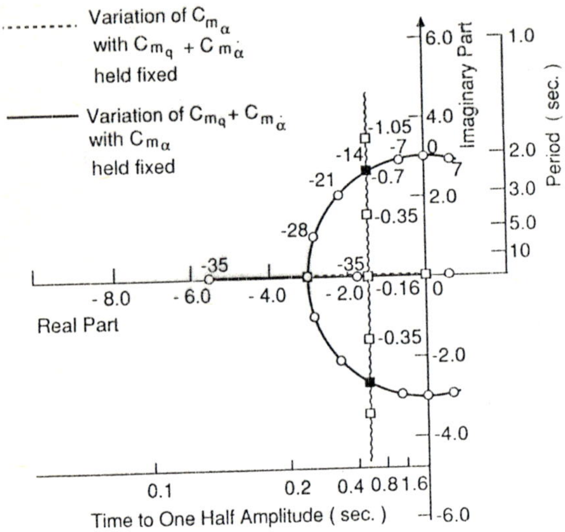

FIGURE 4.7
Variation of the characteristic roots of the pitching motion as a function of the stability coefficients.

ratio ζ. Note that as the system damping is increased the maximum overshoot of the response diminishes.

The influence of the stability coefficients on the roots of the characteristic equation can be seen in Fig. 4.7. The curves show the effect of variations in C_{m_α} and $C_{m_q} + C_{m_{\dot\alpha}}$ on the roots. This type of curve is referred to as a root locus plot. Notice that, as the roots move into the right half-plane, the vehicle will become unstable.

The roots of the characteristic equation tell us what type of response our airplane will have. If the roots are real, the response will be either a pure divergence or a pure subsidence, depending upon whether the root is positive or negative. If the roots are complex, the motion will be either a damped or an undamped sinusoidal oscillation. The period of the oscillation is related to the imaginary part of the root, as follows:

$$\text{Period} = \frac{2\pi}{\omega} \tag{4.46}$$

The rate of growth or decay of the oscillation is determined by the sign of the real part of the complex root. A negative real part produces a decaying oscillation, whereas a positive real part causes the motion to grow. A measure of the rate of growth or decay of the oscillation can be obtained from the time for halving or doubling the initial amplitude of the disturbance. Figure 4.8 shows a damped and undamped oscillation and how the time for halving or doubling of the amplitude can be calculated. The expression for the time for doubling or halving of the amplitude is

$$t_{\text{double}} \text{ or } t_{\text{halve}} = \frac{0.693}{|\eta|} \tag{4.47}$$

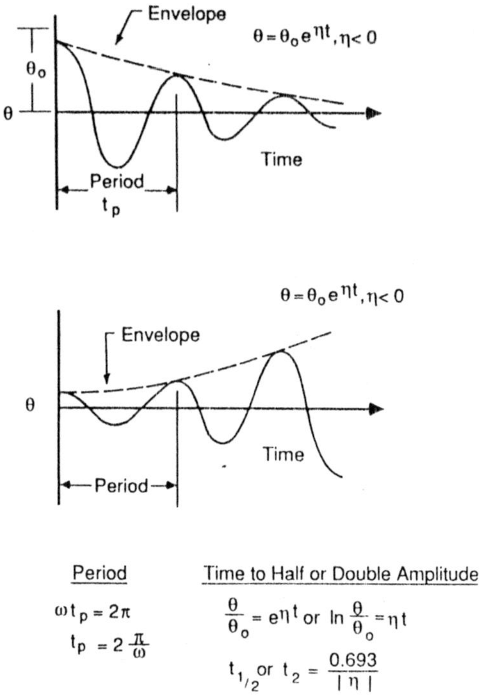

FIGURE 4.8
Relationships for time to halve or double amplitude and the period.

and the number of cycles for doubling or halving of the amplitude is

$$N(\text{cycles})_{\text{double or halve}} = 0.110 \frac{|\omega|}{|\eta|} \tag{4.48}$$

4.4 STICK FIXED LONGITUDINAL MOTION

The motion of an airplane in free flight can be extremely complicated. The airplane has three translational motions (vertical, horizontal, and transverse), three rotational motions (pitch, yaw, and roll) and numerous elastic degrees of freedom. To analyze the response of an elastic airplane is beyond the scope of this book.

The problem we shall address in this section is the solution of the rigid body equations of motion. This may seem to be a formidable task; however, we can make some simplifying assumptions which will reduce the complexity of the problem. First, we shall assume that the aircraft's motion consists of small deviations from its equilibrium flight condition. Second, we shall assume that the motion of the airplane can be analyzed by separating the equations into two groups. The X-force, Z-force and pitching moment equations comprise the longitudinal equations, and the Y-force, rolling and yawing moment

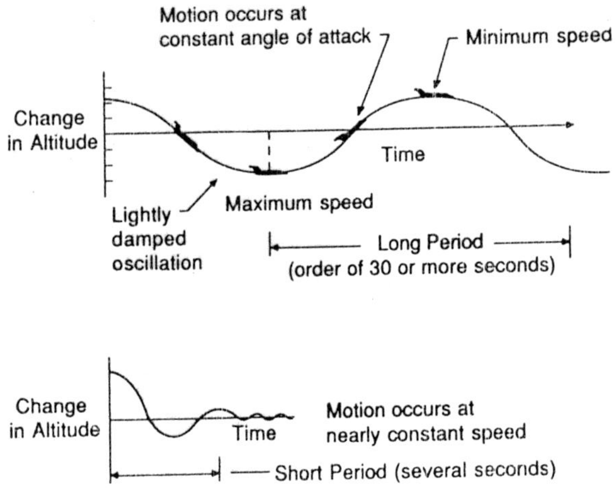

FIGURE 4.9
Sketch of the phugoid and short period motions.

equations are called the lateral equations. To separate the equations in this manner, the longitudinal and lateral equations must not be coupled. These are all reasonable assumptions provided the airplane is not undergoing a large-amplitude or very rapid maneuver.

In aircraft motion studies, one must always be sure that the assumptions made in an analysis are appropriate for the problem at hand. Students are all too eager to use the first equation they can find to solve their homework problems. This type of approach can lead to many incorrect or ridiculous solutions. To avoid such embarrassment, one must always verify that the assumptions used in developing the equations one wishes to use are consistent with the problem one is attempting to solve. This is particularly important when solving problems related to aircraft dynamics.

In the following sections we shall examine the longitudinal motion of an airplane without control input. The longitudinal motion of an airplane (controls fixed) disturbed from its equilibrium flight condition is characterized by two oscillatory modes of motion. Figure 4.9 illustrates these basic modes. We see that one mode is lightly damped and has a long period. This mode is called the long-period or phugoid mode. The second basic mode is heavily damped and has a very short period; it is appropriately called the short-period mode.

STATE VARIABLE REPRESENTATION OF THE EQUATIONS OF MOTION. The linearized longitudinal equations developed in Chapter 3 are simple, ordinary linear differential equations with constant coefficients. The coefficients in the differential equations are made up of the aerodynamic stability derivatives, mass, and inertia characteristics of the airplane. These equations can be written as a set of first-order differential equations. When the equations are written as a system of first-order differential equations they are called the state-space or state variable equations and are represented mathematically as

$$\dot{\mathbf{x}} = \mathbf{A}\mathbf{x} + \mathbf{B}\eta \tag{4.49}$$

where $\mathbf{x}$ is the state vector, η is the control vector and the matrices $\mathbf{A}$ and $\mathbf{B}$ contain the aircraft's dimensional stability derivatives.

The linearized longitudinal equations developed earlier are repeated below.

$$\begin{aligned}
&\left(\frac{d}{dt} - X_u\right)\Delta u - X_w\,\Delta w + (g\cos\theta_0)\,\Delta\theta = X_\delta\,\Delta\delta + X_{\delta_T}\,\Delta\delta_T\\
&-Z_u\,\Delta u + \left((1 - Z_{\dot{w}})\frac{d}{dt} - Z_w\right)\Delta w - \left((u_0 + Z_q)\frac{d}{dt} - g\sin\theta_0\right)\Delta\theta\\
&\qquad\qquad = Z_\delta\,\Delta\delta + Z_{\delta_T}\,\Delta\delta_T\\
&-M_u\,\Delta u - \left(M_{\dot{w}}\frac{d}{dt} + M_w\right)\Delta w + \left(\frac{d^2}{dt^2} - M_q\frac{d}{dt}\right)\Delta\theta = M_\delta\,\Delta\delta + M_{\delta_T}\,\Delta\delta_T
\end{aligned} \tag{4.50}$$

In practice, the force derivatives Z_q and $Z_{\dot{w}}$ are usually neglected because they contribute very little to the aircraft response. Therefore, to simplify our presentation of the equations of motion in the state space form we will neglect both Z_q and $Z_{\dot{w}}$. Rewriting the equations in the state-space form yields

$$\begin{bmatrix}\Delta\dot{u}\\ \Delta\dot{w}\\ \Delta\dot{q}\\ \Delta\dot{\theta}\end{bmatrix} = \begin{bmatrix} X_u & X_w & 0 & -g\\ Z_u & Z_x & u_0 & 0\\ M_u + M_{\dot{w}}Z_u & M_w + M_{\dot{w}}Z_w & M_q + M_{\dot{w}}u_0 & 0\\ 0 & 0 & 1 & 0\end{bmatrix}\begin{bmatrix}\Delta u\\ \Delta w\\ \Delta q\\ \Delta\theta\end{bmatrix}$$
$$+ \begin{bmatrix} X_\delta & X_{\delta_T}\\ Z_\delta & Z_{\delta_T}\\ M_\delta + M_{\dot{w}}Z_\delta & M_{\delta_T} + M_{\dot{w}}Z_{\delta_T}\\ 0 & 0\end{bmatrix}\begin{bmatrix}\Delta\delta\\ \Delta\delta_T\end{bmatrix} \tag{4.51}$$

where the state vector $\mathbf{x}$ and control vector η are given by

$$x = \begin{bmatrix}\Delta u\\ \Delta w\\ \Delta q\\ \Delta\theta\end{bmatrix} \qquad \eta = \begin{bmatrix}\Delta\delta\\ \Delta\delta_T\end{bmatrix} \tag{4.52}$$

and the matrices **A** and **B** are given by

$$\mathbf{A} = \begin{bmatrix} X_u & X_w & 0 & -g \\ Z_u & Z_w & u_0 & 0 \\ M_u + M_{\dot{w}}Z_u & M_w + M_{\dot{w}}Z_w & M_q + M_{\dot{w}}u_0 & 0 \\ 0 & 0 & 1 & 0 \end{bmatrix} \tag{4.53}$$

$$\mathbf{B} = \begin{bmatrix} X_\delta & X_{\delta_T} \\ Z_\delta & Z_{\delta_T} \\ M_\delta + M_{\dot{w}}Z_\delta & M_{\delta_T} + M_{\dot{w}}Z_{\delta_T} \\ 0 & 0 \end{bmatrix} \tag{4.54}$$

The force and moment derivatives in the matrices have been divided by the mass of the airplane or the moment of inertia, respectively as indicated below:

$$X_u = \frac{\partial X/\partial u}{m} \qquad M_u = \frac{\partial M/\partial u}{I_y} \qquad \text{etc.} \tag{4.55}$$

Table 4.2 includes a list of the definitions of the longitudinal stability derivatives. Methods for estimating the stability coefficients were discussed in Chapter 3.

The homogeneous solution to Eq. (4.49) can be obtained by assuming a solution of the form

$$\mathbf{x} = \mathbf{x}_r e^{\lambda_r t} \tag{4.56}$$

TABLE 4.2
Summary of longitudinal derivatives*

$X_u = \dfrac{-(C_{D_u} + 2C_{D0})QS}{mu_0}$	$X_w = \dfrac{-(C_{D_\alpha} - C_{L0})QS}{mu_0}$
$Z_u = \dfrac{-(C_{L_u} + 2C_{L0})QS}{mu_0}$	
$Z_w = \dfrac{-(C_{L\alpha} + C_{D0})QS}{mu_0}$	$Z_{\dot{w}} = -C_{Z\dot{\alpha}}\dfrac{\bar{c}}{2u_0}QS/(u_0 m)$
$Z_\alpha = u_0 Z_w$	$Z_{\dot{\alpha}} = u_0 Z_{\dot{w}}$
$Z_q = -C_{Zq}\dfrac{\bar{c}}{2u_0}QS/m$	$Z_{\delta e} = -C_{Z_{\delta e}}QS/m$
$M_u = C_{m_u}\dfrac{(QS\bar{c})}{u_0 I_y}$	
$M_w = C_{m\alpha}\dfrac{(QS\bar{c})}{u_0 I_y}$	$M_{\dot{w}} = C_{m\dot{\alpha}}\dfrac{\bar{c}}{2u_0}\dfrac{QS\bar{c}}{u_0 I_y}$
$M_\alpha = u_0 M_w$	$M_{\dot{\alpha}} = u_0 M_{\dot{w}}$
$M_q = C_{mq}\dfrac{\bar{c}}{2u_0}(QS\bar{c})/I_y$	$M_{\delta e} = C_{m_{\delta e}}(QS\bar{c})/I_y$

Substituting Eq. (4.56) into Eq. (4.49) yields

$$[\lambda_r \mathbf{I} - \mathbf{A}]\mathbf{x}_r = 0 \tag{4.57}$$

where **I** is the identity matrix

$$\mathbf{I} = \begin{bmatrix} 1 & 0 & 0 & 0 \\ 0 & 1 & 0 & 0 \\ 0 & 0 & 1 & 0 \\ 0 & 0 & 0 & 1 \end{bmatrix} \tag{4.58}$$

For a nontrivial solution to exist, the determinant

$$|\lambda_r \mathbf{I} - \mathbf{A}| = 0 \tag{4.59}$$

must be zero. The roots λ_r of Eq. (4.59) are called the characteristic roots or eigenvalues. The solution of Eq. (4.59) can be accomplished easily using a digital computer. Most computer facilities will have a subroutine package for determining the eigenvalues of a matrix.

The eigenvectors for the system can be determined once the eigenvalues are known from Eq. (4.60).

$$[\lambda_j \mathbf{I} - \mathbf{A}]\mathbf{P}_{ij} = 0 \tag{4.60}$$

where $\mathbf{P}_{ij}$ is the eigenvector corresponding to the jth eigenvalue. The set of equations making up Eq. (4.60) are linearly dependent and homogeneous; therefore, the eigenvectors cannot be unique. A technique for finding these eigenvectors will be presented later in this chapter.

4.5 LONGITUDINAL APPROXIMATIONS

We can think of the long-period or phugoid mode as a gradual interchange of potential and kinetic energy about the equilibrium altitude and airspeed. This is illustrated in Fig. 4.9. Here we see that the long-period mode is characterized by changes in pitch attitude, altitude, and velocity at a nearly constant angle of attack. An approximation to the long-period mode can be obtained by neglecting the pitching moment equation and assuming that the change in angle of attack is zero, i.e.

$$\Delta\alpha = \frac{\Delta w}{u_0} \qquad \Delta\alpha = 0 \rightarrow \Delta w = 0 \tag{4.61}$$

Making these assumptions, the homogeneous longitudinal state equations reduce to the following:

$$\begin{bmatrix} \Delta \dot{u} \\ \Delta \dot{\theta} \end{bmatrix} = \begin{bmatrix} X_u & -g \\ \dfrac{-Z_u}{u_0} & 0 \end{bmatrix} \begin{bmatrix} \Delta u \\ \Delta \theta \end{bmatrix} \tag{4.62}$$

The eigenvalues of the long period approximation are obtained by solving the equation

$$|\lambda \mathbf{I} - \mathbf{A}| = 0 \tag{4.63}$$

or

$$\begin{vmatrix} \lambda - X_u & g \\ \dfrac{Z_u}{u_0} & \lambda \end{vmatrix} = 0 \tag{4.64}$$

Expanding the above determinant yields

$$\lambda^2 - X_u\lambda - \frac{Z_u g}{u_0} = 0 \tag{4.65}$$

or

$$\lambda_p = \left[X_u \pm \sqrt{X_u^2 + 4\frac{Z_u g}{u_0}}\right] \Big/ 2.0 \tag{4.66}$$

The frequency and damping ratio can be expressed as

$$\omega_{n_p} = \sqrt{\frac{-Z_u g}{u_0}} \tag{4.67}$$

$$\zeta_p = \frac{-X_u}{2\omega_{n_p}} \tag{4.68}$$

If we neglect compressibility effects, the frequency and damping ratios for the long-period motion can be approximated by the following equations:

$$\omega_{n_p} = \sqrt{2}\frac{g}{u_0} \tag{4.69}$$

$$\zeta_p = \frac{1}{\sqrt{2}}\frac{1}{L/D} \tag{4.70}$$

Notice that the frequency of oscillation and the damping ratio are inversely proportional to the forward speed and the lift-to-drag ratio, respectively. We see from this approximation that the phugoid damping is degraded as the aerodynamic efficiency (L/D) is increased. When pilots are flying an airplane under visual flight rules, the phugoid damping and frequency can vary over a wide range and they will still find the airplane acceptable to fly. On the other hand, if they are flying the airplane under instrument flight rules, low phugoid damping will become very objectable to them. To improve the damping of the phugoid motion, the designer would have to reduce the lift-to-drag ratio of the airplane. Because this would degrade the performance of the airplane, the designer would find such a choice unacceptable and would look for another alternative, such as an automatic stabilization system to provide the proper damping characteristics.

SHORT-PERIOD APPROXIMATION. An approximation to the short-period mode of motion can be obtained by assuming $\Delta u = 0$ and dropping the X-force equation. The longitudinal state-space equations reduce to the following:

$$\begin{bmatrix} \Delta \dot{w} \\ \Delta \dot{q} \end{bmatrix} = \begin{bmatrix} Z_w & u_0 \\ M_w + M_{\dot{w}} Z_w & M_q + M_{\dot{w}} u_0 \end{bmatrix} \begin{bmatrix} \Delta w \\ \Delta q \end{bmatrix} \tag{4.71}$$

The above equation can be written in terms of the angle of attack by using the relationship

$$\Delta \alpha = \frac{\Delta w}{u_0} \tag{4.72}$$

In addition, one can replace the derivatives due to w and $\dot{w}$ with derivatives due to α and $\dot{\alpha}$ by using the following equations. The definition of the derivative M_α is

$$M_\alpha = \frac{1}{I_y} \frac{\partial M}{\partial \alpha} = \frac{1}{I_y} \frac{\partial M}{\partial (\Delta w / u_0)} = \frac{u_0}{I_y} \frac{\partial M}{\partial w} = u_0 M_w \tag{4.73}$$

In a similar way we can show that

$$Z_\alpha = u_0 Z_w \qquad \text{and} \qquad M_{\dot{\alpha}} = u_0 M_{\dot{w}} \tag{4.74}$$

Using the above expressions, the state equations for the short-period approximation can be rewritten as

$$\begin{bmatrix} \Delta \dot{\alpha} \\ \Delta \dot{q} \end{bmatrix} = \begin{bmatrix} \dfrac{Z_\alpha}{u_0} & 1 \\ M_\alpha + M_{\dot{\alpha}} \dfrac{Z_\alpha}{u_0} & M_q + M_{\dot{\alpha}} \end{bmatrix} \begin{bmatrix} \Delta \alpha \\ \Delta q \end{bmatrix} \tag{4.75}$$

The eigenvalues of the state equation can again be determined by solving the equation

$$|\lambda \mathbf{I} - \mathbf{A}| = 0 \tag{4.76}$$

which yields

$$\begin{vmatrix} \lambda - \dfrac{Z_\alpha}{u_0} & -1 \\ -M_\alpha - M_{\dot{\alpha}} \dfrac{Z_\alpha}{u_0} & \lambda - (M_q + M_{\dot{\alpha}}) \end{vmatrix} = 0 \tag{4.77}$$

The characteristic equation for the above determinant is

$$\lambda^2 - \left(M_q + M_{\dot{\alpha}} + \frac{Z_\alpha}{u_0} \right) \lambda + M_q \frac{Z_\alpha}{u_0} - M_\alpha = 0 \tag{4.78}$$

TABLE 4.3
Summary of longitudinal approximations

	Long-period (phugoid)	Short-period
Frequency	$\omega_{np} = \sqrt{\dfrac{-Z_u g}{u_0}}$	$\omega_{nsp} = \sqrt{\dfrac{Z_\alpha M_q}{u_0} - M_\alpha}$
Damping ratio	$\zeta_p = \dfrac{-X_u}{2\omega_{np}}$	$\zeta_{sp} = -\dfrac{M_q + M_{\dot\alpha} + \dfrac{Z_\alpha}{u_0}}{2\omega_{nsp}}$

The approximate short-period roots can be obtained easily from the characteristic equation and are given below:

$$\lambda_{sp} = \left(M_q + M_{\dot\alpha} + \frac{Z_\alpha}{u_0}\right)\Big/2$$
$$\pm\left[\left(M_q + M_{\dot\alpha} + \frac{Z_\alpha}{u_0}\right)^2 - 4\left(M_q\frac{Z_\alpha}{u_0} - M_\alpha\right)\right]^{1/2}\Big/2 \tag{4.79}$$

or in terms of the damping and frequency

$$\omega_{n_{sp}} = \left[\left(M_q\frac{Z_\alpha}{u_0} - M\right)\right]^{1/2} \tag{4.80}$$

$$\zeta_{sp} = -\left[M_q + M_{\dot\alpha} + \frac{Z_\alpha}{u_0}\right]\Big/(2\omega_{n_{sp}}) \tag{4.81}$$

Equations (4.80) and (4.81) should look familiar to the reader. They are very similar to the equations derived for the case of a constrained pitching motion. If we neglect the Z_α term (i.e. neglect the vertical motion), Eqs (4.80) and (4.81) are identical to Eqs (4.38) and (4.39).

A summary of the approximate formulas is presented in Table 4.3.

To help clarify the preceding analysis, we shall determine the longitudinal characteristics of the general aviation airplane included in Appendix B.

Example Problem 4.1. Find the longitudinal eigenvalues and eigenvectors for the general aviation airplane included in Appendix B and compare these results with the answers obtained by using the phugoid and short period approximations.

Solution. First, we must determine the numerical values of the dimensional longitudinal stability derivatives. The dynamic pressure Q and the terms QS, $QS\bar{c}$ and $\bar{c}/2u_0$ are

$$Q = \tfrac{1}{2}\rho u_0^2 = (0.5)(0.002378)(176)^2$$
$$= 36.8\ \text{lb/ft}^2$$
$$QS = (36.8)(180) = 6624\ \text{lb}$$
$$QS\bar{c} = (6624)(5.7) = 37\,757\ \text{ft·lb}$$
$$(c/2u_0) = (5.7)(2 \times 176) = 0.016\ \text{s}$$

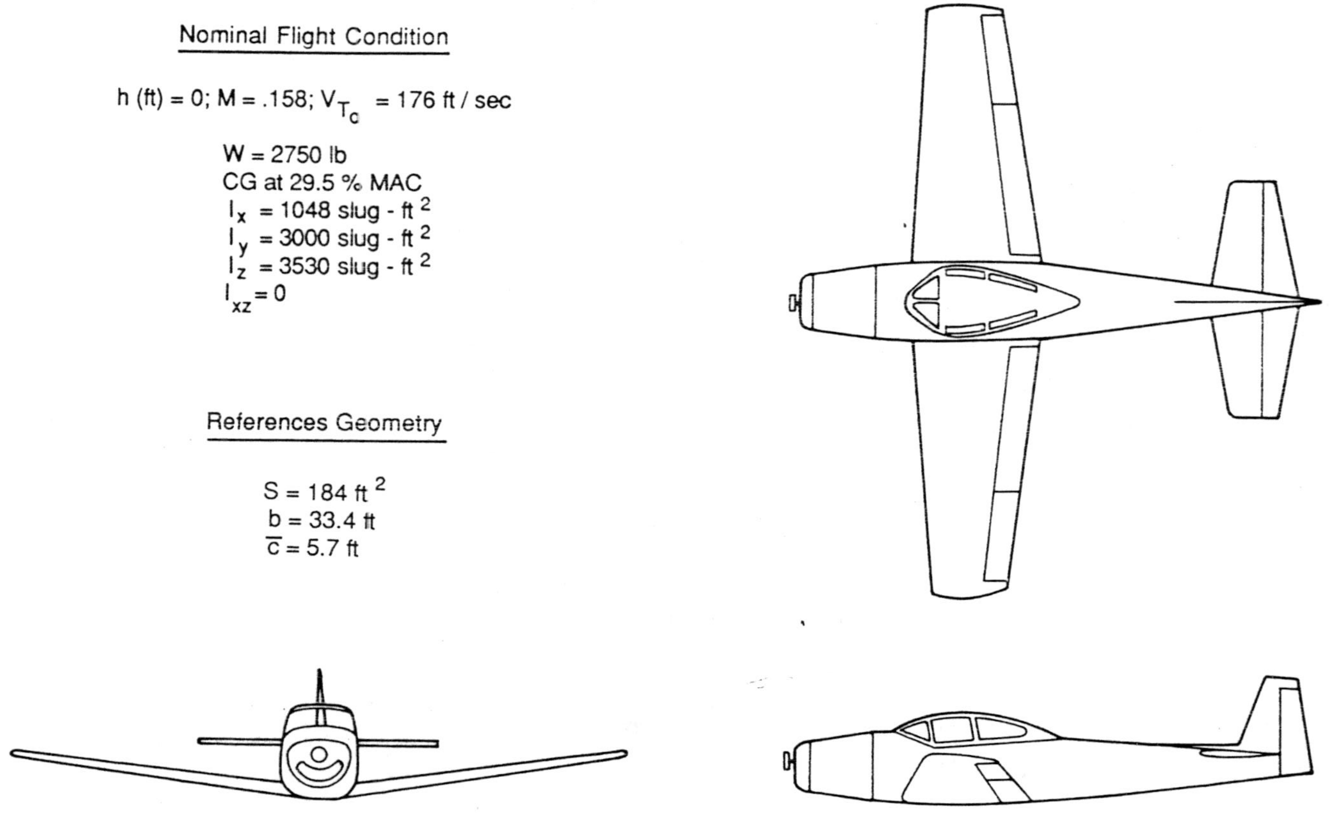

FIGURE 4.10
Geometric, mass, and aerodynamic properties of a general aviation airplane.

The longitudinal derivatives can be estimated from the formulas in Table 4.2.

***U*-derivatives**

$$X_u = -(C_{D_u} + 2C_{D_0})QS/(u_0 m)$$
$$= -(0.0 + 2(0.05))(6624)/(176)(85.5)$$
$$= -0.0446\ (\mathrm{s}^{-1})$$

$$Z_u = -(C_{L_u} + 2C_{L_0})QS/(u_0 m)$$
$$= -(0.0 + 2(0.41))(6624)/(176)(85.5)$$
$$= -0.361\ (\mathrm{s}^{-1})$$

$$M_u = 0$$

***W*-derivatives**

$$X_w = -(C_{D_\alpha} - C_{L_0})QS/(u_0 m)$$
$$= -(0.33 - 0.41)(6624)/(176)(85.5)$$
$$= 0.0357\ (\mathrm{s}^{-1})$$

$$Z_w = -(C_{L_\alpha} + C_{D_0})QS/(u_0 m)$$
$$= -(4.44 + 0.05)(6624)(176)(85.4)$$
$$= -1.978\ (\mathrm{s}^{-1})$$

$$M_w = C_{m_\alpha} QS\bar{c}/(u_0 I_y)$$
$$= (-0.683)(36\,757)/(176)(3000)$$
$$= -0.0488\ (1/(\mathrm{ft}\cdot\mathrm{s}))$$

***W*-derivatives**

$$X_{\dot{w}} = 0$$
$$Z_{\dot{w}} = 0$$
$$M_{\dot{w}} = C_{m_{\dot{\alpha}}} \frac{\bar{c}}{2u_0} QS\bar{c}/(u_0 I_y)$$
$$= (04.36)(0.016)(36\,757)/(176)(3000)$$
$$= -0.0049\ (\mathrm{ft}^{-1})$$

***q*-derivatives**

$$X_q = 0$$
$$Z_q = 0$$
$$M_q = C_{m_q} \frac{\bar{c}}{2u_0} qS\bar{c}/I_y$$
$$= (-9.96)(0.016)(37\,757)/(3000)$$
$$= -2.006\ (\mathrm{s}^{-1})$$

Substituting the numerical values of the stability derivatives into Eq. (4.51), we can obtain the stability matrix:

$$\dot{\mathbf{x}} = \mathbf{A}\mathbf{x}$$

or

$$\begin{bmatrix} \Delta u \\ \Delta w \\ \Delta q \\ \Delta\theta \end{bmatrix} = \begin{bmatrix} -0.0446 & 0.0357 & 0.0000 & -32.2 \\ -0.361 & -1.978 & 168.8 & 0.0000 \\ 0.0018 & -0.0389 & -2.885 & 0.000 \\ 0.0000 & 0.0000 & 1.0000 & 0.0000 \end{bmatrix} \begin{bmatrix} \Delta u \\ \Delta w \\ \Delta q \\ \Delta\theta \end{bmatrix}$$

The eigenvalues can be determined by finding eigenvalues of the matrix **A**:

$$|\lambda \mathbf{I} - \mathbf{A}| = 0$$

The resulting characteristic equation is

$$\lambda^4 + 4.91\lambda^3 + 12.50\lambda^2 + 0.63\lambda + 0.57 = 0$$

The solution of the characteristic equation yields the eigenvalues:

$$\lambda_{1,2} = -0.0165 \pm i(0.2144) \qquad \text{(phugoid)}$$

$$\lambda_{3,4} = -2.436 \pm i(2.521) \qquad \text{(short-period)}$$

The period, time, and number of cycles to half-amplitude are readily obtained once the eigenvalues are known.

Phugoid (long-period)	**Short-period**
$t_{1/2} = 0.69/\lvert\eta\rvert = \dfrac{0.69}{-0.0165}$	$t_{1/2} = 0.69/\lvert\eta\rvert = \dfrac{0.69}{-2.436}$
$t_{1/2} = 41.8$ s	$t_{1/2} = 0.283$ s
Period $= 2\pi/\omega = 2\pi/0.2144$	Period $= 2\pi/\omega = 2\pi/2.521$
Period $= 29.3$ s	Period $= 2.49$ s
Number of cycles to half-amplitude	**Number of cycles to half-amplitude**
$N_{1/2} = \dfrac{t_{1/2}}{P} = 0.110 \dfrac{\omega}{\lvert\eta\rvert}$	$N_{1/2} = 0.110 \dfrac{\omega}{\lvert\eta\rvert}$
$= \dfrac{[0.110][0.2144]}{\lvert -0.0165\rvert}$	$= \dfrac{[0.110][2.521]}{\lvert -2.436\rvert}$
$N_{1/2} = 1.43$ cycles	$N_{1/2} = 0.11$ cycles

Now let us estimate the above parameters by means of the long- and short-period approximations. The damping ratio and undamped natural frequency for the long-period motion was given by Eqs (4.69), (4.70), (4.80) and (4.81).

Phugoid-approximation

$$\omega_{n_p} = \sqrt{\frac{-Z_u g}{u_0}} = \left[\frac{-[-0.361][32.21]}{[176]}\right]^{1/2} = 0.258 \text{ rad/s}$$

$$\zeta_p = \frac{-X_u}{2\omega_{n_p}} = \frac{-[-0.0446]}{2[0.258]} = 0.086$$

$$\lambda_{1,2} = -\zeta_p \omega_{n_p} \pm i\omega_n \sqrt{1 - \zeta^2}$$
$$= -[0.086][0.258] \pm i[0.258]\sqrt{1 - [0.086]^2}$$
$$= -0.022 \pm i0.247$$

$$\text{Period} = \frac{2\pi}{\omega} = \frac{2\pi}{0.247} = 25.54 \text{ s}$$

$$t_{1/2} = \frac{0.69}{\eta} = \frac{0.69}{|-0.022|} = 31.4 \text{ s}$$

$$N_{1/2} = 0.110 \frac{\omega}{|\eta|} = 0.110 \frac{[0.247]}{0.022} = 1.24 \text{ cycles}$$

Short-period approximation

$$\omega_{n_{sp}} = \sqrt{\frac{Z_\alpha M_q}{u_0} - M_\alpha}$$

Recall that $Z_\alpha = u_0 Z_w$, $M_\alpha = u_0 M_w$ and $M_{\dot{\alpha}} = u_0 M_{\dot{w}}$.

$$\omega_{n_{sp}} = ((-1.978)(-2.006) - (-0.0488)(176))^{1/2} = 3.55 \text{ rad/s}$$

$$\zeta_{sp} = \left(M_q + M_{\dot{\alpha}} + \frac{Z_\alpha}{u_0}\right) \Big/ [2\omega_{\omega_{n_{sp}}}]$$
$$= ((-2.006) + (-0.86) + (-1.978))/((2)(3.55))$$
$$= 0.688$$

$$\lambda_{1,2,sp} = -\zeta_{sp}\omega_{n_{sp}} \pm i\omega_n\sqrt{1 - \zeta_{sp}^2}$$
$$= -(0.688)(3.55) \pm i(3.55)\sqrt{1 - (0.688)^2}$$
$$= -2.44 \pm i4.43$$

$$\text{Period} = \frac{2\pi}{\omega} = \frac{2\pi}{2.78} = 2.26 \text{ s}$$

$$t_{1/2} = \frac{0.69}{|\eta|} = \frac{0.69}{|-2.44|} = 0.283 \text{ s}$$

$$N_{1/2} = 0.110 \frac{\omega}{|\eta|} = 0.110 \frac{2.78}{|-2.44|} = 0.125 \text{ cycles}$$

A summary of the results from the exact and approximate analysis is included in Table 4.4. In this analysis, the short-period approximation was found to be in closer agreement with the exact solution than the phugoid approximation. In general, the short-period approximation is the more accurate approximation.

The eigenvectors for this problem can be determined by a variety of techniques; however, we will discuss only one relatively straight forward method. For additional information on other techniques, readers should go to their mathematics library or computer center. Most computer facilities maintain digital computer programs suitable for extracting eigenvalues and eigenvectors of large-order systems.

TABLE 4.4
Comparison of exact and approximate methods

	Exact method	Approximate method	Difference
Phugoid	$t_{1/2} = 41.8$ s	$t_{1/2} = 31.4$ s	24.9%
	$P = 29.3$ s	$P = 25.54$ s	16.2%
Short-period	$t_{1/2} = 0.283$ s	$t_{1/2} = 0.283$ s	0%
	$P = 2.521$ s	$P = 2.25$ s	9.9%

To obtain the longitudinal eigenvectors for this example problem, we will start with Eq. (4.60) which is expanded as follows:

$$(\lambda_j - A_{11})\,\Delta u_j - A_{12}\,\Delta w_j - A_{13}\,\Delta q_j - A_{14}\,\Delta\theta_j = 0$$

$$-A_{21}\,\Delta u_j + (\lambda_j - A_{22})\,\Delta w_j - A_{23} q_j - A_{24}\,\Delta\theta_j = 0$$

$$-A_{31}\,\Delta u_j - A_{32}\,\Delta w_j + [\lambda_j - A_{33}]\,\Delta q_j - A_{34}\,\Delta\theta_j = 0$$

$$-A_{41}\,\Delta u_j - A_{42}\,\Delta w_j - A_{43}\,\Delta q_j + [\lambda_j - A_{44}]\,\Delta\theta_j = 0$$

In this set of equations, the only unknowns are the components of the eigenvector, the eigenvalues λ_j and the elements of the **A** matrix were determined previously. Dividing the preceding equations by any one of the unknowns (for this example we will use $\Delta\theta_j$), we obtain four equations for the three unknown ratios. Any three of the four equations can be used to find the eigenvectors. If we drop the fourth equation, we will have a set of three equations with the three unknown ratios, as follows:

$$(\lambda_j - A_{11})\left(\frac{\Delta u}{\Delta\theta}\right)_j - A_{12}\left(\frac{\Delta w}{\Delta\theta}\right)_j - A_{13}\left(\frac{\Delta q}{\Delta\theta}\right)_j = A_{14}$$

$$-A_{21}\left(\frac{\Delta u}{\Delta\theta}\right)_j + (\lambda_j - A_{22})\left(\frac{\Delta w}{\Delta\theta}\right)_j - A_{23}\left(\frac{\Delta q}{\Delta\theta}\right)_j = A_{24}$$

$$-A_{31}\left(\frac{\Delta u}{\Delta\theta}\right)_j - A_{32}\left(\frac{\Delta w}{\Delta\theta}\right)_j + (\lambda_j - A_{33})\left(\frac{\Delta q}{\Delta\theta}\right)_j = A_{34}$$

This set of equations can easily be solved by conventional techniques to yield the eigenvector $[\Delta u/\Delta\theta, \Delta w/\Delta\theta, \Delta q/\Delta\theta, 1]$.

The nondimensional eigenvectors for the example problem have been computed and are listed in Table 4.5. The longitudinal modes can now be examined by means of a vector or Argand diagram. The magnitude of the eigenvectors are arbitrary so only the relative length of the vectors is important.

Figure 4.11 is an Argand diagram illustrating the long-period and short-period modes. In this diagram the lengths of the vectors are decreasing exponentially with time, while the vectors are rotating with the angular rate ω. The motion of the airplane can be imagined as the projection of the eigenvectors along the real axis.

On close examination of Fig. 4.11, several observations can be made. For the long-period mode, we see the changes in angle of attack and pitch rate are

TABLE 4.5
Longitudinal eigenvectors for general aviation

Eigenvector	Long-period	Short-period
		Eigenvalue
	$\lambda = -0.0165 \pm 0.214i$	$\lambda = -2.436 \pm 2.52i$
$\dfrac{\Delta u/u_0}{\Delta\theta}$	$-0.114 \pm 0.837i$	$0.034 \pm 0.025i$
$\dfrac{\Delta w/u_0}{\Delta\theta} = \dfrac{\Delta\alpha}{\Delta\theta}$	$0.008 \pm 0.05i$	$1.0895 \pm 0.733i$
$\dfrac{\Delta(qc/2u_0)}{\Delta\theta}$	$-0.000027 \pm 0.00347i$	$-0.039 \pm 0.041i$

negligible. The motion is characterized by changes in speed and pitch attitude. Notice that the velocity vector leads the pitch attitude by nearly 90° in phase. In contrast, the short-period mode is characterized by changes in angle of attack and pitch attitude with negligible speed variations. As we can see from the vector diagrams, the assumptions we made earlier in developing the long- and short-period approximations are, indeed, consistent with the exact solution.

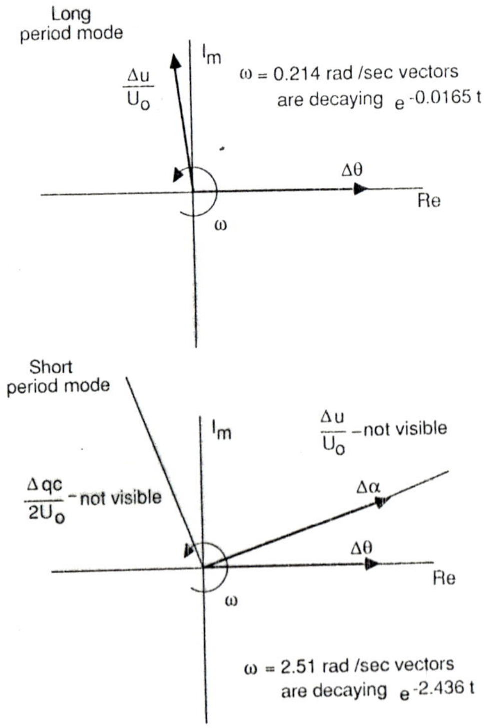

FIGURE 4.11
Eigenvectors for the general aviation airplane in Problem 1.

4.6 THE INFLUENCE OF STABILITY DERIVATIVES ON THE LONGITUDINAL MODES OF MOTION

The type of response we obtain from solving the differential equations of motion depends on the magnitude of the stability coefficients. This can easily be seen by examining the expressions for the damping ratio and frequency of the long- and short-period approximations. Table 4.6 summarizes the effect of each derivative on the longitudinal motion.

Of the two characteristic modes, the short-period mode is the most important. If this mode has a high frequency and is heavily damped, then the airplane will respond rapidly to an elevator input without any undesirable overshoot. When the short-period mode is lightly damped or has a relatively low frequency, the airplane will be difficult to control and, in some cases, may even be dangerous to fly.

The phugoid or long-period mode occurs so slowly that the pilot can easily negate the disturbance by small control movements. Even though the pilot can easily correct for the phugoid mode, it would become extremely fatiguing if the damping were too low.

Figures 4.12 and 4.13 show the effects of varying the center of gravity position and the horizontal tail area size on the long- and short-period response. As the center of gravity is moved rearward, the longitudinal modes become aperiodic and, eventually, unstable.

From a performance standpoint, it would be desirable to move the center of gravity further aft so that trim drags during the cruise portion of the flight could be reduced. Unfortunately, this leads to a less stable airplane. By using an active control stability augmentation system, the requirement of static stability can be relaxed without degrading the airplane's flying qualities.

Recent studies by the commercial aircraft industry have shown that fuel savings of three or four percent are possible if relaxed stability requirements

TABLE 4.6
Influence of stability derivatives on the long and short period motions

Stability derivative	Mode affected	How affected
$M_q + M_{\dot{\alpha}}$	Damping of short-period mode of motion	Increasing $M_q + M_{\dot{\alpha}}$ increases damping
M_α	Frequency of short-period mode of motion	Increasing M_α or static stability increases the frequency
X_u	Damping of the phugoid or long-period mode of motion	Increasing X_u increases damping
Z_u	Frequency of phugoid mode of motion	Increasing Z_u increases the frequency

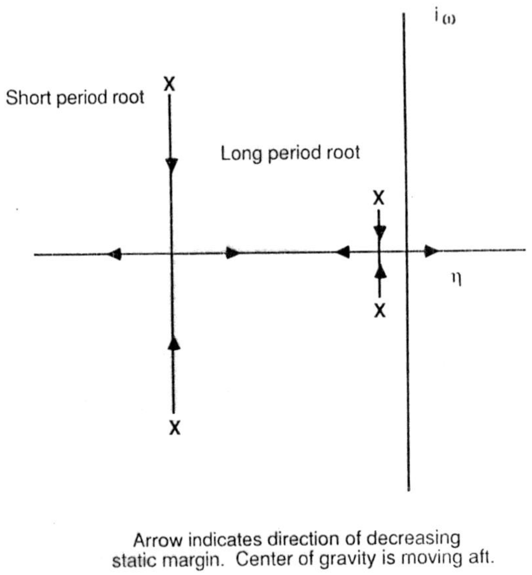

FIGURE 4.12
Influence of center of gravity position on longitudinal response.

and active control stability augmentation are incorporated into the design. With the ever rising costs of jet fuel, this small percentage could mean the savings of many millions of dollars for the commercial airlines.

4.7 FLYING QUALITIES

In the previous sections we examined the stick fixed longitudinal characteristics of an airplane. The damping and frequency of both the short- and long-period motions were determined in terms of the aerodynamic stability derivatives.

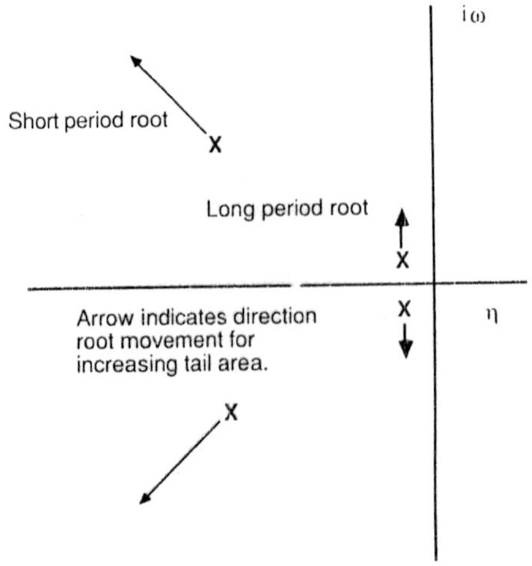

FIGURE 4.13
Influence of horizontal tail area on longitudinal response.

Because the stability derivatives are a function of the geometric and aerodynamic characteristics of the airplane, designers have some control over the longitudinal dynamics by their selection of the vehicle's geometric and aerodynamic characteristics. For example, increasing the tail size would increase both the static stability of the airplane and the damping of the short-period motion.[2] However, the increased tail area would also increase the weight and drag of the airplane and thereby reduce the airplane's performance. The designer is faced with the challenge of providing an airplane with optimum performance that is both safe and easy to fly. To achieve such a goal, the designer needs to know what degree of stability and control is required for the pilot to consider the airplane safe and flyable.

The flying qualities of an airplane are related to the stability and control characteristics and can be defined as those stability and control characteristics that are important in forming the pilot's impression of the airplane. The pilot forms subjective opinions about the ease or difficulty of controlling the airplane in steady and maneuvering flight. In addition to the longitudinal dynamics, the pilot's impression of the airplane is also influenced by the feel of the airplane that is provided to the pilot by the stick force and stick force gradients. The Department of Defense and Federal Aviation Administration has made a list of specifications dealing with airplane flying qualities. These requirements are used by the procuring and regulatory agencies to determine whether an airplane is acceptable for certification. The purpose of these requirements is to ensure that the airplane has flying qualities which do not place any limitation on the vehicle's flight safety or restrict the ability of the airplane to perform its intended mission. The specification of the requirements for airplane flying qualities can be found in Ref. 4.5.

As one might guess, the flying qualities expected by the pilot depend upon the type of aircraft and the flight phase. Aircraft are classified according to size and maneuverability as shown in Table 4.7. The flight phase is divided into three categories as shown in Table 4.8. Category A deals exclusively with military aircraft. Most of the flight phases listed in categories B and C are applicable to either commercial or military aircraft. The flying qualities are specified in terms of three levels:

Level 1 Flying qualities clearly adequate for the mission flight phase.

Level 2 Flying qualities adequate to accomplish the mission flight phase, but some increase in pilot workload or degradation in mission effectiveness, or both, exists.

Level 3 Flying qualities such that the airplane can be controlled safely, but pilot workload is excessive or mission effectiveness is inadequate,

---

[2] Because the aerodynamic derivatives are also a function of Mach number, the designer can only optimize the dynamic characteristics for one flight regime. To provide suitable dynamic characteristics over the entire flight envelope, the designer must provide artificial damping by using stability augmentation.

TABLE 4.7
Classification of airplanes

Class I	Small, light airplanes, such as light utility, primary trainer, and light observation craft
Class II	Medium-weight, low-to-medium maneuverability airplanes, such as heavy utility/search and rescue, light or medium transport/cargo/tanker, reconnaissance, tactical bomber, heavy attack and trainer for Class II
Class III	Large, heavy, low-to-medium maneuverability airplanes, such as heavy transport/cargo/tanker, heavy bomber and trainer for Class III
Class IV	High-maneuverability airplanes, such as fighter/interceptor, attack, tactical reconnaissance, observation and trainer for Class IV

or both. Category A flight phases can be terminated safely, and Category B and C flight phases can be completed.

The levels are determined on the basis of the pilot's opinion of the flying characteristics of the airplane.

Extensive research programs have been conducted by the government and the aviation industry to quantify the stability and control characteristics of the airplane with the pilot's opinion of the airplane's flying qualities. Figure 4.14 is an example of the type of data generated from flying qualities research. This figure shows the relationship between the level of flying qualities and the damping ratio and undamped natural frequency of the short-period mode. This kind of figure is sometimes referred to as a thumbprint plot. Table 4.9 is a summary of the longitudinal specifications for the phugoid and short-period motions which is valid for all classes of aircraft.

TABLE 4.8
Flight phase categories

Nonterminal flight phase	
Category A	Nonterminal flight phases that require rapid maneuvering, precision tracking, or precise flight-path control. Included in the category are air-to-air combat ground attack, weapon delivery/launch, aerial recovery, reconnaissance, in-flight refueling (receiver), terrain-following, antisubmarine search, and close-formation flying
Category B	Nonterminal flight phases that are normally accomplished using gradual maneuvers and without precision tracking, although accurate flight-path control may be required. Included in the category are climb, cruise, loiter, in-flight refueling (tanker), descent, emergency descent, emergency deceleration, and aerial delivery.
Terminal Flight Phases:	
Category C	Terminal flight phases are notmally accomplished using gradual maneuvers and usually require accurate flight-path control. Included in this category are takeoff, catapult takeoff, approach, wave-off/go-around and landing.

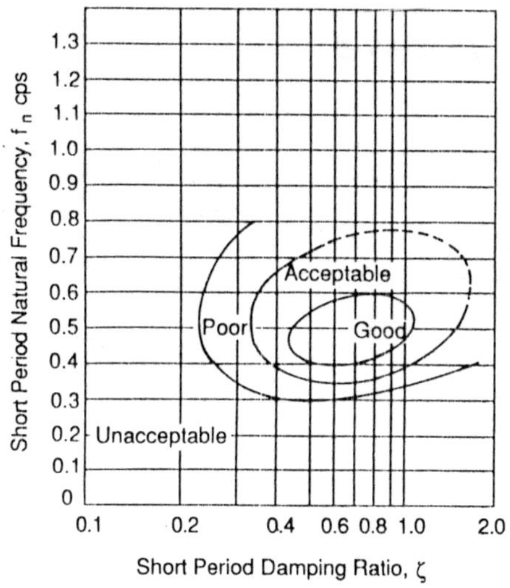

FIGURE 4.14
Short-period flying qualities.

The information provided by Table 4.9 provides the designer with valuable design data. As we showed earlier, the longitudinal response characteristics of an airplane are related to its stability derivatives. Because the stability derivatives are related to the airplane's geometric and aerodynamic characteristics it is possible for the designer to consider flying qualities in the preliminary design phase.

TABLE 4.9
Longitudinal flying qualities

	Phugoid mode			
	Level 1	$\zeta > 0.04$		
	Level 2	$\zeta > 0$		
	Level 3	$T_2 > 55$ s		
	Short-period mode			
	Categories A & C		**Category B**	
Level	ζ_{sp} min	ζ_{sp} max	ζ_{sp} min	ζ_{sp} max
1	0.35	1.30	0.3	2.0
2	0.25	2.00	0.2	2.0
3	0.15	—	0.15	—

Example Problem 4.2. A fighter aircraft has the aerodynamic, mass and geometric characteristics given below. Determine the short-period flying qualities at sea level, at 25 000 ft and at 50 000 ft for a true airspeed of 800 ft/s. How can the designer improve the flying qualities of this airplane?

$$W = 17\,580 \text{ lb} \qquad I_y = 25\,900 \text{ slug}\cdot\text{ft}^2$$

$$S = 260 \text{ ft}^2 \qquad \bar{c} = 10.9 \text{ ft}$$

$$C_{L_\alpha} = 4.0 \text{ rad}^{-1} \qquad C_{m_q} = 04.3 \text{ rad}^{-1} \qquad C_{m_{\dot\alpha}} = -1.7 \text{ rad}^{-1}$$

Solution. The approximate formulas for the short-period damping ratio and frequency are given by Eqs (4.80) and (4.81):

$$\omega_{n_{sp}} = \sqrt{\frac{Z_\alpha M_q}{u_0} - M_\alpha}$$

$$\zeta_{sp} = -\frac{(M_q + M_{\dot\alpha} + Z_g/u_0)}{2\omega_{n_{sp}}}$$

where

$$Z_\alpha = -C_{L_\alpha} QS/m$$

$$M_q = C_{m_q}\left(\frac{\bar{c}}{2u_0}\right)\frac{QS\bar{c}}{I_y}$$

$$M_{\dot\alpha} = C_{m_{\dot\alpha}}\left(\frac{\bar{c}}{2u_0}\right)\frac{QS\bar{c}}{I_y}$$

If we neglect the effect of Mach number changes on the stability coefficient, the damping ratio and frequency can easily be calculated from the preceeding equations. Figure 4.15 is a plot of ζ_{sp} and $\omega_{n_{sp}}$ as functions of the altitude. Comparing the estimated short-period damping ratio and frequency with the pilot opinion contours in Fig. 4.14, we see that this airplane has poor handling qualities at sea level which deteriorate to unacceptable characteristics at altitude.

To improve the flying qualities of this airplane, the designer needs to provide more short-period damping. This could be accomplished by increasing the tail area and/or increasing the tail moment arm. Such geometric changes would increase the stability coefficients C_{m_α}, C_{m_q} and $C_{m_{\dot\alpha}}$. Unfortunately, this can not be accomplished without a penalty in flight performance. The larger tail area results in increased structural weight and empenage drag. For low-speed aircraft, geometric design changes can usually be used to provide suitable flying

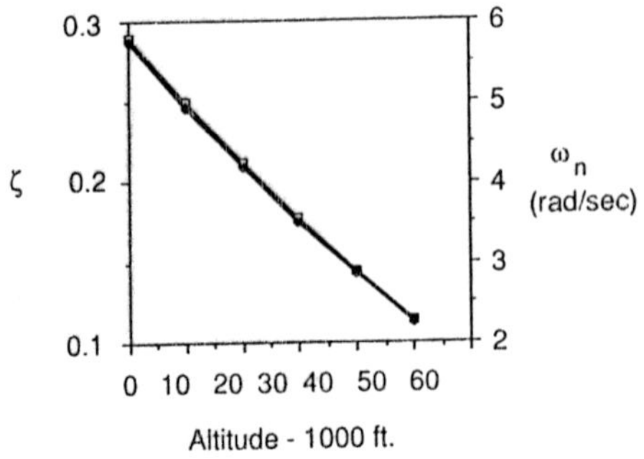

FIGURE 4.15
Variation of ζ_{sp} and $\omega_{n,sp}$ as a function of altitude.

qualities; for aircraft that have an extensive flight envelope, such as fighters, it is not possible to provide good flying qualities over the entire flight regime from geometric considerations alone. This can be accomplished, however, by using a stability augmentation system.

4.8 FLIGHT SIMULATION

To determine the flying quality specifications described in the previous section requires some very elaborate test facilities. Both ground-based and in-flight simulators are used to evaluate pilot opinion on aircraft response characteristics, stick force requirements and human factors data such as instrument design, size, and location.

The ground-based flight simulator provides the pilot with the "feel" of flight by using a combination of simulator motions and visual images. The more sophisticated flight simulators provide six degrees of freedom to the simulator cockpit. Hydraulic servo actuators are attached to the bottom of the simulator cabin and are driven by computers to produce the desired motion. The visual images produced on the windscreen of the simulator are created by projecting images from a camera mounted over a detailed terrain board or by computer-generated images. Figure 4.16 is a sketch of a ground-based simulator.

An example of an in-flight simulator is shown in Fig. 4.17. This figure is a sketch of the U.S. Air Force's Total in Flight Simulator (TIFS) which is a modified C131 transport. By using special force producing control surfaces such as direct lift flaps and side force generators, this airplane can be used to simulate a wide range of larger aircraft. The TIFS has been used to simulate the B-1, C-5 and Space Shuttle among other craft.

The stability characteristics of the simulator can be changed through the computer. This capability permits researchers to establish the relationship between pilot opinion and aircraft stability characteristics. For example, the short-period characteristics of the simulator could be varied and the simulator pilot would be asked to evaluate the case or difficulty of flying the simulator. In this manner, the researcher can establish the pilot's preference for particular airplane response characteristics.

4.9 SUMMARY

In this chapter we have examined the stick fixed longitudinal motion of an airplane using the linearized equations of motion developed in Chapter 3. It was shown that the longitudinal dynamic motion consists of two distinct and separate modes; a long-period oscillation which is lightly damped, and a very short-period but heavily damped oscillation.

Approximate relationships for the long- and short-period modes were developed by assuming that the long-period mode occurred at constant angle of attack and the short-period mode occurred at a constant speed. These

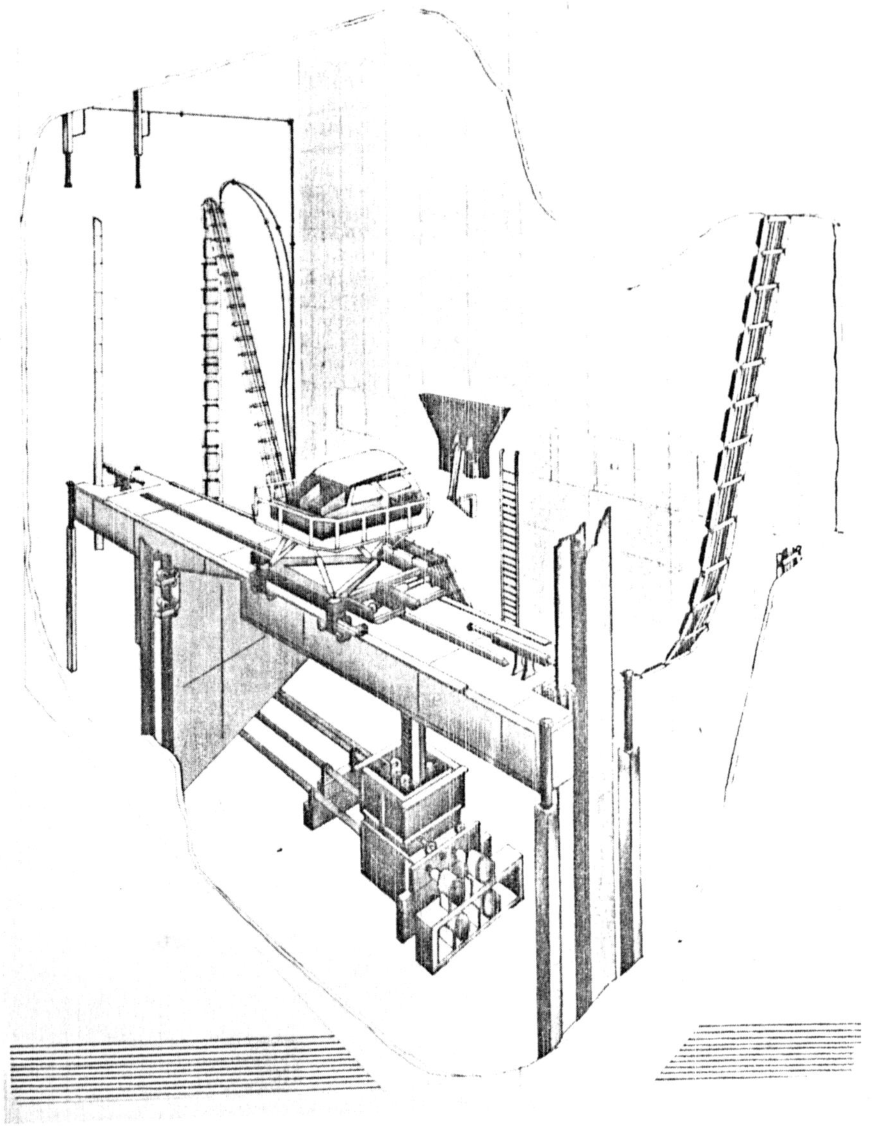

FIGURE 4.16
Sketch of a ground based motion simulator. (Courtesy of SAE Aerospace Engineering magazine).

assumptions were verified by an examination of the exact solution. The approximate formulas permitted us to examine the relationship of the stability derivatives on the longitudinal motion.

Before concluding, it seems appropriate to discuss several areas of research that will affect how we analyze aircraft motions. As mentioned, active control technology in commercial aircraft can be used to improve aerodynamic

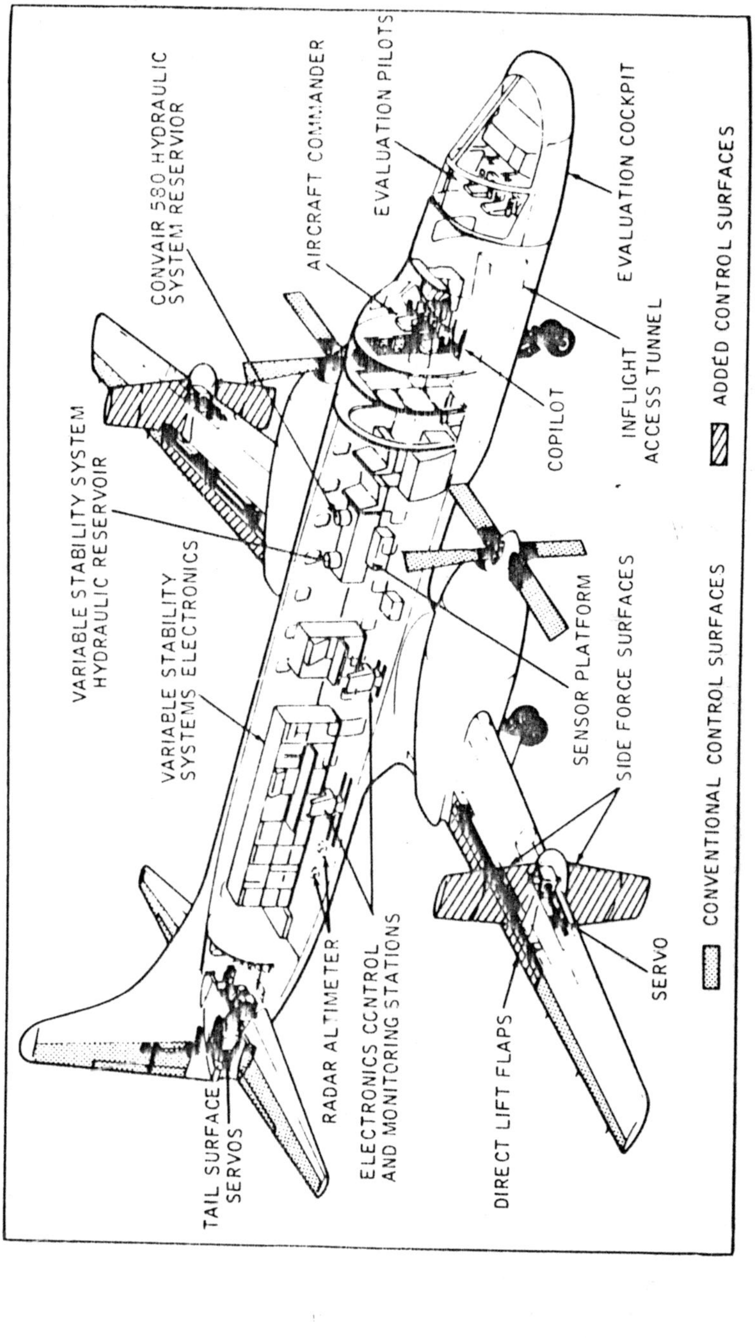

FIGURE 4.17
Airborne flight simulator.

efficiency. With active controls, the aircraft can be flown safely with more aft center of gravity position than would be possible with a standard control system. By shifting the center of gravity further aft, the trim drag can be reduced substantially. This allows for improved fuel economy during the cruise portion of the flight.

Active control technology can also be used to improve ride comfort and reduce wing bending during flight in turbulent air. With active controls located on the wing, a constant load factor can be maintained. This alleviates most of the unwanted response associated with encounters with a vertical gust field. In addition to improving the ride for passengers, the gust alleviation system reduces the wing bending moments, which means the wing can be lighter. Again, this will result in potential fuel savings.

One final topic that has a bearing on the material presented in this chapter is the influence of high angle of attack flight on dynamic stability [4.6, 4.7]. The flight envelopes of high-performance aircraft and missiles have expanded considerably in recent years owing to the increased demands for improved maneuverability. As a result, many of these vehicles encounter higher angles of attack and angular rates than in the past.

As the angle of attack of the airplane increases, the flow around the fuselage separates. The separated flow field can cause non-linear static and dynamic aerodynamic characteristics. An example of the complexity of the leeward wake flows around a slender aircraft and a missile is sketched in Fig. 4.18. Notice that, as the angle of attack becomes large, the separated body vortex flow can become asymmetric. The occurrence of this asymmetry in the flow can give rise to large side forces, yawing, and rolling moments on the airplane or missile even though the vehicle is performing a symmetric maneuver (i.e. sideslip angle equals zero). The asymmetric shedding of the nose vortices is believed to be one of the major contributions to the stall spin departure characteristics of many high-performance airplanes.

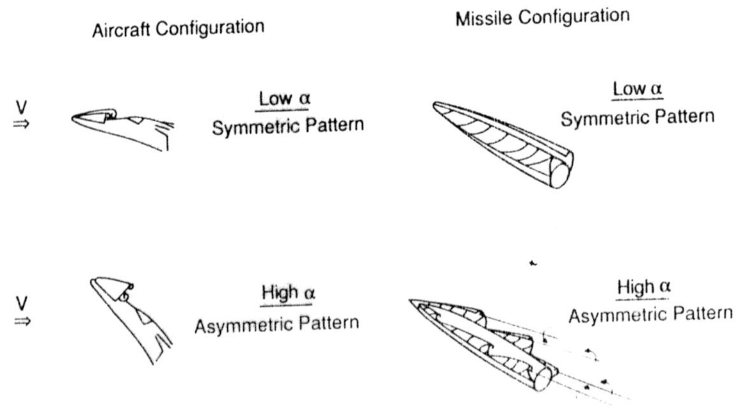

FIGURE 4.18
Vortex flows around an aircraft at large angles of attack.

The asymmetric vortex wake can lead to aerodynamic cross-coupling between the longitudinal and lateral equations of motion. Analyzing these motions requires a much more sophisticated analysis than that presented in this chapter.

4.10 PROBLEMS

4.1. Starting with Newton's second law of motion, develop the equation of motion for the simple torsional pendulum shown in Fig. P4.1. The concept of the torsional pendulum can be used to determine the mass moment of inertia of aerospace vehicles and/or components. Discuss how one could use the torsional pendulum concept to determine experimentally the mass moment of inertia of a test vehicle.

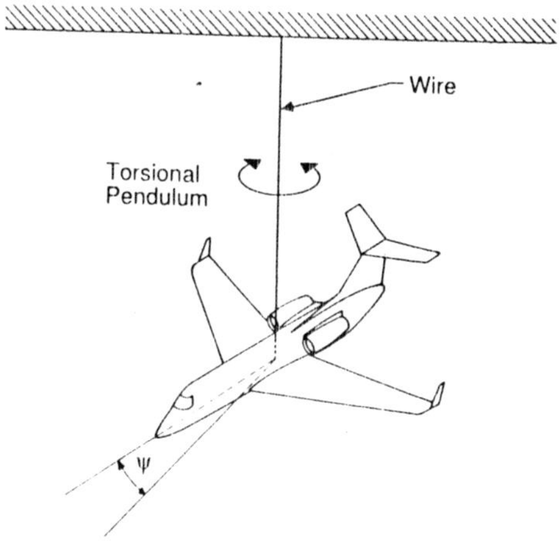

FIGURE P4.1
Sketch of aircraft model swinging as a torsional pendulum.

4.2. A mass weighing 5 lb is attached to a spring as shown in Fig. P4.2. The spring is observed to extend 1 inch when the mass is attached to the spring. Suppose the mass is given an instantaneous velocity of 10 ft/s in the downward direction from the equilibrium position. Determine the displacement of the mass as a function of time.

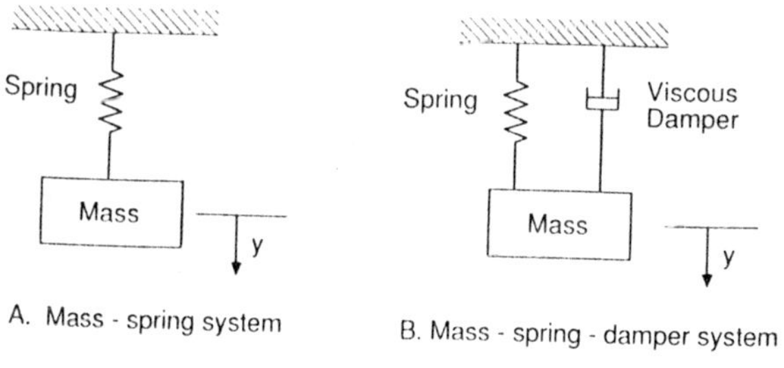

FIGURE P4.2
Spring mass and spring mass damper systems.

4.3. The differential equation for the constrained center of gravity pitching motion of an airplane is computed to be:

$$\ddot{\alpha} + 4\dot{\alpha} + 36\alpha = 0$$

Find the following:

(a) ω_n, natural frequency, rad/s

(b) ζ, damped ratio

(c) ω_d damped natural frequency, rad/s

4.4. Determine the eigenvalues and eigenvectors for the following matrix:

$$\mathbf{A} = \begin{bmatrix} 2 & -3 & 1 \\ 3 & 1 & 3 \\ -5 & 2 & -4 \end{bmatrix}$$

4.5. The characteristic roots of a second-order system are shown in Fig. P4.5. If this system is disturbed from equilibrium, find the time to half-amplitude, the number of cycles to half-amplitude, and the period of motion.

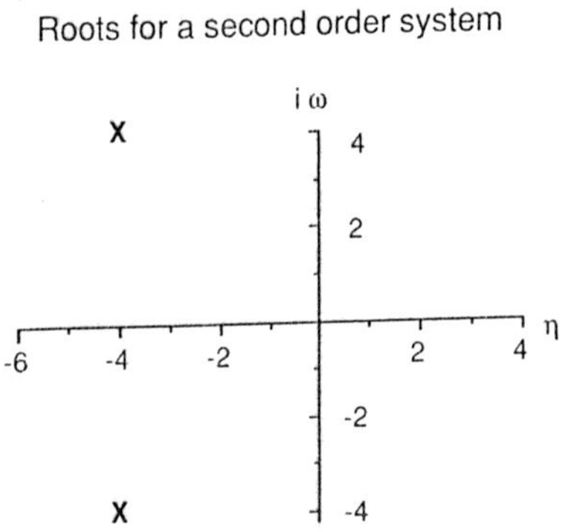

FIGURE P4.5
Second order system roots.

4.6. The missile shown in Fig. P4.6 is constrained so that only a pitching motion is possible. Assume that the aerodynamic damping and static stability come completely from the tail surface (i.e., neglect the body contribution). If the model is displaced 10° from its trim angle of attack ($\alpha_t = 0$) and then released, determine the angle of attack time history. Plot your results. What effect would moving the center of gravity have on the motion of the model?

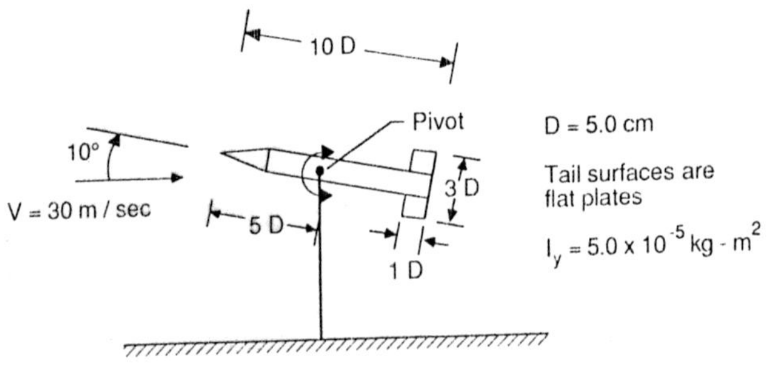

FIGURE P4.6
Pitching wind tunnel model.

4.7. Use the short- and long-period approximations to find the damping ratio for the executive jet airplane described in Appendix B.

4.8. Show that if one neglects compressibility effects, the frequency and damping ratio for the phugoid mode can be expressed as:

$$\omega_{np} = \sqrt{2}\frac{g}{u_0} \qquad \text{and} \qquad \zeta_p = \frac{1}{\sqrt{2}}\frac{1}{L/D}$$

4.9. What effect will increasing altitude have on the short- and long-period modes? Use the approximate formulas in your analysis.

4.10. Develop the equation of motion for an airplane that has freedom only along the flight path, i.e. variations in forward speed. If the airplane is perturbed from its equilibrium state what type of motion would you expect? Clearly state all of your assumptions.

4.11. Develop a computer program to compute the eigenvalues for the longitudinal equations of motion. Use your program to determine the characteristic roots for the executive jet airplane described in Appendix B. Compare your results with those obtained in Problem 4.7.

4.12. An airplane has the following stability and inertial characteristics:

$W = 564\,000$ lb	$C_L = 1.11$
$I_x = 13.7 \times 10$ slug·ft^2	$C_D = 0.102$
$I_y = 30.5 \times 10$ slug·ft^2	$C_{L_\alpha} = 5.7$ rad^{-1}
$I_z = 43.1 \times 10$ slug·ft^2	$C_{D\alpha} = 0.66$ rad^{-1}
$h =$ sea level	$C_{m_\alpha} = -1.26$ rad^{-1}
$s = 5500$ ft^2	$C_{m_{\alpha}} = -3.2$ rad^{-1}
$b = 195.68$ ft	$C_{L_q} = -20.8$ rad^{-1}
$c = 27.3$ ft	$C_{L_\alpha} = -6.7$ rad^{-1}
	$C_{m_q} = -20.8$ rad^{-1}

(a) Find the frequency and damping ratios of the short- and long-period modes.
(b) Find the time to half-amplitude for each mode.
(c) Discuss the influence of the coefficients C_{m_q} and C_{m_α} on the longitudinal motion.

4.13. From data in Fig. P4.13, estimate the time to half amplitude and the number of cycles for both the short- and long-period modes.

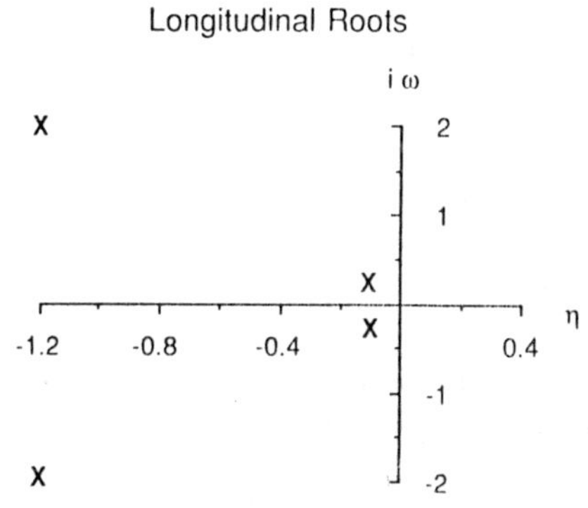

FIGURE P4.13

4.14. A wind-tunnel model is constrained so that only a pitching motion can occur. The model is in equilibrium when the angle of attack is zero. When the model is displaced from its equilibrium state and released, the motion shown in Fig. P4.14 is recorded. Using the data given below determine C_{m_α} and $C_{m_q} + C_{m_{\dot{\alpha}}}$.

$$U_0 = 100 \text{ ft/sec} \qquad \bar{c} = 0.2 \text{ ft}$$
$$Q = 11.9 \text{ lb/ft}^2 \qquad I_y = 0.01 \text{ slug-ft}^2$$
$$S = 0.5 \text{ ft}^2$$

Assume that equation of motion is

$$\theta(t) = \theta_0 e^{\eta t} \cos \omega t$$

where

$$\eta = (M_q + M_{\dot{\alpha}})/2.0$$

and

$$\omega = \sqrt{-M_\alpha}$$

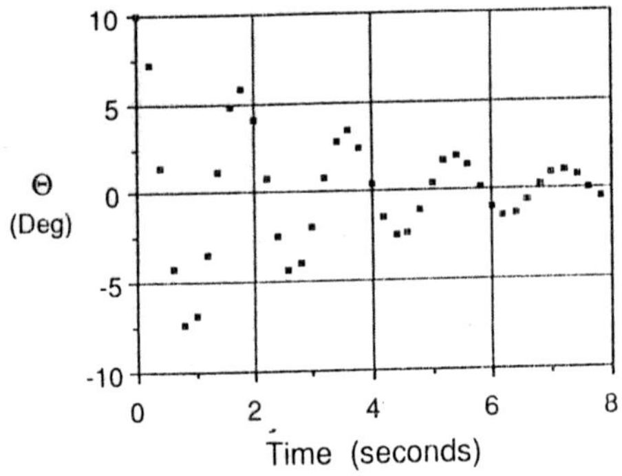

FIGURE P4.14

REFERENCES

4.1. Lanchester, F. W.: *Aerodonetics*, Archibald Constable, London, 1908.
4.2. Perkins, C. D.: "Development of Airplane Stability and Control Technology," *AIAA Journal of Aircraft*, vol. 7, no. 4, 1970 pp. 290–301.
4.3. Bairstow, L.: *Applied Aerodynamics*, 2nd ed., Longmans, Green, New York, 1939.
4.4. Garber, P. E.: 'The Wright Brothers' Contribution to Airplane Design," from *Proceedings of the AIAA Diamond Jubilee of Powered Flight—The Evolution of Aircraft Design*, AIAA, New York, 1978.
4.5. *MIL-F-8785B Military Specifications—Flying Qualities of Piloted Airplanes*, August 1969.
4.6. Curry, W. H., and K. J. Orlick-Ruckermann: "Sensitivity of Aircraft Motion to Aerodynamic Cross-Coupling at high Angles of Attack," paper presented at the AGARD Fluid Dynamics Panel Symposium on Dynamic Stability Parameters held in Athens, Greece, May 22–24, 1978.
4.7. Butler, R. W., and T. F. Langham: "Aircraft Motion Sensitivity to Variations in Dynamic Stability parameters," paper presented at the AGARD Fluid Dynamics Panel Symposium on Dynamic Stability Parameters held in Athens, Greece, May 22–24, 1978.

CHAPTER 5

LATERAL MOTION (STICK FIXED)

"Dutch Roll is a complex oscillating motion of an aircraft involving rolling, yawing and sideslipping. So named for the resemblance to the characteristic rhythm of an ice skater".

[5.1]

5.1 INTRODUCTION

The stick fixed lateral motion of an airplane disturbed from its equilibrium state is a complicated combination of rolling, yawing, and sideslipping motions. As was shown in Chapter 2, an airplane produces both yawing and rolling moments due to sideslip angle. This interaction between roll and yaw produces the coupled motion. There are three potential lateral dynamic instabilities of interest to the airplane designer: they are directional divergence, spiral divergence, and the so-called Dutch roll oscillation.

Directional divergence can occur when the airplane does not possess directional or weathercock stability. If such an airplane is disturbed from its equilibrium state, it will tend to rotate to ever-increasing angles of sideslip. Owing to the side force acting on the airplane, it will fly a curved path at large sideslip angles. For an airplane that has lateral static stability (i.e. dihedral effect) the motion can occur without any significant change in bank angle. Obviously, such a motion cannot be tolerated and can readily be avoided by proper design of the vertical tail surface to ensure directional stability.

Spiral divergency is a nonoscillatory divergent motion which can occur when directional stability is large and lateral stability is small. When disturbed from equilibrium, the airplane enters a gradual spiraling motion. The spiral becomes tighter and steeper as time proceeds and can result in a high-speed spiral dive if corrective action is not taken. This motion normally occurs so gradually that the pilot unconsciously corrects for it.

The Dutch roll oscillation is a coupled lateral-directional oscillation which can be quite objectionable to pilots and passengers. The motion is characterized by a combination of rolling and yawing oscillations which have the same frequency but are out of phase with each other. The period can be of the order of 3 to 15 seconds, so that if the amplitude is appreciable the motion can be very annoying.

Before analyzing the complete set of lateral equations we shall examine several motions with a single degree of freedom. The purpose of examining the single degree of freedom equations is to gain an appreciation of the more complicated motion comprising the stick fixed lateral motion of an airplane.

5.2 PURE ROLLING MOTION

A wind-tunnel model free to roll about its x axis is shown in Fig. 5.1. The equation of motion for this example of a pure rolling motion is

$$\sum \text{Rolling moments} = I_x \ddot{\phi} \tag{5.1}$$

or

$$\frac{\partial L}{\partial \delta_a}\delta_a + \frac{\partial L}{\partial p}p = I_x \ddot{\phi} \tag{5.2}$$

where $(\partial L/\partial \delta_a)\delta_a$ is the roll moment due to the deflection of the ailerons and $(\partial L/\partial p)p$ is the roll-damping moment. Methods for estimating these derivatives were presented in Chapters 2 and 3. The roll angle ϕ is the angle between

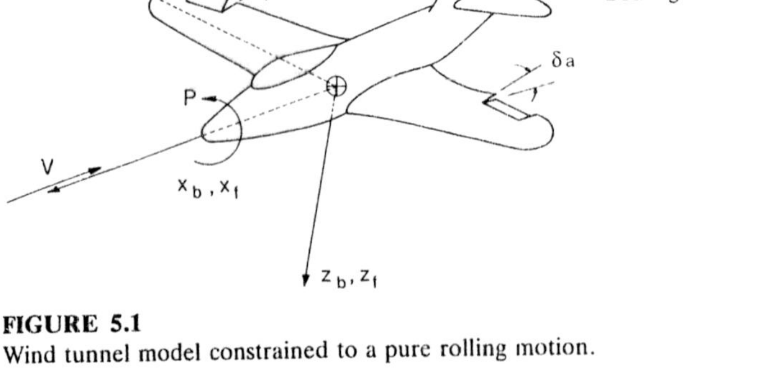

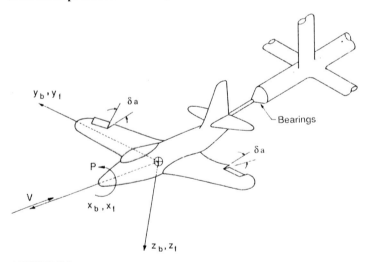

FIGURE 5.1
Wind tunnel model constrained to a pure rolling motion.

z_b of the body axes and z_f of the fixed axis system. The roll rate p is equal to $\dot{\phi}$, which will allow us to re-write Eq. (5.2) as follows:

$$\tau \frac{dp}{dt} + p = \frac{L_\delta \, \Delta\delta_a}{L_p} \tag{5.3}$$

where τ, L_p and $L_{\delta a}$ are defined as follows:

$$\tau = -\frac{1}{L_p} \quad \text{and} \quad L_p = \frac{\partial L/\partial p}{I_x} \quad L_{\delta a} = \frac{\partial L/\partial \delta_a}{I_x} \tag{5.4}$$

The parameter τ is referred to as the time constant of the system. The time constant tells us how fast our system approaches a new steady-state condition after being disturbed. If the time constant is small, the system will respond very rapidly; if the time constant is large, the system will respond very slowly.

The solution of Eq. (5.3) for a step change in the aileron angle is

$$p(t) = -\frac{L_{\delta a}}{L_p}(1 - e^{-t/\tau})\,\Delta\delta_a \tag{5.5}$$

Recall that C_{lp} is negative; therefore, the time constant will be positive. The roll rate time history for this example will be similar to that shown in Fig. 5.2. The steady-state roll rate can be obtained from Eq. (5.5), by assuming that time t is large enough that $e^{-t/\tau}$ is essentially zero:

$$p_{ss} = \frac{-L_{\delta a}}{L_p}\,\Delta\delta_a \tag{5.6}$$

$$p_{ss} = \frac{-C_{l_{\delta a}} QSb/I_x}{C_{l_p}(b/2u_0)QSb/I_x}\,\Delta\delta_a \tag{5.7}$$

$$\frac{p_{ss}b}{2u_0} = -\frac{C_{l_{\delta a}}}{C_{lp}}\,\Delta\delta_a$$

The term $(p_{ss}b/2u_0)$ for full aileron deflection can be used for sizing the aileron. The minimum requirement for this ratio is a function of the class of airplane under consideration:

Cargo or transport airplanes: $pb/2u_0 = 0.07$
Fighter airplanes: $pb/2u_0 = 0.09$

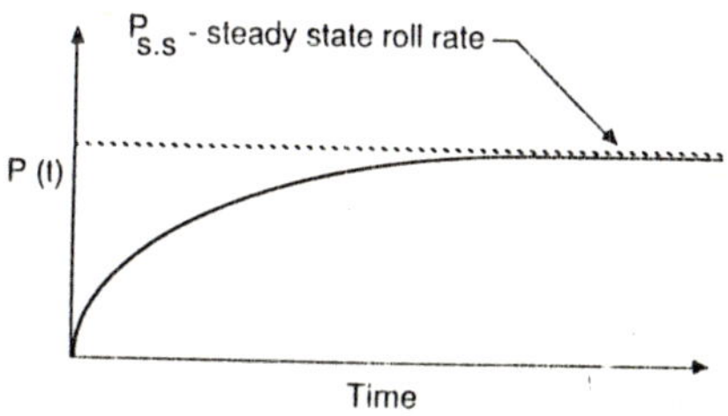

FIGURE 5.2
Typical roll response due to aileron deflection.

Example Problem 5.1. Calculate the roll response of the F104A to a 5° step change in aileron deflection. Assume the airplane is flying at sea level with a velocity of 87 m/s. The F104A has the following aerodynamic and geometric characteristics:

$$C_{l_p} = 0.285\ \text{rad}^{-1} \qquad S = 18\ \text{m}^2$$
$$C_{l_{\delta a}} = 0.039\ \text{rad}^{-1} \qquad b = 6.7\ \text{m}$$
$$I_x = 4676\ \text{kg}\cdot\text{m}^2$$

$$\frac{b}{2u_0} = \frac{6.7}{2(87\ \text{m/s})} = 0.039\ \text{s}$$

$$Q = \tfrac{1}{2}\rho u_0^2 = (0.5)(1.225\ \text{kg/m}^3)(87\ \text{m/s})^2 = 4636\ \text{N/m}^2$$

$$L_p = C_{lp}\frac{b}{2u_0}QSb/I_x$$
$$= (-0.285\ \text{rad}^{-1})(0.039\ \text{s}^{-1})(4636\ \text{N/m}^2)(18\ \text{m}^2)(6.7\ \text{m})(4676\ \text{kg}\cdot\text{m}^2)$$
$$L_p = -1.3\ (\text{s}^{-1})$$
$$\tau = -\frac{1}{L_p} = -\frac{1}{(-1.3\ \text{s}^{-1})} = 0.77\ \text{s}$$

Steady state roll rate

$$p_{ss} = -\frac{L_{\delta a}}{L_p}\Delta\delta_a$$
$$L_{\delta a} = C_{l\delta a}QSb/I_x$$
$$L_{\delta a} = (0.039\ \text{rad}^{-1})(4636\ \text{N/m}^2)(18\ \text{m}^2)(6.7\ \text{m})/(4676\ \text{kg}\cdot\text{m}^2) = 4.66\ (\text{s}^{-2})$$
$$p_{ss} = -(4.66\,1\ \text{s}^{-2})(5\ \text{deg})/(-1.3\ \text{s}^{-1})(57.3\ \text{deg/rad})) = 0.31\ \text{rad/s}$$

Figure 5.4 is a plot of the role rate time history for a step change in aileron deflection.

Let us reconsider this problem. Suppose that Fig. 5.3 is a measured roll

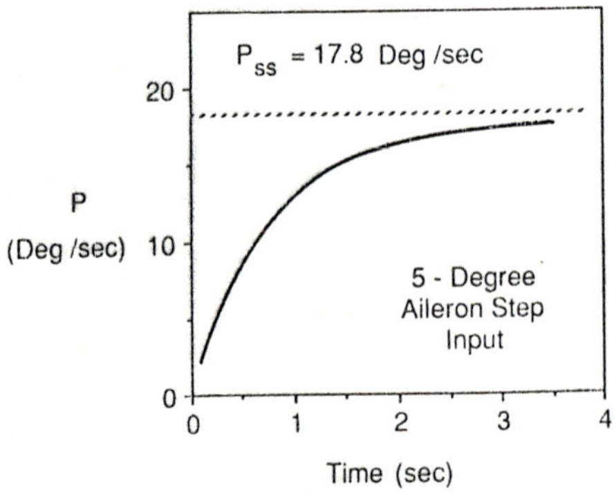

FIGURE 5.3
Roll time history of an F104A to a 5-degree step change in aileron deflection.

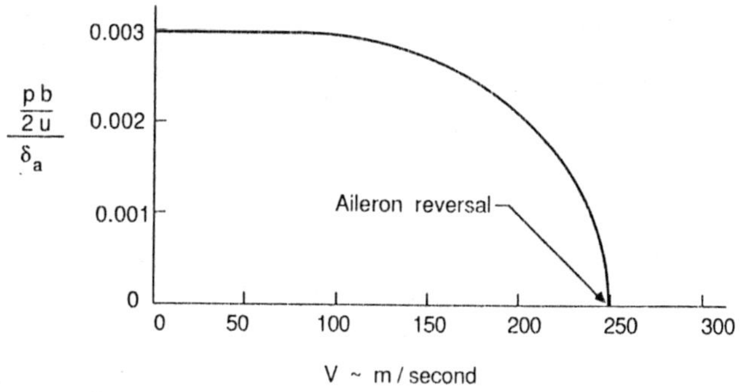

FIGURE 5.4
Aileron control power per degree versus flight velocity.

rate instead of a calculated response. The roll rate of the airplane could be measured by means of a rate gyro appropriately located on the airplane. If we know the mass and geometric properties of the airplane, we can extract the aerodynamic stability coefficients from the measured motion data.

If we fit the solution to the differential equation of motion to the response, we can obtain values for $C_{l_{\delta a}}$ and C_{l_p}. It can be shown that after one time constant, the response of a first-order system to a step input is 63% of its final value. With this in mind, we can obtain the time constant from Fig. 5.3. The steady-state roll rate can also be measured directly from this figure. Knowing τ and p_{ss}, we can compute $L_{\delta a}$ and L_p and, in turn, $C_{l_{\delta a}}$ and C_{l_p}. The technique of extracting aerodynamic data from the measured response is often called in "inverse problem" or "parameter identification".

ROLL CONTROL REVERSAL. The aileron control power per degree, $(pb/2u_0)/\delta_a = 0$ is shown in Fig. 5.4. Note that $(pb/2u_0)/\delta_a$ is essentially a constant, independent of speeds below 140 m/s. However, at high speeds $(pb/2u_0)/\delta_a$ decreases until a point is reached where roll control is lost. The point at which $(pb/2u_0)/\delta_a = 0$ is called the aileron reversal speed. The loss and ultimate reversal of aileron control is due to the elasticity of the wing.

Some understanding of this aeroelastic phenomenon can be obtained from the following simplified analysis. Figure 5.5 is a sketch of a two-dimensional wing with an aileron. As the aileron is deflected downwards, it

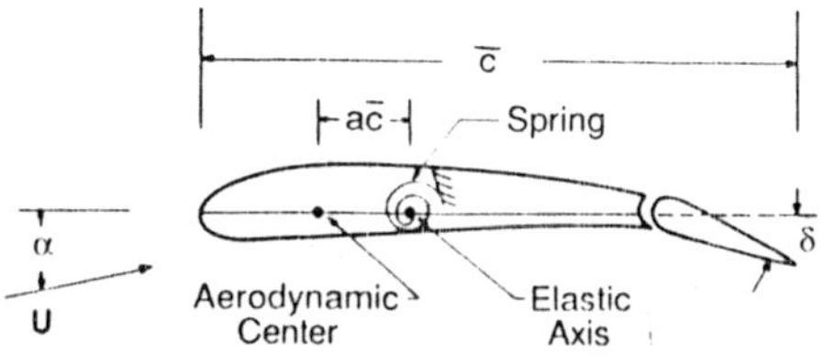

FIGURE 5.5
Two-dimensional wing and aileron.

increases the lift acting on the wing. The increased lift produces a rolling moment. Deflecting the aileron also produces a nose-down aerodynamic pitching moment which tends to twist the wing downward. Such a rotation will reduce the lift and rolling moment. The aerodynamic forces vary with the square of the airplane's velocity, whereas the elastic stiffness of the wing is independent of the flight speed. Thus, it is possible that the wing may twist enough that the ailerons become ineffective. The speed at which the ailerons become ineffective is called the critical aileron reversal speed.

To determine the aileron reversal speed, we shall use the information in Fig. 5.5. The torsional stiffness of the wing will be modeled by the simple torsional spring located at the elastic axis of the wing. The lift and moment coefficients for the two-dimensional airfoil can be expressed as functions of the stability coefficients:

$$C_l = C_{l_\alpha}\alpha + C_{l_\delta}\delta \tag{5.8}$$

$$C_m = C_{m_{\text{ac}}} + C_{m_\delta}\delta \tag{5.9}$$

where δ is the flap angle, i.e. aileron. Aileron reversal occurs when the rate of change of lift with aileron deflection is zero:

$$L = (C_{l_\alpha}\alpha + C_{l_\delta}\delta)Qc \tag{5.10}$$

$$\frac{\mathrm{d}L}{\mathrm{d}\delta} = \left(C_{l_\alpha}\frac{\mathrm{d}\alpha}{\mathrm{d}\delta} + C_{l_\delta}\right)Qc = 0 \tag{5.11}$$

or

$$\frac{\mathrm{d}\alpha}{\mathrm{d}\delta} = -\frac{C_{l_\delta}}{C_{l_\alpha}} \tag{5.12}$$

Note that the angle of attack is a function of the flap angle because the wing can twist. The aerodynamic moment acting about the elastic axis is

$$M[C_{m_{\text{ac}}} + C_{m_\delta}\delta + (C_{l_\alpha}\alpha + C_{l_\delta}\delta)a]Qc^2 \tag{5.13}$$

This moment is balanced by the torsional moment of the wing:

$$k\alpha = [C_{m_{\text{ac}}} + C_{m_\delta}\delta + (C_{l_\alpha}\alpha + C_{l_\delta}\delta)a]Qc^2 \tag{5.14}$$

Differentiating Eq. (5.14) with respect to δ yields

$$k\frac{\mathrm{d}\alpha}{\mathrm{d}\delta} = \left[C_{m_\delta} + \left(C_{l_\alpha}\frac{\mathrm{d}\alpha}{\mathrm{d}\delta} + C_{l_\delta}\right)a\right]Qc^2 \tag{5.15}$$

Substituting Eq. (5.12) into (5.15) and solving for Q yields the critical dynamic pressure:

$$Q_{\text{rev}} = -\frac{kC_{l_\delta}}{c^2 C_{l_\alpha} C_{m_\delta}} \tag{5.16}$$

The reversal speed is given by

$$U_{\text{rev}} = \sqrt{-\frac{2kC_{l_\delta}}{\rho c^2 C_{l_\alpha} C_{m_\delta}}} \tag{5.17}$$

Note that the reversal speed increases with increasing torsional stiffness and increasing altitude.

5.3 PURE YAWING MOTION

As our last example of a motion with a single degree of freedom, we shall examine the motion of an airplane constrained so that it can perform only a simple yawing motion. Figure 5.6 is a sketch of a wind-tunnel model which can only perform yawing motions. The equation of motion can be written as follows:

$$\sum \text{Yawing moments} = I_z \ddot{\psi} \tag{5.18}$$

The yawing moment N and the yaw angle ψ can be expressed as

$$N = N_0 + \Delta N \qquad \psi = \psi_0 + \Delta\psi \tag{5.19}$$

The yawing moment equation reduces to

$$\Delta N = I_z \, \Delta\ddot{\psi} \tag{5.20}$$

where

$$\Delta N = \frac{\partial N}{\partial \beta}\Delta\beta + \frac{\partial N}{\partial \dot{\beta}}\Delta\dot{\beta} + \frac{\partial N}{\partial r}\Delta r + \frac{\partial N}{\partial \delta_r}\Delta\delta_r \tag{5.21}$$

Because the centre of gravity is constrained, the yaw angle ψ and the sideslip

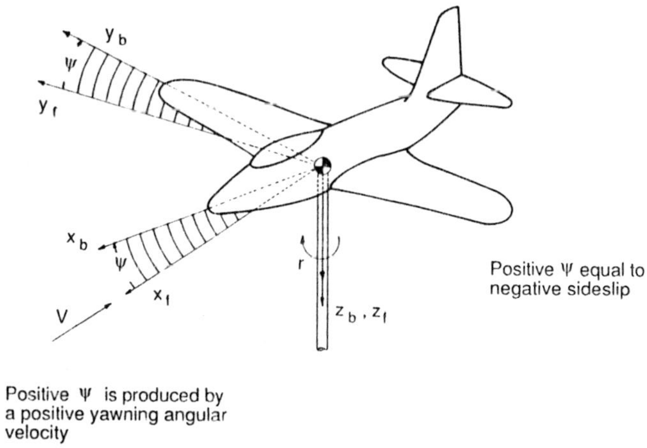

FIGURE 5.6
Wind tunnel model constrained to a pure yawing motion.

angle β are related by the expression

$$\Delta\psi = -\Delta\beta \qquad \Delta\dot{\psi} = -\Delta\dot{\beta} \qquad \Delta\dot{\psi} = r \tag{5.22}$$

Substituting these relationships into Eq. (5.20) and rearranging, yields

$$\Delta\ddot{\psi} - (N_r - N_{\dot{\beta}})\,\Delta\dot{\psi} + N_\beta\,\Delta\psi = N_{\delta r}\,\Delta\delta_r \tag{5.23}$$

where

$$N_r = \frac{\partial N/\partial r}{I_z} \qquad \text{etc.}$$

For airplanes, the term $N_{\dot{\beta}}$ is usually negligible and will be eliminated in future expressions.

The characteristic equation for Eq. (5.23) is

$$\lambda^2 - N_r\lambda + N_\beta = 0 \tag{5.24}$$

The damping ratio ζ and the undamped natural frequency ω_n can be determined directly from Eq. (5.24):

$$\omega_n = \sqrt{N_\beta} \tag{5.25}$$

$$\zeta = -\frac{N_r}{2\sqrt{N_\beta}} \tag{5.26}$$

The solution of Eq. (5.23) for a step change in the rudder control will result in a damped sinusoidal motion, provided the airplane has sufficient aerodynamic damping. As in the case of the pure pitching, we see that the frequency of oscillation is a function of the airplane's static stability (weathercock or directional stability) and the damping ratio is a function of the aerodynamic damping derivative. Figure 5.7 is a sketch of the yawing motion

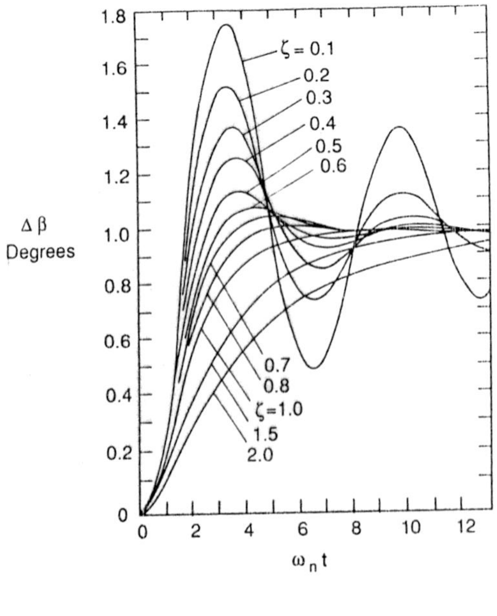

FIGURE 5.7
Yawing motion due to rudder deflection.

due to a step change in rudder deflections for different levels of aerodynamic damping.

5.4 LATERAL-DIRECTIONAL EQUATIONS OF MOTION

The lateral-directional equations of motion consist of the side force, rolling moment and yawing moment equations of motion. The lateral equations of motion can be rearranged into the state space form in the following manner. We start with Eq. (5.27) shown below.

$$\left(\frac{\mathrm{d}}{\mathrm{d}t} - Y_v\right)\Delta v - Y_p\,\Delta p + (u_0 - Y_r)\,\Delta r - g\cos\theta_0\,\Delta\phi = Y_{\delta r}\,\Delta\delta_r$$

$$-L_v\,\Delta v + \left(\frac{\mathrm{d}}{\mathrm{d}t} - L_p\right)\Delta p - \left(\frac{I_{xz}}{I_x}\frac{\mathrm{d}}{\mathrm{d}t} + L_r\right)\Delta r = L_{\delta a}\,\Delta\delta_a + L_{\delta r}\,\Delta\delta_r \tag{5.27}$$

$$-N_v\,\Delta v - \left(\frac{I_{xz}}{I_z}\frac{\mathrm{d}}{\mathrm{d}t} + N_p\right)\Delta p + \left(\frac{\mathrm{d}}{\mathrm{d}t} - N_r\right)\Delta r = N_{\delta a}\,\Delta\delta_a + N_{\delta r}\,\Delta\delta_r$$

Rearranging and collecting terms the above equations can be written in the state variable form:

$$\dot{\mathbf{x}} = \mathbf{A}\mathbf{x} + \mathbf{B}\boldsymbol{\eta} \tag{5.28}$$

The matrices **A** and **B** are defined as follows:

$$\mathbf{A} = \begin{bmatrix} Y_v & Y_p & -(u_0 - Y_r) & g\cos\theta_0 \\ L_v^* + \frac{I_{xz}}{I_x}N_v^* & L_p^* + \frac{I_{xz}}{I_x}N_p^* & L_r^* + \frac{I_{xz}}{I_x}N_r^* & 0 \\ N_v^* + \frac{I_{xz}}{I_z}L_v^* & N_p^* + \frac{I_{xz}}{I_z}L_p^* & N_r^* + \frac{I_{xz}}{I_z}L_r^* & 0 \\ 0 & 1 & 0 & 1 \end{bmatrix} \tag{5.29}$$

$$\mathbf{B} = \begin{bmatrix} 0 & Y_{\delta r} \\ L_{\delta_a}^* + \frac{I_{xz}}{I_x}N_{\delta_a}^* & L_{\delta_r}^* + \frac{I_{xz}}{I_x}N_{\delta_r}^* \\ N_{\delta_a}^* + \frac{I_{xz}}{I_z}L_{\delta_a}^* & N_{\delta_r}^* + \frac{I_{xz}}{I_z}L_{\delta_r}^* \\ 0 & 0 \end{bmatrix} \tag{5.30}$$

$$\mathbf{x} = \begin{bmatrix} \Delta v \\ \delta p \\ \delta r \\ \Delta\phi \end{bmatrix} \quad \text{and} \quad \boldsymbol{\eta} = \begin{bmatrix} \Delta\delta_a \\ \Delta\delta_r \end{bmatrix} \tag{5.31}$$

The starred derivatives are defined as follows:

$$L_v^* = \frac{L_v}{[I - (I_{xz}^2/I_x I_z)]} \qquad N_v^* = \frac{N_v}{[1 - (I_{xz}^2/I_z I_x)]} \qquad \text{etc.} \tag{5.32}$$

If the product of inertia $I_{xz} = 0$, the equations of motion reduce to the following form:

$$\begin{bmatrix} \Delta \dot{v} \\ \Delta \dot{p} \\ \Delta \dot{r} \\ \Delta \dot{\phi} \end{bmatrix} = \begin{bmatrix} Y_v & Y_p & -(u_0 - Y_r) & g \cos \theta_0 \\ L_v & L_p & L_r & 0 \\ N_v & N_p & N_r & 0 \\ 0 & 1 & 0 & 0 \end{bmatrix} \begin{bmatrix} \Delta v \\ \Delta p \\ \Delta r \\ \Delta \phi \end{bmatrix}$$

$$+ \begin{bmatrix} 0 & Y_{\delta r} \\ L_{\delta a} & L_{\delta r} \\ N_{\delta a} & N_{\delta r} \\ 0 & 0 \end{bmatrix} \begin{bmatrix} \Delta \delta_a \\ \Delta \delta_r \end{bmatrix}$$

It is sometimes convenient to use the sideslip angle $\Delta\beta$ instead of the side velocity Δv. These two quantities are related to each other in the following way:

$$\Delta\beta \approx \tan^{-1} \frac{\Delta v}{u_0} = \frac{\Delta v}{u_0} \tag{5.34}$$

Using this relationship, Eqs (5.33) can be expressed in terms of $\Delta\beta$:

$$\begin{bmatrix} \Delta \dot{\beta} \\ \Delta \dot{p} \\ \Delta \dot{r} \\ \Delta \dot{\phi} \end{bmatrix} = \begin{bmatrix} \dfrac{Y_\beta}{u_0} & \dfrac{Y_p}{u_0} & -\left(1 - \dfrac{Y_r}{u_0}\right) & \dfrac{g \cos \theta_0}{u_0} \\ L_\beta & L_p & L_r & 0 \\ N_\beta & N_p & N_r & 0 \\ 0 & 1 & 0 & 0 \end{bmatrix} \begin{bmatrix} \Delta \beta \\ \Delta p \\ \Delta r \\ \Delta \phi \end{bmatrix} + \begin{bmatrix} 0 & \dfrac{Y_{\delta r}}{u_0} \\ L_{\delta_a} & L_{\delta_r} \\ N_{\delta_a} & N_{\delta_r} \\ 0 & 0 \end{bmatrix} \begin{bmatrix} \Delta \delta_a \\ \Delta \delta_r \end{bmatrix} \tag{5.35}$$

The solution of Eq. (5.35) is obtained in the same manner as we solved the state equations in Chapter 4. The characteristic equation is obtained by expanding the following determinant.

$$|\lambda_r \mathbf{I} - \mathbf{A}| = 0 \tag{5.36}$$

where **I** and **A** are the identity and lateral stability matrices, respectively. The characteristic equation determined from the stability matrix **A** yields a quartic equation:

$$A\lambda^4 + B\lambda^3 + C\lambda^2 + D\lambda + E = 0 \tag{5.37}$$

where A, B, C, D, and E are functions of the stability derivatives, mass, and inertia characteristics of the airplane.

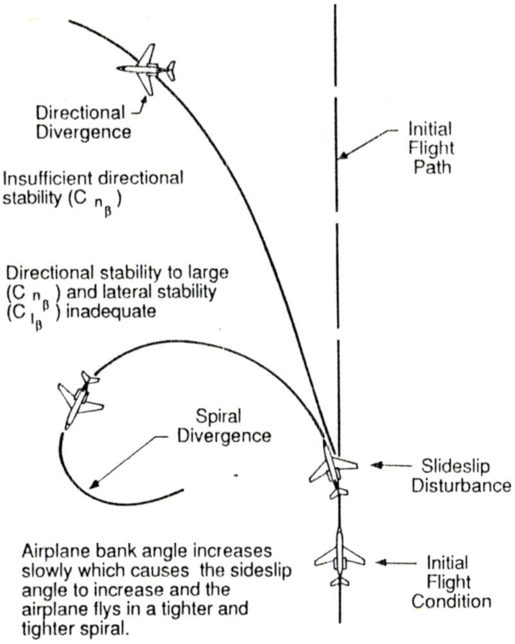
Directional Divergence
Insufficient directional stability (C_{n_β})
Directional stability to large (C_{n_β}) and lateral stability (C_{l_β}) inadequate
Spiral Divergence
Airplane bank angle increases slowly which causes the sideslip angle to increase and the airplane flys in a tighter and tighter spiral.
Initial Flight Path
Slideslip Disturbance
Initial Flight Condition

FIGURE 5.8
Sketch of the spiral motion.

In general, we will find the roots to the lateral-directional characteristic equation to be composed of two real roots and a pair of complex roots. The roots will be such that the airplane response can be characterized by the following motions:

(a) a slowly convergent or divergent motion, called the spiral mode
(b) a highly convergent motion, called the rolling mode
(c) a lightly damped oscillatory motion having a low frequency, called the Dutch roll mode.

Figures 5.8, 5.9 and 5.10 are sketches of the spiral, roll and Dutch roll motions. An unstable spiral mode results in a turning flight trajectory. The airplane's bank angle increases slowly and the airplane flies an ever tightening spiral dive. The rolling motion is usually highly damped and will reach a steady

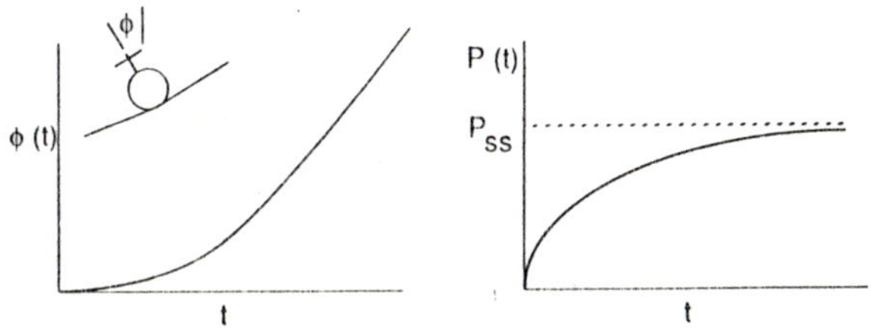
ϕ (t)
t
P (t)
P ss
t

FIGURE 5.9
Sketch of the roll motion.

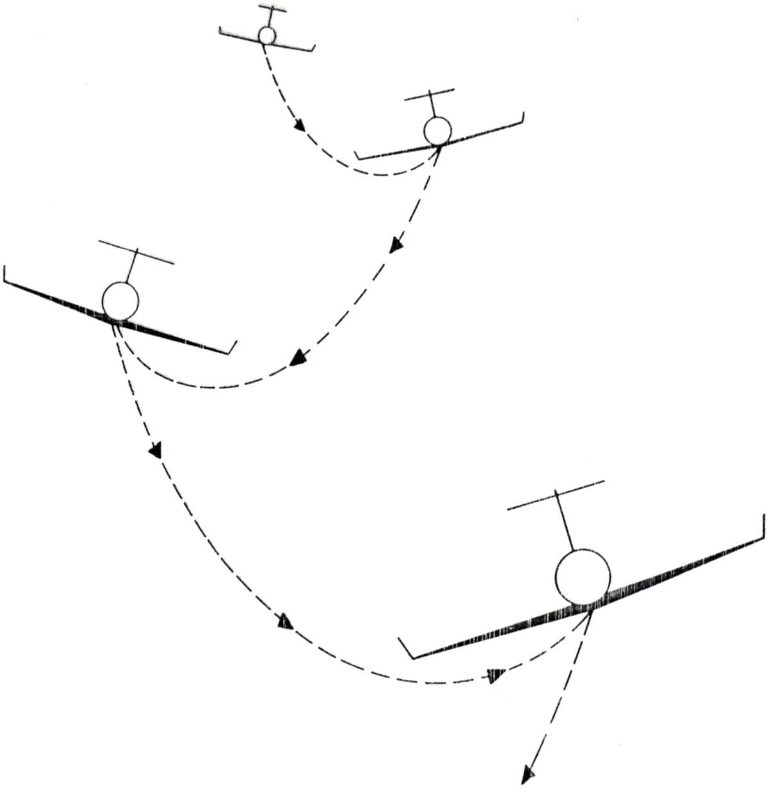

FIGURE 5.10
Sketch of the Dutch roll motion.

state in a very short time. The combination of the yawing and rolling oscillations is called the Dutch roll motion because it resembles the weaving motion of an ice skater.

SPIRAL APPROXIMATION. As indicated in Fig. 5.8, the spiral mode is characterized by changes in bank angle ϕ and heading angle ψ. The sideslip angle is usually quite small but cannot be neglected because the aerodynamic moments do not depend on the roll angle ϕ or the heading angle ψ, but on the sideslip angle β, roll rate p, and yawing rate r.

The aerodynamic contributions due to β and r are usually of the same order of magnitude. Therefore, to obtain an approximation of the spiral mode, we shall neglect the side force equation and $\Delta\phi$. With these assumptions, the equations of motion for the approximation can be obtained from Eqs (5.35):

$$L_\beta \, \Delta\beta + L_r \, \Delta r = 0 \tag{5.38}$$

$$\Delta\dot{r} = N_\beta \, \Delta\beta + N_r \, \Delta r \tag{5.39}$$

or

$$\Delta\dot{r} + \frac{L_r N_\beta - L_\beta N_r}{L_\beta}\Delta r = 0 \tag{5.40}$$

The characteristic root for the above equation is

$$\lambda_{\text{spiral}} = \frac{L_\beta N_r - L_r N_\beta}{L_\beta} \tag{5.41}$$

The stability derivatives L_β (dihedral effect) and N_r (yaw rate damping), are usually negative quantities. On the other hand, N_β (directional stability) and L_r (roll moment due to yaw rate) are generally positive quantities. If the derivatives have the usual sign, then the condition for a stable spiral model is

$$L_\beta N_r - N_\beta L_r > 0 \tag{5.42}$$

or

$$L_\beta N_r > N_\beta L_r \tag{5.43}$$

Increasing the dihedral effect L_β and/or the yaw damping can be used to make the spiral mode stable.

ROLL APPROXIMATION. This motion can be approximated by the single degree of freedom rolling motion which was analyzed earlier in the chapter:

$$\tau\frac{dp}{dt} + p = 0$$

where t is the roll time constant. Therefore

$$\lambda_{\text{roll}} = -\frac{1}{\tau} = L_p \tag{5.44}$$

The magnitude of the roll damping L_p is determined by the wing and tail surfaces.

DUTCH ROLL APPROXIMATION. If we consider the Dutch roll mode to consist primarily of sideslipping and yawing motions, then we can neglect the rolling moment equation. With these assumptions, Eq. (5.35) reduces to

$$\begin{bmatrix}\Delta\dot{\beta} \\ \Delta\dot{r}\end{bmatrix} = \begin{bmatrix}\dfrac{Y_\beta}{u_0} & -\left(1 - \dfrac{Y_r}{u_0}\right) \\ N_\beta & N_r\end{bmatrix}\begin{bmatrix}\Delta\beta \\ \Delta r\end{bmatrix} \tag{5.45}$$

Solving for the characteristic equation yields

$$\lambda^2 - \left(\frac{Y_\beta + u_0 N_r}{u_0}\right)\lambda + \frac{Y_\beta N_r - N_\beta Y_r + u_0 N_\beta}{u_0} = 0 \tag{5.46}$$

From this expression we can determine the undamped natural frequency and the damping ratio as follows:

$$\omega_{n\mathrm{DR}} = \sqrt{\frac{Y_\beta N_r - N_\beta Y_r + u_0 N_\beta}{u_0}} \tag{5.47}$$

$$\zeta_{\mathrm{DR}} = -\frac{1}{2\omega_{n\mathrm{DR}}}\left(\frac{Y_\beta + u_0 N_r}{u_0}\right) \tag{5.48}$$

The approximations developed in this section give, at best, only a rough estimate of the spiral and Dutch roll modes. The approximate formulas should, therefore, be used with caution. The reason for the poor agreement between the approximate and exact solutions is that the Dutch roll motion is truly a three degree of freedom motion with strong coupling between the equations.

Example Problem 5.2. Find the lateral eigenvalues for the general aviation airplane described in Chapter 4 and compare these results with the answers obtained using the lateral approximations. A summary of the aerodynamic and geometric data needed for this analysis is included in the appendices. The stick fixed lateral equations are shown below.

$$\begin{bmatrix} \Delta\dot{\beta} \\ \Delta\dot{p} \\ \Delta\dot{r} \\ \Delta\dot{\phi} \end{bmatrix} = \begin{bmatrix} \frac{Y_\beta}{u_0} & \frac{Y_p}{u_0} & -\left(1 - \frac{Y_r}{u_0}\right) & \frac{g}{u_0}\cos\theta_0 \\ L_\beta & L_p & L_r & 0 \\ N_\beta & N_p & N_r & 0 \\ 0 & 1 & 0 & 0 \end{bmatrix} \begin{bmatrix} \Delta\beta \\ \Delta p \\ \Delta r \\ \Delta\phi \end{bmatrix}$$

Before we can determine the eigenvalues of the stability matrix **A**, we must first calculate the lateral stability derivatives. Table 5.1 is a summary of the lateral stability derivative definitions and Table 5.2 gives a summary of the values of these derivatives for the general aviation airplane.

Substituting the lateral stability derivatives into the stick fixed lateral equations yields

$$\dot{\mathbf{x}} = \mathbf{A}\mathbf{x}$$

or

$$\begin{bmatrix} \Delta\dot{\beta} \\ \Delta\dot{p} \\ \Delta\dot{r} \\ \Delta\dot{\phi} \end{bmatrix} = \begin{bmatrix} -0.254 & 0 & -1.0 & 0.182 \\ -16.02 & -8.40 & 2.19 & 0 \\ 4.488 & -0.350 & -0.760 & 0 \\ 0 & 1 & 0 & 0 \end{bmatrix} \begin{bmatrix} \Delta\beta \\ \Delta p \\ \Delta r \\ \Delta\phi \end{bmatrix}$$

The eigenvalues can be determined by finding the eignevalues of the matrix **A**:

$$|\lambda_r \mathbf{I} - \mathbf{A}| = 0$$

The resulting characteristic equation is

$$\lambda^4 + 9.417\lambda^3 + 13.982\lambda^2 + 48.102\lambda + 0.4205 = 0$$

TABLE 5.1
Summary of lateral directional derivatives

$Y_\beta = \dfrac{QSC_{y\beta}}{m}$ (ft s^2)	$N_\beta = \dfrac{QSbC_{n\beta}}{I_z}$ (s^{-2})	$L_\beta = \dfrac{QSbC_{l\beta}}{I_x}$ (s^{-2})
$Y_p = \dfrac{QSbC_{yp}}{2mu_0}$ (ft/s) or (m/s)	$N_p = \dfrac{QSb^2C_{np}}{2I_z u_0}$ (s^{-1})	
$L_p = \dfrac{QSb^2C_{lp}}{2I_x u_0}$ (s^{-1})		
$Y_r = \dfrac{QSbC_{yr}}{2mu_0}$ (ft/s) or (m/s)	$N_r = \dfrac{QSb^2C_{nr}}{2I_z u_0}$ (s^{-1})	
$L_r = \dfrac{QSb^2C_{lr}}{2I_x u_0}$ (s^{-1})		
$Y_{\delta a} = \dfrac{QSC_{y\delta a}}{m}$ (ft/s^2) or (m/s^2)	$Y_{\delta r} = \dfrac{QSC_{y\delta r}}{m}$ (ft/s^2) or (m/s^2)	
$N_{\delta a} = \dfrac{QSbC_{n\delta a}}{I_z}$ (s^{-2})	$N_{\delta r} = \dfrac{QSbC_{n\delta r}}{I_z}$ (s^{-2})	
$L_{\delta a} = \dfrac{QSbC_{l\delta a}}{I_x}$ (s^{-2})	$L_{\delta r} = \dfrac{QSbC_{l\delta r}}{I_x}$ (s^{-2})	

Solution of the characteristic equation yields the lateral eigenvalues:

$\lambda = -0.00877$ (Spiral mode)

$\lambda = -8.435$ (Roll mode)

$\lambda = -0.487 \pm i(2.335)$ (Dutch roll mode)

$$\lambda_{\text{sprial}} = \frac{L_\beta N_r - L_r N_\beta}{L_\beta}$$

TABLE 5.2
Lateral derivatives for the general aviation airplane

$Y_v = -0.254$ (s^{-1})	$L_v = -0.091$ (ft·s)$^{-1}$
$Y_\beta = -45.72$ (ft/s^2)	$L_\beta = -16.02$ (s^{-2})
$Y_p = 0$	$L_p = -8.4$ (s^{-1})
$Y_r = 0$	$L_r = 2.19$ (s^{-1})
$N_v = 0.025$ (ft·s)$^{-1}$	
$N_\beta = 4.49$ (s^{-2})	
$N_p = -0.35$ (s^{-1})	
$N_r = -0.76$ (s^{-1})	

Substituting in the numerical values for the derivatives yields

$$\lambda_{\text{sprial}} = [(2.193)(4.497) - (-16.02)(-0.761)](-16.02)$$
$$= 0.144$$
$$\lambda_{\text{roll}} = L_p = -8.4$$

The Dutch roll roots are determined from the characteristic equation given by Equation (5.44):

$$\lambda^2 - \frac{(Y_\beta + u_0 N_r)}{u_0}\lambda + \frac{Y_\beta N_r - N_\beta Y_r + u_0 N_\beta}{u_0} = 0$$

or

$$\lambda^2 + 1.102\lambda + 4.71 = 0$$

which yields the following roots

$$\lambda_{\text{DR}} = -0.51 \pm 2.109\text{i}$$

and

$$\omega_{n_{\text{DR}}} = 2.17 \text{ rad/s}$$
$$\zeta_{\text{DR}} = 0.254$$

Table 5.3 compares the results of the exact and approximate analysis. For this example, the roll and Dutch roll roots are in good agreement. On the other hand, the spiral root approximation is very poor.

The relationship between good spiral and Dutch roll characteristics presents a challenge to the airplane designer. In Chapter 2 is was stated that an airplane should possess static stability in both the directional and roll modes. This implies the $C_{n_\beta} > 0$ and $C_{l_\beta} < 0$. However, if we examine the influence of these stability coefficients on the lateral roots by means of a root locus plot, we observe the following. As the dihedral effect is increased, i.e., C_{l_β} becomes more negative, the Dutch roll root moves towards the right half-plane which means the Dutch roll root is becoming less stable, whereas the spiral root is moving in the direction of increased stability. These observations are clearly shown in Figs. 5.11 and 5.12.

Increasing directional stability of the airplane, i.e. C_{n_β} becomes more

TABLE 5.3
Comparison of exact and approximate roots

	Exact			Approximate		
	$T_{1/2}$, s	T_2, s	P, s	$T_{1/2}$, s	T_2, s	P, s
Spiral	78.7	—	—	—	4.79	—
Roll	0.082	—	—	0.082	—	—
Dutch roll	1.42	—	2.69	1.35	—	2.98

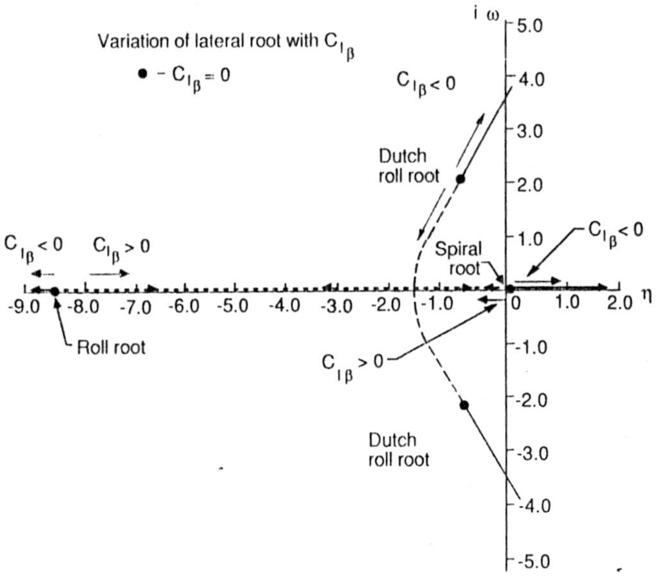

FIGURE 5.11
Variation of lateral roots with C_{l_β}.

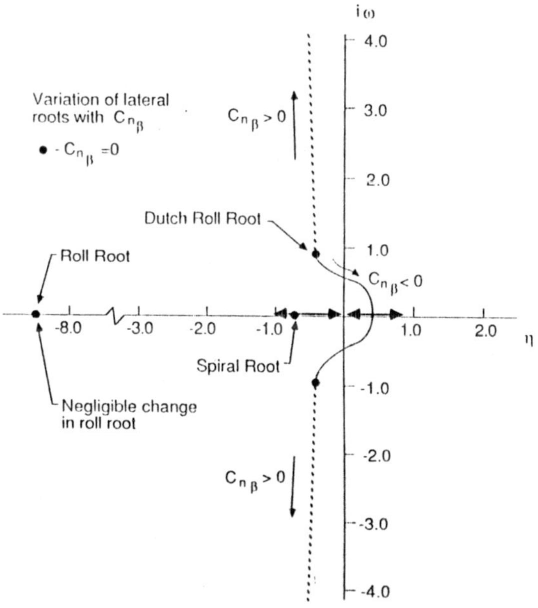

FIGURE 5.12
Variation of lateral roots with C_{n_β}.

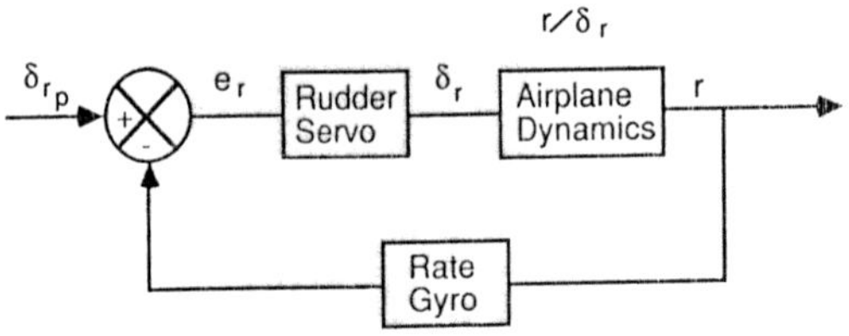

FIGURE 5.13
Sketch of a simple yaw damper.

positive, causing the spiral root to become less stable while the frequency of the Dutch roll root is increased. This is illustrated in the root locus plot of Fig. 5.13. Increasing the yaw damping, i.e. C_{n_r} becomes more negative, will result in better Dutch roll damping. Unfortunately, this is not that easy to achieve simply by geometric design changes. Many airplanes are provided with a rate damper to artificially provide adequate damping in Dutch roll.

5.5 LATERAL FLYING QUALITIES

In this chapter we have examined the lateral direction characteristics of an airplane. The relationship between the aerodynamic stability and control derivatives and the lateral response was discussed. We have developed the necessary equations and analysis procedures to calculate the lateral dynamics. Although these techniques will allow us to determine whether our airplane design is stable or unstable, the analysis does not by itself tell us whether the airplane will be judged by the pilot as having acceptable flying characteristics. To determine whether the airplane will have acceptable flying qualities, the designer needs to know what dynamic characteristics are considered favorable by the pilots who will fly the airplane. This information is available through the lateral-directional flying quality specifications.

The lateral-directional flying quality requirements are listed in Tables 5.4, 5.5 and 5.6. The definition of class and category were presented in Chapter 4. For Example Problem 5.2 the aircraft would be considered as a class 1 vehicle and the flight phase as category B. Using the information from Table 5.4, we find that the aircraft studied here has level 1 flying qualities.

TABLE 5.4
Spiral mode flying qualities

Spiral mode—minunmum time to double amplitude				
Class	**Category**	**Level 1**	**Level 2**	**Level 3**
I & IV	A	12 s	12 s	4 s
	B & C	20 s	12 s	4 s
II & III	All	20 s	12 s	4 s

TABLE 5.5
Roll mode flying qualities

Roll mode—maximum roll time constant, seconds				
Class	**Category**	**Level 1**	**Level 2**	**Level 3**
I, IV	A	1.0	1.4	10
II, III		1.4	3.0	
All	B	1.4	3.0	10
I, IV	C	1.0	1.4	10
II, III		1.4	3.0	

TABLE 5.6
Dutch roll flying qualities

Dutch roll					
Level	**Category**	**Class**	**Min ζ***	**Min $\zeta\omega_n$,* rad/s**	**Min ω_n, rad/s**
	A	I, IV	0.19	0.35	1.0
		II, III	0.19	0.35	0.4
1	B	All	0.08	0.15	0.4
	C	I, II-*C* IV	0.08	0.15	1.0
		II-*L*, III	0.08	0.15	0.4
2	All	All	0.02	0.05	0.4
3	All	All	0.02	—	0.4

Where *C* and *L* denote carrier or land based aircraft.
* The governing damping requirement is that yielding the larger value of ζ.

Example Problem 5.3. As was shown earlier, the Dutch roll motion can be improved by increasing the magnitude of the yaw damping term N_r. One means of increasing N_r is by increasing the vertical tail area. Unfortunately, increasing the vertical tail area will add additional drag to the airplane as well as increasing the directional stability. The increase in directional stability will cause the spiral characteristics of the airplane to be degraded. For most transport and fighter aircraft, increased damping is provided artificially by means of a yaw damper.

In this example we shall examine the basic idea behind a yaw damper. More detailed information on stability augmentation systems and autopilots will be provided in Chapters 7 and 8. To examine how a yaw damper can be used to provide damping for an airplane, let us consider the yawing moment equation developed earlier:

$$\Delta\ddot{\psi} - N_r\,\Delta\dot{\psi} + N_\beta\,\Delta\psi = N_{\delta r}\,\Delta\delta_r$$

Suppose that for a particular airplane the static directional stability, yaw damping and control derivatives were as follows:

$$N_\beta = 1.77\ \mathrm{s}^{-2} \qquad N_r = -0.10\ \mathrm{s}^{-1} \qquad N_{\delta r} = -0.84\ \mathrm{s}^{-1}$$

For this airplane, the damping ratio and undamped natural frequency would be

$$\zeta = -\frac{N_r}{2\sqrt{N_\beta}} = 0.037 \qquad \omega_n = \sqrt{N_\beta} = 1.33\ \mathrm{rad/s}$$

The low damping ratio would result in a free response that would have a large overshoot and poor damping. Such an airplane would be very difficult for the pilot to fly. However, we could design a feedback control system such that the rudder deflection is proportional to the yaw rate, i.e.

$$\delta_r = -k\,\Delta\dot{\psi}$$

Substituting the control deflection expression into the equation of motion and rearranging yields

$$\Delta\ddot{\psi} - (N_r - kN_{\delta r})\,\Delta\dot{\psi} + N_\beta\,\Delta\psi = 0$$

By proper selection of k we can provide the airplane with whatever damping characteristics we desire. For the purpose of this example, let us consider the simple yawing motion to be an approximation of the Dutch roll motion. The flying quality specifications included in Table 5.6 state that a level 1 flying quality rating would be achieved for the landing flight phase if

$$\zeta > 0.08 \qquad \zeta\omega_n > 0.15\ \mathrm{rad/s} \qquad \omega_n > 0.4\ \mathrm{rad/s}$$

A damping ratio of 0.2 and a frequency of 1.33 would be considered acceptable by pilots. The problem now is one of selecting the unknown gain k so that the airplane has the desired damping characteristics. If we compare the yaw moment equation of motion to the standard form for a second-order system, we can establish a relationship for k as follows:

$$2\zeta\omega_n = -(N_r - kN_{\delta r}) \qquad 0.532 = -[-0.1 - k(-0.84)] \qquad k = -0.514$$

Figure 5.14 is a sketch of a simple yaw damper stability augmentation system.

Although we designed a feedback system to provide improved damping, it is possible to control both the damping and the frequency. This can be accomplished by making the rudder deflection proportional to both the yaw rate and yaw angle, i.e.

$$\delta_r = -k_1\,\Delta\dot{\psi} - k_2\,\Delta\psi$$

Substituting this expression back into the differential equation yields

$$\Delta\ddot{\psi} - (N_r - k_1 N_{\delta r})\,\Delta\dot{\psi} + (N_\beta + k_2)\,\Delta\psi = 0$$

The gains k_1 and k_2 are then selected so that the characteristic equation has the desired damping ratio and frequency. The use of feedback control to augment the stability characteristics of an airplane plays an important role in the design of modern aircraft. By using stability augmentation systems, the designer can ensure good flying qualities over the entire flight regime. Furthermore, with the addition of a stability augmentation system, the designer can reduce the inherent aerodynamic static stability of the airplane by reducing the vertical tail size. Thus, the designer can achieve an improvement in performance without compromising the level of flying qualities.

5.6 INERTIAL COUPLING

In the analysis presented in this and the previous chapter, we have treated the longitudinal and lateral equations separately. In so doing we have assumed that there is no coupling between the equations. However, slender high-performance fighter aircraft can experience significant roll coupling which can result in divergence from the desired flight path, causing loss of control or structural failure.

The mechanisms that cause this undesirable behavior can be due to inertial and/or aerodynamic coupling of the equations of motion. To explain how inertial coupling occurs, we examine the non-linearized moment equations developed in Chapter 3. The moment equations are reproduced in Eq. (5.49).

$$\begin{aligned}\sum \text{Roll moments} &= I_x\dot{p} + qr(I_z - I_y) - (\dot{r} + qp)I_{xz}\\ \sum \text{Pitching moments} &= I_y\dot{q} + pr(I_x - I_z) + (p^2 - r^2)I_{xz}\\ \sum \text{Yawing moments} &= I_z\dot{r} + pq(I_y - I_x) + (qr - \dot{p})I_{xz}\end{aligned} \tag{5.49}$$

The first cases of inertial coupling started to appear when fighter aircraft designs were developed for supersonic flight. These aircraft were designed with low aspect ratio wings and long, slender fuselages. In these designs, more of the aircraft's weight was concentrated in the fuselage than in the earlier subsonic fighters. With the weight concentrated in the fuselage, the moments of inertia around the pitch angle yaw axis increased, while the inertia around the roll axis decreased in comparison with subsonic fighter aircraft.

Upon examining Eq. (5.49), we see that the second term in the pitch equation could be significant if the difference in the moments of inertia becomes large. For the case of a slender high performance fighter executing a rapid rolling maneuver the term $pr(I_x - I_z)$ can become large enough to produce an uncontrollable pitching motion.

A similar argument can be made for the product of inertia terms in the equations of motion. The product of inertia I_{xz} is a measure of the uniformity of the distribution of mass about the x axis. For modern fighter aircraft I_{xz} is typically not zero. Again we see that if the airplane is executing a rapid roll maneuver, the term $(p^2 - r^2)I_{xz}$ may be as significant as the other terms in the equation.

Finally, aerodynamic coupling must also be considered when aircraft are maneuvering at high angular rates or at high angles of attack. As was discussed in Chapter 4, high angle of attack flow asymmetries can cause out-of-plane forces and moments even for symmetric flight conditions. Such forces and moments couple the longitudinal and lateral equations of motion.

5.7 SUMMARY

In this chapter we examined the lateral modes of motion. The Dutch roll and spiral motions were shown to be influenced by static directional stability and dihedral effect in an opposing manner. The designer is faced with the dilemma of trying to satisfy the flying quality specifications for both the spiral and Dutch roll modes. This becomes particularly difficult for airplanes that have extended flight envelopes. One way designers have solved this problem is by incorporating a yaw damper in the design. The yaw damper is an automatic system that artificially improves the system damping. The increased damping provided by the yaw damper improves both the spiral and Dutch roll characteristics.

5.8 PROBLEMS

5.1. Determine the response of the A-4D to a 5° step change in aileron deflection. Plot the roll rate versus time. Assume sea-level standard conditions and that the airplane is flying at $M = 0.4$. What is the steady-state roll rate and time constant for this motion?

5.2. For the roll response shown in the Fig. P5.2, estimate the aileron control power L_{δ_a} and the roll damping derivative L_p. Information on the characteristics of the airplane is as follows:

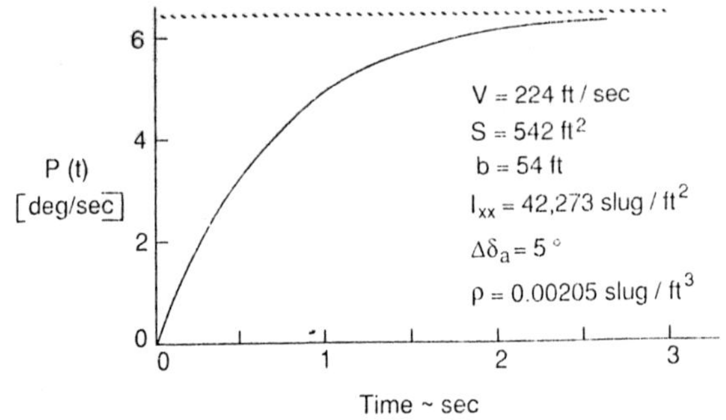

FIGURE P5.2
Roll rate time history.

5.3. A wind-tunnel model free to rotate about its X axis is spun up to 10.5 rad/s by means of a motor drive system. When the motor drive is disengaged, the model spin decays as shown in Fig. P5.3. From the spin time history determine the roll damping derivative L_p.

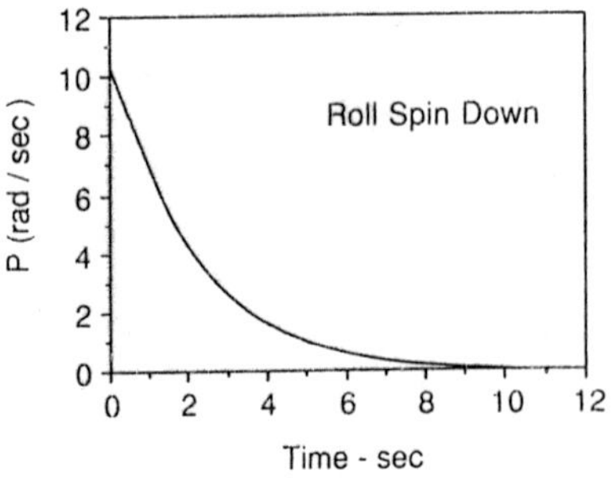

FIGURE P5.3
Roll rate time history.

5.4. Assume the cruciform finned model in Fig. P5.4 is mounted in the wind tunnel so that it is constrained to a pure yawing motion. The model is displaced from its trim position by 10° and then released.

(a) Find the time for the motion to damp to half its initial amplitude.

(b) What is the period of the motion?

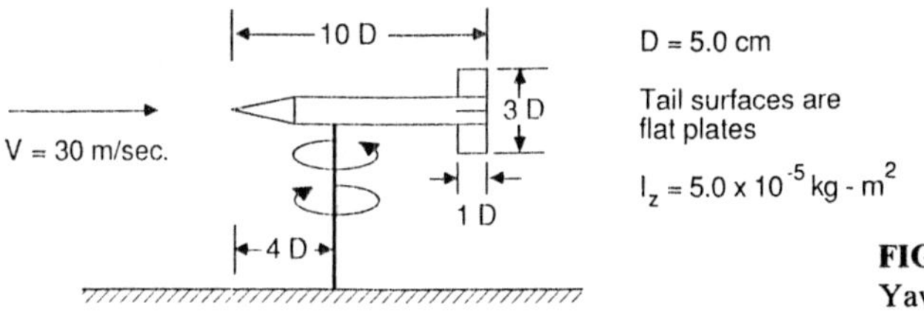

FIGURE P5.4
Yawing Wind Tunnel model.

5.5. Figure P5.5. shows the stick fixed lateral roots of a jet transport airplane. Identify the roots and determine the time for the amplitude and period to halve or double where applicable.

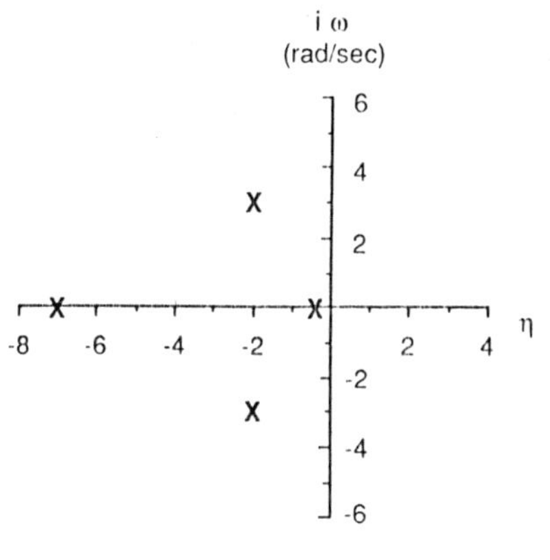

FIGURE P5.5
Lateral roots for a jet transport.

5.6. Develop a computer code to obtain the stick lateral eigenvalues from the lateral stability matrix. Use your computer program to analyze the lateral motion of the 747 jet transport. Estimated aerodynamic, mass and geometric characteristics of the 747 are included in the appendices.

5.7. Using the program developed for Problem 5.5, examine the effect that increasing altitude has on the stick fixed lateral characteristics.

5.8. Using the Dutch roll approximation, determine the state feedback gains so that the damping ratio and frequency of the Dutch roll are 0.3 and 1.0 rad/s, respectively. Assume the airplane has the following characteristics:

$$Y_\beta = -19.5\ \text{ft/s}^2 \qquad Y_r = 1.3\ \text{ft/s}$$

$$N_\beta = 1.5\ \text{s}^{-2} \qquad N_r = -0.21\ \text{s}^{-1}$$

$$Y_{\delta_r} = 4.7\ \text{ft/s}^2 \qquad N_{\delta_r} = -0.082\ \text{s}^{-2}$$

$$u_0 = 400\ \text{ft/s}$$

REFERENCES

5.1. Adams, F. D.: *Aeronautical Dictionary,* National Aeronautics and Space Administration, United States Government Printing Office, Washington, DC., 1959.

5.2. Seckel, E.: *Statility and Control of Airplanes and Helicopters,* Academic Press, New York, 1964.

5.3. Etkin, B.: *Dynamics of Flight,* Wiley, New York, 1972.

5.4. Hage, R. E., and C. D. Perkins: *Airplane Performance, Stability and Control,* Wiley, New York, 1949.

5.5. Rosham, J.: *Flight Dynamics of Rigid and Elastic Airplanes,* University of Kansas Press, 1972.

5.6. Fung, Y. C.: *The Theory of Aeroelasticity,* Wiley, New York, 1955.

5.7. Bisplinghoff, R. L., H. Ashley, and R. L. Halfman: *Aeroelasticity,* Addison Wesley, Reading, MA, 1955.

5.8. Scanlan, R. H., and R. Rosenbaum: *Introduction to the Study of Aircraft Vibration and Flutter,* MacMillan, New York, 1951.

5.9. Abramson, H.: *The Dynamics of Airplanes,* Ronald Press, New York, 1958.

CHAPTER

6

AIRCRAFT RESPONSE TO CONTROL OR ATMOSPHERIC INPUTS

6.1 INTRODUCTION

In the previous chapters we have examined the free response of an airplane as well as several simple examples of single degree of freedom motions with step changes in control input. Another useful input function is the sinusoidal input. The step and sinusoidal input function are important for two reasons. First, the input to many physical systems takes the form of either a step change or sinusoidal signal. Second, an arbitrary function can be represented by a series of step changes or a periodic function can be decomposed by means of Fourier analysis into a series of sinusoidal waves. If we know the response of a linear system to either a step or sinusoidal input, then we can construct the response of the system to an arbitrary input by the principle of superposition.

Of particular importance to the study of aircraft response to control or atmospheric inputs is the steady-state response to a sinusoidal input. If the input to a linear stable system is sinusoidal, then, after the transients have died out, the response of the system will also be a sinusoid of the same frequency. The response of the system is completely described by the ratio of the output to input amplitude and the phase difference, over the frequency range from zero to infinity. The magnitude and phase relationship between the input and output signals is called the frequency response. The frequency response can readily be obtained from the system transfer function by replacing the Laplace variable s by $i\omega$. The frequency response information is usually presented in graphical form using either rectangular, polar, log–log or semi-log plots of the magnitude and phase angle versus the frequency. At first it might appear that the construction of the magnitude and phase plots would be extremely difficult for all but the simplest transfer functions. Fortunately, this is not the case. Consider the factored form of a transfer function, given by

$$G(s) = \frac{k(1+T_a s)(1+T_b s)\cdots}{s^m(1+T_1 s)(1+T_2 s)\cdots\left(1+\dfrac{2\zeta}{\omega_n}s+\dfrac{s^2}{\omega_n^2}\right)} \tag{6.1}$$

The transfer function has been factored into first- and second-order terms. Replacing the Laplace variable s by $i\omega$ and rewriting the transfer function in polar form yields

$$M(\omega) = |G(i\omega)| = \frac{|k| \times |1+T_a i\omega| \times |1+T_b i\omega|\cdots}{m\,|i\omega| \times |1+T_1 i\omega| \times \left|1-\left(\dfrac{\omega}{\omega_n}\right)^2+2\zeta\dfrac{\omega}{\omega_n}i\right|\cdots}$$
$$\times \exp[i(\phi_a+\phi_b\cdots-\phi_1-\phi_2\cdots)] \tag{6.2}$$

Now, if we take the logarithm of the above equation, we obtain

$$\begin{aligned}\log M(\omega) &= \log k + \log|1+T_a i\omega| + \log|1+T_b i\omega|\cdots - m\log\omega\\ &\quad - \log|1+T_1 i\omega| - \log|1+T_2 i\omega|\\ &\quad - \log\left|1-\left(\frac{\omega}{\omega_n}\right)^2+2\zeta\frac{\omega}{\omega_n}i\right| - \cdots\end{aligned} \tag{6.3}$$

or

$$\begin{aligned}\angle G(j\omega) &= \tan^{-1}\omega T_a + \tan^{-1}\omega T_b + \cdots m(90^\circ)\\ &\quad - \tan^{-1}\omega T_1 + \cdots - \tan^{-1}\left(\frac{2\zeta\omega_n\omega}{\omega_n^2-\omega^2}\right)\end{aligned} \tag{6.4}$$

By expressing the magnitude in terms of logarithms, the magnitude of the transfer function is readily obtained by the addition of the individual factors. The contribution of each of the basic factors, i.e. gain, pole at the origin, simple poles and zeros, and complex poles and zeros are presented in an appendix at the end of this book. In practice, the log-magnitude is often expressed in units called decibels (dB). The magnitude in decibels is found by multiplying each term in Eq. (6.3) by 20:

$$\text{Magnitude in dB} = 20\log|G(i\omega)| \tag{6.5}$$

The frequency response information of a transfer function is represented by two graphs, one of the magnitude and the other of the phase angle, both versus the frequency on a logarithmic scale. When the frequency response data are presented in this manner, the plots are referred to as Bode diagrams after

H. W. Bode who made significant contributions to frequency response analysis.

We shall now look at the application of the frequency response techniques to the longitudinal control transfer functions. As the first example, let us consider the longitudinal pitch angle to elevator transfer function developed in Chapter 3. For example, the longitudinal pitch angle to elevator transfer function was found to be

$$\frac{\theta(s)}{\delta_e(s)}=\frac{A_\theta s^2+B_\theta s+C_\theta}{As^4+Bs^3+Cs^2+Ds+E} \tag{6.6}$$

which can be written in the factored form:

$$\frac{\theta(s)}{\delta_e(s)}=\frac{k_{\theta\delta}(T_{\theta 1}s+1)(T_{\theta 2}s+1)}{\left(\dfrac{s^2}{\omega_{n\mathrm{sp}}^2}+\dfrac{2\zeta_{\mathrm{sp}}}{\omega_{n\mathrm{sp}}}s+1\right)\left(\dfrac{s^2}{\omega_{n\mathrm{p}}^2}+\dfrac{2\zeta_p}{\omega_{n\mathrm{p}}}s+1\right)} \tag{6.7}$$

The magnitude and phase angle of the pitch angle of the pitch altitude elevator control transfer function is obtained by replacing s by $j\omega$ as follows:

$$\left|\frac{\theta(i\omega)}{\delta_e(i\omega)}\right|=\frac{|k_{\theta\delta}|\,|T_{\theta 1}i\omega+1|}{\left|\dfrac{(i\omega)^2}{\omega_{n\mathrm{sp}}^2}+\dfrac{2\zeta_{\mathrm{sp}}}{\omega_{n\mathrm{sp}}}i\omega+1\right|}\frac{|T_{\theta 2}i\omega+1|}{\left|\dfrac{(i\omega)^2}{\omega_{np}^2}+\dfrac{2\zeta_p}{\omega_{np}}i\omega+1\right|} \tag{6.8}$$

$$\angle\theta(i\omega)/\delta_e(i\omega)=\tan^{-1}\omega T_{\theta 1}+\tan^{-1}\omega T_{\theta 2}-\tan^{-1}[2\zeta_{\mathrm{sp}}\omega_{n\mathrm{sp}}\omega/(\omega_{n\mathrm{sp}}^2-\omega^2)]$$
$$-\tan^{-1}[2\zeta_p\omega_{np}\omega/(\omega_{np}^2-\omega^2)]$$

The frequency response for the pitch attitude to control deflection for the corporate business jet described in the Appendix is shown in Fig. 6.1. The amplitude ratio at both the phugoid and short period frequency are of comparable magnitude. At very large frequencies the amplitude ratio is very small, which indicates that the elevator has a negligible effect on the pitch attitude in this frequency range.

The frequency response for the change in forward speed and angle of

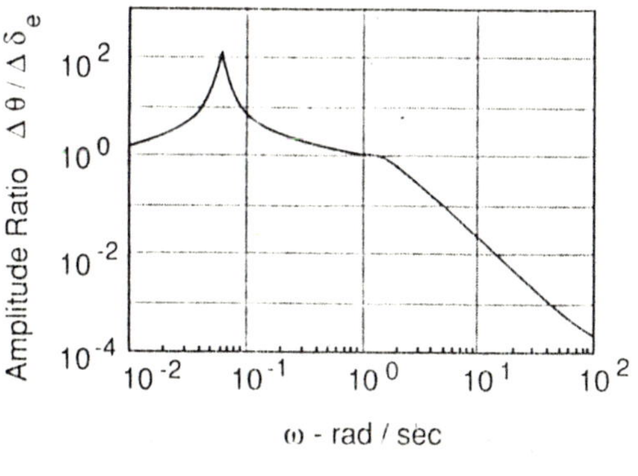

FIGURE 6.1

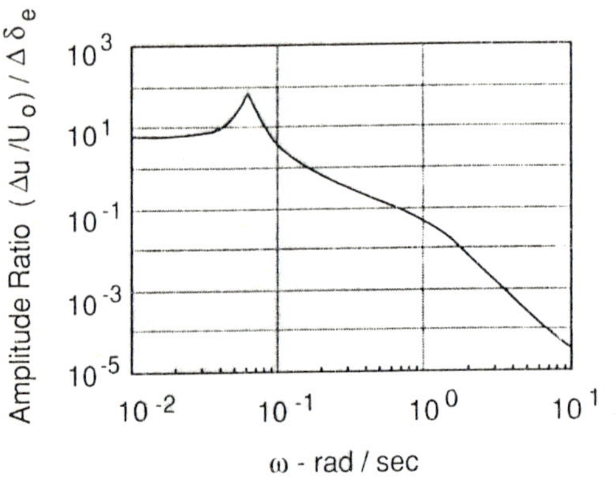

FIGURE 6.2

attack to control input is shown in Figs 6.2 and 6.3 for the same aircraft. For the speed elevator transfer function the amplitude ratio is large at the phugoid frequency and very small at the short-period frequency. Recall that in Chapter 4 we assumed that the short-period motion occurred at essentially constant speed. The frequency response plot confirms the validity of this assumption. Figure 6.3 shows the amplitude ratio of the angle of attack to elevator deflection; here we see that angle of attack is constant at the low frequencies. This again is in keeping with the assumption we made regarding the phugoid approximation. Recall that in the phugoid approximation the angle of attack was assumed to be constant. The phase plots shows that there is a large phase lag in the response of the speed change to elevator inputs. The phase lag for α/δ is much smaller, which means that the angle of attack will respond faster than the change in forward speed to an elevator input.

A similar type of analysis can be conducted for the lateral response to aileron or rudder control input. Several problems dealing with the lateral frequency response are presented at the end of this chapter.

Frequency response techniques are also useful in studying the motion of an aircraft encountering atmosphere turbulence. In Chapter 3, the equations of

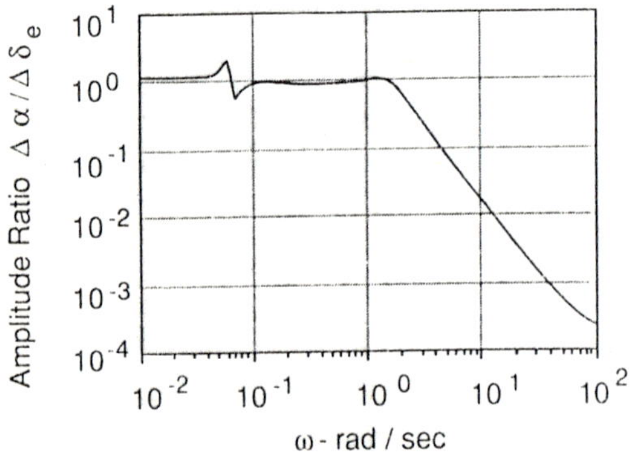

FIGURE 6.3

motion were developed for flight in a stationary atmosphere. In the following sections we shall discuss the influence of wind gusts, i.e. turbulence, on aircraft response.

6.2 EQUATIONS OF MOTION IN A NONUNIFORM ATMOSPHERE

The atmosphere is rarely calm but is usually characterized by winds, gusts, and turbulence. In order to study the influence of atmospheric disturbances on aircraft motions, the equations must be modified. The aerodynamic forces and moments acting on the airplane depend on the relative motion of the airplane to the atmosphere and not on the inertial velocities. Therefore, to account for atmospheric disturbances such as winds, gusts, or turbulence, the forces and moments must be related to the relative motion with respect to the atmosphere. This is accomplished by expressing the velocities used in calculating the aerodynamics in terms of the inertial and gust velocities as shown below.

$$\begin{aligned} \Delta u_a &= \Delta u - u_g \qquad \Delta v_a = \Delta v - v_g \qquad \Delta w_a = \Delta w - w_g \\ \Delta p_a &= \Delta p - p_g \qquad \Delta q_a = \Delta q - q_g \qquad \Delta r_a = \Delta r - r_g \end{aligned} \tag{6.9}$$

where the Δ quantities are the perturbations in the inertial variables and the subscripted variables are the gust velocities. The aerodynamic forces and moments can now be expressed as follows:

$$\begin{aligned} \Delta X &= \frac{\partial X}{\partial u}(\Delta u - u_g) + \frac{\partial X}{\partial w}(\Delta w - w_g) + \frac{\partial X}{\partial \dot{w}}(\Delta \dot{w} - \dot{w}_g) \\ &\quad + \frac{\partial X}{\partial q}(\Delta q - q_g) + \frac{\partial X}{\partial \delta_e}\Delta \delta_e \\ \Delta Z &= \frac{\partial Z}{\partial u}(\Delta u - u_g) + \frac{\partial Z}{\partial w}(\Delta w - w_g) + \frac{\partial Z}{\partial \dot{w}}(\Delta \dot{w} - \dot{w}_g) + \cdots \\ &\;\;\vdots \\ \Delta N &= \frac{\partial N}{\partial v}(\Delta v - v_g) + \frac{\partial N}{\partial r}(\Delta r - r_g) + \frac{\partial N}{\partial p}(\Delta p - p_g) \end{aligned} \tag{6.10}$$

The disturbances in the atmosphere can be described by the spatial and temporal variations in the gust components. The rotational gusts q_g, p_g, etc., included in Eqs (6.10) arise from the variation of u_g, v_g and w_g with position and time.

The rotary gusts p_g, q_g and r_g occur due to the spatial variations of the gust components. For example if the gust field wavelength is large in comparison with the airplane, as shown in Fig. 6.4, the vertical gust produces a spanwise variation of velocity along the span of the wing. The linear variation of velocity across the span is the same as that produced on a rolling wing. The

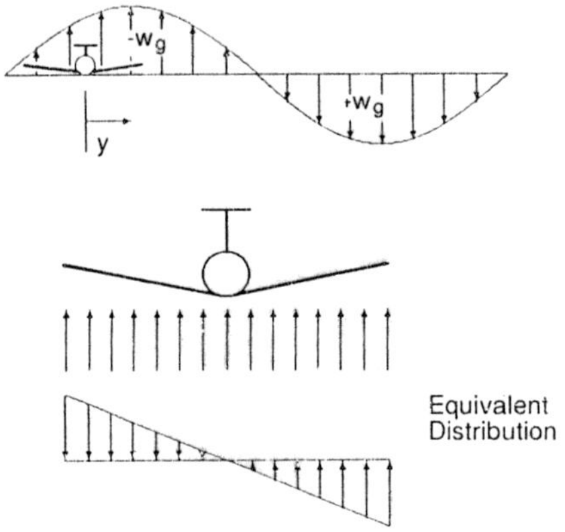

FIGURE 6.4

velocity normal to the wing at some point along the span is given by

$$w = py \tag{6.11}$$

or

$$\frac{\partial w}{\partial y} = p \tag{6.12}$$

Using this analogy, we can express the rotary gust velocity in terms of the gradient in the vertical gust field;

$$p_g = \frac{\partial w_g}{\partial y} \tag{6.13}$$

In a similar manner, the q_g can be developed. The variation of the vertical gust velocity along the X axis of the airplane is similar to the velocity distribution created on a pitching airplane. Figure 6.5 helps to show the origin of rotary

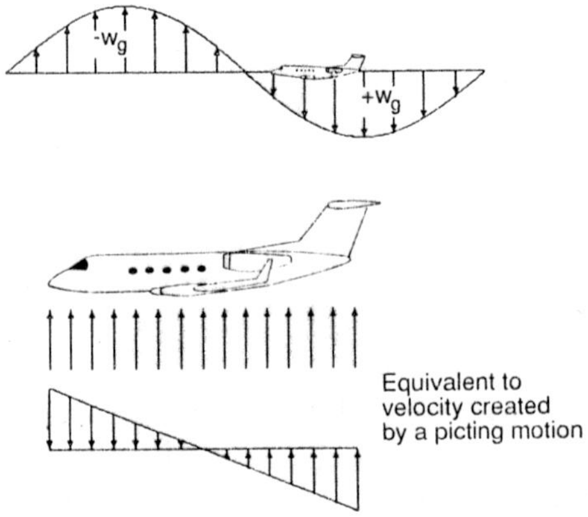

FIGURE 6.5

gust q_g:

$$q_g = \frac{\partial w_g}{\partial x} \tag{6.14}$$

or

$$q_g = -\frac{\partial w_g/\partial t}{\partial x/\partial t} = -\dot{w}_g/u_0 \tag{6.15}$$

The equations of motion, modified to account for atmospheric disturbances, can be written in the state-space form as follows:

$$\dot{\mathbf{x}} = \mathbf{A}\mathbf{x} + \mathbf{B}\eta + \mathbf{C}\boldsymbol{\xi} \tag{6.16}$$

where $\mathbf{x}$, $\boldsymbol{\eta}$ and $\boldsymbol{\xi}$ are the state, control and gust disturbance vectors. The longitudinal equations are

$$\begin{bmatrix} \Delta\dot{u} \\ \Delta\dot{w} \\ \Delta\dot{q} \\ \Delta\dot{\theta} \end{bmatrix} = \begin{bmatrix} X_u & X_w & 0 & -g \\ Z_u & Z_w & U_0 & 0 \\ M_u & M_w & M_q & 0 \\ 0 & 0 & 1 & 0 \end{bmatrix} \begin{bmatrix} \Delta u \\ \Delta w \\ \Delta q \\ \Delta\theta \end{bmatrix} + \begin{bmatrix} X_\delta & X_{\delta t} \\ Z_\delta & Z_{\delta t} \\ M_\delta & M_{\delta t} \\ 0 & 0 \end{bmatrix} \begin{bmatrix} \Delta\delta_e \\ \Delta\delta_t \end{bmatrix}$$
$$+ \begin{bmatrix} -X_u & -X_w & 0 \\ -Z_u & -Z_w & 0 \\ -M_u & -M_w & -M_q \\ 0 & 0 & 0 \end{bmatrix} \begin{bmatrix} u_g \\ w_g \\ q_g \end{bmatrix} \tag{6.17}$$

and the lateral equations are

$$\begin{bmatrix} \Delta\dot{v} \\ \Delta\dot{p} \\ \Delta\dot{r} \\ \Delta\dot{\phi} \end{bmatrix} = \begin{bmatrix} Y_v & 0 & (Y_r - u_0) & g \\ L_v & L_p & L_r & 0 \\ N_v & N_p & N_r & 0 \\ 0 & 1 & 0 & 0 \end{bmatrix} \begin{bmatrix} \Delta v \\ \Delta p \\ \Delta r \\ \Delta\phi \end{bmatrix} + \begin{bmatrix} Y_{\delta r} & Y_{\delta r} \\ L_{\delta a} & L_{\delta r} \\ N_{\delta a} & N_{\delta r} \\ 0 & 0 \end{bmatrix} \begin{bmatrix} \Delta\delta_a \\ \Delta\delta_r \end{bmatrix}$$
$$+ \begin{bmatrix} -Y_v & 0 & 0 \\ -L_v & -L_p & -L_r \\ -N_v & -N_p & -N_r \\ 0 & 0 & 0 \end{bmatrix} \begin{bmatrix} v_g \\ p_g \\ r_g \end{bmatrix} \tag{6.18}$$

The longitudinal and lateral gust transfer functions can be determined by taking the Laplace transform of Eq. (6.17) and (6.18) and then dividing by the gust function. A linear set of algebraic equations in terms of $\Delta\bar{u}/\bar{u}_g$ are obtained. These equations can then be solved for the transfer functions.

To provide some insight into the influence of atmospheric disturbances on aircraft response, we shall examine the vertical motion of an airplane that encounters a vertical gust field.

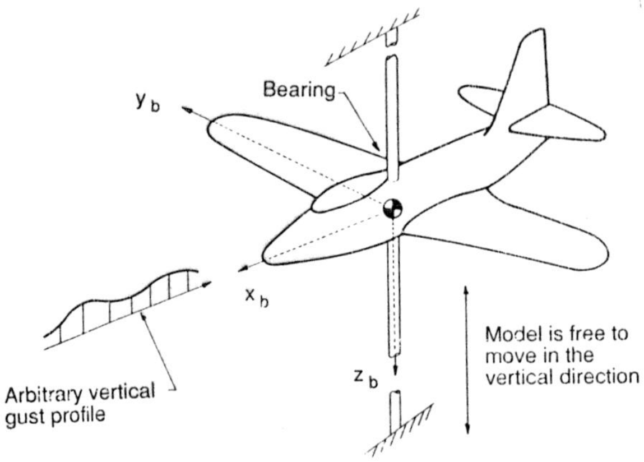

FIGURE 6.6

6.3 PURE VERTICAL OR PLUNGING MOTION

Consider an airplane constrained so that only movement in the vertical direction is possible. This type of motion could be simulated in the wind tunnel using a model constrained by a vertical rod, as illustrated in Fig. 6.6. The model is free to move up or down along the rod, but no other motion is possible.

Now let us examine the response of this constrained airplane subjected to an external disturbance such as a wind gust. The equation of motion for this example is obtained by applying Newton's second law, i.e.

$$\sum \text{Forces in the vertical direction} = m\frac{\mathrm{d}w}{\mathrm{d}t} \tag{6.19}$$

$$Z + WT = m\frac{\mathrm{d}w}{\mathrm{d}t} \tag{6.20}$$

where Z is the aerodynamic force in the z direction and WT is the weight of the airplane model. If we assume the motion of the airplane will be confined to small perturbations from an initial unaccelerated flight condition, then the aerodynamic force and vertical velocity can be expressed as the sum of the reference flight condition plus the perturbation:

$$Z = Z_0 + \Delta Z \qquad w = w_0 + \Delta w \tag{6.21}$$

Substituting Eq. (6.21) into Eq. (6.20) yields

$$Z_0 + \Delta Z + WT = m\frac{\mathrm{d}}{\mathrm{d}t}(w_0 + \Delta w) \tag{6.22}$$

This equation can be simplified by recognizing that in unaccelerated flight the condition for equilibrium is

$$Z_0 + WT = 0 \tag{6.23}$$

Therefore, Eq. (6.22) reduces to

$$\Delta Z/m = \frac{d}{dt}\Delta w \tag{6.24}$$

The aerodynamic force acting on the airplane is a function of the angle of attack and time rate of change of the attack, and can be expressed in terms of the stability derivatives as follows

$$\Delta Z/m = Z_\alpha\,\Delta\alpha + Z_{\dot\alpha}\,\Delta\dot\alpha \tag{6.25}$$

or

$$\Delta Z/m = C_{z_\alpha}\,\Delta\alpha\, QS/m + C_{z_{\dot\alpha}}\frac{\Delta\dot\alpha c}{2u_0}\,QS/m \tag{6.26}$$

where

$$C_{z\alpha} = -C_{L_\alpha} \qquad C_{z\dot\alpha} = -C_{L_{\dot\alpha}}$$

To simplify our analysis we will assume that the lag in lift term, $Z_{\dot\alpha}\,\Delta\dot\alpha$, is negligible in comparison to the $Z_\alpha\,\Delta\alpha$ term.

The change in angle of attack experienced by the airplane is due to its motion in the vertical direction and also to the vertical wind gust. The angle of attack can be written as

$$\Delta\alpha = \frac{\Delta w}{u_0} - \frac{w_g(t)}{u_0} \tag{6.27}$$

Substituting Eqs (6.25) and (6.27) into Eq. (6.24), and rearranging, yields

$$u_0\frac{d\,\Delta w}{dt} - Z_\alpha\,\Delta w = -Z_\alpha w_g(t) \tag{6.28}$$

or

$$-\frac{u_0}{Z_\alpha}\frac{d\,\Delta w}{dt} + \Delta w = w_g(t) \tag{6.29}$$

Equation (6.30) is a first-order differential equation with constant coefficients. Systems characterized by first-order differential equations are referred to as first-order systems. We rewrite Eq. (6.29) to have the form:

$$\tau\frac{d\,\Delta w}{dt} + \Delta w = \left(\tau\frac{d}{dt} + 1\right)\Delta w = w_g(t) \tag{6.30}$$

where

$$\tau = -\frac{u_0}{Z_\alpha} \tag{6.31}$$

and $w_g(t)$ is the gust velocity as a function of time.

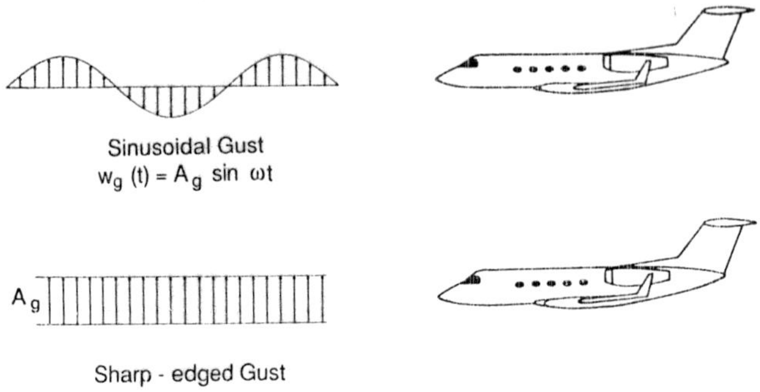

FIGURE 6.7

The solution of Eq. (6.30) for a sharp-edged or sinusoidal gust will now be examined. Figure 6.7 shows an airplane encountering a sharp edged or step gust and a sinusoidal gust profile. The reason for selecting these two types of gust inputs is that they occur quite often in nature. Furthermore, as was mentioned earlier, both the step function and sinusoidal inputs can be used to construct an arbitrary gust profile. For example, Fig. 6.8 shows the construction of an arbitrary gust profile as a series of step changes. Also in the case of an arbitrary-period gust function, the profile can be decomposed into a series of sine waves by Fourier analysis.

The transient response of an airplane to an encounter with a sharp-edged

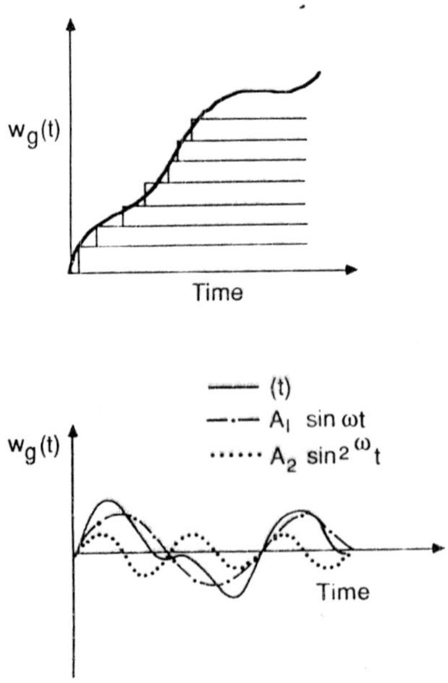

FIGURE 6.8

gust can be modeled by expressing the gust profile as a step function:

$$w_g(t) = \begin{cases} 0 & t = 0^- \\ A_g u(t) & t = 0^+ \end{cases} \tag{6.32}$$

where $u(t)$ is a unit step change and A_g is the magnitude of the gust. The solution of Eq. (6.30) for a step input can be obtained by taking the Laplace transformation of the differential equation

$$\tau s \, \Delta w(s) + \Delta w(s) = w_g(s) \tag{6.33}$$

or solving for the ratio of the output to input yields

$$\frac{\Delta w(s)}{w_g(s)} = \frac{1}{\tau s + 1} \tag{6.34}$$

Equation (6.34) is the transfer function of the change in vertical velocity to the vertical gust input. When the forcing function or input is a step change in the gust velocity, then

$$w_g(s) = \frac{A_g}{s} \tag{6.35}$$

or

$$\Delta w(s) = \frac{A_g}{s(\tau s + 1)} \tag{6.36}$$

Expanding Eq. (6.36) by the method or partial factions, and then taking the inverse Laplace transformation, yields

$$\Delta w(t) = A_g(1 - e^{-t/\tau}) \tag{6.37}$$

The vertical velocity of the airplane grows exponentially from zero to a final value of A_g. The initial slope of the curve at $t = 0$ is given by the derivative

$$\frac{dw}{dt} = \frac{A_g}{\tau} e^{-t/\tau} \qquad \left.\frac{dw}{dt}\right|_{t \to 0} = \frac{A_g}{\tau} \tag{6.38}$$

The parameter τ is referred to as the time constant of the system. The time constant tells us how fast our system approaches a new steady-state condition after being disturbed. If the time constant is small, the system will respond very rapidly; if the time constant is large, the system will respond very slowly. Figure 6.9 shows the response of the airplane to a sharp-edged gust. Notice that the output of the system approaches the final value asymptotically; however, the response is within 2 percent of the final value after only four time constants.

Additional insight into the vehicle's response can be obtained by looking at the maximum acceleration of the airplane. The maximum acceleration

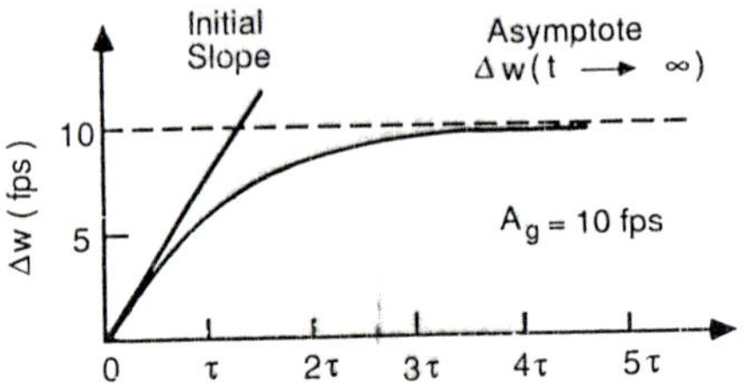

t	$\Delta w/w_g = 1 - e^{-t/\tau}$
0	0
τ	0.632
2τ	0.865
3τ	0.950
4τ	0.982
5τ	0.993
6τ	0.998

FIGURE 6.9

occurs at $t = 0$:

$$\Delta \dot{w} = \frac{C_{L_\alpha} QS}{m u_0} A_g = \frac{A_g}{\tau} \tag{6.39}$$

Dividing Eq. (6.39) by the gravitational constant g we obtain an equation for the change in load factor due to a sharp-edged gust:

$$\frac{\Delta \dot{w}}{g} = \frac{C_{L_\alpha} QS}{m u_0 g} A_g = \frac{\Delta L}{WT} = \Delta n \tag{6.40}$$

or

$$\Delta n = C_{L_\alpha} \frac{\rho u_0}{2} \frac{A_g}{WT/S} \tag{6.41}$$

Equation (6.41) indicates that airplanes having low wing loading WT/S will be much more responsive to the influence of vertical wing gust than airplanes with high wing loadings.

The takeoff and landing performance of an airplane can be shown to be a function of wing loading WT/S. Airplanes having a low wing loading will, in general, have short takeoff and landing field requirements. Airplanes designed for minimum runway requirements such as short-takeoff-and-landing (STOL) aircraft will have low wing loadings compared with conventional transport and fighter airplanes and, therefore, should be more responsive to atmospheric disturbances.

If the gust profile encountered by the airplane is sinusoidal, the response will consist of a transient phase followed by a steady-state sinusoidal oscillation. The steady-state response to a sinusoidal gust can be written as

$$\Delta w(t) = A_g \frac{1}{\sqrt{1+\tau^2\omega^2}} \sin(\omega t - \phi) \tag{6.42}$$

where

$$\phi = -\tan^{-1}(\tau\omega)$$

The steady-state response of the airplane will have the following characteristics:

1. The same frequency as the gust wave
2. The amplitude of the response will be

$$\text{Amplitude} = \frac{A_g}{\sqrt{1+\tau^2\omega^2}}$$

   where the amplitude of the gust is A_g.
3. The phase angle of the response is $\phi = \tan^{-1}(\tau\omega)$; the phase angle of the input gust is zero. The response of the airplane lags the gust wave by the angle ϕ.

Figure 6.10 shows the vertical response of an airplane to a sinusoidal gust encountered for values of $\omega\tau$. Remember that ω is the frequency of the gust and τ is the time constant of the airplane. Notice for small values of $\omega\tau$, i.e. low-frequency gusts or small airplane time constants, the phase angle ϕ is very small and the ratio of the response to gust input amplitudes is near unity. In this situation, the response is in phase with the gust wave and the amplitude of response of the airplane is nearly equal to the amplitude of the gust profile.

For very large values of $\omega\tau$, the response amplitude tends to zero, that is, the airplane is unaffected by the gust profile. These trends are easily observed in the frequency response curve shown in Fig. 6.11. This analysis shows us that the rigid body motion of the airplane is excited by the low frequency or long wave length gusts, and that the high frequency or short wave length gusts have little effect on the airplane's motion. Although the high frequency gusts do not

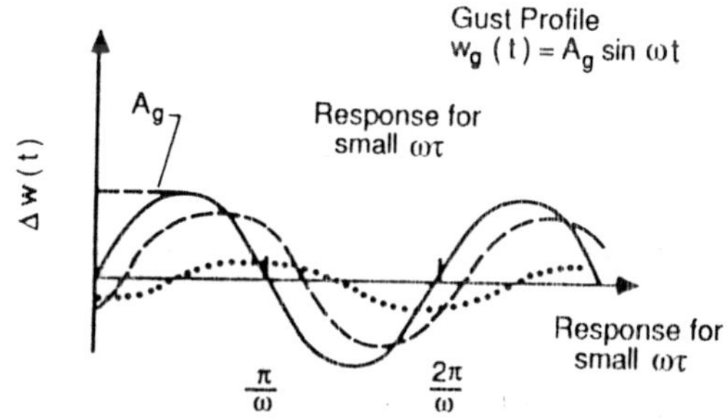

FIGURE 6.10

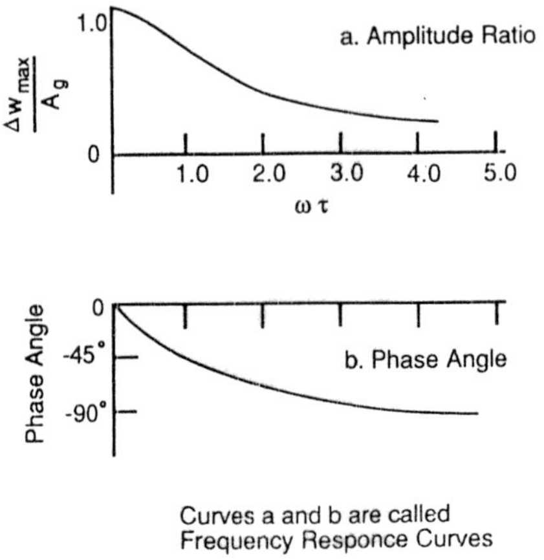

FIGURE 6.11

influence the rigid body motion they will excite the structural modes of the airplane.

Although this example gives us some insight into how atmospheric gusts will affect an airplane, the turbulence in the atmosphere is not deterministic. That is to say, there are no analytical expressions that completely describe atmospheric turbulence. Rather, turbulence is a stochastic process, or random process, and can only be described in a statistical manner.

6.4 ATMOSPHERIC TURBULENCE

The atmosphere is in a continuous state of motion. The winds and wind gusts created by the movement of atmospheric air masses can degrade the performance and flying qualities of an airplane. In addition, the atmospheric gusts impose structural loads that must be accounted for in the structural design of an airplane. The movement of atmospheric air masses is driven by solar heating, the Earth's rotation and various chemical, thermodynamic, and electromagnetic processes.

The velocity field within the atmosphere varies in both space and time in a random manner. This random velocity field is called atmospheric turbulence. The velocity variations in a turbulent flow can be decomposed into a mean and a fluctuating part. Figure 6.12 shows a typical atmospheric turbulence profile. The size or scale of the fluctuations vary from small wavelengths, of the order of centimeters, to wavelengths of the order of kilometers. Because atmospheric turbulence is a random phenomenon, it can only be described in a statistical way.

To predict the effect of atmospheric disturbances on aircraft response, flying qualities, autopilot performance, and structural loads requires a mathematical model of atmospheric turbulence. In the following sections the discussion will include a description of statistical functions used in describing

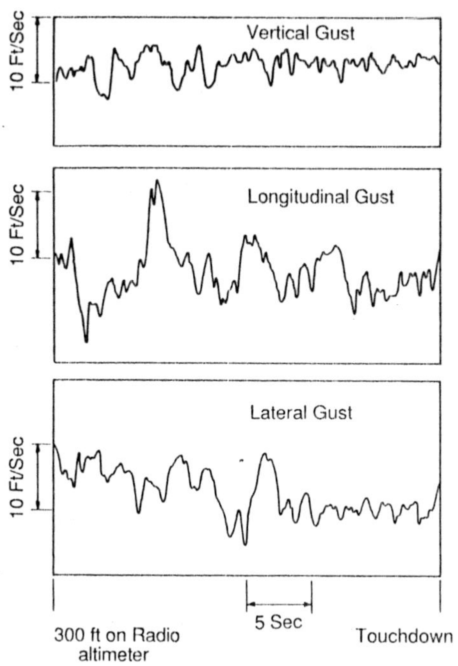

FIGURE 6.12

atmospheric turbulence, a mathematical model of turbulence, and finally an indication of how the turbulence model can be used to determine the response of an airplane to atmospheric disturbances.

Before presenting the mathematical model of turbulence, it is necessary to review some of the basic concepts used to describe turbulence. The discussion will be, at best, only a cursory review of an extremely complicated subject. The reader is referred to Refs 6.2 and 6.3 for a more informative treatment of the subject.

6.5 HARMONIC ANALYSIS

An arbitrary periodic signal having a period T can be represented as an infinite series of cosine and sine functions as follows:

$$f(t) = a_0 + \sum_{n=1}^{\infty} a_n \cos(n\omega t) + \sum_{m=1}^{\infty} b_n \sin(m\omega t) \tag{6.43}$$

where the angular frequency $\omega = 2\pi/T$ and the Fourier coefficients are found from the relationship

$$a_0 = \frac{\omega}{2\pi} \int_{t_0}^{t_0+2\pi/\omega} f(t)\, dt \tag{6.44}$$

$$a_n = \frac{\omega}{\pi} \int_{t_0}^{t_0+2\pi/\omega} f(t) \cos(\omega t) \, \mathrm{d}t \tag{6.45}$$

$$b_n = \frac{\omega}{\pi} \int_{t_0}^{t_0+2\pi/\omega} f(t) \sin(n\omega t) \, \mathrm{d}t \tag{6.46}$$

When the function is not periodic, the technique can still be used by allowing the period T to go to infinity; then the Fourier series becomes a Fourier integral:

$$f(t) = \frac{1}{\pi} \int_{-\infty}^{\infty} \mathrm{e}^{i\omega t} \, \mathrm{d}\omega \int_{-\infty}^{\infty} f(\tau) \mathrm{e}^{-i\omega\tau} \, \mathrm{d}\tau \tag{6.47}$$

If we define the second integral to be

$$G(\omega) = \int_{-\infty}^{\infty} f(\tau) \mathrm{e}^{-i\omega\tau} \, \mathrm{d}\omega \tag{6.48}$$

then

$$f(t) = \frac{1}{2\pi} \int_{-\infty}^{\infty} G(\omega) \mathrm{e}^{i\omega t} \, \mathrm{d}\omega \tag{6.49}$$

where $G(\omega)$ and $f(t)$ are a Fourier transform pair. The integrand $G(\omega)\,\mathrm{d}\omega$ gives the contribution of the harmonic components of $f(t)$ between the frequencies ω and $\omega + \mathrm{d}\omega$. Unfortunately, the harmonic analysis discussed above does not hold for turbulence. In order for the Fourier integral to be applicable, the integrals must be convergent. The nonperiodic turbulence disturbances persist for long periods of time without dying out in time. The persistence of turbulence yields integrals that do not converge.

To obtain a frequency representation for a continuing disturbance requires the use of the theory of random processes. A random process is one which is random by its nature, so that a deterministic description is not practical. For example, we are all familiar with board games we play for relaxation and enjoyment. In most of these games we must roll dice to in order to move around the board. The rolling of the dice constitutes a random experiment. If we denote the sum of the points on the two dice as X, then X is a random variable that can assume integer values between 2 and 12. If we roll the dice a sufficient number of times, we can determine the probabilities of the random variable X assuming any value in the range of X. A function $f(X)$ that yields the probabilities is called the probability or frequency function of a random variable.

Atmospheric turbulence is also a random process, and the magnitude of the gust fields can only be described by statistical parameters. That is, we can conduct experiments to determine the magnitude of a gust component and its probability of occurrence. The properties of atmospheric turbulence include that it is homogeneous and stationary. The property of homogeneity means that the statistical properties of turbulence are the same throughout the region of interest; stationarity implies that the statistical properties are independent of time.

For the case when $f(t)$ is a stationary random process, the mean square $\overline{f^2(t)}$ is defined as

$$\overline{f^2(t)} = \lim_{T \to \infty} \frac{1}{T} \int_0^T [f(t)]^2 \, dt \tag{6.50}$$

where $\overline{f^2(t)}$ represents a measure of the disturbance intensity. The disturbance function $f(t)$ can be thought of as an infinite number of sinusoidal components having frequencies ranging from zero to infinity. That portion of $\overline{f^2(t)}$ that occurs from ω to $d\omega$ is called the power spectral density and is denoted by the symbol $\Phi(\omega)$. The intensity of the random process can be related to the power spectral density.

The response of a physical system such as an airplane to a random disturbance such as atmospheric turbulence can be obtained from the power spectral density of the input function and the system transfer function. If $G(i\omega)$ represents the system frequency response function and $\Phi_i(\omega)$ is the power spectral density of the disturbance input function, then the output $\Phi_0(\omega)$ is given by

$$\Phi_0(\omega) = \Phi_i(\omega)\,|G(i\omega)|^2 \tag{6.51}$$

With Eq. (6.51) we can determine the response of an airplane to atmospheric disturbances. The transfer function G is the gust transfer described earlier. All that remains now is to describe $\Phi_i(\omega)$ for the gust input.

TURBULENCE MODELS. There are two spectral forms of random continuous turbulence used to model atmospheric turbulence for aircraft response studies. They are the mathematical models named after von Karman and Dryden, the scientists who first proposed them. Because the von Karman model is more widely used in practice, it will be the only one described here. The power spectral density for the turbulence velocities is given by

$$\Phi_{ug}(\Omega) = \sigma_u^2 \frac{2L_u}{\pi} \frac{1}{[1 + (1.339L_u\Omega)^2]^{5/6}} \tag{6.52}$$

$$\Phi_{vg}(\Omega) = \sigma_v^2 \frac{2L_v}{\pi} \frac{1 + \frac{8}{3}(1.339L_v\Omega)^2}{[1 + (1.339L_v\Omega)^2]^{11/6}} \tag{6.53}$$

$$\Phi_{wg}(\Omega) = \sigma_w^2 \frac{2L_w}{\pi} \frac{1 + \frac{8}{3}(1.339L_w\Omega)^2}{[1 + (1.339L_w\Omega)^2]^{11/6}} \tag{6.54}$$

where σ is the root mean square intensity of the gust component, Ω is the spatial frequency, defined by $2\pi/\lambda$, where λ is the wavelength of a sinusoidal component, and L is the scale of the turbulence. The subscripts u, v, w refer to the gust components. The scales and intensities of atmospheric turbulence depend on altitude and the type of turbulence, i.e. clear air (high or low altitude) and thunderstorm turbulences.

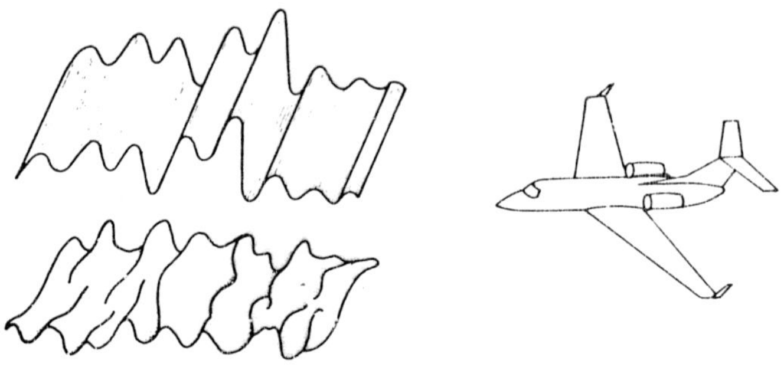

FIGURE 6.13

For an airplane passing through a gust field, it is assumed that the turbulence encountered is independent of time (i.e. the turbulence is stationary). This assumption can be visualized by considering the gust field to be frozen in both time and space, as illustrated in Fig. 6.13. Assuming the frozen-field concept, the turbulence-induced motion is due only to the motion of the airplane relative to the gust field.

The three power spectral densities presented earlier were a function of a spatial frequency; however, as the airplane passes through the frozen turbulent field it senses a temporal frequency. The relationship between the spatial and temporal frequency is given by

$$\Omega = \omega / u_0 \tag{6.55}$$

where ω is in radians/s and u_0 is the velocity of the airplane relative to the air mass it is passing through.

6.6 WIND SHEAR

Wind shear is defined as a local variation of the wind vector. The variations in wind speed and direction are measured in the vertical and horizontal directions. A vertical wind shear is one in which the wind speed and direction varies with changing altitude; horizontal wind shear refers to wind variations along some horizontal distance.

Wind shears are created by the movement of air masses relative to one another or to the earth's surface. Thunderstorms, frontal systems, and the Earth's boundary layer all produce wind shear profiles which, at times, can be hazardous to aircraft flying at low altitudes. The strong gust fronts associated with thunderstorms are created by downdrafts within the storm system. As the downdrafts approach the ground, they turn and move outward along the

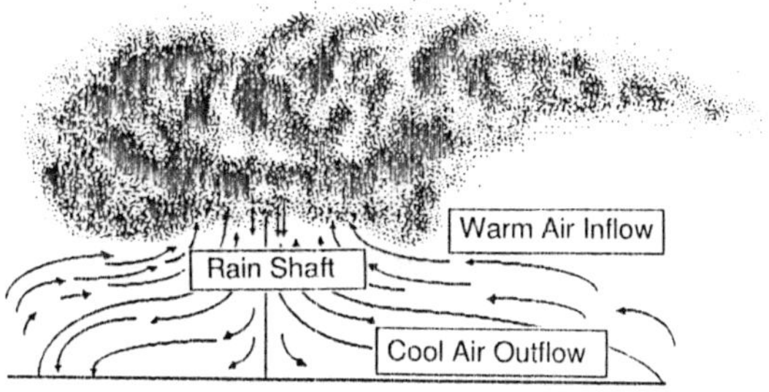

FIGURE 6.14

Earth's surface. The wind shear produced by the gust front can be quite severe.

The wind shear created by a frontal system occurs at the transition zone between two different air masses. The wind shear is created by the interaction of the winds in the two air masses. If the transition zone is gradual, the wind shear will be small. However, if the transition zone is small, the conflicting wind speeds and directions of the air masses can produce a very strong wind shear. Figure 6.14 shows some of the mechanisms that create a wind shear and Fig. 6.15 shows an experimentally measured shear profile near the ground.

There are no simple mathematical formulations to characterize the wind shears produced by the passage of frontal systems or thunderstorms. Generally, these shears are represented in simulation studies by tables of wind speed components with altitude.

The surface boundary layer also produces wind shear. The shape of the profile is determined, primarily, by local terrain and atmospheric conditions. Additional problems arise when there is an abrupt change in surface roughness (which can be expected near airports), resulting in additional internal boundary layers, and when the direction of the wind varies with altitude.

To analyze the influence of wind shear on aircraft motion, the characteristics of wind shear must be known. The magnitude of the shear can be expressed in terms of the change in wind speed with respect to altitude, du/dz, where a positive wind shear is one which increases with increasing altitude. The qualitative criteria for judging the severity of wind shear were proposed to the International Civil Aviation Organization (ICAO). It was suggested that

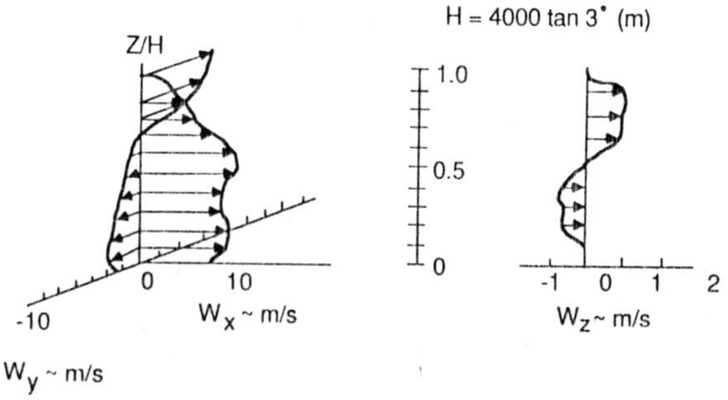

FIGURE 6.15

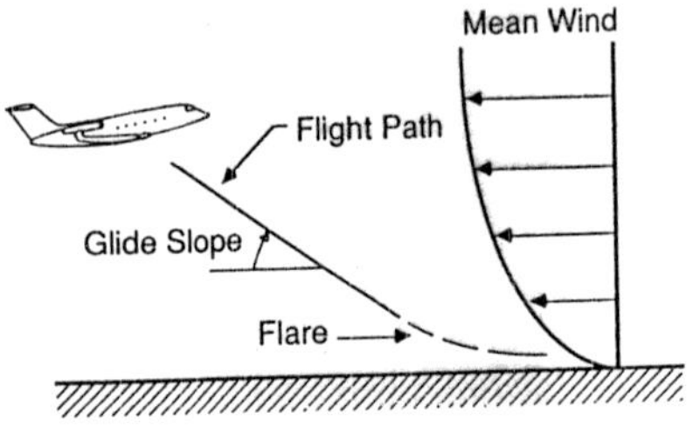

FIGURE 6.16

shear be considered light if du/dh ranged from 0 to $0.08\,\mathrm{s}^{-1}$; moderate for $0.08–0.15\,\mathrm{s}^{-1}$; strong for $0.15–0.20\,\mathrm{s}^{-1}$; and severe if greater than $0.2\,\mathrm{s}^{-1}$. These criteria are useful in giving an idea of the magnitude of wind shear, but the ICAO did not accept them. A shear which is moderate for an airplane with high stall speed may be strong for one with low stall speed, so that universal criteria are impossible owing to differences between aircraft types.

Example Problem 6.1. Consider an airplane on final approach encountering a vertical wind shear, i.e. the variation of horizontal wind velocity with altitude. Figure 6.16 shows an airplane flying into a wind shear. To analyse this problem we can use Eq. (6.17). The change in wind velocity is represented by

$$u_g = \frac{du}{dh}\,dh$$

where du/dh is the velocity gradient and dh is the change in altitude. If we assume that the controls are fixed, Eq. (6.17) reduces to

$$\begin{bmatrix} \Delta\dot{u} \\ \Delta\dot{w} \\ \Delta\dot{q} \\ \Delta\theta \end{bmatrix} = \begin{bmatrix} X_u & X_w & 0 & -g \\ Z_u & Z_w & u_0 & 0 \\ M_u & M_w & M_q & 0 \\ 0 & 0 & 1 & 0 \end{bmatrix} \begin{bmatrix} \Delta u \\ \Delta w \\ \Delta q \\ \Delta\theta \end{bmatrix} + \begin{bmatrix} -X_u \\ -Z_u \\ -M_u \\ 0 \end{bmatrix} [u_g]$$

But u_g is a function of altitude and therefore we must add other equations to the system. The vertical velocity of the airplane can be expressed as the time rate of change of altitude as follows:

$$\Delta\dot{h} = u_0(\Delta\alpha - \Delta\theta)$$

Adding this equation to the state equations, and substituting for u_g, yields

$$\begin{bmatrix} \Delta\dot{u} \\ \Delta\dot{\alpha} \\ \Delta\dot{\alpha} \\ \Delta\dot{\theta} \\ \Delta\dot{h} \end{bmatrix} = \begin{bmatrix} X_u & X_w & 0 & -g & -X_u\dfrac{du}{dh} \\ \dfrac{Z_u}{u_0} & \dfrac{Z_w}{v_0} & 1 & 0 & -Z_w\dfrac{du}{dh} \\ M_u & M_w & M_q & 0 & -M_u\dfrac{du}{dh} \\ 0 & 0 & 1 & 0 & 0 \\ 0 & U_0 & 0 & -u_0 & 0 \end{bmatrix} \begin{bmatrix} \Delta u \\ \Delta\alpha \\ \Delta q \\ \Delta\theta \\ \Delta h \end{bmatrix}$$

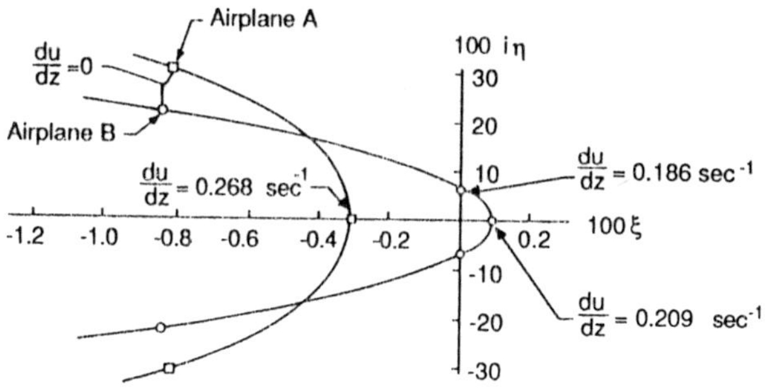

FIGURE 6.17

The solution of this system of equations yields five eigenvalues: two complex pairs representing the phugoid and short-period modes and a fifth, real, root indicating a non-oscillatory motion. The above equations were solved in Ref. 6.6 for STOL aircraft for various magnitudes of the velocity gradient. The results showed that wind shear had very little affect on the short-period motion; however, the phugoid motion was found to be quite sensitive to du/dh. Figure 6.17 shows a root locus plot of the phugoid roots for variations in du/dh. For very large gradients, the phugoid mode can become unstable. An unstable phugoid mode would make the landing approach very difficult for the pilot to control. Therefore, strong wind shears must be avoided for reasons of flight safety.

6.7 SUMMARY

In this chapter we have examined some of the analytical techniques available to the flight control engineers to study the dynamic response of an airplane to control deflection or atmospheric disturbances. Apart from the fact that they create an uncomfortable ride for the pilot and passengers, the loads imposed on the airframe structure by the gust fields must be calculated so that the structure can be properly designed.

Wind shear has been shown recently to be a greater hazard to commercial aviation than had been appreciated. Wind shears created by thunderstorm systems have been identified as the major contributor to several airline crashes. The techniques outlined in this chapter can be used by stability and control engineers to study the effects of atmospheric disturbances on aircraft flight characteristics. Such studies can be used to improve flight safety.

6.8 PROBLEMS

6.1. For the business jet aircraft whose details are included in the Appendix, determine the lateral response curves for an aileron input. Present your results in the form of frequency response curves.

6.2. The vertical motion of an airplane subjected to a sharp-edged gust is described by the equation

$$\Delta w(t) = A_g(1 - e^{-t/\tau})$$

where Δw is in the vertical velocity, A_g is the magnitude of the gust, and τ is the time constant of the airplane. Using the information in Fig. P6.2, determine the maximum vertical acceleration and the time constant of the airplane.

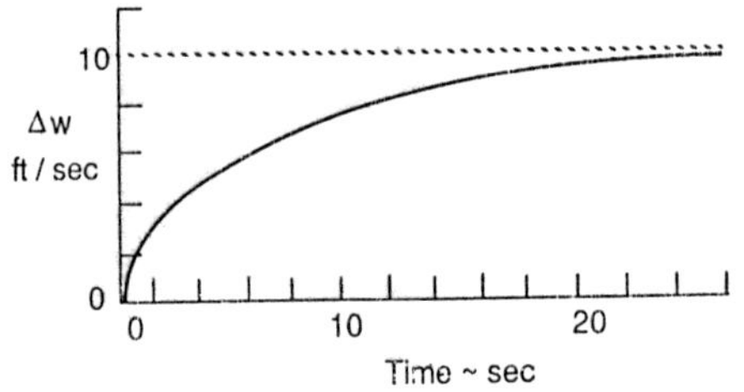

FIGURE P6.2

6.3. For the general aviation airplane whose details are included in the Appendix, determine the vertical response to a sinusoidal gust field. Assume the problem can be modeled by a single degree of freedom vertical equation of motion. Present your results in the form of frequency response curves.

6.4. Discuss how changes in the aerodynamic stability characteristics would effect the response curves obtained in Problem 6.3.

6.5. Assume that an airplane is on final approach and encounters a wind shear which can be represented as

$$u_g = \frac{du}{dh} dh$$

where du/dh is the wind gradient. Assume that the pitch attitude of the airplane is maintained by an automatic control system. Develop the equations of motion governing the vertical and horizontal velocity of the airplane. How does the wind gradient effect the two-dimensional response?

REFERENCES

6.1. McRuer, D., I. Ashkenas, and D. Graham: *Aircraft Dynamics and Automatic Control*, Princeton University Press, Princeton, NJ, 1973.

6.2. Batchelor, G. K.: *The Theory of Homogeneous Turbulence*, Cambridge University Press, London, 1956.

6.3. Lumley, J. L., and H. A. Panofsky: *The Structure of Atmospheric Turbulence*, Interscience, New York, 1964.

6.4. Houbolt, J. C., R. Steiner, and K. G. Pratt: *Dynamic Response of Airplanes to Atmospheric Turbulence Including Flight Data on Input and Response*, NASA TR-R-199, June 1964.

6.5. Press, H., M. T. Meadows, and I. Hadlock: *A Reevaluation of Data on Atmospheric Turbulence and Airplane Gust Loads for Application to Spectral Calculations*, NACA Report 1272, 1956.

6.6. Nelson, R. C., M. M. Curtin, and F. M. Payne: *A Combined Experimental and Analytical Investigation of the Influence of Low Level Wind Shear on the Handling Characteristics of Aircraft*, DOT Report-DOT/RSPA/ADMA-50/83/29, October 1979.

CHAPTER 7

AUTOMATIC CONTROL: APPLICATION OF CONVENTIONAL CONTROL THEORY

"The application of automatic control systems to aircraft promises to bring about the most important new advances in aeronautics in the future". From William Bollay–14th Wright Brothers Lecture[1]

7.1 INTRODUCTION

The development of automatic control systems has played an important role in the growth of civil and military aviation. Modern aircraft include a variety of automatic control systems that aid the flight crew in navigation, flight management and augmenting the stability characteristics of the airplane. In this chapter we shall discuss the application of control theory to the design of simple autopilots that can be used by the flight crew to lessen their work load during cruise and to help them land their aircraft during adverse weather conditions. In addition, we shall also discuss how automatic control systems can be used to provide artificial stability to improve the flying qualities of an airplane.

The development of autopilots closely followed the successful development of a powered man-carrying airplane by the Wright brothers. In 1914 the Sperry brothers demonstrated the first successful autopilot. The autopilot was capable of maintaining pitch, roll, and heading angles. To demonstrate the effectiveness of their design, Lawrence Sperry trimmed his airplane for straight and level flight and then engaged the autopilot. He then proceeded to stand in the cockpit with his hands raised above his head while his mechanic walked out along the wings in an attempt to upset the airplane's equilibrium. Figure 7.1

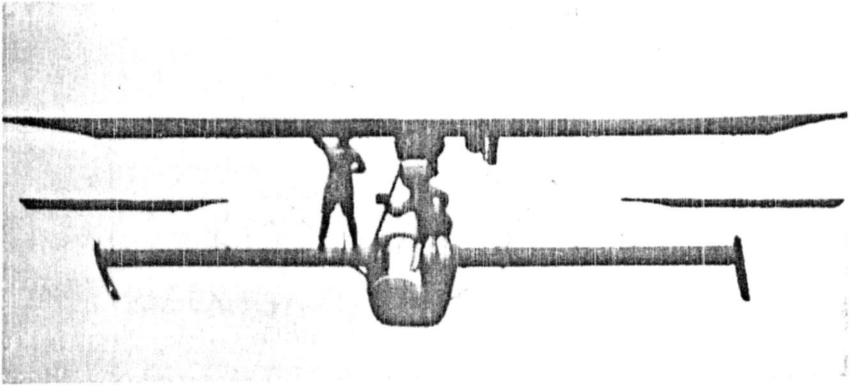

FIGURE 7.1
Photograph of Sperry's flight demonstration of a three axes automatic control system.

shows a photograph of the remarkable flight. The autopilot provided aileron, rudder, and elevator commands to that the airplane remained in a wings-level attitude.

Before examining some simple autopilot designs, we will review some of the basic concepts of control theory. Control systems can be classified as either open-loop or closed-loop systems. An open-loop control system is the simplest and least complex of all control devices. In an open-loop system the control action is independent of the output. A closed-loop system is one in which the control action depends on the output of the system. Closed-loop control systems are called feedback control systems. The advantage of the closed-loop control system is its accuracy.

To obtain a more accurate control system, some form of feedback between the output and input must be established. This can be accomplished by comparing the controlled signal (output) with the commanded or reference input. A feedback system is defined as a system in which one or more feedback loops are used to compare the controlled signal with the command signal so as to generate an error signal. The error signal is then used to drive the output signal into agreement with the desired input signal. The typical closed-loop feedback system shown in Fig. 7.2 is composed of a forward path, a feedback path and an error-detection device called a comparator. Each component of the control system is defined in terms of its transfer function. The transfer function T.F. is defined as the ratio of the Laplace transform of the output to the Laplace transform of the input where the initial conditions are assumed to be zero:

$$\text{T.F.} = \frac{\text{Laplace transform of the output}}{\text{Laplace transform of the input}} \tag{7.1}$$

The transfer function of each element of the control system can be determined

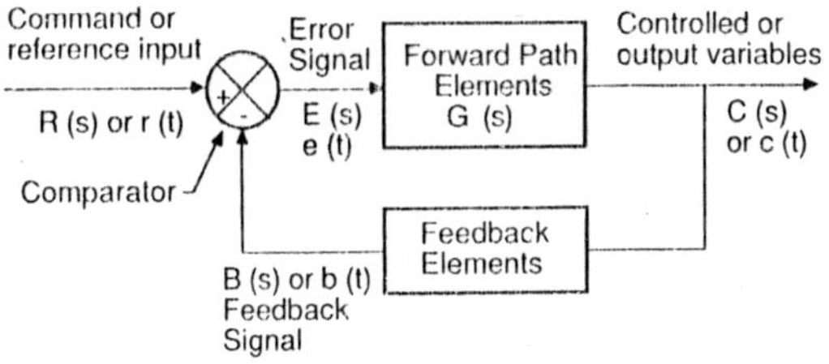

FIGURE 7.2
Block diagram representation of a feedback control system.

from the equations that govern the dynamic characteristics of the element. The aircraft transfer functions were developed in Chapter 3 from the equations of motion.

The closed-loop transfer function for the feedback control system shown in Fig. 7.2 can be developed from the block diagram. The symbols used in the block diagram are defined as follows:

$R(s)$	reference input
$C(s)$	output signal (variable to be controlled)
$B(s)$	feedback signal
$E(s)$	error or actuating signal
$G(s)$	$C(s)/E(s)$ forward path or open-loop transfer function
$M(s)$	$C(s)/R(s)$ the closed-loop transfer function
$H(s)$	feedback transfer function
$G(s)H(s)$	loop transfer function

The closed-loop transfer function $C(s)/R(s)$ can be obtained by simple algebraic manipulation of the block diagram. The actuating or error signal is the difference between the input and feedback signals:

$$E(s) = R(s) - B(s) \tag{7.2}$$

The feedback signal $B(s)$ can be expressed in terms of the feedback transfer function and the output signal:

$$B(s) = H(s)C(s) \tag{7.3}$$

and the output signal $C(s)$ is related to the error signal and forward path transfer function in the following manner:

$$C(s) = G(s)E(s) \tag{7.4}$$

Substituting Eqs (7.2) and (7.3) into Eq. (7.4) yields

$$C(s) = G(s)R(s) - G(s)H(s)C(s) \tag{7.5}$$

Equation (7.5) can be solved for the closed-loop transfer function $C(s)/R(s)$,

$$\frac{C(s)}{R(s)} \quad \frac{G(s)}{1 + G(s)H(s)} \tag{7.6}$$

which is the ratio of the system output to the input. Most control systems are much more complex than the one shown in Fig. 7.2. However, theoretically, the more complex control systems consisting of many feedback elements can be reduced to the simple form described above.

The feedback systems described here can be designed to control accurately the output to some desired tolerance. However, feedback in itself does not ensure that the system will be stable. Therefore, to design a feedback control system one needs analysis tools which permit the designer to select system parameters so that the system will be stable. In addition to determining the absolute stability, the relative stability of the control system must also be determined. A system which is stable in the absolute sense may not be a satisfactory control system. For example, if the system damping is too low, the output will be characterized by large amplitude oscillations about the desired output. The large overshooting of the response may make the system unacceptable.

Autopilots can be designed using either frequency- or time-domain methods developed from servomechanism theory, or by time-domain analysis using state feedback design. In this chapter the techniques from servomechanism theory will be discussed and several simple applications of the design techniques will be demonstrated by applying the techniques to the design of autopilots.

The servomechanism design techniques include the Routh criterion, root locus, Bode, and Nyquist methods. A brief description of these techniques is presented either in the following sections or in the appendices at the end of this book. For a more rigorous treatment of this material, the reader is referred to Refs [7.2–7.5].

7.2 ROUTH'S CRITERION

As noted earlier, the roots of the characteristic equation tell us whether or not the system is dynamically stable. If all the roots of the characteristic equation have negative real parts, the system will be dynamically stable. On the other hand, if any root of the characteristic equation has a positive real part, the system will be unstable. The system is considered to be marginally stable if one or more of the roots is a pure imaginary number. The marginally stable system represents the boundary between a dynamically stable or unstable system. For a closed-loop control system the denominator of Eq. (7.6) is the characteristic equation of the system.

A simple means of determining the absolute stability of a system can be obtained by the Routh stability criterion. The method allows us to determine whether any of the roots of the characteristic equation have positive real parts, without actually solving for the roots. Consider the characteristic equation

$$a_n\lambda^n + a_{n-1}\lambda^{n-1} + a_{n-2}\lambda^{n-2} \cdots a_1\lambda + a_0 = 0 \tag{7.7}$$

In order that there be no roots of Eq. (7.7) with positive real parts, the

TABLE 7.1
Definition of Routh array

Routh table				
λ^n	a_n	a_{n-2}	a_{n-4}	$\cdots$
λ^{n-1}	a_{n-1}	a_{n-3}	a_{n-5}	$\cdots$
λ^{n-2}	b_1	b_2	b_3	$\cdots$
$\vdots$	c_1	c_2	c_3	$\cdots$

where $a_n, a_{n-1}, \ldots a_0$ are the coefficients of the characteristic equation and the coefficients b_1, b_2, b_3, c_1, c_2, etc., are given by:

$$b_1 = \frac{a_{n-1}a_{n-2} - a_n a_{n-3}}{a_{n-1}} \qquad b_2 = \frac{a_{n-1}a_{n-4} - a_n a_{n-5}}{a_{n-1}} \qquad \text{etc.}$$

$$c_1 = \frac{b_1 a_{n-3} - a_{n-1} b_2}{b_1} \qquad c_2 = \frac{b_1 a_{n-5} - a_{n-1} b_3}{b_1} \qquad \text{etc.}$$

$$d_1 = \frac{c_1 b_2 - c_2 b_1}{c_1} \qquad \text{etc.}$$

necessary, but not sufficient, conditions are that:

1. All the coefficients of the characteristic equation must have the same sign.
2. All the coefficients must exist.

To apply the Routh criterion, we must first define the Routh array as in Table 7.1. The Routh array is continued horizontally and vertically until only zeros are obtained. The last step is to investigate the signs of the numbers in the first column of the Routh table. The Routh stability criterion states:

1. If all the numbers of the first column have the same sign, then the roots of the characteristic polynominal have negative real parts. The system is, therefore, stable.
2. If the numbers in the first column change sign, then the number of sign changes indicates the number of roots of the characteristic equation having positive real parts. Therefore, if there is a sign change in the first column, the system will be unstable.

When developing the Routh array, several difficulties may occur. For example, the first number in one of the rows may be zero, but the other numbers in the row may not be. Obviously, if a zero appears in the first position of a row, the elements in the following row will be infinite. In this case, the Routh test breaks down. Another possibility is that all the numbers in a row are zero. Methods for handling these special cases can be found in most textbooks on automatic control theory.

Several examples of applying the Routh stability criterion are shown in Example Problem 7.1.

Example Problem 7.1. Determine whether the characteristic equation

$$\lambda^3 + 6\lambda^2 + 12\lambda + 8 = 0$$

has any roots with positive real parts. The first two rows of the array are written down by inspection and the succeeding rows are obtained by using the relationship for each row element as presented previously:

$$\begin{array}{rrr} 1 & 12 & 0 \\ 6 & 8 & 0 \\ 64 & 0 & \\ 6 & & \\ 8 & & \end{array}$$

There are no sign changes in column 1; therefore, the system is stable.

As another example, consider the characteristic equation

$$2\lambda^3 + 4\lambda^2 + 4\lambda + 12 = 0$$

The Routh array can be written as follows.

$$\begin{array}{rrr} 2 & 4 & 0 \\ 4 & 12 & 0 \\ -2 & 0 & \\ 12 & & \end{array}$$

Note that there are two sign changes in column 1; therefore, the characteristic equation has two roots with positive real parts. The system is unstable.

The Routh stability criterion can be applied to the quartic characteristic equation which describes either the longitudinal or lateral motion of an airplane. The quartic characteristic equation for either the longitudinal or lateral equation of motion can be expressed as

$$A\lambda^4 + B\lambda^3 + C\lambda^2 + D\lambda + E = 0$$

where A, B, C, D and E are functions of the longitudinal or lateral stability derivatives. Forming the Routh array from the characteristic equation yields

$$\begin{array}{lcc} A & C & E \\ B & D & 0 \\ \dfrac{BC - AD}{B} & E & 0 \\ \dfrac{[D(BC - AD)/B] - BE}{(BC - AD)/B} & 0 & \\ E & & \end{array}$$

The condition that the airplane is stable requires that

$$\begin{array}{ll} A, B, C, D, E & > 0 \\ BC - AD & > 0 \\ D(BD - AD) - B^2 & > 0 \end{array}$$

The last two inequalities were obtained by inspection of the first column of the Routh array.

7.3 ROOT LOCUS TECHNIQUE

In designing a control system, it is desirable to be able to investigate the performance of the control system when one or more parameters of the system are varied. As has been shown repeatedly, the characteristic equation plays an important role in the dynamic behavior or aircraft motions. The same is true for linear control systems. In control system design, a powerful tool for analyzing the performance of a system is the root locus technique. Basically, the technique provides graphical information in the s plane on the trajectory of the roots of the characteristic equation for variations in one or more of the system parameters. Typically, most root locus plots consist of only one parametric variation. The root locus technique was introduced by W. R. Evans in 1949. The root locus method allows the control engineer to obtain accurate time-domain response as well as frequency response information of closed-loop control systems.

A brief discussion of the root locus technique will be presented here. Recall that the closed-loop transfer function of a feedback control system can be expressed as

$$\frac{C(s)}{R(s)}=\frac{G(s)}{1+G(s)H(s)} \tag{7.8}$$

The characteristic equation of the closed loop system is found by setting the denominator of the transfer function equal to zero.

$$1+G(s)H(s)=0 \tag{7.9}$$

The loop transfer function $G(s)H(s)$ can be expressed in the factored form as follows:

$$G(s)H(s)=\frac{k(s+z_1)(s+z_2)\cdots(s+z_m)}{(s+p_1)(s+p_2)\cdots(s+p_n)} \tag{7.10}$$

where the z's, p's, and k are the zeros, poles, and gain of the transfer function. The zeros are the roots of the numerator and the poles are the roots of the denominator of the transfer function. As stated earlier, the root locus is a graphical presentation of the trajectory of the roots of the characteristic equation or the poles of the closed-loop transfer function for variation of one of the system parameters. Let us examine the root locus plot for the above equation as the system gain k is varied. The characteristic equation can be rewritten as

$$G(s)H(s)=-1 \tag{7.11}$$

or

$$\frac{k(s+z_1)(s+z_2)\cdots(s+z_m)}{(s+p_1)(s+z_2)\cdots(s+p_n)}=-1 \tag{7.12}$$

For the case when $k=0$, the points on the root locus plot are the poles of

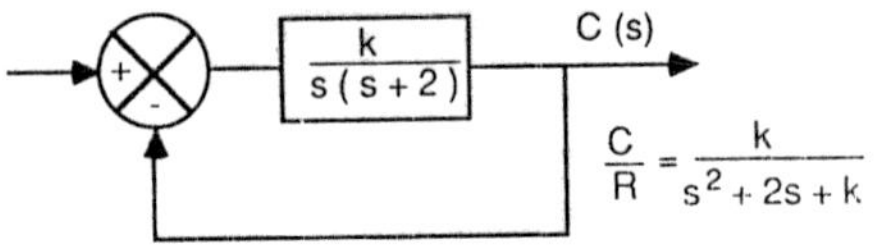

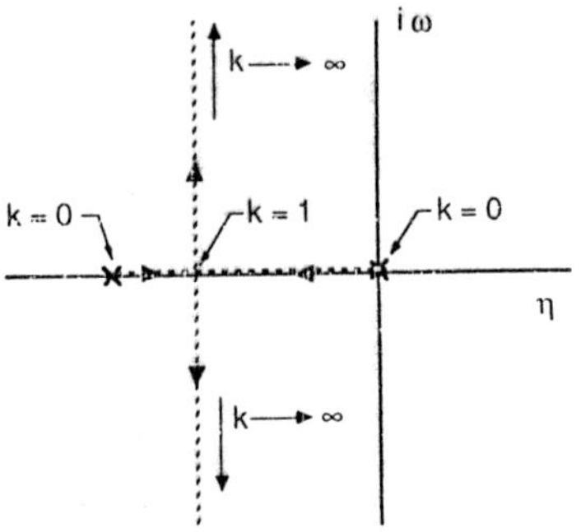

FIGURE 7.3
Root locus sketch for a second order system.

the loop transfer function $G(s)H(s)$. On the other hand, when $k \to 0$ the points on the root locus are the zeros of the loop transfer function. Thus we see that the roots of the closed-loop transfer function migrate from the poles to the zeros of the loop transfer function as k is varied from 0 to ∞. Furthermore, the points on the root locus for intermediate values of k must satisfy the equation

$$\frac{|k|\,|s+z_1|\,|s+z_2|\cdots|s+z_m|}{|s+p_1|\,|s+p_2|\cdots|s+p_n|} = 1 \tag{7.13}$$

and

$$\sum_{i=1}^{m} \underline{/s+z_i} - \sum_{i=1}^{n} \underline{/s+p_i} = (2q+1)\pi \tag{7.14}$$

where $q = 0, \pm 1, \pm 2 \ldots$ all integers.

Figure 7.3 is a sketch of a root locus plot for a second-order control system. The root locus diagram gives the roots of the closed loop characteristic equation as k is varied from 0 to ∞. When $k = 0$ the roots are located at the origin and $s = -2$. As k is increased, the roots move along the real axis towards one another until they meet at $s = -1$. Further increases in k cause the roots to become complex and they move away from the real axis along a line perpendicular to the real axis. When the roots are complex, the system is underdamped and a measure of the system damping ratio is obtained by measuring the angle of a line drawn from the origin to a point on the complex portion of the root locus. The system damping ratio is given by the equation

$$\zeta = \cos^{-1}\theta \tag{7.15}$$

The roots of the closed-loop characteristic equation can be obtained using root-solving algorithms that can be coded on digital computers. Such computer software can be used to obtain root locus contours for variations of system parameters. In addition, there is a simple graphical technique that can be used to rapidly construct the root locus diagram of a control system. An outline of the graphical procedure is presented in one of the appendices.

7.4 FREQUENCY DOMAIN TECHNIQUES

The frequency response of a dynamic system was discussed in Chapter 6. The same techniques can be applied to the design of feedback control systems. The transfer function for a closed-loop feedback system can be written as

$$M(s) = \frac{C(s)}{R(s)} = \frac{G(s)}{1 + G(s)H(s)} \tag{7.16}$$

If we excite the system with a sinusoidal input such as

$$r(t) = A \sin(\omega t) \tag{7.17}$$

the steady-state output of the system will have the form

$$c(t) = B \sin(\omega t + \phi) \tag{7.18}$$

The magnitude and phase relationship between the input and output signal is called the frequency response of the system. The ratio of output to input for a sinusoidal steady state can be obtained by replacing the Laplace transform variable s with $i\omega$.

$$M(i\omega) = \frac{G(i\omega)}{1 + G(i\omega)H(i\omega)} \tag{7.19}$$

Expressing the previous equation in terms of its magnitude and phase angle yields

$$M(i\omega) = M(\omega)\underline{/\phi(\omega)} \tag{7.20}$$

where

$$M(\omega) = \left| \frac{G(i\omega)}{1 + G(i\omega)H(i\omega)} \right| \tag{7.21}$$

and

$$\phi(\omega) = \underline{/G(i\omega)} - \underline{/[1 + G(i\omega)H(i\omega)]} \tag{7.22}$$

The frequency response information can be plotted in rectangular, polar, or logarithmic (Bode) plots. Figure 7.4 is a sketch of the various ways of presenting the frequency response data. The relationship between the frequency- and time-domain performance of a control system is discussed in the next section.

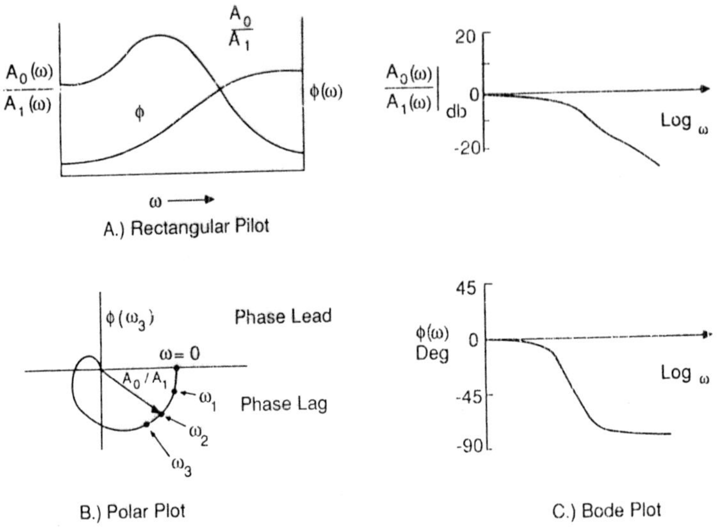

FIGURE 7.4
Various graphical ways of presenting frequency response data.

7.5 TIME-DOMAIN AND FREQUENCY-DOMAIN SPECIFICATIONS

The first step in the design of a feedback control system is to determine a set of specifications of the desired system performance. In the following section we shall present both time- and frequency-domain specifications and their relationship to one another for a second-order system. The transfer function of a second-order system can be expressed as

$$\frac{C(s)}{R(s)} = \frac{\omega_n^2}{s^2 + 2\zeta\omega_n s + \omega_n^2} \tag{7.23}$$

where ζ is the damping ratio and ω_n is the undamped natural frequency of the system. Figure 7.5 is a sketch of the response to a step input of an underdamped second-order system. The performance of the second-order system is characterized by the overshoot, delay time, rise time, and settling time of the transient response to a unit step. The time response of a second-order system to a step input for an underdamped system, i.e. $\zeta < 1$, is given by Eqs (7.24) and (7.25):

$$c(t) = 1 + \frac{e^{-\zeta\omega_n t}}{\sqrt{1-\zeta^2}} \sin(\omega_n \sqrt{1-\zeta^2}\, t - \phi) \tag{7.24}$$

$$\phi = \tan^{-1}\left(\frac{\sqrt{1-\zeta^2}}{-\zeta}\right) \tag{7.25}$$

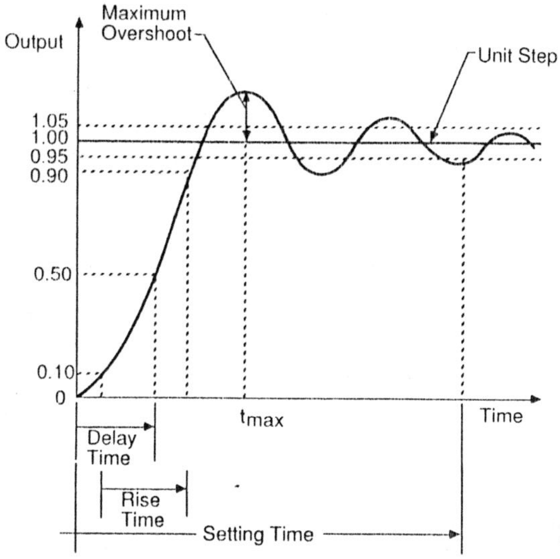

FIGURE 7.5
Time response of a second order system.

The delay and rise time give a measure of how fast the system responds to a step input. Delay time t_d is the time it takes for the response to reach for the first time 50 percent of the final value of the response. The rise time t_r is the time required for the response to rise from 10 percent to 90 percent of the final value. The other two parameters of interest are the settling time and peak overshoot. Settling time t_s is the time it takes for the response to stay within a specified tolerance band of 5 percent of the final value. The peak overshoot is a measure of the oscillations about the final output. From the standpoint of control system design, we would like to have a system that responds rapidly with minimum overshoot. Equations (7.24) and (7.25) can be used to determine the relationships between the time-domain specifications t_d, t_r, etc., and the damping ratio ζ and undamped natural frequency ω_n. Table 7.2 is a summary of these relationships.

Figure 7.6 is a sketch of the typical magnitude and phase characteristics of a feedback control system. As was the case in the time-domain analysis, it is desirable to have a set of specifications to describe the control system performance in the frequency domain. In the frequency domain the design specifications are given in terms of the resonance peak M_r, the resonant frequency ω_r, the system bandwidth ω_B, and the gain and phase margins. The maximum value of $M(\omega)$ is called the resonance peak and is an indication of the relative stability of the control system. If M_r is large, the system will have a large peak overshoot to a step input. The resonant frequency ω_r is the frequency at which the resonance peak occurs and is related to the frequency of the oscillations and speed of the transient response. The band width ω_B is the band of frequencies from zero to the frequency at which the magnitude $M(\omega)$ drops to 70 percent of the zero-frequency magnitude. The bandwidth

TABLE 7.2
Time domain specifications

Rise time t_d $t_d \approx \dfrac{1+0.6\zeta+0.15\zeta^2}{\omega_n}$	**Rise time, t_r** $t_r \approx \dfrac{1+1.1\zeta+1.4\zeta^2}{\omega_n}$
Time to peak amplitude, t_p $t_p = \dfrac{\pi}{\omega_n\sqrt{1-\zeta^2}}$	**Settling time, t_s** $t_s = \dfrac{3.0}{\omega_n\zeta}$
Peak overshoot, M_p $M_p = \dfrac{c(t_p)-c(\infty)}{c(\infty)} \times 100\%$	
For a unit step Percent maximum overshoot $= 100\exp(-\pi\zeta/\sqrt{1-\zeta^2})$	

gives an indication of the transient response of the system. If the bandwidth is large, the system will respond rapidly, whereas a small bandwidth will result in a sluggish control system.

The gain and phase margins are measures of the relative stability of the system and are related to the closeness of the poles of the closed-loop system to the $i\omega$ axis.

For a second-order system the frequency domain characteristics M_r, ω_r, and ω_B can be related to the system damping ratio and the undamped natural frequency ω_n. The relationships will be presented here without proof:

$$M_r = \frac{1}{2\zeta\sqrt{1-\zeta^2}} \tag{7.26}$$

$$\omega_r = \omega_n\sqrt{1-2\zeta^2} \tag{7.27}$$

$$\omega_B = \omega_n[(1-2\zeta^2)+\sqrt{4\zeta^4-4\zeta^2+2}]^{1/2} \tag{7.28}$$

The peak resonance and the peak overshoot of the transient response in the time domain is given by the following approximation:

$$c(t)_{max} \leq 1.17M_r \tag{7.29}$$

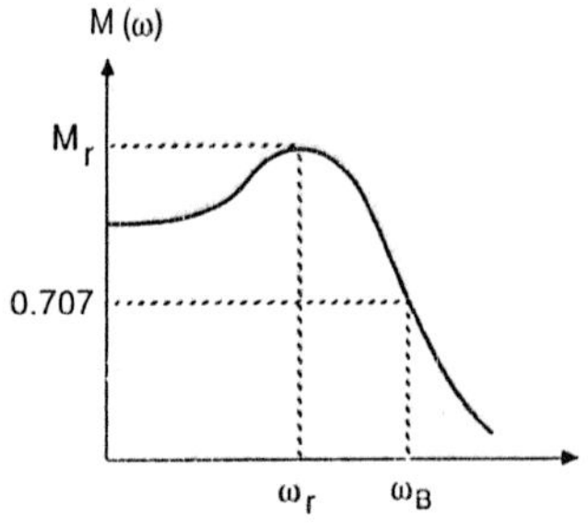

FIGURE 7.6
Frequency response of a closed loop control system.

The phase margin of a second-order system can be related to the system damping ratio as follows:

$$\phi = \tan^{-1}\left[2\zeta\left(\frac{1}{(4\zeta^2+1)^{1/2}-2\zeta^2}\right)^{1/2}\right] \tag{7.30}$$

This very formidable equation can be approximated by the simple relationship

$$\zeta \approx 0.1\phi \qquad \text{for} \qquad \zeta \leq 0.7 \tag{7.31}$$

The phase margin ϕ is in degrees.

From the above relationships developed for the second-order system the following observations can be made.

1. The maximum overshoot for a unit step in the time domain is a function only of ζ.
2. The resonance peak of the closed loop system is only a function of ζ.
3. The maximum peak overshoot and resonance peak are related through the damping ratio.
4. The rise time increases while the bandwidth decreases for increases in system damping for a fixed ω_n. The bandwidth and rise time are inversely proportional to one another.
5. The bandwidth is directly proportional to ω_n.
6. The higher the bandwidth, the larger the resonance peak.

HIGHER-ORDER SYSTEMS. Most feedback control systems are usually of a higher order than the second-order system discussed in the previous sections. However, many higher-order control systems can be analyzed by approximating the system by a second-order system. Obviously, when this can be accomplished, the design and analysis of the equivalent system is greatly simplified.

For a higher-order system to be replaced by an equivalent second-order system, the transient response of the higher-order system must be dominated by a pair of complex conjugate poles. These poles are called the dominant poles or roots and are located closest to the origin in a pole–zero plot. The other poles must be located far to the left of the dominant poles or be located near a zero of the system. The transient response caused by the poles located to the far left of the dominant poles will diminish rapidly in comparison with the dominant root response. On the other hand, if the pole is not located to the far left of the dominant poles, then the poles must be near a zero of the system transfer function. The transient response of a pole located near a zero is characterized by a very small amplitude motion, which can readily be neglected.

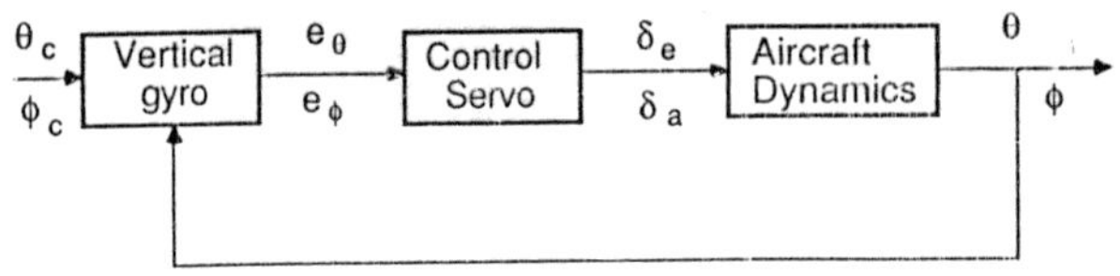

FIGURE 7.7
Block diagram of a roll or pitch displacement autopilot.

7.8 THE DISPLACEMENT AUTOPILOT

One of the earliest autopilots to be used for aircraft control is the so-called displacement autopilot. A displacement type autopilot can be used to control the angular orientation of the airplane. Conceptually, the displacement autopilot works in the following manner. In a pitch attitude displacement autopilot, the pitch angle is sensed by a vertical gyro and compared with the desired pitch angle to create an error angle. The difference or error in pitch attitude is used to produce proportional displacements of the elevator so that the error signal is reduced. Figure 7.7 is a sketch showing the block diagram representation of either a pitch or roll angle displacement autopilot.

The heading angle of the airplane can also be controlled using a similar scheme to that outlined above. The heading angle is sensed by a directional gyro and the error signal is used to displace the rudder so that the error signal is reduced. A sketch of a displacement heading autopilot is shown in Fig. 7.8.

In practise, the displacement autopilot is engaged once the airplane has been trimmed in straight and level flight. To maneuver the airplane while the autopilot is engaged, the pilot adjusts the commanded signals. For example, the airplane can be made to climb or descend by changing the pitch command. Turns can be achieved by introducing the desired bank angle while simultaneously changing the heading command. In the following section we shall examine several displacement autopilot concepts.

PITCH DISPLACEMENT AUTOPILOT. The basic components of a pitch attitude control system are shown in Fig. 7.7. For this design the reference pitch angle is compared with the actual angle measured by a gyro to produce an error signal to activate the control servo. In general, the error signal is amplified and sent to the control surface actuator to deflect the control surface. Movement of the control surface causes the aircraft to achieve a new pitch orientation, which is fed back to close the loop.

To illustrate how such an autopilot would be designed, we will examine this particular pitch displacement autopilot concept for a business jet aircraft.

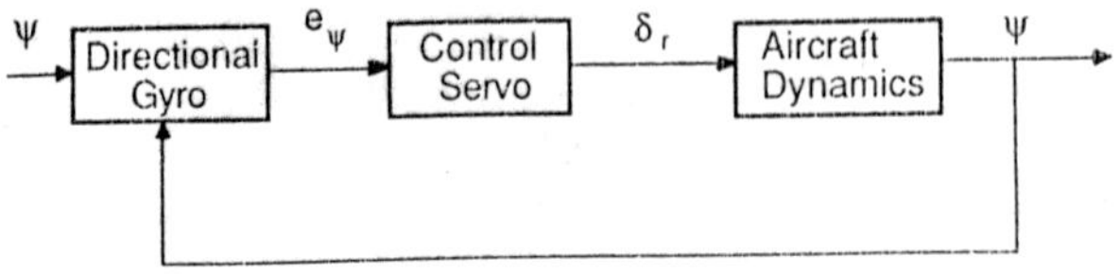

FIGURE 7.8
Block diagram of a leading displacement autopilot.

Once we have decided upon a control concept, our next step must be to evaluate the performance of the control system. To accomplish this we must define the transfer functions for each of the elements in the block diagram describing the system. For the purposes of this discussion we will assume that the transfer functions of both the gyro and amplifier can be represented by simple gains. The elevator servo transfer function can be represented as a first-order system:

$$\frac{\delta_e}{v} = \frac{1}{\tau s + 1} \tag{7.32}$$

where δ_e, v, and τ are the elevator deflection angle, input voltage, and servomotor time constant. Time constants for typical servomotors fall in a range 0.05–0.25 s. For our discussion we assume a time constant of 0.1 s. Finally, we need to specify the transfer function for the airplane. The transfer function relating the pitch attitude to elevator deflection was developed in Chapter 3. To keep the description of this design as simple as possible, we represent the aircraft dynamics by using the short-period approximation. The short-period transfer function for the business can be shown to be

$$\frac{\Delta\theta}{\Delta\delta_e} = \frac{-2.0(s + 0.3)}{s(s^2 + 0.65s + 2.15)} \tag{7.33}$$

Figure 7.9 is the block diagram representation of the autopilot. The problem now is one of determining the gain k so that the control system will have the desired performance. Selection of the gain k can be determined using a root locus plot of the loop transfer function. Figure 7.10 is the root locus plot for the business jet pitch autopilot. As the gain is increased from zero, the system damping decreases rapidly and the system becomes unstable. Even for low values of k, the system damping would be too low for satisfactory dynamic performance. The reason for the poor performance of this design is that the airplane has very little natural damping. To improve the design we could increase the damping of the short-period mode by adding an inner feedback loop. Figure 7.11 is a block diagram representation of a displacement autopilot with pitch rate feedback for improved damping. In the inner loop the pitch

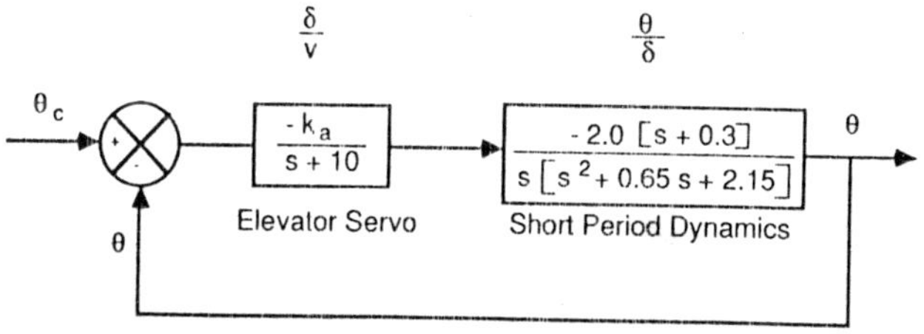

FIGURE 7.9
Block diagram of a pitch displacement autopilot for a business jet.

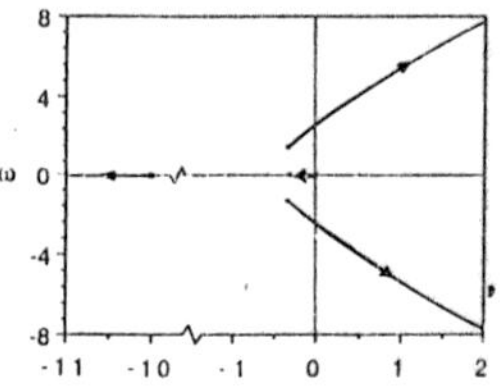

FIGURE 7.10
Root locus plot of the system gain for a pitch displacement autopilot.

rate is measured by a rate gyro and fed back to be added with the error signal generated by the difference in pitch altitude. Figure 7.12 shows the block diagram for the business jet when pitch rate is incorporated into the design. For this problem we now have two parameters to select, namely the gains k and k_{rg}. The root locus method can be used to pick both parameters. The procedure is essentially a trial-and-error method. First, the root locus diagram is determined for the inner loop; a gyro gain is selected, and then the outer root locus plot is constructed. Several iterations may be required until the desired overall system performance is achieved.

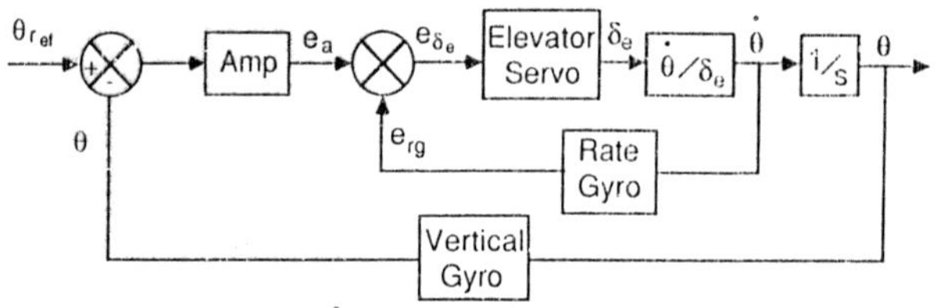

FIGURE 7.11
Block diagram of a pitch altitude control system employing pitch rate feedback.

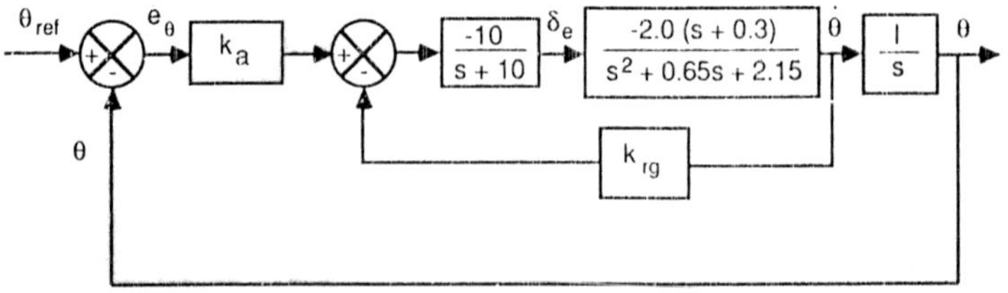

FIGURE 7.12
Block diagram of a business jet pitch attitude control system with pitch rate feedback.

Example Problem 7.2. In this example we examine the performance of a simple roll autopilot for maintaining the roll angle at some specified value. For most applications the autopilot would be used to maintain a wings-level altitude. A conceptual design of a wings-leveling autopilot would be as shown in Fig. 7.7. The aileron servomotor transfer function can again be approximated as a first-order lag term as

$$\frac{\delta_a}{v} = \frac{1}{\tau s + 1}$$

For this example we assume that τ has a value of 0.1 s. The airplane dynamics for this example will be modeled using the single degree of freedom roll equation developed in Chapter 5. The transfer function ϕ/δ can be obtained by taking the Laplace transformation of the roll equation, which yields

$$\frac{\phi}{\delta} = \frac{L_{\delta a}}{s(s - L_p)}$$

The block diagram of the control system for a fighter aircraft having the characteristics $L_{\delta a} = 11.4$, $L_p = -1.4$ is shown in Fig. 7.13.

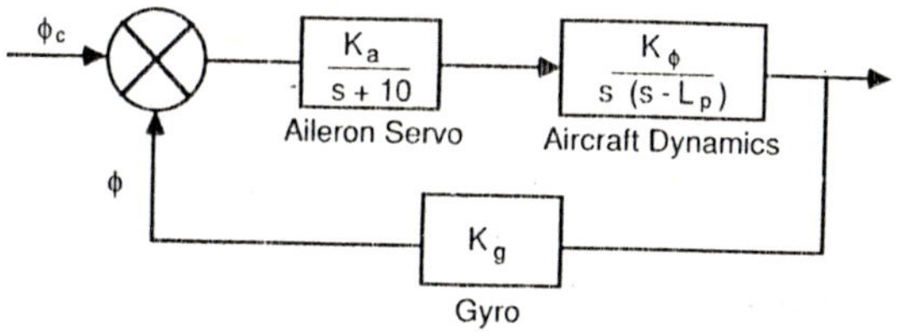

FIGURE 7.13
Block diagram of a roll control autopilot for a jet fighter.

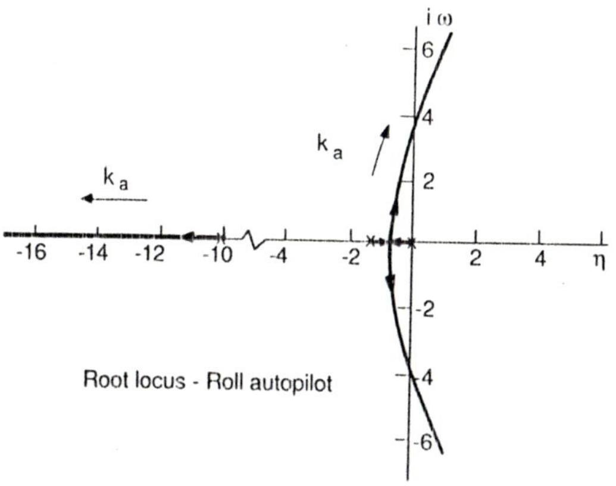

FIGURE 7.14
Root locus plot of a displacement roll control system for a fighter aircraft.

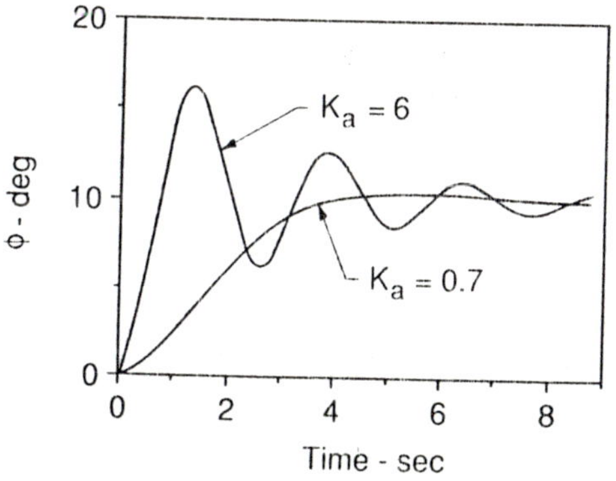

FIGURE 7.15
Roll autopilot response to initial disturbances for several different values of k_a.

The problem is now one of determining the amplifier gain k_a so that the autopilot will perform properly. The root locus analysis technique allows the designer to examine the roots of the characteristic equation of the closed-loop transfer function as one of the system parameters is varied. In this case we will vary k_a from 0 to ∞. Figure 7.14 shows the root locus plot of our autopilot design. As shown earlier, the locus starts as the poles of the open loop transfer function $G(s)H(s)$ when $k_a = 0$, and in this case move towards zeros at infinity as k_a is increase to ∞. The root starting at the servo root moves along the real axis towards $-\infty$ as k_a is increased, while the roots starting at the poles 0 and -1.4 move towards each other and eventually break away from the real axis as complex roots as k_a is increased. The complex portion of the locus intersects the imaginary axis when $k_a = 8$. Further increases in k_a result in an unstable system. Figure 7.15 shows the time-domain response for several different values of k_a.

7.7 STABILITY AUGMENTATION

Another application of automatic devices is to provide artificial stability for an airplane that has undesirable flying characteristics. Such control systems are commonly called stability augmentation systems (SAS).

As we showed earlier, the inherent stability of an airplane depends upon the aerodynamic stability derivatives. The magnitude of the derivatives affects both the damping and frequency of the longitudinal and lateral motions of an airplane. Furthermore, it was shown that the stability derivatives were a function of the airplane's aerodynamic and geometric characteristics. For a particular flight regime it would be possible to design an airplane to possess desirable flying qualities. For example, we know that the longitudinal stability coefficients are a function of the horizontal tail volume ratio. Therefore, we could select a tail size and or location so that $C_{m\alpha}$ and C_{mq} provide the proper damping and frequency for the short-period mode. However, for an airplane that will fly throughout an extended flight envelope, one can expect the stability to vary significantly, owing primarily to changes in the vehicle's configuration (lowering of flaps and landing gear), or Mach and Reynolds number effects on the stability coefficients. Because the stability derivatives vary over the flight envelope, the handling qualities will also change. Obviously, we would like to provide the flight crew with an airplane that has desirable handling qualities over its entire operational envelope. This is accomplished by employing stability augmentation systems.

Example Problem 7.3. To help us understand how a stability augmentation system works, we shall consider the case of an airplane having poor short-period dynamic characteristics. In our analysis we assume that the aircraft has only one degree of freedom—a pitching motion about the center of gravity. The equation of motion for a constrained pitching motion was developed in Chapter 4 and is reproduced below:

$$\ddot{\theta} - (M_q + M_{\dot{\alpha}})\dot{\theta} + M_\alpha \theta = M_\delta \delta$$

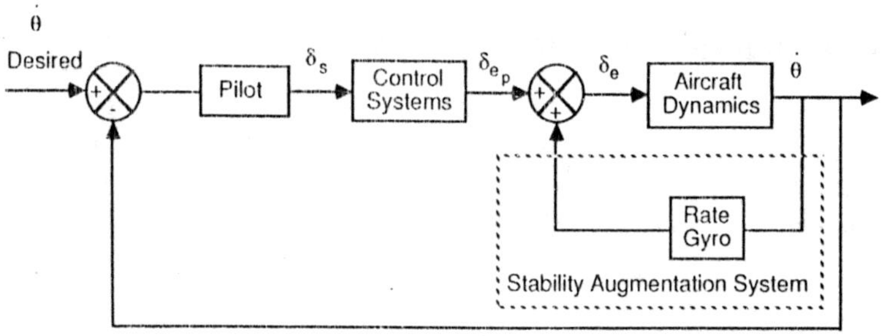

FIGURE 7.16
Stability augmentation system using pitch rate feedback.

The damping ratio and undamped natural frequency are given by

$$\zeta_{\text{sp}} = (C_{mq} + C_{m\dot{\alpha}}) \frac{\rho V S c^2}{4 I_y} / (2\omega_{n\text{sp}})$$

$$\omega_{n\text{sp}}^2 = C_{m\alpha} \frac{\rho V^2 S c}{2 I_y}$$

If the aerodynamic and inertial characteristics of a business jet during cruise are such that the above equations have the numerical values

$$\ddot{\theta} + 0.797\dot{\theta} + 3.27\theta = -2.09\delta_e$$

then the damping ratio and frequency are given by

$$\zeta_{\text{sp}} = 0.22 \qquad \omega_{n\text{sp}} = 1.8 \text{ rad/s}$$

For these short-period characteristics the airplane has level 2 flying qualities. Upon examining the flying quality specification, we see that to provide level 1 flying qualities the short-period damping must be increased so that $\zeta_{\text{sp}} > 0.3$.

One means of improving the damping of the system is to provide rate feedback, as illustrated in Fig. 7.16. This type of system is called a pitch rate damper. The stability augmentation system provides artificial damping without interfering with the pilot's control input. This is accomplished by producing an elevator deflection in proportion to the pitch rate and adding it to the pilot's control input:

$$\delta_e = \delta_{ep} + k\dot{\theta}$$

where δ_{ep} is that part of the elevator deflection created by the pilot. A rate gyro is used to measure the pitch rate and creates an electrical signal that is used to provide elevator deflections. If we substitute the expression for the elevator angle back into the equation of motion, we obtain

$$\ddot{\theta} + (0.342 + 2.09k)\dot{\theta} + 2.07\theta = -2.09\delta_{ep}$$

Comparing this equation with the standard form of a second-order system yields

$$2\zeta\omega_n = (0.342 + 2.09k) \qquad \text{and} \qquad \omega_n^2 = 2.07$$

The short-period damping ratio is now a function of the gyro gain k and can be selected so that the damping ratio will provide level 1 handling qualities. For example, if k is chosen to be 0.25, then the damping ratio $\zeta = 0.3$.

7.8 INSTRUMENT LANDING

With the advent of the instrument landing system (ILS), aircraft were able to operate safely in weather conditions under which visibility was restricted. The instrument landing system is composed of both ground-based signal transmitters and onboard receiving equipment. The ground-based equipment includes radio transmitters for the localizer, glide path, and marker beacons. The equipment on the airplane consists of receivers for detecting the signals and indicators to display the information.

The basic function of the ILS is to provide pilots with information that will permit them to guide the airplane down through the clouds to a point where the pilot re-establishes visual sighting of the runway. In a completely automatic landing, the autopilot guides the airplane all the way down to touchdown and roll out.

Before addressing the autoland system, we briefly review the basic ideas behind the ILS equipment. To guide the airplane down toward the runway, the airplane must be guided laterally and vertically. The localizer beam is used to position the aircraft on a trajectory so that the airplane will intercept the centerline of the runway. The transmitter radiates at a frequency in a band of 108–112 MHz. The purpose of this beam is to locate the airplane relative to a centerline of the runway. This is accomplished by creating azimuth guidance signals which are detected by the onboard localizer receiver. The azimuth guidance signal is created by superimposing a 90-Hz signal that is directed towards the left and a 150-Hz signal that is directed to the right on the carrier signal. Figure 7.17 is a sketch of an instrument landing localizer signal. When the aircraft is flying directly along the projected extension of the runway centerline, both superimposed signals are detected with equal strength.

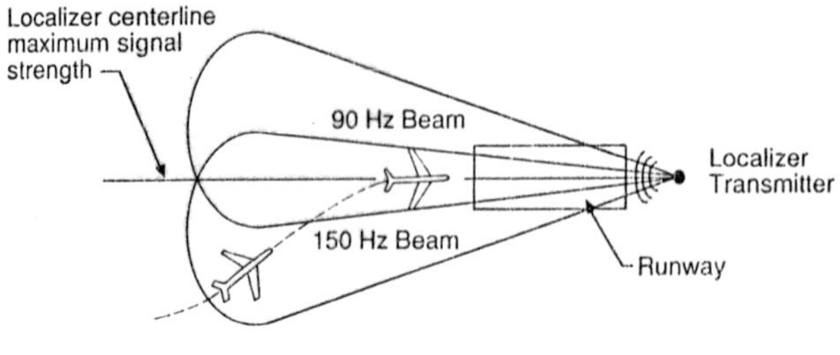

FIGURE 7.17
Sketch of a localizer beam system.

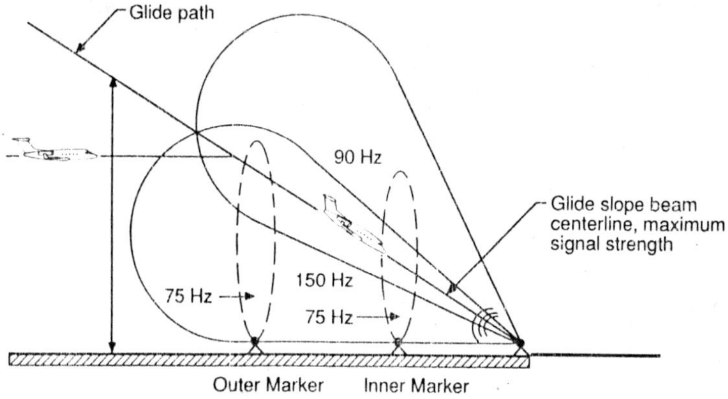

FIGURE 7.18
Sketch of a glide slope beam system.

However, when the aircraft deviates say to the right of centerline, the 150-Hz signal is stronger. The receiver in the cockpit detects the difference and directs the pilot to fly the aircraft to the left by way of a vertical bar on the ILS indicator which shows the airplane to the right of the runway. If the airplane deviates to the left, the indicator will deflect the bar to the left of the runway marker.

The glide path or glide slope beam is located near the runway threshold and radiates at a frequency in the range 329.3–335.0 MHz. Its purpose is to guide the aircraft down a predetermined descent path. The glide slope is typically an angle of 2.5–3° to the horizontal. Figure 7.18 shows a schematic of the glide path beam. As in the case of the localizer, two signals are superimposed on the carrier frequency to create an error signal if the aircraft is either high or low with respect to the glide path. This is usually indicated by a horizontal bar on the ILS indicator that moves up or down with respect to the glide path indicator. The marker beacons are used to locate the aircraft relative to the runway. Two markers are used. One is located four nautical miles from the runway and is called the outer marker. The second, or inner, marker is located 3500 ft from the runway threshold. The beams are directed vertically into the descent path at a frequency of 75 MHz. The signals are coded, and when the airplane flies overhead the signals are detected by an on-board receiver. The pilot is alerted to the passage over a marker beacon by both an audio signal and visual signal. The audio signal is heard over the aircraft's communication system and the visual signal is presented by way of a colored indicator light on the instrument panel.

In flying the airplane in poor visibility, the pilot uses the ILS equipment in the following manner. The pilot descends from cruise altitude under direction of ground control to an altitude of approximately 1200 ft above the ground. The pilot is then vectored so that the aircraft intercepts the localizer at a distance of at least six nautical miles from the runway. The pilot positions the airplane using the localizer display so that the airplane is on a heading towards the runway centerline. When the aircraft approaches the outer marker, the glide path signal is intercepted. The aircraft is placed in its final approach configuration and the pilot flies down the glide path slope. The pilot follows the beams by maneuvering the airplane so that the vertical and horizontal bars on the ILS indicator show zero deviation from the desired flight path. The ILS system does not guide the aircraft all the way to touchdown. At some point during the approach the pilot must look away from the instruments and outside the window to establish a visual reference for the final portion of the landing. The pilot may take five or six seconds to establish an outside visual reference. Obviously the pilot must do this at sufficient altitude and distance from the runway so that if the runway is not visible the pilot can abort the landing. This gives rise to a "decision height" which is a predetermined height above the runway for which the pilot can not go beyond without visually sighting the runway.

The ILS as outlined in the previous paragraphs is an integral part of a fully automatic landing system. To be able to land an airplane without any visual reference to the runway requires an automatic landing system that can intercept the localizer and glide path signals, then guide the airplane down the glide path to some preselected altitude at which the aircraft's descent rate is reduced and the airplane executes a flare maneuver so that it touches down with an acceptable sink rate. The autoland system comprises a number of automatic control systems which include a localizer and glide path coupler, attitude and airspeed control, and an automatic flare control system.

Figure 7.19 shows an airplane descending towards the runway. The airplane shown is below the intended glide path. The deviation d of the airplane from the glide path is the normal distance of the airplane above or below the desired glide path. The angle Γ is the difference between the actual and desired glide path angle and R is the radial distance of the airplane from the glide slope transmitter. To maintain the airplane along the glide path, one must make Γ equal to zero. Figure 7.20 is a conceptual design of an autopilot

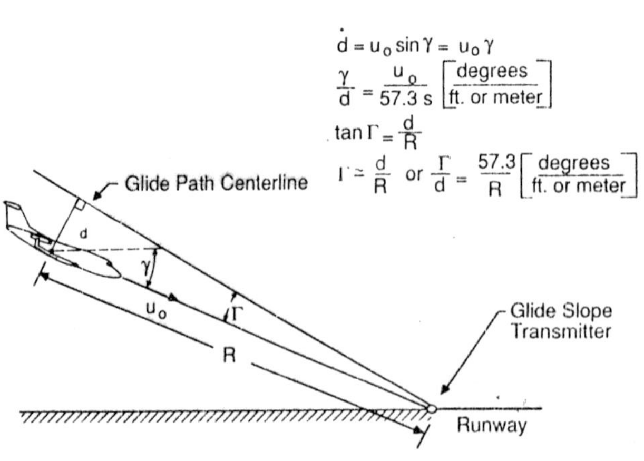

FIGURE 7.19
An airplane displaced from the glide path.

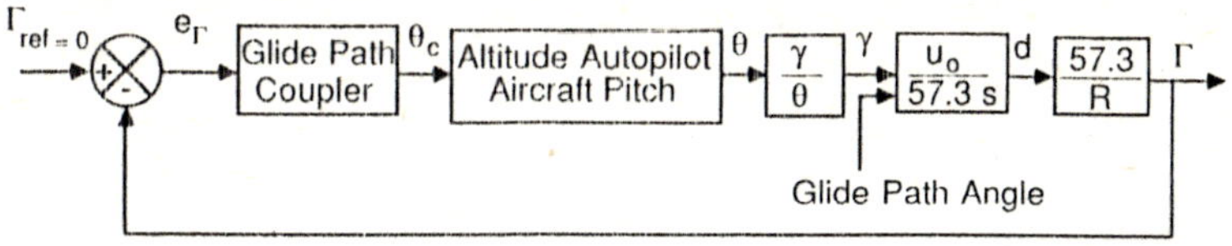

FIGURE 7.20
Conceptual design of an automatic glide path control system.

that will keep the airplane on the glide path. The transfer functions for d and Γ are obtained from the geometry and are noted in Fig. 7.19.

As the airplane descends along the glide path, its pitch attitude, and speed must be controlled. This is again accomplished by means of a pitch displacement and speed control autopilot. The pitch displacement autopilot would be conceptually the same as the one discussed earlier in this chapter. Figure 7.21 is a sketch of an automatic control system that could be used to maintain a constant speed along the flight path. The difference in flight speed is used to produce a proportional displacement of the engine throttle so that the speed difference is reduced. The component of the system labelled compensation is a device that is incorporated into the design so that the closed loop system can meet the desired performance specifications. Finally, as the airplane gets very close to the runway threshold, the glide path control system is disengaged and a flare maneuver is executed. Figure 7.22 illustrates the flare maneuver just prior to touchdown. The flare maneuver is needed to decrease the vertical descent rate to a level consistent with the ability of the landing gear to dissipate the energy of the impact at landing. An automatic flare control system is shown in Figure 7.23. A detailed discussion of the autoland system is provided by Blakelock [7.5].

7.9 SUMMARY

In this chapter we have examined briefly the use of automatic control system that can be used to reduce the pilot's workload, guide the airplane to a safe landing in poor visibility, and provide stability augmentation to improve the flying qualities of airplanes with poor stability characteristics. Additional applications of automatic control technology include load allevation and flutter suppression.

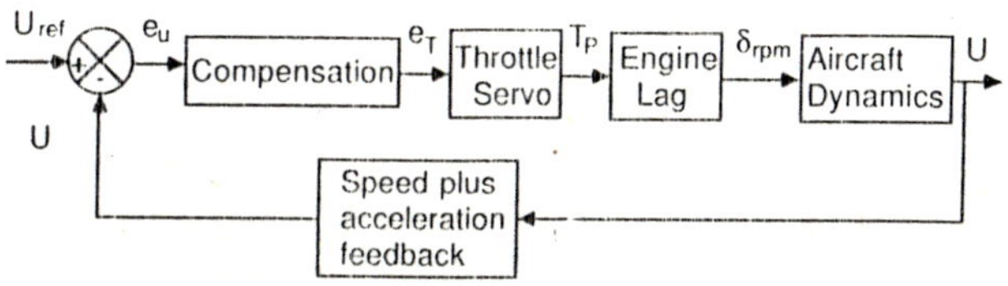

FIGURE 7.21
Conceptual design for an automatic speed control system.

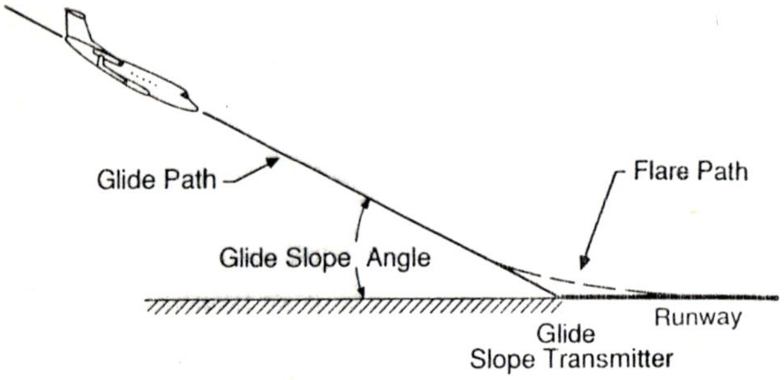

FIGURE 7.22
Sketch of a flare maneuver.

Load alleviation can be achieved by using active wing controls to reduce the wing-bending moments. By reducing the wing design loads through active controls, the designer can increase the wing span or reduce the structural weight of the wing. Increasing the wing span for a given wing area improves the aerodynamic efficiency of the wing, i.e. increases the lift-to-drag ratio. The improvement in aerodynamic efficiency and the potential for lower wing weight result in better cruise fuel efficiency.

Stability augmentation systems can also be used to improve airplane performance without degrading the vehicle's flying qualities. If the horizontal and vertical tail control surfaces are used in an active control system, the tail area can be reduced. Reducing the static stability results in smaller trim drag forces. The combination of smaller tail areas and reduced static stability yields a lower drag contribution from the tail surfaces, which will improve the performance characteristics of the airplane.

Another area in which active control can play an important role is in suppressing flutter. Flutter is an unstable structural motion that can lead to structural failure of any of the major components of an airplane, i.e. wing, tail, fuselage, or control surfaces. Flutter is caused by the interaction between structural vibration and the aerodynamic forces acting on the surface undergoing flutter. During flutter the aerodynamic surface extracts energy from the airstream to feed this undesirable motion. An automatic control system incorporating active controls can be designed to prevent flutter from occurring by controlling the structural vibration.

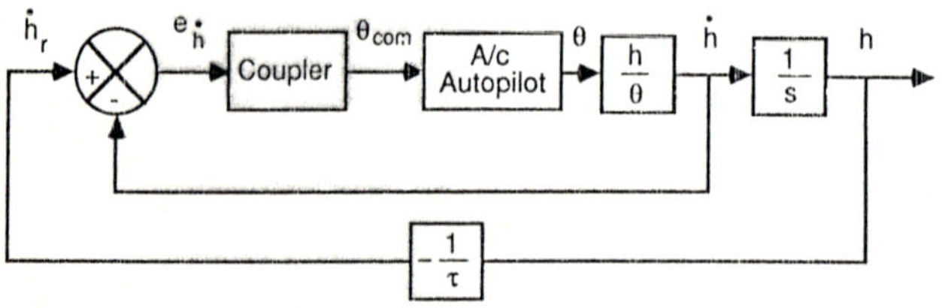

FIGURE 7.23
Conceptual design of an automatic flare control system.

7.10 PROBLEMS

7.1. Given the characteristic equation

$$\lambda^3 + 3\lambda^2 + 3\lambda + 1 + k = 0$$

find the range of values of k for which the system is stable.

7.2. Given the fourth-order characteristic equation

$$\lambda^4 + 6\lambda^3 + 11\lambda^2 + 6\lambda + k = 0$$

for what values of k will the system be stable?

7.3 The single degree of freedom pitching motion of an airplane was shown to be represented by a second order differential equation. If the equation is given as

$$\ddot{\theta} + 0.5\dot{\theta} + 2 = \delta_e$$

where the θ and δ_e are in radians. Estimate the rise time, peak overshoot and settling time for step input of the elevator angle of 0.10 radians.

7.4 Determine the frequency domain characteristic for problem 7.3. In particular estimate the resonance peak, M_r, resonant frequency, ω_r, bandwidth, ω_B, and the phase margin.

7.5. A simple roll control system is shown in Fig. P7.5. Sketch the root locus diagram for this system. Determine the value of the gain k when the system is neutrally stable. Assume that the aircraft characteristics are the same as example problem 7.2.

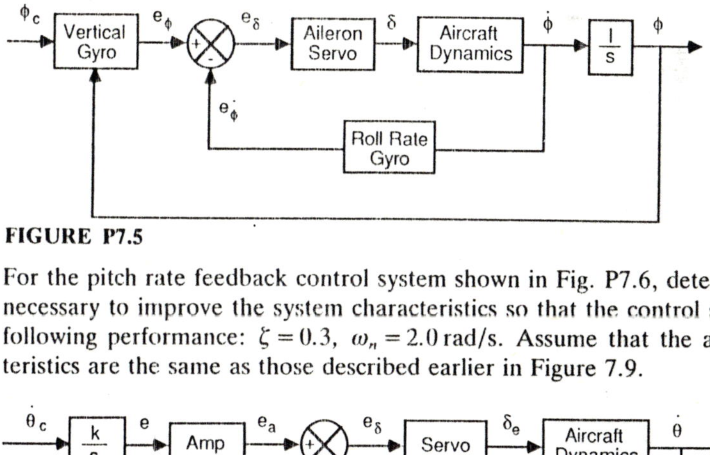

FIGURE P7.5

7.6. For the pitch rate feedback control system shown in Fig. P7.6, determine the gain necessary to improve the system characteristics so that the control system has the following performance: $\zeta = 0.3$, $\omega_n = 2.0$ rad/s. Assume that the aircraft characteristics are the same as those described earlier in Figure 7.9.

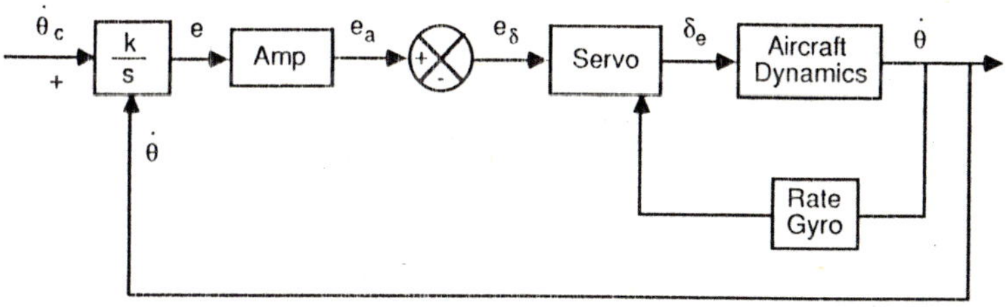

FIGURE P7.6

7.7 The Wright "Flyer" was statically and dynamically unstable. However, because the Wright brothers incorporated sufficient control authority into their design they were able to fly their airplane successfully. Although the airplane was difficult to fly the combination of the pilot and airplane could be made to be a stable system. In reference 7.7 the closed loop pilot is represented as a pure gain, k_p, and the pitch attitude to canard deflection is given as follows.

$$\frac{\theta}{\delta_c} = \frac{11.0(s+0.5)(s+3.0)}{(s^2+0.72s+1.44)(s^2+5.9s-11.9)}$$

Determine the root locus plot of the closed loop system. For what range of pilot gain is the system stable.

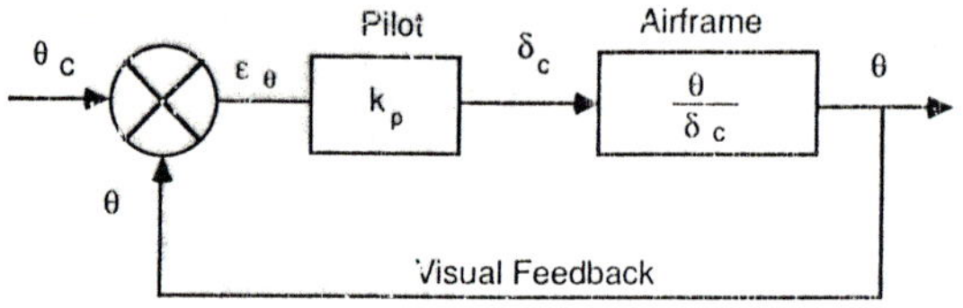

FIGURE P7.7

REFERENCES

7.1. Bollay, W.: "Aerodynamic Stability and Automatic Control," *Journal of the Aeronautical Sciences,* vol. 18, No. 9, pp. 569–617, 1951.
7.2. Raven, F. H.: *Automatic Control Engineering,* McGraw-Hill, New York, 1978.
7.3. Kuo, B. C.: *Automatic Control Systems,* Prentice Hall, Englewood Cliffs, NJ, 1975.
7.4. Shinners, S. M.: *Modern Control System Theory and Application,* Addison Wesley, Reading, MA, 1978.
7.5. Blakelock, J. H.: *Automatic Control of Aircraft and Missiles,* Wiley, New York, 1965.
7.6. Pallett, E. H. J.: *Automatic Flight Control,* Granada Publishing, London, England, 1979.
7.7. Culick, F. E. C.: "Building a 1903 Wright 'Flyer'—by Committee," *AIAA Paper* 88-0094, 1988.

CHAPTER 8

AUTOMATIC CONTROL: APPLICATION OF MODERN CONTROL THEORY

8.1 INTRODUCTION

In Chapter 7, the design of feedback control systems was accomplished using the root locus technique and Bode methods developed by Evans and Bode, respectively. These techniques are very useful in designing many practical control systems. However, the design of a control system using either the root locus or Bode techniques is essentially a trial-and-error procedure. The major advantage of these design procedures is their simplicity and ease of use. This advantage disappears quickly as the complexity of the system increases.

With the rapid development of high-speed computers during the recent decades, a new approach to control system design has evolved. This new approach is commonly called modern control theory. This theory permits a more systematic approach to control system design. In modern control theory, the control system is specified as a system of first-order differential equations. By formulating the problem in this manner, the control system designer can fully exploit the digital computer for solving complex control problems. Another advantage of modern control theory is that optimization techniques can be applied to design optimal control systems. To comprehend this theory fully one needs to have a good understanding of matrix algebra; a brief discussion of matrix algebra is included in the appendices.

It is not possible in a single chapter to present a thorough discussion of modern control theory. Our purpose is to expose the reader to some of the concepts of modern control theory and then apply the procedures to the design of aircraft autopilots. It is hoped that this brief discussion will provide the reader with an appreciation of modern control theory and its application to the design of aircraft flight control systems.

8.2 STATE-SPACE MODELING

The state-space approach to control system analysis and design is a time-domain method. As was shown in Chapters 4 and 5, the equations of motion can easily be written in the state-space form. The application of state variable techniques to control problems is called modern control theory. The state equations are simply first-order differential equations that govern the dynamics of the system being analyzed. It should be noted that any higher-order differential equation can be decomposed into a set of first-order differential equations. This will be shown later by means of an illustration.

In the mathematical sense, the state variables and state equation completely describe the system. The definition of state variables is as follows. The state variables of a system are a minimum set of variables $x_1(t) \cdots x_n(t)$ which, when known at time t_0, and along with the input, are sufficient to determine the state of the system at any other time $t > t_0$. State variables should not be confused with the output of the system. An output variable is one that can be measured, but state variables do not always satisfy this condition. The output, as we will see shortly, is defined as a function of the state variables.

Once a physical system has been reduced to a set of differential equations, the equation can be rewritten in a convenient matrix form:

$$\dot{\mathbf{x}} = \mathbf{A}\mathbf{x} + \mathbf{B}\eta \tag{8.1}$$

The output of the system is expressed in terms of the state and control inputs as follows:

$$\mathbf{y} = \mathbf{C}\mathbf{x} + \mathbf{D}\eta \tag{8.2}$$

The state, control, and output vectors are defined as follows:

$$\mathbf{x} = \begin{bmatrix} x_1(t) \\ x_1(t) \\ \vdots \\ x_n(t) \end{bmatrix} \qquad \text{State vector } (n \times 1) \tag{8.3}$$

$$\eta = \begin{bmatrix} \delta_1(t) \\ \delta_2(t) \\ \vdots \\ \delta_p(t) \end{bmatrix} \qquad \text{Control or input vector } (p \times 1) \tag{8.4}$$

$$\mathbf{y} = \begin{bmatrix} y_1(t) \\ y_2(t) \\ \vdots \\ y_q(t) \end{bmatrix} \qquad \text{Output vector } (q \times 1) \tag{8.5}$$

The matrices **A**, **B**, **C** and **D** are defined in the following manner:

$$\mathbf{A} = \begin{bmatrix} a_{11} & a_{12} & \cdots & a_{1n} \\ a_{21} & & & \vdots \\ \vdots & & & \vdots \\ a_{n1} & a_{n2} & \cdots & a_{nn} \end{bmatrix} \qquad \text{Plant matrix } (n \times n) \tag{8.6}$$

$$\mathbf{B} = \begin{bmatrix} b_{11} & b_{12} & \cdots & b_{1p} \\ b_{21} & & & \vdots \\ \vdots & & & \vdots \\ b_{n1} & b_{n2} & \cdots & b_{np} \end{bmatrix} \qquad \text{Control or input matrix } (n \times p) \tag{8.7}$$

$$\mathbf{C} = \begin{bmatrix} c_{11} & c_{12} & \cdots & c_{1n} \\ c_{21} & & & \vdots \\ \vdots & & & \vdots \\ c_{q1} & & & c_{qn} \end{bmatrix} (q \times n) \tag{8.8}$$

$$\mathbf{D} = \begin{bmatrix} d_{11} & d_{12} & \cdots & d_{1p} \\ d_{21} & & & \vdots \\ \vdots & & & \vdots \\ d_{q1} & & & a_{qp} \end{bmatrix} (q \times p) \tag{8.9}$$

Figure 8.1 is a sketch of the block diagram representation of the state equation given by Eqs (8.1) and (8.2).

The state equations are a set of first-order differential equations. The matrices **A** and **B** may be either constant or functions of time. For the application we are considering, namely aircraft equations of motion, the matrices are composed of an array of constants. The constants making up either the **A** or **B** matrices are the stability and control derivatives of the airplane. It should be noted that, if the governing differential equations are of

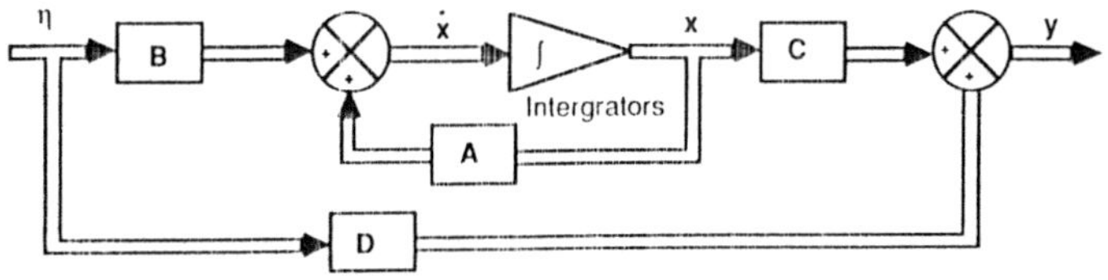

FIGURE 8.1
Block diagram of the linear state equations.

higher order, they can be reduced to a system of first-order differential equations.

For example, suppose that the physical system being modeled can be described by an nth-order differential equation:

$$\frac{d^n c(t)}{dt^n} + a_1 \frac{d^{n-1} c(t)}{dt^{n-1}} + a_2 \frac{d^{n-2} c(t)}{dt^{n-2}} + \cdots + a_{n-1} \frac{dc(t)}{dt} + a_n c(t) = r(t) \quad (8.10)$$

The variables $c(t)$ and $r(t)$ are the output and input variables, respectively. The above differential equation can be reduced to a set of first-order differential equations by defining the state variables as follows:

$$\begin{aligned} x_1(t) &= c(t) \\ x_2(t) &= \frac{dc(t)}{dt} \\ &\vdots \\ x_n(t) &= \frac{d^{n-1}c(t)}{dt^{n-1}} \end{aligned} \quad (8.11)$$

The state equations can now be written as

$$\begin{aligned} \dot{x}_1(t) &= x_2(t) \\ \dot{x}_2(t) &= x_3(t) \\ \\ \dot{x}_n(t) &= -a_n x_1(t) - a_{n-1} x_2(t) - \cdots - a_1 x_n(t) + r(t) \end{aligned} \quad (8.12)$$

The last equation is obtained by solving for the highest-order derivative in the original differential equation. Rewriting the equaton in the state vector form yields

$$\dot{\mathbf{x}} = \mathbf{A}\mathbf{x} + \mathbf{B}\eta \quad (8.13)$$

where

$$\mathbf{A} = \begin{bmatrix} 0 & 1 & 0 & 0 & 0 & \cdot & 0 \\ 0 & 0 & 1 & 0 & 0 & \cdot & 0 \\ 0 & 0 & 0 & 1 & 0 & \cdot & 0 \\ \cdot & \cdot & \cdot & \cdot & \cdot & \cdot & \cdot \\ \cdot & \cdot & \cdot & \cdot & \cdot & \cdot & \cdot \\ 0 & 0 & 0 & 0 & 0 & \cdot & 1 \\ -a_n & -a_{n-1} & -a_{n-2} & -a_{n-3} & -a_{n-4} & \cdot & -a_1 \end{bmatrix} \quad (8.14)$$

$$\mathbf{B} = \begin{bmatrix} 0 \\ 0 \\ 0 \\ 0 \\ \cdot \\ \cdot \\ \cdot \\ 1 \end{bmatrix} \tag{8.15}$$

and the output equation is

$$y = \mathbf{D}\mathbf{x} \tag{8.16}$$

where

$$\mathbf{D} = [1 \ 0 \ 0 \cdot \cdot \ 0] \tag{8.17}$$

STATE TRANSITION MATRIX. The state transition matrix is defined as a matrix that satisfies the linear homogeneous state equation, i.e.

$$\dot{\mathbf{x}} = \mathbf{A}\mathbf{x} \qquad \text{Homogeneous state equation} \tag{8.18}$$

$$\mathbf{x}(0) = \begin{bmatrix} x_1(0) \\ \vdots \\ x_n(0) \end{bmatrix} \qquad \text{Initial state at time } t = 0 \tag{8.19}$$

$$\mathbf{x}(t) = \Phi(\mathbf{t})\,\mathbf{x}(0) \tag{8.20}$$

where $\Phi(t)$ is the state transition matrix.

State transition matrix by Laplace transformation technique

$$\dot{\mathbf{x}} = \mathbf{A}\mathbf{x} \qquad \text{and} \qquad \mathbf{x}(0) = \begin{bmatrix} x_1(0) \\ \vdots \\ x_n(0) \end{bmatrix} \tag{8.21}$$

Taking the Laplace transformation of the above equation yields

$$s\mathbf{x}(t) - \mathbf{x}(0) = \mathbf{A}\mathbf{x}(s) \tag{8.22}$$

or

$$\mathbf{x}(s) = [\mathbf{sI} - \mathbf{A}]^{-1}\mathbf{x}(0) \tag{8.23}$$

The state transition matrix is obtained by taking the inverse Laplace transform of equation 8.23

$$\Phi(t) = \mathcal{L}^{-1}[(\mathbf{sI} - \mathbf{A})^{-1}] \tag{8.24}$$

State transition matrix by classical technique

The state transition matrix can also be found in the following manner

$$\mathbf{x}(t) = e^{\mathbf{A}t}\,\mathbf{x}(0) \tag{8.25}$$

where $e^{\mathbf{A}t}$ is a matrix exponential and $de^{\mathbf{A}t}/dt = \mathbf{A}e^{\mathbf{A}t}$. Substituting the above equation into the homogeneous state equation shows that it is a solution:

$$\mathbf{A}e^{\mathbf{A}t}\,\mathbf{x}(0) = \mathbf{A}e^{\mathbf{A}t}\,\mathbf{x}(0) \tag{8.26}$$

$e^{\mathbf{A}t}$ can be represented by a power series as follows:

$$e^{\mathbf{A}t} = \mathbf{I} + \mathbf{A}t + \tfrac{1}{2}\mathbf{A}^2t^2 + \tfrac{1}{3}\mathbf{A}^3t^3 + \cdots \tag{8.27}$$

The state transition matrix is

$$\Phi(t) = \mathbf{I} + \mathbf{A}t + \tfrac{1}{2}\mathbf{A}^2t^2 + \tfrac{1}{3}\mathbf{A}^3t^3 + \cdots \tag{8.28}$$

Since the state transition matrix satisfies the homogeneous state equation, it represents the free response of the system.

Properties of the state transition matrix

Some of the properties of the state transition matrix are presented below.

1. $\Phi(0) = e^{\mathbf{A}0} = \mathbf{I}$. (8.29)
2. $[\Phi(t)]^{-1} = [\Phi(-t)]$. (8.30)
3. $\Phi(t_1 + t_2) = e^{\mathbf{A}(t_1+t_2)} = e^{\mathbf{A}t_1}e^{\mathbf{A}t_2} = \Phi(t_1)\Phi(t_2) = \Phi(t_2)\Phi(t_1)$. (8.31)
4. $[\Phi(t)]^k = \Phi(kt)$, where k is an integer. (8.32)

Once the state transition matrix has been found, the solution to the nonhomogeneous equation can be determined as follows:

$$\dot{\mathbf{x}} = \mathbf{A}\mathbf{x} + \mathbf{B}\eta \tag{8.33}$$

Taking the Laplace transform of the above equation yields

$$s\{\mathbf{x}(s) - \mathbf{x}(0)\} = \mathbf{A}\mathbf{x}(s) + \mathbf{B}\eta(s) \tag{8.34}$$

solving for $\mathbf{x}(s)$

$$\mathbf{x}(s) = [s\mathbf{I} - \mathbf{A}]^{-1}\mathbf{x}(0) + [s\mathbf{I} - \mathbf{A}]^{-1}\mathbf{B}\eta(s) \tag{8.35}$$

$$\mathbf{x}(t) = \mathscr{L}^{-1}[s\mathbf{I} - \mathbf{A}]^{-1}\mathbf{x}(0) + \mathscr{L}^{-1}[[s\mathbf{I} - \mathbf{A}^{-1}]\mathbf{B}\eta(s)] \tag{8.36}$$

or

$$\mathbf{x}(t) = \Phi(t)\mathbf{x}(0) + \int_0^t \Phi(t-\tau)\mathbf{B}\eta(\tau)\,d\tau \tag{8.37}$$

Example Problem 8.1. Given a system that is described by the following equations,

$$\begin{bmatrix}\dot{x}_1\\ \dot{x}_2\end{bmatrix} = \begin{bmatrix}0 & 1\\ -3 & -4\end{bmatrix}\begin{bmatrix}x_1\\ x_2\end{bmatrix} + \begin{bmatrix}0\\ 1\end{bmatrix}u$$

where

$$\begin{bmatrix} x_1(0) \\ x_2(0) \end{bmatrix} = \begin{bmatrix} 0 \\ 1 \end{bmatrix}$$

and the input function u is a unit step function, determine the response of the system using the state transition matrix, which can be as found as follows:

$$\Phi(s) = [s\mathbf{I} - \mathbf{A}]^{-1} = \begin{bmatrix} s & -1 \\ 3 & s+4 \end{bmatrix}^{-1}$$

$$\Phi(s) = \begin{bmatrix} \dfrac{s+4}{s^2+4s+3} & \dfrac{1}{s^2+4s+3} \\ \dfrac{-3}{s^2+4s+3} & \dfrac{s}{s^2+4s+3} \end{bmatrix}$$

Using the partial fraction expansion, the elements of the transition matrix can be written as

$$\Phi(s) = \begin{bmatrix} \dfrac{3}{2}\dfrac{1}{s+1} - \dfrac{1}{2}\dfrac{1}{s+3} & \dfrac{1}{2}\dfrac{1}{s+1} - \dfrac{1}{2}\dfrac{1}{s+3} \\ \dfrac{-3}{2}\dfrac{1}{s+1} + \dfrac{3}{2}\dfrac{1}{s+3} & \dfrac{-1}{2}\dfrac{1}{s+1} + \dfrac{3}{2}\dfrac{1}{s+3} \end{bmatrix}$$

The state transition matrix $\Phi(t)$ is determined by taking the inverse Laplace transform of $\Phi(s)$:

$$\Phi(t) = \mathscr{L}^{-1}(\Phi(s))$$

$$\Phi(t) = \begin{bmatrix} \frac{3}{2}e^{-t} - \frac{1}{2}e^{-3t} & \frac{1}{2}e^{-t} + \frac{1}{2}e^{-3t} \\ -\frac{3}{2}e^{-t} + \frac{3}{2}e^{-3t} & -\frac{1}{2}e^{-t} + \frac{3}{2}e^{-3t} \end{bmatrix}$$

The state transition matrix can also be obtained by the matrix exponential definition:

$$e^{\mathbf{A}t} = \mathbf{I} + \mathbf{A}t + \frac{1}{2!}\mathbf{A}^2t^2 + \frac{1}{3!}\mathbf{A}^3t^3 + \cdots$$

$$\mathbf{A}^2 = \begin{bmatrix} 0 & 1 \\ -3 & -4 \end{bmatrix}\begin{bmatrix} 0 & 1 \\ -3 & -4 \end{bmatrix} = \begin{bmatrix} -3 & -4 \\ 12 & 13 \end{bmatrix}$$

etc. Then

$$\mathbf{e}^{\mathbf{A}t} = \begin{bmatrix} 1 + 0t - \frac{3}{2}t^2 \cdots & 0 + t - 2t^2 \cdots \\ 0 - 3t + 6t^2 \cdots & 1 - 4t + \frac{13}{2}t^2 \cdots \end{bmatrix}$$

which can be shown to be identical to the equations obtained from the Laplace transform approach. The solution of the state equation is found by

$$\mathbf{x}(t) = \Phi(t)\mathbf{x}(0) + \int_0^t \Phi(t-\tau)\mathbf{B}\eta(s)\,d\tau$$

$$\mathbf{x}(t) = \begin{bmatrix} \frac{3}{2}e^{-t} - \frac{1}{2}e^{-3t} & \frac{1}{2}e^{-t} - \frac{1}{2}e^{-3t} \\ \frac{3}{2}e^{-t} + \frac{3}{2}e^{-3t} & -\frac{1}{2}e^{-t} + \frac{3}{2}e^{-3t} \end{bmatrix} \begin{bmatrix} 0 \\ 1 \end{bmatrix}$$

$$+ \int_0^t \begin{bmatrix} \frac{3}{2}e^{-(t-\tau)} - \frac{1}{2}e^{-3(t-\tau)} & \frac{1}{2}e^{-(t-\tau)} - \frac{1}{2}e^{-3(t-\tau)} \\ -\frac{3}{2}e^{-(t-\tau)} + \frac{3}{2}e^{-3(t-\tau)} & -\frac{1}{2}e^{-(t-\tau)} + \frac{3}{2}e^{-3(t-\tau)} \end{bmatrix} \begin{bmatrix} 0 \\ 1 \end{bmatrix} d\tau$$

CONTROLLABILITY AND OBSERVABILITY. In the following sections we shall examine the application of state feedback design and optimal control theory to aircraft control problems. Two concepts that play an important role in modern control theory are the concepts of controllability and observability. Controllability is concerned with whether the states of the dynamic system are affected by the control input. A system is said to be completely controllable if there exists a control that transfers any initial state $\mathbf{x}_i(t)$ to any final state $\mathbf{x}_f(t)$ in some finite time. If one or more of the states are unaffected by the control, the system is not completely controllable.

A mathematical definition of controllability for a linear dynamical system can be expressed as follows. If the dynamic system can be described by the state equation

$$\dot{\mathbf{x}} = \mathbf{A}\mathbf{x} + \mathbf{B}\eta \tag{8.38}$$

where $\mathbf{x}$ and η are the state and control vectors of order n and m, respectively, then the necessary and sufficient condition for the system to be completely controllable is that the rank of the matrix $\mathbf{P}$ is equal to the number of states. The matrix $\mathbf{P}$ is constructed from the $\mathbf{A}$ and $\mathbf{B}$ matrices in the following way:

$$\mathbf{P} = [\mathbf{B}, \mathbf{AB}, \mathbf{A}^2\mathbf{B}, \ldots \mathbf{A}^{n-1}\mathbf{B}] \tag{8.39}$$

The rank of a matrix is defined as the largest non-zero determinant. Although this definition is abstract, the test for controllability can be easily applied.

Observability deals with whether the states of the system can be identified from the output of the system. A system is said to be completely observable if every state $\mathbf{x}$ can be determined by the measurement of the output $\mathbf{y}(t)$ over a finite time interval. If one or more states cannot be identified from the output of the system, the system is not observable. A mathematical test for observability of an nth-order dynamic system governed by the equations

$$\dot{\mathbf{x}} = \mathbf{A}\mathbf{x} + \mathbf{B}\eta \tag{8.40}$$

$$\mathbf{y} = \mathbf{C}\mathbf{x} + \mathbf{D}\eta \tag{8.41}$$

is given as follows. The necessary and sufficient condition for a system to be

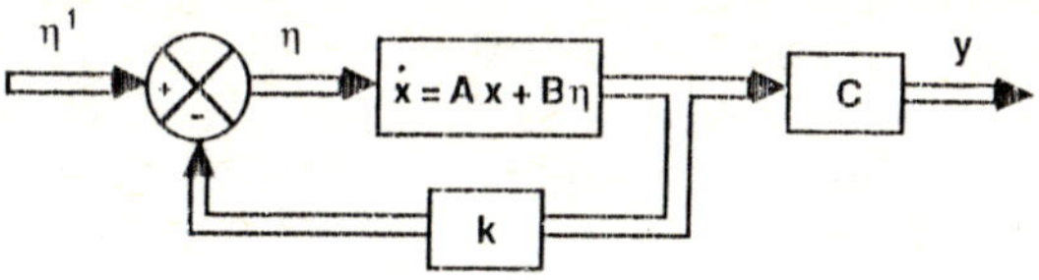

FIGURE 8.2
Block diagram of a linear system with state feedback.

completely observable is that the matrix **U**, defined as

$$\mathbf{U} = [\mathbf{C}^T, \mathbf{A}^T\mathbf{C}^T, \ldots (\mathbf{A}^T)^{n-1}\mathbf{C}^T] \tag{8.42}$$

is of the rank n.

8.3 STATE FEEDBACK DESIGN

State feedback can be used to design a control system with a specific eigenvalue structure. Consider the system represented by the state equations

$$\dot{\mathbf{x}} = \mathbf{A}\mathbf{x} + \mathbf{B}\eta \tag{8.43}$$

$$\mathbf{y} = \mathbf{C}\mathbf{x} \tag{8.44}$$

It can be shown from optimal control theory that if the system is state-controllable, then it is possible to use a linear feedback control law given by

$$\eta = -\mathbf{k}^T\mathbf{x} + \eta' \tag{8.45}$$

to achieve a specific eigenvalue structure. The matrix **k** is called the feedback gain matrix and η' is the original control system input. Figure 8.2 shows the block diagram representation of the system. If we combine Eqs (8.43) and (8.45), the closed loop system is given by

$$\dot{\mathbf{x}} = (\mathbf{A} - \mathbf{B}\mathbf{k}^T)\mathbf{x} + \mathbf{B}\eta' \tag{8.46}$$

or

$$\dot{\mathbf{x}} = \mathbf{A}^*\mathbf{x} + \mathbf{B}\eta' \tag{8.47}$$

where $\mathbf{A}^*$ is the augmented matrix. For the case in which the **A** matrix may have had undesirable eigenvalues, the augmented matrix $\mathbf{A}^*$ can be made to have specific eigenvalues by properly selecting the feedback gains.

The application of state feedback as presented here requires the states be state-controllable. As stated earlier, a system is said to be completely controllable if the control can be used to move the system from its initial state at $t = t_0$ to the desired state at $t = t_1$. Another way of stating this concept is to say that every state is affected by the control input signal.

LONGITUDINAL STABILITY AUGMENTATION. State feedback control can be used to improve the stability characteristics of airplanes that lack good flying qualities. As shown in the previous section, the problem is one of determining the feedback gains that will produce the desired flying characteristics.

Starting with the longitudinal state equations given in Chapter 4, we develop a set of linear algebraic equations in terms of the unknown feedback gains. The state equations for the longitudinal motion have been simplified by neglecting the affect of the control on the X-force equation and the stability derivative $M_{\dot{w}}$. The state equations are given below:

$$\begin{bmatrix} \Delta\dot{u} \\ \Delta\dot{w} \\ \Delta\dot{q} \\ \Delta\dot{\theta} \end{bmatrix} = \begin{bmatrix} X_u & X_w & 0 & -g \\ Z_u & Z_w & u_0 & 0 \\ M_u & M_w & M_q & 0 \\ 0 & 0 & 1 & 0 \end{bmatrix} \begin{bmatrix} \Delta u \\ \Delta w \\ \Delta q \\ \Delta\theta \end{bmatrix} + \begin{bmatrix} 0_\delta \\ Z_\delta \\ M_\delta \\ 0 \end{bmatrix} [\Delta\delta_e] \quad (8.48)$$

or

$$\dot{\mathbf{x}} = \mathbf{A}\mathbf{x} + \mathbf{B}\eta \quad (8.49)$$

where **A** and **B** are the stability and control matrices shown above and **x** and η are the state and control vectors.

The eigenvalues of the **A** matrix are the short- and long-period roots. If these roots are unacceptable to the pilot, a stability augmentation system will be required. State feedback design can be used to provide the stability augmentation system. In state feedback design we assume a linear control law that is proportional to the states, i.e.

$$\eta = -\mathbf{k}^{\mathrm{T}}\mathbf{x} + \eta_p \quad (8.50)$$

where $\mathbf{k}^T$ *is the transpose of the feedback gain vector and* η_p is the pilot input. Substituting the control law into the state equation yields

$$\dot{\mathbf{x}} = (\mathbf{A} - \mathbf{B}\mathbf{k}^{\mathrm{T}})\mathbf{x} + \mathbf{B}\eta_p \quad (8.51)$$

or

$$\dot{\mathbf{x}} = \mathbf{A}^*\mathbf{x} + \mathbf{B}\eta_p \quad (8.52)$$

where $\mathbf{A}^*$ is the augmented matrix and is expressed as

$$\mathbf{A}^* = \mathbf{A} - \mathbf{B}\mathbf{k}^{\mathrm{T}} \quad (8.53)$$

The augmented matrix for the longitudinal system of equations is

$$\mathbf{A}^* = \begin{bmatrix} X_u - X_\delta k_1 & X_w - X_\delta k_2 & -X_\delta k_3 & -g - X_\delta k_4 \\ Z_u - Z_\delta k_1 & Z_w - Z_\delta k_2 & u_0 - Z_\delta k_3 & -Z_\delta k_4 \\ M_u - M_\delta k_1 & M_w - M_\delta k_2 & M_q - M_\delta k_3 & -M_\delta k_4 \\ 0 & 0 & 1 & 0 \end{bmatrix} \quad (8.54)$$

The characteristic equation for the augmented matrix is obtained by solving the equation

$$|\lambda\mathbf{I} - \mathbf{A}^*| = 0 \quad (8.55)$$

which yields a quartic characteristic equation.

$$A\lambda^4 + B\lambda^3 + C\lambda^2 + D\lambda + E = 0 \quad (8.56)$$

where the coefficients are defined below.

$$
\left.
\begin{aligned}
A &= 1.0 \\
B &= Z_\delta k_2 + M_\delta k_3 - (X_u + Z_w + M_q) \\
C &= Z_\delta X_w k_1 + (u_0 M_\delta - X_u Z_\delta - Z_\delta M_q) k_2 \\
&\quad + (Z_\delta M_w - X_u M_\delta - Z_w M_\delta) k_3 + M_\delta k_4 \\
&\quad + X_u M_q + X_u Z_w + Z_w M_q - u_0 M_w - X_w Z_w \\
D &= (u_0 X_w M_\delta - g M_\delta - X_w Z_\delta M_q) k_1 \\
&\quad + (X_u Z_\delta M_q - u_0 X_u M_\delta) k_2 \\
&\quad + (X_u Z_w M_\delta - X_u Z_\delta M_w - X_w Z_u M_\delta + X_w Z_\delta M_u) k_3 \\
&\quad + (Z_\delta M_w - X_u M_\delta - Z_w M_\delta) k_4 \\
&\quad + g M_u - X_u Z_w M_q + u_0 X_u M_w + X_w Z_u M_q - u_0 X_w M_u \\
E &= (g Z_w M_\delta - g Z_\delta M_w) k_1 + (g Z_\delta M_u - g Z_u M_\delta) k_2 \\
&\quad + (X_u Z_w M_\delta - X_u Z_\delta M_w - X_w Z_u M_\delta + X_w Z_\delta M_u) k_4 \\
&\quad + g Z_u M_w - g Z_w M_u
\end{aligned}
\right\} \qquad (8.57)
$$

The characteristic equation of the augmented system is a function of the known stability derivatives and the unknown feedback gains. The feedback gains can be determined once the desired longitudinal characteristics are specified. For example, if the desired characteristic roots are

$$\lambda_{12} = -\zeta_{\text{sp}}\omega_{n\text{sp}} \pm i\omega_{n\text{sp}}\sqrt{1-\zeta_{\text{sp}}^2} \qquad (8.58)$$

and

$$\lambda_{34} = -\zeta_{\text{p}}\omega_{n\text{p}} \pm i\omega_{n\text{p}}\sqrt{1-\zeta_p^2} \qquad (8.59)$$

then the desired characteristic equation is

$$\lambda^4 - [(\lambda_1 + \lambda_2 + \lambda_3 + \lambda_4)]\lambda^3 + [\lambda_1\lambda_2\lambda_3\lambda_4 + (\lambda_1 + \lambda_2)]\lambda^2 - [\lambda_1\lambda_2(\lambda_3 + \lambda_4) + \lambda_3\lambda_4(\lambda_1 + \lambda_2)]\lambda + \lambda_1\lambda_2\lambda_3\lambda_4 = 0 \qquad (8.60)$$

By equating the coefficients of like powers of λ for the augmented and desired characteristic equations, one obtains a set of four linear algebraic equations in terms of the unknown gains. These equations can be solved for the feedback gains.

Example Problem 8.2. An airplane is found to have poor short-period flying qualities in a particular flight regime. To improve the flying qualities, a stability augmentation system using state feedback is to be employed. Determine the feedback gains so that the airplane's short-period characteristics are $\lambda_{\text{sp}} = -2.1 \pm 2.14i$. Assume that the original short-period dynamics are given by

$$\begin{bmatrix} \Delta\dot{\alpha} \\ \Delta\dot{q} \end{bmatrix} = \begin{bmatrix} -0.334 & 1.0 \\ -2.52 & -0.387 \end{bmatrix} \begin{bmatrix} \Delta\alpha \\ \Delta q \end{bmatrix} + \begin{bmatrix} -0.027 \\ -2.6 \end{bmatrix} [\Delta\delta_e]$$

The augmented matrix $\mathbf{A}^*$ can be obtained from Eq. (8.53):

$$\mathbf{A}^* = \mathbf{A} - \mathbf{B}\mathbf{k}^{\mathrm{T}}$$

$$\mathbf{A}^* = \begin{bmatrix} -0.334 + 0.027k_1 & 1.0 + 0.027k_2 \\ -2.52 + 2.6k_1 & -0.387 + 2.6k_2 \end{bmatrix}$$

The eigenvalues of the augmented matrix $\mathbf{A}^*$ are determined from the characteristic equation, which is obtained from

$$|\lambda\mathbf{I} - \mathbf{A}^*| = 0$$

or

$$\begin{vmatrix} \lambda + 0.334 - 0.027k_1 & -1.0 - 0.027k_2 \\ 2.52 - 2.6k_1 & \lambda + 0.387 - 2.6k_2 \end{vmatrix} = 0$$

Expanding the determinant yields the characteristic equation of the augmented system in terms of the unknown feedback gains k_1 and k_2.

$$\lambda^2 + (0.721 - 0.027k_1 - 2.6k_2)\lambda + 2.65 - 2.61k_1 - 0.8k_2 = 0$$

The desired characteristic equation is given as

$$\lambda^2 + 4.2\lambda + 9 = 0$$

Comparing like powers of λ we obtain a set of algebraic equations for the unknown feedback gains:

$$0.721 - 0.027k_1 - 2.6k_2 = 4.2 \qquad 2.65 - 2.61k_1 - 0.8k_2 = 9$$

Solving for the gains yields

$$k_1 = -1.970 \qquad k_2 = -1.318$$

and the state feedback control is given as

$$\Delta\delta_e = 1.970\,\Delta\alpha + 1.318\,\Delta q$$

Figure 8.3 shows the response of the airplane with and without the stability augmentation system. An initial angle of attack disturbance of 5° is used to excite the airplane. Without the stability augmentation, the airplane responds in its natural short period motion. However, when the state feedback stability augmentation system is active the disturbance is quickly damped out.

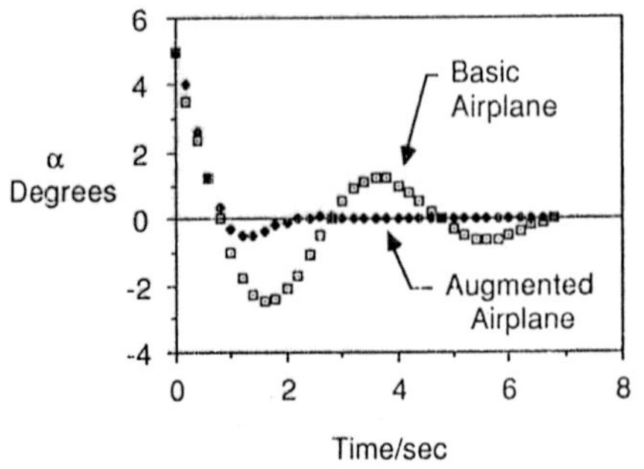

FIGURE 8.3
Longitudinal response of an airplane with and without state feedback.

LATERAL STABILITY AUGMENTATION. The lateral eigenvalues of an airplane can also be modified using state feedback. The lateral state equations were shown to be

$$\begin{bmatrix} \Delta\dot{v} \\ \Delta\dot{p} \\ \Delta\dot{r} \\ \Delta\dot{\phi} \end{bmatrix} = \begin{bmatrix} Y_v & 0 & -u_0 & g \\ L_v & L_p & L_r & 0 \\ N_v & N_p & N_r & 0 \\ 0 & 1 & 0 & 0 \end{bmatrix} \begin{bmatrix} \Delta v \\ \Delta p \\ \Delta r \\ \Delta\phi \end{bmatrix} + \begin{bmatrix} Y_{\delta,a} & Y_{\delta,r} \\ L_{\delta,a} & L_{\delta,r} \\ N_{\delta,a} & N_{\delta,r} \\ 0 & 0 \end{bmatrix} \begin{bmatrix} \Delta\delta_a \\ \Delta\delta_r \end{bmatrix} \tag{8.61}$$

which is of the form

$$\dot{\mathbf{x}} = \mathbf{A}\mathbf{x} + \mathbf{B}\eta \tag{8.62}$$

The state feedback control law can be expressed as

$$\eta = -\mathbf{c}\mathbf{k}^{\mathrm{T}}\mathbf{x} + \eta_p \tag{8.63}$$

which will locate the lateral eigenvalues of the airplane at any desired values. The procedure is identical to that for the longitudinal equations. The constant vector **c** establishes the relationship between the aileron and rudder for augmentation. Either c_1 or c_2 is equal to 1, and the ratio $c_1/c_2 = \Delta\delta a/\Delta\delta r$ is specified by control deflection limits.

Substituting the control vector into the state equation yields

$$\dot{\mathbf{x}} = (\mathbf{A} - \mathbf{B}\mathbf{c}\mathbf{k}^{\mathrm{T}})\mathbf{x} + \mathbf{B}\eta \tag{8.64}$$

or

$$\dot{\mathbf{x}} = \mathbf{A}^*\mathbf{x} + \mathbf{B}\eta \tag{8.65}$$

where $\mathbf{A}^*$ is the augmented matrix and is expressed as

$$\mathbf{A}^* = \mathbf{A} - \mathbf{B}\mathbf{c}\mathbf{k}^{\mathrm{T}} \tag{8.66}$$

The characteristic equation for the augmented system is again solved by expanding the determinant

$$|\lambda\mathbf{I} - \mathbf{A}| = 0 \tag{8.67}$$

The characteristic equation of the augmented matrix is again a quartic equation:

$$A\lambda^4 + B\lambda^3 + C\lambda^2 + D\lambda + E = 0 \tag{8.68}$$

where the coefficients A, B, C, D and E are functions of the stability derivatives and unknown feedback gains. The desired characteristic equation is known once the desired eigenvalues are specified. If the desired lateral roots are given as

$$\lambda_1 = \lambda_{\mathrm{roll}} \qquad \lambda_2 = \lambda_{\mathrm{spiral}} \qquad \lambda_{34} = -\zeta_{\mathrm{DR}}\omega_{n\mathrm{DR}} \pm \mathrm{i}\omega_{n\mathrm{DR}}\sqrt{1-\zeta_{\mathrm{DR}}^2} \tag{8.69}$$

then the desired characteristic equation is given as

$$(\lambda - \lambda_1)(\lambda - \lambda_2)(\lambda - \lambda_3)(\lambda - \lambda_4) = 0 \tag{8.70}$$

or

$$\lambda^4 - [(\lambda_1 + \lambda_2 + \lambda_3 + \lambda_4)]\lambda^3 + [(\lambda_1\lambda_2 + \lambda_3\lambda_4 + (\lambda_1 + \lambda_2)(\lambda_3 + \lambda_4)]\lambda^2 - [\lambda_1\lambda_2(\lambda_3 + \lambda_4) + \lambda_3\lambda_4(\lambda_1 + \lambda_2)]\lambda + \lambda_1\lambda_2\lambda_3\lambda_4 = 0 \quad (8.71)$$

Equating the coefficients of like powers of λ yields a set of algebraic equations in terms of the unknown gains $\mathbf{k}$. Although simple in concept, these equations are quite long in symbolic form, and will not be presented here. However, a simple lateral state feedback augmentation system will be presented to aid the reader in understanding the basic ideas of this section.

The state feedback design described in the previous section requires that all the states be available for feedback. In practise this is not always possible because some states may not be accessible. That is, we may not be able to measure the state with the instrumentation on board the aircraft. If this is the case, it will be necessary to modify the design to include a state observer. The state observer estimates the states of the system from measurements of the input and output signals. A discussion of state observers is to be found in the next section.

Example Problem 8.3. A wind-tunnel is mounted on a bearing system that permits angular rotation in roll and yaw. Use state feedback to position the eigenvalues of the system at λ_1 and λ_2.

For the motion described by the wind-tunnel model, the side force equation and the change in side velocity will be neglected. Furthermore, we will also assume that $I_{xz} = 0$. The equations of motion in state-space form can be written as

$$\begin{bmatrix} \Delta\dot{p} \\ \Delta\dot{r} \end{bmatrix} = \begin{bmatrix} L_p & L_r \\ N_p & N_r \end{bmatrix} \begin{bmatrix} \Delta p \\ \Delta r \end{bmatrix} + \begin{bmatrix} L_{\delta a} & L_{\delta r} \\ N_{\delta a} & N_{\delta r} \end{bmatrix} \begin{bmatrix} \delta_a \\ \delta_r \end{bmatrix}$$

or

$$\dot{\mathbf{x}} = \mathbf{A}\mathbf{x} + \mathbf{B}\eta$$

where

$$\mathbf{A} = \begin{bmatrix} L_p & L_r \\ N_p & N_r \end{bmatrix} \qquad \mathbf{B} = \begin{bmatrix} L_{\delta a} & L_{\delta r} \\ N_{\delta a} & N_{\delta r} \end{bmatrix}$$

If the state feedback is assumed to be

$$\eta = -\mathbf{c}\mathbf{k}^{\mathrm{T}}\mathbf{x}$$

then the augmented matrix $\mathbf{A}^*$ can be expressed as follows.

$$\mathbf{A}^* = \mathbf{A} - \mathbf{B}\mathbf{c}\mathbf{k}^{\mathrm{T}}$$

or

$$\mathbf{A}^* = \begin{bmatrix} L_p - k_1(c_1 L_{\delta a} + c_2 L_{\delta r}) & L_r - k_2(c_1 L_{\delta a} + c_2 L_{\delta r}) \\ N_p - k_1(c_1 N_{\delta a} + c_2 N_{\delta r}) & N_r - k_2(c_1 N_{\delta,a} + c_2 N_{\delta,r}) \end{bmatrix}$$

The eigenvalues of the augmented matrix can be obtained from

$$|\lambda\mathbf{I} - \mathbf{A}^*| = 0$$

which upon expansion yields

$$\lambda^2 + (k_1 L_c + k_2 N_c - L_p - N_r)\lambda + k_1(L_r N_c - N_r L_c) + k_2(N_p L_c - L_p N_c) + N_r L_p - N_p L_r = 0$$

where

$$L_c = C_1 L_{\delta a} + C_2 L_{\delta r} \qquad N_c = C_1 N_{\delta a} + C_2 N_{\delta r}$$

The characteristic equation for desired eigenvalues is given as

$$(\lambda - \lambda_1)(\lambda - \lambda_2) = 0 \qquad \text{or} \qquad \lambda^2 - (\lambda_1 + \lambda_2) + \lambda_1 \lambda_2 = 0$$

If we equate the coefficients of like powers of λ for the desired and augmented characteristic equation, we obtain a set of algebraic equations for the feedback gains k_1 and k_2:

$$k_1 L_c + k_2 N_c - L_p - N_r = -(\lambda_1 + \lambda_2)$$

$$k_1(L_r N_c + N_r L_c) + k_2(N_p L_c - L_p N_c) + N_r L_p - N_p L_r = \lambda_1 \lambda_2$$

The feedback gains are in terms of the known aerodynamic derivatives, the desired eigenvalues and the constants c_1 and c_2 where c_1 and c_2 establish the interconnection between the aileron and rudder. These constants establish the relationship between the aileron and rudder deflections. Once these parameters are selected the feedback gains can readily be obtained.

8.4 STATE VARIABLE RECONSTRUCTION: THE STATE OBSERVER

State feedback design as discussed in the previous section requires the measurement of each state variable. There are systems in which this is not possible, owing either to the complexity of the system or to the expense required to measure certain states. If the states cannot be measured for these reasons, the control law cannot be implemented. An alternate approach for designing the controller when all the states are not available is to use an approximation to the state vector. The approximation to the unavailable states is obtained by a subsystem called an observer. The design of a state feedback control system when some of the states are inaccessible can be divided into two phases. In the first phase, the control system is designed as though all the states were known, for example, the method discussed in the previous section. The second part of the design deals with determining the design of the system that estimates the unavailable states. Figure 8.4 is a sketch of a linear system with state feedback and a state observer.

The designer can select the eigenvalues of the state observer. In choosing the eigenvalues, it should be obvious that one would want the observer to respond faster than the observed system. This means that the eigenvalues of the observer should be more negative than the observed system. In practise, the observer eigenvalues are chosen so that they are only slightly more

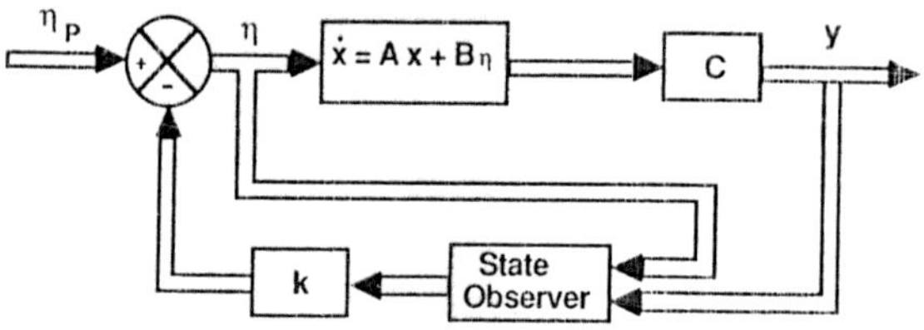

FIGURE 8.4
Block diagram representation of a linear system with state feedback and a state observer.

negative than the observed system eigenvalues. If the observer eigenvalues were chosen to be extremely large negative values, the observer would have extremely rapid response. Such an observer would be highly sensitive to noise. Hence, it has been found that good closed-loop response with an observer is best achieved by selecting eigenvalues of the observer that make the observer only slightly more responsive than the observed system.

There are a number of ways to design a state observer. The basic idea is to make the estimated state $\mathbf{x}_e$ to be very close to the actual state $\mathbf{x}$. Since $\mathbf{x}$ is unknown, there is no direct way of comparing the estimated state to the actual state of the system. However, we do know the output of the system and we can compare it with the estimated output of the observer:

$$\mathbf{y}_e = \mathbf{C}\mathbf{x}_e \tag{8.72}$$

The observer can be constructed as a state feedback problem as illustrated in Fig. 8.5. The problem now is one of determining the observer feedback gains $\mathbf{k}_e$ so that $\mathbf{y}_e$ approaches $\mathbf{y}$ as rapidly as possible. The dynamic characteristics of the observer can be expressed as

$$\dot{\mathbf{x}}_e = (\mathbf{A} - \mathbf{k}_e^T\mathbf{C})\mathbf{x}_e + \mathbf{B}\eta + k_e^T\mathbf{y} \tag{8.73}$$

but

$$\mathbf{y} = \mathbf{C}\mathbf{x} \tag{8.74}$$

or

$$\dot{\mathbf{x}}_e = (\mathbf{A} - \mathbf{k}_e^T\mathbf{C})\mathbf{x}_e + \mathbf{B}\eta + k_e^T\mathbf{C}\mathbf{x} \tag{8.75}$$

If we subtract Eq. (8.75) from the state equation for the actual system, we obtain

$$\dot{\mathbf{x}} - \dot{\mathbf{x}}_e = (\mathbf{A} - \mathbf{k}_e^T\mathbf{C})(\mathbf{x} - \mathbf{x}_e) \tag{8.76}$$

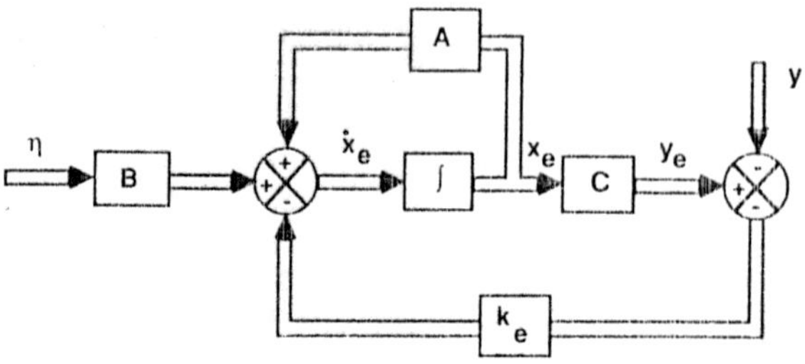

FIGURE 8.5
Design for a state observer.

The characteristic equation for the observer can be determined by solving

$$|\lambda \mathbf{I} - (\mathbf{A} - \mathbf{k}_e^{\mathrm{T}}\mathbf{C})| = 0 \tag{8.77}$$

The gain matrix of the observer is selected so that Eq. (8.76) decays rapidly to zero.

8.5 OPTIMAL STATE SPACE CONTROL SYSTEM DESIGN

The control system can be written in the state-space form

$$\dot{\mathbf{x}} = \mathbf{A}\mathbf{x} + \mathbf{B}\eta \tag{8.78}$$

For the optimal control problem, given an initial state $\mathbf{x}(t_0)$ we want to find a control vector η that drives the state $\mathbf{x}(t_0)$ to the desired final state $\mathbf{x}_d(t_f)$ in such a way that a selected performance index of the form

$$J = \int_{t_0}^{t_f} g(\mathbf{x}, \eta, t)\, \mathrm{d}t \tag{8.79}$$

is minimized. The functional form of the performance index can be expressed in a variety of forms; the most useful form is a quadratic index:

$$J = \int_0^{t_f} \mathbf{x}^{\mathrm{T}}\mathbf{Q}\mathbf{x}\, \mathrm{d}t \tag{8.80}$$

where $\mathbf{Q}$ is a weighting matrix. For many practical control problems, it is desirable to include a penalty for physical constraints such as expenditure of control energy. The performance index can be rewritten as

$$J = \int_0^{t_f} (\mathbf{x}^{\mathrm{T}}\mathbf{Q}\mathbf{x} + \eta^{\mathrm{T}}\mathbf{R}\eta)\, \mathrm{d}t \tag{8.81}$$

Using the quadratic performance index defined above, it can be shown that for a linear feedback control the optimal control law is

$$\eta = -\mathbf{k}^{\mathrm{T}}\mathbf{x} \tag{8.82}$$

where $\mathbf{k}$ is a matrix of unknown gains. This problem is often referred to as the linear regulator problem.

If we apply the principles of the calculus of variations to the minimization of the performance index, we obtain the Riccati equation. A complete development of the Ricatti equation can be found in reference 8.1. The Riccati equation is a set of nonlinear differential equations which must be solved for the Riccati gains $\mathbf{S}(t)$.

$$\frac{\mathrm{d}\mathbf{S}(t)}{\mathrm{d}t} = \mathbf{S}(t)\mathbf{B}\mathbf{R}^{-1}\mathbf{B}^{\mathrm{T}}\mathbf{S}(t) - \mathbf{S}(t)\mathbf{A} - \mathbf{A}^{\mathrm{T}}\mathbf{S}(t) - \mathbf{Q} \tag{8.83}$$

The time-varying gains are related to the Riccati gains in the following

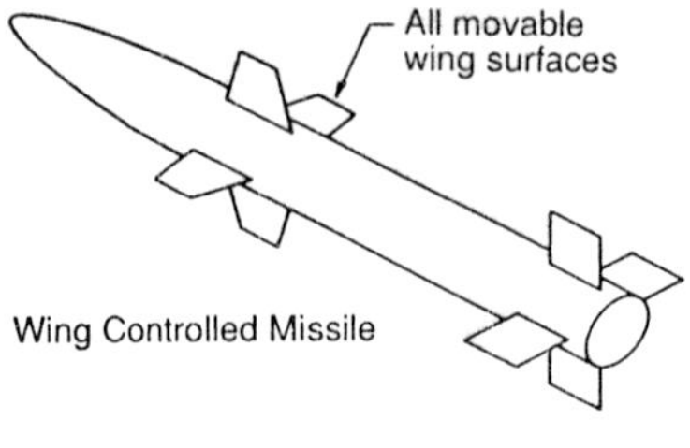

FIGURE 8.6
Sketch of a wing controlled missile.

manner:

$$\mathbf{k}(t) = \mathbf{R}^{-1}\mathbf{B}^{\mathrm{T}}\mathbf{S}(t) \tag{8.84}$$

For the case in which the final time t_f approaches infinity, the Riccati gain matrix becomes a constant matrix and reduces to

$$\mathbf{SBR}^{-1}\mathbf{B}^{\mathrm{T}}\mathbf{S} - \mathbf{SA} - \mathbf{A}^{\mathrm{T}}\mathbf{S} - \mathbf{Q} = 0 \tag{8.85}$$

In this form the Riccati equation is a set of nonlinear algebraic equations in terms of the Riccati gains. Except for the simplest of examples, solution of Eq. (8.85) requires sophisticated computer codes.

Example Problem 8.4. Most guided missiles require that the roll attitude of the missile be kept at a fixed orientation throughout its flight so that the guidance system can function properly. A roll autopilot is needed to maintain the desired roll orientation. Figure 8.6 is a sketch of a wing-controlled missile.

In this example we shall design a feedback control system that will keep the roll orientation near zero, while not exceeding a given limit on the aileron deflection angle. The equations of motion for the rolling motion of the missile were developed in Chapter 5 and are given below:

$$I_x\dot{p} = \frac{\partial L}{\partial P}p + \frac{\partial L}{\partial \delta_a}\delta_a \qquad \dot{\phi} = p$$

Rewriting these equations as

$$\dot{p} = L_p p + L_{\delta a}\delta_a \qquad \dot{\phi} = p$$

where

$$L_p = \frac{\partial L/\partial p}{I_x} \quad \text{and} \quad L_{\delta a} = \frac{\partial L/\partial \delta_a}{I_x}$$

the equations can be easily be written in the state variable form as follows:

$$\begin{bmatrix}\dot{\phi}\\ \dot{p}\end{bmatrix} = \begin{bmatrix}0 & 1\\ 0 & L_p\end{bmatrix}\begin{bmatrix}\phi\\ p\end{bmatrix} + \begin{bmatrix}0\\ L_{\delta a}\end{bmatrix}\{\delta_a\}$$

or

$$\dot{\mathbf{x}} = \mathbf{Ax} + \mathbf{B}\eta$$

where

$$\mathbf{A} = \begin{bmatrix} 0 & 1 \\ 0 & L_p \end{bmatrix} \quad \text{and} \quad \mathbf{B} = \begin{bmatrix} 0 \\ L_{\delta a} \end{bmatrix}$$

The quadratic performance index that is to be minimized is

$$J = \frac{1}{2}\int_0^{\infty} \left[\left(\frac{\phi}{\phi_{max}}\right)^2 + \left(\frac{p}{p_{max}}\right)^2 + \left(\frac{\delta_a}{\delta_{amax}}\right)^2 \right] dt$$

where ϕ_{max} = the maximum desired roll angle
p_{max} = the maximum desired roll rate
δ_{amax} = maximum aileron deflection.

Comparing the performance index given here with the general form allows us to specify the matrices **Q** and **R**:

$$\mathbf{Q} = \begin{bmatrix} \dfrac{1}{\phi_{max}^2} & 0 \\ 0 & \dfrac{1}{p_{max}^2} \end{bmatrix} \quad \text{and} \quad \mathbf{R} = \frac{1}{\delta_{max}^2}$$

The optimum control law is determined by solving the steady-state matrix Reccati equation,

$$\mathbf{SA} + \mathbf{A}^{\mathrm{T}}\mathbf{S} + \mathbf{Q} - \mathbf{SBR}^{-1}\mathbf{BTS} = 0$$

for the values of the **S** matrix. The optimal control law is given by

$$\eta = -\mathbf{kx}(t) \qquad \text{where} \qquad \mathbf{k} = \mathbf{R}^{-1}\mathbf{B}^{\mathrm{T}}\mathbf{S}$$

Substituting the matrices **A**, **B**, **Q**, and **R** into the Ricatti equation yields a set of nonlinear algebraic equations for the unknown elements of the **S** matrix[1]:

$$\frac{1}{\phi_{max}^2} - S_{12}^2 L_{\delta,a}^2 \delta_{max}^2 = 0$$

$$S_{11} + S_{12}L_p - S_{12}S_{22}L_{\delta a}^2 \delta_{max}^2 = 0$$

$$2S_{12} + 2S_{22}L_p + \frac{1}{p_{max}} - S_{22}^2 L_{\delta a}^2 \delta_{max}^2 = 0$$

For the case in which the missile has the aerodynamic characteristics

$$L_p = -2\ \text{rad/s} \qquad L_{\delta a} = 9000\ \text{s}^{-2}$$

and

$$\phi_{max} = 10° = 0.174\ \text{rad} \qquad p_{max} = 300°/\text{s} = 5.23\ \text{rad/s}$$

the control law is found to be $\delta_a = -3.0\phi - 0.103p$.

---

[1] Remember that the S matrix is symmetric, so that it is only necessary to solve the equation generated for the elements along and above the diagonal of the matrix.

8.6 SUMMARY

Modern control theory provides the control systems engineer with a very valuable design tool. Unlike the classical control methods presented in Chapter 7, modern control theory is ideally suited for synthesis of a control system with multiple inputs and for determining optimal control strategies.

State feedback design can be accomplished provided that all the states are controllable. If some of the states are not available for feedback, a state observer can be incorporated into the design to estimate the unknown states provided the states are observable through the system output.

Finally, the state-space representation of the control system lends itself to mathematical techniques that permit the designer to determine optimal control laws consistent with design constraints.

8.7 PROBLEMS

8.1. Given the third-order differential equation

$$\frac{d^3c(t)}{dt^3} + 7\frac{d^2c(t)}{dt^2} 3\frac{d^2c}{dt} + 5c(t) = r(t)$$

write the equation in state vector form.

8.2. Given the second-order differential equation

$$\frac{d^2c(t)}{dt^2} + 3\frac{dc(t)}{dt} + 2ct(t) = r(t)$$

having the initial conditions $c(0) = 1$ and $dc/dt(0) = 0$. Write the equation in state vector form.

(a) Find the state transition matrix.

(b) Determine the solution if $r(t)$ is a unit step function.

8.3. Given the linear time-invariant dynamical system that is governed by the equations

$$\begin{bmatrix} \dot{x}_1 \\ \dot{x}_2 \end{bmatrix} = \begin{bmatrix} 1 & 0 \\ 1 & 1 \end{bmatrix} \begin{bmatrix} x_1 \\ x_2 \end{bmatrix} + \begin{bmatrix} 1 \\ 1 \end{bmatrix} [\eta]$$

where

$$\begin{bmatrix} x_1(0) \\ x_2(0) \end{bmatrix} = \begin{bmatrix} 0 \\ 1 \end{bmatrix}$$

determine the state transition matrix and the response of the system if the input signal is a unit step function.

8.4. Given the state equations

$$\begin{bmatrix} \dot{x}_1 \\ \dot{x}_2 \\ \dot{x}_3 \end{bmatrix} = \begin{bmatrix} 0 & 1 & 0 \\ 0 & 0 & 1 \\ -3 & -6 & -4 \end{bmatrix} \begin{bmatrix} x_1 \\ x_2 \\ x_3 \end{bmatrix} + \begin{bmatrix} 0 \\ 0 \\ 1 \end{bmatrix} [\eta]$$

determine whether the system is completely controllable.

8.5. Given the state equations shown below, use state feedback to position the eigenvalues so closed system has eigenvalues located at $-1 \pm 2i$.

$$\begin{bmatrix} \dot{x}_1 \\ \dot{x}_2 \end{bmatrix} = \begin{bmatrix} 0 & 1 \\ -3 & -1 \end{bmatrix} \begin{bmatrix} x_1 \\ x_2 \end{bmatrix} + \begin{bmatrix} 0 \\ 1 \end{bmatrix} [\eta]$$

8.6. The longitudinal motion of an airplane is approximated by the differential equations

$$\dot{w} = -2.0w + 170\dot{\theta} - 27\delta$$

$$\ddot{\theta} = -0.25w - 15\dot{\theta} - 45\delta$$

(a) Rewrite the equations in state space form

$$\mathbf{x} = \mathbf{A}\mathbf{x} + \mathbf{B}\eta$$

(b) Find the eigenvalues of **A**.

(c) Determine a state feedback control law

$$\eta = -\mathbf{k}^{\mathrm{T}}\mathbf{x}$$

so that the augmented system has a damping ratio $\zeta = 0.5$ and the undamped natural frequency $\omega_n = 20$ rad/s.

8.7. Assume that the vertical velocity for the airplane described in Problem 8.6 cannot be measured. Design a state observer so that the system can achieve the desired performance.

8.8. Design an automatic control system to maintain zero vertical acceleration. The equations of motion governing the aircraft's motion are

$$\Delta\dot{\alpha} = \frac{Z_\alpha}{u_0}\Delta\alpha - \Delta q \qquad \Delta\dot{q} = M_\alpha\,\Delta\alpha + M_\delta\,\Delta\delta_e$$

Find the nonlinear algebraic equations that must be solved to determine the gains for the control law $\eta = -\mathbf{k}^{\mathrm{T}}\mathbf{x}$, that satisfies the performance index

$$J = \int_0^\infty \left[\left(\frac{\delta}{\delta_0}\right)^2 + \left(\frac{\alpha}{\alpha_0}\right)^2 + \left(\frac{q}{q_0}\right)^2 \right] dt$$

where δ_0 = maximum control deflection
α_0 = maximum angle of attack
q_0 = maximum pitch rate

REFERENCES

8.1. Kuo, B. C.: *Automatic Control System,* Prentice-Hall, Englewood Cliffs, NW, 1975.
8.2. Nagrath, I. J., and M. Gopal: *Control Systems Engineering,* Helsted Press, New York, 1975.
8.3. Raven, F. H.: *Automatic Control Engineering,* McGraw-Hill, New York, 1978.
8.4. Blakelock, J. H.: *Automatic Control of Aircraft and Missiles,* Wiley, New York, 1965.
8.5. Bryson, A. E.: "Dryden Lecture: New Concepts in Control Theory 1959–1984", AIAA Paper 84-0161, January 1985.

8.6. Bryson, A. E., and Y. C. Ho: *Applied Optimal Control,* Hemisphere, Washington, DC, 1975.
8.7. Nesline, F. W., and P. Zarchan: "A Classical Look at Modern Control for Missile Autopilot Design," AIAA Paper 82-1512, August 1982.
8.8. Oehman, W. E., and J. H. Suddath: *State-Vector Control Applied to Lateral Stability of High Performance Aircraft,* NASA TN D-2984, July 1965.

APPENDIX A

ATMOSPHERIC TABLES (ICAO STANDARD ATMOSPHERE)

Geometric altitude metric units

H_G, m	H, m	T, K	P, N/m^2	$\frac{P}{P_0}$	ρ, kg/m^{-3}	$\frac{\rho}{\rho_0}$	a, m/s	ν, m^2/s
0	0	288.150	1.01325 +5	1.00000 +0	1.2250 +0	1.0000 +0	340.294	1.4607 −5
1 000	1 000	281.651	8.9876 +4	8.87009 −1	1.1117 +0	9.0748 −1	336.435	1.5813 −5
2 000	1 999	275.154	7.9501 +4	7.84618 −1	1.0066 +0	8.2168 −1	332.532	1.7147 −5
3 000	2 999	268.659	7.0121 +4	6.92042 −1	9.0925 −1	7.4225 −1	328.583	1.8628 −5
4 000	3 997	262.166	6.1660 +4	6.08541 −1	8.1935 −1	6.6885 −1	324.589	2.0275 −5
5 000	4 996	255.676	5.4048 +4	5.33415 −1	7.3643 −1	6.0117 −1	320.545	2.2110 −5
6 000	5 994	249.187	4.7217 +4	4.66001 −1	6.6011 −1	5.3887 −1	316.452	2.4162 −5
7 000	6 992	242.700	4.1105 +4	4.05677 −1	5.9002 −1	4.8165 −1	312.306	2.6461 −5
8 000	7 990	236.215	3.5651 +4	3.51854 −1	5.2579 −1	4.2921 −1	308.105	2.9044 −5
9 000	8 987	229.733	3.0800 +4	3.03979 −1	4.6706 −1	3.8128 −1	303.848	3.1957 −5
10 000	9 984	223.252	2.6500 +4	2.61533 −1	4.1351 −1	3.3756 −1	299.532	3.5251 −5
11 000	10 981	216.774	2.2700 +4	2.24031 −1	3.6480 −1	2.9780 −1	295.154	3.8988 −5
12 000	11 977	216.650	1.9399 +4	1.91457 −1	3.1194 −1	2.5464 −1	295.069	4.5574 −5
13 000	12 973	216.650	1.4170 +4	1.63628 −1	2.6660 −1	2.1763 −1	295.069	5.3325 −5
14 000	13 969	216.650	1.4170 +4	1.39851 −1	2.2786 −1	1.8600 −1	295.069	6.2391 −5
15 000	14 965	216.650	1.2112 +4	1.19534 −1	1.9475 −1	1.5898 −1	205.069	7.2995 −5
16 000	15 960	216.650	1.0353 +4	1.02174 −1	1.6647 −1	1.3589 −1	295.069	8.5397 −5
17 000	16 955	216.650	8.8496 +3	8.73399 −2	1.4230 −1	1.1616 −1	295.069	9.9902 −5
18 000	17 949	216.650	7.5652 +3	7.46629 −2	1.2165 −1	9.9304 −2	295.069	1.1686 −4
19 000	18 943	216.650	6.4674 +3	6.38291 −2	1.0400 −1	8.4894 −2	295.069	1.3670 −4
20 000	19 937	216.650	5.5293 +3	5.45700 −2	8.8910 −2	7.2579 −2	295.069	1.5989 −4
21 000	20 931	217.581	4.7274 +3	4.66709 −2	7.5715 −2	6.1808 −2	295.703	1.8843 −4
22 000	21 924	218.574	4.0420 +3	3.99456 −2	6.4510 −2	5.2661 −2	296.377	2.2201 −4
23 000	22 917	219.567	3.4562 +3	3.42153 −2	5.5006 −2	4.4903 −2	297.049	2.6135 −4
24 000	23 910	220.560	2.9554 +3	2.93288 −2	4.6938 −2	3.8317 −2	297.720	3.0743 −4
25 000	24 902	221.552	2.6077 +3	2.51588 −2	4.0084 −2	3.2722 −2	298.389	3.6135 −4
26 000	25 894	222.544	2.1632 +3	2.15976 −2	3.4257 −2	2.7965 −2	299.056	4.2439 −4
27 000	26 886	223.536	1.8555 +3	1.85539 −2	2.9298 −2	2.3917 −2	299.722	4.9805 −4
28 000	27 877	224.527	1.5949 +3	1.59506 −2	2.5076 −2	2.0470 −2	300.386	5.8405 −4
29 000	28 868	225.518	1.3737 +3	1.37224 −2	2.1478 −2	1.7533 −2	301.048	6.8438 −4
30 000	29 859	226.509	1.1855 +3	1.18138 −2	1.8410 −2	1.5029 −2	301.709	8.0134 −4

Geometric altitude english units

H_G, ft	H, ft	T, °R	P, lb/ft^2	$\frac{P}{P_0}$	ρ, slug/ft^3	$\frac{\rho}{\rho_0}$	a, ft/s	ν, ft^2/s
0	0	518.670	2.1162 +3	1.00000 +0	2.3769 −3	1.0000 +0	1116.45	1.5723 −4
2 500	2 500	509.756	1.9319 +3	9.12910	2.2079 −3	9.2887	1106.81	1.6700
5 000	4 999	500.843	1.7609 +3	8.32085 −1	2.0482 −3	8.6170 −1	1097.10	1.7755 −4
7 500	7 497	491.933	1.6023 +3	7.57172	1.8975 −3	7.9832	1087.29	1.8896
10 000	9 995	483.025	1.4556 +3	6.87832 −1	1.7556 −3	7.3859 −1	1077.40	2.0132 −4
12 500	12 493	474.120	1.3200 +3	6.23741	1.6219 −3	6.8235	1067.43	2.1472
15 000	14 989	465.216	1.1948 +3	5.64587 −1	1.4962 −3	6.2946 −1	1057.36	2.2927 −4
17 500	17 485	456.315	1.0794 +3	5.10072	1.3781 −3	5.7977	1047.19	2.4509
20 000	19 981	447.415	9.7327 +2	4.59912 −1	1.2673 −3	5.3316 −1	1036.93	2.6233 −4
22 500	22 476	438.518	8.7576 +2	4.13834	1.1634 −3	4.8947	1026.57	2.8113
25 000	24 970	429.623	7.8633 +2	3.71577 −1	1.0663 −3	4.4859 −1	1016.10	3.0167 −4
27 500	27 464	420.730	7.0447 +2	3.32892	9.7544 −4	4.1039	1005.53	3.2416
30 000	29 957	411.839	6.2962 +2	2.97544 −1	8.9068 −4	3.7473 −1	994.85	3.4882 −4
32 500	32 449	402.950	5.6144 +2	2.65305	8.1169 −4	3.415	984.05	3.7591
35 000	34 941	394.064	4.9934 +2	2.35962 −1	7.3820 −4	3.1058 −1	973.14	4.0573 −4
37 500	37 432.5	389.970	4.4312 +2	2.093965	6.6196 −4	2.78505	968.08	4.48535
40 000	39 923	389.970	3.9312 +2	1.85769	5.8727 −4	2.4708	968.08	7.70146
42 500	42 413.5	389.970	3.4878 +2	1.64816	5.2103 −4	2.1921	968.08	5.69855
45 000	44 903	389.970	3.0945 +2	1.46227 −1	4.6227 −4	1.9449 −1	968.08	6.4228 −4
47 500	47 392.5	389.970	2.7456 +2	1.29742	4.1015 −4	1.7256	968.08	7.2391
50 000	49 880	389.970	2.4361 +2	1.15116	3.6391 −4	1.5311	968.08	8.1587

52 500	52 368.5	389.970	2.1615 +2	1.02143	3.2290 −4	1.3585	968.08	9.19505
55 000	54 855	389.970	1.9180 +2	9.06336 −2	2.8652 −4	1.2055 −1	968.08	1.0363 −3
57 500	57 341.5	389.970	1.7019 +2	8.042485	2.5424 −4	1.0697	968.08	1.1678
60 000	59 828	389.970	1.5103 +2	7.13664	2.2561 −4	9.4919	968.08	1.3160
62 500	62 313.5	389.970	1.3402 +2	6.33315	2.0021 −4	8.42325	968.08	1.483
65 000	64 798	389.970	1.1893 +2	5.62015 −2	1.7767 −4	7.4749 −2	968.08	1.6711 −3
67 500	67 282.5	390.8835	1.0555 +2	4.988155	1.5767 −4	6.61885	969.21	1.891
70 000	69 766	392.246	9.3672 +1	4.42898	1.3993 −4	5.8565	970.90	2.1434
72 500	72 249	393.6085	8.3134 +1	3.93432	1.2419 −4	5.18435	972.58	2.4283
75 000	74 731	394.971	7.3784 +1	3.49635 −2	1.1022 −4	4.5914 −2	974.26	2.7498 −3
77 500	77 213	396.3325	6.5487 +1	3.10856	9.7829 −5	4.01681	975.94	3.1125
80 000	79 694	397.693	5.8125 +1	2.76491	8.6831 −5	3.6060	977.62	3.5213
82 500	82 174	399.0545	5.1592 +1	2.460355	7.7022 −5	3.19785	979.285	3.98215
85 000	84 655	400.415	4.5827 +1	2.19023 −2	6.7706 −5	2.8371 −2	980.95	4.5012 −3
87 500	87 134.5	401.7755	4.0757 +1	1.95063	5.9598 −5	2.51815	982.62	5.0857
90 000	89 613	403.135	3.6292 +1	1.73793	5.2531 −5	2.2360	984.28	5.7434
92 500	92 091.5	404.495	3.2354 +1	1.549199	4.6362 −5	1.9864	985.94	6.48345
95 000	94 569	405.854	2.8878 +1	1.38133 −2	4.0970 −5	1.7653 −2	987.59	7.3155 −3
97 500	97 046	407.2135	2.5805 +1	1.23226	3.6251 −5	1.56955	989.245	8.25085
100 000	99 523	408.572	2.3085 +1	1.09971	3.2114 −5	1.3960	990.90	9.3017

APPENDIX B

GEOMETRIC, MASS AND AERODYNAMIC CHARACTERISTICS OF SELECTED AIRPLANES

Data on the geometric, mass, and aerodynamic stability and control characteristics are presented for six airplanes. The airplanes include a general aviation airplane, two jet fighters, an executive business jet and two jet transports. The stability coefficients are presented in tabular form for each airplane. Coefficients that were unavailable have been presented with a numerical value of zero in the following tables. The stability coefficients for the A-4D are presented in graphical form as a function of Mach number and altitude. These plots show the large variations in the coefficients due to compressibility effects. The definitions of the stability coefficient and geometric data presented in the figures are given in the following nomenclature list. The information presented in this appendix was taken from the references B1 and B2 given after the nomenclature list.

NOMENCLATURE

b Wing span

$\bar{c}$ Mean chord

$$C_L = \frac{L}{QS}$$

$$C_{L_\alpha} = \frac{\partial C_L}{\partial \alpha} \; (\text{rad}^{-1})$$

$$C_{L_{\dot{\alpha}}} = \frac{\partial C_L}{\partial \left(\frac{\dot{\alpha}\bar{c}}{2u_0} \right)} \; (\text{rad}^{-1})$$

$$C_{L_\mathbf{M}} = \frac{\partial C_L}{\partial \mathbf{M}}$$

$$C_{L_{\delta_e}} = \frac{\partial C_L}{\partial \delta_e} \text{ (rad}^{-1}\text{)}$$

$$C_D = \frac{D}{QS}$$

$$C_{D_\alpha} = \frac{\partial C_D}{\partial \alpha} \text{ (rad}^{-1}\text{)}$$

$$C_{D_\mathbf{M}} = \frac{\partial C_D}{\partial \mathbf{M}}$$

$$C_{D_{\delta_e}} = \frac{\partial C_D}{\partial \delta_e} \text{ (rad}^{-1}\text{)}$$

$$C_m = \frac{M}{QS\bar{c}}$$

$$C_{m_\alpha} = \frac{\partial C_M}{\partial \alpha} \text{ (rad}^{-1}\text{)}$$

$$C_y = \frac{Y}{QS}$$

$$C_{y_\beta} = \frac{\partial C_y}{\partial \beta} \text{ (rad}^{-1}\text{)}$$

$$C_{y_{\delta_r}} = \frac{\partial C_y}{\partial \delta_r} \text{ (rad}^{-1}\text{)}$$

$$C_l = \frac{L}{QSb}$$

$$C_{l_\beta} = \frac{\partial C_l}{\partial \beta} \text{ (rad}^{-1}\text{)}$$

$$C_{l_p} = \frac{\partial C_l}{\partial (pb/2u_0)} \text{ (rad}^{-1}\text{)}$$

$$C_{l_r} = \frac{\partial C_l}{\partial (rb/2u_0)} \text{ (rad}^{-1}\text{)}$$

$$C_{l_{\delta_a}} = \frac{\partial C_l}{\partial \delta_a} \text{ (rad}^{-1}\text{)}$$

$$C_{l_{\delta_r}} = \frac{\partial C_l}{\partial \delta_r} \text{ (rad}^{-1}\text{)}$$

$$C_n = \frac{N}{QSb}$$

$$C_{n_\beta} = \frac{\partial C_n}{\partial \beta} \text{ (rad}^{-1}\text{)}$$

$$C_{n_p} = \frac{\partial C_n}{\partial (pb/2u_0)} \text{ (rad}^{-1}\text{)}$$

$$C_{n_r} = \frac{\partial C_n}{\partial (rb/2u_0)} \text{ (rad}^{-1}\text{)}$$

$$C_{m_{\dot\alpha}} = \frac{\partial C_m}{\partial (\dot\alpha\bar{c}/2u_0)} \text{ (rad}^{-1}\text{)}$$

$$C_{m_\mathbf{M}} = \frac{\partial C_M}{\mathbf{M}}$$

$$C_{m_q} = \frac{\partial C_m}{\partial (q\bar{c}/2u_0)} \text{ (rad}^{-1}\text{)}$$

$$C_{n_{\delta_a}} = \frac{\partial C_n}{\partial \delta_a} \text{ (rad}^{-1}\text{)}$$

$$C_{n_{\delta_r}} = \frac{\partial C_n}{\partial \delta_r} \text{ (rad}^{-1}\text{)}$$

I_x Rolling moment of inertia
I_y Pitching moment of inertia
I_z Yawing moment of inertia
I_{xz} Product of inertia about xz axis
$\mathbf{M}$ Mach number
Q Dynamic pressure
S Wing planform area
u_0 Reference flight speed

REFERENCES

B.1 Teper, G. L., *Aircraft Stability and Control Data*, System Technology, Inc., Hawthorne, California, Technical Report 176-1, April 1969.

B.2 Heffley, R. K., and Jewell, W. F., *Aircraft Handling Qualities Data*, NASA CR-2144, December 1972.

General aviation airplane: NAVION[a]

Longitudinal M = 0.158	C_L	C_D	C_{L_α}	C_{D_α}	C_{m_α}	$C_{L_{\dot\alpha}}$	$C_{m_{\dot\alpha}}$	C_{L_q}	C_{m_q}	C_{L_M}	C_{D_M}	C_{m_M}	$C_{L_{\delta_e}}$	$C_{m_{\delta_e}}$
Sea level	0.41	0.05	4.44	0.33	−0.683	0.0	−4.36	3.8	−9.96	0.0	0.0	0.0	0.355	−0.923

Lateral M = 0.158	C_{y_β}	C_{l_β}	C_{n_β}	C_{l_p}	C_{n_p}	C_{l_r}	C_{n_r}	$C_{l_{\delta a}}$	$C_{n_{\delta a}}$	$C_{y_{\delta r}}$	$C_{l_{\delta r}}$	$C_{n_{\delta r}}$
Sea level	−0.564	−0.074	0.071	−0.410	−0.0575	0.107	−0.125	−0.134	−0.0035	0.157	0.107	−0.072

[a] All derivatives are per radian.

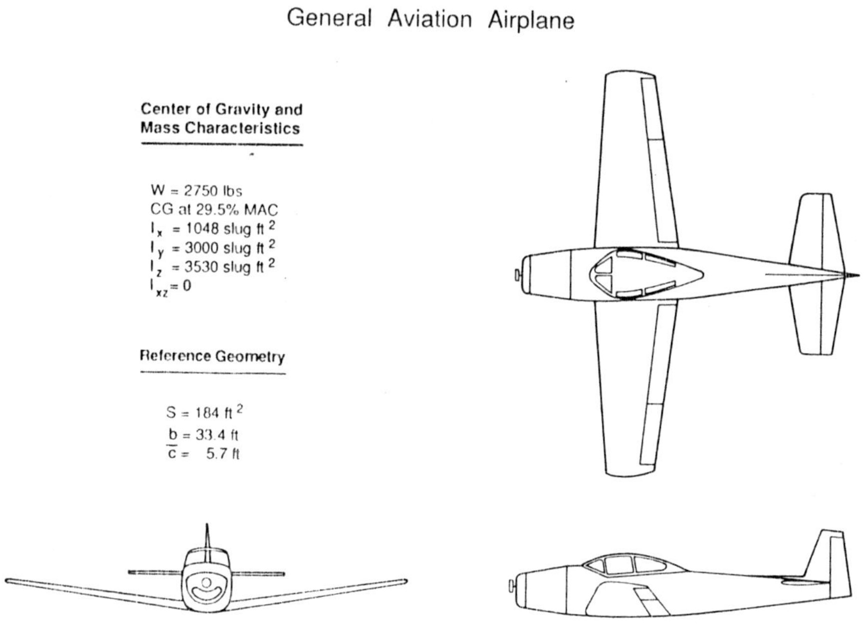

General Aviation Airplane
Center of Gravity and Mass Characteristics
W = 2750 lbs
CG at 29.5% MAC
I_x = 1048 slug ft 2
I_y = 3000 slug ft 2
I_z = 3530 slug ft 2
I_{xz} = 0
Reference Geometry
S = 184 ft 2
b = 33.4 ft
$\bar{c}$ = 5.7 ft

FIGURE B.1
Three-view-sketch and stability data for a general aviation airplane.

Fighter aircraft: F104-A[a]

Longitudinal	C_L	C_D	C_{L_α}	C_{D_α}	C_{m_α}	$C_{L_{\dot\alpha}}$	$C_{m_{\dot\alpha}}$	C_{L_q}	C_{m_q}	C_{L_M}	C_{D_M}	C_{m_M}	$C_{L_{\delta e}}$	$C_{m_{\delta e}}$
M = 0.257 Sea level	0.735	0.263	3.44	0.45	−0.64	0.0	−1.6	0.0	−5.8	0.0	0.0	0.0	0.68	−1.46
M = 1.8 55,000 ft	0.2	0.055	2.0	0.38	−1.30	0.0	−2.0	0.0	−4.8	−0.2	0.0	−0.01	0.52	−0.10

Lateral	C_{y_β}	C_{l_β}	C_{n_β}	C_{l_p}	C_{n_p}	C_{l_r}	C_{n_r}	$C_{l_{\delta a}}$	$C_{n_{\delta a}}$	$C_{y_{\delta r}}$	$C_{l_{\delta r}}$	$C_{n_{\delta r}}$
M = 0.257 Sea level	−1.17	−0.175	0.50	−0.285	−0.14	0.265	−0.75	0.039	0.0042	0.208	0.045	−0.16
M = 1.8 55,000 ft	−1.0	−0.09	0.24	−0.27	−0.09	0.15	−0.65	0.017	0.0025	0.05	0.008	−0.04

[a] All derivatives are per radian.

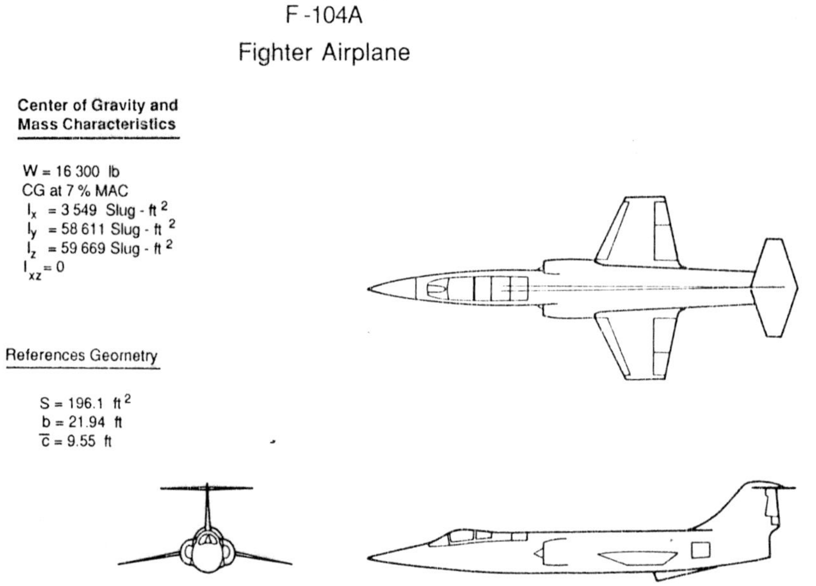

FIGURE B.2
Three-view sketch and stability data for the F-104-A fighter.

Fighter aircraft: A 4D[a]

Longitudinal	C_L	C_D	C_{L_α}	C_{D_α}	C_{m_α}	$C_{L_{\dot\alpha}}$	$C_{m_{\dot\alpha}}$	C_{L_q}	C_{m_q}	C_{L_M}	C_{D_M}	C_{m_M}	$C_{L_{\delta_e}}$	$C_{m_{\delta_e}}$
M = 0.4 Sea level	0.28	0.03	3.45	0.30	−0.38	0.72	−1.1	0.0	−3.6	0.0	0.0	0.0	0.36	−0.50
M = 0.8 35,000 ft	0.30	0.038	4.0	0.56	−0.41	1.12	−1.65	0.0	−4.3	0.15	0.03	−0.05	0.4	−0.60

Lateral	C_{y_β}	C_{l_β}	C_{n_β}	C_{l_p}	C_{n_p}	C_{l_r}	C_{n_r}	$C_{l_{\delta a}}$	$C_{n_{\delta a}}$	$C_{y_{\delta r}}$	$C_{l_{\delta r}}$	$C_{n_{\delta r}}$
M = 0.4 Sea level	−0.98	−0.12	0.25	−0.26	0.022	0.14	−0.35	0.08	0.06	0.17	−0.105	0.032
M = 0.8 35,000 ft	−1.04	−0.14	0.27	−0.24	0.029	0.17	−0.39	0.072	0.04	0.17	−0.105	0.032

[a] All derivatives are per radian.

A4D

Fighter Airplane

Center of Gravity and Mass Characteristics

W = 17,578 lb
CG at 25% MAC
I_x = 8 090 Slug - ft ^{2}
I_y = 25 900 Slug - ft ^{2}
I_z = 29 200 Slug - ft ^{2}
I_{xz} =1 300 Slug - ft ^{2}

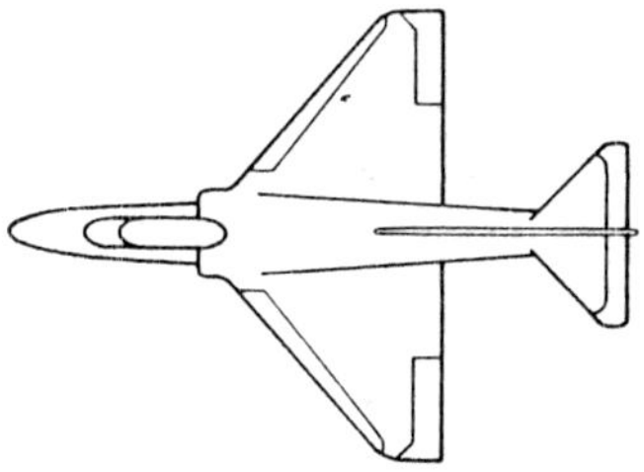

References Geometry

s = 260 ft ^{2}
b = 27.5 ft
$\bar{c}$ = 10.8 ft

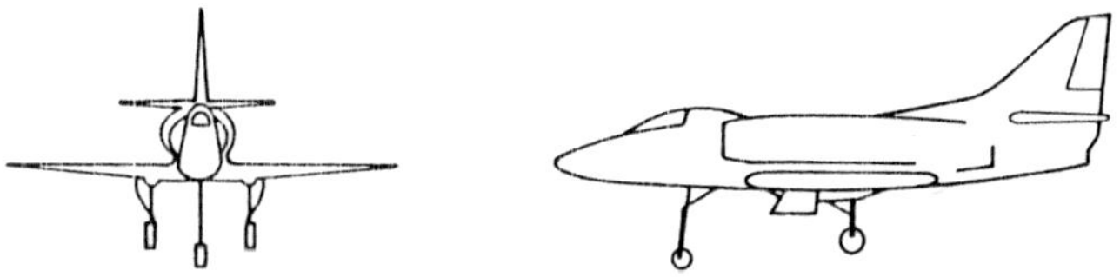

FIGURE B.3
Three-view sketch and stability data for the A-4D fighter.

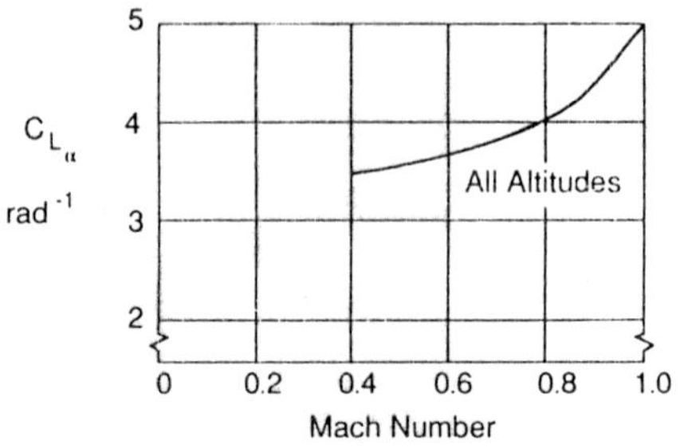

FIGURE B.4
C_{L_α} versus Mach number.

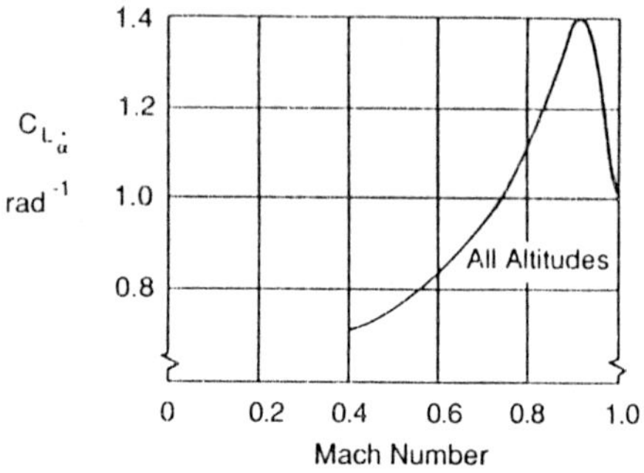

FIGURE B.5
$C_{L_{\dot{\alpha}}}$ versus Mach number.

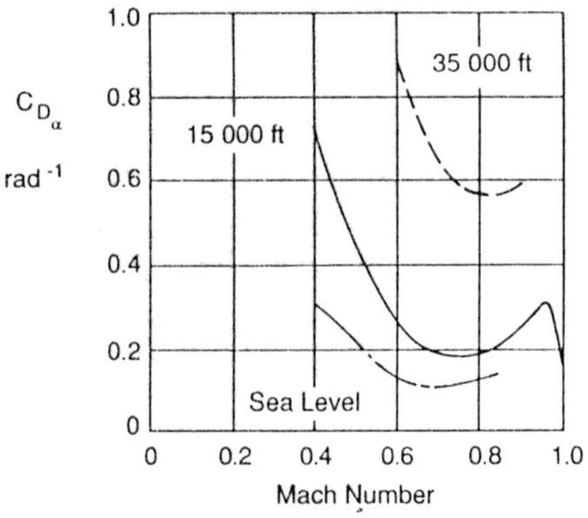

FIGURE B.6
C_{D_α} versus Mach number.

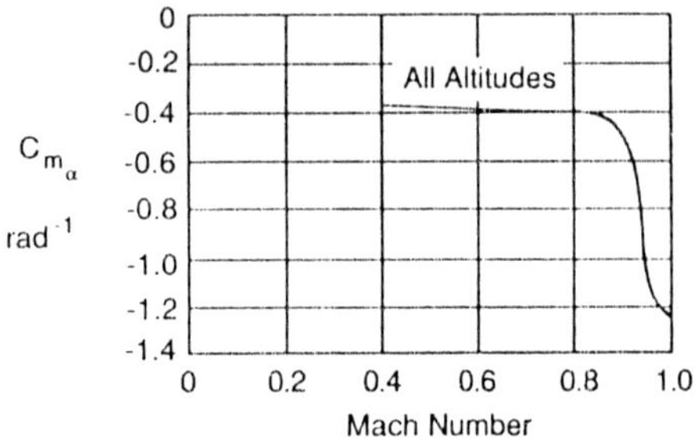

FIGURE B.7
C_{m_α} versus Mach number.

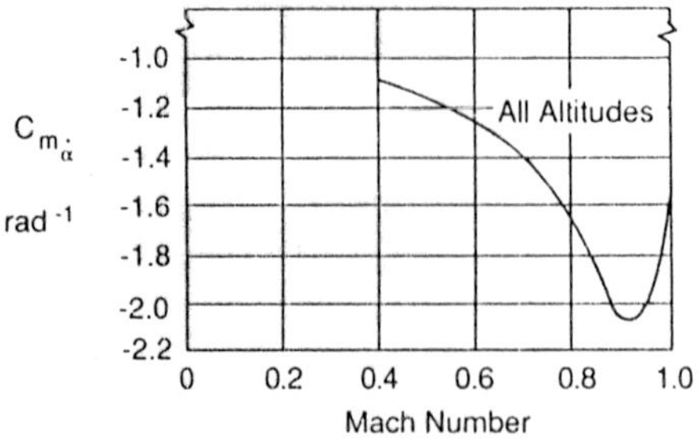

FIGURE B.8
$C_{m_{\dot{\alpha}}}$ versus Mach number.

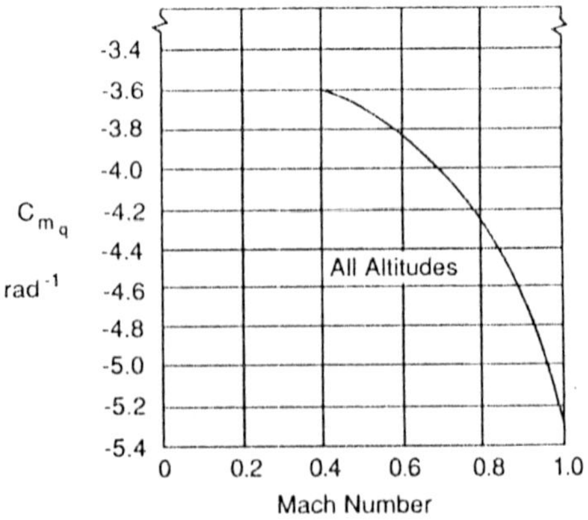

FIGURE B.9
C_{m_q} versus Mach number.

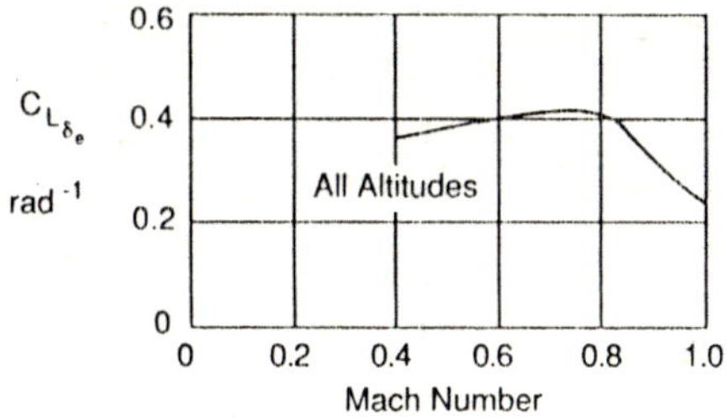

FIGURE B.10
$C_{L\delta_e}$ versus Mach number.

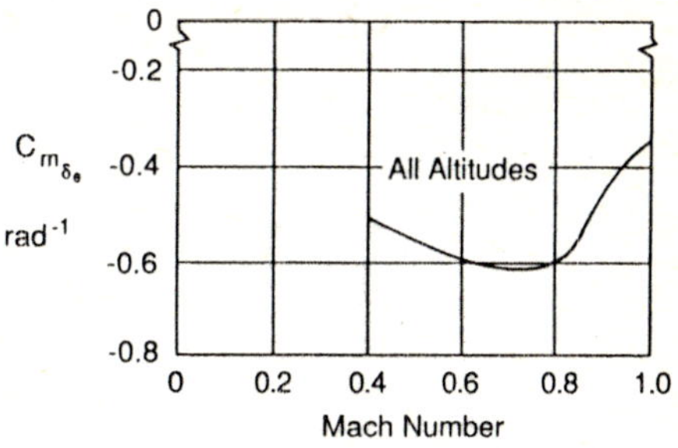

FIGURE B.11
$C_{m\delta_e}$ versus Mach number.

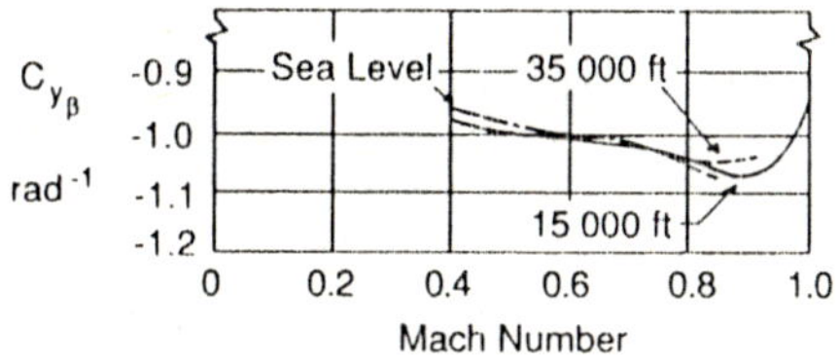

FIGURE B.12
$C_{y\beta}$ versus Mach number.

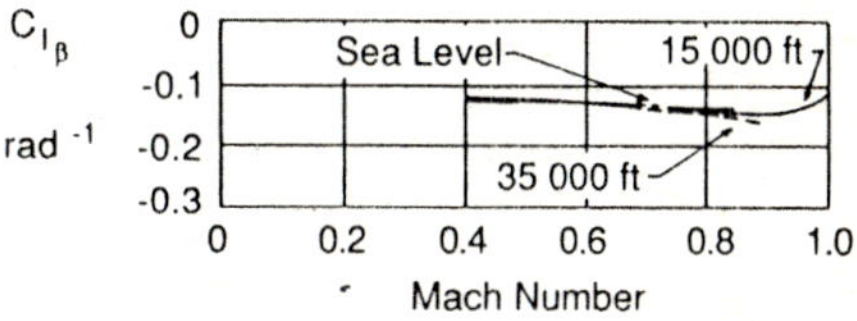

FIGURE B.13
C_{l_β} versus Mach number.

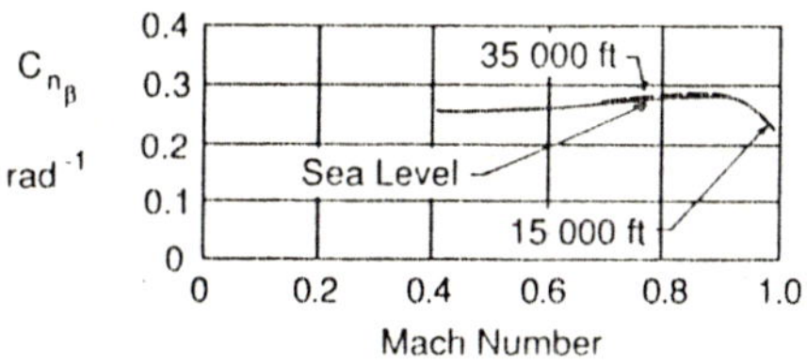

FIGURE B.14
C_{n_β} versus Mach number.

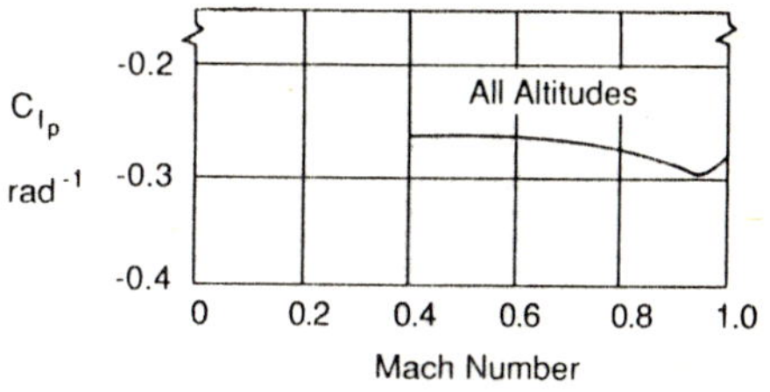

FIGURE B.15
C_{l_p} versus Mach number.

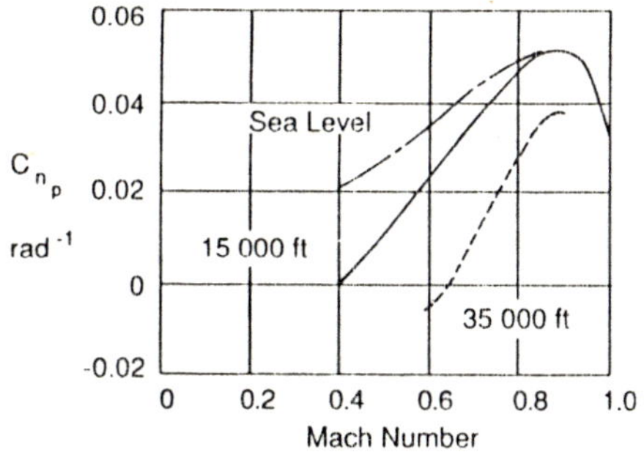

FIGURE B.16
C_{n_p} versus Mach number.

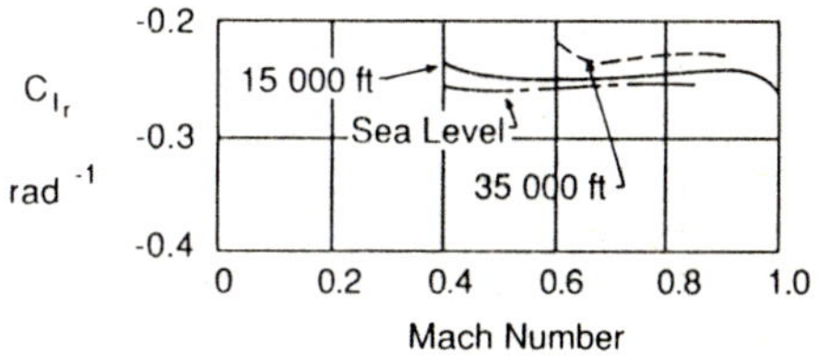

FIGURE B.17
C_{l_r} versus Mach number.

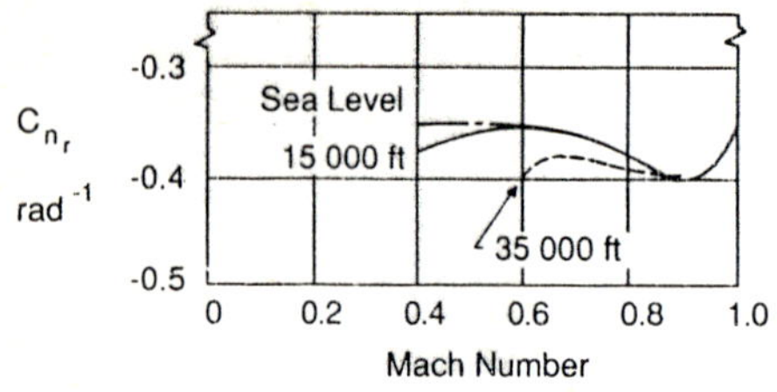

FIGURE B.18
C_{n_r} versus Mach number.

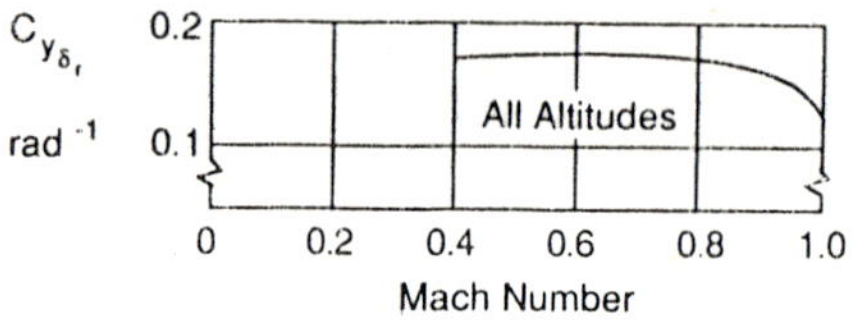

FIGURE B.19
$C_{y\delta_r}$ versus Mach number.

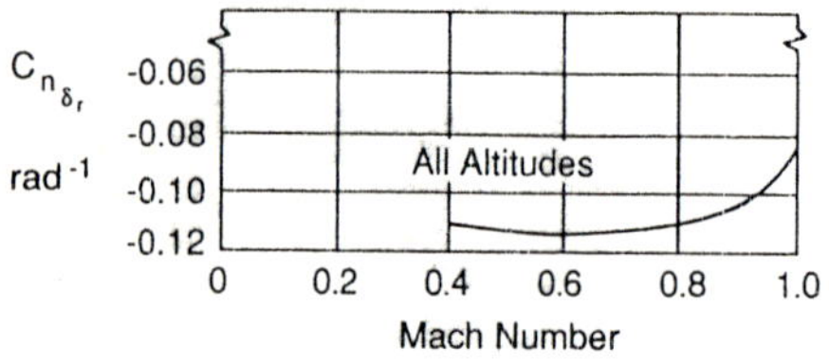

FIGURE B.20
$C_{n\delta_r}$ versus Mach number.

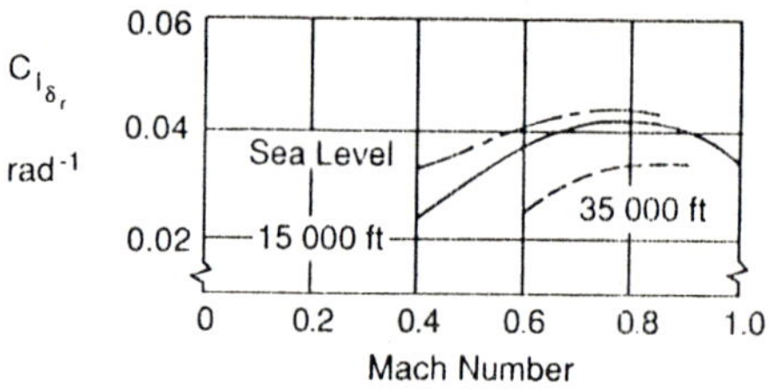

FIGURE B.21
$C_{l\delta_r}$ versus Mach number.

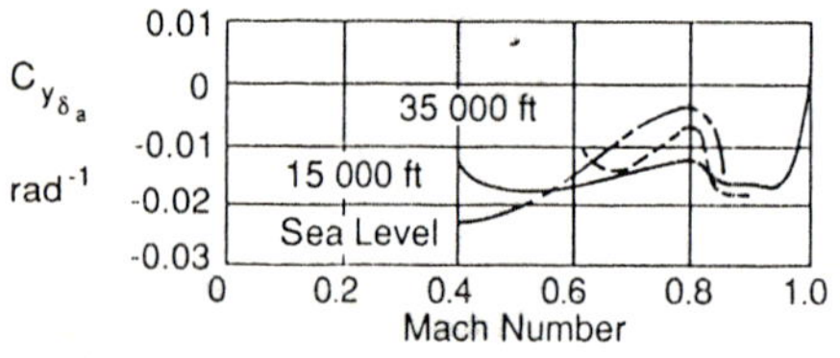

FIGURE B.22
$C_{y\delta_a}$ versus Mach number.

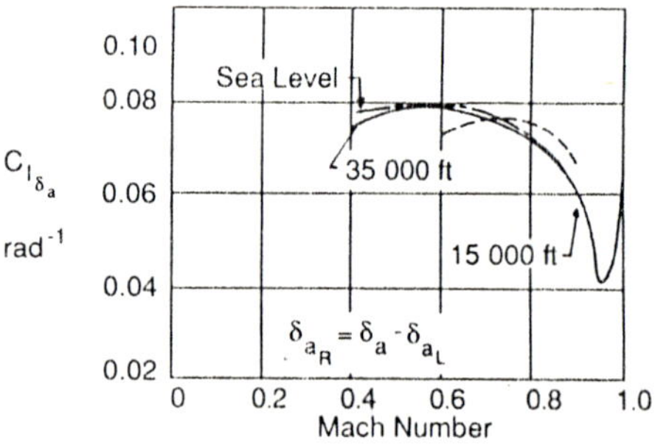

FIGURE B.23
$C_{l\delta_a}$ versus Mach number.

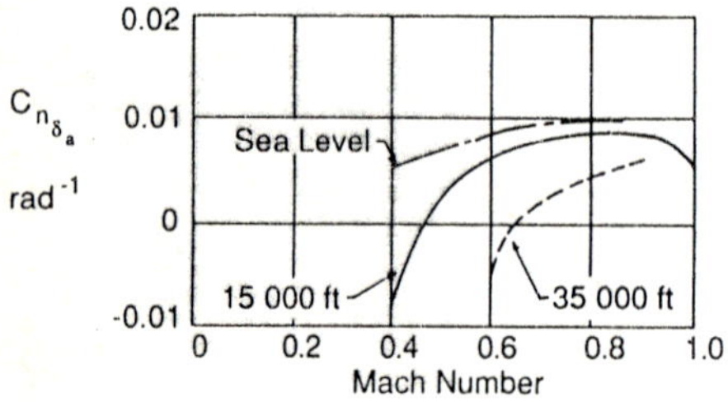

FIGURE B.24
$C_{n\delta_a}$ versus Mach number.

Business Jet: Jetstar[a]

Longitudinal	C_L	C_D	C_{L_α}	C_{D_α}	C_{m_α}	$C_{L_{\dot\alpha}}$	$C_{m_{\dot\alpha}}$	C_{L_q}	C_{m_q}	C_{L_M}	C_{D_M}	C_{m_M}	$C_{L_{\delta_e}}$	$C_{m_{\delta_e}}$
M = 0.20 Sea level	0.737	0.095	5.0	0.75	−0.80	0.0	−3.0	0.0	−8.0	0.0	0.0	−0.05	0.4	−0.81
M = 0.80 40,000 ft	0.4	0.04	6.5	0.60	−0.72	0.0	−0.4	0.0	−0.92	0.0	−0.6	−0.60	0.44	−0.88

Lateral	C_{y_β}	C_{l_β}	C_{n_β}	C_{l_p}	C_{n_p}	C_{l_r}	C_{n_r}	$C_{l_{\delta a}}$	$C_{n_{\delta a}}$	$C_{y_{\delta r}}$	$C_{l_{\delta r}}$	$C_{n_{\delta r}}$
M = 0.20 Sea level	−0.72	−0.103	0.137	−0.37	−0.14	0.11	−0.16	0.054	−0.0075	0.175	0.029	−0.063
M = 0.80 40,000 ft	−0.75	−0.06	0.13	−0.42	−0.756	0.04	−0.16	0.060	−0.06	0.16	0.029	−0.057

[a] All derivatives are per radian.

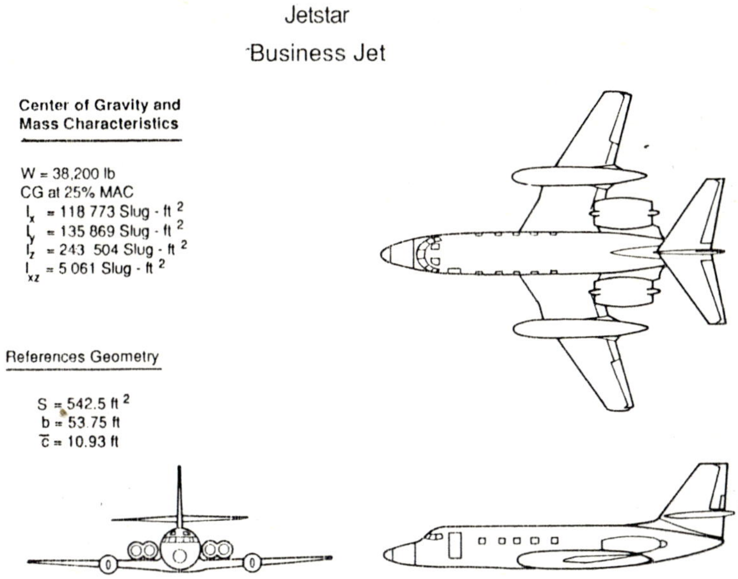

FIGURE B.25
Three-view sketch and stability data for an executive business jet.

Transport aircraft: Convair 880[a]

Longitudinal	C_L	C_D	C_{L_α}	C_{D_α}	C_{m_α}	$C_{L_{\dot\alpha}}$	$C_{m_{\dot\alpha}}$	C_{L_q}	C_{m_q}	C_{L_M}	C_{D_M}	C_{m_M}	$C_{L_{\delta_e}}$	$C_{m_{\delta_e}}$
M = 0.25 Sea level	0.68	0.08	4.52	0.27	−0.903	2.7	−4.13	7.72	−12.1	0.0	0.0	0.0	0.213	−0.637
M = 0.8 35,000 ft	0.347	0.024	4.8	0.15	−0.65	2.7	−4.5	7.5	−4.5	0.0	0.0	0.0	0.190	−0.57

Lateral	C_{y_β}	C_{l_β}	C_{n_β}	C_{l_p}	C_{n_p}	C_{l_r}	C_{n_r}	$C_{l_{\delta a}}$	$C_{n_{\delta a}}$	$C_{y_{\delta r}}$	$C_{l_{\delta r}}$	$C_{n_{\delta r}}$
M = 0.25 Sea level	−0.877	−0.196	0.139	−0.381	−0.049	0.198	−0.185	−0.038	0.017	0.216	0.0226	−0.096
M = 0.8 35,000 ft	−0.812	−0.177	0.129	−0.312	−0.011	0.153	−0.165	−0.050	0.008	0.184	0.019	−0.076

[a] All derivatives are per radian

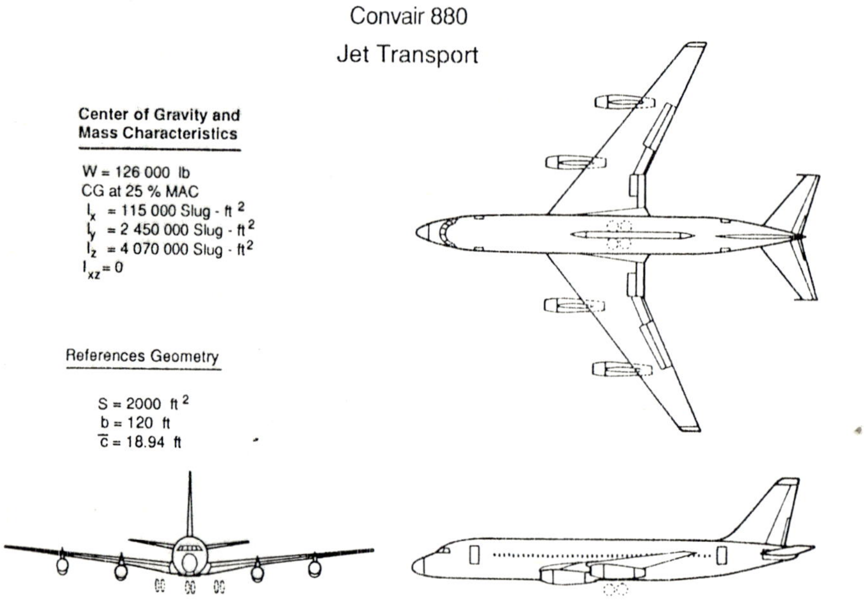

FIGURE B.26
Three-view sketch and stability data for a jet transport.

Transport aircraft: Boeing 747[a]

Longitudinal	C_L	C_D	C_{L_α}	C_{D_α}	C_{m_α}	$C_{L_{\dot\alpha}}$	$C_{m_{\dot\alpha}}$	C_{L_q}	C_{m_q}	C_{L_M}	C_{D_M}	C_{m_M}	$C_{L_{\delta_e}}$	$C_{m_{\delta_e}}$
M = 0.25 Sea level	1.11	0.102	5.70	0.66	−1.26	6.7	−3.2	5.4	−20.8	−0.81	0.0	0.27	0.338	−1.34
M = 0.90 40,000 ft	0.5	0.042	5.5	0.47	−1.6	0.006	−9.0	6.58	−25.0	0.2	0.25	−0.10	0.3	−1.2

Lateral	C_{y_β}	C_{l_β}	C_{n_β}	C_{l_p}	C_{n_p}	C_{l_r}	C_{n_r}	$C_{l_{\delta_a}}$	$C_{n_{\delta_a}}$	$C_{y_{\delta_r}}$	$C_{l_{\delta_r}}$	$C_{n_{\delta_r}}$
M = 0.25 Sea level	−0.96	−0.221	0.150	−0.45	−0.121	0.101	−0.30	0.0461	0.0064	0.175	0.007	−0.109
M = 0.90 40,000 ft	−0.85	−0.10	0.20	−0.30	0.20	0.20	−0.325	0.014	0.003	0.075	0.005	−0.09

[a] All derivatives are per radian.

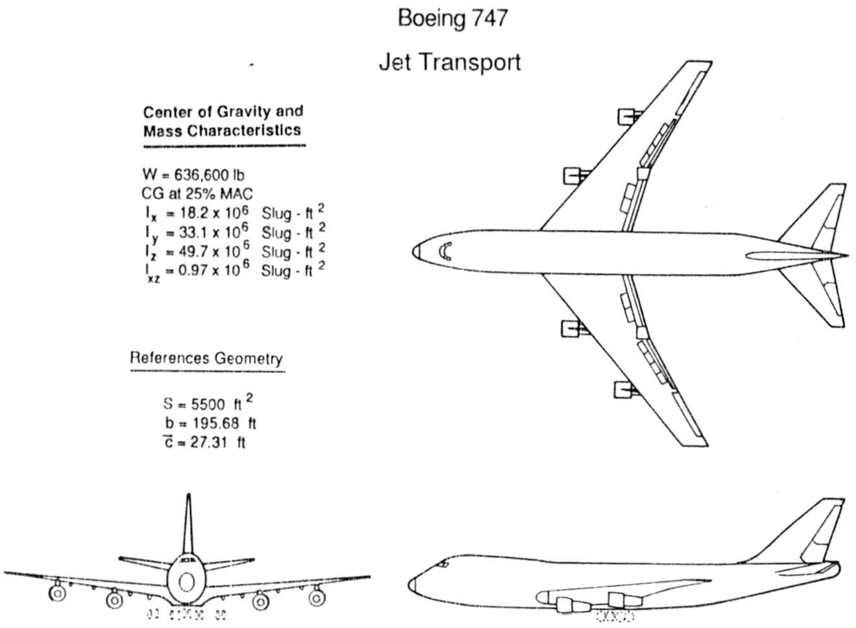

Boeing 747
Jet Transport
Center of Gravity and Mass Characteristics
W = 636,600 lb
CG at 25% MAC
I_x = 18.2 x 10^6 Slug - ft^2
I_y = 33.1 x 10^6 Slug - ft^2
I_z = 49.7 x 10^6 Slug - ft^2
I_{xz} = 0.97 x 10^6 Slug - ft^2
References Geometry
S = 5500 ft^2
b = 195.68 ft
$\bar{c}$ = 27.31 ft

FIGURE B.27
Three-view sketch and stability data for a large jet transport.

APPENDIX C

MATHEMATICAL REVIEW LAPLACE TRANSFORMS MATRIX ALGEBRA

REVIEW OF MATHEMATICAL CONCEPTS

Laplace Transformation

The Laplace transform is a mathematical technique that has been used extensively in control system synthesis. It is a very powerful mathematical tool for solving differential equations. When the Laplace transformation Technique is applied to a differential equation it transforms the differential equation to an algebraic equation. The transformed algebraic equation can be solved for the quantity of interest and then inverted back into the time domain to provide the solution to the differential equation.

The Laplace transformation is a mathematical operation defined by

$$\mathscr{L}[f(t)] = \int_0^\infty f(t)\mathrm{e}^{-st}\,\mathrm{d}t = F(s) \tag{C.1}$$

where $f(t)$ is a function of time. The operator, $\mathscr{L}$, and the complex variable, s, are the Laplace operator and variable, respectively and $F(s)$ is the transform of $f(t)$. The Laplace transformation of various functions $f(t)$ can be obtained by evaluating Eq. (C.1). The process of obtaining $f(t)$ from the Laplace transform $F(s)$ is called the inverse Laplace transformation and is given by

$$f(t) = \mathcal{L}^{-1}[F(s)] \tag{C.2}$$

where the inverse Laplace transformation is given by the following integral relationship.

$$f(t) = \frac{1}{2\pi i}\int_{c-\infty}^{c+\infty} F(s)e^{st}\,ds \tag{C.3}$$

Several examples of Laplace transformations are presented below.

Example Problem C.1. Consider the function $f(t) = e^{-at}$, then the Laplace transform of this expression yields

$$\mathcal{L}[f(t)] = \mathcal{L}[e^{-at}] = \int_0^{\infty} e^{-at}e^{st}\,dt = \int_0^{\infty} e^{-(a+s)t}\,dt$$

and the evaluation of the integral gives the transform $F(s)$

$$F(s) = -\frac{e^{-(a+s)t}}{a+s}\bigg|_0^{\infty} = \frac{1}{s+a}$$

As another example suppose that $f(t) = \sin \omega t$, then upon substituting into the definition of the Laplace transformation one obtains

$$F(s) = \mathcal{L}[\sin \omega t] = \int_0^{\infty} \sin \omega t e^{-st}\,dt = \frac{1}{2i}\int_0^{\infty} (e^{i\omega t} - e^{-i\omega t})e^{-st}dt$$

Evaluating this integral yields

$$F(s) = \frac{\omega}{s^2 + \omega^2}$$

Example Problem C.2. As another example consider the Laplace transformation of operations such as the derivative and definite integral. When $f(t)$ is a derivative, for example $f(t) = dy/dt$, then

$$\mathcal{L}[f(t)] = \int_0^{\infty} \frac{dy}{dt} e^{-st}dt$$

Solution of this integral can be obtained by applying the method of integration by parts. Mathematically integration by parts is given by the following expression.

$$\int_a^b u\,dv = uv\bigg|_a^b - \int_a^b v\,du,$$

Letting u and dv be as follows

$$u = e^{-st}$$

$$dv = \frac{dy}{dt}\,dt$$

then

$$du = -se^{-st}\,dt$$

$$v = y(t)$$

Substituting and integrating by parts yields

$$\mathscr{L}[f(t)] = y(t)e^{-st}\Big|_0^\infty + s\int_0^\infty y(t)e^{-st}\,dt$$

but the integral

$$\int_0^\infty y(t)e^{-st}\,dt = Y(s)$$

$$\mathscr{L}\left[\frac{dy}{dt}\right] = -y(0) + sY(s)$$

In a similar manner the Laplace transformation of higher order derivatives can be shown to be

$$\mathscr{L}\left[\frac{d^n y}{dt^n}\right] = s^n Y(s) - s^{n-1}y(0) - s^{n-2}\frac{dy}{dt}\Big|_{t=0} - \cdots - \frac{d^{n-1}y}{dt^{n-1}}\Big|_{t=0}$$

When all initial conditions are zero the transform simplifies to the following expression.

$$\mathscr{L}\left[\frac{d^n y}{dt^n}\right] = s^n Y(s)$$

Now consider the Laplace transform of a definite integral:

$$\mathscr{L}\left[\int_0^1 y(\tau)\,d\tau\right] = \int_0^\infty e^{-st}\,dt\int_0^t y(\tau)\,d\tau$$

This integral can also be evaluated by the method of integration by parts. Letting u and dv be as follows.

$$u = \int_0^1 y(\tau)\,d\tau$$

$$dv = e^{-st}\,dt$$

then

$$du = y(t)$$

$$v = \frac{1}{s}e^{-st}$$

Substituting and integrating by parts yields

$$\mathscr{L}\left[\int_0^t y(\tau)\,d\tau\right] = \frac{1}{s}e^{-st}\int_0^t y(\tau)\,d\tau\Big|_0^\infty - \frac{1}{s}\int_0^\infty e^{-st}y(t)\,dt$$

or

$$\mathscr{L}\left[\int_0^t y(\tau)\,d\tau\right] = \frac{Y(s)}{s}$$

TABLE C1
Table of Laplace transform pairs

$f(t)$	$F(s)$	$f(t)$	$F(s)$
$u(t)$	$1/s$	$\sin \omega t$	$\omega/(s^2+\omega^2)$
t	$1/s^2$	$\cos \omega t$	$s/(s^2+\omega^2)$
t^n	$n!/s^{n+1}$	$\sinh \omega t$	$\dfrac{\omega}{s^2-\omega^2}$
$\delta(t)$ Unit impulse	1	$\cosh \omega t$	$\dfrac{s}{s^2-\omega^2}$
$\int_{-\varepsilon}^{+\varepsilon} \delta(t)\,dt = 1$		$e^{-at}\sin \omega t$	$\dfrac{\omega}{(s+a)^2+\omega^2}$
e^{-at}	$1/(s+a)$	$t\cos \omega t$	$\dfrac{s^2-\omega^2}{(s^2+\omega^2)^2}$
te^{-at}	$\dfrac{1}{(s+a)^2}$	$t\sin \omega t$	$\dfrac{2\omega s}{(s^2+\omega^2)^2}$
$t^n e^{-at}$	$n!/(s+a)^{n+1}$		

By applying the Laplace Transformation to various functions $f(t)$ one can develop a table of transform pairs as shown in Table C1. This table is a list of some of the most commonly used transform pairs that occur in control system analysis.

Solution of Ordinary Linear Differential Equations

In control system design, a linear differential equation of the form

$$a_n \frac{d^n y}{dt^n} + a_{n-1}\frac{d^{n-1} y}{dt^{n-1}} + \cdots + a_1 \frac{dy}{dt} + a_0 y = f(t) \tag{C.4}$$

is common. This is a nonhomogeneous linear differential equatoin with constants coefficients. The Laplace-transformations of a differential equation results in an algebraic equation in terms of the transform of the derivatives and the Laplace variables. The resulting algebraic equation can be manipulated to solve for the unknown function $Y(s)$. The expression for $Y(s)$ can then be inverted back into the time domain to determine the solution $y(t)$.

Example Problem C.3. Given a second order differential equation

$$\frac{d^2 y}{dt^2} + 2\zeta\omega_n \frac{dy}{dt} + \omega_n^2 y = \omega_n^2 u(t)$$

where $u(t)$ is a unit step function and the initial conditions are as follows

$$y(0) = 0$$

$$\frac{dy(0)}{dt} = 0$$

Taking the Laplace transformation of the differential equation yields

$$[s^2 + 2\zeta\omega_n s + \omega_n^2]Y(s) = \frac{\omega_n^2}{s}$$

Solving for $Y(s)$ yields

$$Y(s) = \frac{\omega_n^2}{s(s^2 + 2\zeta\omega_n s + \omega_n^2)}$$

$y(t)$ can now be obtained by inverting $Y(s)$ back into the time domain.

$$y(t) = 1 + \frac{1}{\sqrt{1-\zeta^2}} e^{-\zeta\omega_n t} \sin(\omega_n \sqrt{1-\zeta^2} t - \phi)$$

where

$$\phi = \tan^{-1}[\sqrt{1-\zeta^2}/-\zeta]$$

Partial Fractions Technique for Finding Inverse Transformations

When solving a differential equation using the Laplace transformation approach, the major difficulty is in inverting the transformation back into the time domain. The dependent variable is found as a rational function of the ratio of two polynomials in the Laplace variable, s. The inverse of this function can be obtained by the inverse Laplace transform defined by Eq. (C.3). However, in practice it is generally not necessary to evaluate the inverse in this manner. If this function can be found in a table of Laplace Transform pairs, the solution in the time domain is easily obtained. On the other hand, if the transform cannot be found in the table, then an alternate approach must be used. The method of partial fractions reduces the rational fraction to a sum of elementary terms which are available in the Laplace Tables.

The Laplace transform of a differential equation typically takes the form of a ratio of polynomials in the Laplace variable s.

$$F(s) = \frac{N(s)}{D(s)}$$

The denominator can be factored as follows.

$$D(s) = (s + p_1)(s + p_2) \cdots (s + p_n)$$

These roots can be either real or complex conjugate pairs and can be of multiple order. For the case when the roots are real and of order one the

Laplace transform can be expanded in the following manner

$$F(s) = \frac{N(s)}{D(s)} = \frac{N(s)}{(s+p_1)(s+p_2)\cdots(s+p_n)}$$

$$= \frac{C_{p1}}{s+p_1} + \frac{C_{p2}}{s+p_2} + \cdots + \frac{C_{pn}}{s+p_n}$$

where the constants C_{p_i}'s are defined as

$$C_{p_1} = \left[(s+p_1)\frac{N(s)}{D(s)}\right]_{s=-p_1}$$

$$C_{p_2} = \left[(s+p_2)\frac{N(s)}{D(s)}\right]_{s=-p_2}$$

$$C_{p_i} = \left[(s+p_i)\frac{N(s)}{D(s)}\right]_{s=-p_i}$$

For the case where some of the roots are repeated, the Laplace transform can be represented as

$$F(s) = \frac{N(s)}{D(s)} = \frac{N(s)}{(s+p_1)(s+p_2)\cdots(s+p_i)^r(s+p_n)}$$

and in expanded form

$$F(s) = \frac{C_{p_i}}{s+p_1} + \frac{C_{p2}}{s+p_2} + \cdots + \frac{k_1}{(s+p_i)} + \frac{k_2}{(s+p_i)^2} + \cdots + \frac{k_r}{(s+p_1)^r}$$

The coefficients for the nonrepeated roots are determined as shown previously and the coefficients for the repeated roots can be obtained from the following expression.

$$k_j = \frac{1}{(r-j)!}\frac{d^{r-j}}{ds^{r-j}}\left[(s+p_i)^r\frac{N(s)}{D(s)}\right]_{s=-p_i}$$

With the partial fraction technique the Laplace transform of the differential equation can be expressed as a sum of elementary transforms that can be easily inverted to the time domain.

Matrix Algebra

In this section we will review some of the properties of matrices. A matrix is a collection of numbers arranged in square or rectangular arrays. Matrices are used in the solution of simultaneous equations and are of great utility as a shorthand notation for large systems of equations. A brief review of some of the basic algebraic properties of matrices are presented in the following section.

A rectangular matrix is a collection of elements that can be arranged in rows and columns as shown below

$$\mathbf{A} = a_{ij} = \begin{bmatrix} a_{11} & a_{12} & \cdots & a_{1j} \\ a_{21} & & & \\ a_{31} & & & \\ a_{41} & & & \\ \vdots & & & \\ a_{i1} & a_{i2} & \cdots & a_{ij} \end{bmatrix}$$

where the index i and j represent the row and column respectively. The rectangular matrix reduces to a square matrix when $i = j$.

A unit matrix or identity matrix is a square matrix with the elements along the diagonal being unity and all other elements of the array are zero. The identity matrix is denoted in the following manner

$$\mathbf{I} = \begin{bmatrix} 1 & 0 & \cdots & 0 \\ 0 & 1 & \cdots & 0 \\ \vdots & \vdots & & \vdots \\ 0 & 0 & \cdots & 1 \end{bmatrix}$$

Addition and Subtraction

Two matrices are equal if they are of the same order, i.e., they have the same number of rows and columns and if the corresponding elements of the matrices are identical. Mathematically this can be stated as follows

$$\mathbf{A} = \mathbf{B}$$

if

$$a_{ij} = b_{ij}$$

Matrices can be added provided they are of the same order. Matrix addition is accomplished by adding corresponding elements together.

$$\mathbf{C} = \mathbf{A} + \mathbf{B}$$

or

$$c_{ij} = a_{ij} + b_{ij}$$

Subtraction of matrices is defined in a similar manner.

$$\mathbf{C} = \mathbf{A} - \mathbf{B}$$

or

$$c_{ij} = a_{ij} - b_{ij}$$

Multiplication of Two Matrices

Two matrices **A** and **B** can be multiplied provided that the number of columns of **A** is equal to the number of rows of **B**. For example, suppose the matrices **A** and **B** are defined as follows

$$\mathbf{A} = [a_{ij}]_{n,p}$$
$$\mathbf{B} = [b_{ij}]_{q,m}$$

These matrices can be multiplied if the number of columns of **A** is equal to the number of rows of **B**, i.e., $p = q$

$$\mathbf{C} = \mathbf{AB} = [a_{ij}]_{n,p}[b_{ij}]_{a,m} = [c_{ij}]_{n,m}$$

where

$$c_{ij} = \sum_{k=1}^{p} a_{ik}b_{kj} \qquad \begin{array}{l} i = 1, 2, \ldots, n \\ j = 1, 2, \ldots, m \end{array}$$

Example Problem C.4. Given the matrices **A** and **B**, determine the product **AB**.

$$\mathbf{A} = \begin{bmatrix} a_{11} & a_{12} & a_{13} \\ a_{21} & a_{22} & a_{23} \\ a_{31} & a_{32} & a_{33} \end{bmatrix} \quad \text{and} \quad \mathbf{B} = \begin{bmatrix} b_{11} & b_{12} \\ b_{21} & b_{22} \\ b_{31} & b_{32} \end{bmatrix}$$

A and **B** can be multiplied together because the number of columns of **A** is equal to the number rows of **B**.

$$\mathbf{C} = \mathbf{AB}$$

$$\mathbf{C} = \begin{bmatrix} (a_{11}b_{11} + a_{12}b_{21} + a_{13}b_{31})(a_{11}b_{12} + a_{12}b_{22} + a_{13}b_{32}) \\ (a_{21}b_{11} + a_{22}b_{21} + a_{23}b_{31})(a_{21}b_{12} + a_{22}b_{22} + a_{23}b_{32}) \\ (a_{31}b_{11} + a_{32}b_{21} + a_{33}b_{31})(a_{31}b_{12} + a_{32}b_{22} + a_{33}b_{32}) \end{bmatrix}$$

Some additional properties of matrix multiplication are included in Table C2. Notice that in general matrix multiplication is not commutative. Multiplication of a matrix **A** by a scalar constant k is equivalent to multiplying each element of the matrix by the scalar k

$$k\mathbf{A} = \begin{bmatrix} ka_{11} & ka_{12} & ka_{13} \\ ka_{21} & ka_{22} & ka_{23} \\ ka_{31} & ka_{32} & ka_{33} \end{bmatrix}$$

TABLE C2
Properties of matrix multiplication

$(\mathbf{AB})\mathbf{C} = \mathbf{A}(\mathbf{BC})$	Associative
$(\mathbf{A} + \mathbf{B})\mathbf{C} = \mathbf{AC} + \mathbf{BC}$	Distributive
$\mathbf{A}(\mathbf{B} + \mathbf{C}) = \mathbf{AB} + \mathbf{AC}$	Distributive
$\mathbf{AB} \neq \mathbf{BA}$	Commutative

TABLE C3
Properties of an inverse matrix

1) $\mathbf{A}\mathbf{A}^{-1} = \mathbf{A}^{-1}\mathbf{A} = \mathbf{I}$
2) $[\mathbf{A}^{-1}]^{-1} = \mathbf{A}$
3) If **A** and **B** are nonsingular and square matrices then $(\mathbf{AB})^{-1} = \mathbf{B}^{-1}\mathbf{A}^{-1}$

Matrix Division (Inverse of a Matrix)

The solution of a system of algebraic equations requires matrix inversion. For example, if a set of algebraic equations can be written in matrix form

$$\mathbf{A}\mathbf{x} = \mathbf{y}$$

then the solution is given as

$$\mathbf{x} = \mathbf{A}^{-1}\mathbf{y}$$

where $\mathbf{A}^{-1}$ is the inverse of the matrix **A**. For the inverse of **A** to exist, the matrix **A** must be square and nonsingular. The condition that **A** be nonsingular means that the determinant of **A** must be a nonzero value. The inverse of a matrix, is defined as follows

$$\mathbf{A}^{-1} = \frac{Adj\mathbf{A}}{|\mathbf{A}|}$$

where Adj **A** is called the adjoint of **A**. The adjoint of a matrix is obtained by taking the transpose of the co-factors of the **A** matrix, where the cofactors are determined as follows

$$C_{ij} = (-1)^{i+j} D_{ij}$$

and D_{ij} is the determinant obtained by eliminating the ith row and jth column of **A**. Some additional properties of the inverse matrix are given in Table C3.

The transpose of a matrix is obtained by interchanging the rows and columns of the matrix. Given the matrix **A**

$$\mathbf{A} = \begin{bmatrix} a_{11} & a_{12} & a_{13} \\ a_{21} & a_{22} & a_{23} \\ a_{31} & a_{32} & a_{33} \end{bmatrix}$$

then the transpose of **A**

$$\mathbf{A}^T = \begin{bmatrix} a_{11} & a_{21} & a_{31} \\ a_{12} & a_{22} & a_{32} \\ a_{13} & a_{23} & a_{33} \end{bmatrix}$$

For additional properties of matrices the reader should consult his or her mathematics library.

APPENDIX D

REVIEW OF CONTROL SYSTEM ANALYSIS TECHNIQUES

BODE DIAGRAMS

The frequency response of a linear system is determined experimentally by applying a sinusoidal input signal and then measuring the sinusoidal response of the system. The frequency response data includes the measurement of the amplitude and phase shift of the sinusoidal output compared to the amplitude and phase of the input signal as the input frequency is varied. The relationship between the output and input to the system can be used by the designer to determine the performance of the system. Furthermore, frequency response data can be used to deduce the performance of a system to an arbitrary input that may or may not be periodic.

The magnitude of the amplitude ratio and phase angle can be presented graphically in a number of ways. However, one of the most useful presentations of the data is in the so called Bode diagram, named after H. W. Bode for his pioneering work in frequency response anlysis. In a Bode diagram the logarithm of the magnitude of the system transfer function, $|G(i\omega)|$, and the phase angle, ϕ, are plotted separately versus the frequency.

The frequency response, output/input amplitude ratio and phase with respect to the input can be determined analytically from the system transfer function written in factored time constant form.

$$G(s) = \frac{k(1 + T_a s)(1 + T_b s)\cdots}{s^r(1 + T_1 s)(1 + T_2 s)\cdots\left(1 + \frac{2\zeta}{\omega_n}s + \frac{s^2}{\omega_n^2}\right)} \tag{D.1}$$

This transfer function has simple zeros at $-1/T_a$, $-1/T_b$ . . . , a pole at the origin of order r, simple poles at $-1/T_1 - 1/T_2, \ldots$, and complex poles at $-\zeta\omega_n \pm i\omega_n\sqrt{1-\delta^2}$. It can be shown that the steady state response can be determined by substituting $i\omega$ for the Laplace variable s, in the system transfer function. Upon substituting $i\omega$ for s, one can express the transfer function in terms of the magnitude of its amplitude ratio and phase angle as follows

$$\begin{aligned} 20 \log |G(i\omega)| &= 20 \log k + 20 \log |1 + i\omega T_a| + 20 \log |1 + i\omega T_b| + \cdots \\ &\quad - 20\, r \log(\omega) - 20 \log |1 + i\omega T_1| - 20 \log |1 + i\omega T_2| \\ &\quad - 20 \log |1 + 2\zeta(\omega/\omega_n) - (\omega/\omega_n)^2| \cdots \end{aligned} \tag{D.2}$$

and the phase angle in degrees

$$\begin{aligned} \angle G(i\omega) &= \tan^{-1}\omega T_a + \tan^{-1}\omega T_b + \cdots - r(90^\circ) - \tan^{-1}\omega T_1 \\ &\quad - \tan^{-1}\omega T_2 \cdots - \tan^{-1}\left(\frac{2\zeta\omega\omega_n}{\omega_n^2 - \omega^2}\right) \end{aligned} \tag{D.3}$$

The magnitude has been expressed in terms of decibels. A magnitude in decibels is defined as follows

$$\text{Magnitude in db} = 20 \log \frac{|\text{Magnitude of Output}|}{|\text{Magnitude of Input}|} \tag{D.4}$$

where the logarithm is to the base 10.

The Bode diagram can now be constructed using a semi-log plot. The magnitude in decibels and phase angle are plotted separately on a linear ordinate versus the frequency on a logarithmic abscissa. Because the Bode diagram is obtained by adding the various factors of $G(i\omega)$ one can construct the Bode diagram quite rapidly.

In the general case the factors that will make up the transfer function are a constant term (system gain), poles at the origin, simple poles and zeros on the real axis and complex conjugate poles and zeros. The graphical representation of each of these individual factors are described in the following section.

System Gain

The log-magnitude of the system gain is given as follows

$$20 \log k = \text{constant db} \tag{D.5}$$

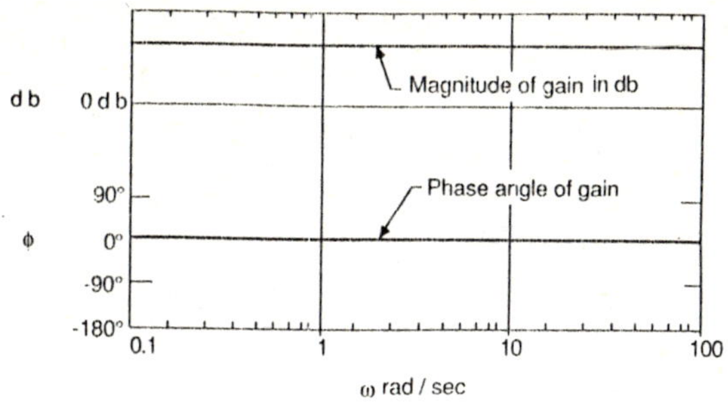

FIGURE D1
Bode representation of the magnitude and phase of the system gain *k*.

and the phase angle by

$$\angle k = \begin{array}{ll} 0^\circ & k > 0 \\ 180^\circ & k < 0 \end{array} \tag{D.5}$$

Figure D1 shows the Bode plot for a positive system gain.

Poles or Zeros at the Origin $(i\omega)^{\pm r}$

The log-magnitude of a pole or zero at the origin of order r can be written as

$$20 \log |(i\omega)^{\pm r}| = \pm 20r \log \omega \text{ db} \tag{D.6}$$

and the phase angle is given by

$$\angle (i\omega)^{\pm r} = \pm 90r \tag{D.7}$$

The log-magnitude is zero db at $\omega = 1.0$ rad/sec and has a slope of 20 db/decade, where a decade is a factor of 10 change in frequency. Figure D2 is a sketch of the log-magnitude and phase angle for a multiple zero or pole.

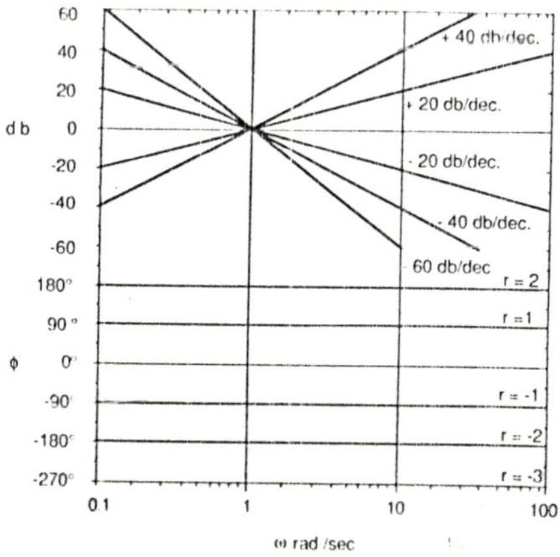

FIGURE D2
Bode representation of the magnitude and phase of a pole or zero at the origin.

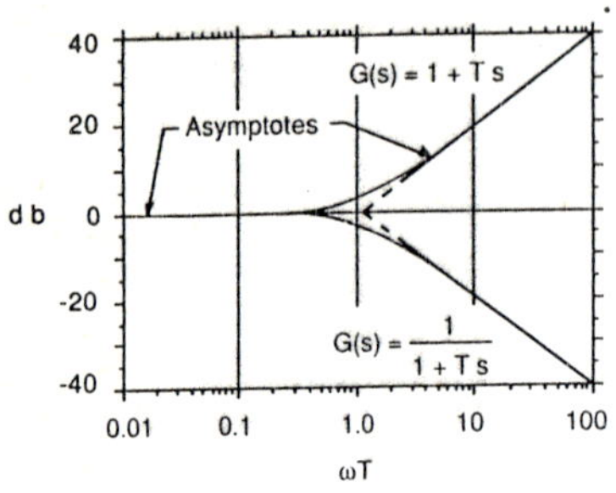

FIGURE D3
Bode representation of the magnitude of a simple pole or zero.

Simple Poles or Zeros $(1+i\omega T)^{\pm 1}$

The log-magnitude of a simple pole or zero can be expressed as

$$\pm 20 \log |1+i\omega T| = \pm 20 \log \sqrt{1+(\omega T)^2} \tag{D.8}$$

For very low values of ωT i.e., $\omega T \ll 1$, then

$$\pm 20 \log \sqrt{1+(\omega T)^2} \cong 0 \tag{D.9}$$

and for very large values of ωT i.e., $\omega T \gg 1$ then

$$\pm 20 \log \sqrt{1+(\omega T)^2} \cong \pm 20 \log \omega T \tag{D.10}$$

From this simple analysis one can approximate the log-magnitude plot of a simple pole or zero by two straight line segments as shown in Fig. D3. One of the asymptotic lines is the 0 db line and the second line segment has a slope of 20 db/decade that intersects the 0 db line at the frequency $\omega = 1/T$. The intersection frequency is called the corner frequency. The actual log-magnitude differs from the asymptotic approximation in the vicinity of the corner frequency.

The phase angle for a simple pole or zero is given by

$$\angle(1+i\omega T)^{\pm 1} = \pm \tan^{-1} \omega T \tag{D.11}$$

Figure D4 is a sketch of the phase angle.

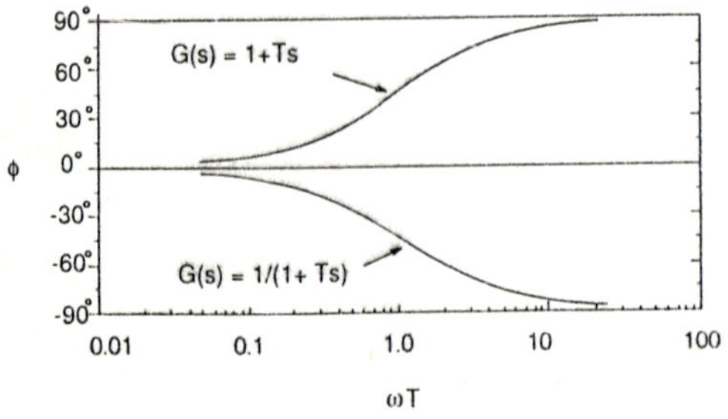

FIGURE D4
Bode representation of the phase angle of a simple pole or zero.

Complex Conjugate Pole or Zero $(1 + i2\zeta\omega/\omega_n - (\omega/\omega_n)^2)^{\pm 1}$

The log-magnitude of the complex pole can be written as

$$20 \log \left| \frac{1}{1 + i2\zeta\omega/\omega_n + (\omega/\omega_n)^2} \right| = -20 \log[(1 - (\omega/\omega_n)^2)^2 + (2\zeta\omega/\omega_n)^2]^{1/2}$$
$$= -10 \log[(1 - (\omega/\omega_n)^2)^2 + (2\zeta\omega/\omega_n)^2] \tag{D.13}$$

The log-magnitude can be approximated by two straight line segments. For example, when $\omega/\omega_n \ll 1$

$$20 \log \left| \frac{1}{1 + i2\omega/\omega_n - (\omega/\omega_n)^2} \right| \cong 0 \tag{D.14}$$

and when $\omega/\omega_n \gg 1$

$$20 \log \left| \frac{1}{1 + i2\omega/\omega_n - (\omega/\omega_n)^2} \right| \cong -40 \log \omega/\omega_n \tag{D.15}$$

The two straight line asymptotes consist of a straight line along the 0 db line for $\omega/\omega_n \ll 1$ and a line having a slope of -40 db/decade for $\omega/\omega_n \gg 1$. The asymptotes intersect at $\omega/\omega_n = 1$ or $\omega = \omega_n$, where ω_n is the corner frequency. Figure D5 shows the asymptotes as well as the actual magnitude plot for various damping ratios for a complex pole.

The phase angle for a complex pole is given by

$$\angle(1 + i2\omega/\omega_n - (\omega/\omega_n)^2)^{-1} = -\tan^{-1}\left[\frac{2\zeta\omega/\omega_n}{1 - (\omega/\omega_n)^2}\right] \tag{D.16}$$

Figure D6 shows the phase angle for a complex pole. Similar curves can be developed for a complex zero.

If the transfer function is expressed in time constant form, then the Bode diagram can be easily constructed from the simple expressions developed in this section.

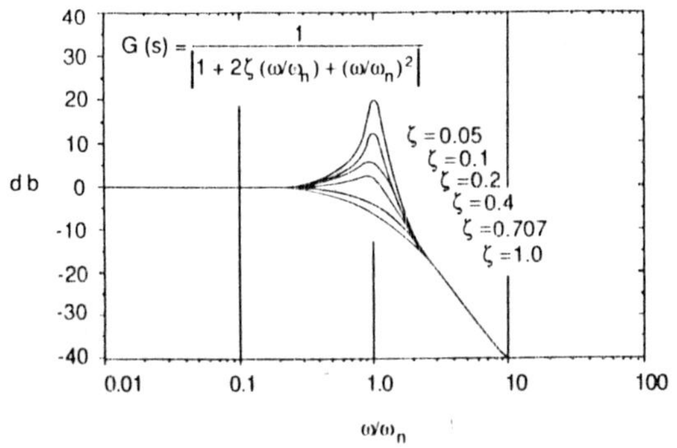

FIGURE D5
Bode representation of the magnitude of a complex conjugate pole.

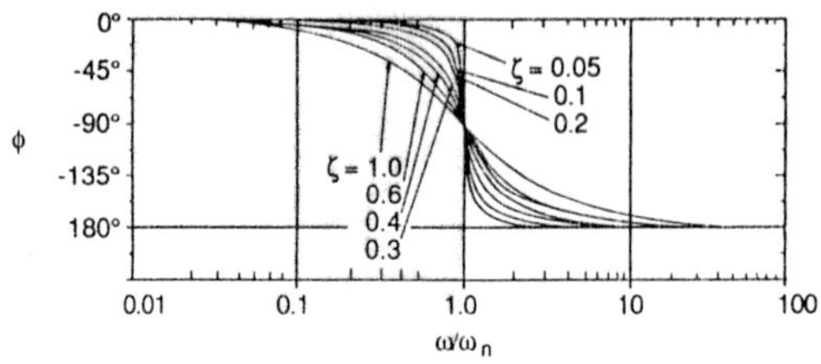

FIGURE D6
Bode representation of the phase angle of a complex conjugate pole.

Root Locus Technique

The transfer function was described earlier as the ratio of the output to the input. Upon examining a transfer function we note that the denominator is the characteristic equation of the system. The roots of the denominator are the eigenvalues that describe the free response of the system, where the free response is the solution of the homogeneous equation. In controls terminology the characteristic roots are called the poles of the transfer function. The numerator of the transfer function governs the particular solution and the roots of the numerator are called zeros.

As was noted earlier in Chapters 4 and 5 the roots of the characteristic equation (or poles) must have negative real parts if the system is to be stable. In control system design, the location of the poles of the closed loop transfer function allows the designer to predict the time domain performance of the system.

However, in designing a control system the designer will typically have a number of system parameters that are unspecified. The root locus technique permits the designer to view the movement of the poles of the closed loop transfer function as one or more unknown system parameters are varied.

Before describing the root locus technique it would be helpful to examine the significance of the root placement in the complex plane and the type of response that can be expected to occur. Figure D7 illustrates some of the

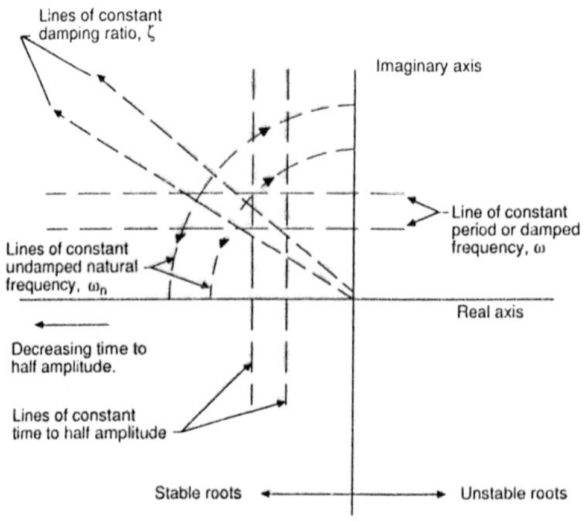

FIGURE D7
Graphical description of characteristic roots.

important features of pole location. First we note that any pole lying in the left half portion of the complex plane is stable, i.e., the response decays with time. Any pole in the right half plane leads to a response that grows with time which will result in an unstable system. The farther the root is to the left of the imaginary axis the faster the response decays. All poles lying along a particular vertical line will have the same time to half amplitude. Poles lying along the same horizontal line have the same damped frequency, ω, and period. The farther the pole is from the real axis the higher the frequency of the response will be. Poles lying along a radial line through the origin have the same damping ratio, and roots lying on the same circular arc around the origin will have the same undamped natural frequency. Finally some comments must be made about the poles lying on the imaginary axis. Poles of the order one on the imaginary axis lead to undamped oscillations, however, multiple order poles result in responses that grow with time.

The closed loop transfer function was shown earlier to be

$$M(s) = \frac{G(s)}{1 + G(s)H(s)} \tag{D17}$$

The characteristic equation of the closed loop system is given by the denominator of Eq. (D.17)

$$1 + G(s)H(s) = 0 \tag{D.18}$$

or

$$G(s)H(s) = -1 \tag{D.19}$$

The open loop transfer function $G(s)H(s)$ can be expressed in factored form as follows

$$G(s)H(s) = \frac{k(s+z_1)(s+z_2)\cdots(s+z_m)}{(s+p_1)(s+p_2)\cdots(s+p_n)} \tag{D.20}$$

where $n > m$ and k is an unknown system parameter. Substituting this equation into the characteristic equation yields

$$\frac{k(s+z_1)(s+z_2)\cdots(s+z_m)}{(s+p_1)(s+p_2)\cdots(s+p_n)} = -1 \tag{D.21}$$

The characteristic equation is complex and can be written in terms of a magnitude and angle as follows

$$\frac{|k|\,|s+z_1|\,|s+z_2|\cdots|s+z_m|}{|s+p_1|\,|s+p_2|\cdots|s+p_n|} = 1 \tag{D.22}$$

$$\sum_{i=1}^{m} \angle(s+z_i) - \sum_{i=1}^{n} \angle(s+p_i) = (2q+1)\pi \tag{D.23}$$

where $q = 0, 1, 2$ etc. Solution of the above equations yields the movement of the roots as a function of the unknown system parameter. These equations can be solved on the computer to determine the root locus contours. There is, however, a simple graphical technique developed by W. R. Evans that can be used to sketch a root locus plot. This graphical procedure is presented in the next section.

It can be easily shown that the root locus contours start at the open loop transfer function $G(s)H(s)$ poles and ends at the open loop zeros as k is varied

TABLE D1

Rules for graphical construction of the root locus plot

1. The root locus contours are symmetrical about the real axis.
2. The number of separate branches of the root locus plot is equal to the number of poles of the open loop transfer function $G(s)H(s)$. Branches of the root locus originate at the open loop poles of $G(s)H(s)$ for $k = 0$, and terminate at either the open loop zeros or at infinity for $k = \infty$. The number of branches that terminate at infinity is equal to the difference between the number of poles and zeros of the open loop transfer function.
3. Segments of the real axis that are part of the root locus can be found in the following manner. Points on the real axis that have an odd number of poles and zeros to their right are part of the real axis portion of the root locus.
4. The root locus branches that approach the open loop zeros at infinity do so along straight line asymptotes that intersect the real axis at the center of gravity of the finite poles and zeros. Mathematically this can be expressed as follows.

$$\sigma = \left[\sum \text{Real Parts of the Poles} - \sum \text{Real Parts of the Zeros}\right] \Big/ (n - m)$$

   where n is the number of poles and m the number of finite zeros.
5. The angle that the asymptotes make with the real axis is given by

$$\phi_a = \frac{180°[2q + 1)}{n - m}$$

   for $q = 0, 1, 2, \ldots, (n - m - 1)$.
6. The angle of departure of the root locus from an open loop pole can be found by the following expression.

$$\phi_p = \pm 180°(2q + 1) + \phi \qquad q = 0, 1, 2, \ldots$$

   where ϕ is the net angle contribution at the pole interest due to all other open loop poles and zeros. The arrival angle at the open loop zero is given by a similar expression

$$\phi_z = \pm 180°(2q + 1) - \phi \qquad q = 0, 1, 2, \ldots$$

   The angle ϕ is determined by drawing straight lines from all the poles and zeros to the pole or zero of interest and then summing the angles made by these lines.
7. If a portion of the real axis is part of the root locus and branch is between two poles, the branch must break away from the real axis and so that the locus ends on a zero as k approaches infinity. The break away points on the real axis are determined by solving

$$1 + kGH = 0$$

   for k and then find the roots of the equation $dk/ds = 0$. Only roots that lie on a branch of the locus are of interest from zero to infinity. For example, if we rearrange the magnitude criteria as

$$\frac{|s + z_1|\,|s + z_2| \cdots |s + z_m|}{|s + p_1|\,|s + p_2| \cdots |s + p_n|} = \frac{1}{|k|} \tag{D.24}$$

then as k goes to zero the function becomes infinite. This implies that the roots approach the poles as k goes to zero. On the other hand, as k goes to infinity the functon goes to zero which implies the roots are at the open loop zeros. Therefore, the root locus plot of the closed loop system starts with a plot of the poles and zeros of the open loop transfer function. Evans developed a series of rules based upon the magnitude and angle criteria for rapidly sketching the root locus branches on a pole zero map. A proof of these rules can be found in most control textbooks and will not be presented here. Table D1 is a summary of the rules for constructing a root locus contour.

INDEX